Carbonate Sedimentology

Maurice E. Tucker
University of Durham

V. Paul Wright
University of Reading

with a chapter by
J.A.D. Dickson
University of Cambridge

b

**Blackwell
Science**

BLACKWELL PUBLISHING
350 Main Street, Malden, MA 02148-5020, USA
9600 Garsington Road, Oxford OX4 2DQ, UK
550 Swanston Street, Carlton, Victoria 3053, Australia

First published 1990

13   2008

Library of Congress Cataloging-in-Publication Data

Tucker, Maurice E.
        Carbonate sedimentology/Maurice E. Tucker &
        V. Paul Wright: with a chapter by J. A. D. Dickson.
            p.   cm.
        ISBN 0-632-01472-5 (pbk)
            1. Rocks, Carbonate.   2. Sedimentation and deposition
        I. Wright, V. Paul, 1953-   II. Dickson, J. A. D.
        III. Title.
        QE471.15.C3T3   1990
        552'.58—dc20

ISBN-13: 978-0-632-01472-9 (pbk)

A catalogue record for this title is available from the British Library.

Set in by Setrite Typesetter Ltd, Hong Kong

For further information on
Blackwell Publishing, visit our website:
www.blackwellpublishing.com

Contents

3 Modern carbonate environments 70

Maurice Tucker

4 Carbonate depositional systems I: marine shallow-water and lacustrine carbonates 101

5 Carbonate depositional systems II: deeper-water facies of pelagic and resedimented limestones 228

Maurice Tucker

6 Carbonate mineralogy and chemistry 284

Tony Dickson

7 Diagenetic processes, products and environments 314

Maurice Tucker

8 Dolomites and dolomitization models 365

Maurice Tucker

9 The geological record of carbonate rocks 401
Maurice Tucker

Preface

The intention of this book is to provide a detailed synthesis of the enormous body of research which has been published on carbonate sediments and rocks. Such rocks are worthy of attention for several reasons. They are volumetrically a most significant part of the geological record and possess much of the fossil record of life on this planet. Most importantly they contain at least 40% of the world's known hydrocarbon reserves. They also play host to base metal deposits and groundwater resources, and are raw materials for the construction and chemical industries. No other rock type is as economically important.

From a scientific viewpoint, carbonates are especially interesting for the diversity of their origins. Most limestones are ultimately biogenic in origin and an appreciation of biological and palaeobiological factors is essential in understanding their formation. Their simple mineralogies, usually monominerallic, belie their petrographic and chemical complexity and highly sophisticated microscopic and analytical techniques are required to decipher their diagenetic histories. Besides biological and geochemical expertise, understanding carbonate deposits also demands an appreciation of physical sedimentology and oceanography.

Carbonate deposition involves a more complex suite of processes than many other sediment types. Facies analysis has been heavily dependent on 'comparative' studies but these have suffered from several drawbacks. The present is not necessarily a key to the past as regards carbonate deposition. Most modern carbonate depositional systems are less than 6000 years old and these will hardly reflect the variety of settings established in the past. Many of the facies models currently in use are based on just several modern environments, and many descriptions are now over 20 years old. The main danger in over-emphasizing the comparative approach is that the present situation has not been the norm because both biological and chemical factors have not remained static. As a result of biological evolution and subtle changes in ocean chemistry, biological and chemical factors have varied through time. This has affected limestone composition, diagenetic potential, and the geochemical pathways limestones have evolved through during their burial histories.

Our view and understanding of carbonate deposition and diagenesis has changed greatly in the last 10 years and this book was conceived to synthesize this new knowledge. It contains coverage of modern depositional systems, facies models, mineralogy, and diagenesis. It is impossible, with the constraints of time and book size, to cover every aspect of the subject; nor is it possible to satisfy every specialist or to include everyone's 'favourite' references. Some topics, such as reefs and carbonate diagenesis, require books of their own. We hope we have provided a thorough introduction, from the basics, such as petrography, to the most up-to-date concepts. In particular, the 'rock' aspects have been emphasized, making the book of direct use to practising geologists. The subject of carbonate sedimentology is dogged by inconsistent terminology and, where possible, we have argued for the use of specific terms and the abandonment of others.

It is an exciting time in carbonate sedimentology. New techniques in microanalysis are providing major opportunities for unravelling detailed diagenetic histories. At the opposite end of the spectrum, seismic data and sequence stratigraphy are enabling us to interpret the evolution of whole carbonate successions. A current theme is the use of computer models to generate sedimentological or diagenetic sequences. To date these have been used to model peritidal sequences and carbonate platform dynamics and shelf-margin diagenesis. An important factor in the drive to develop models is the more widespread acceptance of the role of orbitally-forced climatic cycles, although there is still much argument here. Carbonate sediments are particularly sensitive to environmental changes. Temperature influences biogenic production and most carbonate production is strongly depth dependent, so that even small-scale changes in sea-level can cause a decrease in production as a result of exposure or deepening. The recognition of Milankovitch cyclicity in ancient carbonates will be a major theme over the next few years. Another major trend will be the integration of hydrogeological and diagenetic models. This approach is already proving very successful in understanding platform diagenesis, with the Kohout convection model, and the fascinating discoveries of the 'openness' of the Bahamian Platform groundwater system. Such hydrogeologically-based diagenetic models may, at last, allow carbonate diagenesis to be predictive and not simply descriptive.

The study of carbonate sedimentology is entering a period of transition, from its earlier descriptive phase to one of quantitative models and synthesis. We hope this book will provide both a source and a stimulus for this new work.

Maurice Tucker, Paul Wright and Tony Dickson
September 1989

Acknowledgements

This book has its foundations in many years of research and teaching in carbonate geology by the authors. We acknowledge the support of the universities where we have studied and taught over the years; at home — Bristol, Cambridge, Cardiff, Durham, London, Newcastle, Nottingham, Reading and the Open University — and in the USA — Baton Rouge (Louisiana), Berkeley (California) and Stony Brook (New York). Support for our research has also come from the NERC (Natural Environment Research Council), the British Council, the Royal Society and oil companies BP, Britoil, Chevron, Lasmo, Mobil, Shell (KSEPL), Texaco and Unocal. The book itself derives from courses which have been given to the petroleum industry, and there is also the input from experiences gained in undergraduate and postgraduate teaching, both here in Britain, elsewhere in Europe and in North America.

Many individuals have been involved in the preparation of this book. We are sincerely grateful to Professor Fred Read of Virginia Polytechnic Institute, Blacksburg, for his critical reading and incisive comments on a draft of the book. Fred made many valuable suggestions for the improvement of the text. We are also grateful to Phil Choquette (Marathon Oil Co., Littleton, Colorado), Bill Meyers (State University of New York, Stony Brook), Craig Smalley (BP Sunbury) and Rob Gawthorpe (University of Manchester) for reading some of the chapters and giving their opinions.

Thanks are due to many people who have provided photographs, slides, preprints of papers, permission to use figures and advice: Alfonso Bosellini (Università di Ferrara, Italy), Martin Brasier (University of Oxford), Richard Bromley (University of Copenhagen, Denmark), Barbara Brown and R.P. Dunne (University of Newcastle upon Tyne), Dan Bosence (RHBNC, University of London), Trevor Burchette (BP London), Robert F. Dill (University of South Carolina, Columbia), Peter Ditchfield (University of Liverpool), Y. Druckman (Geological Survey of Israel), Peter Frykman (Geological Survey of Denmark), Ian Goldsmith (University of Durham), Gill Harwood (University of East Anglia, Norwich), Malcolm Hart (Plymouth Polytechnic, Devon), Albert Hine (University of South Florida, St Petersburg), Noel James (Queens University, Ontario), John Kaldi (Arco, Texas), Kerry Kelts (EAWAG, Zurich, Switzerland), Chris Kendall (University of South Carolina, Columbia), Jim Kennedy (University of Oxford), David Kitson (Robertson Research), Andrew Leitch (Agricultural Research Council, Hertfordshire), Jonathan Lewis (Imperial College London), Jean-Paul Loreau (Université de Nancy, France), Roy McGregor (St Andrews University, Scotland), Clyde Moore (Louisiana State University, Baton Rouge), Hank Mullins (Syracuse University, New York), Nigel Platt (University of Bern, Switzerland), John Powell (British Geological Survey), Robert Riding (University of Wales, Cardiff), Art Saller (Unocal, California), Colin Scrutton (University of Newcastle upon Tyne), Gene Shinn (USGS, Fisher Island, Miami), Peter Smart (Bristol University), Denys Smith (University of Durham), Roger Till (BP London), Bill Ward (University of New Orleans) and Nigel Watts (Texaco, Canada).

Permission to use published material was kindly given by the American Association of Petroleum Geologists; Elsevier, Amsterdam; Geological Association of Canada; Geological Society of London; International Association of Sedimentologists; Macmillan Journals Ltd, London; Museum National d'Histoire Naturelle, Antenne de Tahiti; Ocean Drilling Program, College Station, Texas; Society of Economic Paleontologists and Mineralogists; Springer-Verlag, Heidelberg; and Christopher Springman Photographer Inc., Pt Reyes, California.

The authors are most grateful to Carole Blair and Janette Finn, Alison Ruegg, Elizabeth Wyeth and Chris Chester, and Cherri Webre for typing many drafts of this book over the last few years. We are indebted to Karen Gittins and Pam Baldaro for their artwork, and to Gerry Dresser for photography. Thanks are also required for Vivienne, Ashley and Zöe (the Tuckers) for their assistance with the references.

Finally, at Blackwell Scientific Publications we should thank Navin Sullivan for his patience and encouragement and Emmie Williamson for her careful handling of the manuscript and proofs.

1 Carbonate sediments and limestones: constituents

1.1 INTRODUCTION

To a student first looking at limestones in outcrop they may look very boring, for they are usually grey to white in colour and typically lack reassuring points of reference such as sedimentary structures; indeed even the bedding may not be a primary feature. In thin section all the usual familiar polarizing colours are absent and the rock is usually monomineralic. Instead, the rock contains a remarkable variety of grain types, the natures of many of which are difficult to resolve easily. None of this is helped by the jargon which has to be learned with terms such as oobiosparite, packstone, pisoid, neomorphic pseudospar, fascicular optic calcite and baroque dolomite, to name only a few.

The aim of this chapter is to introduce the reader to some of the basic concepts and terminology of carbonate petrography. In each of the following chapters more information is given about sedimentary processes and their products, and finally diagenetic processes and products are discussed in Chapters 6 to 8.

At the simplest level carbonate petrography aims to classify a limestone, enabling comparisons to be made with present-day carbonate sediments. However, limestones differ from carbonate sediments in two main ways. The first is obvious, limestones are lithified but sediments, by definition, are not. A more subtle difference exists in mineralogy. Modern carbonate sediments are composed mainly of two mineral forms (Chapter 6), aragonite and calcite. Two types of calcite are recognized, low magnesian calcite (<4 mole% $MgCO_3$) and high magnesian calcite (>4 mole% $MgCO_3$). Both aragonite and high magnesian calcite are less stable than low magnesian calcite under normal diagenetic conditions and are replaced by it. All three may be replaced by dolomite so that most limestones in the geological record consist of low magnesian calcite and/or dolomite. It may be putting things too strongly to state that looking at limestones is like 'looking through a glass darkly' but it is most important to be aware that significant mineralogical and textural changes have occurred in most limestones. Even making the most basic observations in carbonate petrography requires some interpretation first, or at least an awareness of possible diagenetic overprinting. This is part of the fascination of the subject.

Limestones, despite being mineralogically quite simple, are highly varied in composition. However, three main components can be recognized: grains, matrix and cement. The grain types can be further subdivided into non-skeletal grains and skeletal grains. In limestones the most common cement type is sparite, represented by relatively coarsely crystalline, clear calcite crystals. Cementation is discussed in detail in Chapter 7.

1.2 NON-SKELETAL GRAINS

Non-skeletal grains are those not obviously derived from the skeletal material of micro-organisms, invertebrates or the thalli of calcareous plants. Four main types are recognized (Folk, 1959): coated grains, peloids, aggregates and clasts.

1.2.1 Coated grains

A remarkable variety of carbonate-coated grains occurs and the existing terminology is rife with inconsistencies and problems. Indeed almost every carbonate petrographer seems to have his or her own set of definitions. Coated grains are polygenetic in origin with different processes forming similar types of grains and many of these processes are still very poorly understood. Furthermore, similar coated grains can form in very different environments which makes their use in environmental interpretation difficult.

A bewildering variety of terms exists for describing coated grains and some of the more bizarre names include macro-oncoid, pisovadoid, cyanoid, bryoid, tuberoid, putroid and walnutoid (Peryt, 1983a). Things have gone too far and perhaps the terminological mess can never be cleared up. The most recent review of coated grain terminology is that by Peryt

(1983b) who proposed yet another classification using both a generic and genetic system for classifying these grains. This classification has been criticized by Richter (1983a) and the reader is referred to these two papers for an introduction to the problems which exist in this topic. Many classifications, including Peryt's, have distinguished two broad categories of coated grains: chemically formed (especially ooids) and biogenically formed (oncoids). However, it is often impossible to prove if a coated grain was biogenically formed and many ooids, usually classified as chemically formed, are either directly biogenically formed or their growth may be biochemically influenced. In their classifications, Flügel (1982) and Richter (1983a) take a more descriptive (generic) approach to the terms ooid and oncoid. The following definitions are modified from those given by these authors and stress the nature of the cortical laminae form and continuity (Fig. 1.1).

An ooid (or oolith) is a coated grain with a calcareous cortex and a nucleus which is variable in composition. The cortex is smoothly and evenly laminated especially in its outer parts, but individual laminae may be thinner on points of stronger curvature on the nucleus. They are typically spherical or ellipsoidal in shape with the degree of sphericity increasing outwards. Obvious biogenic structures are lacking or only constitute a minor part of the cortex.

Ooids may be classified on their microfabric or mineralogy. Diagenesis may obliterate many characteristic features and this is especially the case where originally aragonitic ooids have been replaced by calcite. A rock largely made of ooids is an oolite.

An oncoid (or oncolith) is a coated grain with a calcareous cortex of irregular, partially overlapping laminae. They are typically irregular in shape and may exhibit biogenic structures. Some forms lack a clear nucleus.

Oncoids may be classified on the types of biogenic structure they contain, so that for example ones formed by red algal coatings are called rhodoliths (or rhodoids). A rock made of oncoids should be called an oncolite. Some workers have restricted the term to algal nodules but this usage is fraught with problems.

The term pisoid is commonly used in carbonate petrography but again no consensus exists for its definition. Flügel (1982) considered it to mean a non-marine ooid while most workers have stressed it to mean an ooid larger than 2 mm in diameter (Leighton & Pendexter, 1962; Donahue, 1978). As a result of being larger than ooids, they typically have a less regular lamination. It may be advisable to add a

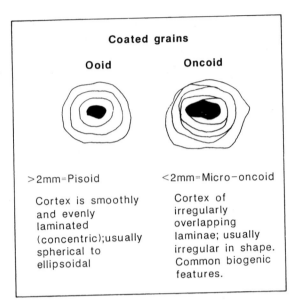

Fig. 1.1 *Classification of coated grains.*

similar size division to the oncoids as shown in Fig. 1.1. While most oncoids are larger than 2 mm a special category for oncoids under 2 mm might be used, the micro-oncoids.

Peryt (1983b) has defined three size categories for coated grains based on their diameters: microid (<2 mm), pisoid (2–10 mm) and macroid (>10 mm). These have been used by Peryt as prefixes (for example to define large oncoids as macro-oncoids), but his system then recognizes genetic, interpretative types which are highly debatable (Richter, 1983a).

A recent addition to this subject has come from Krumbein (1984) who classified ooids and oncoids on the nature of the regularity of shape and continuity of the laminae, and he recognized micro-oncoids as outlined above but further added a genetic terminology based on whether the whole grain assemblage is biogenic or abiogenic. This classification recognizes oolites and oncolites as assemblages of coated grains which are biogenically formed and ooloids and oncoloids as assemblages of abiogenically-formed grains. Since it is impossible at present to tell if many coated grains are biogenic or not, this latest classification system is unusable and hopefully these additional confusing terms will *not* enter the literature.

Cortoids are another type of 'coated grain' recognized by some workers (Flügel, 1982). They are grains covered by a micrite envelope, presumed to have formed by endolithic micro-organisms (Section 1.2.2).

These are not strictly coated grains but represent an alteration of the grain surface. However, many thick micrite envelopes are partly accretionary in origin formed by the encrustation of micro-organisms which are partly endolithic and partly epilithic (Kobluk & Risk, 1977a,b). These would constitute a type of non-laminated coated grain for which the term cortoid might reasonably be used.

1.2.1a Ooids and pisoids

These two grain types will be considered together as there are many similarities in their mineralogies, fabrics and origins.

Mineralogy. The mineralogy of ooids and pisoids influences not only their subsequent diagenesis but also their microfabric. The evidence for, and significance of, the changing mineralogies of marine ooids is discussed in Section 9.3. Recent marine ooids are mainly aragonite in composition (Richter, 1983b), and so also are ooids in saline lakes with a high Mg/Ca ratio (Section 4.4.6b). High magnesian calcite ooids also occur but are much less common (Marshall & Davies, 1975; Milliman & Barretto, 1975; Land *et al.*, 1979), typically having 11−17 mole % $MgCO_3$; low magnesian calcite ooids have been described from present-day lakes, streams, caves and calcareous soils (Wilkinson *et al.*, 1980; Geno & Chafetz, 1982; Richter, 1983b). Biminerallic ooids have been described from hypersaline lagoons of Texas by Land *et al.* (1979) (aragonite and high magnesian calcite), from the Pleistocene of the Florida Shelf by Major *et al.* (1988) (aragonite and high magnesian calcite), and also from the alkaline Pyramid Lake of Nevada by Popp & Wilkinson (1983) (aragonite and low magnesian calcite). Ancient examples of biminerallic ooids have been described by Tucker (1984) and Wilkinson *et al.* (1984).

Pisoids are, generally speaking, less common in recent marine settings but both aragonite and high magnesian calcite pisoids have been recorded from marginal marine hypersaline environments (Purser & Loreau, 1973; Scholle & Kinsman, 1974; Picha, 1978; Ferguson *et al.*, 1982). Low magnesian calcite pisoids have been described from a variety of non-marine environments and may reach quite large sizes (Risacher & Eugster, 1979; Chafetz & Butler, 1980).

Ooid cortical microfabrics. The microfabrics of ooids have attracted considerable interest and a number of detailed studies have been made of both Recent and

ancient ooids. Reviews have been given by Simone (1980), Richter (1983b) and Medwedeff & Wilkinson (1983).

In ooids forming at the present day three main microfabrics occur: tangential, radial and random (Fig. 1.2). *Tangential* microfabrics are the main microfabric in Bahamian ooids (Fig. 3.10) and consist of aragonite 'grains' whose long axes are aligned parallel to the ooid laminae. The grains consist of aragonite rods without crystal terminations, which average 1 μm long (maximum 3 μm) and have diameters of 0.1 to 0.3 μm. Granules (or nannograins) of aragonite also occur with diameters of 0.1 to 0.3 μm (Fabricius, 1977; Gaffey, 1983). *Radial* microfabrics consist of fibrous or bladed crystals of aragonite, low magnesian calcite or high magnesian calcite. As a general rule, radial aragonite fabrics in ooids are less

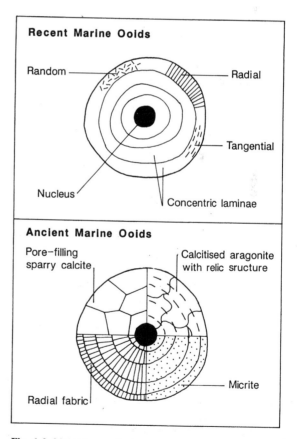

Fig. 1.2 *Major types of microstructure seen in modern and ancient ooids. Variations on these types have been described from ancient ooids by Tucker (1984), Strasser (1986), Chow & James (1987) and Singh (1987).*

common in marine settings than tangential fabrics but they have been described from the Trucial Coast of the Arabian Gulf (Loreau & Purser, 1973) and from the Great Barrier Reef in Australia (Davies & Martin, 1976). Several studies have also been carried out on the radial ooids of the hypersaline Great Salt Lake of Utah (Kahle, 1974; Sandberg, 1975; Halley, 1977).

Radial aragonite coatings in the Trucial Coast consist of needles 10–50 µm long and 2–5 µm in diameter (Loreau & Purser, 1973). In the Great Salt Lake ooids, large aragonite 'rays' occur up to 150 µm long (Halley, 1977) (Fig. 1.3). Low magnesian calcite radial fabrics have recently been described in detail from Pyramid Lake in Nevada (Popp & Wilkinson, 1983) and the calcite occurs as laths or blades 10–20 µm long and 4–8 µm wide. High magnesian calcite radial cortices which have been described from marine settings have component 'crystals' of similar sizes to those in aragonitic and low magnesian calcite ooids (Richter, 1983b). In radial ooids the individual crystals may be arranged in thick radial coats or may have a finer, banded radial structure (Medwedeff & Wilkinson, 1983). Ooids with thick radial coats, but lacking a nucleus (arguably not an ooid), are termed *spherulites*.

Random microfabrics result either because of the random arrangement of the constituents, such as aragonite rods, or because the only crystals present are equant grains. Intense micritization by endolithic micro-organisms can create a random microstructure.

Some ooids exhibit more than one type of microfabric, usually related to differences in mineralogy. The bimineralic ooids described by Land *et al.* (1979) from Baffin Bay, Texas have radial high magnesian calcite laminae and tangential–random aragonite laminae, while those described by Popp & Wilkinson (1983) from Pyramid Lake, Nevada, have radial aragonite laminae and radial to random calcite laminae.

The origin of ooids as discussed above is still open to question so it is not possible to explain all the differences in microfabric. Certainly mineralogy is a major control but whether this ultimately reflects the chemistry of the ambient waters or is due to organic influences, as suggested by Land *et al.* (1979) for the bimineralic (and bifabric) ooids of Baffin Bay, is uncertain. What is now quite well documented is the role of energy level in influencing ooid microfabric. In low-energy environments along the Trucial Coast, such as protected lagoons or troughs between oolite bars, radial aragonite ooids with a loose structure predominate. In higher-energy environments such as

the crests of bars or tidal deltas the ooids outer laminae have a tangential arrangement with tightly packed crystals. This has been interpreted by Loreau & Purser (1973) as partly reflecting the preservation potential of the aragonite crystals. Their studies indicated that the aragonite crystals actively grow on the surface of the ooid usually with a radial or random orientation, especially in the low-energy settings. However, when the ooid is reworked into more agitated environments the crystals become flattened and broken into a tangential orientation. Those crystals already present on the surface with a tangential orientation are more likely to be preserved during grain to grain collisions. These suggestions have been confirmed by Davies *et al.* (1978) in experiments in which tangential fabrics formed on ooids grown under turbulent conditions where abrasion by grain to grain contact was such that any other growth orientation was prevented. Land *et al.* (1979) noted that tangential aragonite ooids were preferentially found in areas of maximum agitation in Baffin Bay, Texas. However, Medwedeff & Wilkinson (1983) have recorded radial ooids from turbulent environments.

A variety of microstructures occur in ancient ooids (Fig. 1.2). Radial microfabrics have been the subject of considerable debate. It was once assumed that all ancient ooids were originally like the aragonitic Bahamian ooids (Fig. 3.10). The radial fabric of ancient ooids was considered to be a diagenetic feature formed during the replacement of aragonite by calcite (Shearman *et al.*, 1970). However, the very delicate radial structure is quite unlike the typical aragonite to calcite replacement fabrics seen in limestones and this has led a number of workers to interpret these fabrics as original calcites (Simone, 1974; Sandberg, 1975; Wilkinson & Landing, 1978). Such primary calcite ooids may reflect different seawater chemistries than those at present (Section 9.3).

While such radial ooids presumably represent low-energy forms, interpreting ancient micritic (random) ooids is more difficult. They may represent high-energy low magnesian calcite ooids, or possibly recrystallized radial high magnesian calcite; however, many high magnesian calcite ooids also retain a fine radial structure (Richter, 1983b). It is quite possible that ancient random fabric ooids may have been biogenically formed in low-energy environments as described by Fabricius (1977) for present-day ooids.

Ancient ooids also exhibit neomorphic spar replacements (Fig. 1.2). In such cases the calcite has either passively filled a void created by the dissolution of aragonite, or has replaced the aragonite on a fine

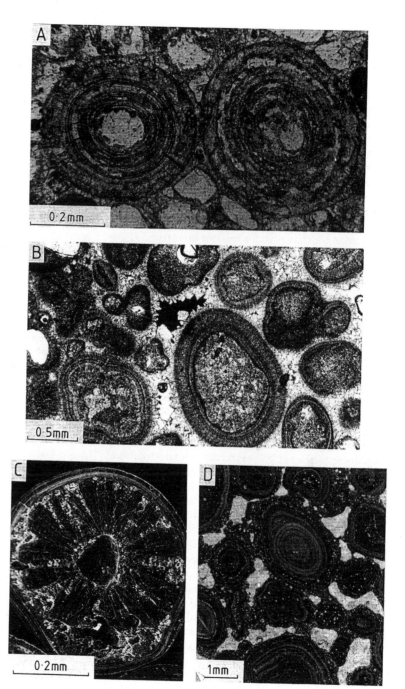

Fig. 1.3 *Types of ooid. (A) Calcitized ooids from the Miami Oolite, Pleistocene, Florida. (B) Radial concentric ooids with finely preserved radial calcite microfabric. Note dead oil-filled pore (dark). Late Precambrian. Huqf Group, central Oman. (C) Great Salt Lake aragonite ooid showing large radial 'club' structures and concentric layers in the outer part of the cortex (see Halley, 1977). (D) Vadose pisolite, Shunda Formation, Lower Carboniferous, near Calgary. The pisoids have a well-preserved radial, mainly isopachous fabric indicating a moderate degree of movement during accretion. These are coated by irregular micritic coatings showing 'bridge' features, typical of a soil profile; thus the rock shows two phases of coated grain growth.*

scale to leave ghosts of the original structure (Richter, 1983b; Tucker, 1984). These two replacement mechanisms are directly analogous to the replacement of aragonitic molluscan walls as discussed in Section 1.3.

A number of examples of ancient limestones have been described in which adjacent ooids exhibit radically different microstructures (or in which different microstructures occur within the same ooid). In such cases these variations have been interpreted as representing differences in original mineralogy (e.g. Tucker, 1984; Wilkinson *et al.*, 1984; Chow & James, 1987b; Singh, 1987).

Any interpretations of the microfabrics of ancient ooids need to consider several factors: energy level, original mineralogy, diagenesis and organic influences. Ooid microfabric has been used as a guide to the chemistry of ancient oceans (Section 9.3) and as a palaeoenvironmental tool. For example, Heller *et al.* (1980) have used changes in ooid microstructure in Cambrian ooids as a hydrodynamic indicator. They noted that ooids in the Cambrian Warrior Formation of Pennsylvania have an inner coarsely radial zone in the cortex, followed by a radial banded zone, inter-laminated with layers with a random orientation. This was interpreted as reflecting a change from suspension load (radial growth) to bed load transport (radial—concentric/banded and random) with more frequent abrasion during bed load transport. A similar sequence of microfabric changes occurs in the Great Salt Lake ooids where the larger ooids have inner cortices with a radial fabric, surrounded by a zone with denser banding with finer radial, random and tangential fabrics. Similar structures occur in the Jurassic Twin Creek ooids of Wyoming described by Medwedeff & Wilkinson (1983). Reijers & Ten Have (1983) found that ooids in the Devonian of northern Spain exhibit a radial fabric in areas where weak agitation occurred but a concentric fabric in areas where there was strong agitation. Strasser (1986) has related ooid microfabrics from the Cretaceous of Switzerland and France to both energy level and salinity. Chow & James (1987b) found physical conditions were the main control on ooid mineralogy and fabric in Cambrian ooids from Newfoundland.

Origin. Ooids and pisoids are a polygenetic group of grains. Some form by biogenic processes while others are purely chemical precipitates. However, by far the most common coated grains are the ooids found in ancient marine limestones, analogues for which are abundant in shallow tropical seas.

Our knowledge of exactly how such marine ooids form is still poor, and considerably more attention has been given of late to the significance of ooid miner-alogy than to its origin. Three sets of processes have been invoked for forming ooids: mechanical, chemical and biological. One of the earliest explanations was that given by Sorby (1879) who suggested that ooids mainly grow by mechanical accretion, the 'snow-ball' model. This seems largely untenable because the highly polished ooids of modern oolite shoals (Fig.

3.10) have no obvious means of accreting fine particles. Certainly this model cannot explain the formation of ooids with prominent radial fabrics. This is not to say that sediment accretion does not occur. Gaffey (1983) has shown that some tangential ooids do contain sediment particles which are trapped in small borings in the ooids. Mechanical accretion by microbial trap-ping and binding is an important process in the growth of oncoids (Section 1.2.1b).

Chemical mechanisms are probably important in forming at least some ooids, with the simple growth of crystals on the ooid surface. This seems the most likely mechanism for the formation of radial ooids which may be later modified by grain collisions and abrasion. Deelman (1978a) has produced Bahamian ooid-like structures in experiments and has suggested that random aragonite needle fabrics may arise by a combination of needle-breeding and collision-breeding mechanisms. These experiments took place in the total absence of any organic compounds.

There is considerable evidence that biological pro-cesses are important in forming many ooids, and this was a view widely held in the late 19th and early 20th centuries. Detailed work on Recent marine tangential ooids has shown that the aragonite rods and nanno-grains, which comprise the cortex, are identical to those associated with endolithic and epilithic algae (cyanobacteria), or with algal mucilaginous films occurring on, or in, the ooids (Fabricius, 1977; Kahle, 1981; Gaffey, 1983). This suggests that such ooids are also formed primarily by the micro-organisms, or by biologically-influenced processes associated with the organic substances. The role of such organic substances in forming ooids has long been suspected (Trichet, 1968; Mitterer, 1972; Suess & Futterer, 1972). Davies *et al.* (1978) and Ferguson *et al.* (1978) conducted a series of experiments and produced radial and tangential aragonitic and high magnesian calcite ooids. They produced *radial* ooids under 'low-energy' conditions in supersaturated seawater solutions con-taining humic acids, but no ooids were formed in the absence of these acids. *Tangential* ooids were pro-duced under agitated conditions with supersaturated seawater, but without the intervention of organic compounds. Their experiments showed that such ooids were inorganic precipitates but that crystal growth poisoning occurred due to Mg^{2+} and H^+. However, organic membranes developed on the ooids provided new, 'unpoisoned' substrates for further growth. Dahanayake *et al.* (1985) have also stressed the for-mation of coated grains by biogenic processes within microbial mats.

So while studies of marine tangential ooids indicate the direct influence of micro-organisms on formation, tangential ooids grown experimentally can form in the absence of such micro-organisms but may require the indirect influence of organic membranes for *continued* growth. Further work is needed on ooids, both by experiment and observation of actual marine ooids, but it seems likely that many are biologically formed, either directly or because of biochemical processes. This calls into question the long-held view that 'ooids' are physicochemically formed while oncoids are biologically produced (Scoffin, 1987).

Environment. Pisoids and ooids form in a wide variety of environments from shallow-marine settings to lagoons, lakes, rivers, caves and even in calcareous soils (Fig. 1.4; Flügel, 1982; Peryt, 1983a). They have no direct environmental significance beyond indicating formation in a setting where calcium carbonate was available and where the encrustation could

occur in a regular and *even* way, to form isopachous concentric layers.

For the correct interpretation of any coated grain it is essential to note the associated features, *for ooids and pisoids themselves are not diagnostic, or even characteristic, of any particular environment.* In the late 1960s and 1970s the problems over interpreting the sites of formation of ooids and pisoids became almost an obsession with some workers. Most of this interest was stimulated by the realization that some coated grains were formed in non-marine settings and were, in some cases, vadose in origin. The focus of attention was on the spectacular pisoids from the Permian Capitan Reef complex of New Mexico which had originally been regarded as algal but were later interpreted as 'vadose' in origin. Some of these coated grains show a variety of features indicating *in situ* formation unlike most pisoids. They occur associated with large tepee structures (Section 4.3.3) and with evidence of extensive early cementation. They are

Fig. 1.4 *Sites of formation and deposition of coated grains.*

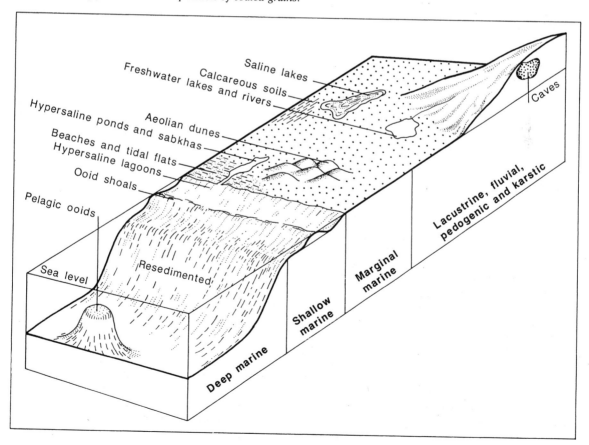

now regarded as having formed within large cavities or within depressions between the tepees and probably precipitated from saturated spring-fed marine or continental waters emerging on supratidal flats (Ferguson *et al.*, 1982; Esteban & Pray, 1983; Handford *et al.*, 1984). Such pisoids are relatively common in peritidal sequences in the geological record (Fig. 1.5; papers in Peryt, 1983a) and typically show two phases of growth. Firstly, they exhibit an inner evenly laminated cortex representing a mobile phase of growth, followed by irregular, commonly elongate downward, laminae (geopetal or gravitational fabric), polygonal fitting of laminae, perched inclusions and bridge-like structures (Figs 1.3D, 1.5). These latter features indicate *in situ* growth (Esteban & Pray, 1983). Calcrete (caliche) pisoids form largely by microbial processes (Calvet & Julia, 1983) within the soil profile. During soil creep they are rotated as they move downslope (Read, 1974a; Arakel, 1982). The microbial coatings are formed mainly of calcified microbial (fungal?) tubes and fungally-precipitated needle-fibre calcite (Fig. 1.5). The coatings on non-mobile grains are typically non-isopachous with filamentous outgrowths formed by fungal mycelia.

The vast majority of ooids in the geological record formed in very shallow (generally under 2 m deep) tropical marine settings like the oolite shoals of the Bahama Banks and Arabian Gulf (Sections 3.21.1b and 3.4.1). Their requirements are warm, shallow waters, saturated or supersaturated with respect to calcium carbonate. Both the Trucial Coast oolite shoal complex and those of the Bahamas are marginal to areas where salinity and water temperatures are high. The waters must also be agitated to form regular concentric coats, perhaps while the ooids are in suspension, and also possibly to promote precipitation by loss of CO_2. An essential factor is for the ooids to be confined in their site of formation.

Ooids may spend only a fraction of their 'lives' actually forming and may remain buried for long periods and it may take hundreds or even thousands of years for a mature ooid to form. During this time the grain must remain near the loci where precipitation occurs. Extensive oolite shoals today occur in tidally-influenced bars, channels and deltas where ooids are retained within the site of precipitation by the tidal regime. Tidal belts marginal to shallow, restricted lagoons or shelf and platform interiors, therefore,

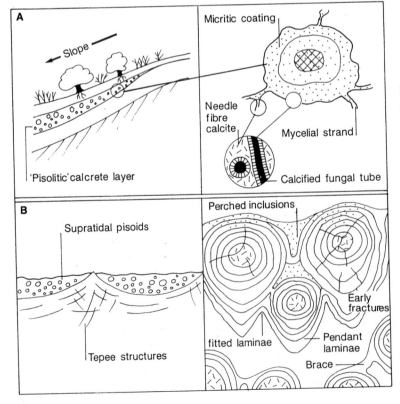

Fig. 1.5 *Vadose and calcrete-coated grains. (A) Calcrete profile with coated grain horizon related to soil creep. The coatings are commonly non-isopachous and may have calcified outgrowths of mycelial bundles. Internally the micritic laminae exhibit a variety of microbial tubes and acicular calcites (e.g. Calvet & Julia, 1983). (B) Supratidal vadose pisoids commonly formed within tepee pools (e.g. Esteban & Pray, 1983). During the early stages of growth the pisoids were free to rotate and have isopachous coatings. In the later stages they were immobile and the coatings exhibit downward (pendant) forms and are 'fitted' (laminae shared by adjacent pisoids). Other features indicating static growth include perched inclusions and braces. Desiccation, it seems, results in early fracturing.*

make ideal settings for ooid formation. However, ooids are easily transported and may constitute important components of aeolianites (Section 4.1.3c) or even deep sea resedimented deposits flanking shelves or platforms (Section 5.12).

Other types of ooid and pisoid. Besides the simple types of ooid and pisoid discussed above a variety of less common types occur. *Superficial ooids* are defined as ooids in which the thickness of the cortex is less than half the radius of the nucleus. Some workers use it for ooids with only one or two laminae in the cortex (Flügel, 1982). Ooids may be joined to form *composite ooids* surrounded by further oolitic coatings. *Cerebroid ooids* have a speckled appearance with an irregular surface with small pits which are probably the result of endolithic organisms (Richter, 1983b). *Eccentric ooids* exhibit alternating concentric oolitic laminae and irregular micritic laminae (Gasiewicz, 1984). They probably represent alternations of normal marine oolitic coatings and micritic layers associated with endolithic or epilithic microorganisms. Other forms of ooid have been recognized which reflect various degrees of destruction such as half moon ooids, broken ooids, distorted ooids and details have been given by Flügel (1982) and Richter (1983b).

1.2.1b Oncoids and micro-oncoids

These have been defined above as coated grains with a calcareous cortex of non-concentric, partially overlapping laminae. If these grains have a diameter of less than 2 mm they are called micro-oncoids, while oncoids have larger diameters. Their terms are purely generic and have no genetic connotations as regards mechanisms of formation, nor for environmental occurrence. At the most basic level the difference between ooids and pisoids, and micro-oncoids and oncoids is that the former have more continuous and regular coatings reflecting a more uniform mechanism for their formation.

By far the most common types of oncoid in the geological record are biogenically-coated grains. These may be formed by coatings of a variety of encrusting organisms such as bryozoans, corals, foraminifers, serpulids and algae, especially the crustose corallines (rhodophytes) and cyanobacteria (blue−green algae). Many biogenically-coated grains consist of several of these organisms.

Algal nodules, including those formed by cyanobacteria, are especially common in shallow-marine

limestones and several major types occur. Micritic oncoids are a prominent component of many Palaeozoic and Mesozoic limestones and typically possess either a dense micritic fabric or a rather spongy fabric, referred to as a spongiostromate fabric by Monty (1981). Other common forms exhibit fine micritic-walled tubules, generally less than 100 µm in diameter which is called a porostromate fabric (Fig. 1.6; Monty, 1981). Riding (1983) has suggested that such oncoids be called cyanoliths or skeletal oncoids. Such tubes are very similar to those found in present-day freshwater or brackish cyanobacterial bioherms or coated grains (Section 4.4.6b). However, most porostromate-coated grains in the pre-Tertiary geological record are marine in origin, but such marine occurrences are very rare in modern seas. In shallow-marine settings microbial mats do form oncoids but these are due to the trapping and binding of sediment by the finely filamentous mats rather like a rolling snowball (Gebelein, 1976; Jones & Goodbody, 1985).

Fig. 1.6 *Porostromate oncoid, Lower Carboniferous, South Wales (Wright, 1983). Porostromate tubes, probably representing calcified cyanobacterial sheaths, are very common in Palaeozoic and Mesozoic shallow-water limestones. They contribute to oncoids, but in this case the use of the term oncoid is problematic. The structure is not a 'coated grain' sensu stricto but is a radial growth of calcareous tubes. Perhaps the term cyanolith (Riding, 1983) or cyanoid is more appropriate.*

This may be the origin of some micritic and spongi-ostromate oncoids in the geological record but the porostromate forms are quite different. Micro-oncoids and oncoids have been described from some deeper-water limestones by Jenkyns (1972) and Massari (1983), and these appear to have formed by the accretion of mud and sediment grains by the snow-ball mechanism. Other micritic coatings can result from calcification of dead endolithic microbial filaments on the surfaces of substrates (Kobluk & Risk, 1977a,b).

Most, but not all, oncoids possess an identifiable nucleus which has become coated. Jones & Goodbody (1985) have described oncoids initially forming as collars around *Halimeda* plants and later becoming mobile on the death of the plant. These oncoids, at least initially, have a hollow core which probably becomes overgrown. Perhaps such hollow oncoids are indicators of macrophyte meadows and this mechanism may explain the occurrences of hollow oncoids in the geological record (e.g. Wright, 1981).

Rhodoliths (or rhodoids) are a particular type of coated grain formed by crustose or branching coralline red algae. It is based on the term rhodolite coined by Bosellini & Ginsburg (1970) for '...nodules and detached branched growths with a nodular form composed principally of coralline algae'. The term rhodolite, strictly speaking, refers to a type of garnet and is not used in this context (Bosence, 1983a). Three main structural types occur with laminar, branched and columnar forms (Bosence, 1983a). The former type represents true oncoids with typically wavy laminae but the latter two forms exhibit digitate growths and may not be oncoids *sensu stricto*. They may also lack a coated nucleus.

Environment of formation. The environmental distribution of each type of biogenic oncoid will be controlled by the tolerances of the organisms or communities involved. This is well established for rhodoliths based on living forms (Bosence, 1983b). They occur from tropical to arctic environments and inhabit a wide variety of depths. The morphology of rhodoliths is a reasonable guide to energy levels and this can be applied to fossil examples (Bosence, 1983b). However, in Palaeozoic and Mesozoic carbonates oncoids assigned to cyanobacteria are abundant. Potentially, cyanobacteria have a very wide environmental range (Monty, 1971) yet most fossil oncoids seem to have inhabited very shallow environments and are common in peritidal carbonates. Their morphology can provide a guide to the prevailing

energy levels during formation. It is commonly thought that oncoids require frequent overturning to form, but Jones & Wilkinson (1978), studying fresh-water oncoids, showed that concentric laminae can form in static oncoids with the cyanobacteria growing and calcifying on the sides, top and underside. However, such static forms show a marked asymmetry to the lamina width (Golubic & Fischer, 1975). If the laminae are regularly concentric, arguably grading to pisoids, some regular degree of turning may occur though how much is required has never been quantified.

1.2.2 Peloids

A peloid is a sand-sized grain with an average size of $100-500 \, \mu m$, composed of microcrystalline carbonate. They are generally rounded or subrounded, spherical, ellipsoidal to irregular in shape and are internally *structureless*. The term is purely descriptive and was coined by McKee & Gutschick (1969). The term pellet is also commonly used but it has connotations for a faecal origin. The term pelletoid has also been suggested (e.g. Milliman, 1974) but is not widely used.

Peloids are an important constituent of shallow-marine carbonate sediments. On the Great Bahama Banks west of Andros Island, the pellet mud facies (Section 3.2.1d) covers 10 000 km². Peloids comprise over 30% of the total sediment and 75% of the sand fraction. The low-energy, lagoonal sediments of the northern Belize Shelf and of the Trucial Coast (Section 3.4.2) are also peloidal, and in general such limestones in the geological record are typical of shallow, low-energy, restricted marine environments.

Peloids are a polygenetic group of grains (Fig. 1.7) and identifying their exact origin is often impossible in limestones. Many are faecal in origin, produced by deposit-feeding animals. Such peloids or pellets commonly are ovoid ellipsoidal in form (Fig. 3.16B), with long diameters 1.5 to 3 times their short diameter, and commonly contain high organic contents (Folk & Robles, 1964). Criteria for their recognition in limestones have been discussed by Flügel (1982), and concentrations of well-sorted peloids, especially in burrow structures, are often used as evidence of a faecal origin. Some organisms can produce very large numbers of pellets, for example the shrimp *Callianasa major* has been estimated to produce some 450 pellets per day. Such pellets, unlike most faecal pellets, have a distinctive internal structure and similar forms called *Favreina* have been recorded in Mesozoic limestones

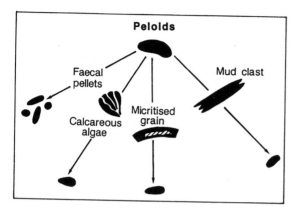

Fig. 1.7 *Origins of peloids.*

(Sellwood, 1971; Senowbari-Daryan, 1978). Faecal pellets are soft and significant compaction can occur during even very shallow burial (Ginsburg, 1957; Shinn & Robbin, 1983). Many ancient, finely mottled lime mudstones, wackestones and packstones probably owe their origin to the compaction of soft faecal pellets. The preservation of recognizable pellets in limestones is clear evidence of early lithification which is a common process in recent pellets. This lithification is poorly understood but may involve both micritization of the pellet rim or whole pellet, or more pervasive cementation (Bathurst, 1975). Lithified and soft faecal pellets have very different hydrodynamic properties (Wanless *et al.*, 1981).

Many peloids represent micritized grains such as abraded shell fragments or ooids (called bahamites by Beales, 1958) (Fig. 1.7). The original grain has been completely micritized by endolithic micro-organisms (Bathurst, 1975) and descriptions of recent peloids formed in this way are given by Kendall & Skipwith (1969), Logan (1974) and Pusey (1975). While endolithic algae play a key role in micritization, many peloids have been regarded as directly algal in origin either as detritus from fine-grained algal remains (Wolf, 1965) or by the calcification of cyanobacteria in algal mats (Friedman *et al.*, 1973). Coniglio & James (1985) have stressed the importance of calcareous algae and microproblematica in producing peloids in Early to Middle Palaeozoic limestones. Many peloids are simply sand-sized intraclasts or lithoclasts (Section 1.2.4) derived from pre-existing micrite substrates. Some peloidal textures in the geological record are purely chemical in origin and represent cements in which the pellet 'clots' are the nucleation sites of small crystals of high magnesian

calcite (Macintyre, 1985). Such pelleted sediments are commonly found associated with modern submarine cements, especially in reefs (Sections 4.5.2d and 7.4.1a). The term 'structure grumeleuse' has been used to describe limestones with a clotted, irregular microstructure in which sparite cement contains small peloids with indistinct margins. A variety of explanations have been offered to explain this distinctive microstructure (Bathurst, 1975). Similar textures occur in irregularly calcified cyanobacterial mats (Monty, 1967) and are termed 'spongiostromate fabrics' (Monty, 1981). The importance of microbial calcification and degradation in the formation of this fabric has also been stressed by Coniglio & James (1985) in a study of Early Palaeozoic limestones from Newfoundland.

1.2.3 Grain aggregates

Grain aggregates are formed when several carbonate particles become bound and cemented together. They usually range in size from 0.5 to 3 mm and have irregular shapes (Fig. 3.15). The constituent grains are generally sand-sized and heavily micritized. A variety of types have been recognized by Illing (1954) from the sediments of the Bahama Banks. *Grapestones* are aggregates of spherical grains (commonly micritized ooids) which resemble small clusters of grapes. *Lumps* are aggregates with a smoother outline which commonly have hollow interiors, and *botryoidal lumps* are grapestones or lumps with a thin oolitic coating. There is no real genetic significance between grapestones and lumps, and both can grade into large peloids or clasts depending on the degree of obliteration of internal detail.

Under the microscope the interstices between the grains consist of micritic carbonate but some organic structures may be visible. Detailed studies using scanning electron microscopes have revealed a distinctive series of growth stages leading to the formation of the aggregates (Fig. 1.8). The carbonate grains, whether peloids, ooids or bioclasts are bound by filamentous micro-organisms (cyanobacteria and algae), or by encrusting foraminifera. Cementation (encrustation) occurs on those binding filaments to create a solid structure. Further encrustation, cementation and possibly recrystallization results in a dense micritic 'cement' to the grains. The micro-organisms within the interstices are termed chasmolithic (living in holes not of their own creation), but they also bore into the grains (endolithic) and so micritize them. The formation of grapestones has been described in detail by

Gebelein (1974a), Winland & Matthews (1974), Fabricius (1977), Kobluk & Risk (1977b) and Cros (1979).

The grapestone belt on the Great Bahama Bank platform covers a very large area transitional from the agitated oolite shoals to the protected pellet mud and mud environments (Section 3.2.1c). The grapestones

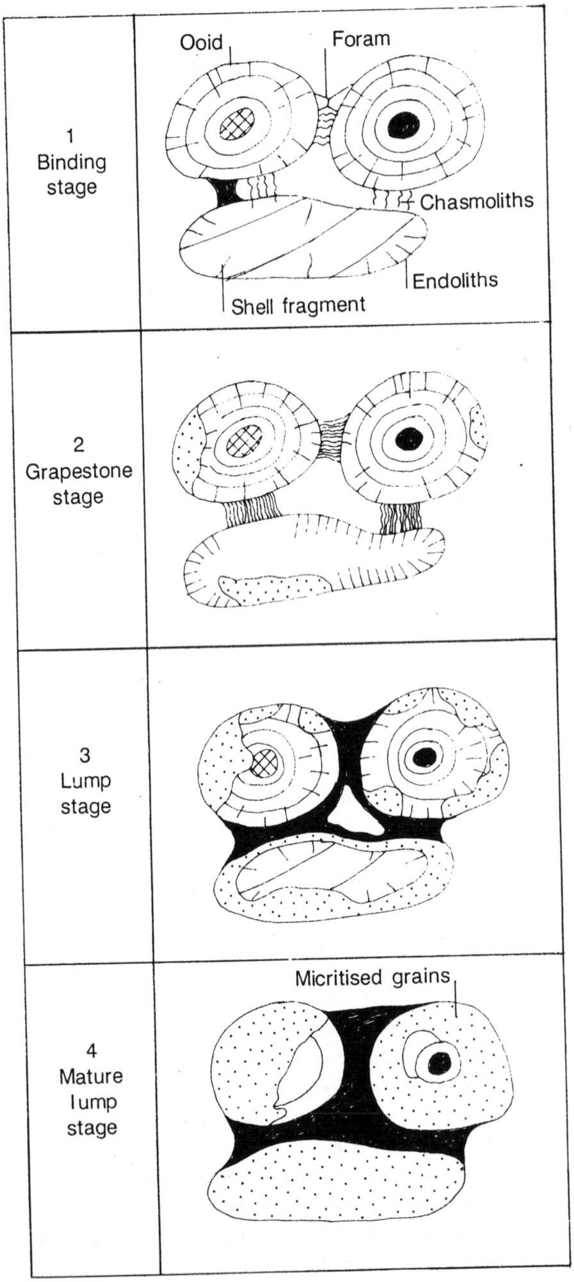

Fig. 1.8 *Formation of grapestones and lumps. Stage 1: sediment grains are bound together by foraminifers, microbial filaments and mucilage. Chasmolithic micro-organisms occur between the grains while endolithic forms bore into the carbonate substrates. Stage 2: calcification of the microbial braces occurs, typically by high magnesian calcite, to form a cemented aggregate (grapestone). Progressive micritization of the constituent grains takes place. Stage 3: increased cementation at grain contacts, by microbially-induced precipitation, fills depressions to create a smoother relief (lump stage). Stage 4: filling of any central cavity to form a dense, heavily micritized and matrix-rich aggregate. Some replacement of the HMC components by aragonite may also occur. Based on Gebelein (1974a), Winland & Matthews (1974) and Fabricius (1977).*

form in areas where waves and currents are sufficient to remove mud and silt, but not sand. These areas are generally very shallow, under 3 m deep, with a maximum depth of 10–15 m. Many of these grapestone belt areas are covered by a surface microbial mat which binds the sediment grains (Fig. 3.13; Bathurst, 1975). The mat has the property of protecting the sediment from currents and waves. Surprisingly such binding and microbial cementation can also occur in higher-energy settings and can even lead to the formation of oolitic hardgrounds (Dravis, 1979). However, grapestones form in lower-energy settings, commonly behind and protected by shoals, where the sediment is moved relatively infrequently allowing microbial binding and cementation and where sand-sized grains may be transported from the shoals during higher-energy events.

1.2.4 Clasts

A fourth category of non-skeletal grains are limestone clasts (lime-clasts). These represent reworked fragments of at least partly consolidated carbonate sediments. Two categories are widely recognized: intraclasts and lithoclasts (extraclasts). Intraclasts (Folk, 1959) are fragments of typically weakly consolidated sediment reworked from within the area of deposition. Common examples are mud flake breccias formed by desiccation and cementation (or dolomitization) on supratidal zones (Section 4.3.3d). Such breccias are also known as flat-pebble or edge-wise breccias or conglomerates. Reworking of subtidal sediments by, for example, storms can also create intraclasts and early lithification surfaces (hardgrounds) can be reworked in this way (Dravis, 1979). Early lithification can result in a variety of other intraclasts also called hiatus concretions (Brown & Farrow, 1978). Resedimentation on slopes can also create intraclasts (Section 5.10). Many limestones appear to contain clasts which are actually mottles caused by bioturbation, and differential dolomitization can also create 'pseudoclasts'.

The second category, lithoclasts or extraclasts, consists of lithologies not represented in the immediate depositional area. An obvious example would be a clast of a trilobite-bearing bioclastic grainstone in a Tertiary limestone. However, these terms rely on interpretation not simply description. If samples are very limited, such as well material, it may be impossible to judge if a clast is 'intra' or 'extra' in origin. The use of the simple descriptive term lime-clast or limestone clast is advisable.

A particularly striking type of lime-clast commonly associated with marginal marine limestones are 'black-pebbles'. These are darkened clasts, usually of lithologies represented in the associated environments (intraclasts) which have a variety of origins ranging from staining by organic solutions (in soils), blackening during fires, or pyritization. In general they are associated with subaerial exposure surfaces (see Section 4.3.3h).

1.3 SKELETAL GRAINS

There seems at first to be a bewildering variety of different types of skeletal grains in limestones. At any one time in the geological past many different types of organisms have been capable of producing calcareous skeletons. These assemblages of biogenic grains changed with time and with environment. A Jurassic shallow-water bioclastic limestone is as different from a Cambrian one as it is from a Jurassic deep-water bioclastic limestone. However, correct identification of the types of skeletal grains occurring in a limestone is critical for its environmental interpretation. In addition to all this variety, the type of preservation seen in biogenic grains varies from phylum to phylum, and within a phylum. To add to the difficulties many biogenic particles are complex three-dimensional structures and their appearance in two-dimensional thin sections can be highly variable, depending on the orientation of the section. However, help is available and there are books which assist in putting a name (and a significance) to skeletal grains (Majewske, 1969; Horowitz & Potter, 1971; Milliman, 1974; Bathurst, 1975; Scholle, 1978; Adams et al., 1984).

Skeletal grains are identified on the basis of their size and shape, microstructure and original miner-

alogy. However, any conclusions on the nature of the original microstructure and mineralogy must be based on a sound understanding of the diagenesis the fossil has subsequently undergone.

As discussed above, the inherent instability of carbonate sediments during diagenesis results from the presence of 'unstable' minerals, such as aragonite and high-Mg calcite (HMC). Much of this instability comes from skeletal grains which, in the case of shallow-marine limestones, are mainly composed of the unstable forms. Figure 1.9 shows the mineralogies of common plants and animals with calcareous skeletons, tests and thalli. The nature of the replacement mechanisms and products of the unstable skeletal mineralogies is complex but some broad points can be made at this stage, which are critical for interpreting even basic features in limestones. Aragonite shells or plant thalli are generally completely dissolved out during diagenesis and the voids filled by sparry calcite cement. Thus, in all organisms which were aragonitic in composition, there has usually been a considerable, if not complete, loss of microstructural detail.

The dissolution of skeletal aragonite is one of the most widespread diagenetic processes. Most aragonite shells undergo complete dissolution and passive mould filling by calcite. Their shell shapes are preserved in limestones because they are defined by micrite envelopes (Fig. 1.10). Skeletal debris lying on the sea floor is attacked by endolithic (boring) micro-organisms such as algae, cyanobacteria and fungi (Section 7.2). The individual borings, which are only a few microns or tens of microns long, become filled with micritic cement when vacated. This zone of borings will eventually form a rim or envelope (Figs 1.10A,B & 1.11). During diagenesis, when the aragonite comes into contact with undersaturated fluids, it is dissolved out with the micrite envelope remaining to define the original shell outline (Fig. 1.10C). The mould can later be filled with sparry calcite cement (Fig. 1.10D) unless collapse of the mould occurs. Although the micrite envelope explanation for the preservation of aragonite shells, proposed by Bathurst (1966), is widely applicable, many originally aragonitic shell fragments have extremely fine micritic dust lines defining their forms which do not appear to represent endolithic envelopes. In an alternative process, the aragonite may be replaced by low-Mg calcite (LMC) via a fine-scale dissolution−precipitation process (calcitization) with the preservation of original shell architectural structures as ghosts (Fig. 1.10E; Section 7.2). The resulting calcites are rich in inclusions of aragonite and are

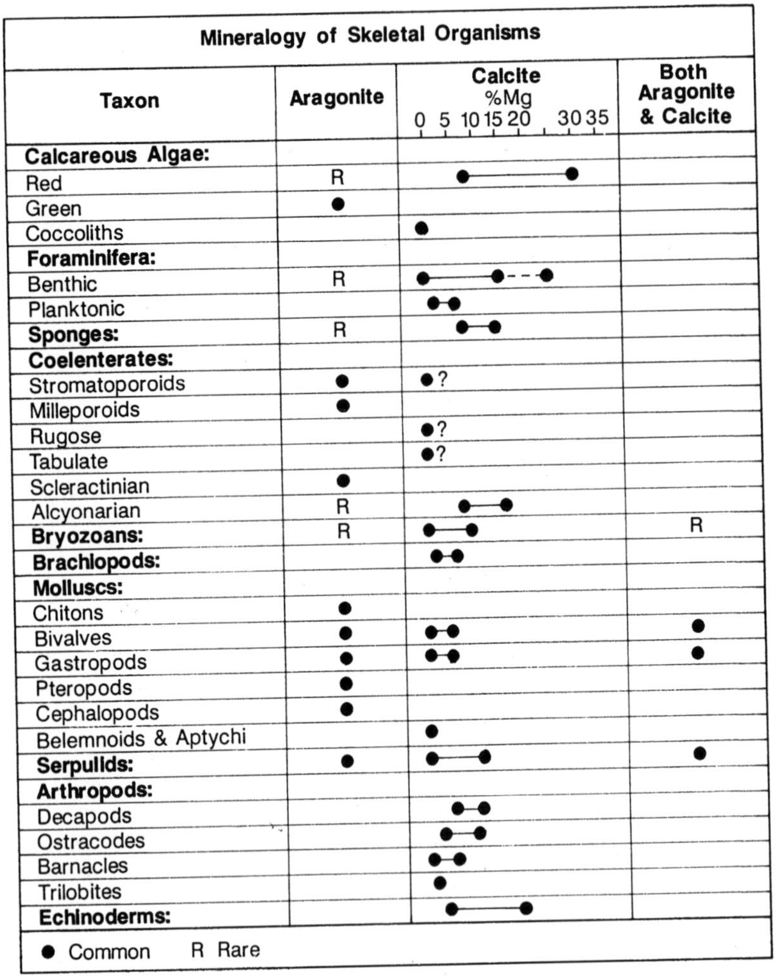

Mineralogy of Skeletal Organisms			
Taxon	Aragonite	Calcite %Mg (0 5 10 15 20 30 35)	Both Aragonite & Calcite
Calcareous Algae:			
Red	R	●———●	
Green	●		
Coccoliths		●	
Foraminifera:			
Benthic	R	●——●--●	
Planktonic		●-●	
Sponges:	R	●-●	
Coelenterates:			
Stromatoporoids	●	●?	
Milleporoids	●		
Rugose		●?	
Tabulate		●?	
Scleractinian	●		
Alcyonarian	R	●———●	
Bryozoans:	R	●—●	R
Brachiopods:		●-●	
Molluscs:			
Chitons	●		
Bivalves	●	●-●	●
Gastropods	●	●-●	●
Pteropods	●		
Cephalopods	●		
Belemnoids & Aptychi		●	
Serpulids:	●	●—●	●
Arthropods:			
Decapods		●—●	
Ostracodes		●-●	
Barnacles		●-●	
Trilobites		●	
Echinoderms:		●———●	
● Common R Rare			

Fig. 1.9 *Mineralogical composition of skeletal organisms. Modified from Scholle (1978).*

typically pseudopleochroic (Sandberg & Hudson, 1983).

Forms composed of high-Mg calcite undergo little visible alteration at the light microscope level when converted to low-Mg calcite, but using a scanning electron microscope significant ultrastructural alteration can be observed (Sandberg, 1975). However, some workers have found that little alteration may occur during this transformation (Towe & Hembleben, 1976). During the transformation of high- to low-Mg calcites, ferrous iron may be incorporated into the calcite to form ferroan calcite, and the presence of this form of iron provides an indicator of original mineralogy in extinct forms (Richter & Füchtbauer,

1978). Biogenic particles originally composed of low-Mg calcite are very stable, such as brachiopods and coccoliths.

Secular variation in skeletal composition is discussed in Section 9.4.

1.4 MATRIX

The matrix of most limestones consists of dense, fine-grained calcite crystals usually referred to as micrite. The crystal size is commonly less than 4 μm but a variety of terms have been used to describe the different crystal size populations which occur (Flügel, 1982). For most sedimentological purposes the term

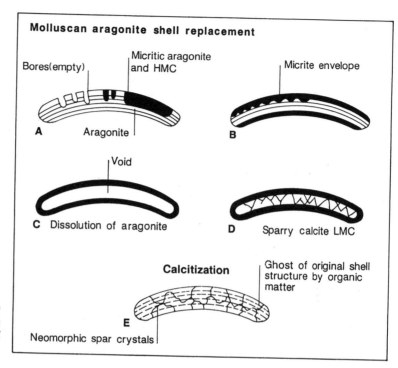

Fig. 1.10 *Molluscan aragonite shell replacement. (A−D) By a dissolution−cementation process. (E) By calcitization with the preservation of relic microarchitecture (see text and Section 7.2).*

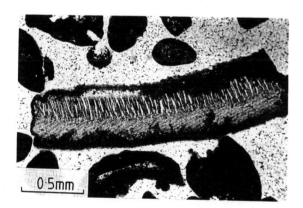

Fig. 1.11 *Micrite envelope around a bivalve fragment. The internal structure of the bivalve shell is clearly shown and is still aragonite. Modern beach sand, Trucial Coast.*

matrix is used to describe the component under 62 μm in size which corresponds to the usage of mud in descriptions of modern sediments. One of the problems in classifying micritic matrices is that micrite is not homogeneous but is poorly 'sorted' with areas of finer and coarser crystals (Flügel *et al.*, 1968; Steinen, 1978). The useful term *microspar* was introduced by Folk (1959) to describe crystal sizes between 5 and 15 μm. However, various authors have used different size definitions and others have restricted the term for recrystallized micrites. Micrite is prone to aggrading recrystallization, that is, neomorphism (Section 7.2), and studies of micrites have revealed the common occurrence of curved and sutured crystal boundaries indicating that recrystallization is a common process. It seems that most limestone matrices have undergone recrystallization and cementation, and it may be more reasonable to qualify descriptions of matrices by using a standardized crystal size classification such as one of those in Fig. 1.12. Thus micrites would be synonymous with aphanocrystalline matrices and microspar with very finely crystalline matrices, and pseudospar (commonly taken as up to 50 μm in size) would correspond to finely crystalline calcite.

Many limestone matrices were originally composed of aragonite and high-Mg calcite but were replaced by low-Mg calcite during diagenesis. It has been suggested that the original Mg content of the mud was

A		
Extremely Coarsely Crystalline	>4mm	
Very Coarsely Crystalline	1–4mm	
Coarsely Crystalline	1–250µm	
Medium Crystalline	62–250µm	
Finely Crystalline	16–62µm	
Very Finely Crystalline	16–4µm	
Aphanocrystalline or Cryptocrystalline	1–4µm	

B	
Micron–sized	0–10µm
Decimicron–sized	10–100µm
Centimicron–sized	100–1000µm
Millimetre–sized	1–10mm
Centimetre–sized	10–100mm

Fig. 1.12 *Terminology for crystal sizes in limestones and dolomites. (A) After Folk (1962). (B) After Friedman (1965).*

able to exert a control over the crystal size of the resulting low-Mg calcite. The Mg^{2+} formed 'cages' around the calcite crystals preventing them from growing beyond 2–3 µm. Removal of the cage by freshwater flushing or by the absorption of Mg^{2+} by clays (Longman, 1977) may allow further crystal growth. Interesting case studies of the recrystallization of micrites are given by Steinen (1978) and Longman & Mench (1978). Lime mud diagenesis is further discussed in Section 7.6.4.

The origin of micrites remains a major problem in carbonate sedimentology. Like other limestone components micrite is polygenetic and after diagenesis has taken its toll it is usually impossible to ascertain its origin. Some micrite is a chemical precipitate associated with high temperatures and salinities or changes in partial pressure of CO_2. This is certainly the case in many lakes (Sections 4.4.5b,c). Categorical proof for such a process in the present-day marine realm is still lacking (Ellis & Milliman, 1986). Such precipitation may be occurring in the lagoons along the Trucial Coast of the Arabian Gulf where the aragonite needle

muds contain strontium values close to the theoretical value for direct precipitation from lagoon waters (Kinsman & Holland, 1969). However, similar geochemical studies of aragonite mud in the Bahamas raised some doubts over direct inorganic precipitation (review in Bathurst, 1975). Clouds of carbonate mud known as whitings, which occur in both the Arabian Gulf and Bahama Banks, have been interpreted as due to precipitation events (Shinn et al., 1985), but some may in fact be due to shoals of bottom-feeding fish stirring up the mud. The most widely held view on the origin of the Bahamian muds (Fig. 3.18) is that they are produced by the disintegration of calcareous green algae, especially the codiaceans such as *Halimeda* and *Penicillus*. Studies of the algal production rate for the Bight of Abaco on the Little Bahama Bank revealed that more mud is produced by algal disintegration than can be accommodated on the platform top and that mud is being transported both on to the adjacent tidal belt and into deeper waters around the platform (Neumann & Land, 1975). This budget is shown diagrammatically in Fig. 1.13 (Section 3.2.1d).

Direct biogenic formation of micrite also occurs with the accumulation of skeletal fragments of calcareous plants such as coccolithophorids (Section 5.2), and similar mud-grade bioclasts occur in lakes (Section 4.4.5c). The breakdown of skeletal organisms, other than plants, is also important in forming micritic muds (Matthews, 1966; Turmel & Swanson, 1976; Bosence et al., 1985) (Section 4.5.8). Bioerosion by organisms which rasp or crunch or bore into carbonate substrates also produces carbonate muds (also see Section 4.5.2).

Biogenically-induced precipitation is a common source of micrite in lakes where algal photosynthesis is the main trigger for carbonate precipitation (Section 4.4.5c). In marginal marine and freshwater algal marshes, huge amounts of micrite may be produced by precipitation mainly by cyanobacteria (Gleason, 1975; Monty & Hardie, 1976).

In all the above cases the micrite is a true matrix, actually being deposited on a sediment surface. However, some micrites are diagenetic and represent cements (Section 7.2). Many calcretes (secondary terrestrial carbonates, mainly of soil origin) possess micrite matrices which are purely secondary, being partly cement and partly replacive of original grains (Read, 1974a; Seminiuk & Meagher, 1981). Such cements also occur in marine settings. The deeper fore-reef and slope sediments of Jamaica contain extensive micrites and pelleted micrites which are the

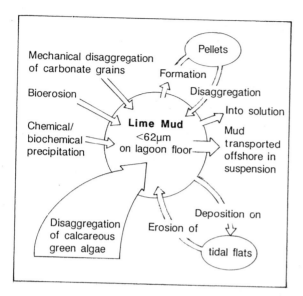

Fig. 1.13 *Lime mud budget for the Bight of Abaco, Bahamas. Based on Neumann & Land (1975) and Tucker (1981).*

products of biological erosion and micritic high-Mg calcite cementation (Land & Moore, 1980). Similar extensive micrite cements may be common in ancient reefs (Section 4.5.2d). Micrite cements occur in microborings to form micrite envelopes (Section 1.2.2; Hook *et al.*, 1984).

The term micrite should be used as a generic term for microcrystalline calcite and should not be restricted to refer to micrite-grade matrix. Milliman *et al.* (1985) have discussed the terminological problems of micrite matrix and cement.

1.5 GRANULOMETRIC AND MORPHOMETRIC PROPERTIES

The granulometric and morphometric properties of a sediment or rock, such as grain size, sorting, roundness and sphericity, are given less significance in carbonate sedimentology than they are in the study of siliciclastic deposits. It is the actual grain types which are generally more useful for environmental interpretation in limestones. These properties are still important and should be noted, but the use of such data must be made with many provisos.

Grain size measurements of carbonate sediments and limestones are usually classified using the Udden–Wentworth system, and measurements on sediments can be made using, for example, pipettes, sieves and settling chambers. In limestones the grains may be measured from thin sections or peels, but a variety of errors can occur by this method and, in addition, a conversion is needed (Blatt *et al.*, 1980; Flügel, 1982; Gutteridge, 1985; Tucker, 1988). The grain size of a rock may provide a guide to energy levels in the environment of deposition, but a number of factors must be borne in mind; for example, the types of biological destruction a grain undergoes are particularly important. Some organisms, such as fish, actually crush hard skeletons and produce carbonate mud and silt. Skeletal grains also have unique hydrodynamic properties and are commonly porous or hollow or contain organic matter (Braithwaite, 1973b). In studying any carbonate grain size population it is important to remember that calcareous organisms have different architectures and therefore disintegrate in different ways; for example, codiacean algae such as *Halimeda* decay readily to form sand- to mud-grade carbonate while associated corals such as *Acropora* predominantly form gravel- to sand-sized fragments (Folk & Robles, 1964).

Statistical analyses of grain size distributions, sorting, skewness, etc. must also be used with care in carbonates. Sorting can quickly be assessed for carbonates in thin section using a comparison chart. The degree of sorting is dependent on the transport and depositional regime and also on the nature of the source material, which in carbonates may as easily reflect the size range of local organisms as any hydrodynamic property.

Morphometric properties include grain shape and roundness and both may reflect the amount of transport and abrasion a grain has undergone. However, these properties must be viewed critically. As Folk (1962) has pointed out, when regarding grain roundness, only some bioclasts provide meaningful results, for grains such as ooids form as well-rounded, spherical grains, and faecal pellets are rounded by their passage through an organism's waste disposal system. Some bioclastic grains are, of course, already rounded and/or spherical such as some foraminifera. Furthermore, the degree of rounding shown by a biogenic grain will depend on a number of factors including the original microstructure of the particle. Pilkey *et al.* (1967) have provided a comparison chart for describing roundness of bioclasts in thin section.

When interpreting the energy levels that a limestone was deposited in using these grain properties, it is usually assumed that the sediment surface is in equilibrium with the hydrodynamic regime. This may not always be the case, for, as pointed out by Bathurst (1975), large areas of the Bahama Banks are covered in subtidal microbial mats, which have stabilized the sediment enabling it to withstand current velocities as much as five times as high as those eroding nearby sediments lacking a microbial cover.

1.6 GRAIN ORIENTATION AND PACKING

The most important factors controlling grain orientation (alignment) are the transport regime and compaction. Flake breccias (intraclast limestones) are common in peritidal sequences where elongate mud flakes become imbricated in tidal channels. In some limestones small grains are aligned in a circular pattern representing burrow linings, while in general bioturbation results in the obliteration of any original grain alignments.

Grain packing in limestones can be described using the systems applied to sandstones (Pettijohn *et al.*, 1972) although measurement errors can occur using thin sections (Harrell, 1981). Compaction can also radically alter grain packing and carbonate grains are particularly susceptible to pressure dissolution. A variety of packing types can result, as described in Section 7.6.3.

1.7 LIMESTONE CLASSIFICATION

A variety of properties are available for classifying limestones such as colour, grain or crystal size, composition and texture–fabric. Any classification may have a generic or a genetic base. A generic classification simply involves defining certain properties and allocating a name to them. A genetic classification is one in which the basis for the classification uses some fundamental property which relates directly to the origin of the item being classified. A system based on colour, which is a highly variable property, would largely be a genetic one. A system based on grain size is widely used recognizing three categories: calcilutite (grains <62 μm), calcarenite (62 μm to 2 mm) and calcirudite (>2 mm). Such a classification, though having some genetic connotations as regards the possible energy level of grain transport and deposition, is limited. The two most important features of a limestone seen in thin section are the actual grain properties (including composition) and the rock fabric, that is the relationships of the grains to one another and to any groundmass. These grain properties will include granulometrics and morphometrics, but grain composition is probably the most important property.

Since most simple petrographic analyses of lime-stones are for the purpose of environmental interpretation, the most useful classification would be one which relates grain properties and fabric to some environmental property such as energy level. In fact the classifications which are most widely used are based on the concept of textural (fabric) maturity, where the fabric is believed to relate to the energy level during the deposition of the limestone. This is the basis of the classifications given by Folk (1959, 1962), Dunham (1962), Leighton & Pendexter (1962), Bissell & Chilingar (1967) and Füchtbauer (1974). The most widely used classifications are those of Folk and Dunham. Folk (1959, 1962) recognized three main constituents in limestones: allochem (grains), matrix (micrite) and sparite (cement). He recognized four categories of allochems: peloids, ooids, bioclasts and intraclasts. Three main 'families' of limestones were recognized (Fig. 1.14): sparry allochemical limestones (allochem cemented by sparry calcite), microcrystalline allochemical limestones (allochems but with micrite matrix), and microcrystalline limestones lacking allochems or showing small sparry patches. The last actually represent either partly recrystallized or fenestral micrites (Section 4.3.3d). In addition, limestones showing coherent, *in situ* organic structures are termed biolithites.

Based on the types of allochem present eight subdivisions can be made (Fig. 1.14). Folk also provided a system for classifying the allochemical limestones based on the volumetric proportions of each allochem type. The classification included a system of naming each rock with sparites and micrites and the abbreviated allochem name. Thus such terms as pelsparite (peloidal sparite) came into use. A grain size term could be added such as a biosparrudite, to describe a coarse-grained bioclastic sparite (Folk took 1.0 mm as the lower boundary for rudites).

A further development was to subdivide the main limestone types into eight groups reflecting a 'textural spectrum' (Fig. 1.15) (Folk, 1962). This system is clearly a genetic one and its use not only involves classifying a rock, but at the same time provides an idea of the energy levels in the environment of deposition. A micrite to packed biomicrite reflects deposition in a setting where current or wave energy was insufficient to winnow away the fine matrix. The unsorted biosparites to rounded biosparites reflect an increasing energy gradient of abrasion and sorting.

However, the most widely used and simplest classification is that of Dunham (1962; Fig. 1.16). The classification is based on the rock or sediment fabric, and the presence of any biological binding. The three

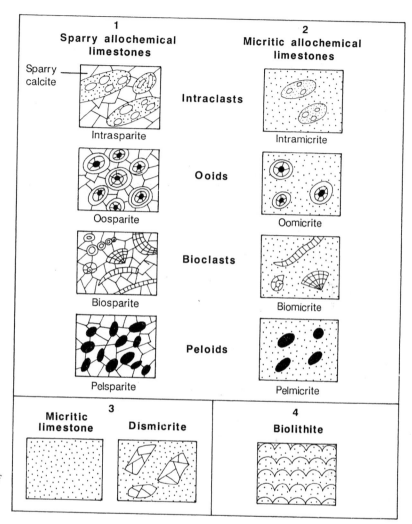

Fig. 1.14 *Basic limestone types of Folk's (1959) classification. Four groups are recognized (see text).*

main divisions are between limestones which are matrix-supported (lime mudstones and wackestones, micrite taken as <20 μm by Dunham), grain-supported (packstones and grainstones) and biologically bound (boundstones). A fourth category, crystalline limestones, is also recognized. Figure 1.17 shows such a selection of limestone types. This classification was extended by Embry & Klovan (1971) to include the types of rocks encountered in reef deposits (Fig. 4.94).

In Dunham's classification the significance of each rock type as regards energy level is relatively clear. However, unlike siliciclastic rocks where all the grains, and most of the matrix, are usually allochthonous, in limestones both the matrix and the grains can be locally produced. A lime mudstone or wackestone with large skeletal fragments is not a contradiction in terms of energy level, but reflects mud accumulation and the original presence of a benthos. The limestone fabrics reflect the interaction of hydraulic processes and biological production. Thus a lime mudstone reflects deposition in a low-energy setting where grain production, especially by organisms, was reduced, perhaps because of restricted conditions. Not all sediments are in equilibrium with their environment of deposition and benthic microbial mats are capable of

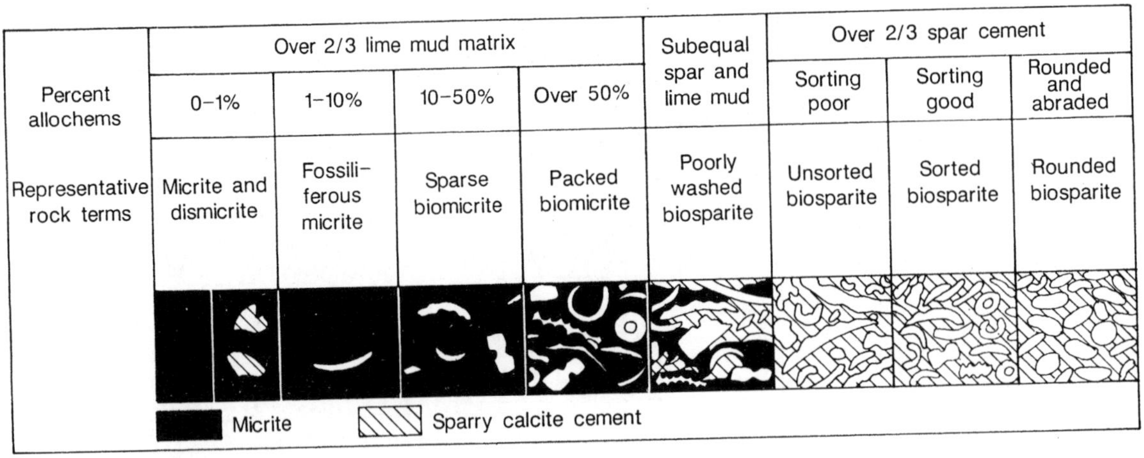

Percent allochems	Over 2/3 lime mud matrix				Subequal spar and lime mud	Over 2/3 spar cement		
	0–1%	1–10%	10–50%	Over 50%		Sorting poor	Sorting good	Rounded and abraded
Representative rock terms	Micrite and dismicrite	Fossili-ferous micrite	Sparse biomicrite	Packed biomicrite	Poorly washed biosparite	Unsorted biosparite	Sorted biosparite	Rounded biosparite

Micrite   Sparry calcite cement

Fig. 1.15 *Textural maturity classification of Folk (1962).*

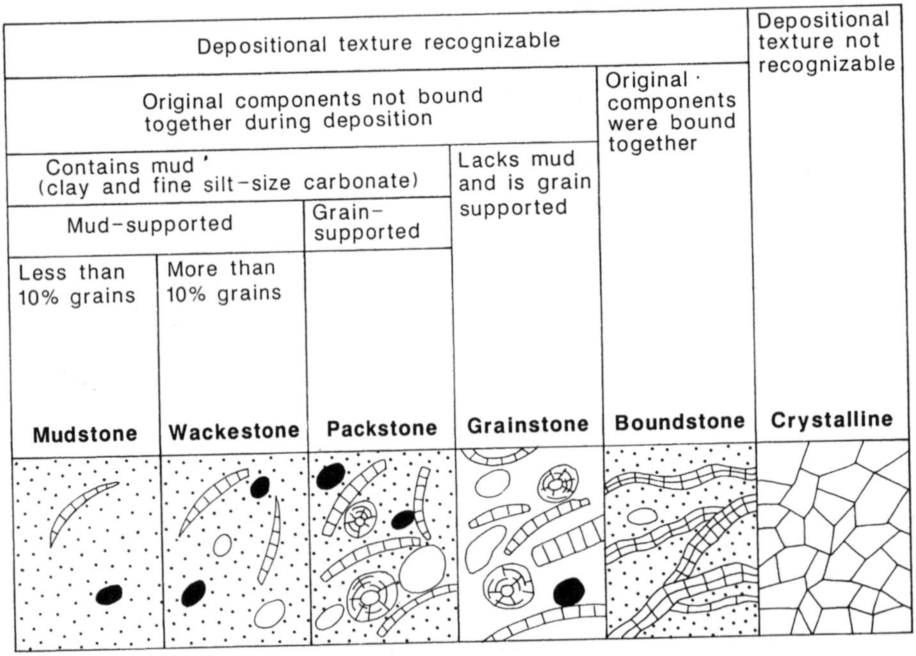

Depositional texture recognizable					Depositional texture not recognizable
Original components not bound together during deposition				Original components were bound together	
Contains mud ' (clay and fine silt–size carbonate)			Lacks mud and is grain supported		
Mud-supported		Grain-supported			
Less than 10% grains	More than 10% grains				
Mudstone	**Wackestone**	**Packstone**	**Grainstone**	**Boundstone**	**Crystalline**

Fig. 1.16 *Dunham's (1962) classification of limestones with schematic diagrams of each rock type.*

stabilizing grain populations which would otherwise be transported out of that site (Section 3.2.1c).

In describing a rock type using these classification systems grain type terms can also be used; for example, a bioclastic grainstone is the equivalent of a biosparite in Folk's classification. However, strict grain class types do not exist in the Dunham classification.

Recently, Smosna (1987) has introduced another concept to limestone classification, compositional maturity. This is defined as 'the extent to which a sediment approaches a constituent end member (intraclasts, ooids, fossils, peloids, micrite matrix and terrigenous materials) to which it is driven by the environmental processes operating upon it'. Smosna

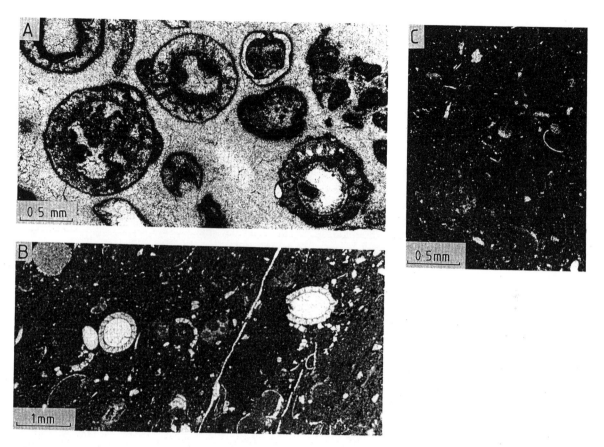

Fig. 1.17 *Selection of Dunham limestone types. (A)* Mizzia *(dasycladacean alga) grainstone, Permian, Dark Canyon, Carlsbad, New Mexico. The grains are separated by cement. (B) Charophyte wackestone. Gyrogonites and molluscan fragments are matrix supported. Upper Jurassic 'Cabacos Beds', Cabo Mondego, Portugal. (C) Lime mudstone with ostracode fragments, Lower Carboniferous, Avon Gorge, England.*

envisaged compositionally immature (varied) lime-stones as being produced in environments where many biological, physical and chemical processes operate simultaneously. As the number of operating processes decreases so the sediment will progress to an advanced stage.

The Dunham classification, like that of Folk, pro-vides a clue to the energy levels during deposition; however, there are both problems in recognizing some categories and also in their interpretation. One of the most common problems in first studying limestones is that in thin section many grains are not in contact but appear to float. This may occur in a grainstone where no supporting matrix is present. Such a property makes distinguishing a wackestone from a packstone difficult. This problem is an artefact of the two-dimensional view of the three-dimensional rock fabric (Dunham, 1962). Grains in a grain-supported fabric are not in contact along every surface and sections through such a fabric will show some grains apparently floating in matrix, or cement, or pore space. The shape of the grains is very important and highly irregular shapes may form a self-supporting frame-work with only 20–30% of the volume actually being *grain*, with such grains having apparently few points of contact. This three-dimensional aspect should be borne in mind in all carbonate petrography. True grain-supported rocks have a packing index (Sander, 1951) of less than 2, and usually around 1.2 (Flügel, 1982), but care needs to be taken in assessing this (Harrell, 1981).

The origins of lime mudstone, wackestone and

grainstone are clear enough as regards energy level. The first two types are deposited in low-energy settings where mud matrix is not winnowed away. But what about packstones? These are a very common rock type, yet if the fabric is grain supported, how is the mud introduced? Some packstones certainly have had mud washed in, for geopetal sediments occur; many calcretes are diagenetic packstones because micrite has been precipitated within intergranular pores. However, most muddy limestones have undergone compaction caused by dewatering. In this way many sediments deposited as wackestones become converted to packstones (Shinn & Robbin, 1983). During pressure dissolution many uncemented grains become welded together to form fitted and condensed fabrics (Section 7.6.3) which, lacking clear sparry cement, are commonly classified as packstones even though they lack any matrix.

These features indicate the failing in the existing classification systems which were designed to describe primary, depositional fabrics. Since these systems were introduced, carbonate sedimentologists have begun to appreciate how susceptible carbonate sediments and limestones are to diagenesis. It is now known that the matrix in limestone is prone to recrystallize to more coarsely crystalline calcite (sparite), and that micrite can be matrix or cement in origin (Section 1.4). Processes such as compaction and pressure dissolution can radically alter the primary rock fabric (Section 7.6.3). Furthermore, many fabrics occur in limestones which are *not* covered in either classification.

In limestones, the final fabric has both a depositional component and one relating to its subsequent diagenesis. It seems unlikely that any convenient classification system can weld together both these aspects, so any description of a limestone must be binary, by being a name which describes its fabric but with diagenetic provisos. Perhaps the most commonly seen example is due to bioturbation, for many limestones display mottling with complex mixtures of various packing arrangements caused by burrowing and sediment ingestion by the infauna.

Limestones may also be impure, with admixtures of siliciclastics. Such rocks present further problems in classification and a new scheme has been proposed by Mount (1985).

1.8 DATA COLLECTION AND PROCESSING

As described in earlier sections, a variety of compari-

son charts are available for estimating sorting, roundness and sphericity in limestones, but as discussed earlier it is the composition of the grains which is probably the most useful single property of a limestone. Calculating the percentages of grains to matrix, or the porosity, or the relative percentages of various grain types, is most reliably done in thin section by point counting. This technique is reviewed by Flügel (1982). A semi-quantitative method is to use comparison charts for visual estimates. A variety of charts are available which include different grain shapes (Emrich & Wobber, 1963; Bacelle & Bosellini, 1965; Schafer, 1969).

Having collected data on the limestone composition it may then be processed into statistically meaningful groups by such techniques as cluster and multivariate analysis. A relatively simple technique for testing the changes in different grain populations, optimized similarity matrices, has been introduced by Hennebert & Lees (1985). By such methods lithofacies types may be defined quantitatively, which are at least reasonably objective. A brief review of these techniques as applied to carbonates is given by Flügel (1982), and Smosna & Warshauer (1979) provided an interesting case study.

However, most facies are defined and interpreted on a purely qualitative basis, usually on the presence of one or more significant features. Interpretation is usually a very subjective exercise, with the interpreter selecting important features and giving less 'weight' to unimportant ones. A correct interpretation needs to take into account rock texture, grain types, fossil biota, sedimentary structures, diagenetic features and also the sedimentary associations of the lithofacies in question. A 'short-hand' method of studying limestones is to compare them with standard microfacies (SMF) types. This is a system devised by Wilson (1975) from a concept of Flügel's (1972), which recognizes 24 standard microfacies assignable to nine standard facies belts (Table 1.1). Each facies has a distinctive composition which should be diagnostic for a particular environment. This is often the case but there are many exceptions and some SMF types can occur in more than one facies belt (Wilson, 1975; Flügel, 1982). In studies where outcrops are available, the microfacies approach can be checked against outcrop data, but if material is limited, such as in well cuttings, solely relying on a microfacies approach can lead to significant errors. Two cases from the author's experience may illustrate this problem. A wildcat well drilled in western Portugal in the late 1970s encountered over 250 m of oolitic and peloidal grainstones

Table 1.1 Standard microfacies types based on Wilson (1975) and Flügel (1972, 1982)

SMF type	Lithology	Environment
1	Spiculite. Dark clayey mudstone or wackestone, organic rich, or siliceous spiculitic calcisiltite. Spicules usually oriented, generally siliceous monaxons, commonly replaced by calcite	Basinal, deep-water environment with slow sedimentation
2	Microbioclastic calcisiltite. Small bioclasts and peloids in very fine-grained grainstone or packstone; small-scale ripple cross-lamination	Open sea shelf near the lower slope, deeper shelf margin
3	Pelagic mudstone and wackestone. Micritic matrix containing scattered pelagic microfossils (e.g. radiolarians or globigerinids) or megafauna (e.g. graptolites or thin-shelled bivalve fragments)	As 1 and 2
4	Microbreccia or bioclastic–lithoclastic packstone. Rounded grains, often graded. Polymict or monomict in origin. Also quartz, cherts, and carbonate detritus	Fore-slope talus; resedimented limestones
5	Grainstone–packstone or floatstone with bioclasts of reef. Geopetal sediments	Reef flank facies
6	Reef rudstone; no matrix material	Fore-reef slope, debris from the reef; commonly in high-energy zone
7	Boundstone. Subtypes of framestone, bindstone, or bafflestone	Reef, often found on platform margins
8	Wackestone with whole organisms. Well-preserved infauna and epifauna	Shelf lagoon with circulation; low-energy water below normal wave-base.
9	Bioclastic wackestone or bioclastic micrite. Fragments of diverse organisms, bioturbated. Bioclasts may be micritized	Shallow waters with open circulation close to wave-base
10	Packstone–wackestone with coated and abraded bioclasts	Textural inversion; dominant particles from high-energy environment have moved down local slopes to low-energy settings
11	Grainstones with coated bioclasts, in sparry cement	Winnowed platform edge sands; areas with constant wave action at or above wave-base

Table 1.1 (cont'd)

SMF type	Lithology	Environment
12	Coquina, bioclastic packstone, grainstone or rudstone with concentrations of organisms. Certain types of organisms dominate (e.g. dasyclads, shells, or crinoids)	Slopes and shelf edges
13	Oncoid (biosparite) grainstone	Moderately high-energy areas, very shallow water
14	Lags. Coated and rounded particles, in places mixed with ooids and peloids. May be blackened and iron-stained with phosphate; lithoclasts; usually thin beds	Slow accumulation of coarse material in zone of winnowing
15	Oolites of well-sorted ooids; fabric usually overpacked; always cross-bedded	High-energy environment on oolite shoals, beaches, and tidal bars
16	Grainstone with peloids. Probably faecal pellets, in places admixed with concentrated ostracode tests or foraminifera	Shallow water with only moderate water circulation
17	Grapestone, pelsparite or grainstone with aggregate grains, isolated and agglutinated peloids, some coated grains	Shelf with restricted water circulation and tidal flats
18	Foraminiferal or dasycladacean grainstones	Tidal bars and channels of lagoons
19	Fenestral, laminated mudstone–wackestone, grading occasionally into pelsparite with fenestral fabrics. Ostracodes and peloids, sporadic foraminifera, gastropods and algae	Restricted bays and ponds
20	Microbial stromatolite mudstone	Commonest in the intertidal zone
21	Spongiostrome mudstone. Convolute microbial fabric in fine-grained micrite lime mud sediment	Tidal ponds
22	Micrite with large oncoids, wackestone or floatstone	Low-energy environments, shallow water, back-reef; often on the edges of ponds or channels

Table 1.1 (cont'd)

SMF type	Lithology	Environment
23	Unlaminated, homogeneous unfossiliferous pure micrite; evaporitic minerals may occur	Hypersaline tidal ponds
24	Rudstone or floatstone with coarse lithoclasts and bioclasts. Clasts usually consist of unfossiliferous micrite; may be imbricated and cross-bedded; matrix sparse	Lag deposit in tidal channels ('intraformaional breccia')

corresponding to SMF types 15 and 16. The well cuttings were interpreted as a thick shelf limestone sequence of oolite shoal and back shoal associations. However, on closer examination, it was found that mixed in with these cuttings were small numbers of cuttings of SMF type 3, argillaceous lime mudstones with thin-shelled pelagic bivalves (*Bositra*). At outcrop the explanation was simple, the grainstones were turbidites separated by hemipelagic lime mudstones. The thick sequence of grainstones actually represented a deep-water, prograding, carbonate submarine fan (Section 5.13.2 and Fig. 5.57), i.e. resedimented shelf deposits. In another example, a set of well cuttings, this time from the Upper Palaeozoic of Canada, consisted of argillaceous lime mudstones with sponge spicules and chert fragments. This lithofacies corresponds to SMF type 1, a basinal limestone, and the well interval was interpreted as such. However, some doubts developed and, on further examination of associated samples, calcrete textures and stromatolitic fragments were found. In fact, the spicule-bearing limestones were probably of lagoonal origin, while the chert turned out to contain minute anhydrite laths. The interval was reinterpreted as a peritidal deposit. This is not to say that the use of SMF types is not a very useful technique, but there will be a tendency to fit the data to the model, instead of the other way round.

1.9 POROSITY

As interesting as grains, matrix and cements are, it is the holes in limestones which make them so important in hydrocarbon exploration. The porosity of a rock is the ratio of the total pore space to the total volume of the rock, and this is usually given as a percentage. The importance of a carbonate reservoir really depends more on its permeability, which controls the recovery of hydrocarbons, than its simple porosity. Some rocks are porous but have low permeabilities and so it is the *effective* porosity which is important.

Various techniques can be used to estimate porosity in limestones, but a commonly used technique involves point counting. However, estimates of pore volume by this technique are subject to error (Halley, 1978).

Porosity in limestones is rather different from that in sandstones. It is much more erratic in type and distribution within a reservoir and is generally much lower than in sandstone reservoirs (Choquette & Pray, 1970). Carbonate reservoirs with as low as 5–10% porosity are known, while most sandstone reservoirs have values of 15–30%. Most porosity in sandstone reservoirs is primary, depositional, interparticle porosity, although the importance of diagenetic (secondary) porosity is now better appreciated. Broadly speaking, most porosity in limestones is diagenetic in origin, and as a result it is more difficult to predict carbonate reservoir quality, which will be controlled by the original facies types and later diagenetic processes. Reviews of carbonate porosity have been given by Longman (1981) and Moore (1979, 1980).

A number of classifications of porosity are available (reviewed in Flügel, 1982) but the most widely used is that by Choquette & Pray (1970) (Fig. 1.18). This classification has four elements: basic porosity types, and three sets of modifying terms covering: (1) time, origin of the porosity. (2) the pore size and shape, and (3) abundance.

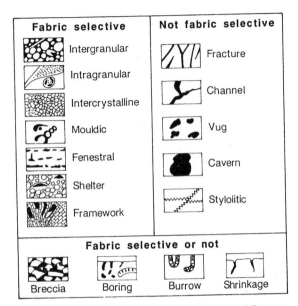

Fabric selective	Not fabric selective
Intergranular	Fracture
Intragranular	Channel
Intercrystalline	Vug
Mouldic	Cavern
Fenestral	Stylolitic
Shelter	
Framework	

Fabric selective or not
Breccia Boring Burrow Shrinkage

Fig. 1.18 *Porosity types. Based on Choquette and Pray (1970).*

1.9.1 Porosity types

There are a variety of porosity types in limestones and Choquette & Pray have divided them into three groups (Fig. 1.18). The fabric-selective types have their pores defined by fabric elements of the rock, such as grains or crystals. The non-fabric-selective porosity types cross-cut the actual fabric of the rock, for example fracture porosity. The third group may display a fabric control or not. A review of porosity types is given by Longman (1982).

Intergranular (interparticle) porosity. This is the original, primary depositional porosity of the sediment, and the types of packing fabric will be important in controlling the types of pore space found. Intergranular porosity also occurs in some very fine-grained sediments and porosities of as high as 40% have been found in ancient chalks (Section 5.7). Not all intergranular porosity is strictly primary in origin for dissolution of matrix, or even cement, may have occurred creating secondary intergranular porosity.

Intragranular porosity. This is porosity within the grains, especially in skeletal material. Such porosity is commonly very localized and its effectiveness will depend on the micropermeability of the grain, and on the overall fabric of the rock. A striking example is

rudists, which are a group of extinct, aberrant large bivalves of Mesozoic age. Some families had complex, highly porous skeletons. Commonly, rudist buildups developed in low-energy, below wave-base settings where the porous rudists became isolated in a fine-grained matrix. However, rudist grainstones display both high intergranular and intragranular porosity and many reservoirs in rudist limestones are found in such grainstone facies not in the actual rudist buildups.

Intercrystalline porosity. This is porosity between crystals. This most commonly occurs in replacive dolomites and represents a secondary porosity (Wardlaw, 1979). It also occurs in evaporite deposits and in recrystallized limestones.

Mouldic porosity. The bulk of shallow-water carbonate grains in recent sediments consists of aragonite and high magnesian calcite. Aragonite is highly susceptible to dissolution in undersaturated waters and, as a result, highly fabric-selective porosity can result. Similar aragonitic mouldic porosity is abundant in carbonate reservoirs. The original depositional fabric is very important for if mouldic porosity occurs in a matrix-supported rock such as a wackestone, the mouldic pores may become isolated as the matrix compacts and cements. However, in a grain-supported rock, high porosities and permeabilities can occur. Aragonitic dissolution of bioclasts and ooids creates *biomoulds* and *oomoulds* respectively. Evaporite minerals are also prone to dissolution and during dedolomitization dolomoulds can form.

Fenestral porosity. Fenestrae are small pores which are common in peritidal carbonates and typically form because of desiccation and gas generation (Section 4.3.3d). While local porosities may be high, fenestrae form in a very narrow range of environments and occur in thin, discontinuous horizons. They are prone to early cementation and often contain geopetal sediments. They are, at best, only a minor porosity type complementing the more typical porosity types in peritidal sequences, such as intercrystalline and mouldic porosity.

Shelter porosity. Shelter pores, also called umbrella pores, are cavities formed beneath larger particles such as convex-up shells. This is a minor porosity type but may complement other porosities.

Growth or framework porosity. This type is created by the skeletal growth of frame-building organisms such as corals, stromatoporoids or calcareous algae. The varied branching and foliose growth of *in situ* reef organisms can create numerous cavities (Section 4.5.2a). This type is often quoted as important in reef reservoirs, and is certainly conspicuous in present-day reefs, but as pointed out by Longman (1982) it is rarely preserved in ancient reefs because of high sedimentation rates around reefs and early cementation.

Fracture porosity. Fracture porosity, like channel, vuggy and cavern porosity, is non-fabric selective, and cuts across the fabric elements of the rock. Fractures commonly result from tectonic deformation, slumping or solution collapse associated with evaporites or limestone dissolution. Fracture porosity is very common and can greatly increase the effective permeability of a limestone by many times (Watts, 1983).

Channel porosity. Limestones are prone to dissolution in undersaturated waters, and a common product is channel porosity, commonly developed along fractures. Near-surface karstic solution pipes are a common example in the geological record. By definition a channel is an elongate pore with a length of ten times its diameter.

Vuggy porosity. Vugs are pores with a diameter greater than 1/16 mm, and so are just visible to the naked eye. They are roughly equant in shape. The term *pin-point* porosity is sometimes used for microvuggy porosity, especially in dolomites, which may encompass forms of intercrystalline porosity (Longman, 1982). Most vugs represent the solutional enhancement (enlargement) of intergranular or intercrystalline pores and this may involve the dissolution of early cements. Some vuggy porosity is thought to be created by CO_2-rich groundwaters generated during hydrocarbon maturation. The migration of these fluids creates porosity ahead of migrating hydrocarbons.

Cavern porosity. Cavern pores are person-sized pores or larger, of channel or vug shape. They are mainly solutional in origin and are associated with palaeokarstic processes (Section 7.5).

The remaining pore types may be fabric selective or not.

Breccia porosity. This is a continuation of fracture porosity where the fragments have their own interparticle porosity, and like fracture porosity they may have a tectonic or dissolutional origin. Examples of breccia porosity reservoirs occur in many salt basin sequences such as the Permian Auk Field of the North Sea (Brennand & van Veen, 1975). Recently Ijirigho & Schreiber (1988) have offered a classification for both fracture and breccia porosity.

Minor types include boring and burrow porosity resulting from biological activity and shrinkage porosity resulting from desiccation.

An additional type not included by Choquette & Pray is stylolitic porosity. While in many limestones stylolites represent zones of very low or zero porosity, they can act both as porosity and as important conduits for fluid migration (Longman, 1982).

The second part of the Choquette & Pray system involves the use of modifying terms to describe the processes affecting the pore (dissolution, cementation or internal sedimentation), the stage that process is in (enlarging, reducing or filling the pore) and the timing (primary or secondary). A series of terms is also used for describing the environment of formation of secondary porosity. '*Eogenetic*' describes diagenesis which occurred during the time interval between final deposition and burial of the newly deposited sediment below the depth of significant influence of processes operating at, or dependent upon proximity to, the surface. Meteoric diagenesis is an example. '*Mesogenetic*' diagenetic processes take place in the time interval in which sediments or rocks are buried at depths below the major influences related to the surface. Mesogenetic diagenesis is also called burial diagenesis. '*Telogenetic*' diagenetic processes take place during the time interval when sediments or rocks, long buried, are affected by processes associated with the formation of an unconformity. These terms are not widely used.

Other modifiers are available to cover the size and abundance of the pores and Choquette & Pray also provided a standardized system for describing porosity.

The origins of different porosity types are also discussed in Chapter 7, and are briefly reviewed by Longman (1982). The descriptive system of Choquette & Pray only deals with the types of pores, but it is the nature of the pore system which influences the recovery efficiency of a carbonate reservoir. Basic work on this subject is to be found in Wardlaw & Cassan (1978) and Wardlaw (1979, 1980).

2 Geological background to carbonate sedimentation

2.1 INTRODUCTION

This chapter discusses the geological background to carbonate environments and facies. There are many factors which determine whether carbonates will be deposited, and their facies and sequences, but the two overriding controls are geotectonics and climate. Much of our understanding of limestones has come from studies of modern carbonates, and the sediments and facies of three classic areas, the Bahama Platform, the Florida Shelf, and Trucial Coast are reviewed in Chapter 3. However, modern carbonates do not provide all the answers and much information has to come from the limestones themselves. There are three considerations which have to be borne in mind when interpreting ancient limestones from modern carbonates:

1 As a consequence of the Pleistocene glaciation and associated major fluctuations in sea-level over the last one million years or so, in most areas where carbonates are forming today, sedimentation only began 4000–5000 years ago. Thus relic topographies exert a strong control on sedimentation and in some areas (mostly low-latitude, deeper-water, mid to outer continental shelves) relic carbonate sediments abound. A steady-state situation with an equilibrium between sediments and environments has not been attained in many locations. Modern carbonate sediment thicknesses are mostly only a few metres, although in areas of active reef growth, such as along the Florida Reef Tract, up to 14 m have accumulated since around 5000–6000 years BP.

2 Sea-level at the present time is relatively low compared with much of the geological record. This means that shallow-water carbonate environments are not as widespread today as they were at certain times during the Phanerozoic. In particular, there are now no extensive, low-latitude shallow seas (epeiric or epicontinental seas) where carbonates are accumulating, comparable with instances in the past of large platforms being covered by knee-deep marine waters.

3 Modern carbonate sediments are almost entirely produced by biogenic processes, apart from ooids and some lime muds. The organism types contributing their skeletons to limestones have varied drastically throughout the Phanerozoic due to changing fortunes: the evolution of new groups and demise of others. The roles played by organisms have also changed through time; this is particularly important when considering reef limestones. In addition, the dominant mineralogical composition of organism skeletons and of inorganic $CaCO_3$ precipitates has varied through time, in response to fluctuations in seawater–atmosphere chemistry (see Chapter 9).

As a result of the above three points, limestones cannot be interpreted entirely from studies of modern carbonate sedimentary processes and products; essential information has to come from the rock record. In addition, as alluded to above, consideration has to be given to the evolutionary pattern of carbonate-secreting organisms through time, and of changes in seawater chemistry, which affect marine precipitates and patterns of early diagenesis.

2.2 MAJOR CONTROLS ON CARBONATE SEDIMENTATION

Although there are many factors which determine the nature of a carbonate formation, there are two overriding controls on carbonate sedimentation: (1) geotectonics, and (2) climate, and these together control the other important variable, sea-level. The geotectonic context is of paramount importance. It controls one of the prime requisites for carbonate sedimentation, the lack of siliciclastic material, by determining hinterland topography and river drainage. Apart from simply diluting the carbonate component of a sediment, terrigenous material also has a detrimental effect on carbonate production, particularly where coral reefs are concerned. Shallow-water carbonates are extensively developed on the Bahama Platform, since the deep Florida Straits effectively prevent terrigenous mud from reaching the platform. Carbonate sedimentation is inhibited around much of the northern and western shelves of the Gulf of Mexico by mud emanating from the Mississippi River and being transported around the coast by longshore drift. In the Far East, the equatorial Sunda Shelf off Indonesia is blanketed by river-borne mud and coral reef development is restricted to distant shelf margins. Along the Queensland coast of Australia, clastic sediments are trapped on the inner shelf in a major depression, permitting the luxuriant growth of the Great Barrier Reef along the mud-free shelf edge. In the Red Sea, reefs occur in close proximity to clastic sediments derived from coastal alluvial fans and fan deltas. However, for the most part, the terrigenous debris is relatively coarse and so quickly sedimented. In the extensional regime of the Red Sea, narrow graben structures also trap clastics (Purser *et al.*, 1987).

Geotectonics determine the depositional setting of carbonate sedimentation and as discussed in a later section (2.4), five major categories of carbonate platform are recognized: shelf, ramp, epeiric platform, isolated platform and drowned platform (Fig. 2.1). Each has a distinctive pattern of carbonate facies, generalized facies models can be constructed and various types of each platform category can be distinguished. The settings can be modified and changed during sedimentation by tectonic effects, sea-level changes or carbonate sedimentation alone.

Global sea-level stand is a major factor in carbonate sedimentation, with more widespread and thicker sequences being deposited at times of high stand (see Fig. 9.1A). The broad position of sea-level is determined by ocean-basin volumes and glacial ice volumes, both controlled by geotectonics and climate. Rises and falls in sea-level greatly affect carbonate sedimentation, and cycles of sea-level change are recognized on five orders of magnitude, i.e. a first order cycle of hundreds of millions of years (10^8 years) down to a fifth order cycle of tens of thousands of years (10^4 years). There are two first order cycles through the Phanerozoic (see Fig. 9.1B), the result of supercontinent fragmentation and construction, and opening and closing of major oceans, and the second order cycles (10^7 years duration) are largely the result of passive margin subsidence. Third order cycles of sea-level rise and fall (10^6 years) are responsible for whole carbonate formations (i.e. carbonate platforms), but the origin of these cycles is a matter of much discussion. They are attributed to global eustatic sea-level changes in the Vail *et al.* (1977) and Haq *et al.* (1987) schemes, but another school of thought relates the megasequences deposited during third order scale sea-level changes to tectonic extension and subsequent thermal subsidence (e.g.

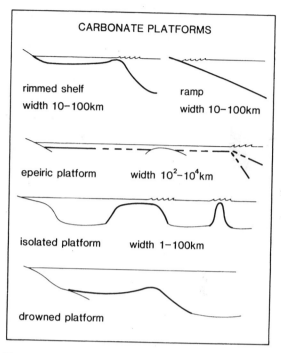

Fig. 2.1 *Carbonate platforms, sketches of the main categories.*

Hubbard, 1988). Fourth and fifth order cycles (10^5–10^4 years) control the development and repetition of metre-scale shallowing-up limestone units, which are a feature of many carbonate platform sequences. The 10^4–10^5 year sea-level changes *may* be the result of orbital forcing, i.e. an astronomic control on sea-level via polar ice-caps due to the Milankovitch rhythms (see Section 2.10). The latter also control the rhythmic layering of deeper-water, pelagic facies (see Section 5.7.1).

In the concepts of *sequence stratigraphy*, derived largely from the work of seismic stratigraphers (papers in Wilgus *et al.*, 1988), the major control on deposition is the relative changes of sea-level, and this is determined by eustatic sea-level changes and tectonic subsidence (see Fig. 2.2). Particular *depositional systems tracts* are developed during specific time intervals of the sea-level curve (third order) (Fig. 2.2). Thick carbonate sequences are deposited principally in the highstand systems tract, but also in the transgressive systems tract (Sarg, 1988). Two types of sequence are recognized (types 1 and 2), depending on whether sea-level falls below the shelf break, giving lowstand fans and wedges, or not, when a shelf-margin wedge

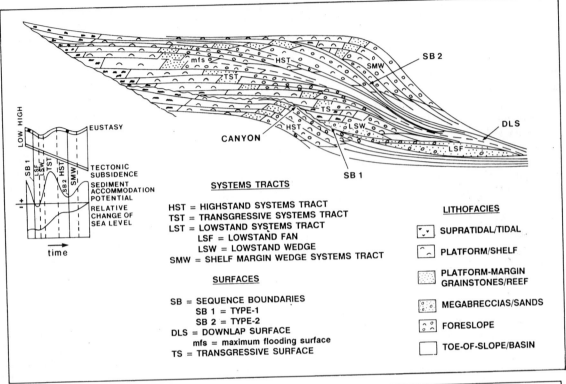

SYSTEMS TRACTS

HST = HIGHSTAND SYSTEMS TRACT
TST = TRANSGRESSIVE SYSTEMS TRACT
LST = LOWSTAND SYSTEMS TRACT
LSF = LOWSTAND FAN
LSW = LOWSTAND WEDGE
SMW = SHELF MARGIN WEDGE SYSTEMS TRACT

LITHOFACIES

SUPRATIDAL/TIDAL
PLATFORM/SHELF
PLATFORM-MARGIN GRAINSTONES/REEF
MEGABRECCIAS/SANDS
FORESLOPE
TOE-OF-SLOPE/BASIN

SURFACES

SB = SEQUENCE BOUNDARIES
SB 1 = TYPE-1
SB 2 = TYPE-2
DLS = DOWNLAP SURFACE
mfs = maximum flooding surface
TS = TRANSGRESSIVE SURFACE

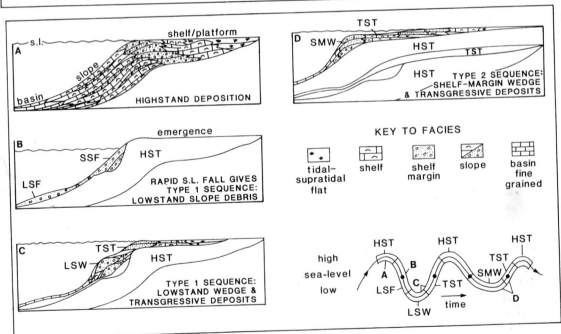

A HIGHSTAND DEPOSITION

B RAPID S.L. FALL GIVES TYPE 1 SEQUENCE: LOWSTAND SLOPE DEBRIS

C TYPE 1 SEQUENCE: LOWSTAND WEDGE & TRANSGRESSIVE DEPOSITS

D TYPE 2 SEQUENCE: SHELF-MARGIN WEDGE & TRANSGRESSIVE DEPOSITS

KEY TO FACIES

tidal-supratidal flat   shelf   shelf margin   slope   basin fine grained may form (see Fig. 2.2). If there is an overriding, third order, eustatic control on deposition, with near-regular, near-symmetrical, rises and falls in sea-level, then this also has implications for the diagenesis of carbonates, especially in the degree of subaerial exposure and meteoric diagenesis. Studies applying sequence stratigraphy to carbonate formations can be found in Wilgus *et al.* (1988), notably Sarg (1988), Crevello *et al.* (1989) and Tucker *et al.* (1990).

Computer modelling of the carbonate platform–basin system, depositional sequences and parasequences (metre-scale cycles) is very fashionable (see papers in Crevello *et al.*, 1989; Tucker *et al.*, 1990). Bice (1988) for example has demonstrated the importance of sea-level fluctuations and subsidence rate, as well as the higher-order, short-period sea-level cycles, in controlling the cross-sectional geometry of carbonate platforms.

Climate, along with geotectonics, is important in determining water circulation patterns, temperature, salinity, nutrient supply, turbulence, storm and tidal current strengths and wave activity. Shallow-water carbonates can accumulate wherever there is a paucity of terrigenous material, but the highest rates of organic productivity (Section 2.2.1) take place in low latitudes, chiefly 30° either side of the equator. Many important carbonate-producing organisms, such as corals and codiacean green algae, can only exist in warm tropical waters. Molluscs and calcareous red algae are much more tolerant and can exist, and indeed do form substantial deposits, up to very high latitudes (76° N in arctic Norway for example).

The two main factors affecting shallow-marine carbonate-secreting organisms are *temperature* and *salinity* (Lees, 1975), and on this basis, Lees & Buller (1972) recognized three principal skeletal grain as-sociations: chlorozoan, foramol and chloralgal as-sociations (Fig. 2.3A). In low latitudes, biogenic carbonate sediments are produced by the *chlorozoan* association, that is, the characteristic skeletal carbonate producers are hermatypic corals and calcareous green algae, along with many other organisms. The chlorozoan association does not exist where minimum surface temperatures fall below 15°C, and it can only tolerate a quite narrow salinity range (32–40‰). Hence coral reefs are vulnerable to cold (and very warm) waters, especially if they are of low salinity. Some coral reefs along the Florida Reef Tract have died off over the last few thousand years because of the detrimental effects of Florida Bay water draining on to the shelf (Section 3.3.1e). At elevated salinities, corals do not survive but green algae do, giving their name to the *chloralgal* association. Where seawater temperatures are mostly below 15°C, even down to 0°C, the *foramol* association is present, and sediments are dominated by benthic foraminifera and molluscs, with carbonate also coming from echinoderms, barnacles, bryozoans, calcareous red algae and ostracods. The foramol association comprises *temperate carbonates* currently being deposited off western Ireland (Bosence, 1980), off northwestern and northern Scotland (Farrow *et al.*, 1984) and New Zealand (Nelson *et al.*, 1982). Ancient examples of temperate carbonates are rare in comparison with low-latitude limestones, but they have been described from the Permian of Australia (Rao & Green, 1982), and Cainozoic of Tasmania and New Zealand (Nelson, 1978; McGregor, 1983).

Non-skeletal grains, ooids, aggregates such as grapestones and pellets (Section 1.2.1) occur with the chlorozoan association, and pellets extend into the foramol regions. Water temperature and salinity

Fig. 2.2 *The sequence stratigraphy approach to carbonate deposition (after Sarg, 1988). The rationale is that stratal patterns are controlled by the relative change of sea-level (determined by eustasy and tectonic subsidence) and this controls the sediment accommodation potential (+ = deposition and − = erosion), see graph figure. During specific time intervals of the sea-level curve, particular systems tracts are established: HST, TST, LST and SMW. Thick aggradational-to-progradational carbonate sequences develop especially during sea-level highstands (A). Two types of sequence boundary are recognized. In a type 1 boundary, the rate of eustatic sea-level fall is greater than subsidence, so that sea-level drops below the platform margin. Carbonate debris sheets and turbidites may be deposited at this time from slope front erosion to form a lowstand fan (LSF) apron at the toe-of-slope (B). Debris may also fill erosional scars and canyons on the slope front itself to give slope scar fills (SSF, Fig. B). During the sea-level low stand, a bank may develop (lowstand wedge, LSW), and then a retrogradational, onlapping transgressive systems tract as sea-level rises (C). In a type 2 sequence boundary, the rate of sea-level fall is less than the subsidence, so that the shoreline moves towards the edge of the shelf but not over the break, and a shelf-margin wedge (SMW) systems tract is established during the sea-level low stand; this onlaps during the subsequent sea-level rise (TST). The LSW and SMW systems tracts will be better developed on ramps and the LSF at steep platform margins. The mfs separates the TST from the HST, and the TS separates the LSW from the TST.*

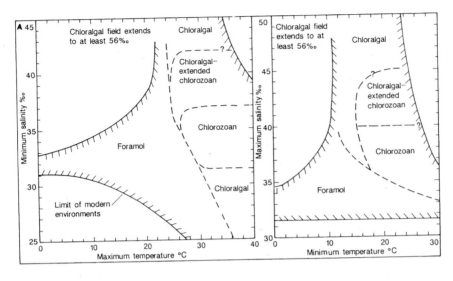

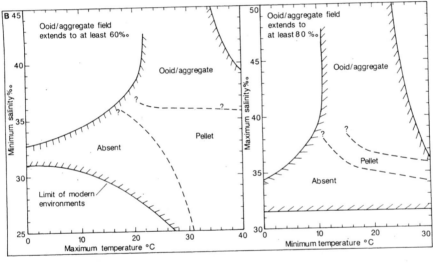

Fig. 2.3 *Salinity–temperature annual ranges and occurrence of skeletal grain associations (A) and non-skeletal grain associations (B) in modern shelf carbonate sediments. After Lees (1975).*

again appear to be the controlling factors (Fig. 2.3B). Ooids and aggregates form in areas where mean temperatures exceed 18°C. They also appear to be restricted to the subtropical belts, 15–25° latitude (rather than equatorial regions), where evaporation rates exceed precipitation and seawater salinity is slightly higher (Lees, 1975). Inorganic precipitation of lime mud is likely to be restricted to the subtropical belts too.

Many organisms cannot tolerate great fluctuations in salinity and/or temperature, so that the number of species generally decreases where these occur, as in protected lagoons and on tidal flats. Although diversity is generally low in such situations, frequency can be

very high. Cerithid gastropods, for example, dominate evaporitic tidal flats in many parts of the world and occur in vast numbers.

In the shallow-marine environment, *water depth* and *turbidity* are important. Most organic productivity takes place in less than 10–15 m of water. As discussed in Chapter 3, codiacean green algae such as *Penicillus* and *Halimeda* produce vast quantities of sediment on the Bahama Platform and Florida Shelf, and although they exist into quite deep water (*Halimeda* recorded from >50 m) highest production rates are in water depths less than 15 m. Coral growth on barrier reefs also goes down to more than 50 m depth, but most skeletal carbonate is produced in water depths less than 5 m. Turtle grass (*Thalassia*) is also mostly a feature of the shallower parts of the Florida Shelf (<8 m); it traps sediment and is host to many organisms producing sediment so that higher sedimentation rates prevail in these shallower, grass-covered areas. Since the formation of carbonate sediments is most intense at depths less than 15 m, this shallow subtidal environment is frequently referred to as the *carbonate factory*.

The lower limit of the *photic zone* is an important depth in carbonate deposition. Above this, photosynthetic organisms such as algae and marine grasses can grow, as well as metazoans which have a symbiotic relationship with algae, such as the hermatypic corals. Endolithic algae, which play such a major role in bioerosion by producing micrite envelopes around bioclasts and micritized grains (see Section 1.3), also only occur within the photic zone. The latter varies in its depth, depending on turbidity and sunlight intensity, but it may reach down to 100 m or more in clear waters.

Turbidity can be brought about by suspended clay from rivers (etc.), but even on siliciclastic-free shelves, lime mud is frequently put into suspension by waves, storms and fish. Turbidity inhibits carbonate production by cutting down the amount of light reaching the seafloor and thus discouraging the growth of calcareous algae and sea-grass which rely on light for photosynthesis. Also many benthic organisms with carbonate skeletons cannot tolerate suspended mud as it interferes with their feeding mechanisms.

The distribution of carbonate-secreting organisms also depends on *water circulation* and *current regime*. Some organisms, exemplified by the corals, flourish in turbulent areas and so are concentrated along shelf margins; others prefer the quieter waters of the inner shelf. The wave–current regime is responsible for reworking and sorting sediment and transporting it from its site of formation. In fact, apart from close to the shelf margin where offshelf and onshelf transport of sediment is significant as a result of the concentration of wave and storm action there, most carbonate sediments have formed *in situ* and have only been subjected to minimal transport.

Seawater in low latitudes is supersaturated *re* $CaCO_3$ in the upper few hundred metres, but in mid and high latitudes it is undersaturated. Precipitation of $CaCO_3$ can thus take place in the tropics and dissolution in temperature shallow seas. From considerations of carbonate equilibria (reviewed in Bathurst, 1975; Krauskopf, 1979), processes removing CO_2 from seawater can lead to $CaCO_3$ precipitation. These include

temperature increases (solubility of CO_2 decreases with increasing temperature), evaporation, pressure changes in seawater such as where deep water is brought up to the surface, turbulence and photosynthesis. These processes account for the precipitation of cement in reefs and lime sands, formation of ooids and precipitation of lime mud on protected platforms and in lagoons. Below 200–300 m in the tropics, carbonate grains are liable to suffer dissolution (especially if sedimentation rates are low), but this is not an important process until depths of several thousand metres when the rate of dissolution increases drastically (the lysocline) and the carbonate compensation depth (CCD) is reached, where the rate of sedimentation equals the rate of dissolution (see Section 5.3 and Fig. 5.1).

2.2.1 Organic productivity and sedimentation rates

Carbonate sediments can accumulate rapidly compared with other sedimentary rock types, but this topic can be considered on three levels: the growth rates of modern carbonate skeletons, which give a maximum value for limestone deposition; rates of sedimentation of Holocene carbonate sediments, which allow for variations in growth rates over 5000 years, and rates derived from thick limestone sequences of the geological record which are basically reflections of subsidence rates and include breaks in sedimentation due to sea-level fluctuations and periods of emergence, and compaction and pressure dissolution of the limestone mass.

Where optimum conditions exist for the growth of organisms with carbonate skeletons, then it appears that the carbonate production rate is fairly constant, regardless of the types of organism involved (Smith, 1973; Hallock, 1981). Gross production rates determined for benthic foraminifera, corals and coralline algae on seaward reef flats are around 1.5–4.5 kg $CaCO_3$ m^{-2} yr^{-1}, equivalent to a carbonate deposition rate of 0.5–1.5 mm yr^{-1} or 0.5–1.5 m 1000 yr^{-1}. Rates are somewhat lower in back-reef lagoons (0.1–0.5 m 1000 yr^{-1}) but on the reef front, coral growth rates of 6 m 1000 yr^{-1} have been recorded (Longman, 1981). It appears from studies of carbonate skeletons that carbonate production rates are determined by ocean physicochemistry (i.e. temperature, agitation, P_{CO_2}, nutrient supply, etc.) rather than organic–biological factors.

At the present time, only around 10% of carbonate production takes place in the shallow-marine realm.

Ninety per cent of $CaCO_3$ production today is accounted for by the mostly calcitic plankton (coccolithophorida and foraminifera), and their deposits of pelagic carbonate ooze on the deep seafloor (100–4500 m) account for 80% of the global deposition of marine $CaCO_3$. Some planktonic carbonate is lost by dissolution where the seafloor is below the carbonate compensation depth (Section 5.3). Reefs are major sites of carbonate sedimentation, being responsible for 6% of production. Aragonitic lime muds total 3% of primary production and the remaining 1% $CaCO_3$ production is mostly by benthic organisms on shelves.

Holocene carbonate sediments in the Bahama–Florida region have only been accumulating for 3000–6000 years, since the post-glacial sea-level rise covered the area. From the Florida Reef Tract, a maximum thickness of 14 m is recorded, in 6000 years (2.3 m 1000 yr^{-1}), but the average rate for the outer reef is more like 1 m 1000 yr^{-1} (e.g. Carysfort Reef and Molasses Reef, where 6–8 m have been deposited in 6700 years; Enos, 1977a). Sedimentation rates on the inner shelf are generally much lower (0.2–0.5 m 1000 yr^{-1}) except where mud banks have developed (Section 3.3.1g) and then rates are comparable with the shelf-edge reefs (e.g. Rodriguez Bank: 5 m in <5000 years, Turmel & Swanson, 1976; Tavernier: 4 m in 5000 years, Bosence et al., 1985; and Crane Key mud bank in Florida Bay: 3 m in 3000 years, Stockman et al., 1967).

It is instructive to compare the typical sedimentation rate of modern carbonates, 1 m 1000 yr^{-1}, with rates of subsidence and sea-level change. Measurements of modern subsidence rates in sedimentary basins give a range of 0.01–2.5 m 1000 yr^{-1}, depending on the type of subsidence involved, flexural, thermal, or fault-controlled. Typical subsidence rates of passive continental margins, where many ancient carbonate platforms developed, are of the order of 0.01–0.1 m 1000 yr^{-1}. These values suggest that carbonate production should be able to keep up with an average amount of subsidence. Eustatic sea-level changes can be brought about by fluctuations in ocean-basin volume through plate tectonic processes, and fluctuations in ocean-water volumes through changes in the mass of glacial ice at the poles. However, the former process typically gives a rate of sea-level change of 0.01 m 1000 yr^{-1}, while glacial eustatic changes are up to three orders of magnitude faster (e.g. Flandrian transgression up to 10 m 1000 yr^{-1}, Donovan & Jones, 1979). Thus on the whole, shallow-water (<15 m) carbonate sedimentation is relatively fast and is able to keep up with moderate ocean-level changes. It cannot keep up with rapid subsidence, generally fault-induced, or with major sea-level rises through glacial melting (see Schlager, 1981 and Kendall & Schlager, 1981 for further discussion).

Many ancient carbonate sequences are measured in thicknesses of kilometres, but the net sedimentation rates are very low compared with the figures given above from the Holocene. For example, some 4 km of limestone, with some dolomite and evaporite, were deposited on the Bahama Platform from the Upper Cretaceous (100 Ma ago). This gives a net sedimentation rate of 0.04 m 1000 yr^{-1}. The Urgonian (Barremian–Aptian) Platform carbonates of the Subalpine Chains, France, are up to 500 m thick and accumulated in 10 Ma, giving a rate of 0.05 m 1000 yr^{-1}. The Upper Permian strata of west Texas, 800 m thick and deposited in 15 Ma, also give a rate of 0.05 m 1000 yr^{-1}. These figures illustrate the well-known fact that only a fraction of time is represented by the rock record. Breaks in sedimentation, mostly through emergence in shallow-water carbonates, may represent long periods of time. However, since many carbonate sequences consist of shallow-water sediments deposited close to sea-level, the sedimentation rates are really a reflection of the *accommodation*, i.e. the amount of space available to be filled by sediment, and this is dependent on subsidence rate and any eustatic sea-level change.

2.3 CARBONATE FACIES AND FACIES SEQUENCES

The environmental interpretation of ancient limestones mostly involves the same facies approach as is used for siliciclastics (e.g. Walker, 1984; Reading, 1986), along with more detailed laboratory studies to identify microfacies. A *facies* is defined by a particular set of sediment attributes: a characteristic lithology, texture, suite of sedimentary structures, fossil content, colour, etc. *Lithofacies*, defined on the basis of sedimentary characteristics, are most often used by sedimentologists (and simply called facies). *Biofacies* are occasionally identified and rely on palaeontological differences; *ichnofacies* are distinguished on their trace fossil suites. Within a carbonate sequence, there may be many different facies present and careful fieldwork logging and collecting are needed to identify these. A particular facies will often be found to recur several times within a sequence and one facies may pass vertically or laterally into another facies by a change in one or several of its characteristic features, such as fossil content or grain size. In sedimentary sequences, it is often found that groups of facies occur together to form *facies associations*. The facies comprising an association were deposited in the same general environment, but local differences existed or several depositional processes were involved.

Carbonate facies are usually referred to in one of three ways: (1) in a purely descriptive sense using a few pertinent adjectives, such as cross-bedded oolitic grainstone facies or fenestral peloidal mudstone facies, (2) in terms of their environment of deposition, such as oolite shoal facies or tidal flat facies, or (3) in terms of their depositional process, such as wave-dominated oolite sand facies or storm-deposited skeletal mudstone facies. The first is purely descriptive; the second and third are both interpretative and may need to be modified or changed in the light of further work.

With carbonates, it is usually necessary to go much further than fieldwork data to define facies. Often it is difficult to identify a limestone's composition in the field, so that microscopic study of thin sections and peels is necessary. Sedimentary structures may be poorly developed or unclear so that polished slabs of hand specimens are needed. Diagenetic effects such as stylolitization and dolomitization may make the original limestone type unrecognizable in the field. Laboratory studies of carbonates are thus necessary to confirm field identifications, to classify according to the Folk and Dunham schemes (Section 1.7), to make modal analyses, and to elucidate the rock's diagenesis. From thin-section studies, the *microfacies* of the carbonate sequence are identified. Like lithofacies, a microfacies is defined by a particular set of characteristics, but in this case, the data are obtained from thin sections, peels and polished slabs. Microfacies identification mainly involves determination of the limestone's composition and texture. This may simply be a question of distinguishing between various types of grainstone, such as peloidal versus skeletal versus oolitic types, but frequently it is necessary to look more closely at the skeletal contribution, and carry out modal analyses by point counting to separate the various types of skeletal grainstone. Factor analysis or multivariate cluster analysis can be used to separate out the various microfacies. Examples where this has been applied to carbonate sediments include Ekdale *et al.* (1976), Rao & Naqvi (1977), Smosna & Warshauer (1979) and Bosence *et al.* (1985).

The microfacies concept is discussed in Wilson (1975) and Flügel (1982). From a review of Phane-rozoic limestones, Wilson (1975) identified 24 standard microfacies types (Table 1.1) and produced a generalized model of carbonate sedimentation involving nine standard facies belts. The model is basically that of carbonate shelf sedimentation which has been known for many decades. Since Wilson's (1975) synthesis, the concept of the ramp model has become an alternative to the shelf model, so that one single generalized facies model is not applicable to all ancient carbonate sequences.

Carbonate facies interpretation is often facilitated by considering the vertical *facies sequence*, rather than taking each individual facies (or microfacies) and attempting to interpret that in isolation. However, it must be said that with carbonates, much more than with clastics, an individual facies may contain sufficient diagnostic information to pinpoint its depositional environment very precisely. Certain sedimentary structures, birdseyes for example, are restricted to a specific environment, and particular skeletal components, if not transported any distance, may indicate quite specific depositional conditions. Nevertheless, consideration of the vertical sequence of facies can help identify the larger-scale depositional processes that were operating. Where there is a conformable *vertical* succession of facies, with no erosive contacts or non-sequences, the facies sequence may be the product of *laterally* adjacent environments. This concept is frequently referred to as Walther's Law, which holds that the vertical succession of facies is produced by the progradation (building out) or lateral migration of one environment over another. With carbonates, tidal flat progradation and basin-ward, shelf-margin reef growth are two depositional processes which can give distinctive vertical facies sequences (Section 2.3.1). However, there are many exceptions to Walther's Law, particularly with carbonate facies where there is a strong biological control on facies development. Upward facies changes can simply result from *in situ* organic growth, as in the formation of patch reefs for example. In addition, changes in the position of sea-level, which can drastically affect platform carbonate sedimentation, can induce major vertical facies changes, without any lateral migration of environments. Sedimentary breaks in a vertical sequence of carbonate facies commonly take the form of erosion surfaces, overlain by an intraformational conglomerate or shell lag, emergence horizons including paleosols and paleo-karstic surfaces, or hardground surfaces. Lengthy periods of time may be represented by the break, and quite different environments may have existed there, leaving no trace through non-deposition or subsequent erosion.

The arrangement of facies within a succession can be completely random, but in many instances there is a regular repetition of the various facies to give *sedimentary cycles*. Carbonate cycles are mostly of the rhythmic ABC, ABC, ABC type. With many carbonate cycles there is a distinct boundary at the top of each cycle, and this is commonly an emergence horizon with paleokarstic surface, paleosol or sabkha evaporite. There are usually systematic changes in the character of the sediments up through a cycle, such as in the grain size, lime mud content or skeletal composition, and in the majority of cases, these reflect a shallowing of the depositional environment (see Section 2.10).

In many cases, a cyclic character is obvious in the field through weathering effects, or it becomes apparent from logging the section and petrographic studies. In other instances a cyclic pattern may be more subjective or not apparent at all. Various techniques are available to test a sequence for facies relationships, randomness and repetitions. Simple methods involve counting the number of times each facies is overlain by the others and noting the type of boundary (sharp or gradational) between facies. A facies relationship diagram can be constructed from the data or the data can be tabulated (a 'data array') and compared with a table of values for a random arrangement of the facies (for details see Walker, 1984). For a more rigorous statistical treatment of the data, Markov chain analysis and other techniques

such as power spectral analysis can be used (e.g. Jones & Dixon, 1976; Powers & Easterling, 1982; Weedon, 1986).

Carbonate facies should be studied at several scales to obtain maximum information. An individual bed can be examined for its mechanism of deposition and microfacies. A group of beds and microfacies may constitute a sequence, and consideration of this package will reveal the nature of changing environments. The sequence may be a cycle, repeated many times, so that the mechanisms causing the repetition need to be sought. Looking at the cycles up through a formation may show subtle but systematic changes in bed thickness, microfacies, grain size, etc., which reflect much longer-term changes in the relative position of sea-level through tectonic or eustatic mechanisms. Then on the scale of formations and basins, sequence stratigraphy analysis may uncover much larger-scale relationships reflecting regional or even global tectonic/eustatic events. Figure 2.4 shows how facies analysis contributes on these various scales and comes from the work of Aigner (1984) on the Muschelkalk ramp carbonates of Germany, discussed in Section 2.6.3.

2.3.1 Depositional processes and facies sequences in carbonate rocks

The facies sequences observed in carbonate formations are the result of changes in the environment through time, but these changes may be brought about by natural processes operating within the en-

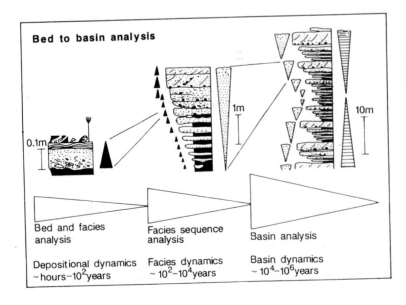

Bed to basin analysis

0.1m   1m   10m

Bed and facies analysis

Facies sequence analysis

Basin analysis

Depositional dynamics ~hours–10^2 years

Facies dynamics ~10^2–10^4 years

Basin dynamics ~10^4–10^6 years

Fig. 2.4 *Diagram illustrating the levels of description and interpretation for bed to basin analysis. Consideration of a single bed gives information on its depositional environment and process, in this case, deposition from a waning storm current on a carbonate ramp. Examining the vertical sequence of beds reveals a unit (cycle) showing an upward thickening and coarsening of the beds, reflecting ramp shoreline progradation. On a larger scale still, the thickness of the cycles shows a regular upward changing pattern, as a result of major, long-term, relative sea-level fluctuations. After Aigner (1984).*

vironment or by fluctuations in the external factors controlling sedimentation, such as a rise in sea-level. Consideration of the various depositional processes that operate in carbonate environments shows that there are five principal processes and these lead to the formation of characteristic vertical and lateral facies sequences (Fig. 2.5), *without any change in the position of sea-level.*

1 *Tidal flat progradation* results largely from the redeposition of shallow subtidal sediment on to tidal flats and beach ridges during major storms. Trapping and some precipitation of sediment in microbial mats on the flats is important. Some carbonate (and other minerals) can be precipitated inorganically on tidal flats in an arid climate. The net result is a shallowing-upward sequence (discussed in Sections 2.10 and 4.3) of intertidal sediments overlying subtidal sediments. In detail there are variations in the microfacies of these shallowing-upward sequences, depending on the type of tidal flat, energy level and climate (etc.).

2 *Reef progradation* is important at rimmed shelf margins and mostly involves seaward growth of the reef over its storm-produced talus on the fore-reef slope (discussed further in Sections 2.5 and 4.5).

3 *Vertical accretion of subtidal carbonates* can take place when sediment production rates are high. Shallowing-upward sequences are produced, of deeper subtidal facies giving way to shallower subtidal facies (and of course intertidal facies could follow naturally).

4 *Migration of carbonate sand bodies* is significant in relatively high-energy locations, giving beach barrier—tidal delta complexes, especially on ramps, and sand shoals, especially at shelf margins. Under stable sea-level, beach barrier—tidal delta complexes will prograde offshore if there is a good supply of sediment (i.e. high organic productivity in the shoreface zone or much ooid formation) (see Section 4.1.4). With sand shoals at shelf rims, their shoreward migration into the shelf lagoon is important in windward locations, giving rise to quiet-water, below fairweather wave-base packstones and wackestones passing up into above wave-base storm- or tide-dominated grainstones. On leeward margins, offshore, basinward transport of skeletal sands is significant and can lead to progradation of the margin itself (see Section 4.2).

5 *Offshore storm transport and deposition* of shore-

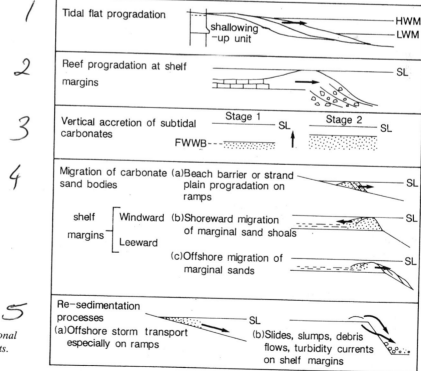

Fig. 2.5 *The principal depositional processes of carbonate sediments. After Tucker (1985a).*

face carbonate sediment is very important on ramps (see Section 4.1.3b). Other *resedimentation processes*, slides, slumps, debris flows and turbidity currents, all of which may be storm or seismically induced, are important at shelf margins. These are discussed in Chapter 5.

The vertical facies sequences produced by the above processes have been described from many ancient carbonate formations and have been documented from coring in Holocene carbonate sediments (Chapters 3 and 4). When external factors change for one reason or another, then many more facies patterns are produced in response to the changing conditions. The position of sea-level is one major factor affecting carbonates and this can change through glacio-eustatic tectono-eustatic or regional tectonic effects. At the extremes, such as major regression giving subaerial exposure, major hinterland uplift causing clastic influx, or major transgression causing drowning of the carbonate platform, carbonate sedimentation may cease altogether. Smaller-scale sea-level changes will promote new carbonate facies developments, and organism communities and habitats will adjust, evolve and change to accommodate the new conditions. A change in climate, circulation pattern, salinity, water temperature and other factors affecting carbonate environments will also result in facies changes in shallow-marine limestones. Examples are given in Chapters 4 and 5 where the various carbonate facies are discussed in more detail.

2.4 CARBONATE PLATFORMS

Although terms for the large-scale depositional environments of carbonates have been used in different ways by different people over the years (e.g. Ahr, 1973; Ginsburg & James, 1974; Wilson, 1975; Kendall & Schlager, 1981; Read, 1982, 1985; Tucker, 1985a) a consensus has emerged in recent years. *Carbonate platform* is used as a very general and loose term for a thick sequence of mostly shallow-water carbonates. Carbonate platforms develop in a whole range of geotectonic settings, but particularly along passive continental margins, in intracratonic basins to failed rifts, and back-arc basins to foreland basins. Various types of carbonate platform are recognized and the broad categories are: rimmed shelf, ramp, epeiric platform, isolated platform and drowned platform (Fig. 2.1). More specific platform types occur within each category.

A *rimmed shelf* is a shallow-water platform with a pronounced break of slope into deep water. Along this wave-agitated shelf margin there is a near continuous rim of barrier reefs and/or skeletal—oolitic sand shoals. These restrict circulation in the shelf lagoon located landward of the shelf margin. Some shelves may have quite deep basins (intrashelf basins) developed behind the rimmed margin. Widths of rimmed shelves are typically a few to 100 km. Rimmed shelves can be divided into accretionary, bypass and erosional types. Modern examples are the Queensland Shelf off eastern Australia with the Great Barrier Reef, the South Florida Shelf and the Belize Shelf. Rimmed shelves are discussed in detail in the next section (Section 2.5).

A *carbonate ramp* is a gently sloping surface (generally less than 1°) on which nearshore wave-agitated sandy facies pass offshore into deeper-water, more muddy facies. Barrier reefs are generally absent, but mud mounds and pinnacle reefs are not uncommon on ramps. On the basis of ramp profile, homoclinal and distally-steepened types are distinguished. In terms of depositional processes and geomorphology, a carbonate ramp is really analogous to the open shelf of the siliciclastic literature. The Trucial Coast of the Arabian Gulf, the eastern Yucatan coast of Mexico and Shark Bay in Western Australia are modern examples of carbonate ramps. Ramps are discussed in detail in Section 2.6.

Epeiric platforms are very extensive (100—10000 km wide), quite flat, cratonic areas covered by a shallow sea. Oceanwards, an epeiric platform may be bound by a margin which has a gentle (ramp-like) or steep (shelf-like) slope. The margin could be rimmed by barrier reefs and lime sand shoals. However, the margin is not an integral feature of the epeiric platform, which has its own particular set of depositional conditions. For the most part, epeiric platforms are dominated by shallow subtidal—intertidal low-energy facies and shallowing-upward, tidal flat cycles are typical. There are no good modern examples of epeiric platforms of the size which existed in the past, but the interior of the Great Bahama Bank and Florida Bay could be close analogues. Like shelves, epeiric platforms may have deep-water basins developed within them, surrounded by ramps and rimmed shelves. Epeiric platforms are discussed in Section 2.7.

Isolated platforms are shallow-water platforms, usually with steep sides, which are surrounded by deep water. A feature of these platforms is that the facies distribution is much affected by prevailing wind and storm directions. The Bahamas would be a large

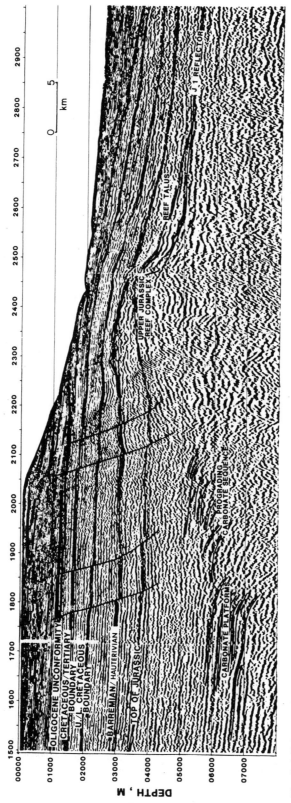

Fig. 2.6 Seismic reflection profile from the continental margin off northern USA in the region of Baltimore Canyon (offshore from Atlantic City), showing carbonate platforms with prograding ramps and an Upper Jurassic rimmed shelf reef complex with associated reef talus and back-reef, lagoonal deposits. The succeeding Cretaceous—Tertiary sediments are mostly clastics. After Gamboa et al. (1985), courtesy of American Association of Petroleum Geologists.

modern example, and smaller ones occur off the Belize Shelf. Atolls are a more particular type of isolated platform with a reefal rim and central, deeper lagoon, although true oceanic atolls have a volcanic foundation. Isolated platforms are discussed in Section 2.8.

Drowned platforms are ramps, rimmed shelves, isolated and epeiric platforms which have been subjected to a rapid relative sea-level rise so that deeper-water facies are deposited over the shallow-water limestones. Many pelagic limestones were deposited on drowned platforms (see Chapter 5). *Incipiently drowned platforms* are ones where the water depth is not too great after the sea-level rise, so that if production rates are high, the system may recover. Drowned platforms are discussed in Section 2.9.

Carbonate platform type is principally determined by geotectonics and relative sea-level, but one setting may evolve into another, either through the natural processes of carbonate sedimentation or through tectonic effects and sea-level changes. Common patterns are: (1) a ramp may develop into a rimmed shelf, especially through reef growth, (2) a shelf may develop into a ramp through differential subsidence along a hinge-line, (3) an epeiric platform may develop into local shelves, ramps and basins through contemporaneous fault movements, and (4) any of the shallow-water carbonate platform types may become drowned or incipiently drowned through rapid relative sea-level rise. An example of carbonate platform development is shown in the seismic profile in Fig. 2.6, from the Jurassic of the eastern North American continental margin (Gamboa *et al.*, 1985, also see Jansa, 1981). A system of prograding carbonate ramps gave way to a rimmed shelf with marginal reefs.

2.5 CARBONATE RIMMED SHELVES

Rimmed shelves are characterized by the development of reefs and carbonate sand bodies along the shelf margin (Fig. 2.7). Depths are shallow adjacent to the shelf-break, even subaerial if islands have formed. The shelf margin is a turbulent, high-energy zone where ocean waves (swell), storm waves, and possibly tidal currents, impinge on the sea floor. Organic productivity is highest under these conditions, especially if the sea is fertile through up welling. Much precipitation of $CaCO_3$ in the form of ooids and cements occurs along shelf margins. Behind the rim, there is usually a shelf lagoon. This will vary in its degree of protection from the marginal turbulent zone, depending on how well the reefs and sand shoals of the margin act as a barrier. At one extreme, a true shelf lagoon will exist, being a very quiet environment with poor circulation and perhaps hypersalinities during dry seasons. It will only be affected by major storm events. At the other, a more open lagoon will exist, subject to continuous wave and tidal motions.

The shoreline at the inner margin of a rimmed

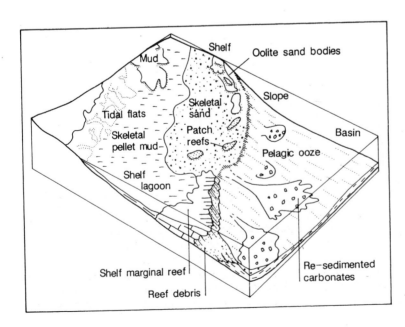

Fig. 2.7 *The carbonate rimmed shelf depositional model. After Tucker (1985a).*

shelf will generally be dominated by tidal flats, especially if there is a significant tidal range (mesomacrotidal). A beach barrier–tidal delta coastline may form if wave energy is substantial (determined by prevailing climate and coastline orientation) and tidal range is low (micro- to meso-tidal). This type of shoreline, however, is more typical of carbonate ramps (Section 2.6).

2.5.1 Modern rimmed shelves

Modern rimmed shelf carbonates are accumulating in many areas in the Caribbean. They have been studied extensively off Belize (James & Ginsburg, 1979) where a fringing reef is developed along the shelf margin and a quiet-water shelf lagoon occurs to shoreward (Fig. 2.8). On the Bahama Platform, a typical rimmed shelf facies pattern is seen to the east of Andros Island, where reefs and sand shoals are developed along the platform margin and there is a narrow (1–5 km) lagoon behind, before the shoreline of the island. The sediments of this area are described in Section 3.2.1. The 300 km long South Florida Shelf is a 6–35 km wide area of shallow-water carbonate sedimentation, with marginal reefs and sand shoals along the shelf margin (see Section 3.3.1). The Florida Shelf has a facies distribution similar to many ancient rimmed shelf carbonates: shelf-marginal reefs (framestones and boundstones) and carbonate sands (oolitic and skeletal grainstones), giving way to skeletal packstones and wackestones of the protected inner shelf lagoon. Skeletal mudstones are accumulating in plant-stabilized mud banks, and mangrove swamps and tidal flats occur along the shoreline. Another well-documented rimmed shelf occurs off Queensland where the Great Barrier Reef is a narrow (5 km), but long (2000 km), zone of active coral growth at the shelf margin, with a broad (10–50 km) shelf lagoon behind, where carbonates and clastics are being deposited (Hopley, 1982). The modern reefs are commonly developed upon older reefs and there is much relic Pleistocene sediment on the shelf too.

Along some rimmed shelf margins, carbonate sedimentation can lead to the development of islands, composed of reef rubble and carbonate sand. This has happened along parts of the southern Florida Reef Tract (Section 3.3.1) and at Joulter's Cay in the Bahamas (3.2.1b.).

Studies of the Recent have provided a well-defined facies model for shelf-marginal reefs: the fore-reef zone of reef slope and proximal and distal talus,

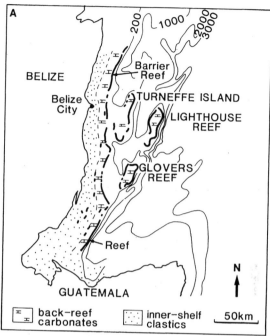

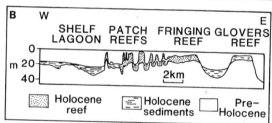

Fig. 2.8 *The Belize carbonate rimmed shelf. (A) Map showing shelf-margin barrier reefs and reefs around isolated platforms (Lighthouse and Glovers Reefs and Turneffe Island). (B) Sketch cross-section from a seismic profile showing development of modern reefs upon pre-existing topographic highs of Pleistocene limestone. After James & Ginsburg (1979).*

the reef itself (or reef core) of reef framework and reef crest, and the back-reef zone of reef flat and back-reef sand (see Sections 3.2.1a, 3.3.1e and 4.5.5, and Longman, 1981). This facies model can be applied to many ancient rimmed shelf–reef complexes, but in detail there are usually departures.

Carbonate sands are usually generated in abundance along rimmed shelf margins. Much of the sand is of skeletal origin, derived from shelf-break reefs

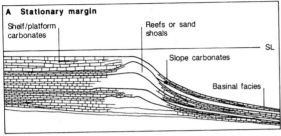

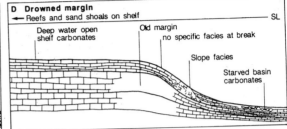

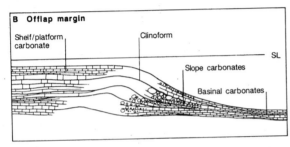

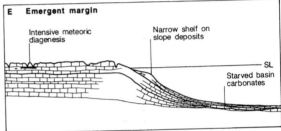

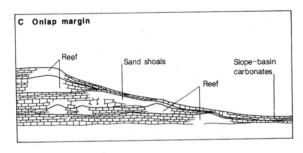

Fig. 2.9 *General models for rimmed carbonate shelves responding to sea-level changes. (A) Stationary. (B) Offlap. (C) Onlap. (D) Drowned. (E) Emergent. After James & Mountjoy (1983).*

(if present) and from the skeletons of organisms which live in the shelf-margin area. The sudden barrier that the shelf slope makes to open ocean and storm waves ensure continuous turbulence along the shelf margin and constant reworking of sediment and erosion of reefs. Ooids are commonly an important component of shelf-margin sands since $CaCO_3$ precipitation is promoted by the active movement of suitable nuclei (fine skeletal grains) and the warming and CO_2 degassing of ocean water as it comes on to the shallow shelf. Modern rimmed shelf sand bodies are well represented on the Bahama Platform (Hine, 1977; Hine *et al.*, 1981; Handford, 1988), and several types exist, dependent on the orientation of the shelf-break relative to the prevailing wind direction, and on wave–storm versus tidal effects (see Section 4.2).

The distribution of carbonate sediments on many modern rimmed shelves is very much controlled by relict topographies, produced during the Pleistocene (e.g. Purdy, 1974a). Reefs and sand bodies formed during high sea-level stands over the past million years, especially during the Sangamon high of +120 000 years ago, were modified by erosion and karstification during subsequent low sea-level stands. Many modern reefs at shelf margins are thin veneers over Pleistocene reefs, and topographic highs on shelves are sites of patch reefs and lime sand deposition. The effect of this relict topography is well seen on the Belize Shelf (Fig. 2.8) and off Queensland along the Great Barrier Reef.

2.5.2 Rimmed shelves and sea-level changes

Rimmed shelf margins respond in different ways to sea-level change and James & Mountjoy (1983) identified five types: (1) stationary, (2) offlap, (3) onlap, (4) drowned, and (5) emergent (Fig. 2.9). In the *stationary mode*, the position of the rimmed shelf margin does not change much through time, so that vertical accretion takes place. Since carbonate production is much higher at the shelf edge, compared to the slope and basin, the relief between rimmed shelf and basin gradually increases. The stationary mode typically develops where the relative rate of sea-level rise is more or less balanced by the sedimentation rate at the shelf margin. Since sedimentation rates are higher at the rimmed shelf margin, a shelf lagoon will commonly be well developed behind the margin, with the sediments very much determined by the degree of connection to the open ocean and the climate. In the *offlap mode*, the rimmed shelf margin progrades basinwards since there is an overproduction of carbonate at the shelf edge and sea-level is not rising fast. In this case, the whole shelf will be close to sea-level so that connection with the open ocean will often be poor or non-existent, leading to hypersaline lagoons with evaporites and/or exposure with a whole range of emergence phenomena possible, depending on climate. When sea-level does rise quickly, then either onlap or drowning takes place. *Onlap* can occur in steps or more gradually through a shelfwards migration of facies belts. The latter will mostly occur when the shelf margin is dominated by carbonate sands and the sea-level rise is not too fast. *Backstepping* of the margin is more likely to occur when relative sea-level rises in stages and/or where reefs are present at the shelf margin. When the sea-level rise is really too fast for carbonate sedimentation to keep pace, then the shelf and slope are drowned and pelagic carbonates and other deeper-water sediments are usually deposited over the whole region. In the *emergent type* of margin, sea-level falls below the shelf-break and the shelf is subject to subaerial exposure. A narrow zone of shallow-water carbonate production may be established upon the former slope. These various types of shelf-margin response to relative sea-level change have been well documented in ancient carbonate formations (discussed in Section 2.5.4).

2.5.3 Rimmed shelf types

Consideration of modern and ancient rimmed shelf margins reveals that there are three main types: accretionary, bypass and erosional shelf margins (McIlreath & James, 1984; Read, 1982, 1985).

Rimmed shelf margins can also be divided on whether they are reef or lime sand dominated.

Accretionary rimmed shelf margins are ones exhibiting lateral migration, with some or little vertical aggradation, and the shallow shelf-margin reefs and sand bodies build out over fore-reef and slope deposits (Fig. 2.10). The shelf slope consists of a prograding apron or talus. There is an interdigitation of shallow-water rimmed shelf, resedimented slope and deeper-water basinal facies. Frequently a well-defined *clinoform* bedding is seen, which consists of large-scale dipping surfaces, formed largely of the shallow-water debris shed off the shelf margin. Clinoforms generally have a significant dip, $5-15°$, and a height of several tens to hundreds of metres (next section and 5.13.1). Accretionary shelf margins mostly form when sea-level is stable, or only rising slowly, and carbonate productivity is high at the shelf margin. Modern accretionary margins occur along parts of the Bahama Platform (e.g. Hine *et al.*, 1981; Eberli & Ginsburg, 1987), and clinoform-like structures are revealed in seismic reflection data (Fig. 2.11). Similar features are seen along parts of the Florida Shelf margin (Enos, 1977a).

Bypass rimmed shelf margins are ones where little sediment is deposited on the shelf slope. They occur in areas of rapid vertical accretion where shelf-margin sedimentation is able to keep pace with rising sea-level, but insufficient sediment is deposited on the slope for any significant lateral accretion. Sediment shed off the shelf edge is deposited at the toe of the shelf slope or in the basin. A steep, shelf-edge cliff may be developed through rapid upbuilding in the 'escarpment bypass' type of Read (1982), and in the 'gullied-slope bypass' type debris is taken down submarine channels to fans and aprons at the base of the shelf slope (Fig. 2.10). Modern examples of bypass shelf margins are seen off Queensland where vertical reef growth has dominated over any lateral migration (see Johnson & Searle, 1984).

Erosional rimmed shelf margins occur in areas of strong tidal or ocean currents and cliffs and escarpments characterize the shelf slope (Fig. 2.10). Debris from the erosional margin may accumulate in fans and aprons at the toe of the shelf slope. Some modern rimmed shelf margins are erosional, and in most cases it is Pleistocene shallow-water limestones that are exposed and they are being eroded both by physical and biological processes. Examples include parts of the shelf margin off Jamaica and Belize (Land & Goreau, 1970; James & Ginsburg, 1979).

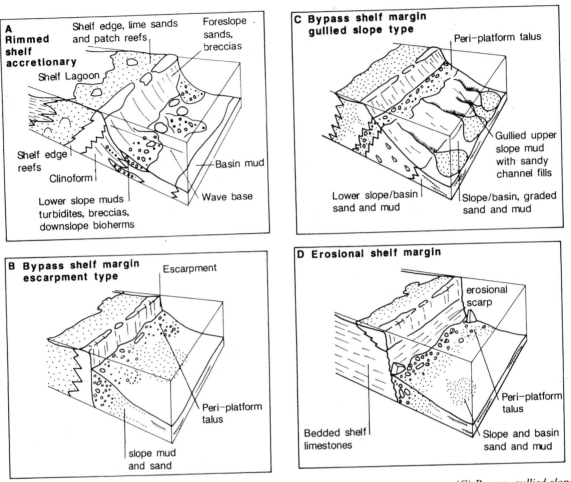

Fig. 2.10 *Rimmed carbonate shelf types. (A) Accretionary type. (B) Bypass, escarpment type. (C) Bypass, gullied-slope type. (D) Erosional rimmed shelf margin. After Road (1982).*

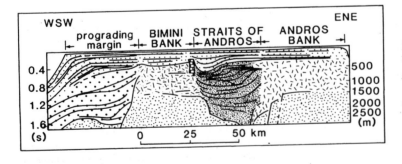

Fig. 2.11 *Schematic section across the Great Bahama Bank from a seismic profile showing two buried banks (Andros and Bimini), lateral progradation of the western margin of Bimini and complex filling of the Straits of Andros basin. After Eberli & Ginsburg (1987).*

2.5.4 Ancient carbonate rimmed shelves

There are many well-documented ancient rimmed shelf limestone formations which can be compared with modern rimmed shelf carbonates. Classic shelf carbonates occur in the Alpine Chain and were particularly well developed during the Triassic. Shelf margins were sites of reef growth with corals, sponges, *Tubiphytes* and red algae the dominant organisms. *The*

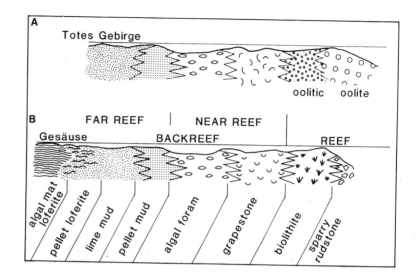

Fig. 2.12 *Upper Triassic rimmed shelf facies patterns in the Totes Gebirge and Gesäuse ranges of the northern calcareous Alps, Austria. After Flügel (1981).*

Late Triassic Dachstein Limestone of the northern calcareous Alps, Austria contains many shelf-marginal reef complexes (Fig. 2.12; Flügel, 1981), with smaller patch reefs on the shelf (e.g. Schäfer & Senowbari-Daryan, 1981). In some regions, oolite sand bodies developed at the shelf-break, rather than reefs. Shelf-lagoonal facies are typically grapestone grainstones and algal–foram grainstones and packstones (Fig. 2.12). Tidal flat-protected lagoonal facies of the inner shelf are fenestral, pelletal lime mudstones of the Lofer cycles (Section 2.10).

In the *Dolomites of the southern Alps, Italy, Middle Triassic carbonates* were deposited on rimmed shelves and isolated platforms (Leonardi, 1967; Bosellini & Rossi, 1974; Bosellini, 1984, 1989. A complex platform–basin topography was produced by block-faulting and an early phase of carbonate platform sedimentation in the Ladinian was terminated by basaltic lavas filling the basinal areas when the platforms were subaerially exposed and karstified in an emergent mode (Fig. 2.13). The succeeding Carnian platforms prograded rapidly into basins where San Cassian Formation deep-water marls were deposited. Although extensively dolomitized, reefs have been recognized at shelf margins, with bedded lagoonal–tidal flat facies behind. Shelf margins commonly show spectacular clinoforms of reef debris (Fig. 2.13), and these accretionary margins, many of the offlap type, can be shown to have prograded many kilometres into the basin. In whole mountain-side exposures, Bosellini (1984) has recognized several types of clinoform (discussed further in Section 5.13.1), many of which are

comparable to the dipping planes seen in seismic profiles (as in Fig. 2.5) and used for sequence analysis in seismic stratigraphy.

The *Middle to Upper Devonian* was a time of extensive shallow-water carbonate sedimentation, and shelf-margin reefs were well developed in some instances, such as western Canada, western Germany and Western Australia. Various rimmed shelf margins existed, including accretionary and bypass types in the Canning Basin, showing offlap, onlap, stationary and drowned sequences (see Fig. 4.98; Playford, 1980). The 'classic face', Windjana Gorge, Canning Basin (Fig. 2.14) clearly shows bedded shelf-lagoon facies, shelf-margin reef and reefal-slope clinoforms. A vertical platform-margin unconformity was produced by erosion during a short period of sea-level fall and emergence. A variety of rimmed shelf types is well seen in Upper Cambrian through Devonian strata of Nevada (Fig. 2.15).

Jurassic–Lower Cretaceous carbonate platforms formed a discontinuous belt over 6000 km long off the eastern USA and Canada, from the Grand Banks to the Bahamas, and the rimmed shelves here were of various types, but particularly accretional–offlap and stationary as deduced mainly from seismic reflection studies (e.g. Fig. 2.6; Eliuk, 1978; Jansa, 1981; Gamboa *et al.*, 1985). Jansa (1981) also recognized two destructional kinds of margin where either erosion or growth faulting took place. The erosion of the carbonate platform margins is thought to have taken place during the Early Tertiary by North Atlantic contour current, which in places cut the margin back 30 km.

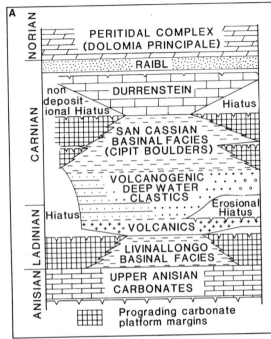

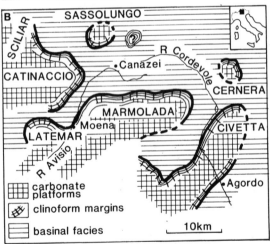

Fig. 2.13 *Middle Triassic carbonate platform margins in the Dolomites, northern Italy. (A) Stratigraphy showing platform and basin facies. (B) Early Ladinian palaeogeography showing carbonate platforms, margins with clinoforms and basinal Livinallongo Formation. After Bosellini (1984). (C) Clinoforms of the Carnian Lagazuoi Platform margin prograding in an oblique–tangential pattern on to basinal mudrocks and turbidites of the San Cassian Formation. Photo courtesy of Alfonso Bosellini.*

In the *Permian Basin of Texas and New Mexico*, reefs occur at shelf margins in some areas (including the classic Capitan Reef), and oolite and skeletal sand belts developed in others (e.g. Lower Clear Fork and Wichita Formations; Mazzullo, 1982). In fact, there has been much discussion over the precise nature of the Capitan Reef, whether it was a barrier-reef complex, shelf-marginal mound or a deeper-water reef rim (see Wilson, 1975). The reef consists of much bioclastic grainstone and wackestone, and

boundstone of sessile benthic and encrusting organisms, including bryozoans, red algae, sponges, *Tubiphytes*, foraminifera, brachiopods and crinoids. A well-developed fore-reef slope facies consists of reef debris arranged in inclined beds (clinoforms, Fig. 2.16). Lateral progradation of 20–30 km can be demonstrated, making the Capitan complex an example of an accretionary shelf margin. However, it does appear as if the reef did not grow at a shallow shelf margin, but at depths from 30 m down to 200 m. Grainstones with a large variety of bioclasts and ooids appear to have formed at a shallow rimmed shelf margin a few kilometres back from this slightly deeper-water reef. Onshelf 'Carlsbad facies' are skeletal grainstones–packstones, stromatolitic lime mudstones and calcisphere and ostracod wackestones, with tepee structures, vadose pisolites and fenestrae (Fig. 2.16). Much of this shelf-lagoon limestone is dolomitized, contrasting with the mostly non-dolomitic 'Capitan facies'. Along the inner shelf margin, there occur evaporites and red-bed clastics ('Chalk Bluff' and 'Bernal' facies).

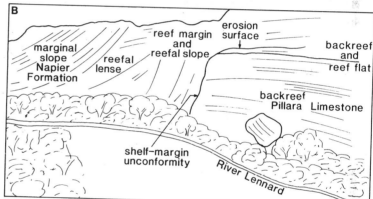

Fig. 2.14 *The 'classic face' of Windjana Gorge, Canning Basin, Western Australia, in Upper Devonian reefal limestones (for stratigraphy see Fig. 4.98). (A) Field photo courtesy of Colin Scrutton. (B) Interpretation after Playford (1980).*

Back-reef facies in ancient rimmed shelf sequences are commonly cyclic, generally of the shallowing-upward type. They are particularly well developed in Devonian rimmed shelf complexes, such as those of western Europe (Burchette, 1981), western Canada (e.g. the Snipe Lake and Kaybob Reefs, Havard & Oldershaw, 1976; Wong & Oldershaw, 1980) and Western Australia (e.g. the Pillara Limestone of the Canning Basin, Read, 1973; Playford, 1980). They also occur in the Carlsbad facies of the Permian Capitan Reef complex of west Texas (Fig. 2.16). The Devonian cycles commonly consist of coral–stromatoporoid biostromes passing up into fenestral, planar stromatolitic limestones with evidence of emergence such as desiccation cracks, evaporite pseudomorphs, vadose cements and paleokarsts. Where well developed, a vertical zonation of stromatoporoid growth forms is seen through the biostrome, from domal and tabular at the base through bulbous to branched forms at the top. Towards the shelf interior cycles become thinner and restricted facies more prominent, with even fluvial sandstones and conglomerates at the top of some cycles, cutting down into the carbonates. The back-reef cycles reflect an upward decrease in turbulence and water depth and increase in restriction of the environment as tidal flats were established. Carbonate cycles are discussed further in Section 2.10.

2.6 CARBONATE RAMPS

A carbonate ramp is a gently sloping surface, gradients of the order of a few metres per kilometre, contrasting markedly with the steep slope up to a carbonate shelf. On a ramp, shallow-water carbonates pass gradually offshore into deeper water and then into basinal sediments (Fig. 2.17). There is no major

BASIN
SLOPE
SHELF MARGIN
SHELF
DEVONIAN
U
M
L
SILURIAN
CAMBRO-ORD
PILOT
PILOT
DEVILS GATE
GUILMETTE
DEVILS GATE
DENAY
RABBIT HILL
McCOLLEY C.
OXYOKE
"NEVADA"
LONE MTN.
BEACON PK.
ROBERTS MTN.
LONE MTN.
LAKETOWN
HANSON CRK.
ELY SPRINGS
EUREKA
EUREKA
"POGONIP"
GOODWIN
DUNDERBERG
HALES
SWARBRICK
Stationary
Offlap margin
Gradual onlap
Onlap backstepping
Unconformity and stratigraphic gap

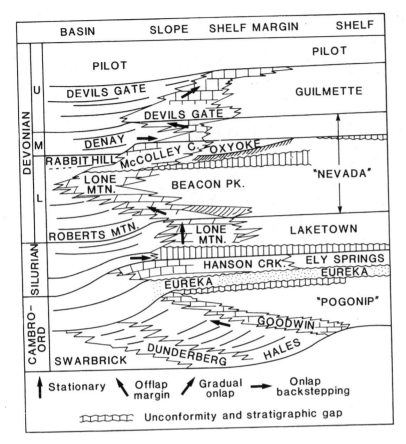

Fig. 2.15 *Upper Cambrian through Devonian carbonates of central Nevada, with different types of platform margin sequence. The vertical sequence shown is of the order of 4–6 km thick and the platform width some 400–500 km. After Cook et al. (1983).*

Fig. 2.16 *Sedimentary facies of the Capitan–Carlsbad complex (Upper Guadalupian, Permian) of the Guadalupe Mountains, west Texas and New Mexico. After Matthews (1984) based on work of Dunham.*

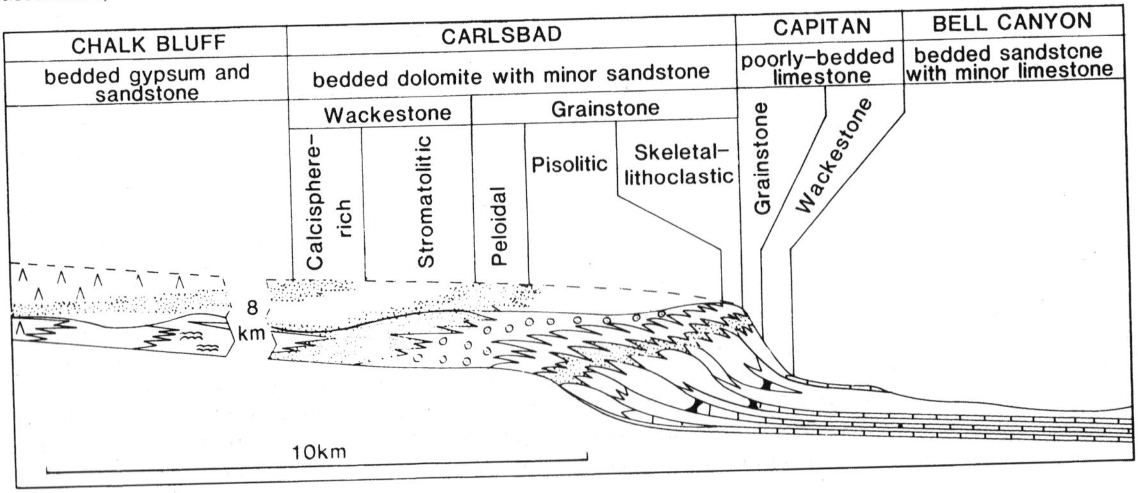

CHALK BLUFF
bedded gypsum and sandstone
CARLSBAD
bedded dolomite with minor sandstone
Wackestone
Grainstone
Calcisphere-rich
Stromatolitic
Peloidal
Pisolitic
Skeletal-lithoclastic
CAPITAN
poorly-bedded limestone
Grainstone
Wackestone
BELL CANYON
bedded sandstone with minor limestone
8 km
10km

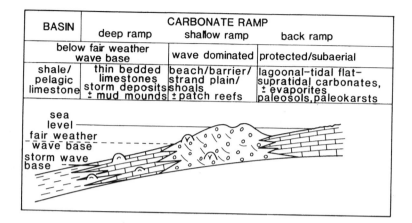

BASIN	CARBONATE RAMP		
	deep ramp	shallow ramp	back ramp
below fair weather wave base		wave dominated	protected/subaerial
shale/ pelagic limestone	thin bedded limestones storm deposits ± mud mounds	beach/barrier/ strand plain/ shoals ± patch reefs	lagoonal–tidal flat– supratidal carbonates, ± evaporites paleosols, paleokarsts

Fig. 2.17 *The carbonate ramp depositional model.*

break in slope which is the characteristic feature of a rimmed shelf to basin transition.

The distinctive sediments of the inner ramp are carbonate sands formed in the agitated shallow subtidal shoreface zone (above fairweather wave-base) and low intertidal. On a ramp, wave energy is not as intense as along a shelf margin where oceanic swell and storm waves are suddenly confronted with a shallow steep slope. Nevertheless, the gradual shoaling of a ramp does result in relatively strong wave action in the shoreface–intertidal zone and this permits the formation of shoreline carbonate sand bodies. Storm events are generally very important on ramps and along with normal wind–wave activity give rise to nearshore shoals and beaches through shoreward movement of sand. Offshore storm surges are important in transporting shoreface sands to the outer, deeper ramp. Along the inner ramp, there are three end-members of the ramp model: (1) a beach barrier–tidal delta complex with lagoons and tidal flats behind, (2) fringing sand banks and shoal complexes with intertidal and supratidal flats behind (but no lagoons), and (3) a strandplain of linear beach ridges with depressions (swales) between.

The best developed and described modern carbonate ramp is off the Trucial Coast of the Arabian Gulf (Loreau & Purser, 1973; Wagner & Van der Togt, 1973), where the seafloor gradually slopes down from sea-level to a depth of 90 m in the axis of the gulf. The ramp is not a smooth surface; there are many local shoal areas. Along the inner ramp, there occur barrier–tidal delta complexes with lagoons and tidal flats behind, and fringing shoals with tidal flats to landward. The carbonate facies of the Trucial Coast are described in Section 3.4. The northeast Yucatan coast of Mexico is an example of a ramp

with an inner strandplain zone of beach ridges (Ward *et al.*, 1985); these are described in Section 4.1.2. Shark Bay in Western Australia (Logan *et al.*, 1970; Logan *et al.*, 1974) is also considered an example of a ramp, and the western Florida Platform (Mullins *et al.*, 1988) is a somewhat deeper-water ramp.

2.6.1 Carbonate ramp facies

The nearshore sand belt of a ramp consists of skeletal and oolitic grainstone. Peloids may be important too. Apart from compositional differences, the facies developed will be identical to those of a siliciclastic beach barrier or strandplain system. The beach-barrier/beach-ridge carbonates will show bedding dipping at a low angle offshore (surf-swash deposit) and onshore from deposition on the backsides of beach berms. Onshore-directed cross-bedding will be produced by shoreface megaripples and wave–ripple cross-lamination will also occur. Aeolian cross-bedding is likely through backshore wind-blown dune migration (also shoreward). Rootlets and vadose diagenetic fabrics are possible in the backshore sediments and burrows may occur in the low intertidal–shoreface part. In the strandplain ramp model, beach sands dominate the upper part of the ramp sequence, with aeolian sands above. Some muddy sediments may accumulate in the swales between the beach ridges. In a beach barrier–lagoon ramp model, tidal channels will cut through the barrier and tidal deltas may occur at the ends of the channels. Muddy sands and sandy muds will accumulate in the lagoon and on tidal flats around, with climate playing a major role in controlling the biota, water salinity and mineral precipitation. Tidal deltas, which could be largely oolitic, would give rise to offshore and/or onshore directed cross-bedding from sand wave and dune migration. Herring-bone cross-bedding is possible from tidal current reversals. Tidal inlet migration, resulting from longshore currents, will rework beach-barrier sediments and give rise to a sharp-based shell lag of the channel floor, overlain by variously cross-bedded, cross-laminated and flat-bedded sands of the deep to shallow channel and migrating spit. These inner-ramp environments and facies are described in detail in Section 4.1.3a.

At depths a little greater than fairweather wave-base, organic productivity is still high so that bio-clastic limestones will be dominant. Shoals of skeletal debris (grainstones) are likely to be formed through reworking by storm waves. These would have medium- to small-scale cross-stratification and contain win-nowed horizons of coarser debris (rudstones). Hummocky and swaley cross-stratification is possible here too. Migration of these skeletal sand shoals would take place during major storms and result in coarsening-up, thickening-up units, a few metres thick, of skeletal wackestones passing up into skeletal grainstones and rudstones.

In slightly deeper areas, below storm wave-base, skeletal packstones and particularly skeletal wacke-stones will dominate. Fine carbonate will be derived largely from shallow-water areas where fragmentation of skeletal debris takes place; some lime mud will be formed *in situ*. Thin, graded bioclastic grainstones and packstones will be common in the below storm wave-base areas of the ramp, deposited from seaward flowing storm-surge currents. Scoured bases, grooves, even flutes, can be expected on the bases of these storm beds, as well as a sequence of internal struc-tures indicating deposition from waning flow (flat-bedding to ripple cross-lamination especially). Burrows may occur on the base and within the storm bed. These outer-ramp, deeper-water environments and facies are described in detail in Section 4.1.3b.

Major reef developments are rare on carbonate ramps since there is no major break of slope in shallow water for colonization by reef-building or-ganisms. However, small patch reefs or reef mounds are common, occurring seaward of the beach-barrier system, in the back-barrier lagoon, or on any topo-graphic highs on the deep ramp. Pinnacle reefs (Section 4.5.6) may develop in these deeper-water settings. One particular type of buildup which devel-oped on outer ramps (also on slopes up to shelves), especially during the Palaeozoic, is the mud mound (also called reef knoll or mud bank). Composed largely of micrite, usually clotted and pelleted, they

usually had some form of seafloor relief, as shown by flank beds which slope off the mound structure with dips up to 20°. Mud mounds do not appear to contain any metazoan frame-builders, and this has given rise to much discussion, posing two questions: how did mound structure form, and where did all the lime mud come from? (see Section 4.5.8).

2.6.2 Carbonate ramp types

Two categories of ramp are distinguished (Fig. 2.18): *homoclinal ramps* where slopes are relatively uniform and *distally-steepened ramps* where there is an in-crease in gradient in the outer, deep ramp region (Read, 1982, 1985). In the first type, there are very few slumps, debris flow deposits or turbidites in the deeper-water ramp facies, but these sediment gravity flow deposits are common in the second type. In this respect, the facies of distally-steepened ramps are similar to accretionary shelf margins, but one import-ant difference is that the break in slope comes in deeper water on the ramp, so that the resedimented deposits consist of outer-ramp and upper-slope sands and muds, and clasts of this material, rather than shallow-water debris as is typical of slope and basin facies adjacent to a rimmed shelf margin. The Trucial Coast of the Arabian Gulf and Shark Bay of Western Australia are examples of homoclinal ramps, whereas northeastern Yucatan and western Florida are distally-steepened ramps.

Sea-level changes and subsidence rates will be important in determining the thicknesses of ramp facies and the lateral migration of sand bodies. Small sea-level changes will mostly affect the inner-ramp environments where water is shallow. On the deeper ramp, sea-level changes will affect the position of wave-base and result in an increase or decrease in the amount of reworking of bottom sediments. On the inner ramp, during a still-stand or slight sea-level fall, a thick sand body can develop through seaward progradation of the shoreline beach-barrier or strand-plain system, if there is a continuous supply of sand. A slow sea-level rise can lead to the shoreward mi-gration of the sand belt, and the generation of a transgressive sequence. As with siliciclastic shorelines, rapid transgression may leave little record of the shoreface, other than a disconformity (a ravinement) and a basal conglomerate, produced by surf-zone erosion. Alternatively, inner-ramp sand bodies may be drowned as sea-level rises, leaving the sand bodies abandoned on the outer ramp to be covered by muds or reworked by storms (see Sections 4.1.4 and 4.1.5).

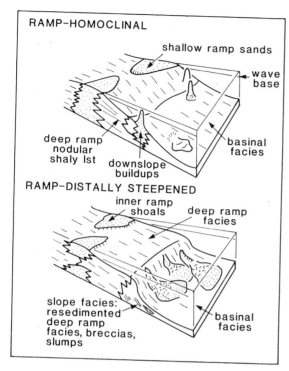

Fig. 2.18 *Two types of ramp: homoclinal and distally steepened. After Read (1982).*

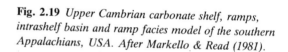

Fig. 2.19 *Upper Cambrian carbonate shelf, ramps, intrashelf basin and ramp facies model of the southern Appalachians, USA. After Markello & Read (1981).*

2.6.3 Ancient carbonate ramps

One of the first carbonate formations to be interpreted in terms of the ramp model was the *Jurassic Smackover Formation of the Texas–Louisiana–Arkansas subsurface* (Ahr, 1973; Section 4.1.6, see Fig. 4.20). It consists of a seawards-prograding wedge, 30–80 km wide and 100 m thick, of oolitic and peloidal grainstone. The deeper-ramp facies are peloidal skeletal wackestones and mudstones which pass into basinal, pelagic organic-rich lime mudstones. Behind the oolitic sand belt, lime mudstones, evaporites and red beds of the Buckner Formation developed in lagoons and supratidal and subaerial environments. Ramp carbonates are well represented in the *Upper Cambrian to Middle Ordovician of the Appalachians in Virginia* (Read, 1980; Markello & Read, 1981). They were deposited around an intrashelf basin and characteristically consist of oolites in the shallow ramp and ribbon carbonates in the deeper ramp (Fig. 2.19). The latter are thin-bedded storm deposits.

In the UK, parts of the *Lower Carboniferous succession in south Wales* are ramp in character (Fig. 2.20). In the southernmost part, deeper-water mud-stones and packstones locally contain mud mounds and give way northwards to inner-ramp skeletal, oolitic packstones and grainstones. Several major phases of shallowing result from oolite shoal and beach progradation in a strandplain type of ramp model (also see Section 4.1.6 and Wright, 1986a). During strandplain progradation, palaeokarstic surfaces and soils developed upon the subaerially exposed beach ridges. In the Bowland Basin of northern England (Gawthorpe, 1986), Dinantian lagoonal lime mudstones pass laterally into inner-ramp grainstones and then outer-ramp wackestones and basinal mudstones. In vertical sequence, the ramp deposits pass up into the slope deposits, with slumps, slides and debris flow deposits, derived from a now fault-bounded rimmed carbonate shelf.

The examples cited above show facies distributions consistent with the homoclinal ramp model. Read (1985) has suggested that distally-steepened carbonate ramps existed along the western USA shelf margin during the Upper Cambrian and Lower Ordovician (Cook & Taylor, 1977), and Ruppell & Walker (1984) have described a distally-steepened ramp sequence

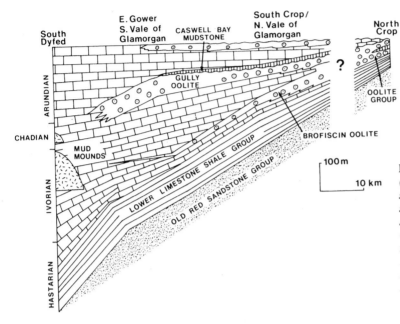

Fig. 2.20 *Lower Carboniferous (Dinantian) ramp sequences in southwest UK. Stratigraphic cross-section shows three strandplain–shoal oolite complexes, Brofiscin, Gully and Cefn Hendy (un-named at top) which prograded southwards. Beach-barrier complexes occur within the Lower Limestone Shale Group, see Section 4.1.6. After Wright (1986a) and Hird & Tucker (1988).*

from the mid-Ordovician of Tennessee, deposited in the Appalachian foreland basin.

Ramp carbonates are well developed in the *Triassic Muschelkalk of western Europe*, in the South German Basin and the Catalan Basin, Spain. The Upper Muschelkalk in the intracratonic German Basin consists of many shallowing-upward cycles (Fig. 2.4, 2.21). Two types of cycle are recognized: oolite grainstone cycles of the more marginal parts of the ramp system and thickening-upward cycles of the more open marine, outer ramp (Aigner, 1984). In the first type, 2–6 m thick, nodular (through bioturbation) lime mudstones pass up into skeletal wackestones and packstones and then into oolitic grainstones. The top surface of the cycle is a submarine hardground in some cases, and in others there is evidence of subaerial exposure in the form of vadose cements. The deeper-ramp cycles, 1–7 m thick, begin with widespread marlstone horizons, used as marker beds in the stratigraphy, which pass up into thin- and then thicker-bedded limestones, showing much evidence of deposition from storm-induced flows (tempestites). Some oolite cycles can be traced into the thickening-upward cycles, showing that position upon the ramp determines the character of these cycles. Aigner (1984) showed that the Upper Muschelkalk ramp consisted of a back-bank lagoonal area, where dolomites and oncolitic packstones–wackestones were deposited, a zone of shelly ooid banks, forming the inner shallow ramp, and then a deeper-water outer ramp, where

skeletal packstones, lime mudstones and marlstones were deposited (Fig. 2.21). The cycles formed as a result of rapid transgressive events followed by still-stand or slow regression, when the oolite sand bodies prograded basinwards, and the thin- to thick-bedded limestones were deposited by storms. See Section 2.10 for discussion of shallowing-upward cycles.

Aigner (1984, 1985) has shown how the vertical sequence of the small-scale cycles can be used to infer longer time-scale changes in relative sea-level within the basin, which may reflect eustatic and/or tectonic controls. On this larger scale, the Muschelkalk ramp carbonates show two major developments of lime sands: the first, more crinoidal, is part of a long-term transgressive motif (i.e. the small-scale cycles show a progressive upward trend towards less sand), and the second, more oolitic, is part of a long-term regressive motif (i.e. the small-scale cycles show a progressive upward trend towards more sand). Thus in the South German Basin, facies analysis of the Upper Muschelkalk ramp cycles reveals short-term rapid transgressive events, followed by still-stand/slight regression and a much longer-term transgression–regression on the scale of the whole Upper Muschelkalk (Aigner, 1984).

Cyclic ramp carbonates in the *Upper Muschelkalk of the Catalan Basin, eastern Spain* (Calvet & Tucker, 1988) are of the outer-ramp type. Marlstone passes up into thin-, then thicker-bedded limestones, but there are few discrete storm beds. The latter could be a

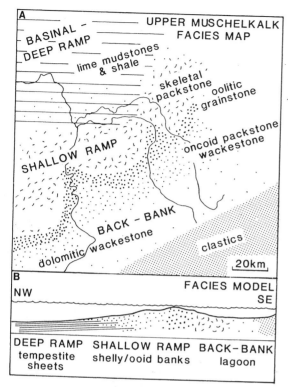

Fig. 2.21 *Carbonate ramp facies of the Upper Muschelkalk (Triassic) of southern West Germany. The sequence consists of storm beds arranged into many shallowing-upward cycles, each consisting of thickening-up and coarsening-up beds (shown in Fig. 2.3). (A) Schematic palaeogeography and facies pattern. (B) Facies model. After Aigner (1984).*

consequence of the orientation of the Catalan ramp relative to prevailing winds and storms. The ramp sequences of the Upper Muschelkalk in the Catalan Basin include reefal mounds which grew in a mid- to outer-ramp location. A sharp sea-level drop exposed the mounds to some karstification and a subsequent sea-level rise resulted in the establishment of a stratified sea and organic-rich carbonate sedimentation. Pinnacle reefs occur in a somewhat similar ramp setting in the Silurian (Niagaran) of the Michigan Basin (Fig. 2.22; Shaver, 1977; Sears & Lucia, 1980).

2.7 EPEIRIC CARBONATE PLATFORMS

Epeiric carbonate platforms are very extensive areas of negligible topography. Although non-existent today, shallow epeiric seas covered extensive areas of cratons at times during the geological record. Examples include the Late Precambrian to Ordovician of west-central China, the Cambro-Ordovician of North America, the Upper Dinantian (Mississippian) and Triassic—Jurassic of parts of western Europe, and some of the Tertiary of the Middle East. Water depths on these platforms were generally less than 10 m, so that shallow subtidal to intertidal environments dominated. The intertidal areas consisted of tidal flats many kilometres to tens of kilometres wide. These would have developed extensively in the epeiric platform interiors, along a broad shoreline, with supratidal flats beyond, giving way to the peneplained land surface where subaerial processes such as pedogenesis and karstification would have operated. Tidal flats would also have developed around slightly more positive areas upon the epeiric platform. Apart from local shoals of skeletal sand, the subtidal zone would also be a near-flat surface but probably with slightly deeper and slightly shallower areas reflecting pre-existing topography on the craton or the effects of differential subsidence. Deep, intraplatform basins can occur within an epeiric sea, and may be surrounded by ramps or rimmed shelves.

The traditional view of epeiric seas is that they had a low tidal range, with the tides being damped out by frictional effects over the very extensive shallow seafloor. In this model, tidal currents would have been insignificant on the open epeiric platform, but perhaps quite strong in any channels of the broad intertidal zone. For much of the time, epeiric platforms would have been very quiet, low-energy environments, with only wind—wave activity. Fairweather wave-base would have been quite shallow, less than 5 m. In the general and much-cited model of Irwin (1965) for epi-continental seas, a low-energy, below wave-base zone X of the open sea, gives way to a relatively narrow zone Y of higher energy where waves impinge on the seafloor and there are strong tidal currents. Beyond this, in zone Z up to hundreds of kilometres wide, there is restricted circulation, tidal effects are minimal, and storm-wave action is only periodically significant. Hypo- and/or hypersalinities are likely in zone Z.

The dominant processes affecting epeiric platform sedimentation in tideless epeiric seas would have been storms, their frequency, direction and magnitude controlled by climatic factors. Severe storms can raise sea-level by several metres and give rise to currents reaching 1 m s^{-1}. On an epeiric platform several hundred kilometres across, storm winds blowing per-

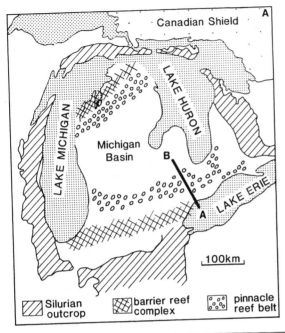

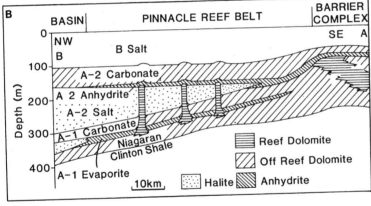

Fig. 2.22 *Silurian barrier reef and pinnacle reef belt in the Michigan Basin, northeastern USA. (A) Location of reefs, with the pinnacles developed on the slope into the basin. (B) Cross-section along line A–B. Barrier reef–pinnacle reef growth was terminated by evaporative drawdown in the basin when the A-1 evaporite was deposited. Further flooding and evaporation led to other carbonate–evaporite sequences. After Gardner & Bray (1984).*

sistently from one quadrant will pile up water in a downwind direction. Where normally quite shallow water exists (<2 m), the epeiric platform floor itself could even be exposed as the sea is blown off it. Strong surges would cross the epeiric platform after the storm subsided and the sea returned to its normal level. During storms, the epeiric platform interior tidal flats would be flooded and much shallow subtidal sediment deposited upon them. In the subtidal, skeletal debris would be transported and sorted during storms and post-storm surges, and deposited to give grainstone beds. Winnowed shell lags (rudstones)

would be left after the passage of storm currents and waves.

The classic, storm-dominated, tideless epeiric sea model has recently been challenged by Pratt & James (1986). They asserted that tides would only be damped in epeiric seas if there was a highly rugged seafloor and quoted studies (e.g. Klein & Ryer, 1978; Cram, 1979) showing that tidal range increases with the width of the continental shelf, to postulate that epeiric seas would have been subject to a tidal regime. For a clastic epeiric sea, Slingerland (1986) computed a numerical model for the Upper Devonian Catskill Sea of eastern North America, which showed that it was likely tide-dominated, with regions of high mesotidal to low macrotidal ranges. From a consideration of facies relationships in the Lower Ordovician St George Group carbonates of Newfoundland, Pratt & James (1986) proposed that many epeiric seas consisted of low-relief supratidal islands and intertidal banks, surrounded by open water (Fig. 2.23). The evidence for a tidal regime in the St George Group sediments is the following *suite* of features: (1) herring-bone cross-stratification, (2) flaser, lenticular and wavy bedding, (3) mudcracks, (4) local bioturbation, (5) stromatolites, (6) patch reefs and oolites, (7) linear channels and elongate patch reefs, (8) rare evaporites, and (9) lateral facies variations such as planar stromatolites ('cryptalgal laminites') passing into wavy and lenticular-bedded carbonates and thin-bedded lime mudstones—grainstones. Although many of these features could reflect non-tidal mechanisms, taken together, a tidal regime is considered most likely. The main evidence for tidal islands and banks is the lateral impersistence of facies. Storm waves would still have been important depositional processes within this tidal environment. It remains to be demonstrated how applicable the tidal island model is to other epeiric platform carbonate sequences in preference to the storm-dominated, non-tidal model for epeiric sea carbonates.

Under constant sea-level, apart from some aggrading of the shallow subtidal sediments through simple skeletal carbonate production, the dominant depositional process would be progradation of the tidal flats. In the traditional storm-dominated epeiric sea model, the progradation will mostly take place along the platform interior shoreline. In the tidal island model, banks and islands will occur over much of the platform and will accrete vertically and migrate laterally. In both models, shallowing-upward sequences are the typical product, but there are differences. These are explored in Section 2.10 discussing shallowing-upward cycles in general. The movement of shallow subtidal sand shoals by tidal currents and/or storm waves could be important in both models.

Although there are no modern examples of the very extensive platforms of the past, an indication of what sedimentation may have been like, at least under the storm-dominated, non-tidal, epeiric sea model, can be gleaned from the studies of the interior of the Great Bahama Bank (Shinn *et al.*, 1969; Gebelein, 1974a; Hardie, 1977; Multer, 1977). To the west of Andros Island, there occur protected tidal flats along a broad shoreline and a shallow subtidal platform. Tidal range is very low and wind—wave activity is also weak. Sedimentation is largely controlled by the rare storm events. The facies and sequences of western Andros are described in Sections 3.2.2b and 4.3. Florida Bay, too, is an extensive area of shallow-water carbonate deposition, with insignificant tidal currents (see Section 3.3).

2.8 ISOLATED CARBONATE PLATFORMS

Isolated platform refers to shallow-water carbonate accumulations surrounded by deep water. No size restriction is implied, although if the platform is very

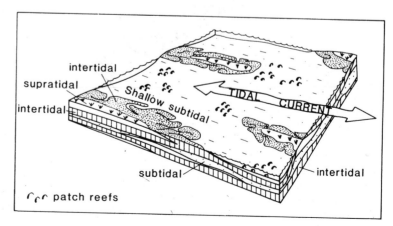

Fig. 2.23 *Epeiric sea carbonate platform depositional model involving low-relief islands oriented parallel to tidal currents. The sequences produced as a result of this type of deposition are laterally discontinuous subtidal, intertidal and supratidal deposits commonly arranged into shallowing-upward units. After Pratt & James (1986), devised for the Cambro-Ordovician St George Group of Newfoundland.*

large, then it can be looked at in terms of the shelf, ramp and epeiric platform models discussed in previous sections. It is the smaller isolated platforms that do have distinctive facies patterns, since the various margins of the platform will have been subjected to different wave and storm regimes, depending on their orientation with respect to the direction of prevailing winds and major storms. Most isolated platforms have steep margins and slopes into deep or very deep water. Isolated platforms will have marginal reefs and sand bodies, with quieter-water, sandy muds and muds in the platform interior, and possibly sandy islands with tidal flats around. If there is a high rate of carbonate production around the margins against a background of steady subsidence, then a rimmed isolated platform may develop, with a deep lagoon in the centre. This variant can be termed an atoll, but true oceanic atolls have formed upon extinct, subsiding volcanoes.

2.8.1 Modern isolated platforms

The Bahama Banks form a very large isolated platform (area of 96 000 km²), cut off from the Florida Shelf by the deep Florida Straits. The facies around the margin of the Bahama Platform are very much dependent on whether the margin is leeward or windward, open or protected, or tide-dominated (see Sections 3.2, 4.2 and 5.11; Hine *et al.*, 1981). Reefs are particularly well developed along windward margins, and the type of sand body varies considerably in geometry, structure and grain type depending on location. The central parts of the Bahama Platform are occupied by shallow subtidal quiet-water skeletal muds and sandy muds, and tidal flats with peloidal muds, algal mats and mangroves (etc.) occur around islands (Section 3.2.1). Because the Bahama Platform is very large, the facies developments there can be used in the facies models for the rimmed shelf and epeiric platform, described in previous sections. Much smaller isolated platforms include those on the Coral Sea Plateau (Orme, 1977; Read, 1985), and some of the reefs off the Belize Shelf. In addition, shallow-water carbonates are being deposited upon fault-blocks in the Red Sea (Fig. 2.24; Purser *et al.*, 1987), and in the Arabian Gulf they are accumulating upon topographic highs caused by salt diapirism (Section 3.4.1; Purser, 1973b).

Glovers Reef is an oval-shaped isolated platform, 28 × 10 km, situated 15 km east of the major barrier reef rimming the Belize Shelf (James & Ginsburg, 1979). Glovers atoll appears to be a fault-block upon a rifted continental margin, trailing edge (Fig. 2.8). This isolated platform has a nearly continuous marginal reef, but it is narrower on the western, leeward side, and *Acropora palmata*, the robust frame-building coral of many Caribbean reefs, is less common there. The reef rim provides protection for a central lagoon, 6–18 m deep, where patch reefs are numerous. Poorly-sorted coarse sands occur around the patch reefs and muddy fine sands cover the lagoon floor.

Off the Trucial Coast of the Arabian Gulf, topographic highs with coralgal reefs rise above the deep, muddy ramp, and are a few square kilometres in area. In fact, there is quite a variation in the carbonate facies upon these highs, so that a range of facies models can be produced to illustrate the progressive development of carbonate sedimentation (Purser, 1973b; reviewed in Tucker, 1985a). Four stages are

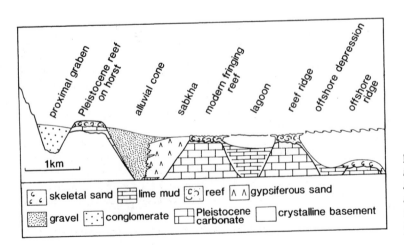

Fig. 2.24 *Schematic cross-section of the western Red Sea shoreline area showing development of reefs on structural blocks and close association with fluvial clastics. After Purser* et al. *(1987).*

recognized. The initial stage is the development of carbonate sands on the high through enhanced carbonate skeletal production on this shallow agitated area and winnowing out of lime mud. Sand distribution is asymmetric, sedimentation taking place mainly in a downwind direction. Stage 2 is the establishment of a reef on the windward side of the high, giving rise to more rapid production and accumulation of skeletal sand. In stage 3, reef growth is well advanced, producing more of a barrier to waves and currents, so that downwind tails of reef debris from the margins of the reef produce a leeside lagoon. Island formation is likely from storm piling-up of reef talus. In stage 4, the reef has grown most or all the way around the topographic high to enclose a lagoon. Lime muds may accumulate here in this atoll-like stage.

In the Gulf of Suez, carbonates are being deposited upon tilted fault-blocks formed during rifting, and the facies pattern here is one of reefs developed along the gulf-ward side of the block, with lagoonal sediments behind (Fig. 2.24; Purser *et al.*, 1987). Where several blocks occur, those nearer to the rift margin are subject to periodic influxes of clastic sediment and freshwater from alluvial fans, and this can kill off some of the nearshore reefs.

2.8.2 Ancient isolated platforms

Isolated carbonate platforms occur in the *Devonian of western Canada*, where reef development is more extensive on the windward side. The subsurface Golden Spike Reef complex of the Frasnian Leduc Formation is a good example (Walls, 1983). The reefs are mainly massive stromatoporoid framestones to bindstones, and stromatoporoid rudstones. A reef talus surrounds the isolated platform and passes rapidly into basinal shale. The platform interior facies consist of interbedded skeletal, sand flat grainstones and packstones and peritidal planar stromatolites. Small patch reefs of massive and branching stromatoporoids also occur. The Golden Spike Reef is a prolific oil producer, with the highest porosities occurring in the marginal reef facies and the skeletal sands.

Mesozoic isolated platforms are well represented in the Tethyan region. They occur in the *Triassic of the Dolomites and Jurassic of the Venetian Alps* in northern Italy (Bosellini *et al.*, 1981; Bosellini, 1984; Blendinger, 1986), and in the Cretaceous of the Apennines of central Italy. The *Golden Lane 'atoll'* of mid-Cretaceous age in Mexico (Enos, 1977b, 1988) is another good example.

2.9 DROWNED CARBONATE PLATFORMS

Rapid relative sea-level rises are required to drown a carbonate platform, since, as explained in Section 2.2.1, shallow-water carbonate production rates are generally more than sufficient to keep pace with moderate rates of subsidence and sea-level rise. Rapid relative sea-level rises can be brought about by fault-induced subsidence and glacio-eustatic sea-level rises. One other possibility for drowning is a drastic reduction in carbonate productivity as a result of some environmental stress, so that shallow-water carbonate deposition is unable to keep up with even moderate relative sea-level rises. The development of poorly-oxygenated and/or low nutrient waters on a platform or climatic changes could have this effect.

Drowned platforms typically have deep-water carbonate facies overlying the shallow-water facies. Where the sea-level rise was sufficient to take the platform below the photic zone, then many benthic organisms, but especially algae, will be excluded, and the carbonates will be dominated by planktonic–nektonic fossils (such as coccoliths, planktonic foraminifera, pteropods, certain thin-shelled bivalves and ammonoids). These pelagic carbonates are usually fine-grained, nodular to thin-bedded limestones, with shaley partings or a clay-poor and clay-rich bedding (see Chapter 5). Hardgrounds may occur, and stratigraphic thickness is generally reduced relative to contemporaneous shallow-water facies. Mineralized surfaces may be present, with Fe and Mn oxides and phosphate. Disconformities and stratigraphic breaks can also occur in drowned platform sequences.

During drowning of a platform, reefs with their higher production rate than surrounding areas may be able to keep up with the rapid sea-level rise. This can generate a high-relief barrier or a pinnacle morphology if starting from a patch reef. The latter commonly develop on drowned ramps. Backstepping of a rimmed shelf may take place when sea-level rise is rapid and reef growth resumes to landward during an ensuing still-stand. Drowning of a rimmed shelf can lead to the formation of a distally-steepened ramp.

When the sea-level rise is not sufficient to terminate most benthic carbonate production, then an *incipiently drowned platform* is produced. Depths will still be within the photic zone in these instances and if there are no further sea-level rises, the platform may aggrade into very shallow water again.

Modern drowned platforms are common in the

Pacific and Indian Oceans, where atolls and isolated platforms have sunk below the photic zone, largely as a result of thermal subsidence following a cessation of volcanicity. In the Caribbean, Cay Sal Bank (shown in Fig. 3.2) is a small drowned platform, and the Blake Plateau north of the Bahamas is a much larger one. The latter was a shallow-water platform in the Cretaceous which subsequently underwent rapid subsidence. It is now at 2000–4000 m depth receiving pelagic sediments and Fe and Mn oxide precipitates. Modern distally-steepened ramps such as eastern Yucatan and west Florida are the result of the rapid Holocene post-glacial sea-level rise. The Holocene Queensland Shelf has been termed an incipiently-drowned rimmed shelf, where the shelf-edge reefs were able to keep growing and so form the Great Barrier Reef, with a deep and broad lagoon behind.

There are numerous ancient examples of drowned platforms (see Kendall & Schlager, 1981; Schlager, 1981). They are particularly well developed in the Jurassic of the Tethyan area and in the Devonian of western Europe; in both instances, shallow-water carbonates are overlain by ammonoid-bearing pelagic limestones (see Chapter 5). Drowned platforms also occur widely in the Cretaceous, as in Texas, Mexico and the Middle East. Ancient incipiently-drowned platforms also occur in the Devonian, in western Canada and the Canning Basin for example. In the latter case, rapid sea-level rise caused backstepping of reefs at the rimmed shelf margin (see Fig. 4.98). Pinnacle reefs formed by continuous sea-level rise are common on the slope into the Silurian Michigan Basin (see Fig. 2.22).

2.10 CARBONATE CYCLES

Through normal depositional processes, such as progradation of tidal flats, lateral migration of tidal islands, progradation of carbonate sand bodies, and vertical accretion of shallow subtidal sediments into shallower depths, carbonate platforms will build up to sea-level and just above. The typical sequence produced is a shallowing-upward unit of subtidal through to intertidal–supratidal deposits. Such a sequence is revealed by coring through Holocene sediments on the western side of Andros Island (see Fig. 3.22, Section 3.2.2). In view of the deposition close to sea-level, a relative drop in sea-level will expose the platform to supratidal–subaerial processes, such as sabkha evaporite precipitation if the climate is arid and there is still a source of seawater, soil for-

mation, like calcretization if a semi-arid climate prevails, or karstification if more humid. With a relative rise in sea-level, subtidal environments are widely established over a platform, with tidal flats at the distant shoreline.

2.10.1 Shallowing-upward carbonate cycles

Repetitions of small-scale shallowing-upward units are common in carbonate formations, demonstrating the periodic flooding of platforms through transgressive events. For examples see Laporte (1967), Coogan (1969), Ginsburg (1975), Wilson (1975), Lohmann (1976), Somerville (1979), Bridges (1982), Enos (1983), Grotzinger (1986a,b), Hardie (1986a,b), Hardie *et al.* (1986) and Goldhammer *et al.* (1987). In detail, the cycles are very variable. Microfacies analysis reveals differences within one cycle when traced laterally across a platform, and between cycles in a vertical sequence. Frequently, the cycles of one particular stage or substage of a geological period have features in common, which are different from those in the cycles of a succeeding stage. Such is the case with cycles in the British Dinantian (Lower Carboniferous), reviewed in Walkden (1987).

Shallowing-upward cycles are developed in the *Asbian and Brigantian stages of the Upper Dinantian in Wales* (Fig. 2.25 Somerville, 1979; Gray, 1981) When cycles are traced towards the platform interior, gradual but distinct changes are observed in addition to a general shorewards thinning of each cycle (Fig. 2.26). Away from the open platform, the transgressive phase (a below fairweather wave-base, thin argillaceous packstone–wackestone facies), which forms the lower part of each cycle, gradually reduces in thickness. Sequences of more proximal areas tend to have shallow subtidal (above fairweather wave-base) sediments in their lower parts. The 'regressive' phase tidal flat sediments (fenestral, stromatolitic, peloidal wackestones) increase in thickness as cycles are traced shorewards and may comprise the whole cycle in very proximal areas. Lateral variations are also seen in the nature of emergence horizons at the top of each cycle: in proximal areas, paleokarstic surfaces are usually developed (possibly above a calcrete). These pass distally into 'sutured discontinuity surfaces', interpreted as the product of intertidal, rather than wholly subaerial, dissolution and erosion (cf. Read & Grover, 1977).

The lateral variations in these Dinantian cycles are primarily a function of the gradient of the platform. The transgressions appear to have been relatively

Fig. 2.25 *Shallowing-upward cycles in the Lower Carboniferous (Asbian—Brigantian) near Llangollen, Wales. Each major limestone cliff is one cycle. Courtesy of Ian Somerville.*

rapid, and during the transgression, a basal bed was developed in distal to medial areas. Sedimentation after the initial transgression was determined by depth, especially relative to wave-base (see Fig. 2.26A). Differences between cycles of different stages relate to the magnitude of the transgressions. For example, compared with Asbian cycles, those of the Brigantian are dominated by thin-bedded below wave-base packstones—wackestones. This indicates that the transgressions were more extensive, resulting in a greater depth of water over the platform.

Within the Upper Dinantian sequence of mid-Wales, and elsewhere in the UK, it is not uncommon to find cycles 'missing' in proximal areas or unrecognizable in more distal areas. Problems of stratigraphic correlation can result. The absence of sediments of a particular cycle in proximal areas results from weak transgressions that did not extend landwards as far as earlier and later ones (see Fig. 2.26B). Where this happened, the subaerial processes operated for a longer time (over several cycles) and very marked palaeokarstic horizons can develop (multiple palaeokarsts). In more distal platform areas, it could happen that the prograding tidal flats did not arrive, so that subtidal conditions were maintained throughout. Fluctuations from below to above wave-base may be recognized by careful study of microfacies (e.g. Jefferson, 1980), or it may be that sea-level changes were not substantial enough on the open platform to cause any modification to the sedimentary facies.

In the *Early Proterozoic Rocknest carbonates of northwest Canada*, four cycle types occur, reflecting

their palaeogeographic position on the platform (Fig. 2.27; Grotzinger, 1986a,b). Shale-based cycles of the inner platform, pass westwards into dolosiltite-based cycles and these into tufa-based cycles of a shoal complex. The latter give way to grainstone-based cycles developed in a back-reef sands environment near the platform margin. The upper parts of the inner platform cycles consist of tufa and planar stromatolites, sharply overlain by an intraclast packstone bed, up to 0.3 m thick, which is the transgressive basal unit of the succeeding cycle. Cycle boundaries are erosion surfaces which become more pronounced towards the platform margin shoal complex. The tufas are beds several metres thick of cement laminae in smooth, undulatory, colloform and columnar arrangements, along with tepees, desiccation cracks and intraclasts. Much of the tufa was precipitated as aragonite (Grotzinger & Read, 1983).

Metre-scale cycles are prominent in the platform carbonates of the *Middle and Upper Triassic of the Dolomites in northern Italy*, especially in the Latemar Limestone (Ladinian), Durrenstein (Carnian) and Dolomia Principale (Norian) (Hardie, 1986a,b; Hardie *et al.*, 1986; Goldhammer *et al.*, 1987). In the Latemar Limestone, the cycles are mostly less than 1 m thick (0.65 m average) and more than 500 have been recognized. Many are limestone—dolomite couplets, consisting of shallow subtidal packstones—grainstones with a thin cap rock (the dolomite) containing vadose cements, pisoids and tepee structures (Fig. 2.28). At certain horizons, complex tepee structures on a larger scale disrupt many metres of the

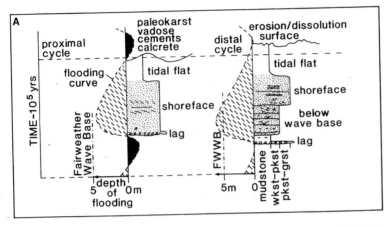

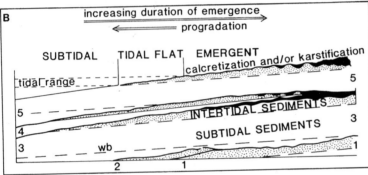

Fig. 2.26 *Shallowing-upward cycles of the Lower Carboniferous, near Llangollen, Wales. (A) Two typical cycles with the lateral variation interpreted in terms of depth of platform flooding, from a distal cycle to a proximal cycle of the platform interior. (B) Model for development of shallowing-upward cycles in a broadly transgressive sequence. The lateral variation of individual cycles and the vertical differences between cycles are reflections of the degree of flooding of each transgression. After Gray (1981).*

thin cycles. Thus in the Latemar Limestone the metre-thick cycles (which have been called '*diagenetic cycles*') reflect frequent periods of subaerial exposure on a short time-scale, whereas the superimposed tepee horizons reflect much longer periods of exposure. Of particular note is the absence of peritidal deposits, which rules out tidal flat progradation as the cause of the exposure. Studies of the thickness variation of the thin Latemar cycles up through the formation show that they commonly group into packets of five ('pentacycles') with the thickness of each cycle in a packet decreasing upwards (Goldhammer *et al.*, 1987; Goldhammer, 1988). Durrenstein cycles average 2 m in thickness and are also subtidal facies with a tepee cap. In the Dolomia Principale, cycles of subtidal sediments with thin pisolitic breccia caps are again present, although there are peritidal units of shallowing-upward cycles capped by mudcracked laminites.

Cycles are also prominent in the *Norian Dachstein*

Limestone of the northern calcareous Alps in Austria (Fischer, 1964), where they have been referred to as Lofer cycles. These are also largely subtidal deposits, but they do contain intertidal fenestral, stromatolitic and dolomitic limestones. The latter is a distinctive rock type referred to as *Loferite*. Some of these Lofer cycles appear to have an intertidal facies at the base, passing up into a thick subtidal facies, with evidence of exposure at the top (Section 4.3.6).

Shallowing-upward carbonate cycles may be part of much thicker cycles of sedimentation, and in the Late Precambrian through Early Palaeozoic of North America these have been termed *Grand Cycles* (Aitken, 1966, 1978; Grotzinger, 1986a,b; Chow & James, 1987a). These sequences, of the order of 100–1000 m in thickness, are bound by abrupt stratigraphic contacts (sequence boundaries), and some show shale-based or shale-dominated cycles in the lower part, passing up into limestone–dolomite cycles in the upper part (Fig. 2.29).

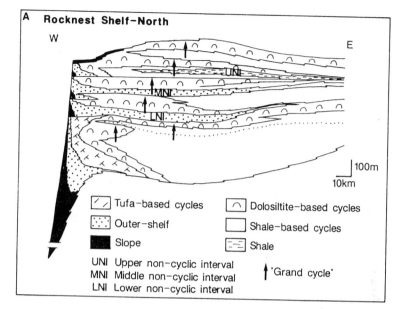

Fig. 2.27 *Shallowing-upward cycles of the Rocknest Platform Early Proterozoic, Canada. (A) Rocknest stratigraphy showing facies zonation from platform margin in the west to platform interior in the east. (B) Shallowing-upward cycles of the Rocknest Platform. After Grotzinger (1986a,b).*

2.10.2 Carbonate–evaporite and carbonate–clastic cycles

The cycles described in the preceding section consist mostly of limestone, or limestone–dolomite, but there are carbonate cycles where other lithologies are involved. One particular class of shallowing-upward cycle is where gypsum–anhydrite occurs at the top of the cycle. In many cases, this was deposited in a supratidal sabkha environment and occurs above a dolostone or limestone with intertidal sedimentary structures (planar stromatolites and birdseyes) (see Sections 3.4.4 and 4.3.3g, and Fig. 4.48).

Thicknesses of sulphate units are generally in the range of 1 to 10 m. Examples of sulphate-capped shallowing-upward carbonate cycles occur in the Permian San Andres Formation of Texas (Handford, 1982), Upper Permian Bellerophon Formation of the southern Alps, Italy (Bosellini & Hardie, 1973), Carboniferous Windsor Group of the Maritime Provinces, eastern Canada (Schenk, 1969), Jurassic Arab Dharb Formation, Saudi Arabia (Wood & Wolfe, 1969) and the Lower Fars/Gachsaran Formation of Iraq and Iran (Gill & Ala, 1972; Shawkat & Tucker, 1978).

Some sulphate–carbonate cycles have gypsum–anhydrite in the lower part, passing up into limestone. Such is the case in the Cretaceous Ferry Lake Anhydrite, Texas subsurface, where the sulphate was pre-

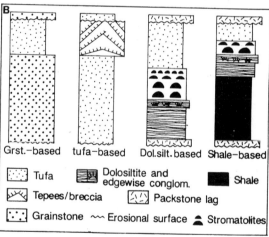

cipitated in a lagoon as selenitic gypsum, and then covered by carbonate from prograding tidal flats (Loucks & Longman, 1982).

Cyclic carbonate deposits may include siliciclastic members. This is particularly the case towards the interiors of platforms where there are significant fluvial inputs, or where there were siliciclastic shorelines and delta fronts, and carbonates accumulated in the outer shoreface and below wave-base. Carbonate–clastic cycles are well seen in the Yoredale Series of the Upper Dinantian–Namurian of northern England and equivalent Pennsylvanian strata of eastern USA, where below wave-base limestones

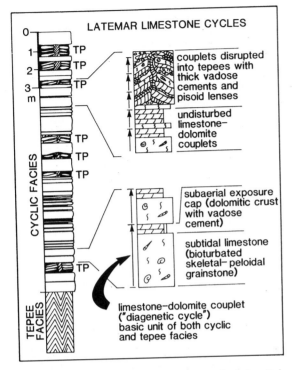

Fig. 2.28 *Cycles in the Latemar Limestone, Ladinian, Italy (see Fig. 2.13 for stratigraphy and location). The sequence consists of many thin cycles of limestone–dolomite couplets (diagenetic cycles), with tepee horizons (TP) that affect several cycles. Much thicker, complex tepee horizons disrupt many cycles. After Hardie* et al. *(1986).*

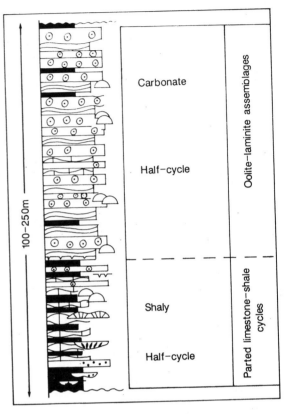

Fig. 2.29 *Schematic Grand Cycle from the Cambrian of eastern North America consisting of many shale–limestone, shallowing-upward cycles in the lower part and oolite–stromatolite cycles in the upper part. After Chow & James (1987a).*

(mostly skeletal packstones and wackestones) pass up into mudrocks and then sandstones as deltaic shorelines prograded across platforms. Delta abandonment phases resulted in extensive swamp development and the subsequent formation of coal. For these Yoredale cycles, Leeder & Strudwick (1987) proposed a tectono-sedimentary model, whereby the carbonates were deposited following a transgression brought about by compaction of underlying deltaic sediments. Subsequent deltaic advance was encouraged by movements on basin-margin listric normal faults.

2.10.3 Causes of carbonate cyclicity: orbital forcing or not?

The cause of the repetition of the carbonate shallowing-upward cycles which are a feature of many platform limestone sequences has frequently been discussed in the literature and is an emotive topic. The repetition of the cycles is generally of the order of 20 000 to 500 000 years. Cycles in Ladinian through Norian strata (Triassic) of northern Italy have average durations of 10 000 to 40 000 years (Hardie, 1986b) and the Upper Triassic Lofer cycles, Austria of 40 000 to 50 000 years (Fischer, 1964). Goodwin & Anderson (1985) calculated the average recurrence interval of Devonian cycles as 80 000 years or less, and for the Early Proterozoic Rocknest cycles, Grotzinger (1986a) estimated 18 000 to 30 000 years. For the 'minor' cycles of the British Late Dinantian, Walkden (1987) estimated 250 000–500 000 years and Leeder (1988) 190 000–240 000. There are basically three categories of explanation for the repetition of the shallowing-upward cycle: (1) eustatic, (2) sedimentary, and (3) tectonic.

A *eustatic control* involves changes in ocean level and this has been the most popular explanation for cycle repetition for many years. Ocean level is controlled by both climate and tectonics: climate controls the volume of ocean water by determining glacial ice volumes, while tectonic processes lead to vertical and horizontal plate movements which affect the volume of ocean basins. Thus glacio-eustasy can be distinguished from tectono-eustasy. However, sea-level fluctuations through changes in ocean-basin volume, which are controlled by rates of seafloor spreading, are on a long time-scale (0.01 m 1000 yr^{-1}) and cannot lead to rapid sea-level rises as required for many shallowing-upward cycles. A glacio-eustatic mechanism was advocated by Goodwin & Anderson (1985) for Devonian cycles in eastern USA, for which they earlier introduced the term punctuated aggradational cycle (PAC). In their hypothesis, sea-level rises relatively rapidly, and then sedimentation builds up to sea-level during a relatively long period of base-level stability. The mechanism invoked for the short-lived transgressive events is glacio-eustasy, driven by orbital perturbations (Milankovitch cycles). The rationale is that the astronomic cycles of precession of the equinoxes (periodicity 23 000 and 19 000 years), obliquity of the ecliptic (41 000 years) and eccentricity (100 000 years) (see Fig. 2.30) give rise to sea-level changes and these are stepwise, with rapid rises followed by periods of stability (Fig. 2.31). Against a high subsidence rate, Goodwin & Anderson (1985) reasoned that five PACs could result in each 100 000 year cycle. A similar explanation has been put forward by Goldhammer *et al.* (1987) for the Middle Triassic cycles in northern Italy (see Fig. 2.28), where they are apparently grouped into asymmetric packages of five cycles, with a thick basal cycle overlain by four progressively thinner cycles. This pattern could reflect the 20 000 year precession periodicity superimposed on the 100 000 year eccentricity cycle.

The underlying mechanism in this glacio-eustatic hypothesis is that the orbital perturbations affect the solar insolation and that variations in this cause fluctuations in the extent of polar ice-caps, which result in changes in sea-level. That this mechanism does work is well known from studies of Pleistocene deep sea cores where Milankovitch rhythms have been detected in the oxygen isotope record (see Section 9.6.2, Fig. 9.20B and Hays *et al.*, 1976; Imbrie & Imbrie, 1980; Chapell & Shackleton, 1986). These data show that the 100 000 year cycle is the major influence and that the sea-level rises were much more rapid than the sea-level falls. This is the result of ice-cap melting being

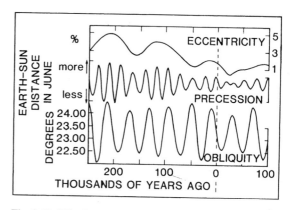

Fig. 2.30 *Milankovitch rhythms of the last 250 000 years, and into the future. After Imbrie & Imbrie (1980).*

faster than ice-cap growth. Milankovitch cyclicities are also seen in ages of Pleistocene coral reef terraces of Barbados (Bloom *et al.*, 1974), where a 20 000 year periodicity is superimposed on the longer 100 000 year cycle (Fig. 2.32). Such a pattern is not seen in the Bahamas or Florida since sea-level was below the shelf-break in these areas from the high stand 120 000 years ago until the recent flood of the shelves some 7000–10 000 years ago (one eccentricity cycle). The cycle periodicities and patterns of cycle thickness (the pentacycles) in the Triassic Latemar Limestone of northern Italy (Fig. 2.28) are consistent with Milankovitch rhythms of precession and eccentricity (Goldhammer *et al.*, 1987); the thick units of tepee structures can be accounted for by extended periods of exposure, when sea-level dropped below the platform margin, perhaps for one whole eccentricity cycle (or more) (see Fig. 2.33). Loss of cycles in this way accounts for the often longer, non-Milankovitch, periodicities recorded in cyclic sequences. The apparently longer the UK period Dinantian minor cycles of could reflect an eccentricity cycle of control on sea-level, rather than the shorter-term precession.

Computer modelling of glacio-eustatic sea-level changes and sedimentation has been used to simulate shallowing-upward cycles. Read *et al.* (1986) modelled Ordovician cycles of the Appalachians and Grotzinger (1986a) did likewise for Rocknest cycles (Fig. 2.34). One important factor used in the program is the concept of lag-time, the delay in carbonate production during a sea-level rise as a platform was flooded. Dinantian 'minor cycles' of the UK described in Section 2.10.1 (Figs 2.25, 2.26) and clastic–carbon-

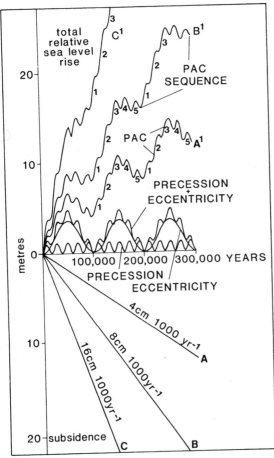

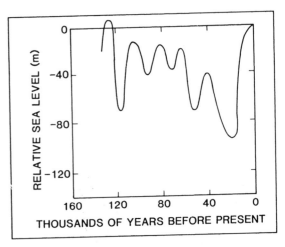

Fig. 2.32 *Sea-level curve for the last 130 000 years showing fluctuations on a 20 000 year and 100 000 year scale, based on the dating of coral reef terraces on Barbados. After Bloom* et al. *(1974).*

Fig. 2.31 *Model for episodic stratigraphic accumulation produced by eustatic responses to orbital perturbations superimposed on continuous subsidence. The precession (20 000 year) and eccentricity (100 000 year) cycles are combined into one curve to represent the general form of eustatic response and this is then added to each of the three subsidence curves (A, B and C) to give three patterns of total relative sea-level rise (lines A¹, B¹ and C¹). The resulting patterns could produce shallowing-upward or punctuated aggradational cycles (PACs) and PAC sequences. The PAC sequence could be truncated by erosion in regions of low subsidence (line A¹) or be more complete in areas of high subsidence (line C¹). After Goodwin & Anderson (1985).*

ate Yoredale cycles have been modelled by Walkden & Walkden (1990).

The orbital-forcing, Milankovitch-rhythm mechanism for sea-level change is a most attractive hypothesis since it produces sea-level changes of the right rate, periodicity and asymmetry. However, it

does require the waxing and waning of polar ice-caps, and it is difficult to see how the mechanism could operate at times when these were absent. This may well have been the case during the Triassic to Late Cretaceous, when global temperatures were relatively high, and continental plates were not located at the poles. It is always possible that small ice-caps existed on microplates or Alpine glaciers were present in mountain ranges during this time. Insolation variations alone are unlikely to affect the position of sea-level, although there is the possibility of global temperature changes causing ocean-water volume changes. The rate is apparently of the order of 1 m per °C (Donovan & Jones, 1979), but the efficiency of this mechanism is open to debate. Thus the role of orbital forcing as a general cause of cyclicity throughout the geological record still needs to be evaluated. Orbital forcing certainly does affect sedimentation, as is revealed in studies of limestone–shale rhythms and pelagic limestone bed thickness variations (e.g. Schwarzacher & Fischer, 1982; Weedon, 1986), but in these cases it is fluctuations in either clastic input or biological productivity, which is brought about by the variations in insolation caused by the orbital perturbations.

One feature of cycles produced in a eustatic mechanism is that the cycles should be laterally very extensive, regionally, even globally. However, in many instances it remains to be demonstrated convincingly that individual cycles can be correlated over a large area; indeed, the study of Pratt & James

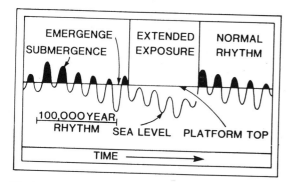

Fig. 2.33 *Model of deposition and exposure during aperiodic, 100 000 year eccentricity cycles of uneven amplitude with superimposed 20 000 year precession cycles. After Hardie et al. (1986).*

(1986) showed that some of the St George Group (Ordovician) cycles were very impersistent laterally. In the Upper Triassic Dolomia Principale of northern Italy, cycles vary in thickness and facies from locality to locality, and differential subsidence of adjacent fault-blocks appears to have exerted a strong control on platform cyclicity there (Hardie, 1986b). In one of the few well-documented cases, Grotzinger (1986a) was able to trace Rocknest cycles of northwest Canada for more than 200 km along strike and more than 120 km across strike. Variable subsidence rates of different carbonate platforms, local tectonic movements and the simple difficulties of correlating cycles which are similar, account for the apparent regional rather than global development of carbonate cycles.

Fig. 2.34 *Synthetic models for shallowing-upward cycles on the Rocknest Platform, Proterozoic, Canada (see Fig. 2.27 for stratigraphy and cycles). (A) General model, with a symmetrical sea-level fluctuation of period 50 000 years and amplitude of 5 m. A lag-time of 5000 years is used, subsidence rate of 10 cm 1000 yr^{-1}, tidal range of 1 m, subtidal sedimentation rate of 100 cm 1000 yr^{-1} and intertidal sedimentation rate of 50 cm 1000 yr^{-1}. With time, sea-level rises according to the sine curve and the sediment surface subsides because of lag-time. After 5000 years, water depths are near those of the low-tide zone, and sedimentation begins (dotted line marks sediment surface). Sea-level then peaks and starts to fall, but sedimentation continues to high tide (follow path of sediment surface). As sea-level falls, the platform is exposed because rate of sea-level fall is greater than rate of subsidence; a vadose zone is formed. Subsequent cycles have well-developed subtidal bases because the subsiding platform is not submerged until nearly one-third of the sea-level rise has been attained. (B) Model for the dolosiltite-based cycles of the Rocknest inner platform (see Fig. 2.27 B), with asymmetric sea-level fluctuations (20% of cycle period in sea-level rise), period of 24 000 years, amplitude 10 m, lag-time 1000 years, subsidence rate 35 cm 1000 yr^{-1}, shallow subtidal dolosiltite sedimentation rate of 80 cm 1000 yr^{-1} at water depth 5 to 3 m, stromatolite sedimentation rate of 100 cm 1000 yr^{-1} at depth 3 to 1 m, and tufa sedimentation rate of 50 cm 1000 yr^{-1} at 1 to 0 m depth. In this model, the sediment surface quickly builds up, intersecting the sea-level curve during the fall, so that the tidal flat, tufa facies is relatively thin. The platform is exposed during low stands forming the erosional surfaces. After Grotzinger (1986a).*

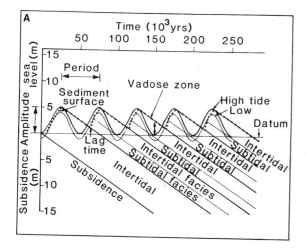

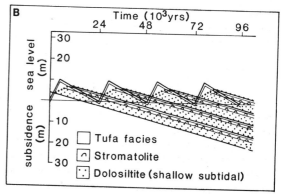

One important but rarely considered aspect of eustasy and sea-level is the effect of changes in the geoid, that is the equipotential surface of the Earth's gravity field (Morner, 1976, 1983). On a global scale, the real ocean surface is not a smooth surface but it has highs and lows of several to tens of metres amplitude. This geodetic sea-level is not fixed, but varies through time as a result of changes in the Earth's gravity field, brought about by such processes as mantle/core interactions, mantle density changes and horizontal and vertical crustal movements. The significant point about this geoidal eustasy is that the position of sea-level may go up in one area, but down in another. Such a mechanism could thus also account for cycles not being of global extent. Whether geoidal changes could actually cause the cycles, however, is unknown.

Two types of *sedimentary explanation* are currently in vogue for shallowing-upward cycles, again generally taken as acting against a background of continuous subsidence and/or sea-level rise. The first, after R.N. Ginsburg, involves a gradual decrease in sediment supply as a tidal flat progrades across a carbonate platform (Fig. 2.35; reviewed in Wilkinson, 1982; James, 1984a; Hardie, 1986b). The rationale here is that most carbonate sediment is generated in the shallow subtidal area (the carbonate factory) and carried on to tidal flats by storms, waves and tidal currents. As the area of sediment generation decreases, so tidal flat progradation ceases. With the continuing subsidence of the platform, there is an ensuing transgression across the tidal flat to re-establish the subtidal carbonate factory and permit progradation of a tidal flat of the next cycle. This autocyclic mechanism would allow tidal flat progradation in one area and tidal flat subsidence in another, so that shallowing-up units need not be correlatable over large areas. This mechanism is reminiscent of that frequently applied to small-scale cycles in deltaic sequences, where delta lobes prograde and then are abandoned and subside, although the sediment is coming from a different direction.

The second autocyclic mechanism is the tidal island model of Pratt & James (1986), referred to earlier (Fig. 2.23). In this scenario, the platform is never completely exposed or submerged, but is dotted with tidal flat islands which accrete and migrate laterally with time at variable rates, but keeping pace with rising sea-level through eustasy or subsidence. The tidal banks will build up towards sea-level, but then the rate of accretion will decrease once the supratidal zone is reached. Lateral progradation of tidal islands

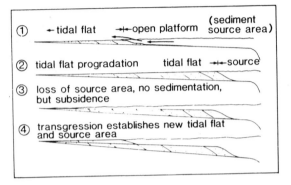

Fig. 2.35 *Model for formation and repetition of shallowing-upward cycles by tidal flat progradation, loss of the subtidal source area (carbonate factory), subsidence and ensuing transgression. After James (1984a).*

will take place until the adjoining subtidal regions, where the sediment is produced, have been reduced in area and are unable to provide sufficient sediment to maintain island growth. Hydrographic forces will then shift the site of sediment accretion to a nearby area, a new bank will develop and lead to island growth, while the former areas of sediment buildup subside below sea-level. Small-scale cycles are produced as the location of sedimentation moves around on the platform, against relatively constant sea-level rise, but the cycles themselves are of local rather than regional extent. The model was devised for the St George Group (Lower Ordovician of Newfoundland) where a lack of lateral continuity of cycles required a new hypothesis for cycle development (Pratt & James, 1986).

Tectonic explanations for sedimentary cycles mostly invoke episodic subsidence as the cause of the periodic rise in sea-level across the platform. The problem with a tectonic control on cycle repetition is that until recently there has not been a suitable mechanism to operate on the relatively short time-scale of several thousand years. Subsidence along fault-lines can be rapid and will be important in extensional regimes. Cisne (1986) did suggest synsedimentary strike-slip faulting as a cause of the Lofer cycles in the Dachstein Limestone of Austria (Fischer, 1964), but in reality, faults are unlikely to be the major control of cycle repetition on extensive carbonate platforms. Synsedimentary faulting may interrupt a normal sequence of cycles or give rise to different cycle stratigraphies on adjacent carbonate platforms. Such is the case with the Dolomia Principale cycles of northern Italy (Hardie, 1986b). Flexural and thermal subsidence of the crust after rifting is relatively slow, and although it does account for thick sedimentary sequences (megasequences or super-cycles), it cannot explain the small-scale cycles with a periodicity of 20 000 years. The much larger-scale *Grand Cycles* (Fig. 2.29) could well be the result of long-term tectonic subsidence (or a third order slow eustatic sea-level rise), the rate decreasing through time accounting for the change in the nature of the minor cycles.

The concept of plate tectonics explains many fea-tures of sedimentary geology, but particularly those of a large scale or long time-scale. The first order cycles of the global sea-level curve, for example, are ex-plained through opening and closing oceans, and the second order cycles through passive margin subsid-ence. However, plate tectonics is a global continuum, in space and time, so that plate tectonic processes will operate on the small scale and short time-scale. One mechanism supposedly able to cause rapid sea-level changes is *in-plane stress* (Cloetingh *et al.*, 1985; Karner, 1986). The lithospheric plates are mostly under stress, with the continental plates being under compression. The in-plane stress in a plate is changed by reconfiguration of plates. Large-scale events, such as plate collisions, oceanic subduction and obduction, continental rifting and new ocean formation, through to small-scale events, such as major earthquakes, local basin formation and filling, and volcanic out-pourings, will cause in-plane stress variations. Calcu-lations by Karner (1986) for a lithospheric plate of typical elastic thickness have shown that if there is an increase in the in-plane stress of 1 kbar, then this leads to uplift in the basin margin of around 20 m, and to subsidence in the basin centre of around 75 m (Fig. 2.36). If there is a decrease in the in-plane stress, then the reverse occurs, basin margin subsidence and basin centre uplift. An important consideration will be the position of global sea-level relative to the basin, since this will determine where the shoreline is located. The continental plates are generally in compression, and increases in this in-plane stress would result in basin-edge uplift, leading to sediment erosion or karstification, and a relatively rapid transgressive event within the basin centre. Such stress changes could thus account for the repetition of cycles. The predicted order of magnitude of the sea-level changes as a result of in-plane stress changes is 10 m in 5000 years. A sea-level rise of this magnitude and time-scale would be sufficient to drown tidal flats on a carbonate platform and re-establish shallow subtidal conditions.

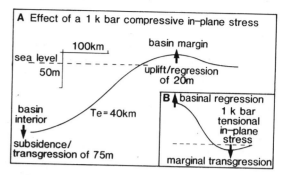

Fig. 2.36 *Sketch to show the effects of a 1 kbar change in in-plane stress on a continental plate of elastic thickness (Te) 40 km. (A) If compressive, there is uplift of 20 m in marginal areas, causing a regression, and subsidence in basinal areas of 75 m, leading to a relative sea-level rise. (B) The effect of a tensional in-plane stress giving marginal transgression and basinal regression. Based on work of Karner (1986).*

The study of carbonate cycles is clearly very topical and the next few years should see a clarification of the roles of the eustatic, sedimentary and tectonic mech-anisms discussed in this section. As with many prob-lems in sedimentary geology, the answer in the end is likely to be a compromise between existing models.

2.11 TECTONIC CONTEXT OF CARBONATE PLATFORMS

Although carbonate sediments can accumulate in most depositional environments, major shallow-marine car-bonate sequences tend to occur in particular tectonic settings. Low-latitude passive continental margins, at times of relatively high sea-level so that clastic input is reduced, are one of the main tectonic environments for the development of carbonate platforms. Ramps, rimmed shelves, epeiric and isolated platforms can all develop along drifting continental margins. The subsurface Jurassic–Lower Cretaceous of the eastern North America continental margin is a good example of this. Carbonate ramps and rimmed shelves with lagoons behind developed along this margin and have been documented from core and seismic data (see Fig. 2.6; Eliuk, 1978; Jansa, 1981; Gamboa *et al.*, 1985). Similar platforms formed along the eastern Atlantic during this time (Ellis *et al.*, 1989). During the Lower Palaeozoic, thick carbonate sequences were deposited along the western continental margin of North America, in ramp and rimmed shelf setting (see Fig. 2.15).

Where initial rifting caused much faulting of the continental margin, or where later growth faulting occurred, then isolated platforms may be created. Modern isolated reefal platforms produced through extension in the early stages of continental rifting occur in the Red Sea (Fig. 2.24; Purser *et al.*, 1987) and small Miocene platforms occur there too (Burchette, 1988; James *et al.*, 1988). The Glovers Reef, Turneffe Islands and Lighthouse Reef off the Belize Shelf (Fig. 2.7) are examples of fault-blocks developed during Tertiary plate separation where carbonate sedimentation has been able to keep up with subsidence. Thus 1000 m of shallow-water limestone occur beneath the Turneffe Islands and 570 m beneath Glovers Reef.

Carbonate platforms were extensively developed during the Mesozoic along the Tethyan continental margins, now preserved in southern Europe, North Africa and the Middle East. Many Triassic–Jurassic shallow-water platforms in the Mediterranean area were fragmented, tilted and drowned during the extension associated with the opening of Tethys (see Section 5.7.1 and Fig. 5.11). In some areas, such as the Swiss and French Alps (e.g. Eberli, 1987, see Fig. 5.56), half-graben–tilt block topographies resulted in reefs and sand bodies on footwall uplifts and small rimmed shelves there. Resedimented carbonates accumulated in the deeps. Carbonate ramps may form on the hanging-wall sides of half-graben structures. In the Jurassic of the Venetian Alps, Italy, the Trento and Friuli Platforms were two fault-bounded platforms, with the Belluno Trough between, on an extending continental margin. The shallow-water Trento Platform was drowned in the Upper Liassic and pelagic limestones (ammonitico rosso type) accumulated there; the Friuli Platform on the other hand, became a site of massive ooid production along its western, windward margin, supplying the adjacent Belluno Trough with lime sand (Vajont Limestone). In the Late Jurassic, into the Cretaceous, reefs flourished along the margin, prograding the platform westwards and supplying skeletal sand into the trough (Bosellini *et al.*, 1981).

Extensive, epeiric carbonate platforms develop over cratons when sea-level stand is relatively high and there is tectonic stability. This tends to occur during the lengthy drift phase after plate separation. The Cambro-Ordovician of North America was a time of widespread shallow-water carbonate deposition, including many cyclic peritidal sequences (e.g. Pratt & James, 1986; see Section 2.10). This was a time of relative tectonic stability, following the major

plate movements and realignments of the latest Precambrian/Early Cambrian, which included the opening of Iapetus. A similar picture of epeiric platforms is seen for the Lower Palaeozoic in western China and the Jurassic of the Middle East. Within these platforms, small basins may occur, surrounded by rimmed shelves and ramps. Such is the case in the Appalachians (see Fig. 2.19) and in the Middle East during the Late Precambrian–Cambrian time, and the Jurassic–Cretaceous (see Murris, 1980; Wright *et al.*, 1989). Hydrocarbon source rocks, or even evaporites, may be deposited in these intrashelf basins.

Carbonate deposition around intracratonic basins is mainly of the ramp type, unless there is much faulting during the stretching to produce distinct breaks of slope or reef growth is sufficiently prolific to form a barrier when a rimmed shelf may be created. Intracratonic basins with marginal carbonates are well developed in North America, and include the Michigan, Williston-Elk Point, Paradox and Delaware Basins. Several of these, at certain times, show a fringing reef belt along a shelf margin, and then a gentle slope or deep ramp into the basin with numerous pinnacle reefs. The Michigan Basin is a useful example here, with its Middle Silurian (Niagaran) barrier reef up to 180 m thick, pinnacle reefs up to 100 m high on the deep ramp, and only 25 m of fine-grained, commonly organic-rich limestone in the starved basin centre (Fig. 2.22). The Delaware Basin in Pennsylvanian–Permian time has a complicated geometry, with shelves along some margins and ramps along others, and these margins evolved through time as a result of fringing reef growth, sand body progradation and tectonic movements. The Devonian–Carboniferous Williston-Elk Point Basin contains well-developed shallowing-upward cycles in the Red River and Madison–Mission Canyon Groups.

Foreland basins resulting from loading and bending of the lithosphere by thrust sheets and nappes generally contain much clastic material derived from the hinterland. However, carbonates may be well developed towards the foreland and here ramps are likely to be the platform type rather than a rimmed shelf, unless reefs are prominent. Carbonate platforms developed in the foreland basin to the south of the Pyrenees during the Eocene, and also in the Appalachian foreland basin, in the Lower Palaeozoic (Read, 1980).

In recent years, a more quantitative approach has been taken with sedimentation and subsidence, particularly of passive margins, but also foreland basins. One aim has been to separate the role of tectonic subsidence from eustatic sea-level change in producing sedimentary sequences (see for example, Bond & Kominz, 1984; Thorne & Watts, 1989; papers in Crevello *et al.*, 1989, and Wilgus *et al.*, 1989). These approaches, involving thermal subsidence modelling, backstripping and geohistory curves, are likely over the next years to improve our understanding of the larger-scale controls on carbonate deposition.

3 Modern carbonate environments

3.1 INTRODUCTION

Carbonate sediments are being deposited in many low-latitude areas at the present time, but three locations warrant an extended discussion since they possess a wide range of environments and lithofacies and provide good analogues to many ancient carbonate facies. These are the Bahama Platform, the South Florida Shelf and the Trucial Coast of the Arabian Gulf. These three carbonate platforms also provide modern examples of an isolated carbonate platform, rimmed shelf and ramp (see Section 2.4).

3.2 THE BAHAMA PLATFORM

The Bahama Platform is a classic area for carbonate sedimentation which has provided many modern analogues for ancient limestones. The size of the platform, 700 km north−south and 300 km east−west, is comparable with the areal distribution of many ancient carbonate sequences. The platform is close to sea-level and is divided into several banks by the deep channels, Exuma Sound, Tongue of the Ocean and Providence Channel (Figs 3.1 and 3.2). It is separated from the USA mainland by the deep Florida Straits and from Cuba by the Old Bahama Channel. These deep channels effectively cut the Bahama Platform off from any siliciclastic material, permitting very pure carbonate sediments to accumulate. The Bahamas is a large isolated carbonate platform (see Sections 2.4, 2.8).

The Bahama Platform has been an area of shallow-water carbonate sedimentation since the Jurassic, and around 5 km of limestone and dolomite, with some evaporite, rest on continental basement. The platform is immediately underlain by Pleistocene limestone and outcrops of this give rise to islands and cays, mostly located close to the platform margin. The distribution of the Pleistocene outcrops and the local topography on the platform are partly the result of karstic weathering during the pre-Holocene sea-level low when the platform was subaerially exposed (Purdy, 1974a).

Shallow waters cover much of the Bahama Platform and have an average depth of around 7 m; these give way very quickly to deeper waters at the platform margin where the seafloor descends rapidly to depths of several hundred metres in only a few kilometres, with local slopes reaching 40°.

The Bahama Platform lies within the Trade Wind Belt so that during the summer, March through August, persistent winds blow from the east and southeast, and during the winter, winds are more from the northeast (see Fig. 3.3). Waves generated by these winds are particularly strong along east-facing platform margins. The tidal range is low, around 0.8 m at the platform margin, decreasing on to the platform. Tidal currents are only significant in channels between islands, reefs and sand shoals, and where the platform margin configuration enhances the tidal effects, such as at the heads of the deep Tongue of the Ocean and Exuma Sound. Flood tidal currents are generally stronger than the ebb currents. Storms and hurricanes produce extreme currents and temporary sea-level rises. Winter storms are mostly from the northwest; hurricanes on the other hand are not so predictable, although many originate in the Atlantic. On the Great Bahama Bank, hurricanes may blow water off the platform from east to west, or pile up water against the Pleistocene-founded islands and reefs of the eastern margin. The great lateral extent of the platform, together with the marginal rocky shoals, considerably reduce the exchange of on-platform water with the surrounding open ocean. This protection has led to the development of an extensive region of relatively sheltered water, only affected by major storms. There are some cross-bank currents, mostly occurring where marginal shoals and islands are absent. The energy regime of the platform is such that the daily normal wind and current activity is capable of moving sand-sized particles only near the platform margin. This is where active bed forms occur. Sand and bed forms on the platform itself are only moved during major storms and hurricanes. Mud-grade carbonate can be moved routinely (Gebelein, 1974a).

Rainfall is heavy over the Bahama Platform at

Fig. 3.1 *Aerial photo mosaic of the Great Bahama Bank (GBB) and Little Bahama Bank (LBB). Features clearly visible include the banks and deep channels (Providence, Tongue of the Ocean and Exuma), the oolite shoals and reefs along the northern margin of LBB, Andros Island with tidal flats on the western side, and tidal oolite bars at the head of Tongue of the Ocean. Compare with Fig. 3.2. After Gebelein (1974a).*

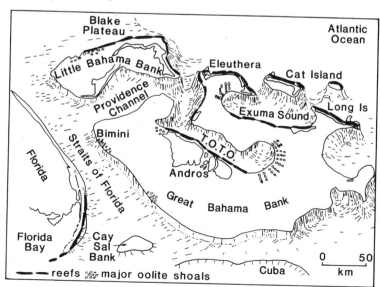

Fig. 3.2 *Banks, channels (Providence, Tongue of the Ocean, Exuma Sound and Straits of Florida), reefs and oolite shoals of the Bahamas and Florida. After Gebelein (1974a).*

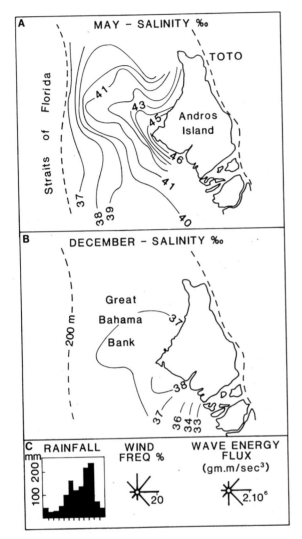

Fig. 3.3 *Salinity, rainfall, wind frequency and wave-energy flux for the Great Bahama Bank. After Gebelein (1974a).*

3.2.1 Subtidal carbonate sediments of the Bahamas

Research on the sediments of the Bahama Platform has identified four major shallow-marine lithofacies and these reflect, to varying degrees, eight habitats and eight organism communities (Table 3.1; Newell *et al.*, 1959; Purdy, 1963a,b; reviews in Milliman, 1974 and Bathurst, 1975). Important attributes which distinguish the lithofacies from each other are grain composition, size and sorting, and lime mud content. Habitat refers to the environment where a particular group of organisms (a community) lives and of the many factors involved in determining the nature of a particular habitat (topography and depth, salinity, temperature, turbulence, turbidity, substrate, nutrient supply, etc.); the two main factors controlling the Bahamian habitats are topography and substrate. The interrelationships between organisms and their substrates, and other environmental parameters, are discussed in Newell *et al.* (1959). The Bahama—Florida region has also been discussed in terms of its biofacies (Coogan in Multer, 1977), defined as a group of organisms living together in a particular area which can be mapped. In effect, biofacies correspond to the organism communities of Newell *et al.* (1959). Biofacies distributions closely follow those of the lithofacies (Fig. 3.4), since much of the sediment is of biogenic origin; however, there are departures, since some sediment is strongly affected by physical processes and the biofacies respond to these and the several other factors which affect the habitat.

The four major shallow subtidal lithofacies are: (1) coralgal, (2) oolite, (3) oolitic and grapestone, and (4)

1200 mm yr^{-1}, but it is seasonal, most falling in the summer months, June through October (Fig. 3.3). Water temperatures generally vary from 22 to 31°C, but higher values are reached locally during the summer. Salinity has a normal oceanic value of 36‰ near the margins but it does reach 46‰ on the platform at times of intense evaporation as a result of the relatively poor circulation (Fig. 3.3).

The Bahama Platform is covered with only a thin veneer of modern carbonate sediments, generally less than 5 m thick, and these have accumulated in the last 4000 years.

Table 3.1 Lithofacies, habitats and communities of the Great Bahama Bank. After Bathurst (1975)

Lithofacies	Habitat	Community
Coralgal	Reef	*Acropora palmata*
	Rock pavement	Plexaurid (sea whips)
	Rocky shore	Littorine
	Rocky ledges and prominences	*Millepora*
	Unstable sand	*Strombus samba*
Oolitic and grapestone	Stable sand	*Strombus costatus*
Oolite	Mobile oolite	*Tivela abaconis*
Mud and pellet mud	Mud and muddy sand	*Didemnum candidum/ Cerithidea costata*

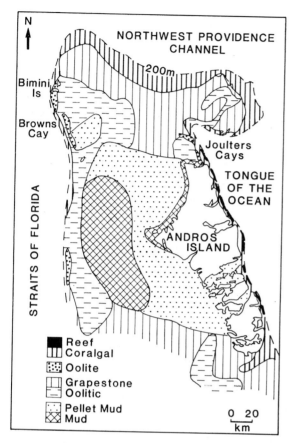

Fig. 3.4 *Lithofacies distribution on the Great Bahama Bank. After Newell* et al. *(1959) and Gebelein (1974a).*

sands and muddy sands, and fore-reef platform-margin skeletal sands (both unstable sand habitat). Pleistocene limestone outcrops on the seafloor or along cliffed shorelines form the rocky-shore and rock pavement habitats where coralgal facies sediments also accumulate.

The *platform-margin reefs* are concentrated on windward margins (see Fig. 3.2) where wave action, turbulence and oxygenation are high, turbidity is low and where water temperature and salinity are similar to those of the open ocean. Such reefs are well developed along the eastern side of the Great Bahama Bank, forming a near continuous barrier opposite Andros Island with a back-reef lagoon 0.5–4 km wide between it and Andros (Newell & Rigby, 1957; Newell *et al.*, 1959; Gebelein, 1974a; Multer, 1977). They are also well developed along the northeast side of the Little Bahama Bank and off Eleuthera Island. Reefs are rare along the western side of the platform, since the dominantly easterly winds push warm hypersaline turbid water of the platform interior over the platform margin, and this is not conducive to prolific coral growth.

The reefs are constructed mainly of coral frame-builders with calcareous algae playing a binding and encrusting role. Five subenvironments are recognized in modern shelf-margin reefs (Fig. 2.8; Longman, 1981). From the ocean side to the lagoon these are: the *reef slope* in deep water (30–100 m) varying from steep to gently sloping with patches of bare reef rock (or Pleistocene limestone outcrops) colonized by various hard-substrate organisms (corals, sponges, gorgonians, etc.) and areas of reef talus; the *reef framework* and *reef crest* (may be just emergent at low spring tide), where most coral growth takes place; the *reef flat* just behind the reef and close to low tide level and consisting mostly of dead, *in situ* coral and other skeletons; and the *back-reef coralgal sand belt*, consisting of debris derived from the reef and washed over during storms, and sloping down to the lagoon floor at a depth of 2–6 m.

Some 30 species of coral are common and these frequently show a zoned distribution across the reef. The elkhorn coral (*Acropora palmata*), a strong branching coral, is a major frame-builder; luxuriant growth of this coral occurs in the reef crest area where wave action is intense (Fig. 3.5A,B). *A. palmata* is commonly oriented into or away from the wave surge on the reef crest, whereas on the reef flat it is unoriented, and much of it is dead and encrusted. Thinner-branched *A. palmata* occurs on the upper fore-reef slope. *Acropora cervicornis* (staghorn coral)

lime mud and pellet mud lithofacies. The first two are principally platform-margin facies, whereas (3) and (4) are platform-interior facies. A fifth category is also described: (5) the deeper-water, periplatform ooze lithofacies.

3.2.1a Coralgal lithofacies

Coralgal lithofacies sediments consist of material from corals and calcareous red algae, along with skeletal debris from many other organisms. This facies occurs along the platform margins in depths ranging from intertidal (reef crest and beach) to around 50 m at the platform edge (Fig. 3.4). It occurs in areas of high wave energy, strong currents and much turbulence. Within this lithofacies group are included the important platform-margin reefs and lagoonal patch reefs (reef habitat, Table 3.1), back-reef (lagoonal) skeletal

Fig. 3.5 *Bahamian–Caribbean reef corals. (A)* Acropora palmata, *the elkhorn coral. (B)* Acropora palmata *with branches oriented oceanwards, also soft coral (gorgonian) in foreground. (C)* Acropora cervicornis, *the staghorn coral. (D)* Diploria, *brain coral, with soft corals close by. From slides courtesy of R.P. Dunne and Barbara Brown.*

is a more delicately branching coral (Fig. 3.5C) that is common in less-exposed parts of the reef crest and on the upper fore-reef slope. Massive and domal corals, such as *Montastrea annularis*, *Diploria* (brain coral) (Fig. 3.5D) and *Siderastrea* are abundant on the upper fore-reef slope, with some growing to depths of 60 m. Small head corals, such as *Diploria*, *Siderastrea* and *Porites*, and delicately branching corals, *Porites*, *A. cervicornis* and *Agaricia* occur on the back-reef slope. Within and close by the reef live a variety of other organisms including molluscs, echinoids, foraminifera, gorgonians and sponges. The hydrocoral *Millepora*, is abundant on the reef crest. The bushy calcareous green (codiacean) alga *Halimeda opuntia* (see Fig. 3.7) is especially common on the reef and on death breaks up to give much lime sand.

Sheet-like calcareous red algae, such as *Litho-thamnium* and *Goniolithon*, encrust corals and coral rubble to consolidate the reef framework. Some accessory organisms such as calcareous worms, bryozoans, foraminifera and bivalves also perform an encrusting role.

Many organisms are involved in bioerosion of the reef and this can also produce carbonate detritus. Lithophagid bivalves, clionid sponges, echinoids and endolithic algae bore into corals and other skeletons, making them more susceptible to damage and fracture by waves. Parrotfish eat coral polyps and in so doing rasp the coral skeleton to generate much lime sand and mud.

The platform-margin reefs are not continuous; they are traversed by major tidal channels every few kilometres which connect the back-reef lagoon with the open ocean. Also, along many of the reefs there are zones which are basically dead, and are undergoing erosion (biological and physical), contrasting with zones of active reef growth. In some areas a distinctive spur and groove morphology is developed, with lime sand derived from the reef occupying the grooves and the spurs forming buttresses 5–50 m across, of luxuriant coral growth, which descend down the reef slope into deep water. Studies of spurs and grooves of the Florida Reef Tract (see Section 3.3.1e) show that they are constructional in origin, not related to any antecedent topography in the underlying Pleistocene limestone (Shinn *et al.*, 1981).

Although forming a massive limestone, the reef has an open structure and there is much internal circulation of water. Fine sediment is carried into reef crevices and the pumping action of seawater promotes much precipitation of cement. Internal sediment is lithified, especially by micritic and peloidal high-Mg calcite, and acicular aragonite and bladed high-Mg calcite are precipitated as isopachous linings on cavity walls (see Section 7.4.1a).

Patch reefs occur in lagoons behind barrier reefs (e.g. Fig. 3.6) and are well seen in the windward lagoon adjacent to Andros Island (Gebelein, 1974a). They are several metres to tens of metres across, with 3–4 m of relief above the seafloor. They initiate on a hard substrate, which is mostly a Pleistocene limestone outcrop. Most luxuriant coral growth takes place around the margins of the patch reefs, where currents and turbulence are at a maximum. Sand accumulates in the depression on the reef top generated by marginal growth, and a halo of reef rubble and skeletal sand occurs around the patch reef. A similar organism community to the barrier reefs exists on the patch reefs, but *A. palmata* only occurs on the seaward side of the patch reefs where wave action is most intense. *A. cervicornis*, *Montastrea*, *Diploria*, *Siderastrea* and *Porites* dominate. Low mounds composed largely of calcareous red algae, with few corals, have formed in some back-reef areas.

The unstable sands of the coralgal facies occur close to the reefs and around much of the Bahama Platform margin as a belt generally 5 km in width, but reaching 40 km where exposed to much wave and storm action (e.g. northern margin of the Great Bahama Bank and south of Andros, see Fig. 3.4). These sand bodies display a hierarchy of bed forms, including sand waves, dunes and ripples. Various types of sand body can be distinguished depending on the relative roles of tidal currents, storms and waves

Fig. 3.6 *Patch reefs in lagoon behind shelf-margin reef. Rippled sand occurs between the patch reefs which show some zonation of corals. From slide courtesy of R.P. Dunne and Barbara Brown.*

Fig. 3.7 *The green codiacean alga* Halimeda *which on death disintegrates into sand-sized particles of aragonite.*
H. incrassata *on the left with conspicuous holdfast is common on the Bahama Platform and Florida Shelf.* H. opuntia *on the right is a more bushy form, typical of reef habitats of the shelf margin.*

in their formation (Ball, 1967; Hine *et al.*, 1981; Section 4.2.1).

Common organisms of the unstable sand habitat are molluscs, echinoids and some foraminifera, and in areas not constantly affected by currents, calcareous green algae such as *Halimeda incrassata* (Fig. 3.7), and *Thalassia* (turtle grass) grows in the sand.

In the more protected back-reef lagoon, areas of *Thalassia* are common and many rooted calcareous algae are associated, including *Halimeda incrassata*,

Fig. 3.8 *The green codiacean alga* Penicillus *with sediment attached to holdfast. On death this alga breaks down into micron-sized aragonite needles.*

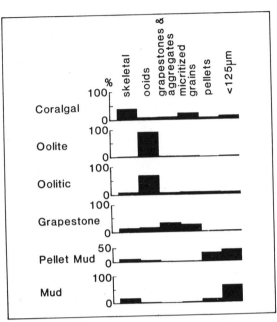

Fig. 3.9 *Compositions of Bahamian lithofacies. After Newell et al. (1959).*

Penicillus (Fig. 3.8), *Udotea* and *Rhipocephalus*. Molluscs, echinoids, crustaceans, foraminifera, holothurians and some small corals also occur. Sediments are lime sands and muddy sands, rippled where there is no grass and bioturbated. The effects of boring endolithic algae have commonly micritized grains, making identification difficult or impossible. A typical coralgal sand consists of 40% skeletal grains, 30% micritic grains, many of which will be altered bioclasts, 15% peloids and aggregates, and 15% others (Fig. 3.9).

3.2.1b Oolite lithofacies

The oolite lithofacies is rather localized in its distribution, occurring in some of the highest-energy shallow water (less than 3 m) locations close to the platform margin (Figs 3.2 and 3.4). Examples include Joulter's oolite shoal just north of Andros (Harris, 1979), Cat and Brown Cay south of Bimini, Lily Bank on the northeast side of Little Bahama Bank (Hine, 1977), and oolite ridges at the heads of Tongue of the Ocean and Exuma Sound. In these areas, many grains are in near constant motion and aragonite is being precipitated around them to produce the ooids (Fig. 3.10). Ooid shoals, like coralgal sand bodies, have a variable geometry depending on the importance of tidal currents and waves. Where on-platform wave and storm action is strong, linear, platform-margin parallel sand bodies develop, cut through by tidal channels with spillover lobes on their lagoonward sides (Fig. 3.11; e.g. Hine, 1977). Where tidal currents dominate sand

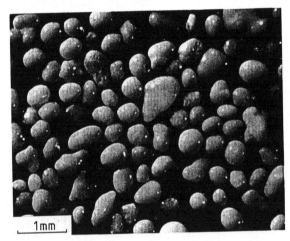

Fig. 3.10 *Bahamian ooids with high polish. Many have probably nucleated upon peloids and are spherical, whereas others probably have less regular bioclasts as nuclei. From slide courtesy of Roger Till.*

bodies are oriented normal to the platform margins, as linear ridges up to 50 km in length, separated by troughs occupied by *Thalassia* and muddy sand. Ooid shoals may be exposed at low tide and with high rates of ooid formation they may give rise to islands. This has happened at Joulter's, where a narrow zone of

Fig. 3.11 *Oolite shoal along Bahamian Platform margin. The open ocean is to the top right; the platform lower left is covered by sea-grass and hence appears dark. Ebb tidal channels cut through the oolite shoal. Sand waves are well developed on the shoal. Lily Bank, view to NW, Little Bahama Bank. From negative courtesy of Albert Hine.*

Fig. 3.12 *Large stromatolite columns, rising 1–2 m from the seafloor, in a tidal channel near Exuma Islands. Photo courtesy of Robert Dill, Gene Shinn and* Nature.

active ooid precipitation close to the platform margin led to the on-platform development of extensive sand flats cut by tidal channels, tidal bars and cays (islands), as a 400 km² ooid shoal complex (Harris, 1979; see Section 4.2.2).

Local seafloor cementation of ooids by acicular aragonite is giving rise to cemented crusts and hardgrounds (e.g. Dravis, 1979). These are broken up by storms to give intraclasts. The biota is limited in the active ooid sand facies by the unstable substrate. Filter feeders dominate and include the burrowing bivalve *Tivela* sp.

A recent discovery of much interest is the occurrence of giant stromatolites within tidal channels close to the Exuma Islands (Fig. 3.12; Dill *et al.*, 1986). The stromatolites, up to 2 m high, are growing in 7–8 m of water and occur in an area of oolitic–peloidal sand waves, where currents reach 1 m s⁻¹. They are individual columns and large, coalesced bioherms, elongated perpendicular to tidal flow. Many are asymmetric, growing into the incoming tide. They have smooth to pustular surfaces and an internal structure of convex-up laminae which either define small columns (10–40 mm across) or encompass the whole structure. The sediment trapped by the microbial film, which includes blue–green algae and diatoms, is ooid and pelletal sand. The stromatolites are being cemented by acicular aragonite and this has produced a tough rock some 0.3 m below the structure's surface.

3.2.1c Oolitic and grapestone lithofacies

The oolitic and grapestone lithofacies covers large areas of the platform from water depths of around 9 m to just exposed at low tide. Currents and wave action are generally sufficient to prevent much deposition of lime mud, although the lime sand itself is moved only during storms. This is the stable sand habitat since much of the sediment is covered by a thin scummy layer of algae and diatoms (Fig. 3.13) or by *Thalassia* grass (Fig. 3.14A). The surficial algal mat (Bathurst, 1967) reaches 10 mm in thickness and is dominated by the blue–green alga *Schizothrix*. The mat contains within it many minute organisms, such as small molluscs, foraminifera, annelids, nematodes, ostracods and diatoms, as well as trapped sedimentary particles carried on to the mat during storms. The surficial mat, turtle grass and rooted codiacean algae all have a stabilizing effect on the sediment and are able to prevent erosion by quite strong currents (Neumann *et al.*, 1970; Scoffin, 1970). Sea-grass protects the seafloor from tidal currents up to 0.7 m s⁻¹ by being bent over and laid flat, but extensive sediment erosion begins when currents exceed 1.5 m s⁻¹ (without grass, extensive erosion takes place at velocities of 0.5 m s⁻¹). Erosion takes place more easily in wave-driven, oscillatory current regimes, since the grass blades are then swept back and forth (Scoffin, 1970).

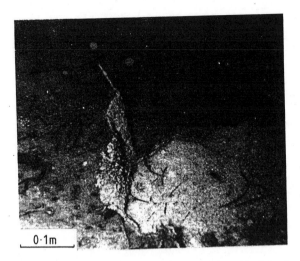

Fig. 3.13 *Surficial algal mat from a lagoonal area in 3 m of water. Part of the mat has lifted off the sediment surface. Microbial mats like this protect the seafloor from erosion during storms.*

A rich and diverse fauna occurs in areas of this lithofacies, with skeletal grains coming from bivalves, gastropods, echinoids, starfish, foraminifera and small corals. The aragonitic *Halimeda incrassata* (Fig. 3.7)

and related forms disintegrate to give sand-sized particles, and *Penicillus* (Fig. 3.8), *Udotea* and *Rhipocephalus* supply lime mud. Many small organisms, such as foraminifera, serpulids and melobesoid algae (epibionts) live upon the *Thalassia* blades (Fig. 3.14B) and contribute carbonate to the sediment. Ooids are washed in from platform-margin ooid shoals during storms. Grapestones are aggregates of grains cemented on the seafloor by micritic aragonite (Fig. 3.15; also see Section 1.2.3, Fig. 1.8). Faecal pellets are common. Much of the sediment consists of sand-sized micritic grains (Fig. 3.9) which have been produced by the boring activities of endolithic algae on skeletal fragments (Bathurst, 1966). An early stage in this micritization is the development of a micrite envelope (see Section 1.3, Fig. 1.11).

Sand waves and ripples occur in some areas of the oolitic and grapestone facies but they are mostly only active during severe storms. Bioturbation is prevalent, especially by crustaceans such as *Callianassa*.

3.2.1d Pellet mud and mud lithofacies

The pellet mud and mud lithofacies occurs in the most protected part of the platform, depths usually less than 4 m, such as to the west of Andros Island and in the Bight of Abaco. It occurs in areas where tidal

Fig. 3.14 *(A)* Thalassia *sea-grass from protected lagoonal area. Calcareous and other algae occur within these grassy areas, which protect the seafloor from erosion during stroms. (B) Blades of* Thalassia *with epibionts, mainly melobesoid (red) calcareous algae.*

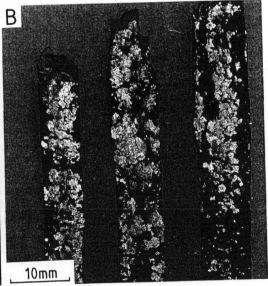

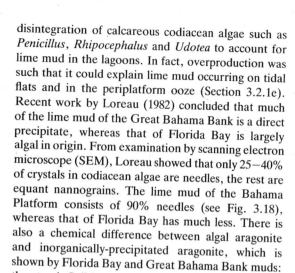

Fig. 3.15 *Grapestones: aggregate grains consisting of bioclasts and peloids cemented by micritic aragonite. From slide courtesy of Roger Till.*

currents and waves are extremely weak, and there are no cross-bank currents, so that sediments are only affected during major storms. The sediment largely consists of aragonite mud and faecal pellets (50–200 μm long) composed of this mud (Fig. 3.16), produced mainly by polychaete worms, but also gastropods. The lime mud itself consists of aragonite needles, a few microns in length. There are few ripples and much of the sediment is bioturbated, especially by the crustacean *Callianassa*, which produces distinctive conical mounds on the seafloor and a simple branching burrow system (Fig. 3.17). Turtle grass is widely distributed but generally sparse, and a surficial algal mat may cover the sediment surface. On the whole, the fauna is sparse and of low diversity, with molluscs the most important group. Echinoids, a soft bottom-living coral (*Manicena*), sponges and tunicates also occur.

There has been much discussion over the origin of Bahamian lime mud, with the controversy centring on an inorganic versus algal origin (see Bathurst, 1975 for a review). Cloud (1962) argued from chemical evidence and the low standing crop of calcareous algae in the mud facies area that the aragonite needles were a direct precipitate from seawater as a result of evaporation. However, later studies in the Bight of Abaco by Neumann & Land (1975), and also in south Florida by Stockman *et al.* (1967), were able to show that more than enough aragonite is produced by the

disintegration of calcareous codiacean algae such as *Penicillus*, *Rhipocephalus* and *Udotea* to account for lime mud in the lagoons. In fact, overproduction was such that it could explain lime mud occurring on tidal flats and in the periplatform ooze (Section 3.2.1e). Recent work by Loreau (1982) concluded that much of the lime mud of the Great Bahama Bank is a direct precipitate, whereas that of Florida Bay is largely algal in origin. From examination by scanning electron microscope (SEM), Loreau showed that only 25–40% of crystals in codiacean algae are needles, the rest are equant nannograins. The lime mud of the Bahama Platform consists of 90% needles (see Fig. 3.18), whereas that of Florida Bay has much less. There is also a chemical difference between algal aragonite and inorganically-precipitated aragonite, which is shown by Florida Bay and Great Bahama Bank muds: the atomic Sr/Mg ratio of aglae is less than 2, whereas it is more than 4 in inorganic aragonite.

Quite commonly on the Bahama Platform (and in Florida Bay), the sea takes on a milkiness due to suspended material, mostly aragonite needles. These *whitings* appear to be the actual inorganic precipitation of aragonite taking place (Shinn *et al.*, 1989), although stirring up of bottom sediments by shoals of fish can produce the same effect.

3.2.1e Periplatform ooze

Periplatform ooze occurs on the slopes around the Bahama Platform and consists of platform-derived shallow-water mud and sand mixed with pelagic carbonate, mostly planktonic foraminifera and coccoliths, with some pteropods (Schlager & James, 1978; Mullins *et al.*, 1984, 1985). The shallow-water material is mostly taken off the platform during storms, but resedimentation processes of slumping, debris flows and turbidity currents are important, especially in moving sediment from the upper part of the slope into deeper water. The platform component of the ooze is mostly composed of aragonite and high-Mg calcite, contrasting with the dominantly low-Mg calcite mineralogy of pelagic carbonate. Monitoring the composition of the periplatform ooze shows that there is a linear decrease in the contribution from the platform with increasing distance from thc shallow-water source (Heath & Mullins, 1984). Periplatform oozes are being lithified on the seafloor and in the shallow subsurface on the slopes around the Bahama Platform (Schlager & James, 1978; Dix & Mullins, 1988), and in some areas this is generating nodular structures (Mullins *et al.*, 1980, 1985). This slope

Fig. 3.16 *Peloidal skeletal sand grains washed out of sandy mud. (A) Surface view. (B) Photomicrograph of peloids showing homogeneous micritic nature. From slides courtesy of Roger Till.*

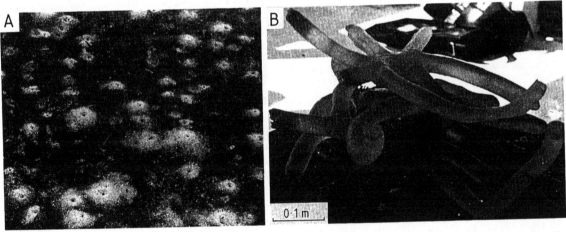

Fig. 3.17 *(A) Conical mounds, 0.1–0.2 m across, produced by the crustacean* Callianassa, *in 0.5 m water depth. The mounds are composed of white muddy sand brought up by the crustaceans on to a lagoon floor covered by a thin, dark surficial microbial mat. (B) Burrow system of* Callianassa *preserved in resin.*

lithofacies and the associated resedimented carbonates are considered further in Section 5.11.

The distribution of subtidal lithofacies on the Bahama Platform largely reflects water movement high-energy platform margins with reefs, associated skeletal sands and oolite shoals, contrast with stabil-ized sands, pelleted muds and lime muds of the quieter-water platform interior. Most sediments are deposited close to where constituent grains were formed, and long-distance sediment transport is only important at the platform margins where material may be taken on to the adjoining slope to basin by storms, or transported deep into the basin by turbidity currents and other resedimentation processes. Micro-bial micritization of skeletal debris is widespread and

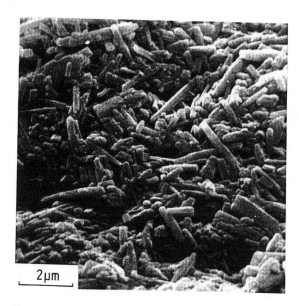

Fig. 3.18 *Bahamian mud. SEM view showing predominance of aragonite needles. Photo courtesy of Jean-Paul Loreau.*

inorganic processes of precipitation are important, giving aragonite needle muds, ooids and seafloor cementation of sands to form grapestones and hardgrounds. In spite of this, biogenic processes of carbonate formation still contribute most material to the sedimentary package.

3.2.2 Intertidal–supratidal carbonate sediments of the Bahamas

A wide variety of lithofacies are deposited in intertidal and supratidal settings on the Bahama Platform and they fall into two broad groups: those of relatively high-energy shorelines where narrow sandy beaches are backed by aeolian dunes, and those of low-energy shorelines where broad, laterally extensive muddy tidal flats are backed by supratidal freshwater marshes.

3.2.2a High-energy shorelines

High-energy shorelines are developed along windward coasts such as the eastern sides of Andros, the Berry Islands, Great Abaco and Eleuthera. Lime sand comprising the beach is derived from the shoreface and most consists of skeletal debris and ooids. Sedimentary structures are identical to those of clastic beaches: flat-bedding in truncated sets dipping seaward at a low angle, with some wave ripple cross-lamination and

crustacean burrows. Shoreface (low tide to wave-base) sands are also rippled and bioturbated. Syn-sedimentary cementation in the intertidal zone gives rise to beachrock (Scoffin & Stoddart, 1983), well known from Bimini. In the backshore, beach berms and dunes are the result of storm waves and onshore winds blowing sand off the beach. Anchoring of sand by vegetation is common and soils may develop, giving rise to crusts, local cementation and rhizo-cretions. Shoreline carbonate sands are discussed at length in Section 4.1.

3.2.2b Tidal flats on west side of Andros

The tidal flats on the west side of Andros Island have been the focus of numerous studies (for example, Black, 1933; Shinn *et al.*, 1965; Shinn *et al.*, 1969; Gebelein, 1974a; Hardie, 1977; Multer, 1977; Gebelein *et al.*, 1980; Shinn, 1983a). In this area, the tidal range is very low (0.46 m) and wind–wave activity is weak since Andros Island acts as a barrier to the dominant easterly winds. Occasional winter storms from the west to north produce strong waves in spite of the shallowness of the platform.

The Andros tidal flats are complex with many subenvironments including tidal channels, beach ridges (hammocks), levees, ponds, intertidal flats themselves, areas of surficial crusts, areas of algal mats, mangrove clumps and swamps, and freshwater algal marsh (Figs 3.19 and 3.20). Parts of the tidal flat are permanently subaqueous, the ponds and channels for instance, whereas other areas are exposed for some of the tidal cycle or for certain seasons of the year (Fig. 3.23). To describe the fluctuations in water cover, an exposure index was introduced by Ginsburg *et al.* (1977) to indicate the percentage exposure of a subenvironment over a year (Fig. 3.21, also see Section 4.3).

Two distinct types of tidal flat occur on the west side of Andros Island (Gebelein, 1974a; Hardie, 1977): to the northwest of Williams Island, the tidal flats are 5 km wide and are dissected by many tidal channels (comprising 15% of the flat complex) which drain ponds, flats and algal marshes (Figs 3.19, 3.20A and 3.22); to the southwest of Williams Island, the tidal flats are up to 35 km wide but they have few tidal channels, and consist instead of broad depressions separated by former beach ridges rising 1–2 m above normal high water (Fig. 3.20B). The depressions are variably occupied by water to form ponds which are surrounded by intertidal flats, algal marshes and areas of surficial crust.

Fig. 3.19 *Aerial view of the intertidal flats on the northwest side of Andros Island, Great Bahama Bank, showing tidal channels, levees, patchy areas of mangroves and algal mats (dark) and depressions, ponds and flats (grey areas). Photo courtesy of Ian Goldsmith.*

At their seaward margins, the Andros tidal flats have a low beach ridge which is constructed of sediment thrown up from the shallow subtidal zone during storms. The sediment of the present and former beach ridges is largely skeletal–peloidal sand with fine laminae and small irregular and laminoid fenestrae. Levee sediments along channel banks are similar.

Sedimentation on the tidal flats mostly takes place during major storms, which are very sporadic. Waters move on to the tidal flats via the tidal channels every tidal cycle, but these waters are clear, carrying little suspended sediment. During storms, sediment from the adjacent nearshore is put into suspension and carried on to the tidal flats via the channels, to be deposited as a thin pelletal lime mud blanket. In the ponds and low intertidal parts of the flats, this thin layer is mixed into the sediment packet by burrowing and grazing animals (gastropods, annelids and crustaceans). In the high intertidal and supratidal areas, the storm layer is incorporated into algal mats.

Channels meander across the tidal flat and rework the sediments. Erosion takes place on the outside of meander bends and deposition occurs on point bars and levees. Intraclasts occur in channel bottoms along with many gastropods. Pelleted lime muds occur in quieter reaches of the channels. The ponds and intertidal areas are composed of pelletal lime muds, with polychaetes and gastropods providing the pellets. Birdseye vugs are common in higher tidal flat sedi-

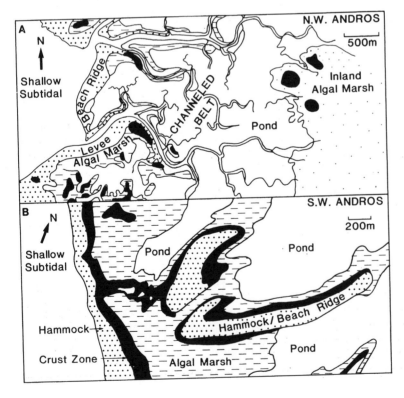

Fig. 3.20 *Tidal flat environments of western Andros, Great Bahama Bank. (A) Northwest Andros, with well-developed channelled belt. After Hardie (1977). (B) Southwest Andros, with beach ridges. After Gebelein et al. (1980).*

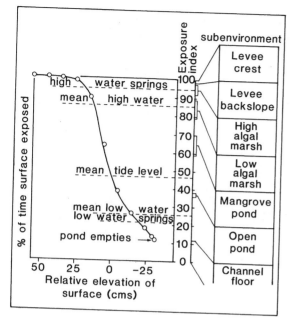

Fig. 3.21 *Exposure index (percentage of time tidal flat surface exposed) and subenvironments of western Andros. After Ginsburg* et al. *(1977).*

ments along with root-moulds and burrows (Fig. 3.23; Shinn, 1968a, 1983b). Faunal diversity is low on the tidal flats; apart from numerous gastropods, benthic foraminifera and polychaete worms are common. Microbial (algal–bacterial–diatom) mats (Fig. 3.24) are widespread in the upper intertidal and supratidal zones of the channelled belt of Andros and these give rise to stromatolitic laminae by the trapping of sediment within the mat. Laminoid fenestrae are common between microbial layers. The organic mats are often broken up or disrupted into polygonal structures

by desiccation; small domes may develop through buckling up of the mat. Thin graded storm beds are common in the supratidal sediments, where there are few burrowing organisms to destroy them.

In the freshwater, inland algal marsh (Black, 1933; Monty, 1972; Monty & Hardie, 1976), an area 4–8 km wide and 50 km long, the microbial mats are dominated by the tufted blue–green alga *Scytonema*, which can form a peat several centimetres thick. A range of mat types is again developed, from flat sheets to polygons and domes, with exposure and desiccation determining gross morphology. Periodically, storms bring marine sediment on to the freshwater marsh, covering the microbial mat with a peloidal–foraminiferal layer. Another alga *Schizothrix* then colonizes the surface forming a thin sheet before the *Scytonema* mat is re-established. An important feature of the freshwater marsh is that calcification of algal filaments takes place here (Monty & Hardie, 1976). Micritic calcite (low Mg) is precipitated around the algal filaments and within the mats through evaporation and biochemical effects. This 'algal tufa' does not form in microbial mats of areas frequently inundated by seawater.

In the high intertidal and supratidal zones, aragonite and dolomite are being precipitated to cement the surficial sediments and form crusts (Fig. 3.25; Shinn *et al.*, 1965; Shinn, 1983a). This is happening on the sides of beach ridges and levees and around ponds. Frequently crusts are broken up to produce intraclasts and these can be reworked to form edge-wise conglomerates or flakestones. Dolomitization of surficial sediments is probably an evaporitic process, caused by porewaters with increased Mg/Ca ratio resulting from precipitation of aragonite and possibly gypsum in the sediment. The dolomitic crusts form just above high tide mark where marine groundwaters are drawn up to the surface by capillary action and

Fig. 3.22 *Schematic cross-section of the tidal flats of northwestern Andros. After Hardie (1977).*

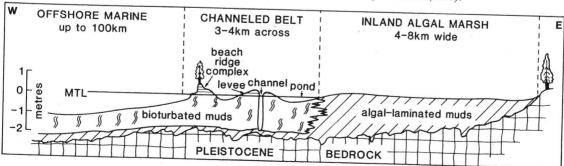

Fig. 3.23 *Tidal flat sediment showing crumbly, peloidal muddy sediment with small irregular fenestrae (birdseyes), some horizontal, laminoid fenestrae, and prominent burrows.*

Fig. 3.25 *Cemented dolomite crust and surface covered with small intraclasts. Andros tidal flat. Lens cap 6 cm diameter. Photo courtesy of Ian Goldsmith.*

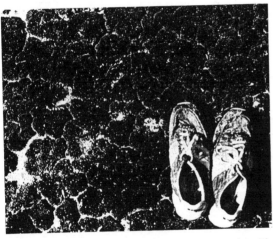

Fig. 3.24 *Algal mats with polygonal cracks from Andros tidal flat. Photo courtesy of Ian Goldsmith.*

evaporated. The dolomite is Ca-rich, poorly ordered and very fine grained (crystals 2–4 μm). A poorly-ordered dolomite is also forming in the subsurface of the southwestern Andros tidal flats (Gebelein *et al.*, 1980). It is possible that the dolomite is forming where marine groundwaters are mixing with fresh-water beneath porous beach-ridge sediments (also see Section 8.7.3).

Cores taken through the tidal flats on the northwest coast of Andros reveal a thin (3 m) seaward-thickening sediment wedge over Pleistocene limestone (Fig. 3.22; Shinn *et al.*, 1969; Gebelein, 1974a; Shinn, 1983a). Immediately overlying the Pleistocene is a marsh deposit (peat) laid down prior to and during the Holocene sea-level rise across the platform, and this passes into the modern algal marsh deposits at the back of the present-day tidal flat. Above the basal transgressive marsh deposit, there occurs a shallowing-upward sequence of bioturbated pelleted muds with a good marine fauna (offshore subtidal) passing up into pelleted muds with a restricted fauna and vertical burrows (low intertidal) and then algal laminated sediments of the upper intertidal. The Holocene sequence upon the Pleistocene thus records the initial transgression (marsh deposits, onlapping), followed by depositional regression as the tidal flats began to prograde seawards. It is thought that progradation took place until about 1000 years BP and that now the tidal flats are being eroded (Gebelein, 1974a). It has been suggested that this erosional state accounts for the abundance of tidal channels in this northwestern area of Andros. In fact, the orientation of this north-western coastline is such that it receives the full force of storm winds and waves, i.e. the meteorological tidal range can be high (Gebelein, 1974a; Hardie, 1977). It also appears that the adjacent subtidal area has a low sediment production rate. By way of con-trast, the tidal flats on the southwest side of Andros face away from the direction of major storms and have continued to prograde until the present time, although the rate appears to have slowed down over the last 1000 years (Gebelein, 1974a). The few tidal channels may reflect the more protected nature of this shoreline. The tidal flats here appear to have pro-graded seaward in discrete jumps, with the old shore-lines marked by hammocks (former beach ridges). The depressions between hammocks are being filled with pond and bay sediments (low intertidal) and then by algal laminites (high intertidal) and supratidal sediments (marsh with algal tufa).

3.3 RECENT CARBONATES OF THE FLORIDA SHELF

The Florida carbonate shelf extends some 300 km south and southwest from Miami curving towards the west past Key West to Dry Tortugas (Figs 3.26 and 3.27). A discontinuous string of elongate islands, the Florida Keys, delineates the inner shelf margin. Behind these islands and connected to the shelf by tidal channels is Florida Bay, a very shallow region with numerous mud banks and mangrove-covered islands. The shelf itself is 5 to 10 km wide with a shelf-break in 8–18 m of water where a seaward slope of 1–10° descends and then gradually flattens into the Straits of Florida with a water depth of 800–1000 m. Along some parts of the shelf margin there is a deeper shelf in 200–400 m of water (the Pourtales Terrace and the Miami Terrace) with a generally steeper slope (8–18°) up to the modern shelf. The deeper terraces are thought to be remnants of a Miocene carbonate platform with marginal reefs. The Florida Shelf–Florida Bay region is underlain by Pleistocene lime-stones which were deposited during the last high stand of sea-level (the Sangamon), some 120 000 years ago (Perkins, 1977). The present topography of south Florida broadly reflects the Pleistocene facies distri-bution, and on a local scale, the karstic weathering which affected the Pleistocene limestones after their deposition, when sea-level fell to around −100 m. The antecedent topographic control is well seen around Miami, where the small hills and troughs (glades) reflect an oolite beach-barrier and tidal channel system in which the Pleistocene Miami Oolite was deposited (see Section 4.2.3; Evans, 1984). The elongate Middle and Upper Florida Keys, which run parallel to the shelf margin, are formed upon a Pleistocene patch reef complex (the Key Largo Lime-stone) and in the Lower Keys (Big Pine to Key West), the more shelf-normal arrangement of islands (well seen in Fig. 3.27) is due to the underlying Pleistocene oolite having been deposited in a more tide-dominated oolite complex (Section 4.2.3). Small lagoons and

bays occur within the Keys belt, particularly in the Lower Keys, where the configuration of the rocky islands provides protection from waves and tides. Strongly affecting Holocene carbonate sedimentation, the modern shallow shelf-break is the former site of a Pleistocene shelf-marginal reef.

The prevailing winds in southeast Florida (Fig. 3.26) are the same as affect the Bahamas; southeast trade winds during the summer, and winds from the northeast during the winter. These generate onshore waves and currents which can be strong along the shelf margin (0.5 m s^{-1} bottom currents), but are much weaker on the inner shelf. For much of the time, wave-base on the shelf is very shallow, at less than 3 m. The Gulf Stream moves northwards in the Florida Straits at a mean velocity of around 1.3 m s^{-1} and this gives rise to a southward-flowing counter current along the shelf margin. The tidal range along the Florida Shelf is very low, at 0.7 m, so that tidal currents are only significant within and near the channels which connect Florida Bay with the shelf. Of great importance to sedimentation along the Florida Shelf, and in the Caribbean generally, are the effects of hurricanes. These short-lived intense but rare storms give rise to extreme waves and currents, which can break large coral stands, transport much coarse sediment, cut channels through barrier islands and mud banks, and put vast quantities of lime mud into suspension (Ball *et al.*, 1967; Perkins & Enos, 1968).

Water temperatures on the Florida Shelf generally range from 18 to 30°C, but occasionally much lower values pertain and these can have a detrimental effect on the corals. Salinities on the shelf are generally the normal marine values of 35–38‰. However, in Florida Bay salinity ranges from 6 to 58‰ with the low values reflecting extreme freshwater run-off from the Everglades during a very wet summer. Lowered salinities can then occur on the shelf, especially in the vicinity of the tidal channels between the Keys, connecting the bay to the shelf. Reviews of the Florida Shelf carbonates are given in Ginsburg (1956, 1964), Bathurst (1975), Multer (1977) and Enos (1977a).

3.3.1 Subtidal carbonate sediments of the Florida Shelf

The carbonate sediments of the Florida Shelf are almost entirely biogenic in origin. There are no oolites and much of the lime mud is probably of codiacean algal origin. Pleistocene limestone lithoclasts make up a small percentage of the carbonate sediments, and a

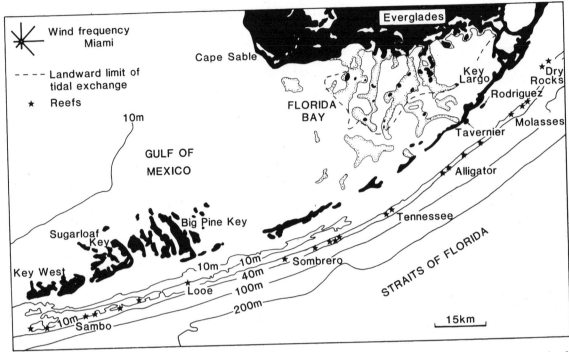

Fig. 3.26 *Map of southern Florida showing the Florida Keys located along a line of Pleistocene reefs and oolite shoals, the modern reefs along the shelf-break, the two mud banks at Tavernier and Rodriquez, and Florida Bay with network of banks and 'lakes'.*

Fig. 3.27 *Satellite photo of southern Florida showing Florida Bay with network of 'lakes' and banks, the linear Upper and Middle Florida Keys which are present-day islands developed upon a line of Pleistocene patch reefs (Key Largo Limestone) and the Lower Florida Keys which are modern islands developed upon a Pleistocene tide-dominated oolite shoal complex (the Miami Oolite).*

little quartz is present, increasing northwards to form a substantial part of the sediment in the Key Biscayne region. As for the Bahamas, the sediments of the Florida Shelf have been discussed in terms of their lithofacies, and organism communities and habitats have been defined (Enos, 1977a; Multer, 1977). Enos recognized eight habitat communities: (1) rock and dead reef, (2) lime mud, grass-covered or bare, (3) lime sand, grass-covered or bare, (4) patch reef, (5) outer reef, (6) fore-reef muddy sand, (7) shoal fringe (mound), and (8) reef rubble. Each habitat has a distinctive group of organisms living there, although many organisms can live in a number of habitats. The habitats are broadly zoned across the Florida Shelf (Fig. 3.28) and there is an associated variation in sediment composition across the shelf and into Florida Bay (Fig. 3.29).

3.3.1a The rock and dead reef habitat

The rock and dead reef habitat occurs along the inner shelf margin where Pleistocene limestone crops out, from the shoreline to up to 3 km seawards. Many

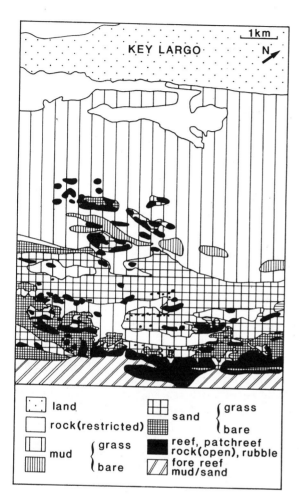

land

rock(restricted)

mud { grass
      { bare

sand { grass
       { bare

reef, patchreef
rock(open), rubble

fore reef
mud/sand

Fig. 3.28 *Lithofacies distribution across the Florida Shelf off Key Largo. After Enos (1977a).*

and water draining out from the bay on to the shelf through gaps in the chain of islands (the Florida Keys) had a detrimental effect on the outer reefs.

3.3.1b Lime mud and sandy mud habitat

Areas of lime mud and sandy mud habitat mostly occur along the inner shelf margin where water circulation and wave action is at a minimum. Where water depths are less than about 8 m, there is a dense cover of turtle grass. Many rooted algae grow here, such as *Penicillus*, *Halimeda incrassata* and *Udotea*, and they contribute vast quantities of aragonite needles and sand-sized grains to the sediment. Molluscs and echinoids are common, along with some soft sediment-living corals and red algae such as *Goniolithon*. Crustacean burrows are abundant. Small bays and lagoons in the Lower Florida Keys area (e.g. Fig. 3.30) such as Coupon Bight, are also locations of lime mud deposition (see Section 4.1.5) but the faunal diversity is lower.

As on the Bahama Platform, the dense sea-grass carpets stabilize mud and sand, and baffle waves and currents; they also trap sediment. The stabilization effect of sea-grass is clearly demonstrated when hurricanes strike the Florida Shelf (Ball *et al.*, 1967). Little erosion takes place in the sea-grass areas, whereas areas of lime sand and the outer reefs are considerably affected. Storms can generate 'blowouts' in grass-covered areas (Wanless, 1981), and the subsequent migration and recolonization of these flute-shaped hollows gives rise to fining-upwards units, around 1 m thick. A shell lag begins the unit, formed in the base of the erosional hollow, and this is overlain by skeletal sand and then sandy mud trapped by the grass-covered advancing leeside of the hollow.

3.3.1c Lime sand

Lime sand occurs in a belt several kilometres wide inshore from the shelf margin. Sea-grass colonizes areas of lower wave energy; elsewhere sand is rippled and sand waves occur. The latter are well seen on White Bank, 3–5 km offshore from Key Largo, and 1–3 km in from the shelf edge (Fig. 3.31). Sediment is entirely skeletal in origin, being derived from the outer reef tract, and from molluscs and algae particularly, but also from echinoids and foraminifera, which live in the bare sand habitat. Grains are commonly micritized but there are no ooids forming on the Florida Shelf.

Lime sand also occurs in tidal channels and associ-

organisms live in this area, particularly those liking a hard substrate and able to tolerate the low energy of the inner shelf, such as some corals (*Siderastrea, Porites*), sponges, echinoids, gorgonians and some algae. Patches of sand and muddy sand are common.

Dead reef is common along the outer shelf margin and forms a similar hard substrate for encrusting organisms. Red algal sheets are particularly common over dead coral and play a major role in generating reef rock. Bioerosion of dead coral by boring bivalves, sponges and algae is intensive. Some areas of dead reef along the shelf edge are situated opposite major channels draining from Florida Bay. These reefs, such as Alligator, were living until about 4000 years BP. Florida Bay came into existence from about that time,

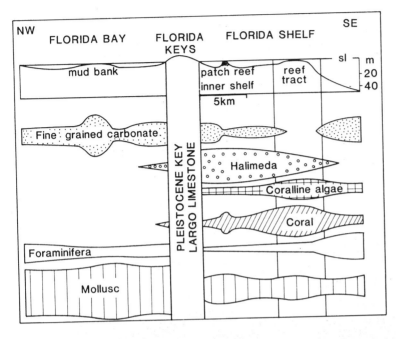

Fig. 3.29 *Distribution of sediment grain size and type across the Florida Shelf and Florida Bay. After Ginsburg (1956).*

Fig. 3.30 *Shallow, protected lagoonal area of the southern Florida Keys showing Callianassa mounds in the foreground, shallows and channels in the middle part, and low islands and barriers colonized by mangroves in the distance.*

Fig. 3.31 *Aerial view of White Bank, ridges of skeletal sand near the shelf margin off Key Largo, Florida. Areas between the sand waves are covered in sea-grass. Photo courtesy of Ian Goldsmith.*

ated tidal deltas developed between islands of the Florida Keys, where there is substantial drainage from the bay and there are strong tidal currents (e.g. the sand complex between Lower and Upper Matecumbe Keys, Ebanks & Bubb, 1975; and in Bluefish Channel in the Lower Keys, Jindrich, 1969). Beaches of lime sand are not well developed along the inner shelf

margin (on the east side of the Florida Keys) since it is mostly a low-energy shoreline. This is a function of shelf dynamics (Section 2.5), where maximum wave action is concentrated at the shelf margin and wave energy is progressively damped across the shelf. Along most of the shoreline, there occur mangrove swamps against and upon the Pleistocene limestone which forms the Florida Keys. Sandy beaches occur close to major tidal channels cutting through the Keys. Examples include Bahia Honda in the Lower Keys and the southern part of Long Key (Multer, 1977). Even so, these beaches are narrow, with a 1–3 m wide foreshore and 1–10 m wide backshore before vegetation, because of the low tidal range.

3.3.1d Patch reefs

Patch reefs occur upon the Florida Shelf, several kilometres back from the shelf margin. They are generally elongate and stand several metres above the surrounding seafloor, which is usually bare lime sand close to the reefs. The patch reefs, such as Hens and Chickens, Matecumbe Coral Gardens and Mosquito Banks, are similar to those described in Section 3.2.1a from the Bahamas. *Porites*, *Montastrea*, *Siderastrea* and *Diploria* are the dominant corals, along with the hydrocoral *Millepora*. Gorgonians and *Halimeda opuntia* are abundant, and molluscs and echinoderms are common.

3.3.1e Outer reef tract

The outer reef tract is a belt up to 1 km wide extending along the shelf margin from off Key Biscayne for 200 km south-west to the Dry Tortugas area. Flourishing reef growth is only taking place where islands occur along the inner shelf margin, for example along the shelf margin seaward of Key Largo, where Molasses Reef, Grecian Rocks, Key Largo Dry Rocks and Carysfort Reef occur. Where there are no Keys and Florida Bay water drains through on to the shelf, then reefs are impoverished or dead along the shelf margin (e.g. Alligator Reef, see Fig. 3.26). Some of the Florida outer reefs, such as Looe Key and Molasses, show a well-developed spur and groove topography (Section 3.2.1a), and this is ascribed to a constructional process, being related to zones of coral growth and strength of prevailing wave energy (Shinn *et al.*, 1981).

Subenvironments of the outer reefs are similar to those described from the Bahamas (Section 3.2.1a). As an example, Shinn (1980) has described five ecologic zones across the reef at Grecian Rocks, which occurs 1 km back from the shelf margin (Fig. 3.32): (1) a deep seaward coral rubble zone at depths of 6–8 m where there is still some strong coral growth and many soft corals (alcyonarians), (2) an area of spurs and grooves in 3–7 m of water where massive coral heads, especially *Montastrea* and the hydrocoral *Millepora* are common, (3) a zone of oriented *A.*

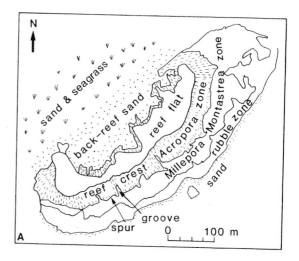

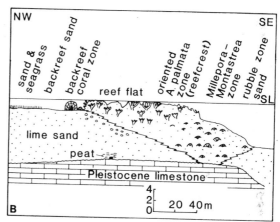

Fig. 3.32 *The reef at Grecian Rocks, near Dry Rocks, off Key Largo. (A) Zonation of the reef. (B) Cross-section of the reef based on shallow coring and radiocarbon dating. After Shinn (1980).*

palmata, in water depths of 0.5–4 m, which receives the brunt of waves and oceanic swells. Branches of this coral are mostly oriented landward, away from incoming waves, (4) the reef flat, composed of un-oriented *A. palmata* which has grown to spring low tide level. Much of this coral is dead and there is rubble around, but active growth does occur on the leeward side of the reef flat, by *A. palmata* and many other corals, (5) the back-reef zone, consisting of scattered colonies of *A. palmata*, many thickets of *A. cervicornis*, large heads of *Montastrea annularis* and *Diploria*, with rubble and lime sand between. Some debris in this back-reef area is transported from the fore reef during hurricanes (Ball *et al.*, 1967). Reef debris extends back on to the shelf for up to 1 km in places as back-reef talus lobes (Enos, 1977a) and rubble islands are formed locally (e.g. the Sambo reefs and rubble islands 7 km south of Boca Chica Key, Fig. 3.26). Seaward transport of reef debris occurs down chutes between coral buttresses. Coring through Grecian Rocks Reef (Shinn, 1980) has shown that the reef is around 5 m thick and founded on 5 m of lime sand which rests on a flat surface of Pleistocene lime-stone (Fig. 3.32B). Reef development at Grecian Rocks appears to have involved the seaward growth of the oriented *A. palmata* (reef crest) zone over the *Montastrea—Millepora* zone, and it also appears that the reef has been extending leewards, mainly by corals colonizing the storm-derived rubble (Fig. 3.32B). A similar trend has been documented from other Florida shelf-margin reefs (Shinn *et al.*, 1977) where it can be shown that underlying Pleistocene bedrock topo-graphy controlled the location and trend of Holocene reef development. The Florida Shelf was flooded by the post-glacial transgression 6000−7000 years BP and coral reef growth followed soon after. However, it appears that more tolerant massive corals such as *Montastrea* and *Porites* were the dominant forms until 3000−4000 years BP, and then the more sensitive *Acropora* genera were able to grow when more open oceanic conditions were established.

3.3.1f Fore-reef muddy sand belt

An extensive fore-reef muddy sand belt occurs sea-ward of the outer reefs on the gentle fore-reef slope at depths greater than around 20 m. Sediment is derived from molluscs, green algae (*Halimeda* can exist to depths in excess of 50 m), echinoids, foraminifera and corals. Coarse reef debris and rippled sand, with patches of live coral, occur in the fore-reef zone between this muddy sand blanket and the outer reef itself.

3.3.1g Mud banks

One important feature of the inner Florida Shelf is the presence of mud banks or *mounds* (Fig. 3.33) and the associated *shoal fringe habitat*. The best documented are Rodriguez Bank (Turmel & Swanson, 1976) and Tavernier (Bosence *et al.*, 1985), location shown on Fig. 3.26. The banks are mostly composed of sandy mud but there is a distinct zonation of lithofacies and habitats around the banks (Fig. 3.34). The banks are emergent and covered by red and black mangroves,

which baffle and trap sediment during storm flooding, and give a peaty soil. Around the banks in the shallow subtidal there is a grass and green algal zone where much lime mud is produced by *Penicillus* and lime sand by *Halimeda*. Sea-grass stabilizes the sediment and prevents erosion and acts as a baffle to trap suspended sediment. On the windward margin there next occurs a *Neogoniolithon* zone where this red branching alga grows profusely. Seawards is the *Porites* zone and the branching finger coral *Porites divaricata* forms a dense intergrowth of colonies (Fig. 3.33B), with many other animals, grass and algae associated, which is a wave-resistant hedge. The red algae and finger corals are brittle and quite easily broken to form coarse sand and gravel. In the deeper water around the bank, grass-covered muddy sand is present.

The mounds have developed through sediment trapping by mangroves, rooted green algae, sea-grass, red algae and branching corals to form a buildup which rises from water depths of around 3−5 m to just above HWM.

In the subsurface of Tavernier mound, a molluscan-rich gravelly mud facies dominates (Fig. 3.34B) with minor foraminifera, ostracods, *Halimeda* and sponge debris, and some 40% cryptocrystalline grains (Bosence *et al.*, 1985). Vertical *Thalassia* roots are common in this facies which is thought to represent the develop-ment of the mound by sea-grass trapping of sediment produced by the local molluscan, green algal com-munity and epibionts upon the grass blades. In fact, the mound originated in a valley within the Pleistocene surface as sea-level was rising from 8000 years ago, and once the whole shelf was flooded, this area con-tinued to be a grass-dominated site of high organic productivity. Aragonite and Sr contents and SEM study (Fig. 3.33C) show that the mud in the mound is derived from mixing *Penicillus*, *Halimeda* and *Porites* on the one hand with *Thalassia* epibionts and *Neo-goniolithon* on the other, with the subsurface mol-luscan mud facies having a higher green algal input than the surface muds (Fig. 3.34C). Breakdown of molluscs contributes little to the mud fraction. More open marine conditions in the last few thousand years have resulted in growth of the windward *Neogoniolithon* and *Porites* zones and the seaward (eastwards) growth of the mound as a whole.

Mud banks also occur in *Florida Bay*. This is a large, rock-floored lagoon occurring behind the Florida Keys and south of the Everglades, connected to the open Atlantic Ocean through tidal channels between the Keys. Florida Bay consists of a network

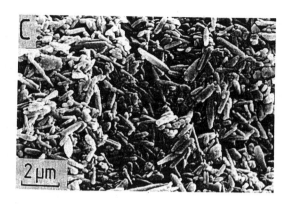

Fig. 3.33 *The carbonate mounds of the inner Florida Shelf. (A) Aerial view of Rodriguez Key showing subaerial part vegetated by mangroves and surrounding shallow subaqueous part where corals, calcareous algae and sea-grass cover the seafloor. The Florida Keys are in the distance behind the mound with Florida Bay in the far distance. Photo from slide courtesy of Jon Lewis.*
(B) Underwater photo of colonies of the coral Porites *and sea-grass in a water depth of 1–2 m on the seaward side of the mound. (C) SEM photo of lime mud from Tavernier mound molluscan mud facies composed of crystals and grains of a variety of shapes and sizes, derived from the breakdown of codiacean and red algae, corals and* Thalassia *epibionts. The composition is 30% HMC, 60% aragonite, 10% LMC. Photo courtesy of Dan Bosence.*

of mud banks and mangrove keys which are interconnected to divide the bay into many shallow basins, commonly referred to as 'lakes' (Enos & Perkins, 1979). The area of Florida Bay was flooded about 4000 years ago and since then sediments have been produced largely by the breakdown of carbonate skeletons, especially the algae *Penicillus* and *Halimeda*, molluscs which dominate the bottom fauna, and foraminifera. Close to the channels connecting with the shelf, a more diverse fauna occurs, including some corals. Much of the muddy bay floor is covered with turtle grass and this has helped construct the banks (Enos & Perkins, 1979). Although vertical accretion has generally been assumed for the Florida Bay mud banks, recent coring has shown that in some banks there are important differences in the sediments of leeward and windward sides. Also, a crude, large-scale cross-stratification suggests that the banks in effect may be bed forms, which migrate during major storms (Bosence, 1989).

3.3.2 Intertidal and supratidal carbonates of the inner Florida Shelf

Intertidal–supratidal flats are poorly represented along the inner Florida Shelf; much of the shoreline is occupied by mangroves, sandy beaches are rare (see Section 3.3.1c), and there are a few rocky shores (rock habitat, features noted earlier, Section 3.3.1a). Mangroves (Fig. 3.35) contribute greatly to the development of new land by their baffling effect on sediment in transit. The black mangrove *Avicennia nitida* occurs in the highest intertidal and supratidal zones and is distinguished by its aerial roots. The red mangrove *Rhizophora mangle* with prop roots occurs more in the upper intertidal. The latter produces more peat than the former. The thick tangle of roots traps sediment carried into the swamp during high tides and storms. Many organisms are associated with the mangrove swamps: molluscs and barnacles on the stems and branches, and crustaceans between the

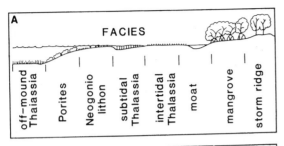

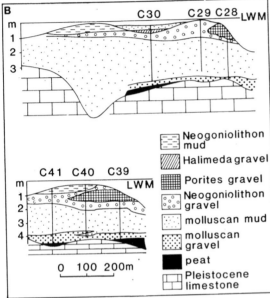

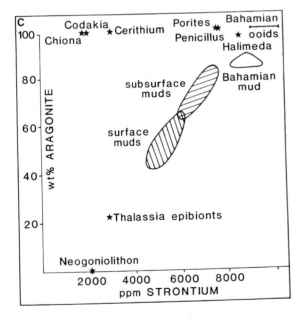

roots and in burrows. Mangrove swamps are well developed on the Florida Bay side of the Florida Keys and they cap some mud banks in Florida Bay. The main development, however, is along the southern and western coasts of mainland south Florida in the Everglades National Park. These swamps are considered analogues for ancient coal-forming environments (e.g. Cohen & Spackman, 1980).

Tidal flats are not as well developed in the Florida Keys area as on the Bahamas. They occur mostly in the Lower Keys, such as Sugarloaf and Big Pine Keys, and then generally on the bay side, where there is most protection from the waves and storms which come dominantly from the east. They also occur on some of the larger islands in Florida Bay, such as Crane Key and Cluett Key. The tidal flats are quite small, being only a few square kilometres in area, and they do not have tidal channels. Depressions permanently occupied with water are common. Clumps of small mangroves are widely distributed. Microbial (algal) mats are the main surface feature of these tidal flats and they vary from extensive carpets where the tidal flat is permanently wet, to desiccated and buckled mats where they are exposed for long periods of time. These mats give rise to stromatolitically-laminated pelleted mud with laminoid fenestrae. Birdseyes (irregular fenestrae) are common in unlaminated pelleted lime muds (Shinn, 1968a). Storms, especially hurricanes, transport much sediment on to the flats to give a distinctive layering. A 50 mm bed was deposited on Crane Key supratidal flat during Hurricane Donna in 1960 (Ball *et al.*, 1967; Perkins & Enos, 1968).

The supratidal flat sediments of Sugarloaf contain up to 80% dolomite, and this mostly occurs within surficial crusts which are cemented storm layers (Shinn, 1968b). Dolomitic intraclasts are common. The origin of this dolomite has been put down to evaporation of seawater and an increase in Mg/Ca ratio, perhaps induced by the precipitation of aragonite. Recent work by Carballo *et al.* (1987) has suggested that the dolomite is formed by tidal pumping of Florida Bay water through the thin Holocene sediment package during spring tides (see Section 8.7.5).

Fig. 3.34 *The carbonate mound at Tavernier.*
(A) Subenvironments across the mound. (B) Two cross-sections showing distribution of sediment types.
(C) Strontium—aragonite contents of mound muds and of sediment-producing organisms. Also shown are Bahamian ooids and muds. After Bosence et al. *(1985).*

Fig. 3.35 *Mangrove swamps along the Florida Shelf shoreline. (A) Tidal channels through the swamp. (B) Dense network of subaerial roots and branches of the mangrove.*

3.4 CARBONATE SEDIMENTS OF THE TRUCIAL COAST

The Trucial Coast of the Arabian Gulf is an area of extensive subtidal and intertidal carbonate sedimentation, and supratidal carbonate, dolomite and evaporite precipitation (papers in Purser, 1973a, and reviews in Bathurst, 1975 and Schreiber *et al.*, 1986). It is a modern example of a carbonate ramp, where the seafloor gradually slopes from sea-level to many tens of metres, without any major break of slope (Fig. 3.36; see Section 2.6). This ramp, in fact, is not a smooth surface; there are many local positive areas, shoals and islands, which are structurally controlled, some being due to movement of salt (halokinesis). The Trucial Coast is a mesotidal area with a tidal range of 2.1 m along the shoreline, dropping to 1.2 m within lagoons. The northeast—southwest oriented coast directly faces strong winds (Shamals) coming from the north—northwest. Because of a very arid climate, and the partly enclosed nature of the gulf (Fig. 3.36), salinity (40–45‰) is a little higher than in the Indian Ocean (35–37‰), and in the lagoons it may reach 70‰.

The offshore outer ramp, in deep water below wave-base where fine sediments accumulate, gives way to a complex of nearshore inner-ramp sedimentary environments of sand shoals, beach-barrier islands and coral reefs (Fig. 3.37). Behind the beach-barrier island system of the Trucial Coast there occur lagoons which are connected to the open gulf via tidal channels through the barriers. Ebb and flood tidal deltas occur at the gulf and lagoonward ends of the channels. Extensive intertidal flats dissected by tidal creeks occur on the landward side of the lagoons and these are partly covered by microbial mats. Still further landwards there occur the broad supratidal flats or sabkhas, wherein gypsum—anhydrite is precipitating. Dolomite is being precipitated in high intertidal—supratidal zones. The back-barrier environments are particularly well developed along the Trucial Coast because this area is a tectonic depression.

3.4.1 Shoals, barriers and reefs of the inner ramp

In the Arabian Gulf, skeletal sandy muds of the deeper, outer ramp give way to carbonate sands and reefs in the shallow subtidal to intertidal zones of the inner ramp along the Trucial Coast (Wagner & Van der Togt, 1973). The outer-ramp skeletal sandy mud lithofacies consists largely of bivalve and foraminiferal debris, often unbroken and unabraded, with mud which has come largely from the Tigris and Euphrates

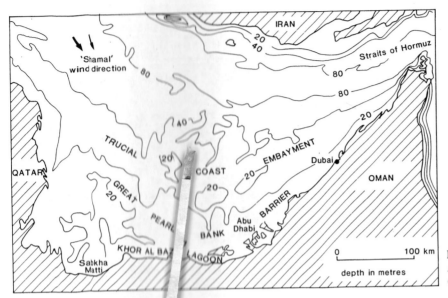

Fig. 3.36 *The Trucial Coast Embayment of the Arabian Gulf. After Purser (1973a).*

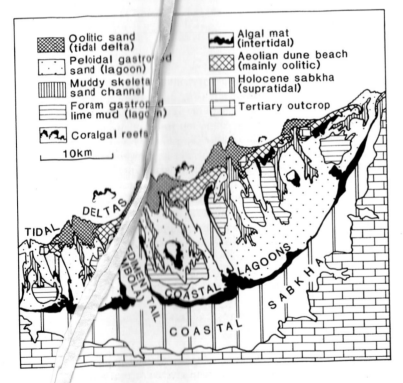

Fig. 3.37 *Schematic map of Abu Dhabi region of the central part of the Trucial Coast showing depositional environments and sediments. After Purser (1973a).*

River at the head of the Arabian Gulf. The inner-ramp sands form an east–west oriented barrier, whose location is structurally determined, with a major lagoon, the Khor al Bazm, located behind (Fig. 3.38; Purser & Evans, 1973). In the west, the sand shoal is submerged and forms the Great Pearl Bank; towards the east it becomes emergent to form a beach-barrier island system, capped by aeolian dunes and dissected by tidal channels. Outcrops of Pleistocene limestone along the structural high are foundations to some of

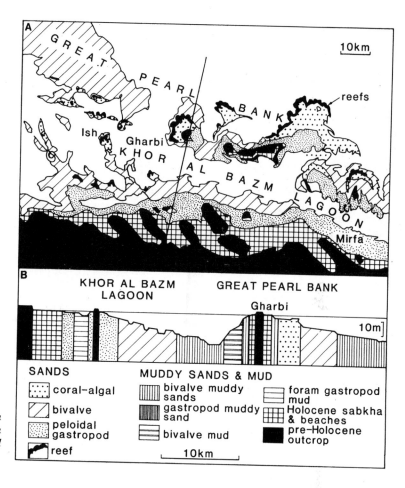

Fig. 3.38 *Sediment distribution and schematic cross-section of the western part of the Trucial Coast, in the region of the Great Pearl Bank and Khor al Bazm Lagoon. After Purser (1973a).*

the islands. The asymmetric Great Pearl Bank is steeper on the lagoon side as a result of cross-bank transport during storms.

Coral patch reefs are developed on the lagoonward side of the bank on shoals and around islands rising up from the deeper parts of the ramp in the central part of the gulf (Purser, 1973b). Prolific coral growth is inhibited by the slight hypersalinity of gulf waters and occasional low (<20°C) water temperatures. Three coral types are common: the branching *Acropora*, the massive *Porites* and the brain coral *Platygyra*. Encrusting red melobesoid calcareous algae are important reef binders, occurring especially on dead corals. Sponges, serpulids and oysters are also common. Various gastropods, regular echinoids and fish graze on the reefs and along with wave action they generate much sand- and mud-sized skeletal debris.

Towards the east, the Great Pearl Bank widens to 20 km and consists of many sand banks and islands. Once the latter reach several kilometres in size, they develop tidal flats on their lagoonward sides. The barrier is traversed by tidal channels, some of which have tidal deltas building into the lagoon. Further east around Abu Dhabi, the coastline becomes a complex of barrier islands, peninsulas and small lagoons (Figs 3.37 and 3.39). Island growth around Pleistocene outcrops has taken place by the lagoonward growth of tails of carbonate sand and lateral accretion of sand spits by longshore currents. Channels between islands have spectacular oolite tidal deltas at their seaward ends, with minor bars and deltas developed within the channels and at their lagoonward ends (Fig. 3.39). The channels are the sites of ooid precipitation and the high rates of production are the main reasons for

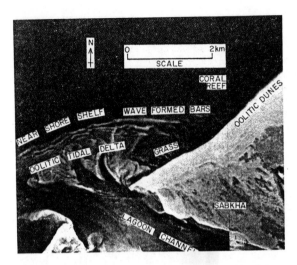

Fig. 3.39 *Tidal channel with bars and shoals, ebb-tidal delta with sand waves and spillover lobes and beach-dune ridges with sabkha behind. In protected areas of the open marine environment, sea-grass is developed and in high-energy areas, coral reefs are present. After Schreiber* et al. *(1986).*

the rapid barrier growth (Loreau & Purser, 1973). Fringing coral reefs just offshore from the barrier islands also supply some sand. Sea-grass occurs in some quieter-water areas seaward of the barriers.

In many shallow-water areas of the inner ramp, carbonate sediments are being cemented on and just below the seafloor to form surficial crusts and hardgrounds. Off Qatar especially, seafloor cementation by acicular aragonite and high-Mg calcite has produced pavements with polygonal crack patterns and pseudo-anticlinal structures (*tepees*) where the crust has fractured and expanded through cement precipitation (Shinn, 1969). Cemented sands form hard substrates (hardgrounds) for encrusting organisms and they are frequently penetrated by borings.

From the foregoing, three principal lithofacies can be distinguished for the Trucial Coast inner-ramp environments:
1 Skeletal and oolitic grainstones of the relatively high-energy shoals, beach barriers, tidal deltas and tidal channels. Skeletal material is generally well abraded and is derived from molluscs, echinoids, foraminifera and corals. Sedimentary structures are common, mainly various types of cross-stratification and some burrows, and these together with data on lithofacies geometry and sequence, palaeocurrents

and subtleties of the sediment composition, would enable the various environments to be distinguished in a fossil situation.
2 More muddy skeletal sands of the quieter-water locally grass-covered areas seaward of the barrier.
3 Coral−algal bindstones and framestones of the fringing and patch reefs.

3.4.2 Lagoons

The major Khor al Bazm Lagoon behind the Great Pearl Bank decreases in width and depth towards the east and salinity increases from 40‰ to 50‰. Along with this increasing restriction, corals, echinoids and algae decrease in importance and gastropods and foraminifera dominate. The sediments are mostly skeletal−pelletal sands, with little lime mud. The small protected lagoons at the eastern end of the Trucial Coast (e.g. Fig. 3.40) have salinities up to 60‰ and sediments are pelleted lime muds rich in imperforate foraminifera and gastropods. Locally, swamps occur around the edges of the lagoons, colonized by the black mangrove *Avicennia marina* and other halophytic plants. Here lime muds and pelleted muds accumulate and they are extensively bioturbated by crustaceans.

The lime mud and pellets of the lagoons are composed of aragonite needles and chemical evidence suggests that these are a direct precipitate (Kinsman & Holland, 1969). The needles are $1-4$ μm long and have a strontium content of around 9400 ppm. Calcareous green algae, the source of much lime mud in the Caribbean (Section 3.2.1d), are rare in the gulf and the main aragonite-producing organisms, the molluscs, have $1000-4000$ ppm Sr. Corals have high Sr but they do not contribute much sediment to the lagoon floor. Furthermore, around 9400 ppm Sr is close to the expected value for aragonite precipitated from seawater at the known temperature, and inorganically-precipitated, aragonitic ooids in the gulf have a similar Sr content (9600 ppm). The occurrence of 'whitings' (Section 3.2.1d) in the open gulf and lagoons could be the inorganic precipitation taking place, although Ellis & Milliman (1986) documented evidence against this. Milkiness in gulf waters is also produced by shoals of fish stirring up the bottom sediment and discharge from oil tankers.

The dominant lithofacies of the lagoons are thus skeletal-pelletal sands in areas of moderate circulation and lime mud and pelletal mud in the most protected parts. Faunal diversity is low compared with seaward lithofacies but certain species of gastropods and

Fig. 3.40 *View over lagoon and tidal flats behind the Trucial Coast barrier showing well-developed polygonal mud cracks in the lime mud. Photo from slide courtesy of John Powell.*

foraminifera are abundant, and burrow structures produced by annelids and crustaceans are common, giving a mottled appearance to the sediments. Mangrove swamps give rise to a lime mud and pelleted mud lithofacies with characteristic root structures and possibly some peat.

3.4.3 Tidal flats

The intertidal zone of the Trucial Coast is up to 5 km across and is dominated by microbial mats in the upper part. Tidal creeks dissect the flats and patches of halophytic plants occur at their lagoonward ends. Although the flats themselves are mostly composed of lime mud and pelleted mud, low beach ridges, strandline deposits and some tidal creeks consist of skeletal—pelletal sand, dominated by cerithid gastropods. As in the Bahamas, the tidal flat sediments are laminated as a result of storm and spring tide deposition, and laminoid, irregular and burrow fenestrae are common. In some parts of the tidal flat, active precipitation of aragonite is leading to the formation of cemented crusts. In a similar way to the subtidal hardgrounds, polygonal cracks and tepee structures are generated (Evamy, 1973). Fracture and breakage of the crusts gives rise to intraclasts which can be reworked into edge-wise conglomerates or flakestones.

A distinctive feature of the Trucial Coast tidal flats is the development of a microbial (algal) mat belt, up to 2 km wide, along protected lagoon margins where

surface slopes are very low. In the low intertidal and lagoonal areas, mat growth is mostly precluded by the grazing activities of cerithid gastropods. However, the latter cannot tolerate high salinities, so where these occur, small domal stromatolites up to 60 mm high and 180 mm across form, in waters up to 3 m deep.

The Trucial Coast algal–bacterial mats are dominated by two basic algal communities, those of *Microcoleus* and *Schizothrix* (Park, 1976). The microbial mat types, however, are determined by environmental factors, rather than the structure of the algal–bacterial community. Four principal types of mat are recognized and the major control on their distribution is the frequency of flooding by tidal and storm-driven lagoon waters (Fig. 3.41). The most important type is the *smooth mat* of Kinsman & Park (1976), also called polygonal mat (Kendall & Skipwith, 1968) and flat mat (Park, 1977). It covers 30–60% of the mat belt and is best developed in shallow pool and channel sites (Fig. 3.42). The dominant alga is the filamentous *Microcoleus chthonoplastes*. Smooth mat gives rise to stromatolites consisting of alternations of organic and sediment laminae (Fig. 3.42B). The thicknesses of the carbonate laminae are generally a millimetre or two, but they increase to 10 mm in the direction of the lagoon. In areas of lower flooding frequency, in the upper parts of the tidal flat, the smooth mat is more of an algal peat, with little sediment, and the mat is broken into polygonal structures through desiccation (Fig. 3.42A). Smooth mat has a high preservation potential. *Pustular mat* of Kinsman & Park (1976), equivalent to cinder mat of Kendall & Skipwith (1968), is the initial mat type in low to mid intertidal areas and is dominated by the coccoid alga *Entophysalis major*. The fabric produced by sediment trapping on a pustular mat has a clotted or thrombolitic appearance. *Pinnacle mat* occurs in the higher, usually well-drained parts of the tidal flat and has small tufts, formed by the large filamentous *Lyngbya aestuarii*. *Blister mat* (crinkle zone of Kendall & Skipwith, 1968) occurring in the highest intertidal zone, consists of a leathery, domed mat with little trapped sediment, but it has a low preservation potential.

Although a little sediment is deposited on the microbial mats during every diurnal tidal flooding, the thicker and thus more significant laminae are invariably the product of major storms (Park, 1976). Growth rates of living mats appear to be of the order of 2–2.5 mm yr^{-1}, but buried mat horizons suggest that after desiccation and compaction, the accretion rate is of the order of 0.2 mm yr^{-1}.

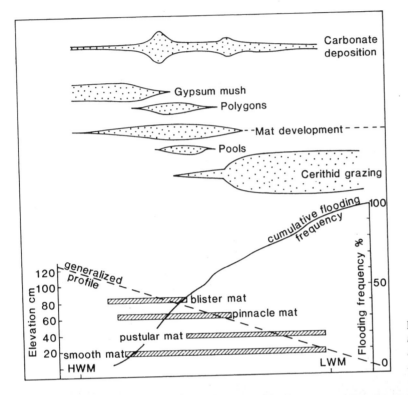

Fig. 3.41 *Distribution of algal mat types and sedimentary features across the tidal flats of the Trucial Coast. After Park (1977).*

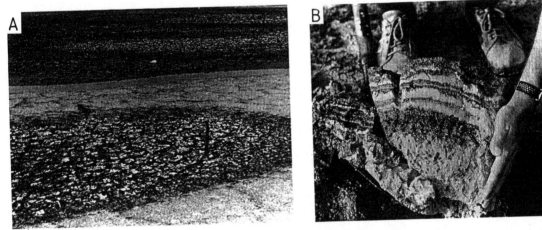

Fig. 3.42 *Algal mats of the Abu Dhabi tidal flats. (A) View showing desiccated polygonal mats, and larger polygonal structures close to tidal channel. Knife 0.2 m long. (B) Section through tidal flat sediment showing algal mat layers alternating with sediment layers. Photos from slides courtesy of John Powell.*

In the higher parts of the tidal flat, high evaporation results in the precipitation of gypsum crystals within and beneath the microbial mats. Seaward progradation of the supratidal sabkha surface progressively buries the mats so that an organic horizon occurs beneath the sabkha (Fig. 3.43). Within this buried algal bed and in the sabkha sediments above, more gypsum and anhydrite are precipitated. The precipitation of these evaporite minerals has a disruptive effect upon the mat laminae, and may destroy the

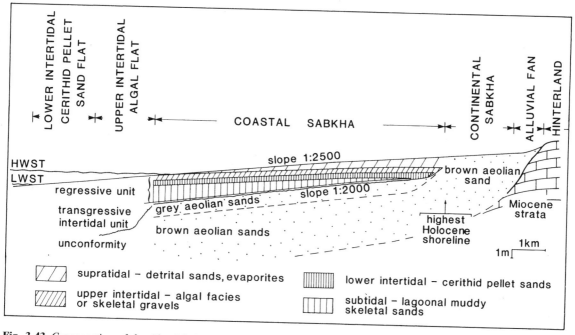

Fig. 3.43 *Cross-section of the Abu Dhabi tidal flat–sabkha showing sediment types. After Kinsman & Park (1976).*

Fig. 3.44 *Large polygonal structures in supratidal sediments. Abu Dhabi sabkha. Photo from slide courtesy of John Powell.*

horizon altogether (Park, 1977).

The two characteristic tidal flat lithofacies of the Trucial Coast are pellet mud and lime mud with storm laminae and fenestrae, and stromatolitically-laminated lime mud with a variety of fabrics reflecting the microbial mat type. Minor lithofacies are skeletal, mostly gastropod, sands and gravels and skeletal–pelletal sands of tidal creeks and channels and former beach ridges.

3.4.4 Supratidal flats and sabkhas

Landwards from the intertidal zone of the Trucial Coast occurs an extensive supratidal area known as the sabkha. The sabkha extends for up to 10 km inland and has an imperceptible seaward slope of around 1:2500. The sabkha surface has formed in the last 5000 years as a result of the gradual seaward migration of the shoreline. Progradation was mostly brought about by the filling of lagoons and sedimentation in the supratidal–intertidal zones, but in addition to this sedimentary offlap, there has been a relative fall in sea-level of 1.2 m. The offlap has proceeded at an average rate of 2 m yr^{-1}, so that a broad supratidal zone (the sabkha) has been generated (Patterson & Kinsman, 1981).

Seawater flooding of the supratidal flat occurs during high tides and storms and transports much lime mud and sand on to the sabkha. Carbonate laminae are deposited but these are mostly disrupted by desiccation and evaporite mineral growth and large polygons develop on the sabkha surface (Fig. 3.44).

Flooding frequency decreases quickly across the extensive supratidal flat, so that seawater may only reach the most landward parts of the sabkha once every few years. As a result of the intense evaporation (1.5 m yr^{-1}), a suite of minerals is successively precipitated from the high intertidal zone across the sabkha and pore-fluid chemistry changes systematically (see Section 8.7.1 and Fig. 8.20). In the upper intertidal zone, aragonite is precipitated as a cement to generate surficial crusts and lithified subsurface layers. Dolomite is also being precipitated within the intertidal sediments (Illing *et al.*, 1965; McKenzie, 1981; Patterson & Kinsman, 1982), and it appears to be concentrated near remnant channels close to and above the present strandline. The dolomite itself is poorly ordered and Ca-rich, with an oxygen isotopic composition ($\delta^{18}O + 2‰$) indicating an evaporative origin (see Section 8.6). As a result of the sabkha progradation over the intertidal sediments, a zone of dolomite is present beneath the sabkha surface, especially in the buried upper intertidal sediments. The absence of aragonite in these sediments suggests that the dolomite has formed by replacing this mineral. Analyses of porewaters have shown that dolomitization is taking place from fluids with a high Mg/Ca ratio (>6), pH of $6.3-6.9$ and temperature of $25-40°C$ (Patterson & Kinsman, 1982).

Gypsum is also being precipitated in the upper intertidal sediments. The crystals are mostly less than 1 mm long, lens-shaped and flattened normal to the c-axis. A little further inland, in the outer supratidal zone where flooding occurs at intervals of a month or more, gypsum crystals up to several centimetres long form a surface mush which reaches 0.3 m in thickness. The gypsum is precipitated displacively within the sediment, but frequently includes carbonate grains and even microbial lamination. Replacement of aragonitic shells by gypsum also occurs.

In the mid sabkha, where flooding is more frequent than once a month, the gypsum mush gives way to anhydrite, which occurs as nodules of lath-shaped crystals, $1-100 \text{ µm}$ in length (Butler, 1970). As a result of displacive growth, the original sediment occurs as stringers between the nodules to give a net or chicken-wire texture. Anhydrite also forms seams of coalesced nodules in the sabkha sediment above the former gypsum mush horizon, and these beds usually show contortions in the form of ptygmatic and disharmonic folds, referred to as enterolithic

structures (Butler, 1970; Butler *et al.*, 1982). Lower in the sediment profile, in the buried algal and lagoonal sediments, lenticular gypsum crystals, up to 0.25 m long, are common.

Apart from dolomite, gypsum, anhydrite and halite, two other minerals, which are precipitated in minor quantities, are celestite and magnesite (Bush, 1973). Although some celestite ($SrCO_3$) is precipitated through the evaporative concentration of seawater, most is precipitated as a result of the release of Sr during the dolomitization of aragonite and during the secondary gypsification of anhydrite. Magnesite ($MgCO_3$) occurs within the sediments of the mid sabkha, above the buried microbial mats. In this area, brines have a very high Mg content; there is no aragonite in the sediments but some dolomite. The magnesite probably forms after dolomitization has removed all aragonite. Although much of the Ca^{2+} for the gypsum—anhydrite comes from seawater, another source is dolomitization. During this process, Ca^{2+} is released according to the equation:

$$2CaCO_3 + Mg^{2+} \rightarrow CaMg(CO_3)_2 + Ca^{2+}$$

and the Ca^{2+} combines with SO_4^{2-} in the porewaters to give more $CaSO_4$.

The typical sabkha lithofacies, as being formed in the Trucial Coast, is thus a nodular anhydrite and gypsum crystal mush in lime mud and muddy sand, with some dolomite. It is a distinctive facies of arid—semi-arid supratidal flats.

As a result of sedimentation upon the intertidal flats and sabkhas, and the early diagenetic precipitation of evaporites within the sediment, this coastal belt has prograded up to 10 km in the last 5000 years. The product of sabkha progradation is a definite sequence of sediments: the evaporitic supratidal sediments come to overlie the tidal flat sediments, and these in turn overlie lagoonal deposits. A pit dug in the outer to mid sabkha of the Trucial Coast reveals this sequence beneath the surface (Fig. 4.48), with the tidal flat part around 1.2 m thick, reflecting the tidal range, and the supratidal nodular and enterolithic anhydrite reaching 2 m. In the geological record, there are now many carbonate—evaporite formations interpreted as the product of sabkha sedimentation and precipitation, and most of these consist of repeated sabkha cycles (see Sections 2.10.2, 4.3.3g and 4.3.6).

4 Carbonate depositional systems I: marine shallow-water and lacustrine carbonates

4.1 COASTAL AND OFFSHORE ENVIRONMENTS: beaches, barriers, lagoons, tidal deltas, the shoreface and offshore

4.1.1 Introduction

Where carbonate shorelines are subjected to moderate to high wave action, lime sands are generated in abundance, especially in the shoreface zone. Beaches and backshore dunes are supplied with sand, particularly during storms. Depending on a number of factors (detailed later), there are two end-member types of carbonate sand shoreline: (1) beach-barrier island complexes with tidal channels connecting through to back-barrier lagoons usually surrounded by tidal flats, and (2) strandplain complexes consisting of a series of parallel beach ridges, with depressions (swales) between (Fig. 4.1). Strandplain and beach-barrier island–lagoon complexes can give rise to thick limestone sequences and are typical of inner-ramp settings (Section 2.6). Low beach ridges border the seaward margins of some tidal flats but sediments deposited there are usually a minor component of the tidal flat sequence (Section 4.3). Frequently, they are reworked by tidal channels and incorporated into the tidal flat package during tidal flat progradation. Carbonate beaches and aeolianites may also develop along platform and shelf margins where high rates of lime sand production lead to the development of sand flats and islands.

Carbonate sand shorelines are directly comparable to siliciclastic shorelines and show similar lateral and vertical distributions of sedimentary facies. For information on siliciclastic beaches and barriers see Komar (1976), Davis (1978), Reinson (1984) and Elliott (1986). Naturally there are differences, mainly in the more biological control on sedimentation, the more local formation of grains and contemporaneous cementation in carbonate environments.

4.1.2 Modern shoreline carbonate sand systems

Modern carbonate sand shoreface–foreshore–backshore environments are widely distributed in the Caribbean–Bahamas area, occurring along exposed shorelines where wave action is high. Sediments are mostly skeletal grains derived from carbonate-secreting organisms living in the shoreface area, and from patch reefs if these occur nearshore. In some instances ooids are an important component of beaches and dunes, such as at Joulter's Cay, in the Bahamas. Along the northeast coast of the Yucatan Peninsula, Quintana Roo, Mexico (Fig. 4.2A), the modern beach-barrier islands of Mujeres, Contoy, Blanca and Cancun consist of oolitic sand which has formed in nearby shallow shoreface zones. The islands are capped by prominent aeolian dune ridges which reach 15–20 m in height. On some islands, washover fans occur on the lagoonward side, through over-barrier storm transport of sand. Tidal flats occur on the landward side of some barriers and evaporitic ponds fill depressions between dune ridges. Lime mud is being deposited behind Islas Cancun and Blanca, where protected embayments exist, and submarine sand shoals occur in the strait between Isla Mujeres and the mainland (Fig. 4.3). The location of these islands is controlled by underlying topography (Fig. 4.2B): the islands occur upon Pleistocene aeolianites which capped barrier islands formed along a break of slope at −10 m during a lower stand of sea-level. Along the southeastern Yucatan mainland, a strandplain of beach ridges and swales developed during the Upper Pleistocene and this is clearly visible from aerial photographs (Fig. 4.4). At its widest part, some 20 ridges are present, rising 1–5 m above intervening swales, and spaced at intervals of 50–200 m (Ward et al., 1985). The strandplain developed behind a barrier reef, located along the seaward margin of a shallow

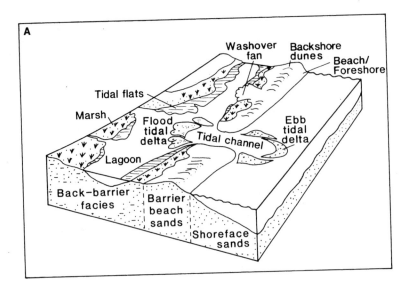

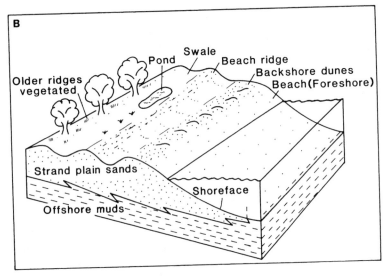

Fig. 4.1 *Shoreline facies models. (A) Beach-barrier island—lagoon model. After Reinson (1984). (B) Beach—strandplain model.*

and narrow shelf (Fig. 4.5).

A well-developed carbonate beach-barrier shoreline with extensive lagoon and tidal flats behind occurs along the Trucial Coast of the Arabian Gulf and has been described in Section 3.4 (also see Evans *et al.*, 1973; Loreau & Purser, 1973).

Sandy islands occur out on carbonate shelves where there are local topographic highs. Ambergris Cay on the Belize Shelf is an example of a shelf island (see Fig. 2.7; Ebanks, 1975). It has an eastern (windward) marginal beach ridge, a less well-developed leeward beach, supratidal flats behind the beaches and an intra-island lagoon.

4.1.3 Environments and facies of shoreline sands

Carbonate beach-barrier island systems develop in regions of moderate to high wave energy, where the tidal range is generally less than 3 m and carbonate sand production rate is high. A moderately stable low gradient coastal region of an inner ramp or open shelf shoreline favours barrier island formation, although in the clastics literature there has been much discussion over how the barrier is actually initiated. With many modern carbonate examples, the development of sand shoals and barriers is related to underlying topographic

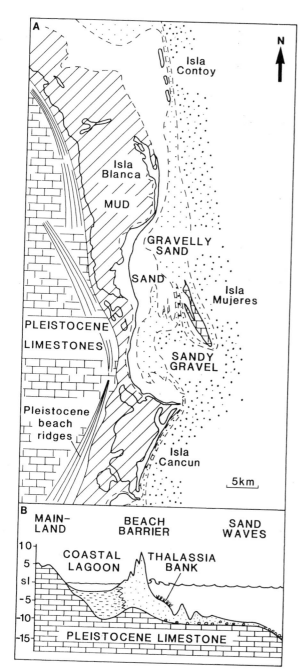

Fig. 4.2 *Modern carbonates of northeast Yucatan, Mexico.* *(A) Generalized map of grain size distribution of Recent carbonates. (B) Schematic cross-section of modern environments and sediments occurring upon Pleistocene limestone. After Ward & Brady (1979) and Ward et al. (1985).*

highs in the Pleistocene limestone bedrock. Lagoons are generally located behind a barrier island and connected to the open sea via tidal inlets (Fig. 4.1A). In microtidal (<2 m) areas, tidal inlets are widely spaced along the barrier, but in mesotidal locations (2–4 m), tidal inlets are prominent and tidal deltas are usually developed at the ends of the inlets. Tidal flats, freshwater marshes, salt ponds or sabkhas may occur around the lagoon, particularly on the landward side, and their nature very much depends on the climate. Where tidal inlets are sparse, then sediment is transported over the barrier during storms, to be deposited in back-barrier washover fans.

Strandplain systems also develop in regions of moderate to high wave energy, where the tidal range is low and carbonate sand production rate is high. They develop during still-stands or slight sea-level falls along shorelines of low to moderate gradient. The swales may be sites of temporary lakes but mostly the water is meteoric. Older ridges may be vegetated, and swales occupied by marsh or swamp (Fig. 4.1B). Many modern carbonate beaches along shorelines without lagoons behind are associated with rocky coasts, usually of Pleistocene limestone.

4.1.3a Shoreface–foreshore facies

Sediments of the shoreface (fairweather wave-base to low tide) and foreshore (beach, low to high tide marks) show a sequence of bed forms reflecting the changes in wave-form characteristics as water depth decreases (see Fig. 4.6). Offshore, sinusoidal waves (swell) approaching the shoreline become steeper when the depth is less than half the wavelength (wave-base). In this *buildup zone*, flow is still oscillatory, although there is a net shoreward movement of water. Fairweather wave-base generally occurs at depths of around 10–20 m (varies with fetch, orientation of ramp, latitude, etc.) and shallower than this the sea-floor is affected by normal waves. Sediment is moved to and fro as each wave passes and symmetrical ripples, becoming more asymmetric shorewards, are the common bed form (λ's 0.1–0.5 m typically). Ripple wavelength is related to water depth and sediment grain size, with longer wavelengths occurring in deeper water and coarser sediments (see Allen, 1982 for a review). Wave-ripple cross-lamination, which can show much variation (e.g. de Raaf et al., 1977), is the characteristic structure of buildup-zone sediments. In high-energy shorefaces, larger-scale bed forms may occur, such as onshore-directed lunate dunes (mega-ripples) with a wavelength of a metre or more, giving

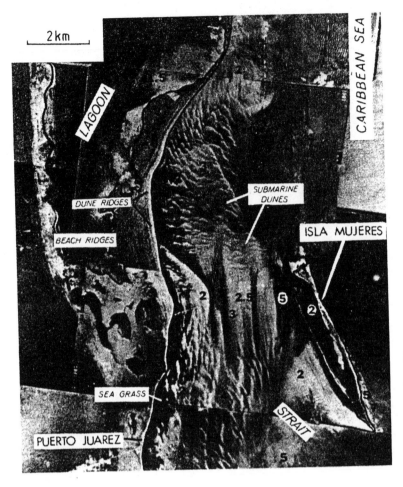

Fig. 4.3 *Aerial photo of the northeast Yucatan coast (compare with Fig. 4.2A) showing sand waves in the strait between Isla Mujeres and the mainland, and Holocene beach and dune ridges of the Isla Blanca Peninsula. Depths given in fathoms (nearly 2 m). After Ward* et al. *(1985) with permission of Bill Ward.*

trough cross-bedding. Eventually, as water depth decreases and wave steepness increases the waves break (*breaker zone*), giving rise to fast shoreward-directed flows in the *surf zone*, and thin sheet flows of the *swash-backwash zone*. The turbulent surf zone is wide on low gradient beaches and absent on steep beaches where the swash develops immediately after the waves have broken. Asymmetric ripples and dunes occur in the breaker and surf zones, and larger-scale bars may also form there, giving rise to onshore-directed planar cross-bedding. Swaley cross-stratification (SCS, next section) may be formed in the shoreface zone. Upper flow regime sheet flows in the swash-backwash zone give planar surfaces with current lineation in the foreshore zone (Fig. 4.7A) and deposit flat-bedded well-sorted sand with a low angle, offshore dip. Summer–winter changes in beach profile result in subtle, mostly planar erosion surfaces

between packets of flat-bedded sand (see Fig. 4.7B). Also along a shoreline, offshore-directed rip currents may occur in localized areas and generate shallow channels (e.g. Fig. 4.7A) with offshore-directed bed forms. Rip currents thus produce seaward-directed cross-bedding and shallow channel fills and erosion surfaces. Towards the upper foreshore and backshore, low beach berms occur, and washover sedimentation leads to onshore-directed planar cross-bedding. In the lower foreshore (Fig. 4.7A), low ridges covered in dunes and ripples, and runnels (shallow channels) floored by ripples result in cross-bedding, cross-lamination and shallow scours.

Carbonate sand bodies also form in the shallow subtidal, shoreface areas of inner ramps and open shelves. These *mobile submarine shoals* are well developed along parts of the northeast Yucatan coast, especially in the strait between Isla Mujeres and the

Fig. 4.4 *Aerial photo of beach ridges in the Upper Pleistocene strandplain of eastern Yucatan. After Ward* et al. *(1985) with permission of Bill Ward.*

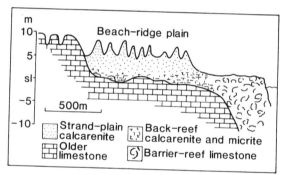

Fig. 4.5 *Diagrammatic cross-section through the Upper Pleistocene of northeast Yucatan. After Ward & Brady (1979).*

mainland (Fig. 4.3; Harms *et al.* in Ward *et al.*, 1985). The shoals are elongate ridges and patches of sand, parallel to the strong tidal currents, with north-ward-moving, strongly asymmetric sinuous to linguoid crested sand waves (Fig. 4.3). The latter have a spacing of 10−400 m and height of 1−3 m. Water depth is around 5−6 m, with sand wave crests reaching to within 3−4 m of the sea surface. Smaller-scale dunes and current ripples occur upon the sand waves. Avalanche faces of the sand waves dip at the angle of repose (32−35°) and are planar, or have poorly-developed wave ripples upon them. Toes of foresets

are mostly sharp and angular, rather than tangential. Areas of inactive sand are stabilized by algae and *Thalassia*. Sediment of the mobile shoals is skeletal and oolitic, with some aggregates. Internally, these Yucatan submarine shoals are likely to show large-scale tabular, planar cross-bedding from sand wave migration, with sets up to 4.5 m high. Individual cross-beds would coarsen downward, with the highest percentage of coarse grains occurring at the toe of the foreset. Although mostly unidirectional, some reversals of cross-bed dip are likely from storm winds, and reactivation surfaces could also be developed (Ward *et al.*, 1985). This facies is likely to show many similarities to clastic, tide-dominated, shelf sands, which have been well studied recently and commonly have sigmoidal cross-bedding, tidal bundles, mud drapes and reactivation surfaces.

Longshore drift is a feature of many sandy shore-lines where dominant winds strike the coast at an angle. This can lead to the formation of sand spits and sand shoals at the ends of barrier islands and tombolos connecting the islands with the mainland. Examples of these features are well seen along the Trucial Coast (Section 3.4.1) and off northeast Yucatan.

Burrows are common in shoreface sediments, generally less so in beach sediments. Crustaceans, echinoids, sea anenomes, annelids and bivalves all make distinctive burrows in modern shallow shoreface sands. In modern shoreface environments, *Thalassia*

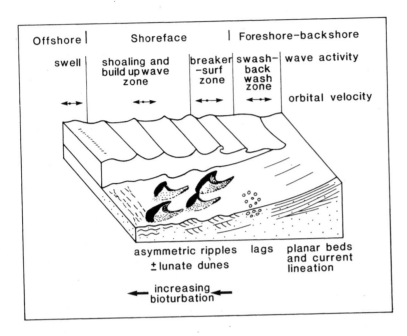

Fig. 4.6 *Environments, wave zones and bed forms of a non-barred, high wave energy shoreline. After Elliott (1986), based on work of Clifton et al.*

sea-grass is sparsely developed, and algae such as *Halimeda* with holdfasts are common. Various solitary corals, such as *Porites* and *Manicenia*, occur in the sands, along with gastropods, bivalves, crustaceans, holothurians and echinoids.

Small coral–algal patch and fringing reefs may develop in the shoreface zone and be a source of sand. Such reefs occur along the seaward side of barrier beaches in the central region of the Trucial Coast, and these can give rise to protected areas and small lagoons between them and the barrier.

With regard to the sediment itself, skeletal grain-stones dominate in most modern carbonate shoreface–foreshore–backshore environments, but in some regions there is a high content of ooids (e.g. northeast Yucatan and Trucial Coast). Aggregates such as grapestones are not common but peloids may be abundant. Many will be micritized skeletal grains. Grain size usually increases from the shoreface to the foreshore, but mostly this trend is not as well defined as along siliciclastic shorelines since skeletal grains start out with a wide range of sizes, shapes and mechanical strengths. Concentrations of coarse shell debris are common. There is usually a shoreward decrease in lime mud content as reworking and win-nowing of fine sediment become more intense in that direction. Sorting generally improves shorewards too; this is especially evident where skeletal grains domi-nate the sediment (rather than ooids), and then

the degree of rounding also improves as abrasion increases. Skeletal content will depend on the organ-isms living in the shoreface, but a wide variety of grains is typical, reflecting the normal salinity and good circulation of the environment. Typical grains will be of mollusc, echinoid, coral, foraminifera and algal origin.

A common feature of the foreshore in low latitudes is the presence of beachrock (Scoffin & Stoddart, 1983, and others). Carbonate sands are locally being cemented by acicular aragonite and micritic high-Mg calcite and it is likely that this is taking place just below the beach surface (Section 7.4.1c). Cemented sands also occur in the shoreface zone and are well documented from off Qatar (Section 3.4.1, Shinn, 1969) and on the Bahama Platform (Section 3.2.1b, Dravis, 1979).

4.1.3b Below fairweather wave-base facies, HCS and storm beds (tempestites)

Beyond fairweather wave-base, the seafloor is only affected by currents and waves during storms and so packstones–wackestones–mudstones are typical of these deeper-water locations, usually with much bioturbation. Two storm processes should be dis-tinguished here although they commonly operate together. Storms themselves can set up waves which affect the seafloor and cause reworking of sediment

Fig. 4.7 *The foreshore environment and facies. (A) The intertidal region of a Gulf Coast beach, Fort Myers, Florida, showing asymmetric (onshore-directed) wave ripples on a low ridge cut through by a rip channel in the lower foreshore, a shallow, shoreline-parallel runnel of the mid foreshore and a flat, gently seaward-dipping swash-backwash surface of the upper foreshore. (B) Swash-backwash bedding, sets of planar, seaward-dipping (to right) laminae with slight disconformities and truncations between sets, reflecting changes in beach profile. Raised beach (Pleistocene), Mallorca, Spain.*

and the development of bed forms. A storm wave-base can thus be recognized for an area, occurring in the region of around 30−50 m, a depth down to which major storms will interact with the seafloor. Storms can also generate currents; this is especially common where storms are directed onshore, leading to a buildup of water in the nearshore region or in the lagoon and, when the storm abates, an offshore-directed bottom current (storm surge) (Fig. 4.8).

Shoreface sediment put into suspension can be transported considerable distances offshore in a density current (like a turbidity current), to beyond storm wave-base.

Storm waves and currents deposit beds and packets of grainstone−packstone with distinctive sedimentary structures. In the area between fairweather wave-base and storm wave-base, grainstones−packstones may show an undulating bedding referred to as hummocky cross-stratification (HCS) (Dott & Bourgeois, 1982; Duke, 1985). It is characterized by gently curved, low-angle cross-lamination (Fig. 4.9A), and within individual beds curvature of laminae typically is both convex-upward (a 'hummock') and concave-upward (a 'swale'). Cross-strata dip at maximum angles less than about 10−15°. The intersections between laminae vary from erosional truncations of overlying against underlying laminae, to non-erosional terminations of overlying on underlying laminae. The spacing of hummocks is usually between 1 to 6 m and in plan view they are three dimensional and radially symmetrical. There is normally no preferred orientation of cross-strata, and vertical sections through HCS appear very similar, regardless of orientation. In many cases, hummocky cross-stratified grainstones (and sandstones) are mantled with wave-formed ripples.

Hummocky cross-stratified grainstones occur in a spectrum from units 0.1 to 2 m thick interbedded with shale or lime mudstone through to a sequence of amalgamated grainstones, where each bed is in erosional contact with the one below, and there is little mudstone between (Fig. 4.10). Some hummocky grainstones show a sequence of divisions, designated from the base up B (basal lag), P (parallel-laminated), H (hummocky), F (flat-laminated), X (cross-laminated) and M (mudstone) (see Fig. 4.9A; Dott & Bourgeois, 1982; Walker *et al.*, 1983).

Grainstones (and sandstones) with HCS are interpreted as the product of wave-generated oscillatory flows produced by the passage of storms. There has been much discussion, however, over the role of unidirectional currents with some authors suggesting that HCS is produced by combined flows (e.g. Allen, 1985). Where a complete BPHFXM sequence is present then this is interpreted as indicating initial deposition from a powerful unidirectional current (divisions B and P), with higher divisions (PHFX) being deposited from oscillatory-dominant flow, as the current subsides (Walker *et al.*, 1983). The mudstone division (M) is deposited after the storm and is usually bioturbated. The spectrum from interbedded grainstone−mudstone to amalgamated grainstones

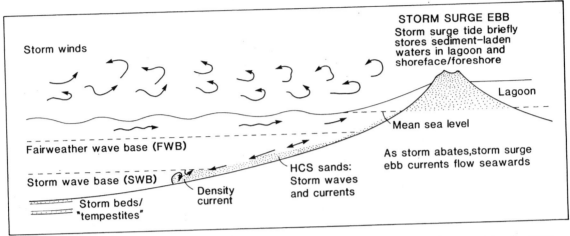

Fig. 4.8 *Model for development of HCS sands and storm beds. After Elliott (1986), based on a variety of sources.*

(Fig. 4.10) is interpreted as reflecting depth, frequency of storm events and proximity to source area.

Related to HCS is swaley cross-stratification (SCS, Fig. 4.9B), in which swales are preferentially preserved, and hummocks are rare. Flat-bedding is usually associated with SCS, and may form substanital thicknesses. Sedimentation is apparently nearly continuous, rather than episodic as in HCS deposits, although a storm-wave origin is still likely since all known SCS deposits are closely associated with beds containing HCS. A shallower-water, probably shore-face origin is thus advanced for limestones (and sand-stones) with SCS (Leckie & Walker, 1982; Duke, 1985).

Reviewing the effects of storms on shallow-water environments, Duke (1985) concluded that severe tropical cyclones (hurricanes) and mid-latitude winter wave cyclones (intense winter storms) are the only types of storm capable of producing HCS. In addition, a survey of all known occurrences of HCS showed that palaeolatitude and palaeogeography were consistent with a direct storm influence (Duke, 1985). However, these views have generated much discussion (Klein *et al.*, 1987).

Hummocks and swales have yet to be observed on the seafloor although probable HCS is recorded from vibrocores from the North Sea (Aigner & Reineck, 1982). However, there are many occurrences of HCS in shallow-marine limestones (e.g. Fig. 4.11); examples include the Triassic Muschelkalk of Germany (Aigner, 1982), the Upper Cambrian Nolichucky Formation of Virginia (Markello & Read, 1981), the Middle–Upper Ordovician of Virginia

(Kreisa, 1981), and the Lower Carboniferous of south Wales (Wu, 1982; Wright, 1986a).

Limestone (and sandstone) beds deposited by storms are being increasingly referred to as *tempestites* and these show much variation in thickness, grain size and internal structures, depending on proximity to area of storm and the intensity of the storm (e.g. Aigner, 1982). As with HCS, because of difficulties of sampling modern shelf sediments, especially to show larger-scale sedimentary structures, our knowledge of storm beds has mostly come from the geological record, an exception being the work of Aigner & Reineck (1982) in the North Sea and Snedden & Nummedal (1990) in the Gulf of Mexico. *Storm beds* grade from amalgamated sandy sequences, with HCS, as described above, through to centimetre-thick graded units within mudstones (Figs 4.10 and 4.12). The base of storm beds is always sharp and erosional and a variety of sole structures occur. Broad scours may be several metres across and 0.1–0.2 m deep. Gutter casts are common and tool marks may be abundant. The sole marks give an indication of palaeo-current direction and although this is mostly uni-directional, bipolar patterns are not uncommon, reflecting the passage of storm waves. Internally, storm beds are variable in their structures. Graded bedding is common, and there may be concentrations of bioclasts as a basal layer. These may have formed through storm reworking of the seafloor, to produce a lag, but with no great transport of shells involved (Kreisa & Bambach, 1982). Infiltration fabrics are common in the shell layers, produced by the percolation of finer-grained sediment down into

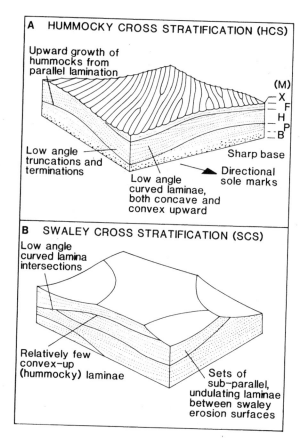

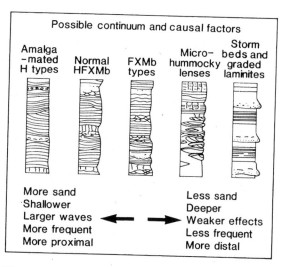

Fig. 4.10 *The spectrum of storm deposits. After Dott & Bourgeois (1982).*

Fig. 4.9 *Hummocky and swaley cross-stratification. (A) HCS as part of a 'complete' storm wave deposit, consisting of basal lag (B), parallel-laminated (P), hummocky (H), flat-laminated (F), cross-laminated (X) and mudstone (M) divisions. After Duke (1985). (B) SCS, which has similar appearance regardless of orientation of vertical section and tends to occur in grainstones (and sandstones) several metres thick with few muddy interbeds.*

the lag, as the storm subsides and material in suspension settles out. Flat-bedding, hummocky cross-stratification and cross-lamination occur within storm beds, and commonly a sequence of structures reflecting waning flow is seen (e.g. flat-bedding with parting lineation to cross-lamination, Fig. 4.12). The tops of storm beds are commonly rippled, with symmetrical wave-generated forms or current ripples. Other storm beds have gradational upper boundaries into overlying mudstones or are hardground surfaces. Burrows are common, on the undersides of beds, going through beds (escape burrows) and in the upper part due to organisms colonizing the seafloor after the storm.

Within a storm bed there is commonly evidence for both unidirectional currents and waves, and palaeocurrent markers may indicate variations in the flow direction. For example, storm beds in the German Muschelkalk show gutter casts parallel to the inferred shoreline, oscillation ripples oriented onshore–offshore and internal structures indicating offshore flow (Aigner, 1982).

Storm beds commonly show marked changes in character with increasing distance from the shoreline and increasing water depth (Aigner, 1982; Aigner & Reineck, 1982). 'Proximal' storm beds are relatively thick-bedded, bioclast-dominated and coarse-grained, with many composite and amalgamated beds. Distal equivalents are mud-dominated and thinner one-event beds. This pattern is a function of decreasing strength of storm waves and currents away from the shoreline. However, a simple proximal-to-distal model with increasing distance from the shoreline is complicated by shelf topography and variations in strength and frequency of the storms. In addition, storms may be centred on the shelf itself, rather than along the shoreline.

The orientation of ramps and open shelves relative to storms is an important consideration. Where a ramp faces the direction of oncoming storms, the offshore, bottom-hugging storm surges are likely to be much more frequent than on a ramp where the dominant storm direction is offshore. Regardless of the orientation of the ramp, the seafloor can be affected by storm waves, but it is on ramps in windward

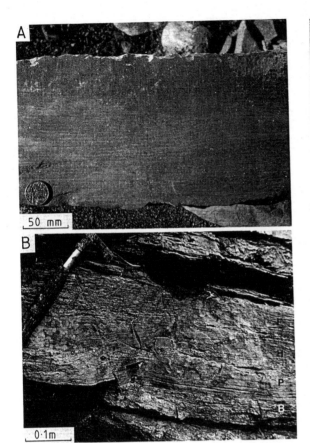

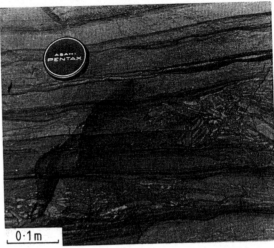

Fig. 4.11 *Hummocky cross-stratification. (A) Small-scale HCS in an oolitic sandy limestone, Etina Formation, Late Precambrian, South Australia. (B) HCS as part of a storm-wave deposited bed with basal lag (B), overlain by parallel-laminated unit (P), erosive-based hummocky division (H) and flat-laminated top (F). Shipway Limestone, Lower Carboniferous, south Wales. From Wright (1986a).*

Fig. 4.12 *Storm beds ('tempestites') consisting of graded units with flat-bedding passing up into cross-lamination and ripples. Also, intraclastic beds of dolomicrite formed by storm erosion of lime muds precipitated as a background sediment, between storm events. Note passage of thin carbonate bed in centre into breccia to the right. This indicates seafloor cementation of the carbonate beds too. Biri Formation, Late Precambrian, southern Norway.*

4.1.3c Backshore dunes and aeolianites

In the backshore area, sands accumulate through wind deflation of the beach and storms carrying sediment beyond normal high tide levels. Both beach barriers and beach ridges are capped by dunes. A variety of aeolian dune forms develop, but transverse ridges with steep landward faces are most common. They can reach several tens of metres, or more, in height, and may cover an area a kilometre or two in width. On Abu Dhabi Island, Trucial Coast, barchan and parabolic dunes occur inland from transverse dunes. Typical aeolian cross-bedding, tabular-planar and wedge-planar with high angles of dip (up to 34°), is common in these coastal dune ridges and it is invariably directed onshore. Trough cross-bedding is also recorded, along with contorted bedding. Aeolian carbonate dunes of the eastern Yucatan shoreline have leeward (to the west) cross-bed dips in the range 28–32°, with a maximum of 39°. Foresets are planar, concave upward or slightly convex upward. Windward cross-beds dip at low angles mostly (<10°).

A feature of wind-blown lime sands is that they are commonly cemented to form *aeolianite*. Such deposits are well developed in Bermuda, northeast Yucatan

settings that storm beds will be most common, from the generation of sediment-laden storm surges and currents in the nearshore area and their travel to the offshore, outer ramp. A case in point is the Upper Muschelkalk of the Catalan Basin, Spain, where an easterly-dipping carbonate ramp was the site of marl-stone, and nodular to bedded limestone deposition, with few discrete storm beds. The prevailing wind and storm direction of the time was towards the east, so that the Catalan Basin was in a leeward location and storm surges would have been very infrequent (Calvet & Tucker, 1988).

(Fig. 4.13A; Ward *et al.*, 1985), Mallorca and Ibiza, Spain, Bahamas (Ball, 1967) and the Trucial Coast (where they are termed miliolite). Cementation mostly takes place in the meteoric vadose zone so that grain-contact (meniscus) and microstalactitic calcite (low Mg) cements are the common types. Syntaxial overgrowths occur upon echinoderm fragments. One further type of cement which is common in the Yucatan aeolianites is needle-fibre cement, consisting of straight fibres of calcite, up to 200 µm long and 4 µm wide. Also grains are coated with a micritic rim cement superficially resembling a micrite envelope. Needle-fibre cement and grain coats usually occur in close association with rootlet and paleosol horizons. A characteristic feature of aeolianites is the presence of rhizocretions (Ward, 1975; McKee & Ward, 1983). These are calcified root systems of dune plants (Fig. 4.13B). Many are vertically oriented, show branching and are several centimetres to less than a millimetre across. Paleosols, especially calcretes and laminated crusts, are common within and upon aeolianites. *Microcodium* is a spherical structure produced by the calcification of mycorrhizal associations and consisting of radiating calcite prisms. It is common in aeolianites, as well as calcretes (Klappa, 1978).

Where Pleistocene and Holocene aeolianites occur in the same region, then a progressive change in the mineralogy of the aeolianites can be demonstrated (Ward, 1975). Young Holocene dune ridges have an aragonite–high-Mg calcite content similar to local beach and shoreface sands from which they were derived. Contact with meteoric water lowers the content of metastable carbonates in older ridges.

4.1.3d Tidal channels and tidal deltas

Tidal channels cutting through beach-barrier islands commonly have tidal deltas developed at their seaward and/or lagoonward ends. Along the Trucial Coast (see Section 3.4), it is ebb-tidal deltas which are most prominent (Fig. 3.39). These are shoal areas of sand, several kilometres across, dissected by channels, where much wave energy is expended as waves from the open gulf break along the seaward margin of the delta. Large bed forms, breaker bars, sand waves and dunes, as well as smaller-scale ripples, cover the tidal delta. Levees, just exposed at low tide, border the channels that cross the sand shoal. Tidal channels are generally floored by sand waves and dunes, and in quieter reaches sea-grass and coral patch reefs occur. Internal structures of tidal delta sand bodies are likely to be cross-stratification, on various scales, with

Fig. 4.13 *Holocene aeolianites from northeast Yucatan, Mexico. (A) Large-scale cross-bedded (onshore-directed) oolitic aeolianite at back of modern beach. Cliff is 6 m high. (B) Weathered-out rhizocretions in uppermost part of aeolianite, formed by preferential cementation around plant roots.*

orientations mostly normal to the shoreline, lagoonward and seaward, the proportions of each direction determined by the dominant flow direction. The Trucial Coast ebb-tidal deltas are dominated by seaward-flowing currents so that over much of the shoals bed forms are directed offshore and spillover lobes of sand occur adjacent to channels. Near the margins of the shoals and away from channel influences, sand waves are directed landwards in response to onshore waves and currents.

The tidal delta is the main site of ooid precipitation in the Trucial Coast and maximum size (>2 mm) is attained close to the axial channel, decreasing away to the outer parts of the delta where skeletal grains rapidly dominate the sediment where depths exceed 2 m. The channel itself is occupied by a mixed pelletal–ooid–skeletal sand, which also forms the beaches and dunes.

Tidal channels and tidal deltas also occur between islands in the Florida Keys (Jindrich, 1969; Basan, 1973; Ebanks & Bubb, 1975). Although the Keys are composed of Pleistocene bedrock and are not a Holocene beach-barrier–dune system, the gross sedimentological features of the tidal channels and deltas are similar to those of northeast Yucatan and Trucial Coast. In the Bahamas, too, tidal deltas have formed at the ends of tidal passes between Pleistocene-cored islands (Halley *et al.*, 1983), e.g. in the Exuma Islands of Great Bahama Bank, and in Carter and Strangers Cays of northeast Little Bahama Bank.

In the Florida tidal deltas and banks associated with the channels between the Matecumbe Keys, there are strong differences in the sediments of the seaward and bay-side tidal sand banks (Ebanks & Bubb, 1975). Sediments on seaward shoals are coralgal packstones and grainstones of *Halimeda*, coralline algae, corals and many other skeletal fragments, with relatively little lime mud. These sediments, as well as the flora and fauna, reflect the strong tidal currents, waves and regular storms on the seaward side. Tidal banks on the bay side have a greater abundance of molluscan grains, usually whole and unabraded, a matrix-support fabric and a high percentage of lime mud. Sea-grass covers less agitated parts of all tidal banks.

4.1.3e Carbonate lagoons

Carbonate lagoons behind beach barriers (Fig. 4.14) generally are the sites of accumulation of fine-grained sediments, in some cases with restricted faunas. However, there is quite a variation in sediment type depending on the circulation within the lagoon and this is largely controlled by the frequency of tidal channels and by the climate. Most lagoons are protected from ocean swell and storms by the beach-barrier system so that packstones through to lime mudstones are the typical sediments. Where there is a good connection with the open sea, normal salinities will occur within the lagoon and a diverse and abundant fauna can be expected, along with much bioturbation by infaunal organisms. Small patch reefs may occur in open lagoons. Pellets are abundant in most lagoonal sediments, produced by annelids, molluscs

Fig. 4.14 *Carbonate lagoon of the southern Florida Keys, with mangroves in the nearshore part and crustacean burrows (small mounds of white lime sediment). Coupon Bight, Big Pine Key.*

and crustaceans. Open lagoons occur along the Trucial Coast, the western end of the Khor al Bazm Lagoon for instance, behind the Great Pearl Bank (Fig. 3.38). In northeast Yucatan, lagoons occur behind Islas Blanca and Cancun (Fig. 4.2).

Where circulation is poor within a lagoon, then the fauna typically is impoverished and species diversity is low, although the few species may be present in very large numbers. In arid regions, evaporation will elevate salinities in the lagoons and may even lead to shallow subaqueous evaporite precipitation. In restricted parts of the Khor al Bazm Lagoon of the Trucial Coast, salinities reach 60‰ and imperforate foraminifera and cerithid gastropods are extremely abundant. Hard pelletal sands are common too. Precipitation of aragonite on the lagoon floor has produced aggregates and crusts, as well as lime mud.

High salinities are reached in the partly enclosed Freycinet and Hamelin Basins of *Shark Bay, Western Australia* (Fig. 4.15), and oceanic (35–40‰), metahaline (40–53‰) and hypersaline (53–70‰) waters are distinguished (Logan *et al.*, 1970; Logan *et al.*, 1974). Circulation is greatly restricted by shallow barriers and banks of skeletal sand and mud, partly covered by sea-grass and cut through by tidal channels, located at the entrances to the basins (the Fork Flat and Faure Sills, Fig. 4.15A). Three major lagoonal environments are: the *intertidal–supratidal zones* of beaches, rocky shorelines and notably tidal flats with algal mats (see Section 4.3.3j), the *sublittoral platform* from low tide to 6–9 m depth fringing the basins and merging with the barrier sills, and the *embayment plain*, a flat surface in the basin centre at a depth of 9–10 m (Fig. 4.15B). The sublittoral platform and sills are wedges and banks of grainstone through wackestone largely formed by the trapping of sediment by sea-grass and wave action over the last few thousand years. At the present time, the platform in metahaline areas is occupied by a burrowing mollusc–sea-grass community, and in hypersaline areas the gastropod *Fragum* dominates, with little grass, but stromatolites in some areas of Hamelin Basin. The embayment plain is mainly an area of skeletal lime mud accumulation, with molluscs, foraminifera and the green alga *Penicillus*.

In a more humid region, *Florida Bay* (see Section 3.3.1g) is a lagoonal-type environment located behind a barrier made by the rocky Florida Keys. The bay is triangular in shape, covering 1500 km², located to the south of the Everglades. Circulation is restricted through the presence of the near-continuous Florida Keys along the southeast side and shallow mud banks along the southwest side, which reduce tidal exchange

with the Gulf of Mexico. Mud banks, mostly linear features, occur within the bay too, and divide it into smaller basins or 'lakes'. Salinity and temperature vary drastically over the year. During the wet season, run-off from the Everglades reduces salinity in the bay to 10–15‰. During the dry season, intense evaporation results in salinities up to 70‰. Temperature varies from 15 to 40°C. Tidal range is mostly less than 0.3 m and average depth is 1.5 m. Tidal currents in the bay and wind-driven waves and currents are weak through the baffling effects of the mud banks and the shallowness of the bay. However, Florida Bay is the site of extensive lime mud deposition and much of this is accumulating in the plant-stabilized banks (see Section 3.3.1g, Stockman *et al.*, 1967; Enos & Perkins, 1979). The bottom fauna is dominated by molluscs with at least 100 genera (Multer, 1977), and to a lesser extent by foraminifera. Turtle grass is widespread, as are rooted calcified green algae. Much of the lime mud is formed by the disintegration of the green algae, especially *Penicillus*, but breakdown of other skeletal grains, especially molluscs, by crabs, holothu-

Fig. 4.15 *Shark Bay, Western Australia. (A) Location, salinity zones and sills. (B) Major depositional environments in Hamelin Basin. After Logan* et al. *(1974).*

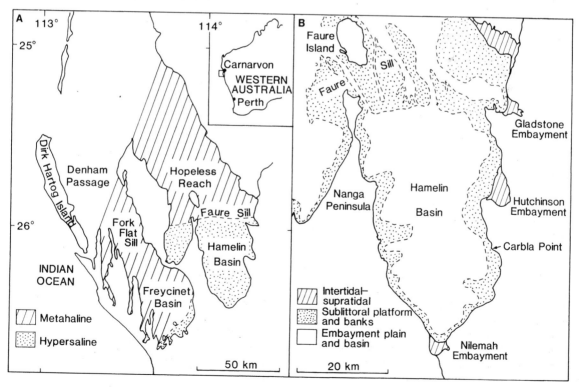

rians, boring sponges and algae contributes much fine sediment. There is sufficient depth and fetch in the basins for waves to winnow fines, leaving skeletal lags, and the mud is deposited on the leeward (south-west) sides of the banks. Storms and hurricanes are particularly important, dumping decimetres of sediment on the banks and taking fines out of the bay. There is evidence that some mud banks migrate during storms (Bosence, 1989a).

Smaller lagoons occur elsewhere in south Florida (e.g. Coupon Bight, Big Pine Key, Fig. 4.14, described in Multer, 1977), and on the Bahama Platform, although again these are not associated with modern barrier islands, but exist because of the configuration of Pleistocene limestone outcrops. Bimini Lagoon on the Great Bahama Bank is of particular interest since it is virtually mud-free, and very thin oolitic coatings are forming around the sand grains, in what is a low-energy environment (Bathurst, 1967; Multer, 1977). The lagoon is enclosed on three sides by islands and is an area of around 20 km² with depths mostly less than 2 m. Tidal channels between islands drain the lagoon, and with a tidal range of 0.7 to 1 m, much of the lagoon floor is exposed at low tide. The sandy sediment of Bimini Lagoon is mostly skeletal, derived from the rich and varied *Strombus costatus* community which inhabits the lagoon; peloids of faecal and micritized skeletal grain origin are common. *Thalassia* seagrass and a subtidal algal mat cover much of the sediment surface, binding sediment and inhibiting grain movement (Scoffin, 1970). The thin oolitic films on sand grains are only a few microns thick, but their origin is unclear. The environment contrasts with the turbulent, constantly agitated shallow bars and shoals of the platform margin where ooids are forming in abundance (Section 4.3.1).

4.1.3f Intertidal back-barrier environments

These are diverse, ranging from bare tidal flats to algal marshes and freshwater ponds to dense mangrove swamps. Around the lagoons behind the Trucial Coast barriers, tidal flats vary considerably. Some areas are covered with blue–green algal mats or have an aragonite-cemented crust, which is commonly brecciated. Other areas are bare, or covered in ripples. Thick stands of mangroves also occur, with pelletal lime mud intensely burrowed by crabs. Tidal creeks drain the mangrove swamps and tidal flats. Intertidal areas on the lagoonward side of the barriers are composed of oolitic–skeletal sand derived from barrier-top dunes and locally produced pelletal and

skeletal sand and mud. Intertidal areas along the mainland shore of the lagoon are more muddy and extensive, and are covered in a variety of algal mat types which reflect the degree of exposure and desiccation (see Section 3.4.3).

Lagoons in Florida and the Bahamas are also surrounded by tidal flats but mangrove swamps are the dominant feature, with many thousands of square kilometres occurring around Florida Bay. Beyond the reach of seawater in the Everglades, calcite (low Mg) is being precipitated in freshwater ponds (Enos & Perkins, 1979), and algal mats occur there too.

Where beach barriers are narrow and there are few tidal inlets, then sand is often transported over the barriers during storms to form washover fans on the lagoonward sides. A temporary channel may be cut through the barrier during the storm. Washover fans prograde into the lagoon generating sets of tabular cross-bedding. Trough cross-bedding and cross-lamination may form from dunes and ripples upon the fan surface. Washover fans occur on the lagoonward side of Isla Blanca, Yucatan (Ward, 1975), where most of the back-barrier shores are lined with mangroves. Small lagoons occur between discontinuous older aeolianite ridges.

4.1.4 Beach-barrier island–lagoonal sequences

Thick sedimentary packages are deposited along barrier–lagoonal carbonate shorelines and on a broad scale the main factor determining the facies sequence is the pattern of relative sea-level change. A number of scenarios are possible, and our understanding of these has largely come from studies of Recent and Pleistocene siliciclastic barrier island systems.

Where the position of sea-level changes little or falls only slightly, then seaward progradation of the beach-barrier island system can take place. This generates a coarsening-upward sequence (Fig. 4.16) from the offshore muds with storm beds, through amalgamated storm beds with HCS to cross-bedded shoreface sands and then to low-angle flat-bedding of the foreshore. Large-scale aeolian cross-bedding could occur at the top of the unit from barrier-top dunes. The thickness of a progradational or regressive barrier sequence would depend on the energy level and tidal range of the system, which determine the depth to wave-base and the height of the beach, and also on the subsidence rate. A typical thickness would be around 10–30 m from the shoreface through foreshore facies.

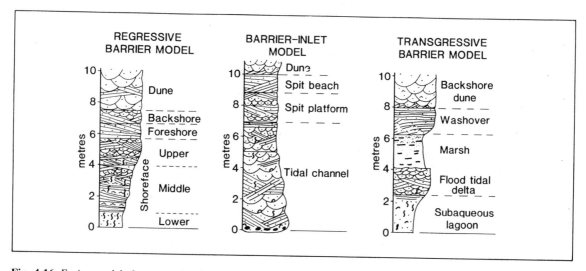

Fig. 4.16 *Facies models for regressive, barrier inlet and transgressive barrier island sequences. After Reinson (1984).*

If there was sufficient subsidence, then lagoonal mudstones could come to overlie the barrier-top facies.

A variation on the facies model for a progradational barrier is necessary where tidal channels are common, typically in mesotidal regions (Fig. 4.16). The channels migrate along the barrier in response to longshore drift and development of spits. Consequently, a barrier island sand body with many tidal inlets will consist largely of channel-fill grainstones (Kumar & Sanders, 1974). A unit with sharp base, some fining-upwards and cross-bedding (probably bimodal, onshore−offshore) would thus constitute much of the sequence. The unit would be capped by the foreshore and barrier-top facies and rest upon the lower shoreface sands and offshore muds.

Where there is a relative rise in sea-level there are several possible outcomes for the beach-barrier island system (Fig. 4.17). If sea-level rises quickly, then a barrier can be drowned and abandoned in the deeper waters of the outer ramp or open shelf (Sanders & Kumar, 1975). The barrier sands could be reworked by a variety of currents and waves to produce an offshore sand bar complex as is well documented for some ancient siliciclastic shelf sand bodies (e.g. the Cretaceous Duffy Mountain Sandstone of Colorado, Boyles & Scott (1982), or Sussex Sandstone of Wyoming, Hobson *et al.* (1982)). Alternatively, the barrier could be blanketed with mud and buried. A new beach-barrier island system could develop at

the new shoreline when sea-level stabilized once again.

In other situations, probably when sea-level is rising quickly, but not too rapidly so as to drown the barrier, the barrier can migrate landwards by erosion in the surf zone and washover of sand into the lagoon (Fig. 4.17B; Fischer, 1961; Swift, 1968). Some sand is transported offshore into deeper water by storms (e.g. Swift, 1975). The landward migration of the barrier in this shoreface retreat mode leaves behind an erosion surface, known as a *surf-zone ravinement*, generated in the upper shoreface/low foreshore region. The surface will be underlain by back-barrier and lagoonal facies, and upon the surface there may occur a thin intraclastic breccia/conglomerate of slabs of lagoonal mudstone, derived from erosion of lagoonal deposits exposed in the surf zone of the migrating barrier. Clasts of cemented beach sand (beachrock), calcrete and black-pebbles, and skeletal debris may also occur in the lag deposit. The latter passes up into offshore muds, deposited on the erosion surface after the barrier has migrated. The intraformational disconformity is an important piece of evidence for postulating the former presence of a barrier, along with the occurrence of lagoonal deposits below, which would have needed the barrier for their protected depositional environment. Thus sequences of lagoonal facies overlain by offshore limestones should be examined carefully for evidence of a surf-zone ravinement.

Where sea-level does not rise too fast, there is a high production rate of carbonate and sufficient subsidence, then a *transgressive barrier sequence* can be produced (Fig. 4.16). Lagoonal sediments are buried by the landward migrating barrier, as washovers build into the lagoon. With slowly rising sea-level and subsidence, the back-barrier sands are preserved and barrier-top facies (pond sediments, soils and aeolian dunes) could occur towards the upper part of the sequence. With sufficient subsidence and sea-level rise, marine muds could be deposited over the former barrier sands, while the active barrier is further landward. Tidal channels and deltas could also be preserved in this transgressive barrier.

In the later section on ancient shoreline carbonate sands, examples of progradational, transgressive and drowned barriers are given, as well as surf-zone ravinements, where barriers have been and gone.

4.1.5 Beach ridge–strandplain sequences

Sedimentation along a strandplain shoreline is similar to that along a microtidal beach barrier, but of course with no associated lagoon or tidal channels. Seaward progradation of the beach gives rises to a coarsening-upward sequence, as noted in the last section, and this process appears to be episodic, generating successive beach ridges. In some models of strandplain development, it is considered that the ridges grow from longshore bars in the shoreface zone which migrate towards the shoreline under conditions of high sediment supply. An alternative is that the ridges develop in the foreshore zone during storms and extra high tides, with wind playing a major part in developing the higher backshore part of the ridge. Quarry sections through the Upper Pleistocene strandplain of the eastern Yucatan show the type of sequence produced (Fig. 4.18; Ward & Brady, 1979; Ward *et al.*, 1985). Three units are recognized: (1) a lower unit of burrowed, low-angle and hummocky (?) cross-bedded calcarenite, (2) a middle unit of multidirectional high-angle cross-bedded calcarenite and calcirudite, and (3) an upper unit of parallel-laminated calcarenite, with low dips, mainly to the east (offshore). The sequence represents the progradation of beach over shoreface environments. The sedimentary package rests on a bored calcreted surface capping the underlying Middle Pleistocene limestone. There is a basal conglomerate with reworked bioclasts and calcrete pebbles, formed during the transgression which initiated the strandplain. The Upper Pleistocene sequence is capped by a calcrete with rhizocretions

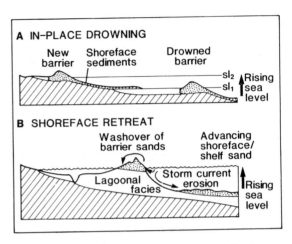

Fig. 4.17 *Sea-level rise and barrier islands. (A) In-place drowning of barrier through rapid sea-level rise. (B) Barrier migration through shoreface retreat and washover sedimentation. After Elliott (1986) based on various sources.*

formed in the backshore, ridge-top dune sands.

The overall geometry of a strandplain package is that of a blanket or sheet sand with an undulating upper surface. Fine-grained sediments may subsequently be deposited in the swales if ponds develop. Carbonate strandplains are likely to be affected by subaerial diagenetic processes, although the extent of these does depend on climate. Karstification, meteoric cementation, calcrete development and mixing-zone dolomitization are all possible processes.

4.1.6 Ancient shoreline carbonates

Although there are many ancient carbonate grainstone formations, few detailed descriptions exist of those deposited in beach-barrier island–dune complexes, with all the attendant back-barrier facies, or in beach ridge–strandplain complexes. Many grainstone sequences appear to have accumulated in shallow subtidal shoals and offshore bars in mid-shelf and inner-ramp settings. In many instances back-shoal facies are not distinctive and there is no indication of island formation. Commonly the only evidence of emergence is in the nature of the early cements.

Beach-barrier island–lagoonal sequences are clearly a complex array of facies and subfacies, but one of the distinguishing features is that the barrier facies separate lagoonal from offshore marine facies. As noted earlier, barriers may prograde and give rise to thick grainstone packages, but they may also be

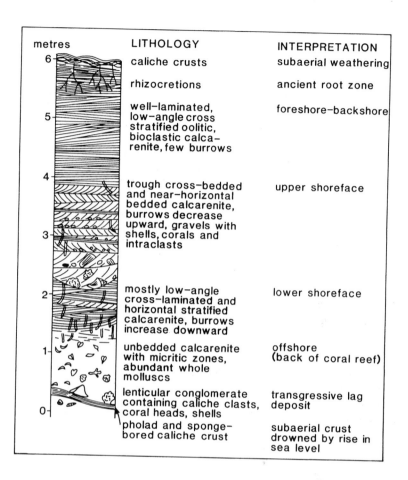

metres	LITHOLOGY	INTERPRETATION
6	caliche crusts	subaerial weathering
	rhizocretions	ancient root zone
5	well-laminated, low-angle cross stratified oolitic, bioclastic calcarenite, few burrows	foreshore-backshore
4	trough cross-bedded and near-horizontal bedded calcarenite, burrows decrease upward, gravels with shells, corals and intraclasts	upper shoreface
3		
2	mostly low-angle cross-laminated and horizontal stratified calcarenite, burrows increase downward	lower shoreface
1	unbedded calcarenite with micritic zones, abundant whole molluscs	offshore (back of coral reef)
0	lenticular conglomerate containing caliche clasts, coral heads, shells	transgressive lag deposit
	pholad and sponge-bored caliche crust	subaerial crust drowned by rise in sea level

Fig. 4.18 *Diagrammatic facies sequence through the Upper Pleistocene strandplain of northeast Yucatan, Mexico. After Ward & Brady (1979).*

transgressive, and shoreface erosion during barrier migration may leave little record of the barrier itself.

Middle Jurassic of the UK. One well-documented ancient carbonate barrier island–lagoonal complex is the mid-Jurassic *Lincolnshire Limestone Formation, Bajocian of eastern England* (Ashton, 1977, reviewed in Tucker, 1985). The Lower Lincolnshire Limestone consists of tidal flat and lagoonal carbonates in the Sproxton, Greetwell and Leadenham members. It is inferred that a north–south barrier existed during these times, with the quiet-water sediments accumulating to the west in a lagoon with a minimum width of 8 km. The lagoonal sediments are a complex array of facies, but a more inshore (westerly) low-energy zone dominated by mud-supported skeletal–peloidal packstone–wackestone can be distinguished from a higher energy, grain-supported belt of outer lagoon, back-barrier environments. In areas of maximum shelter behind the barrier, lagoonal mudstones with stable *in situ* benthic faunas accumulated. All lagoonal carbonates contain a normal marine fauna showing that there was good circulation and exchange with the open sea through tidal inlets in the barrier. In some areas there is evidence for emergence of the barrier in the form of local supratidal algal laminites and packstones. Elsewhere, the barrier appears to have been discontinuous and submergent. Complex bars and spillover lobes existed, and these prograded into the lagoon to give large-scale cross-bedded units. Although there is a predominance of landward (west) directed structures, some seaward-directed cross-bedding indicates strong ebb currents, and a southerly longshore drift component can be recognized too. Initial deposits of the barrier are thin coarsening-upward cycles reflecting oscillations of the barrier's inner margin, and lagoonward barrier advance through washover fan deposition. An ebb-tidal delta has been identified, developed at the mouth of a tidal inlet. Unfortunately, the nature of the sediments seaward of the barrier is not known; later Jurassic uplift removed Upper Bajocian strata. The Great Pearl Bank of the Trucial Coast is regarded by Ashton as a modern analogue of the Lincolnshire Limestone barrier complex. The stratigraphic development of the Lincolnshire Limestone shows that the barrier was transgressive; it migrated landwards through time, building into the lagoon (fig. 16 in Tucker, 1985a). Clearly subsidence was sufficient to accommodate the lagoonal and barrier sediments during the sea-level rise, with little loss of the facies through surf-zone erosion.

Also in the Middle Jurassic of the UK, a carbonate ramp to the south of the London–Brabant Massif was the site of extensive oolite accumulation, particularly in the *Great Oolite of Bathonian* age, well exposed *in the Cotswolds* (Fig. 4.19A; Sellwood *et al.*, 1985; Sellwood, 1986). Large-scale cross-bedding was formed by sand waves, and bimodal and polymodal palaeocurrent patterns suggest tidal and longshore currents. In the subsurface, dipmeter data record the cross-bedded nature of the oolite and increasing-upward and decreasing-upward dip patterns reflect sand wave migration patterns (Fig. 4.19B). Reactivation surfaces are common and major breaks in sedimentation are recorded by hardground surfaces, encrusted by oysters and bored. Small coral–bryozoan–brachiopod patches also occur on the hardground surfaces. The oolite itself contains few fossils, but notably thick-shelled gastropods and mobile bivalves; there are bioturbated and burrow-mottled horizons, and discrete burrows such as *Diplocraterion*. Marine cements occur within the oolite and locally there is evidence for emergence from meteoric diagenetic fabrics. Skeletal–oolitic packstones through to skeletal mudstones shoreward of the mobile sand belt were deposited in stabilized muddy sand and protected lime mud environments. A diverse fauna is present and there is much bioturbation, especially from crustaceans. In the subsurface of Hampshire, the Great Oolite is a hydrocarbon reservoir, with porosity enhanced by burial dolomitization. In the Wealden Basin, best porosities are found where early freshwater diagenesis in the Great Oolite resulted in a more mineralogically stable limestone, which was better able to withstand later burial compaction and pressure dissolution (McLimans & Videtich, 1986).

The Smackover Formation. Upper Jurassic ooid grainstones of the Smackover Formation occur in an arcuate belt in the *subsurface of the Gulf Rim from Florida through to Texas* (Fig. 4.20A). They have been the subject of many studies since they contain important hydrocarbon reservoirs, especially in east Texas and Arkansas (Collins, 1980; papers in Ventress *et al.*, 1984; especially Moore, 1984). The updip grainstone belt passes southwards, offshore, into peloidal bioclastic wackestones and foraminiferal lime mudstones, and then into organic-rich laminated mudstones. The depositional setting is generally taken to have been a ramp (Ahr, 1973; Budd & Loucks, 1981; Moore, 1984), but in some areas oolite formation appears to have been localized near breaks of slope, so that a shelf-margin sand body concept is more appropriate there (e.g. McGillis, 1984), and there are cases where the oolite has developed upon topographic highs on the ramp, caused by salt diapirism or structure. The Smackover sequence in most areas consists of offlapping oolite bodies, prograding southwards over the outer-ramp facies. Northwards, the Smackover is overlain by the Buckner Formation which consists of massive, nodular and laminated anhydrite, together with some halite. There are different views over the correlation of the Buckner and Smackover sequences. Budd & Loucks (1981), and others, believed that the Buckner evaporites were deposited in lagoons and sabkhas behind barriers and shoals of Upper Smackover oolite. Moore (1984), on the other hand, considered that the Buckner was not chronostratigraphically equivalent to the Upper Smackover, but was deposited after Smackover deposition, in an evaporite lagoon located behind a shelf-margin barrier, the Gilmer Member (see Fig. 4.20B). Red-bed clastics which were being deposited updip from the Smackover in Moore's model, spread southwards over the Buckner Lagoon, to terminate carbonate–evaporite deposition. Smackover to Haynesville sedimentation is seen as a ramp to shelf sequence.

The model for Upper Smackover deposition put forward for south Texas (Budd & Loucks, 1981) recognizes beach-barrier islands, washover fans, tidal inlets and deltas and shoreface shoals with sand waves. In Arkansas, one major shallowing-up unit is seen in cores through the Smackover (McGraw, 1984). The Upper Smackover is 20 m thick and shows organic-rich and laminated fine pelletal grainstones–packstones passing up into a similar but coarser facies with cross-bedding and burrows. Overlying algal boundstones with oncolites and skeletal–pelletal grainstones are then capped by a 10 m oolite with bimodal cross-beds and flat-bedding. The whole system is interpreted as an upward-shallowing package of deeper subtidal through shoreface to intertidal facies. Emergence and island formation is indicated by early

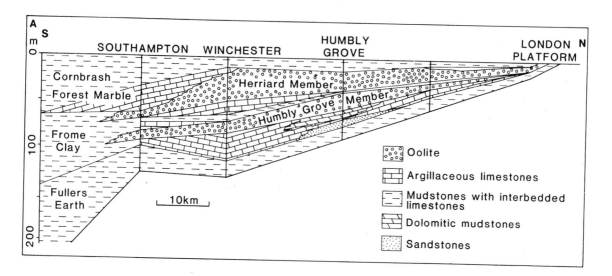

Fig. 4.19 *The Great Oolite Limestone Formation, Middle Jurassic, Wessex Basin, southern England. (A) Lithostratigraphy. After Sellwood* et al. *(1985). (B) Dipmeter data showing upward-decreasing and upward-increasing dip patterns.*

meteoric leaching of the ooids, which accounts for the high porosities and good reservoir characteristics of this formation. The ooid shoals were developed over pre-Smackover structural highs, with maximum wave energy expended on the southeastern flanks of these highs. Domal stromatolites formed in slightly deeper water on the seaward side of the shoals, and stromatolites were deposited in the protected areas behind the shoals.

Smackover–Gilmer carbonates of the East Texas Basin contain up to five shallowing-up cycles (Harwood & Moore, 1984), and where complete a cycle shows mudstone–wackestone passing up into pelletal–ooid packstone, pelletal–ooid grainstone and then ooid grainstone. Cross-bedding and flat-bedding are present but such structures are rarely seen because of the well-sorted nature of the grainstone. Skeletal grains are rare in the Smackover of this region, but micritized grains (peloids) are common. *Favreina* crustacean pellets are abundant. The cycles show below fairweather wave-base through shoreface to intertidal facies and reflect the upward growth of sand bars into beach-barrier bars and their lateral migration. The diagenesis of the East Texas Basin grainstones is complex, but includes early circumgranular meteoric or mixed phreatic calcite cements, dolomi-

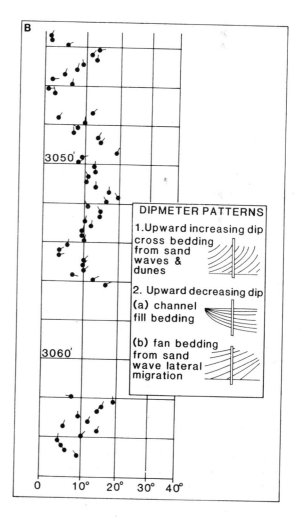

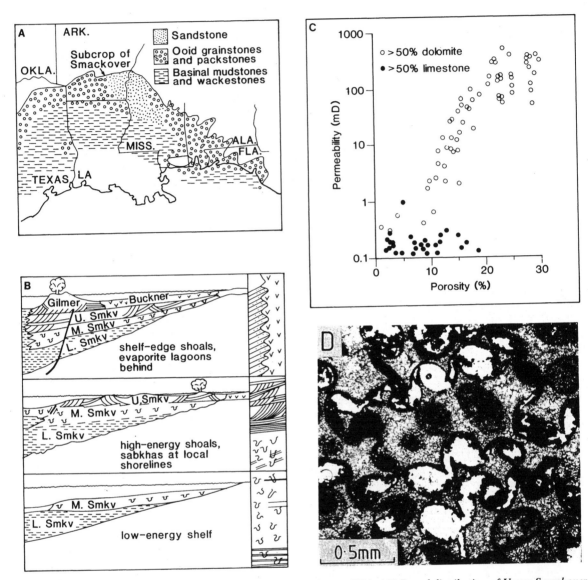

Fig. 4.20 *The Upper Jurassic Smackover Formation of the Gulf Rim, USA. (A) Broad distribution of Upper Smackover lithofacies. (B) Development of the Smackover, Buckner and Gilmer Formations. After Moore (1984). (C) Porosity–permeability plot for Upper Smackover limestones and dolomites. After Harwood & Moore (1984). (D) Oomoldic porosity in formerly aragonitic ooids. Photo courtesy of Gill Harwood.*

tization, ooid dissolution, calcite spar precipitation and replacive and void-filling anhydrite. Highest porosities are developed in zones of dolomitization (Fig. 4.20C) and there is a facies control in that it is the grainstones which are preferentially dolomitized. The dolomitization is attributed to mixed meteoric/marine waters and/or refluxed Buckner brines.

A different interpretation for the Smackover has been advanced for the northeast Texas area (McGillis, 1984). A shelf-break is postulated for this area (contrasting with the ramp setting of other parts of the Gulf Rim) and ooid shoals and bars prograded landwards by spillover lobe sedimentation into the shelf lagoon, where pelleted muds were accumulating. A broad ooid belt was formed, with shoals becoming inactive when they reached sea-level.

Two major influences on Smackover–Haynesville deposition and diagenesis are sea-level changes and basement structures. The gross stratigraphy was determined by sea-level changes: rising sea-level during the Lower Smackover, still-stand or slight fall during Upper Smackover, and rapid rise during the Haynesville (Moore, 1984). Oomouldic porosity (Fig. 4.20D) in updip areas was probably the result of freshwater influx into the sediments during the still-stand. However, studies of the original mineralogy of the Smackover ooids (Chowdhury & Moore, 1986; Swirydczuk, 1988) have revealed that ooids of older, more northern oolite ridges in Arkansas were composed of aragonite (Fig. 4.20D), but that younger, more southern ridges in southern Arkansas and northern Louisiana were (and still are) composed of calcitic ooids (e.g. Fig. 7.14). The reasons for the change in mineralogy are not clear, but it is an important consideration for the diagenesis. Oomouldic porosities are more common in more northern oil fields and intergranular porosities in more southern oil fields. Burial diagenetic effects of compaction and spar precipitation are more important to the south.

In the *Haynesville*, Faucette & Ahr (1984) described tidal oolite bars developed over topographic highs caused by salt movements. Shallowing-up cycles were generated, similar to those in the Upper Smackover. Local uplift produced by salt diapirism exposed the ooid bars and led to meteoric leaching.

Gulf Rim Cretaceous grainstones. Well-developed ooid grainstones are seen in the Early Cretaceous of the Gulf Coast Rim, particularly in the *Sligo Formation*. A shallow, open shelf existed around the Gulf Coast basin during this time. Wiggins & Harris (1984) documented three shallowing-up limestone sequences in 36 m of strata, where each consists of skeletal packstones and wackestones, coarsening up into ooid–skeletal grainstones. The former are interpreted as on–shelf, quieter-water deposits and the latter represent an ooid–sand shoal which built up to near sea-level. The grainstones have porosities up to 13% (reduced intergranular porosity). This is a considerable reduction on the original porosity of ooid lime sand (up to 50%) through two stages of cementation: an early meteoric phreatic isopachous calcite fringe and a later burial (post-compaction) calcite spar. Preferential cementation towards the top of grainstone units reduces the porosity considerably. Oil migrated into the grainstones after calcite spar precipitation.

In the *Black Lake Field of central Louisiana* (Harbour & Mathis, 1984), Sligo carbonate grain-stones were deposited in an inner shelf setting, closely associated with a rudist mud mound. The shelf margin, located some kilometres further south, had a major reef rim. Grainstones were deposited on the seaward side of a caprinid-dominated mud mound which built up to wave-base and may have become emergent. A sea-level rise permitted the development of a large oolitic sand shoal complex to the northwest and its subsequent progradation to the southeast (Fig. 4.21A). Behind the shoal, oncolitic–oolitic grainstones and packstones were deposited in a stabilized sand flat environment. Skeletal miliolid–peloidal grainstones were deposited seaward of the oolite shoals. When carbonate sedimentation was unable to keep up with rising sea-level, deep-water mudrocks (the Pine Island Shale) were deposited over the area. The grainstones are again important hydrocarbon reservoirs, with good intergranular porosities (average 16%) and high permeabilities (average 110 mD) (see Fig. 4.21B). Cementation of the grainstones has been minor and by calcite fringes interpreted as meteoric phreatic in origin, suggesting the development of ephemeral islands with associated floating groundwater lenses. The lack of later cements is attributed to tight updip lagoonal lime mudstones and the overlying Pine Island Shale, which acted as diagenetic seals and prevented the later downdip migration of calcite-spar-precipitating meteoric waters.

In the *Pearsall Formation of central and south Texas*, which overlies the Pine Island Shale, Stricklin & Smith (1973) and Loucks & Bebout (1984) have described a northeast–southwest oriented shoreface–beach grainstone complex (30 m maximum thickness) in the Cow Creek Limestone which prograded southwestwards in response to longshore currents and wave approach from the southeast. Grainstones are composed of echinoid and mollusc debris and ooids, and were deposited in beach, tidal channel, spit, tidal bar, shoreface and sand flat environments. Beachrock cements occur and calcretes indicate local subaerial exposure on islands. Seaward of the shoal complex, in what would have been lower-energy deeper water, occurs a thinner echinoid–mollusc packstone facies and then an oncolitic wackestone–packstone facies of the open shelf. Deeper-water outer shelf deposits are mixed terrigenous mudstones–lime wackestones. Three shallowing-up cycles are recognized in the Upper Cow Creek Limestone, of the wackestones–packstones coarsening up into grainstones of the sand shoal. In the region of a salient on the shelf, the third cycle is capped by a coral–stromatoporoid–rudist bindstone–framestone patch reef complex. Rep-

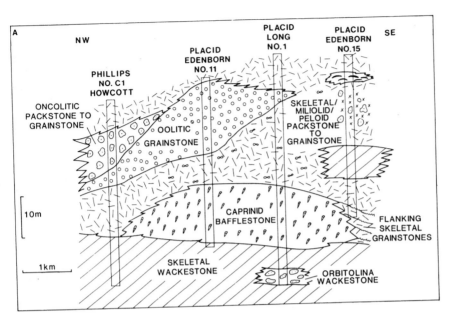

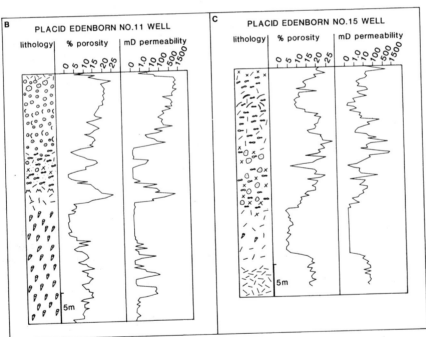

Fig. 4.21 *Lower Cretaceous rudist bioherm and grainstones of the Black Lake Field, Louisiana. (A) Cross-section showing lithofacies. (B) Lithology, porosity and permeability data for two wells. After Harbour & Mathis (1984).*

etition of the cycles is attributed to a slight relative sea-level rise or decrease in sediment production rate, allowing the deeper-water wackestone–packstone facies to be deposited, and then shoaling and sand bar–beach formation, as sedimentation rate caught up with sea-level. The best reservoirs occur in the grainstone and boundstone facies, where again inter-granular and growth cavities have been incompletely filled with cement.

Carboniferous grainstones, Illinois Basin. Oolite grainstones deposited in shoals and barriers, which occasionally became emergent, are well developed on the carbonate ramp which sloped gently southwards into the Illinois Basin, during the Mississippian (Choquette & Steinen, 1980; Cluff, 1984). These grainstones are also important hydrocarbon reservoirs and targets of current exploration programmes. In the *Salem Limestone*, oolites occur within several clearly defined shallowing-up cycles. Offshore, open-marine outer-ramp carbonates (skeletal wackestones) pass up into mobile shoal sands (oolitic–skeletal grainstones) which are overlain by platform-interior muddy sands (bioturbated oolitic–skeletal packstones). The various facies have distinctive wireline log characteristics, enabling correlations and deductions of clay content and porosity to be made (Cluff, 1984). These cycles are notably different from those described earlier, where oolites cap the shallowing-up units. The promi-nent grainstone cycles grade into thick, open-marine skeletal wackestones–mudstones with no apparent cyclicity, towards the south, and into fine-grained restricted marine tidal flat facies towards the north. The geometry of the oolite grainstones is variable; in some areas they are oriented roughly perpendicular to the depositional strike, suggesting that they may be tidally-influenced bars and channels. Elsewhere they are broad linear sand belts, parallel to the shoreline. These have been compared with the inner-ramp barrier sands of the Trucial Coast (Section 3.4.1; Cluff, 1984).

In the *Ste Genevieve*, cycles are not so apparent and two major facies types are recognized: an oolitic bar facies consisting of clean, well-sorted ooid grain-stone, and an interbar facies of bioturbated skeletal wackestones and lime mudstones. The oolitic bar facies was deposited on high-energy shoals and in the Lower Ste Genevieve they are of variable shape and size. In the Upper Ste Genevieve (Fig. 4.22), bars are more linear, 1–5 km long, 0.5–2.5 km across and up to 9 m thick. They are oriented north–south and northeast–southwest, perpendicular to the regional palaeoslope, suggesting a strong tidal influence. These bars were probably emergent at times. Lime mud-stones and wackestones beneath the oolite bodies are dolomitized (Choquette & Steinen, 1980). The oolite sands are thought to have acted as conduits for meteoric water, with dolomitization taking place in mixing zones (see Section 8.7.3).

Lower Carboniferous barrier, lagoonal and strandplain sequences in south Wales. The Dinantian of south Wales (see Fig. 2.20) contains several beach-barrier grainstone sequences, some of which are associated with back-barrier lagoonal facies. Both transgressive and regressive (progradational) barriers can be rec-ognized, as well as strandplain deposits. During the Early Carboniferous, a major sea-level rise led to the development of an extensive carbonate ramp in south Wales (Wright, 1986a). During deposition of the first stratigraphic unit, the Lower Limestone Shale Group, Burchette (1987) recognized three cycles, each rep-resenting the evolution and termination of a barrier–lagoon complex, in the Forest of Dean–southeast Glamorgan region (Fig. 4.23). Facies analysis reveals three lithofacies associations: 1, barrier/shoal cross-stratified skeletal and oolitic grainstones; 2, hyper-saline lagoonal limestones with stromatolites, vermetid gastropods, *Modiolus* and gypsum pseudo-morphs, deposited to landward of I; and 3, offshore shelf/embayment mudstones with thin, graded skeletal packstones of storm origin, deposited seawards (south) of I.

The first lagoon–barrier system (the Tongwynlais Formation) consists of 4 m of lagoonal lime mudstone overlain by a 3 m oolite. A channelled erosion sur-face separates these two units, upon which occurs a conglomerate with *Trypanites*-bored pebbles of the lagoonal lime mudstone. The grainstones pass up through amalgamated and graded shell beds into mudstone (III), containing numerous laminated or bioturbated silt–sand grade, thin limestone and sand-stone beds, and shell layers. This sequence (Fig. 4.23C) is an example of a transgressive barrier shore-line. The lagoonal sediments accumulated behind an oolite barrier and this transgressed landwards (north) by shoreface retreat. The erosion surface is inter-preted as a ravinement, produced by the transgressing barrier. The barrier sand was partly reworked on to the nearshore shelf, and lagoonal deposits exposed at the foot of the migrating barrier were eroded to form the pebbles above the channelled surface. The upper part of the Tongwynlais Formation records the tran-sition to a muddy outer shelf environment affected by periodic storms.

The second barrier, about 25 m of cross-bedded oolite (Castell Coch Limestone), is interpreted as a regressive barrier, but one which was largely subtidal. An extensive tide- and wave-dominated shoal area is envisaged, and the occurrence of outer shelf sedi-ments above suggests that it was drowned 'in place'.

The third barrier (Stowe Oolite) consists of two

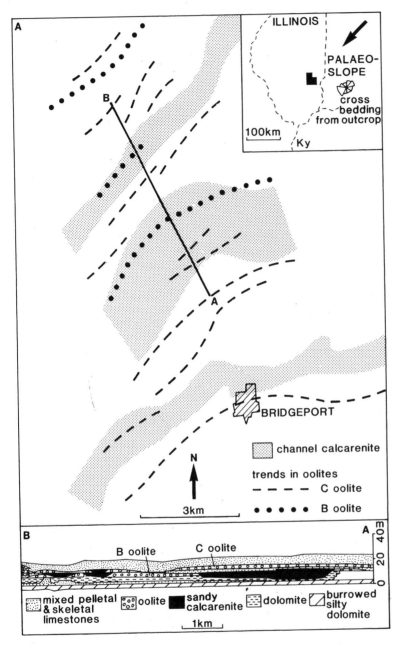

Fig. 4.22 *Lower Carboniferous (Mississippian) Ste Genevieve Limestones of Illinois. (A) Distribution of channel calcarenites and trends (zones of maximum thickness) of B and C oolite bodies. These sand bodies have a northeast–southwest orientation, down the palaeoslope, suggesting a strong tidal influence (see inset for cross-bedding rose from outcrop). In fact, these carbonate grainstones are developed upon a northwest–southeast palaeoshoal, parallel to the strike of the palaeoslope. (B) Cross-section along line A–B, across the trend of the channels and oolites. The sediments beneath the channel calcarenites are preferentially dolomitized (see Section 8.7.3 for further information on this). After Choquette & Steinen (1980).*

coarsening-up and shallowing-up shoreface to beach sequences (Fig. 4.23D). The lower one contains channels and trough cross-bedding interpreted as the product of rip currents. Both sequences have flat-bedded grainstones of beach origin towards the top. The uppermost part of the upper cycle shows evidence of subaerial exposure in the form of vadose cements and calcrete nodules. Subsequent transgression re-moved any supratidal barrier-top deposits and ter-minated barrier growth by drowning. Behind the Stowe Oolite barrier (to the north), two shoaling, restricted lagoonal sequences can be identified, cor-responding to the two barrier units in the Stowe Oolite.

In the *Viséan of mid and south Glamorgan* (Riding & Wright, 1981; Wright, 1986a), carbonate sand

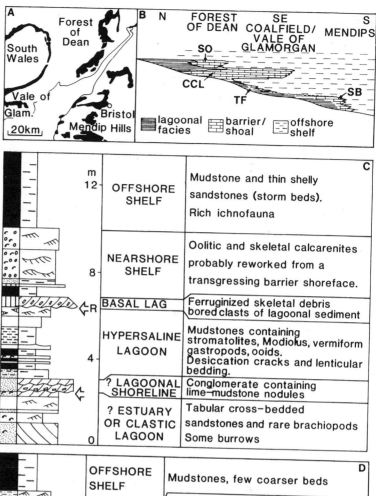

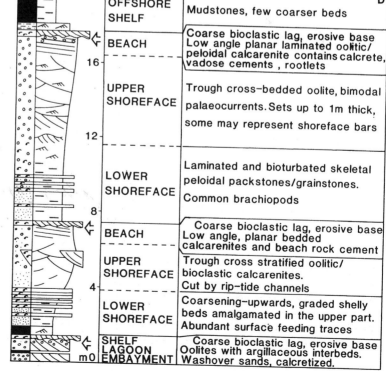

Fig. 4.23 *Beach-barrier–lagoon complexes in the Lower Carboniferous (Tournasian) of southwest UK. (A) Location and outcrop map. (B) Stratigraphic cross-section showing three barrier developments: (1) Shirehampton Beds–Tongwynlais Formation (SB and TF), a transgressive barrier, (2) Castell Coch Limestone (CCL), a regressive barrier, and (3) the Stowe Oolite (SO), of two regressive shoreface to beach sequences with lagoonal sediments behind. (C) Simplified log of lower Tongwynlais Formation. Arrows indicate a ravinement (R) and strong erosion surface. (D) Simplified log of Stowe Oolite showing two stacked shoaling sequences from shoreline progradation. After Burchette (1987).*

bodies are more of the strandplain type, without closely associated lagoonal deposits. In the Gower, three progradational shoreface−beach sequences are seen (see Fig. 2.20). In the Shipway Limestone−Brofiscin Oolite package (Fig. 4.24), thin-bedded graded bioclastic limestones, with intercalated black lime mudstones and shales, give way to more thickly-bedded, graded, laminated and hummocky cross-stratified packstones. Beds commonly show sharp, erosive bases and bioturbated tops. BPHF sequences (Fig. 4.11B) have been recognized, and locally thick units of amalgamated BPH layers occur (Wu, 1982; Wright, 1986a; Faulkner, 1988). This part of the sequence is interpreted as below fairweather wave-base deposits affected by storms. The succeeding cross-bedded oolites and the low-angle accretion surfaces of the Brofiscin, record shoreface and beach deposition. Meteoric phreatic and vadose cements in the Brofiscin Oolite indicate subaerial exposure (see Fig. 9.13A; Hird & Tucker, 1988). There is a sharp, non-karsted, contact with succeeding outer-ramp carbonates.

The Gully (or Caswell Bay) Oolite in the Gower contains two more prograding shoreface to beach sequences with the upper one capped by a paleokarst (Fig. 4.24; Ramsay, 1987). Again, below FWWB, storm-influenced sediments are recognized, as well as rip channel and longshore current deposits. The palaeokarst is of the Deckenkarren type (Wright, 1982), and this is overlain by a thin soilstone crust with alveolar fabrics. Both the Brofiscin and Gully Oolites developed during still-stands or slight sea-level falls as progradational, regressive, strandplain complexes, after rather rapid transgressions.

Overlying the Gully Oolite is the Caswell Bay Mudstone, a lagoonal−tidal flat sequence (laminated peloidal micrites, dolomites, breccias and planar stromatolites, see Fig. 8.8), of protected back-barrier origin (Riding & Wright, 1981; Wright, 1986a). It was deposited during another major, but probably less rapid, transgression. Overlying the mudstone with a sharp erosive contact is the High Tor Limestone, very coarse crinoidal grainstones with intraclasts of mudstone lithologies in the basal part. The erosion surface is interpreted as a surf-zone ravinement produced by a transgressive barrier. In west Gower, the High Tor

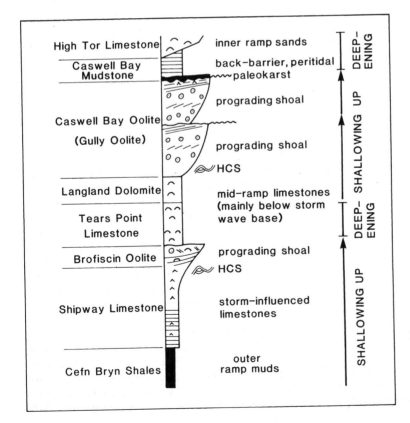

Fig. 4.24 *Simplified log through the mid-Dinantian limestones of Three Cliffs Bay, Gower, south Wales, showing three shallowing-upward sequences. The stratigraphy is given in Fig. 2.20. After Wright (unpubl.).*

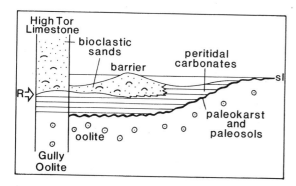

Fig. 4.25 *Interpretation of the Gully Oolite–High Tor Limestone sequence, at the top of the log in Fig. 4.24, as the result of landward migration of a barrier. R is a ravinement surface. After Wright (1986a).*

rests directly on the Gully, as a result of complete shoreface erosion. The High Tor Limestones are inner shelf bioclastic sands formed partly of reworked barrier sands (see Fig. 4.25, and Riding & Wright, 1981). Thus in the south Wales Lower Carboniferous, a range of beach-barrier and strandplain sequences was developed, reflecting rates of change of sea-level and some variations in subsidence (Wright, 1986a).

4.2 SHELF-MARGIN SAND BODIES

4.2.1 Modern shelf-margin sands

Carbonate sands are an important component of shallow shelf-margin sedimentary facies as noted in Chapter 2 (2.5) and studies of modern sand bodies in the Bahamas reveal that there are several types. Ball (1967) distinguished between *tidal bar belts*, occurring in areas of strong tidal currents, and *marine sand belts*, forming where wave action is strong and tidal effects weaker. The latter type can be divided further (Hine *et al.*, 1981a), when consideration is taken of the orientation of the shelf-break relative to prevailing winds. Distinct differences exist between the sand bodies of windward and leeward shelf margins, and those of open and protected aspects (Fig. 4.26). Where there are high rates of lime sand production, new environments such as islands and sand flats may develop. The formation of islands in particular can be very significant; it may allow meteoric diagenesis to affect the sands and this can modify the sediment's porosity–permeability considerably.

Open windward shelf margins are generally the most turbulent and marine sand belts are usually well developed there. On the northeastern side of Little

Bahama Bank an active oolite shoal, Lily Bank (Hine, 1977), is a classic example of this type of marine sand belt. A similar but less well-defined sand shoal occurs along the exposed northern margin of the Great Bahama Bank (Hine *et al.*, 1981a). Along the windward shelf margin of the Lily Bank area, reefs occur a little seaward of the sand belt and contribute skeletal debris to the sand shoal. The near-constant agitation promotes the precipitation of ooids so that at Lily Bank the sediment is an oolitic–skeletal grainstone. The dominant easterly winds result in a net westward wave-energy flux. Some two-thirds of the tropical storms and hurricanes in the region also have a principal bankward component. Strong tidal currents occur in the area and contribute to the local geometry of the sand body. Antecedent topography in the Pleistocene bedrock was probably instrumental in initiating the sand belt in this area. Lily Bank is a typical marine sand belt in being a narrow, linear shoal feature, parallel to the shelf margin; this contrasts with tidal bars which are mostly normal to the shelf-break.

Lily Bank is covered by a hierarchy of bed forms, from ripples up to sand waves, and cutting through the shoal are deep, grass-floored channels, frequently terminating in spillover lobes (Fig. 4.27A). Sand waves (wavelength 90–300 m, height 1–2 m) on the oceanward side of the shoal are mostly flood (onshelf) oriented. They migrate towards a crestal zone on the lagoonward side of the shoal where symmetrical and steep sand waves occur, affected by both ebb- and flood-tidal currents. The crestal region protects the slightly deeper flood-dominated oceanward side of the shoal from ebb-tidal currents, which are then mostly confined to the channels. The ebb currents generate spillover lobes covered in sand waves and dunes at the oceanward ends of the channels. Lagoonward migration of the sand belt probably takes place during major storms, when the protective crestal shield is broken down by storm surges, and flood-oriented sand waves migrate into the lagoon (Fig. 4.27B). There is a sharp boundary along the lagoonward side of the shoal, where the lime sands are prograding over grass-covered muddy peloidal sands of the quieter-water lagoon. This generates a coarsening-up sequence, with an upward trend of better sorting and decreasing lime mud content.

Between Lily Bank and the shelf-edge reefs, there is a series of linear sand bars which are oriented normal to the shelf margin. They are mostly stabilized by sea-grass. These are relic tidal sand bars which formed as sea-level was rising several thousand years

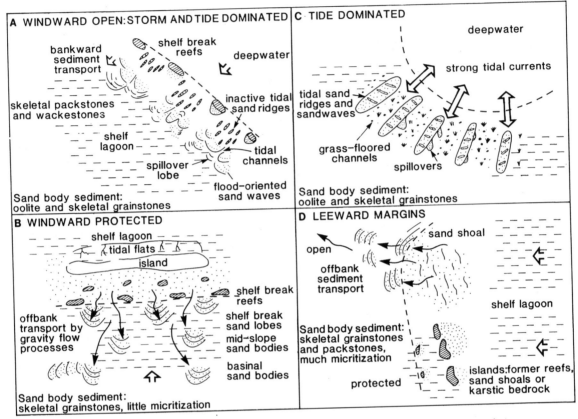

Fig. 4.26 *Facies models for shelf-marginal carbonate sand bodies, where shelf-break orientation relative to waves, storms and tidal currents is the controlling factor. Based on studies in the Bahamas, including Hine* et al. *(1981b). After Tucker (1985a).*

ago, when tidal currents were stronger than today because of the local configuration of the seafloor.

On *windward protected* shelf margins an island or barrier reef affords some protection to the shelf lagoon (see Fig. 4.26). A shelf margin of this type occurs along the south side of the Little Bahama Bank, fronting Grand Bahama Island (Hine *et al.*, 1981a,b). Strong southeast winds and storms affect the narrow shelf. Sand is produced in abundance, from wave and storm attack on the reefs, and this is deposited between the reefs and island, or just offbank from the reefs. However, the important feature of this shelf-margin type is that the dominant onshore wave and storm currents are reflected offshore to produce strong downslope bottom currents. Lobes of sand are thus deposited at the shelf-break, on the shelf slope and at the toe of the slope. Resedimentation is important as sand is transported off the shelf by grain flows, debris

flows and turbidity currents. The high-energy location gives a rapid turnover of sediment so that little is micritized by algae and few ooids are produced.

On *leeward* shelf margins, the dominant direction of water movement is offshelf, and the storms and waves have moved over the shelf itself before reaching the shelf-break (Fig. 4.26). Sand bodies are generally not well developed in this situation. On open leeward margins, sand is produced by reworking of the sandy mud of the lagoon, so that grains are extensively micritized and peloids are an important constituent. Aggregates, including grapestones, are common and molluscan and algal bioclasts dominate. Sand waves may develop at the shelf edge and there is much offshelf transport of sediment. Indeed, in areas of high carbonate production, this process can lead to basinward progradation of the shelf margin (Mullins & Neumann, 1979; Hine *et al.*, 1981a,b). Leeward

A

Fig. 4.27 *Lily Bank oolite shoal, Little Bahama Bank. (A) Aerial photo showing well-developed tidal channels and spillover lobes, migrating on to the shelf, to the right (appearing dark since here grass-covered). Sand waves are prominent on the shoal. From negative courtesy of Albert Hine. (B) Schematic cross-section through a windward, open shelf-margin sand body, based on Lily Bank, Little Bahama Bank. After Hine (1977).*

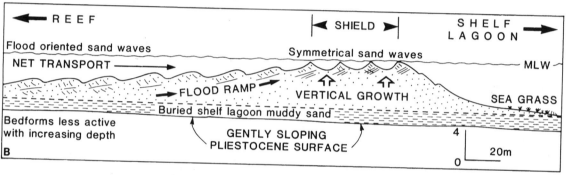

open margins occur on the western side of the Little Bahama Bank (Hine & Neumann, 1977) and on the northwest and western parts of the Great Bahama Bank (Hine *et al.*, 1981), and on seismic reflection profiles from these margins, prominent dipping reflectors, some from hardgrounds, indicate extensive platform margin migration (Fig. 4.28).

Tidal bar belts are typically linear sand bars oriented normal to the shelf-break (Fig. 4.26). Classic examples occur on the Great Bahama Bank at the southern end of the Tongue of the Ocean (Fig. 4.29) and at the northern end of Exuma Sound, in the region of Schooner Cay (shown in Fig. 3.2). Tidal currents are enhanced in these deep cul-de-sacs into the Great Bahama Bank, so that peak tidal currents exceed 1 m s^{-1}.

Tidal bar belts consist of many linear, parallel sand bars, ranging from 0.5−1.5 km across and 12−20 km long. They are separated from each other by broad channels, 1−3 km wide and up to 7 m deep. The channels are occupied by sand, but this is mostly stabilized by *Thalassia* and algae. Upon the sand ridges, there occur large sand waves; many are symmetrical but others are flood or ebb oriented, depending on which currents dominate. Sand waves

vary in their orientation from transverse, to oblique, to near-parallel to the ridge axes. The smaller-scale bed forms, dunes and ripples, are common too. Locally there are small channels cutting through the sand ridges and it is likely that these have been cut during storms. Spillover lobes occur at the ends of these channels, on the flanks of the tidal ridges.

Elongate tidal bars with channels between also occur on the southern margin of the Little Bahama Bank between Grand Bahama and Great Abaco Islands (Hine *et al.*, 1981a). Sand lobes occur along the lagoonward side of the bar belt. At the southeast end of Grand Bahama Island, the tidal sand shoals have become emergent through beach ridge and spit growth. Flood-tidal deltas occur at the lagoonward ends of active tidal channels between islands, while other channels are being abandoned and filled with lime mud.

4.2.2 Marine sand belt to sand flat: Joulter's Cay, Bahamas

One shelf-margin sand body in the Bahamas has developed from a mobile sand belt to an oolite shoal−sand flat−tidal channel−island complex. Joulter's

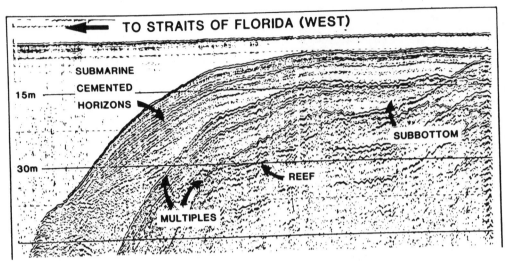

Fig. 4.28 *Seismic profile across the western margin of the Little Bahama Bank showing progradation through off-platform sand transport, subsurface cemented horizons, and buried earlier Holocene reefs. From Hine* et al. *(1981) with permission of Elsevier Pub. Co., Amsterdam.*

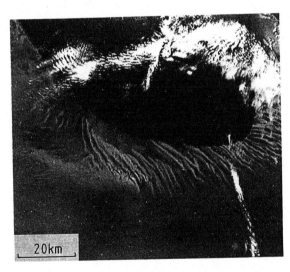

Fig. 4.29 *Satellite view of the tidal oolite ridges from the platform margin around the head of the Tongue of the Ocean, Bahamas.*

Ooid Shoal, north of Andros Island (Harris, 1979, 1984) was initiated on a Pleistocene topographic high (a submerged extension of Andros) as sea-level rose 3000–4000 years ago (Figs 4.30 and 4.31). Between 3000 and 1000 years BP, the ooid shoal developed into an elongate marine sand belt with some spit-like tidal bars at its northern end. Tidal channels cut through the shoals and ooids were transported through these to mix with platform-interior muddy peloidal sand. Ooids were deposited in lobe-shaped fans developed at the ends of the channels and at bank spillovers. An extensive oolitic–peloidal sand flat developed bankward of the ooid shoal and burrowing organisms contributed to sediment mixing. In time, tidal channels were abandoned and filled. The high rates of ooid production in the mobile zone along the seaward edge of the shoal, coupled with waves and storm action, led to islands being formed along the shoal's eastern margin from about 1000 years BP. Beach ridges and dunes developed but the situation was complicated by ridge erosion, filling of tidal channels and cutting of new ones, and accretion of sand spits through longshore drift (Harris, 1979). The islands and sand flats, now close to and just above sea-level, prevented platform water from reaching the open sea and this allowed patch reefs to develop seaward of the ooid shoals.

Three main facies are now seen in the Joulter's area:

1 *ooid grainstones* occurring in (a) the shallow shoreface 'shoal mobile fringe' where the ooids are mostly being precipitated, (b) the spit, beach and dune systems which make up the islands, and (c) the tidal channels;

2 *mixed ooid–peloidal sands* (mostly packstones) of the sand flat environment; and

3 *mixed fine peloidal muddy sand* (wackestones and packstones) of the platform interior.

The significance of the Joulter's region is that it shows that with high rates of lime sand production, sand shoals can build up to sea-level, and above, and give rise to very extensive sand flats, although the zone of ooid precipitation in a marine sand belt may be quite narrow. Cores taken through the Joulter's Ooid Shoal reveal a sequence of facies deposited in progressively shallow water: the platform-interior muddy sands pass up into the mixed ooid–peloidal packstone sand flat facies, and then the ooid grainstone (Harris, 1984).

Ooid grainstones which comprise the islands of Joulter's Cay show the effects of meteoric diagenesis (Section 7.5; Fig. 7.28). Ooids above the water table have meniscus calcite cements at grain contacts; those in the phreatic zone show equant calcite crystals in irregular to isopachous coats. Dissolution of aragonite grains may also take place.

Fig. 4.30 *Aerial view looking west over Joulter's Cays, northwest of Andros Island, Great Bahama Bank. Ooids are forming along the oceanward side of the cays (bottom of picture) in shoal areas. Beach ridges are just visible on the vegetated cays. A major tidal channel is shown between two islands, and platformward, the extensive sand flat is developed. Photo courtesy of Ian Goldsmith.*

Fig. 4.31 *Diagrammatic evolution of the oolite shoal–sand flat complex of Joulter's Cays, Great Bahama Bank. (A) Flooding of bank and then development of oolite shoal around 3000–4000 years* BP. *(B) Enlargement of the shoal and more flooding of the Andros area, around 1000–3000 years* BP. *(C) Extension of the mobile ooid belt bankwards with bars and channels; spit-like tidal bars formed on north side of shoal and mixed ooid–peloid sand flat began to develop on the bankward side of the mobile belt. (D) Over the last 1000 years restricted movement of water across the shoal, extension of sand flat area and development of islands (Joulter's Cays) along bankward side of ooid mobile belt. Patch reefs now flourishing on the shelf as exchange of water across the ooid shoal increasingly limited. After Harris (1979).*

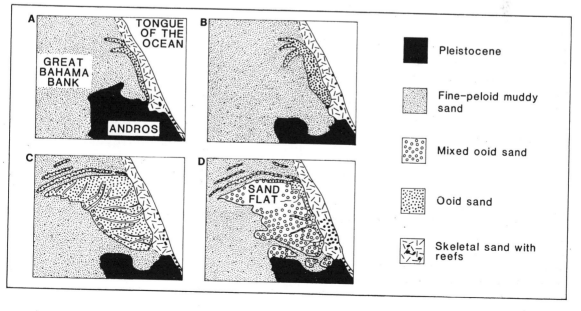

4.2.3 Ancient shelf-margin grainstones

Oolitic and skeletal grainstones are developed along most ancient carbonate shelf margins and many show features comparable with the modern occurrences described above.

The Miami Oolite of Upper Pleistocene age is a useful example of a tide-dominated system. It is developed in two areas, around Miami and in the Lower Florida Keys, from Big Pine to Key West (Fig. 4.32). A line of patch reefs, forming the Key Largo Limestone existed between the two areas of oolite and now crops out in the Upper Florida Keys (Hoffmeister *et al.*, 1967). In the Miami area, a shoal and channel system oriented with long axes perpendicular to the shelf-break gave way through time to a shelf-parallel (north−south) barrier bar, developed on the seaward side through coalescence of ebb-tidal deltas (Halley & Evans, 1983; Evans, 1984). The barrier forms the 'Atlantic Coastal Ridge' in the Miami area and this is cut through by valleys or 'glades', the former site of tidal channels. The Miami Oolite thus shows an evolution from a tidal bar belt to a marine sand belt. Migration of sand waves on the shoals and barrier bar gave rise to planar cross-beds with tangential bases. Cross-beds are graded and the upper, finer parts are usually preferentially cemented. Dip angles are commonly high (>20°), close to the angle of repose, from avalanching of sand from the sand wave crest. The cross-beds have a bimodal distribution, but the eastward direction dominates, indicating stronger ebb-tidal flow. Three orders of bounding surface are recognized in the Miami Oolite (Fig. 4.33, Halley & Evans, 1983): a first order bounding surface which is either a burrowed surface, representing a period of little sediment movement and much infaunal activity, or a major truncation surface which records a phase of erosion. A coarse, shelly lag typically occurs upon the latter type. Between first order bounding surfaces, which are mostly planar, there usually occur several cross-bed sets, each of which is separated by a second order surface. Many sets are tabular, 0.05−1 m thick, so that most second order surfaces are planar and near-parallel. These cross-beds formed from straight-crested sand waves. Wedge-shaped sets also occur and probably result from more sinuous or lobate dunes. Second order surfaces are thus surfaces across which trains of sand waves migrated, eroding underlying cross-beds. Burrows occur within these oolites, but rarely originate on second order surfaces, suggesting that periods of non-deposition were not

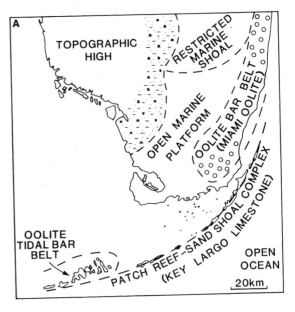

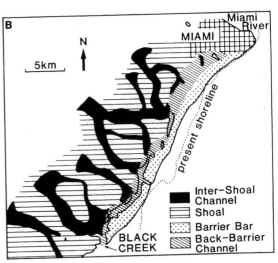

Fig. 4.32 *The Pleistocene Miami Oolite of southern Florida. (A) Broad facies pattern, with oolite bar belts in the Miami and Lower Florida Keys area (the shape of the present islands in the Lower Keys is a relict of the Pleistocene tide-dominated bars) and patch reefs and sand shoals of the Key Largo Limestone forming the present Middle and Upper Florida Keys. After Perkins (1977). (B) Subenvironments of the oolite shoal complex in the Miami area. After Halley & Evans (1983).*

long-lived. Third order bounding surfaces are on a smaller scale, occurring within a cross-bed set as a reactivation surface or a change in angle of dip of

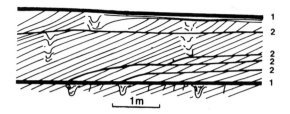

Fig. 4.33 *First and second order bounding surfaces in the Miami Oolite, Pleistocene, Florida, as sketched in the field. After Halley & Evans (1983).*

cross-strata. They represent short-term changes in the shape of the sand wave, through tidal current reversals and/or storm erosion. Development of the barrier caused restriction on the platform behind, and bryozoan-rich peloidal sands accumulated there in waters 3–5 m deep. To seaward of the barrier, mottled oolite is a bioturbated grainstone deposited in a probably grass-stabilized, low-energy sand flat. Evidence for occasional and local emergence of the barrier to form sandy islands is provided by vadose and phreatic meteoric cements (Halley & Evans, 1983), but mostly the barrier was shallow subtidal (0–3 m depth).

In the Lower Florida Keys, the oolite ridges are oriented north-northwest–south-southeast and form the present islands (see Figs 3.27 and 4.32). This orientation, normal to the shelf margin, indicates a strong tidal influence in the generation of the oolite shoal complex.

In the *Alpine Triassic*, lime sands occur in the *Dachstein Limestone of Austria* along some of the carbonate platform margins and around banks and isolated paltforms (e.g. the Totes Gebirge, see Fig. 2.12A, Piller, 1976), whereas in other areas, reefs are developed (Fig. 2.12B; Flügel, 1981). On the Totes Gebirge Shelf, oolite facies forms shoals at the shelf-break and consists mostly of normal and superficial ooids, with aggregates, peloids and a few bioclasts. This facies gives way lagoonwards to an oolitic facies, where peloids and micritized bioclasts are an important constituent, along with some lime mud. Aggregate grains, peloids and bioclasts dominate the next facies belt, the grapestone facies, which was deposited in the shallow subtidal to intertidal protected central portion of the shelf lagoon. Algal–foram, pellet lime mud and lime mud facies, with birdseyes and microbial laminites (loferites) occur in the inner shelf environment (Piller, 1976).

In the *Venetian Alps of northern Italy*, major oolitic grainstones are developed in the Lower–Middle Jurassic (Bosellini *et al.*, 1981). The Belluno Trough was a narrow elongate basin, likened to the modern Tongue of the Ocean in the Bahamas. The Friuli Platform to the east was the site of prolific ooid production, particularly along the western windward margin which was facing an ocean perhaps 1000 km wide. The thick Ternowaner Oolite (500–700 m) of Upper Liassic–Dogger age is cross-bedded with bi-modal palaeocurrents, oriented north-northeast–south-southwest, which were near-parallel to the shelf margin. Much oolite was carried over the shelf-break and transported into the adjacent Belluno Trough by turbidity currents and debris flows, to form the 800–1000 m thick Vajont Limestone. The ooid factory was closed by a short eustatic sea-level drop at the end of the Callovian, and the Ternowaner Oolite was then cemented during the subaerial exposure. Subsequently shelf-margin reefs were extensively developed after a eustatic sea-level rise (the 'Oxfordian transgression'), when the cemented oolite provided a suitable substrate for the corals and hydrozoans. Debris shed into the Belluno Trough was now entirely skeletal.

Another Jurassic example is provided by the *Abenaki Formation* which occurs in the *subsurface off Nova Scotia* (Eliuk, 1978). During the Jurassic and Lower Cretaceous, carbonate shelves and banks formed a discontinuous belt from the Grand Banks down to the Bahamas. Various types of carbonate sequence have been recognized, depending on depositional, palaeo-oceanographic and tectonic processes (Jansa, 1981). From seismic reflection data off Nova Scotia, a distinct break of slope can be identified along part of the shelf margin in the Jurassic–Cretaceous carbonates, but elsewhere the transition to deep water sediments is more ramp in character (see Fig. 4.34; Eliuk, 1978). Massive oolite accumulated along the shelf margin in the Scatarie Limestone Member and just back from the shelf-break, several oolite cycles are developed, ranging from 10 to 60 m in thickness. In these, above a thin basal sandstone, oolite grainstones and packstones pass up into oncolitic skeletal beds and then fossiliferous wackestones and mudstones cap each cycle. The oolites are interpreted as shallow shoal sands and the overlying oncolitic beds as deeper, quieter-water deposits, either seaward or landward of the shoals. A deeper, more open marine environment is favoured for the wackestones and mudstones towards the tops of the cycles, so that overall each sequence is deepening up or transgressive. These Scatarie cycles, contrasting with shallowing-up cycles which characterize many carbonate formations, are accounted for by eustatic sea-level rises, possibly related to the onset of sea floor spreading in the North Atlantic (Eliuk, 1978). Some support for deepening upwards comes from the diagenesis and porosity of the oolites, which mostly show calcite spar cements and little porosity. This suggests that they were not subaerially exposed, but well cemented during burial diagenesis. This contrasts with many other ancient oolite shoal facies described in this and the previous section, which have petrographic evidence for emergence.

In the higher Baccaro Limestone Member of the Abenaki, reefs as well as oolitic and skeletal grainstones occur along the shelf margin, forming a carbonate bank more than 500 m thick. Exposure of the shelf margin and the development of freshwater lenses are indicated by karstic dissolution phenomena (linear dissolution channels, breccia zones and reddening) and leaching of bioclasts. Local dolomitization is attributed to mixing zones at the base of the meteoric-water lenses.

On the *Lower Cretaceous* shelf margin of the *Gulf Coast Basin*, thick, massive skeletal grainstones occur predominantly on the outer shelf, with much debris coming from shelf-marginal rudist reefs (Griffith *et al.*, 1969). This facies pattern is well displayed in the outcropping El Abra Limestone of Mexico, where shelf-edge sands pass rapidly into lagoonal peloidal and muddy facies (Fig. 4.35). A similar facies arrangement is seen in the equivalent Edwards and Stuart City Formations of the south Texas subsurface (see Fig. 8.24).

Grainstones, mostly oolitic, are common in shelf-margin locations in the *Permian Basin of Texas*. During Guadalupian time, oolitic grainstones developed along the shelf edge in the Carlsbad facies, a little shelfward from the sponge–algal reefs of the Capitan facies (see Fig. 2.16). In the subsurface of the Midland Basin, oolites developed at the shelf margin during Wichita and Lower Clear Fork times of Wolfcampian–Leonardian age (Mazzullo, 1982). High ooid production rates and offshelf sediment transport resulted in basinward progradation of the shelf margin, which can be seen on seismic reflection lines (Mazzullo, 1982, fig. 4). The oolite bodies have been dolomitized and are potential stratigraphic hydrocarbon reservoirs.

Also in the subsurface Permian of Texas, a well-documented oolite sand body, comparable in many respects to modern Bahamian shoals, occurs in the Wolfcampian of Ochiltree County, in the northwest

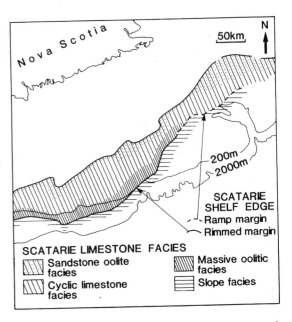

Fig. 4.34 *Upper Jurassic shelf margin off Nova Scotia and facies pattern in the Scatarie Limestone. Oolites occur along the rimmed margin and this passes northeastwards into a margin more ramp in character. After Eliuk (1978).*

corner of the Anadarko Basin (Asquith, 1979). In the Lower Permian Council Grove B zone, bioturbated oolitic, crinoidal, bryozoan wackestones representing a relatively deep, quiet-water facies, pass up into cross-bedded oolites (Fig. 4.36A). The Council Grove B oolite shoal is a hydrocarbon reservoir and from well data, an isopach map reveals the elongate nature of the sand body (Fig. 4.36B). Maximum thickness is 20 m, but there are distinct north–south oriented thinner zones which are interpreted as the sites of tidal channels crossing the oolite shoal. In addition, the shoal has a distinct lobate southern margin which could represent spillover lobes. Electric log correlations across the sand body and reference to a distinct marker horizon below show that the base of the shoal rises to the south. This is taken to indicate southward progradation of the sand body in response to dominant southward-directed tidal and/or storm currents. Wackestones above the oolite indicate a return to quiet-water sedimentation. However, the grainstones have an oomouldic porosity which could indicate emergence of the shoal and freshwater leaching before the wackestones were deposited.

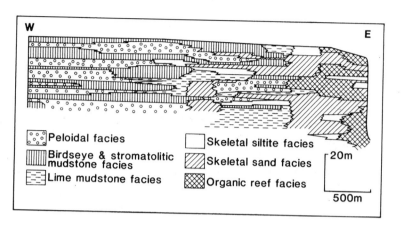

Fig. 4.35 *Cross-section of the Lower Cretaceous shelf margin El Abra Limestone, central Mexico (equivalent to the Edwards and Stuart City Formations of the Texas subsurface), showing facies pattern of shelf-margin rudistid reefs and skeletal sands and back-reef peloidal and lime mudstone lagoonal facies with interbedded tidal flat birdseye and stromatolitic limestones. After Griffith et al. (1969).*

Another well-developed, and described, Permian oolite sand body occurs in the first cycle *Zechstein Cadeby Formation of Yorkshire* (Kaldi, 1986). The thick oolite is characterized by large-scale cross-bedding (sets reaching 12 m) which defines linear, asymmetric bed forms, 2−50 m in wavelength and 0.6−15 m high. These structures are interpreted as sand waves but the palaeocurrent patterns are complex, with unimodal, bimodal and polymodal patterns recorded. The long axes of the sand waves trend northwest−southeast and a predominant current to the southwest is suggested by the onlapping arrangement of southwest-dipping cross-beds. A fan-bedding (Fig. 4.37A) is thought to result from lateral migration of sand waves. Also present are sand lobes, with convex-up cross-bedding, oriented mainly towards the northwest. These are interpreted as spillover lobes and are attributed to storm waves and surges. Hummocky cross-stratification (HCS) occurs at several horizons (Fig. 4.37B) and also indicates

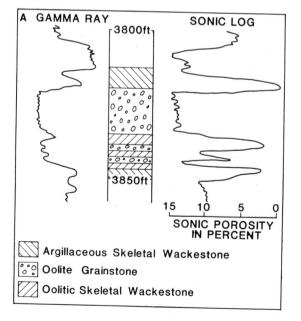

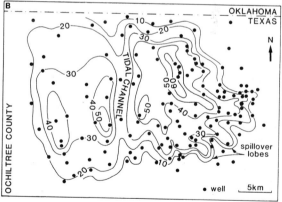

Fig. 4.36 *Subsurface Lower Permian Council Grove Oolite Shoal, northern Texas. (A) Lithology, gamma ray and sonic logs through the oolite which is an oil reservoir. (B) Isopachyte map based on much well data. The shelf margin was located to the north. Note inferred tidal channel and lobate southern margin to shoal, interpreted as spillover lobes prograding in to shelf lagoon. Contours in feet! After Asquith (1979).*

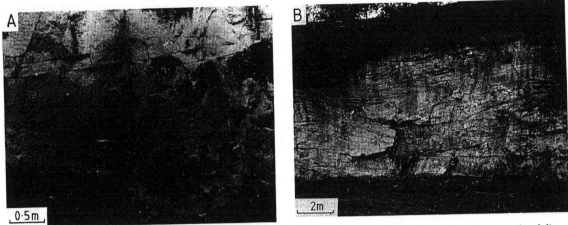

Fig. 4.37 *Cross-stratification in the Cadeby Formation oolites of Yorkshire. (A) Fan-bedding, where the angle of dip of cross-beds decreases upwards and individual beds thicken downwards. Photo courtesy of John Kaldi. (B) Large-scale hummocky cross-stratification. Photo courtesy of Denys Smith.*

Fig. 4.38 *Shelf-margin carbonates of the Upper Cambrian of western Maryland. (A) Facies model for a rimmed shelf stage with marginal reefs and oolite shoals, and intertidal–supratidal flats located just onshelf. (B) Typical shallowing-upward cycle (2–10 m thick) of Conococheague Limestone formed by progradation of the tidal flats over carbonate sands. After Demicco (1985).*

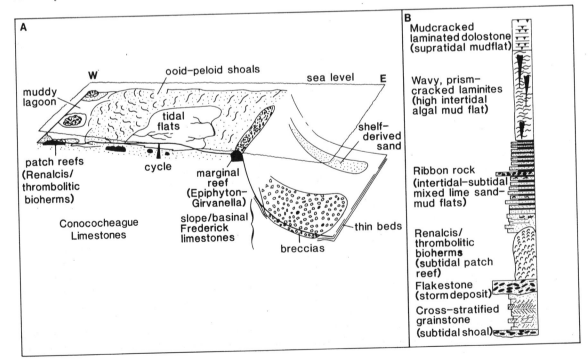

storm-wave activity. Trace fossils, including feeding trails and dwelling burrows, are common. The development of this oolite shoal complex is attributed to seafloor topography, with ooid precipitation and sand wave formation occurring on a shelf close to a break in slope, either due to the nature of the pre-Permian surface, or to reefs in an underlying Zechstein Formation. Shorewards (to the west) of the oolite sand body, skeletal−peloidal packstones to mudstones accumulated in a lagoon fronting tidal flats and sabkhas. The oolites, and most other sediments of the Cadeby Formation, are dolomitized with much fabric destruction.

There are a number of ancient shelf margin sand bodies which have developed into shoals and islands giving sand flats and tidal mud flats towards the lagoon, similar to present-day Joulter's (Section 4.2.2). During the Late Cambrian−Early Ordovician a vast epeiric sea existed over North America and tidal flat sedimentation was dominant. Oolite shoals were developed along the platform margin on the east, in eastern New York, southwest Vermont and western Maryland for example (Mazzullo & Friedman, 1975; Demicco, 1985), bordering the Iapetus Ocean (Fig. 4.38A). Tidal flats were locally developed, especially on the platform side of the oolite shoals. Further west, there occur wackestones and lime mudstones of the platform lagoon with patch reefs, before the more extensive tidal flats of the platform-interior shoreline. In western Maryland, progradation of tidal flats over the oolite shoals led to shallowing-up cycles in the platform margin *Conococheague Limestone* (Fig. 4.38B; Demicco, 1985). Resedimentation off the platform margin resulted in breccias and turbidites in the Frederick Limestone (Fig. 4.38A).

4.3 PERITIDAL CARBONATES

4.3.1 Introduction

Peritidal carbonates (carbonate sediments formed 'around the tides') represent a spectrum of marine environments from shallow subtidal areas adjacent to tidal flats, to the tidal flats themselves, and to supratidal zones including sabkhas, mangrove forests, marshes and coastal 'lakes'. Peritidal deposits are very common in the geological record and some of their modern counterparts, which are readily accessible, have been studied in considerable detail (Ginsburg, 1975). Ancient peritidal deposits have considerable economic importance for they act as hosts for metalliferous and non-metalliferous mineral

deposits, and as sources, seals and reservoirs for hydrocarbons. In addition, peritidal carbonates commonly form stratigraphic traps for hydrocarbons as a result of onlap and offlap geometries, creating pinch-out structures (Shinn, 1983a). Peritidal sediments, having been deposited in marginal settings, commonly undergo both early meteoric diagenesis, resulting in leaching and dolomitization, and they are commonly associated with the development of significant secondary porosity. Several case studies of peritidal reservoirs are to be found in Roehl & Choquette (1985).

4.3.2 **Peritidal environments**

The term peritidal (Folk, 1973) is generally used to describe a variety of carbonate environments associated with low-energy tidal zones, especially tidal flats. Tidal flat environments develop in a variety of settings (Fig. 4.39) on ramps or shelves and platforms. They may accrete from the shorelines of land areas, and from around islands (Ebanks, 1975; Pratt & James 1986); they also occur on the leeward, protected sides of barriers on ramps, and behind shoals and reefs on rimmed shelves.

Traditionally three main zones are recognized.

Subtidal zone. This is the permanently submerged area seaward of the tidal flats. It may be strongly influenced by wave action and tidal currents (shoreface environments), but most tidal flats are associated with protected low-energy lagoons and restricted bays. The biota may be abundant and diverse depending mainly on the salinity and temperature. Tidal channels and ponds represent extensions of the subtidal zone into the intertidal belt.

Intertidal zone. This lies between the normal low-tide and high-tide levels, and is alternately flooded by marine water and exposed. A variety of sedimentary, diagenetic and biological features develop in such settings which are reviewed below. Meteorological and climatic effects are very important in these settings and influence the types of features formed.

Supratidal zone. This zone is infrequently flooded, usually during high spring tides and by storms, and can be very wide on low-relief prograding coastlines. It is characterized by prolonged exposure but its nature is ultimately controlled by the prevailing climate.

The use of these terms has been found to be unsatisfactory in many cases because of the high de-

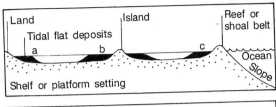

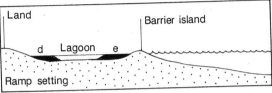

Fig. 4.39 *Sites of tidal flat deposition on shelves/platforms and ramps. Examples: (a) Florida Bay; (b) shelf islands of Belize (Ebanks, 1975) and Lower Ordovician of Newfoundland (Pratt & James, 1986); (c) Andros Island, Bahamas (Chapter 3); (d) and (e) Trucial Coast of Arabian Gulf.*

gree of variability within each zone. Whether a particular area is subtidal, intertidal or supratidal depends on the tidal range, mean sea-level and on its topographic position relative to sea-level. Both mean sea-level and tidal ranges are highly variable and meteorological factors such as wind can affect the degree of exposure. During onshore winds and storms the tidal range in Shark Bay, Western Australia rises to 3 m (normally 0.6–0.9 m) while strong offshore winds along the coast of northwest Andros Island retard the tides and can keep parts of the tidal flats dry for days on end (Hardie & Garrett, 1977). Recognizing discrete tidal zones also becomes difficult where topography is complex. The channelled tidal zone of northwest Andros (Chapter 3) is topographically divided into ponds, levees, beach ridges and channels, each of which has a particular exposure regime, and the tidal zones are highly complex.

To overcome such problems, Ginsburg *et al.* (1977) proposed the use of the exposure index. This is a quantitative measure which enables any point on the tidal zone to be assigned an exposure index (a percentage of the time that area is exposed) (Fig. 3.21). The levee crest zones of the tidal channels of northwest Andros, for example, are exposed 98% of the time. The lower intertidal zone recognized at Cape Sable,

Florida is exposed 12–75% of the time (Gebelein, 1977a). A similar approach has also been used to define the flooding frequencies of different microbial mat types on the arid tidal flats of the Trucial Coast by Kinsman & Park (1976; Fig. 3.41), and to define the different tidal environments of Spencer Gulf, South Australia (Burne & Colwell, 1982).

This technique also allows specific exposure indices to be assigned to sedimentary structures; for example Ginsburg *et al.* (1977) found that irregular fenestrae (see below) and small mudcracks (under 0.2 m diameter) develop with greater than 60% and 50% exposure respectively (Fig. 4.40; Ginsburg & Hardie, 1975). Such information can be easily applied to the geological record. Smosna & Warshauer (1981) used cluster analysis techniques on Silurian peritidal carbonates to recognize a number of lithofacies, to which they assigned specific exposure indices by analogy with modern peritidal facies. While such a semi-quantitative approach is desirable, simply for ease of communication it is usually necessary to use the terms subtidal, intertidal and supratidal.

Invariably intertidal limestones are only reliably recognized in the geological record because they possess features indicating exposure, such as mudcracks or fenestrae or evaporites. However, none of these criteria are diagnostic of intertidal environments, and such features merely indicate exposure, not necessarily on a tidal flat. Examples of coast marginal carbonate environments which possess exposure features but are not peritidal *sensu stricto* will be discussed in a later section. It is often very difficult to distinguish true *peritidal* lithofacies from other types of coast marginal deposits, and the terms perimarine or perilittoral may be more acceptable to cover the wide spectrum of coast marginal carbonates.

4.3.3 Sedimentary processes and products

A characteristic set of depositional, diagenetic and biological processes acts upon peritidal environments. Few of these processes are unique to the setting but together they create a suite of sedimentary features which enables similar environments to be recognized in the geological record.

Sedimentary processes in peritidal settings can be divided into two broad categories. Firstly, there are larger-scale processes, such as progradation and channel migration, which influence the types of facies sequences, and secondly there are smaller-scale depositional and diagenetic processes which control the individual lithofacies types.

Type of Feature

Bedding
1 Smooth flat lamination
2 Disrupted flat lamination (including intraclast sands)
3 Lamination with palisade structure (including lithified LLH and SH protostromatolites and protodolomite crusts)
4 Crinkled fenestral lamination (including LLH 'raised disc' protostromatolites)
5 Smooth domal lamination (SH protostromatolites)
6 Algal tufa–peloidal mud thin beds
7 Disrupted fenestral thin beds
8 Flat–pebble gravels
9 Cross–bedded coarse skeletal sands
10 Bioturbated pelleted mud thick beds

Desiccation Features
1 Small mudcracks (1–5cm polygons)
2 Wide shallow mudcracks (5–15cm poygons)
3 Deep prism cracks (20–30cm polygons)
4 Sheet cracks

Fenestral Pores (Excluding open burrows)

Burrows
1 Cardisoma crab burrows
2 Oligochaete worm burrows
3 *Uca* crab burrows
4 *Polychaete* worm burrows
5 *Alpheus* shrimp burrows
6 *Callianassa* shrimp burrows

Plants
1 Algal mats dominated by *Scytonema*
2 Algal mats dominated by *Schizothrix*
3 Halophytic grasses and shrubs
4 Mangroves
5 Sea–grasses
6 Green algae (*Penicillus*)

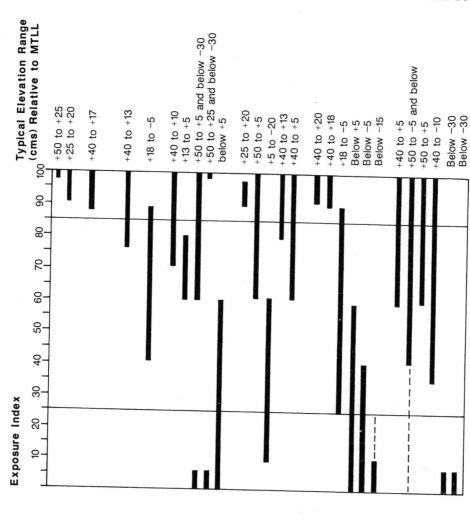

Fig. 4.40 *Relationships of various features to exposure index, southwest Andros Island. The solid lines are measured ranges, while the dashed lines are estimated ranges. After Hardie (1977).*

4.3.3a Progradation

This is one of the most important processes influencing peritidal environments (see Section 2.3.1). It represents the seaward migration of shallower-water environments over deeper-water ones as a depositional basin (e.g. a lagoon) becomes filled by sediment. The result is a sedimentary regression, that is, a migration of the shoreline seawards. Two processes are involved: firstly, the seaward subtidal area must become filled with sediment, and, secondly, the tidal flats build upwards and outwards as sediment accumulates on them. The net result is that subtidal deposits become overlain by intertidal and eventually supratidal or terrestrial deposits. For progradation to occur the rate of sediment supplied to the subtidal zone and tidal flat must be greater than either the rate of landward, wave-induced erosion and/or sea-level rise. The rate of sediment supply on tidal flats can be rapid, especially during storms, and microbial mats have the ability to trap and bind this sediment. These processes lead to the rapid vertical accretion of the flats, and average rates as high as 5 mm yr^{-1} have been recorded at Cape Sable, Florida (Gebelein, 1977a). The rate of seaward prograded can also be very rapid, for example 5 km in 4000 years in the Arabian Gulf (Evans *et al.*, 1969), and a similar amount in Spencer Gulf, South Australia (Burne, 1982). The tidal flats of southwest Andros have prograded at the rate of 5–20 km 1000 yr^{-1} (Hardie, 1986b). By this process, wide supratidal flats may develop which, in arid or semi-arid areas, become the sites of evaporite formation and are called sabkhas, playas or salinas. Prograded has usually been regarded as a purely sedimentary process operating under rates of low sea-level rise or still-stand, and high rates of sediment supply. However, it seems that prograded has been triggered by relative sea-level falls in Spencer Gulf (Burne, 1982) and in Shark Bay, Western Australia (Hagan & Logan, 1974). The classic example of tidal flat prograded, that of the sabkhas of Abu Dhabi, is also associated with a fall in sea-level of 1–2 m (Evans *et al.*, 1969; Schneider, 1975; Kinsman & Park, 1976). This relative fall may have been a genuine one or may reflect changes in the tidal range or wave height, caused by the growth of an offshore, protective barrier island (Evans *et al.*, 1969).

Whatever the exact cause of prograded, the result is the typical shallowing-upward sequence so commonly encountered in the geological record (see Section 2.10). Many of the examples cited below are of this type.

4.3.3b Channel migration

Tidal channels (or creeks) represent a distinctive environment on tidal flats. Typically on low-energy tidal flats, these channels are shallow, 3 m or less, but they may be up to 100 m wide as on northwest Andros (Section 3.3.2). Each channel becomes progressively shallower and narrower as it is traced landward into the upper intertidal zone. They migrate laterally in a similar way to fluvial channels and have point-bar deposits on the inner banks of meanders. Very little information is available on the rates of lateral migration of these channels, but it is widely believed that their migration leaves a blanket layer of channel-fill sediment in the same way as high sinuosity fluvial channels or siliciclastic tidal channels.

The channel floor is usually covered with a basal lag containing intraformational clasts, derived by bank erosion, together with coarse skeletal debris. The intraclasts commonly consist of lithified tidal flat sediments including dolomite crusts. The channel walls typically undergo early cementation and may be reworked as clasts into the channel lag (Davies, 1970b). The bulk of the channel deposits consists of bioturbated muddy sand. The biota may be restricted in nature if the tidal flats have hypersaline waters, or are affected by freshwater run-off. The biota may be exotic, and in some tidal channels shell debris of open marine faunas can be swept landward up the channels for up to 5 km (Barwis, 1978). One distinctive feature of carbonate and siliciclastic tidal channels is the development of channel bars. These are prominent in the tidal channels of northwest Andros Island (Shinn *et al.*, 1969; Hardie & Garrett, 1977). The bars on Andros consist of bioturbated, peloidal muds with scattered skeletal grains, and they possess crude inclined bedding (Fig. 4.41) which may survive bioturbation. The bars developed on the inner banks of meanders are generally coarser grained and may display these inclined surfaces, whereas bars developed in straighter channels are much finer grained and lack any primary sedimentary structures (Shinn *et al.*, 1969; Hardie & Garrett, 1977; Shinn, 1986). Channel bars in siliciclastic tidal channels have been described in detail by Barwis (1978), who recognized the importance of bar types in generating different tidal channel migration sequences. A similar rigorous approach to studying such bars may lead to a better understanding of carbonate tidal channel processes. As the channels on Andros migrate the channel deposits become overlain by facies with higher exposure indices, developed on the inner bank of the meander. Mangroves are common along these creeks, as they are in other tidal channels (Fig. 4.41). The upper part of the Andros channel-fill sequence consists of levee deposits, which have very high exposure indices (85–98% on Andros). As a consequence, these are not heavily bioturbated and possess fine lamination with fenestral fabrics. Such levee deposits are commonly recognized in ancient sequences and are used as evidence for deposition from tidal channels (Loucks & Anderson, 1980; Ramsay, 1987), yet levee deposits are generally poorly developed on tidal channel banks because bank full discharge in these channels occurs when current velocities are low and little sediment is being carried. The prominent, commonly graded, laminae developed on the Andros levees, so readily used as indicators of ancient levee deposits, are, like the levees themselves, products of storm flooding (Hardie & Ginsburg, 1977) and hurricanes (Bourrouilh-le-Jan, 1982; Wanless *et al.*, 1988).

The real significance of this type of channel migration is that it results in a sequence which shows progressive shallowing upwards, closely resembling prograding interchannel deposits (Fig. 4.41). Each sequence begins with a lag containing intraclasts, perhaps with debris of a marine biota, and a lack of exposure features. This could easily be interpreted as a non-channel, subtidal deposit. This is overlain by horizons showing a progressive increase in exposure from intertidal to supratidal. Thus it is possible to confuse prograding tidal flat deposits with those of migrating tidal channels. With the absence of obvious channel-related features, such as tidal bar lamination or erosive-based channel forms, it is very difficult to distinguish these two types of sequence, and, despite being used in the interpretation of ancient carbonates, the model has many flaws (Section 4.3.5).

Small channel sequences are much easier to recognize in the field and Fig. 4.42 shows tidal channel deposits from the Late Precambrian of Oman. The channel fill contains platy, laminated dolomite clasts interpreted as supratidal dolomite crusts broken up by desiccation and storms, comparable to those developed on the tidal flats of Andros Island (Fig. 3.25).

4.3.3c Lamination

One of the most characteristic features of peritidal carbonates is fine-scale lamination, especially in the intertidal zone. This is usually on a millimetre scale, and is commonly associated with fenestral fabrics. Some of the lamination results from deposition by semi-diurnal tides, but on most low-energy tidal flats

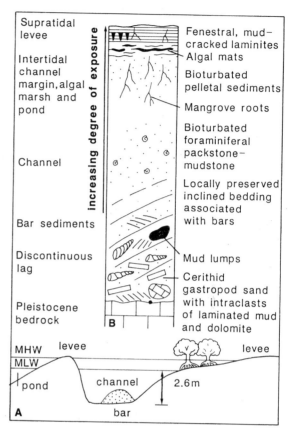

Fig. 4.41 *Andros tidal channels (modified from Wright, 1984). (A) Major environments on the Andros channelled belt. (B) Schematic section through a channel migration sequence. The section represents a transition from channel to channel margin as the channel migrates across the tidal zone. Such a sequence shows increasing exposure upwards and is very similar to shoaling, non-channel deposits. The only clue to the channel origin for such a sequence might be the presence of inclined surfaces formed on channel bars. Such bars are only well developed on the inner banks of meanders whereas the bars of straighter channels are much finer grained and lack sedimentary structures. Based on data in Shinn* et al. *(1969), Hardie & Garrett (1977) and Shinn (1983a). See Figs 3.19, 3.20 and 3.22.*

very little sediment is actually deposited by normal tides. Sediment may also be deposited on tidal flats by wind action, especially in arid and semi-arid areas, and especially by storms and hurricanes (Hardie & Ginsburg, 1977; Wanless *et al.*, 1988).

Sediment deposited on tidal flats may be trapped and bound by microbial mats which consist of layers of filamentous and unicellular micro-organisms, mainly blue–green algae (cyanobacteria), but red and green algae and bacteria also occur. The alternation of layers of microbial mats and sediment gives rise to a prominent lamination on many tidal flats, but diagenetic processes, especially compaction, result in the obliteration of this very fine-scale lamination (Park 1976, 1977). In most ancient peritidal carbonates the lamination consists of millimetre-thick micritic layers, with thin organic seams interlayered with thicker (up to several centimetres) fine grainstone or packstone laminae. These laminae, which usually contain peloids, small intraclasts and may be graded, represent storm deposits (Davies, 1970b; Hardie & Ginsburg, 1977; Park, 1977; Wanless *et al.*, 1988). Thus a micritic laminae may represent many days or even months of normal sediment accumulation (details of which are now destroyed because of diagenesis), while the typically thicker, and more prominent, storm layer represents a very brief depositional episode.

Many laminites in peritidal deposits have an irregular, crinkled appearance and typically contain fenestrae (Fig. 4.43). These irregularities are caused by microbial mats which occurred on, and just below, the sediment surface. The growth and clumping of cyanobacterial communities results in a variety of surface forms (Fig. 4.44) which create irregular surface relief on the tidal flat surface. Irregularities are also caused by desiccation, by mineral (evaporite) growth and by gas generation, especially from the decomposition of buried organic matter (Fig. 4.44). These relief features, usually only a few millimetres or centimetres in height, enable microbially-colonized tidal flat surfaces to be distinguished from purely sedimentary laminated sediments. Microbial layering may also contain calcified microbial filaments or filament moulds (Monty, 1976).

Layering is not evenly distributed on tidal flats. In subtidal and lower intertidal areas the lamination may be destroyed by bioturbation. In areas more frequently exposed, and so unsuitable for marine burrowers, lamination may be preserved. However, in the supratidal zone, desiccation and evaporite mineral growth

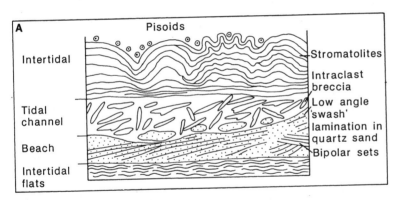

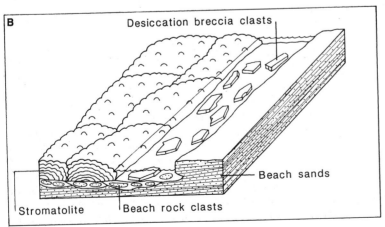

Fig. 4.42 *Tidal channels from Late Precambrian Khufai Formation, central Oman. (A) Schematic of tidal channel sequence, which ranges from 0.3 to 1.0 m in thickness. Cross-laminated beach deposits (siliciclastic sands) are erosively truncated by channel forms filled by dolomicrite flake breccias. Stromatolites cap these channel fills. (B) Reconstruction of tidal channel. After Wright* et al. *(1990).*

Fig. 4.43 *Microbial laminated fenestral intertidal sediment, Shark Bay, Western Australia. Note vertical desiccation cracks and fenestrae. Courtesy of A.R. McGregor.*

may disrupt the lamination. In hypersaline settings, with reduced faunal activity, lamination may be preserved even into the subtidal zone. Detailed discussions of the significance and origins of lamination in peritidal carbonates have been given by Park (1976), Hardie & Ginsburg (1977), Shinn (1983a) and Wanless *et al.* (1988).

4.3.3d Desiccation processes

Strictly speaking these processes are diagenetic and not depositional, but since exposure is an integral part of the tidal flat regime it is here included with depositional processes. The most obvious products of desiccation are mudcracks, which can take a variety of forms depending on many factors such as the presence or absence of a microbial mat cover, the rate of drying, exposure time and lamina or bed thickness (Plummer & Gostin, 1981; Shinn, 1983a).

Desiccated mud polygons, which commonly litter the upper intertidal and supratidal zone, can be reworked, especially during storms, to form intraclasts. A special form of polygonal structure is the tepee. These are antiformal structures with sharp apices whose crests intersect to form irregular, and commonly large, polygons (Fig. 4.45). Although desiccation and related processes can result in small buckled polygonal cracks (Fig. 4.45C), large peritidal tepee structures have more complex histories involving fluctuating groundwater conditions (Section 4.3.4d). Their origins have been discussed by Assereto & Kendall (1977) and Kendall & Warren (1987). Other processes responsible for polygon formation include thermal contraction, the crystallization of minerals such as evaporites, and upward-flowing groundwaters (Ferguson *et al.*, 1982). Further expansion may be caused by the precipitation of minerals along fractures, thermal expansion and further groundwater flow.

One of the most common features of peritidal deposits are fenestrae (or birdseyes). The term fenestra, coined by Tebbut *et al.* (1965), refers essentially to voids in the rock, larger than intergranular pores, which may be partially or completely filled by sediment, or by some cementing material (usually calcite). The term has also been used to refer to tubular voids formed by root-moulds or burrows (Grover & Read, 1978). Shinn (1983a,b) has recommended that the term fenestrae should be confined to the irregular and laminoid types commonly encountered in peritidal sequences. Irregular fenestrae (Fig. 4.46) are approximately equidimensional in form, in the range 1–5 mm in diameter (Logan, 1974; Grover & Read, 1978). In beach sands similar structures form and are called *keystone* vugs (Dunham, 1970). Fenestrae are mainly formed by desiccation and shrinkage or air and gas bubble formation (Shinn, 1968a), but they also commonly occur with stromatolites and form by the irregular growth processes of the microbial mats (Monty, 1976; Playford & Cockbain, 1976). Laminoid fenestrae are associated with tidal flat laminites, and especially with microbial mats (Fig. 4.43). They are flattened, planar or curved, parallel or subparallel to lamination, 1–5 mm high or rarely larger, and less than 20 mm in length. They mainly

Fig. 4.44 *Microbial mat morphologies. (A) Tufted mat formed by* Lyngbya. *(B) Pustular mat from upper intertidal zone; the pustules contain gypsum crystals. Smaller-scale divisions are in centimetres. From tidal flats of Qatar. Photos courtesy of David Kitson.*

result from desiccation, and parting of the laminae, but they may also form by the oxidation of microbial layers (Logan, 1974). Both irregular and laminoid fenestrae are characteristic of the upper intertidal and supratidal zones, with the distribution of laminoid fenestrae being closely related to that of microbial mats. On the complex tidal zone of northwest Andros Island such fenestrae occur where the exposure period is 60–100% (Fig. 4.40; Ginsburg *et al.*, 1977).

Root-moulds and burrows form tubular open-space structures. Worm and small arthropod burrows are common in subtidal settings and root-moulds of

aquatic plants also occur. Similar structures occur in the supratidal zone formed by soil animals and land plants. Thus such structures may occur in association with supratidal or subtidal features and are not characteristic of any setting (Read, 1973; Grover & Read, 1978; Shinn, 1983b).

A variety of other open-space structures are associated with tepee structures and desiccation polygons (Fig. 4.47). Desiccation can result in extensive sheet cracks or vertical mudcracks and both types have been well documented and illustrated in the Triassic Lofer cycle intertidal–supratidal facies (Fischer, 1964). Desiccation also results in the formation of a very fine irregular system of cracks which are common in soils.

Shinn (1983b) has described irregular fenestrae from subtidal grainstones, associated with early cementation and hardgrounds. These fenestrae are large intergranular pores with irregular shapes, formed between compound grains made up of loosely packed and cemented smaller grains. They are a type of compound-packing void but can easily be mistaken for peritidal fenestrae in grainstones. These various open-space structures can fill with sediments. Aissaoui & Purser (1983) have described a variety of internal pore-filling sediments which are characteristic of different environments, and by using these it may be possible to distinguish subtidal from intertidal, or supratidal pores. Fenestrae and related features must be used with care and the absence of fenestrae may not reflect a lack of exposure but may be due to their loss by compaction (Shinn & Robbin, 1983).

Various diagenetic processes operate in peritidal environments, but the main ones are cementation, dolomitization, evaporite precipitation and weathering and soil processes.

4.3.3e Cementation

Cementation, by calcite, aragonite or dolomite, is a common feature of peritidal carbonate environments and the cementation of carbonate sands to give beachrock is particularly common (see Section 7.4.1c). It may occur on sand flats and in shallow subtidal areas adjacent to the intertidal zone, and also along tidal channel banks to form wall-rock (Davies, 1970b). Beachrock and wall-rock may be eroded and provide intraclasts which are common features of peritidal lithologies. This intertidal cementation is important in the formation of stromatolites in Shark Bay (see Section 4.3.3i).

Another type of cementation, which is very com-

Fig. 4.45 *Tepees. (A) Groundwater tepee from strandline of Deep Lake, South Australia (see Warren, 1982a). The tepees define the margins of megapolygons up to 60 m in diameter. The tepee is 0.3 m high. Note stromatolites. (B) Large peritidal tepees, Permian, Carlsbad Group, Dark Canyon, New Mexico. These formed on the lee side of barrier islands. Height of outcrop is 4 m. (C) Intraclasts (flakes), formed by desiccation, arranged as small tepee structures. From Late Precambrian, Finnmark, Norway.*

Fig. 4.46 *Irregular fenestrae, still unfilled, from a back shoal sand flat sequence, Khufai Formation, Late Precambrian, central Oman. After Wright et al. (1990).*

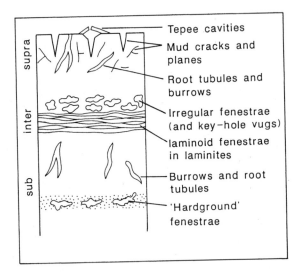

Fig. 4.47 *Open-space structures in peritidal deposits (see text). Planes are fractures, and is a term used by soil scientists for open cracks.*

mon, consists of calcite or aragonite encrustations on the intertidal and supratidal surface. These encrustations may occur on substrates affected by surf spray to give extensive irregular crusts called coniatolites (Purser & Loreau, 1973) or they may occur as encrustations on loose grains to form coated grains (pisoids) (Purser & Loreau, 1973; Scholle & Kinsman, 1974; Picha, 1978). Other encrustations and pisoids occur around groundwater springs such as in Spencer Gulf, South Australia (Ferguson *et al.*, 1982), where waters emerging on to the supratidal zone have high P_{CO_2}, total carbonate and Ca concentrations, and are close to saturation with respect to aragonite. As these waters reach the surface, and lose CO_2, aragonite is precipitated to form cavernous limestones associated with megapolygons, tepees, pisoids and crusts. Unusual as this setting may seem, strikingly similar deposits have been widely recognized in the geological record (see Section 4.3.4d). A common feature of many of these encrustations is that they possess fabrics analogous to meteoric vadose cements, such as gravitational (microstalactitic) and meniscus geometries (Section 7.3). In some of the coastal lakes of South Australia, which are fed by marine groundwater seeping through barriers, the groundwaters also precipitate a variety of similar crusts (Warren, 1982a).

4.3.3f Dolomitization

This topic is discussed in more detail in Chapter 8. Dolomite is common as a surface deposit on many modern supratidal zones such as those of the Trucial Coast, Andros (Fig. 3.25) and Florida (Illing *et al.*, 1965; Hardie, 1977; McKenzie *et al.*, 1980). The evaporation of capillary waters and surface flood water leads to the formation of brines from which the selective removal of Ca, in aragonite and gypsum, raises the Mg/Ca ratio and dolomite is formed. This dolomite is very fine grained and poorly ordered. Similar dolomites can be recognized in ancient peritidal carbonates by the association of fine-grained dolomite with such features as evaporites. Dolomite crusts on the supratidal zone are readily broken up forming intraclasts. Mazzullo & Reid (1988) have recently described dolomites from the peritidal flats of Belize which, anomalously, lack evidence of peritidal deposition and could easily be mistaken for dolomitized subtidal sediments.

4.3.3g Evaporite formation

A variety of evaporite minerals may form in association with peritidal carbonates but the most common are gypsum, anhydrite and halite (Schreiber *et al.*, 1986). They more readily occur in arid or semi-arid areas, and may also form during dry seasons in more humid areas but are redissolved during the wet season.

The most thoroughly documented area of peritidal evaporite formation is the Trucial Coast of the Arabian Gulf (Section 3.4.4). As a result of relative sea-level fall and rapid progradation during the last 4000 to 5000 years, a broad supratidal zone (sabkha) has developed (Kinsman, 1964; Patterson & Kinsman, 1981) (Fig. 3.43). The zone is locally over 10 km wide and dips gently seaward with slopes of 1:1000 to 1:3000. The salinity of the groundwaters increases rapidly landwards (see Fig. 8.20) with the result that evaporite minerals are precipitated.

Waters are supplied to the sabkha in three forms: firstly, by flood recharge during storms (Butler, 1969) when thin sheets of water (20 mm deep) are blown inland over the sabkha; secondly, by capillary evaporation where waters are drawn up through the sabkha by capillary action; and thirdly by a process known as evaporitive pumping (Hsü & Schneider, 1973), which involves the upward flow of groundwater to replace waters lost by capillary evaporation. The source for this groundwater is continental in the Trucial Coast sabkhas and not marine as was once thought (Hsü & Schneider, 1973; Patterson & Kinsman, 1981). Marine waters may also flow into sabkhas through barrier systems (Gavish, 1974; Horodyski & Von der Haar, 1975; Warren, 1982a).

In the upper intertidal zone of the Arabian Gulf sabkhas, gypsum first appears as lens-shaped crystals up to 2 mm long, which have grown displacively in microbially-laminated sediments. In the lower supratidal zone the gypsum occurs as a mush up to 0.3 m thick. Further landward the gypsum is replaced by anhydrite nodules.

These nodules initially form by the dehydration of gypsum followed by later growth. The anhydrite rarely occurs as isolated nodules but more commonly in layers up to 2.4 m thick. As a result of the displacive growth the anhydrite nodules typically display a chicken-wire texture in which the nodules are separated by films of sediment. Internally these nodules consist of a mass of laths 2–100 μm long (Butler et al., 1982). The anhydrite may also occur as highly irregular (ptygmatic) folds showing 'enterolithic structure'. In the most landward zones the anhydrite may be hydrated by continental waters to gypsum.

As a result of progradation of the tidal flats a sequence develops as shown in Fig. 4.48. In this setting the nodular evaporites are composed of anhydrite, but in other sabkhas, nodular gypsum occurs (West et al., 1979; Ali & West, 1983).

In the Trucial Coast sabkhas, halite is limited to ephemeral surface crusts or to localized areas of halite cementation, whereas hopper crystal growth occurs around zones of continental groundwater seepage (Butler et al., 1982). In other supratidal settings halite may precipitate out in salt pans with the formation of chevron halite (Shearman, 1978; Kendall, 1979).

There are many occurrences of sabkha deposits in the geological record and examples include works by Wood & Wolfe, 1969, Bosellini & Hardie, 1973, Gill, 1977, Leeder & Zeidan, 1977 and Presley & McGillis, 1982. The recognition of sabkha facies sequences has been reviewed by Warren & Kendall (1985).

Evaporites are commonly replaced by silica with a complex variety of silica fabrics in ancient carbonates (Milliken, 1979).

4.3.3h Weathering and soil formation

Weathering processes affect sediments deposited in the supratidal and terrestrial zone. Many ancient peritidal sequences contain horizons showing evidence of prolonged subaerial exposure, with evidence of dissolution and pedogenesis (for example, the Triassic

'Lofer' cycles of Europe (Sections 2.10.1 and 4.3.6)). Under humid conditions extensive dissolution may occur in the supratidal zone and karstic surfaces may form. Pedogenic processes also modify the sediment and under suitable conditions result in the formation of secondary carbonate accumulations known as calcretes (or caliches). One distinctive feature associated with exposed limestones is darkened limestone clasts called 'black-pebbles' (Strasser & Davaud, 1983; Strasser, 1984; Shinn & Lidz, 1988).

4.3.3i Microbial activity: microbial mats and stromatolites

Microbial mats are ubiquitous on many tidal flats. They are mainly composed of filamentous and unicellular cyanobacteria although they may contain a wide variety of other micro-organisms. Cyanobacterial mats, particularly those composed of motile, filamentous forms, have the ability to both trap and bind

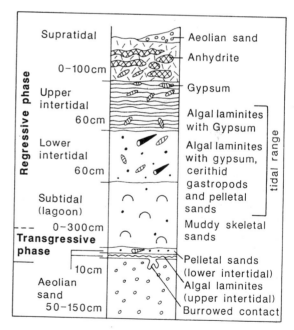

Fig. 4.48 *Stratigraphic sequence from the sabkhas of Abu Dhabi (modified from Wright, 1984). The transgressive phase deposits are thin, a result of low sediment supply. The regressive intertidal unit is much thicker, its thickness being controlled by the tidal range. Sediment supply was much greater during this phase resulting in progradation (see text). Based on data in Kinsman & Park (1976).*

sediment washed on to the tidal flats. As a result laminated sediments are commonly composed of organic and sediment laminae (Fig. 3.42B).

The distribution of these mats depends on a number of factors and they are not limited to tidal flat areas. Subtidal mats are also common but usually leave little trace in the sedimentary record, as is the case in the Bahamas grapestone belt (Section 3.2.1c). A major factor controlling mat distribution on present-day tidal flats is grazing by invertebrates, especially cerithid gastropods. In subtidal and lower intertidal areas the sediment surface is heavily grazed by cerithids and microbial mats are not well developed (Kendall & Skipwith, 1968). In areas with restricted tidal exchange, salinities may become abnormally high, such as in the hypersaline bays of Shark Bay (Fig. 4.15) and the Laguna Madre. Under such conditions grazing invertebrates may be absent and microbial mats may extend into the subtidal zone (Birke, 1974; Playford & Cockbain, 1976). Grazing is limited in the upper intertidal and supratidal zone because of the prolonged exposure, and microbial mats are better developed. The upper limit of their distribution is also controlled by desiccation. However, large stromatolites, built by microbial communities, have been found in tidal channels bordering Lee Stocking Island, Exuma Cays, Bahamas (Fig. 3.12; Dill *et al.*, 1986).

Within the microbial mat zone a variety of different mat communities occur, whose distributions and forms are controlled primarily by the degree of wetting and the sedimentation rate. Each community has a distinctive mat morphology and these have been recognized in many peritidal areas world-wide (Fig. 3.41; Logan *et al.*, 1974; Kinsman & Park, 1976). Such mat forms are potentially very useful for recognizing different environments but the preservation potential for most is extremely low (Park, 1977), and there are few records of similar mat types being identified in the geological record (Leeder, 1975; Wright & Mayall, 1981). The recognition of microbially-laminated sediments has been discussed above and the term crypt-algal laminite is often given to such sediments. Since they most commonly form on the more frequently exposed parts of tidal flats, they may contain mud-cracks, fenestral fabrics and evaporites.

The term stromatolite is generally used for discrete laminated structures with some degree of relief on the lamination and a useful classification is given by Logan *et al.* (1964). The term stromatolite has, like so many other terms in carbonate sedimentology, been the centre of much discussion. Stromatolites are very varied structures and many misconceptions have arisen as to their origins and significance (Awramik, 1984). A widely used definition for stromatolites has been given by Awramik & Margulis (quoted in Walter, 1976, p. 1): 'a stromatolite is an organosedimentary structure produced by the sediment trapping, binding, and/or precipitation activity of micro-organisms, primarily by cyanobacteria'. Recently, Burne & Moore (1987) have introduced the term *microbialite* to describe organosedimentary deposits formed as a result of microbial processes. There is an extensive literature on these remarkable structures, which is well reviewed in Walter (1976). One of the most common misconceptions is that stromatolites indicate intertidal environments. This is not true, and many stromatolites, commonly with considerable relief, are subtidal in origin (Gebelein, 1976; Playford & Cockbain, 1976; Dill *et al.*, 1986).

Stromatolite morphology can be used in conjunction with other criteria, as an indicator of energy conditions (Figs 4.49 and 4.50). In Shark Bay, Western Australia, a variety of stromatolite morphologies occur which reflect the degree of wave exposure and tidal scour. Columnar stromatolites occur on headlands exposed to wave attack, whereas those on protected shorelines have low-relief forms, that is, planar, microbial laminites (Hoffman, 1976). There are, however, exceptions to this rule even within Shark Bay (Playford & Cockbain, 1976).

The geological history of stromatolites is discussed in Section 9.2.3.

4.3.3j Bioturbation

The churning of sediment by burrowing and sediment-ingesting organisms is a common process in peritidal carbonates. Subtidal lagoons adjacent to tidal flats are often highly productive environments, and the sediments are extensively modified by organisms. The distribution of bioturbation is also influenced by the degree of exposure, with few organisms occurring in the more-exposed upper intertidal and supratidal zones. Subtidal sediments are typically heavily bioturbated and so are the lower parts of tidal channels (subtidal areas within the tidal flats). The lower intertidal zone is also commonly heavily bioturbated (Fig. 4.51), and the degree of burrowing can be related to exposure index in the same way as other sedimentary features (Fig. 4.40; Ginsburg *et al.*, 1977; Burne, 1982). Burrowing animals are not the only organisms causing bioturbation and in the subtidal zone the roots of various plants, such as sea-grasses, may cause

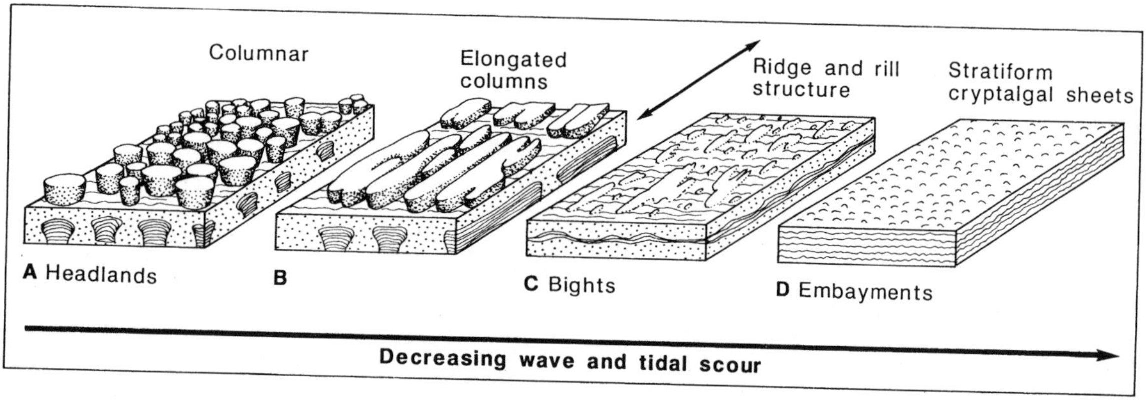

Fig. 4.49 *Schematic diagram of the variations in stromatolite form related to wave and tidal scour (modified from Wright, 1984). Discrete columnar forms (A) occur on headlands fully exposed to waves. The relief of the columns is proportional to the intensity of wave action. Elongation of the columns occurs parallel to the direction of wave attack (B) and occurs in less-exposed bights near headlands. In areas partially protected from wave attack, ridge and rill structures develop (C) with relief of 0.1–0.3 m. In small embayments, completely protected from waves, stratiform sheets occur with relief of less than 40 mm (D). These four all represent pustular mat forms from Shark Bay. Based on data in Hoffman (1976). However, Burne & James (1986) have interpreted the intertidal columnar forms as subtidal forms exposed by a drop in sea-level. Similar subtidal forms have been described from tidal channels in the Bahamas by Dill et al. (1986; Fig. 3.12).*

bioturbation, as do mangroves in intertidal zones and land plants such as halophytes in the supratidal zone. Burrowing by terrestrial organisms is also another feature of supratidal environments.

4.3.4 **Controls on the deposition of peritidal carbonates**

A group of factors influence deposition in peritidal systems, the main ones being tidal range, wave energy, climate, hydrology, sediment supply, biology, topography and sea-level changes (Fig. 4.52). The role of some of these is quite well understood but others, such as sediment supply, have received less attention. These factors interact in complex ways to create a wide spectrum of peritidal depositional systems. The facies models devised from modern tidal flats are few in number, and all these areas have developed during the complex sea-level fluctuations of the Late Pleistocene and Holocene. These few models are unlikely to be representative of the spectrum possible and many other types of peritidal carbonate models await to be recognized from the geological record (Wright, 1984). Reviews of peritidal facies models have been given by James (1984a) and Hardie (1986a).

4.3.4a Tidal influences

While tidal influences should, by definition, affect peritidal environments, there are nevertheless significant variations in the influence that tides have. The majority of modern peritidal carbonates develop in areas with a microtidal regime (tidal range is less than 2 m), e.g. 1.2 m in Trucial Coast lagoons, 0.46 m mean spring range along western Andros, and 0.6–0.9 m in parts of Shark Bay, Western Australia. Variations in the tidal ranges within one area can result in different depositional responses. Such an example occurs on the Florida Shelf where tidal range decreases from 0.7 m around the Florida Keys to only 0.15 m over much of the eastern half of the Florida Bay 'shelf lagoon' (Section 3.3), whereas there is no tidal exchange in the northeastern area. As a result of these changes marked differences occur in the types of peritidal facies sequences which form. Cape Sable, which is on the northwest part of Florida Bay (Gebelein, 1977a), is open to tidal influences and the tidal flats have a well-defined sequence of lithofacies whose characteristics can be related to the flooding frequency. The tidal range (taken as the height of the 88% to 25% flooding frequency interval) is 0.6 m (Gebelein, 1977a). As a result thorough tidal flushing

Fig. 4.50 *Stromatolite morphology, Shark Bay. (A) Subtidal stromatolites, 0.4 m high. (B) Intertidal forms (see Fig. 4.49A) from exposed headland. (C) Elongate columns (see Fig. 4.49B). (D) Ridge and rill structures from less-exposed bight area (see Fig. 4.49C). Photos courtesy of A.R. McGregor.*

occurs, with normal marine salinities in the subtidal zone and a clearly defined tidal zone. However, further east along Florida Bay, tidal exchange is reduced because of friction with the sea bottom (a complex system of channels, basins and shallow mudbanks, section 3.3.1g) and because the main outer banks act as a barrier. A consequence of this reduction is that there is very little tidal exchange, and salinities can fluctuate from 10 to 40‰, reduced by freshwater input from the Everglades. Such an environment is not strictly peritidal and the sedimentary record would contain a subtidal unit with a restricted fauna, overlain by supratidal deposits with a thin or absent intertidal interval.

Fig. 4.51 *Ghost crab burrows on lower tidal flat, Film, east central Oman. Knife is 0.12 m long.*

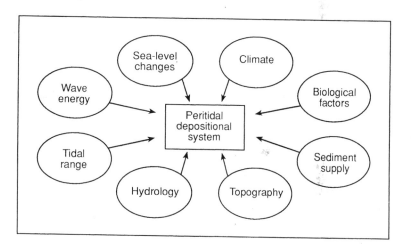

Fig. 4.52 *Major controlling factors on the peritidal depositional system.*

An example of a sequence where tidal range varied across an ancient ramp has been described from the Middle Jurassic (Bathonian) White Limestone Formation of central England (Fig. 4.53; Palmer, 1979). A broad trend has been recognized from the outer margin in the southwest across to a landmass (London Landmass) in the northeast. In the southwest, the White Limestone Formation is predominantly composed of bioclastic and locally of oolitic limestones which are cross-bedded. However, northeastwards the dominant lithofacies types are muddy limestones and marls. These are thought to represent deposition in shallower water, where tidal influences and wave fetch were reduced by the shallowness of the water and by the baffling action of sediment shoals. In the nearshore areas in the northeast, there are lagoonal clays with clear evidence of freshwater input with brackish faunas, lignites and rootlet beds, and these deposits have been compared to the present-day Everglades. Whereas subtidal (lagoonal) and supratidal (swamp) sediments can be recognized, unequivocal intertidal deposits (fenestral laminites) are absent or poorly developed. This lack of tidal flushing resulted in abnormal salinities and in the absence of true intertidal lithofacies. Superimposed on this palaeogeographic trend there are shallowing-up cycles within the formation which also reflect the overall trend (Fig. 4.53).

In the above cases the use of the term peritidal must be questioned. If sedimentary features indicating a moderate frequency of exposure are absent then true 'intertidal' lithofacies cannot be recognized. The reverse problem also arises, that is, features indicating a moderate frequency of exposure can also form in non-tidally-influenced environments. In the back-barrier Laguna Madre of Texas the tidal activity is locally non-existent, but water levels fluctuate because of wind activity (Miller, 1975). As a result, wind flats develop on the lagoon margins which possess features indicative of tidal exposure. Similarly, lakes isolated behind barriers may receive marine waters by seepage through the barrier. These coastal lake sediments commonly possess typical peritidal features such as fenestrae, evaporites, desiccation polygons, tepees and stromatolites (Horodyski et al., 1977; Warren, 1982a) yet are not strictly tidally influenced. As mentioned above, the terms perimarine, perilittoral or paralic may be better general terms to use when the exact setting of sequences with such features cannot be determined (Wright and Wilson, 1987).

4.3.4b Wave action

The characteristic peritidal facies which occur in the Recent and in the fossil record are generally the products of low-energy, protected lagoons, tidal flats and associated environments. Under higher-energy conditions beaches, shoreface deposits and barrier systems develop and these are discussed in Section 4.1.3. At the simplest level this division is between protected and wave-dominated environments. In modern peritidal settings, wave protection is afforded by such features as coral reefs, sand cays and shoals around rimmed shelves and platforms (Florida Shelf and Bahamas Banks) and by nearshore shoal belts on ramps (Trucial Coast). Larger physiographic features may also act to protect coastlines from wind action, such as along the western parts of the Trucial Coast, where the Qatar Peninsula affords protection from the Shamal winds (Fig. 4.54; Section 3.4). Along this coast there are marked changes both in the types of peritidal carbonates and in the diagenesis they undergo, related to their proximity to barriers (Purser & Evans, 1973). In the eastern zone, from Ras Ghanada to the Masandam (Fig. 4.54) the coastline is exposed to waves with the full fetch of the Arabian Gulf behind them. As a result the coastline is wave and storm dominated with storm ridges and

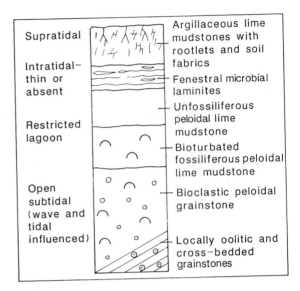

Fig. 4.53 *Shallowing-up sequence from the White Limestone Formation (Middle Jurassic) of central England. The intertidal units are very thin or absent, being locally replaced by brackish to freshwater clays. These cycles are 2 to 3 m thick (see Fig. 4.55C). Such a facies signature both indicates a low tidal range and a non-arid climate. Based on data in Palmer (1979).*

spits. In the west, in areas less protected from the Shamal winds by the Qatar Peninsula, open sand flats have developed such as those at Sabkha Matti (Fig. 4.54). The coastline is accreting by storm beach progradation over the sand flats. The resulting sedimentary record shows coarse storm beach sands with low-angle cross-bedding (oriented mainly seawards), grading down into finer intertidal and subtidal cross-laminated sands (Fig. 4.54B). In other areas in the western region, longshore processes are dominant and spits have migrated over the tidal flats, forming coarsening-up sequences topped by coarse sands with landward-oriented steep cross-sets. In these grain-dominated sequences beachrock is common and is broken up to form intraclast breccias. Microbial mats only form on the high intertidal flats in the more protected areas along this exposed sand flat zone.

In the central region, e.g. Abu Dhabi, offshore barriers give protection from waves and behind these, sheltered lagoons 2–5 m deep occur, with high salinities. Most of the lagoon floor is covered in a thin layer of hard pellets and compound grains showing

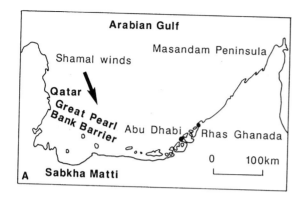

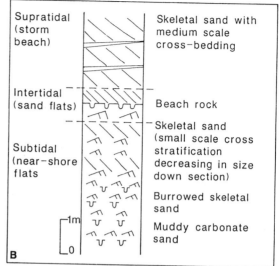

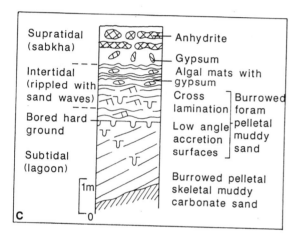

Fig. 4.54 *Contrasting peritidal deposits along the Trucial Coast of the southern Arabian Gulf. (A) Simplified map of the region (see also Fig. 3.36) showing the orientation of the Shamal wind. (B) Prograding sequence developed on the exposed embayment of the Sabkha Matti. (C) Prograding sequence developed in the protected back-shoal zone of central Abu Dhabi. Modified from Wright (1984); based on data in Purser & Evans (1973).*

extensive cementation. Muds, however, accumulate in only the most sheltered areas. The adjacent tidal flats have an extensive microbial mat cover and are rapidly prograding into the lagoon. As a result of the high salinities sabkhas occur in the supratidal zone (Section 3.4.4).

Similar deposits have been recognized by Purser (1975) from the Middle Jurassic (Bathonian) of France. 'Grainy' beach facies represent wave-influenced zones while muddy tidal flat sequences are also recognized; the latter are associated with offshore oolite barrier deposits.

The contrast between wave-influenced and protected shorelines is also seen along the present coasts of Andros Island. The windward east coast has a narrow 'grainy' beach zone whereas the sheltered leeward western coast has tidal flats (Section 3.2.2).

If shorelines are exposed to even a limited amount of wave or storm erosion, and under conditions of low sediment supply, the coastline may undergo an erosional transgression. This is the case along the northwest coast of Andros Island while the more protected southwest coast (with a higher rate of sediment supply) is prograding seawards (Section 3.2.2b; Shinn *et al.*, 1969). This is an important example and shows that parts of the same coastline may simultaneously undergo a regression (progradation) and an erosive transgression. These changes are not the result of any change in sea-level but reflect differences in exposure to wave action and sediment supply. Strasser (1988), based on detailed correlation of Cretaceous 'Purbeckian' peritidal cycles in the Jura Mountains, has been able to recognize areas simultaneously undergoing progradation and local erosion.

A spectrum of environments can occur from wave- and tidally-influenced peritidal carbonates (beaches), to wave-protected low-energy tidal flats, to tidally-restricted lagoonal marsh areas (Fig. 4.55). The recognition of such transitions in ancient limestones would provide useful information on palaeogeography and possibly on other factors such as wind patterns.

4.3.4c Climate

Under this category, both normal climatic factors, and rarer meteorological events such as storms, should be considered. The rainfall—temperature regime, in particular, is the major influence on the types of supratidal deposits which form. In humid or subhumid areas, the supratidal zone may have lowered salinities. Extensive mangrove swamps or marshes may occur, and mangrove peats may develop such as in Florida (Wanless, 1969). One of the most widespread humid, supratidal environments today is the coastal algal marsh, which develops in seasonally-flooded marsh zones, dominated by cyanobacterial mats. Two types occur, referred to as coastal and interior algal marshes, both of which can be represented in shallow-marine sequences as a result of sea-level changes (Monty & Hardie, 1976). Interior algal marshes are situated landward of coastal marshes in low-lying areas, generally with a karsted carbonate bedrock. These areas are very rarely inundated by marine waters and are arguably, strictly speaking, not peritidal or perilittoral deposits. However, as a consequence of their association with low-lying areas, such deposits can become incorporated into peritidal lithosomes (Fig. 4.56; Halley & Rose, 1977; Enos, 1983). The major process operating in such areas is the precipitation of calcite from carbonate-rich, shallow, fresh-waters by cyanobacteria (as benthic mats, floating masses and as periphyton) (Gleason & Spackman, 1974). A fine carbonate mud is produced which, after compaction, results in amorphous, thickly-bedded lime mudstones containing non-marine biotas, with poorly-preserved cyanobacterial remnants, local calcified stromatolites and numerous desiccation features (Monty & Hardie, 1976). Root-moulds and peats may also occur.

Coastal 'algal' marshes are marine influenced, especially during storms, and episodic sedimentation results in a characteristic layering of calcified cyanobacteria, and marine-derived sediment layers, associated with aragonite laminae. Stromatolites may develop, and the detailed petrographic and sedimentological features of this environment have been documented from the Andros coastal marshes by Monty & Hardie (1976).

Even though these marshes are widespread in the humid to subhumid carbonate province of southern Florida—Bahamas, there are very few records from the pre-Pleistocene. Halley & Rose (1977) have described marsh deposits from the Cretaceous of Texas (Fig. 4.56C), Andrews (1986) from the Middle Jurassic of Scotland and Wright (1985) from delta-plain environments, from the Upper Jurassic of Portugal.

In addition, under humid conditions extensive dissolution of carbonate substrates will also occur with the development of soils such as 'terra rossa' (Esteban & Klappa, 1983). In semi-arid or arid areas evaporite minerals may form in the supratidal facies resulting in sabkhas (Sections 3.4.4 and 4.3.3g).

These two settings, the subhumid and semi-arid to arid, are almost at opposite ends of the spectrum and there are many variations between. In some peritidal environments salinities can vary dramatically, often seasonally, because of freshwater input during the wet season, either as run-off or from groundwater. One striking example of this is the Upper Jurassic Purbeck evaporite facies of southern UK (West, 1975, 1979). These limestones consist of a varied sequence of pelletal, bioclastic, stromatolitic and evaporite-bearing limestones deposited in hypersaline subtidal to supratidal environments (West, 1975). Paleosols contain well-preserved plant remains (Francis, 1983). Shallowing-upward cycles have been described passing from intertidal or subtidal to subaerial (Fig. 4.57). While such a setting might readily be interpreted as having formed under arid or semi-arid conditions, two pieces of additional evidence warrant consideration. The trees in the Purbeck paleosols are dominantly conifers, whose growth rings show evidence of a strongly seasonal climate which was probably of the semi-arid Mediterranean type (warm wet winters, hot dry summers) (Francis, 1983). The Purbeck Limestones also contain freshwater molluscs, and the seemingly bizarre association of evaporites and charophyte remains (West, 1975). More light was shed on these unusual deposits when Burne *et al.* (1980) described a variety of coastal lagoons and saline lakes from South Australia which possess the unusual association of charophytes and evaporites, as well as stromatolites. They described a variety of settings but all were influenced by continental groundwater resurgences, with salinities varying markedly as a result of seasonal changes under a Mediterranean-type climate. The Purbeck Formation may have been deposited in coastal lagoons similar to those described by Burne *et al.* (1980), and this type of facies seems to have been widespread in the Late Jurassic—Early Cretaceous of Europe (Strasser & Davaud, 1983). It

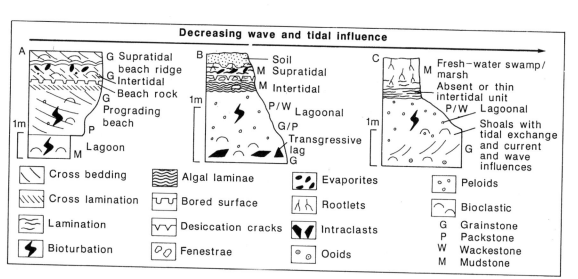

Fig. 4.55 *Spectrum of peritidal shoaling units reflecting differing degrees of wave and tidal influence. (A) High-energy type based on Khor Duwahine, extreme western part of Abu Dhabi; although this area is protected from the Shamal winds it still lacks protection by a barrier system and is wave influenced. Microbial mats only occur on the upper part of the intertidal zone and are associated with evaporites (gypsum). These sand flats are prograding seawards with an accretion slope showing seawards-inclined bedding. Based on data in Purser & Evans (1973). (B) Protected, lower-energy sequence. The initial coarse transgressive lag horizon is followed by a shoaling phase showing well-defined intertidal facies. Based on Lower Carboniferous peritidal deposits from south Wales (Wright, 1986a); similar sequences are shown by James (1984a). (C) Highly restricted sequence; shoaling from a grainstone facies, with wave and tidal influences, into lagoonal and finally a freshwater facies (supratidal−terrestrial). Intertidal deposits are thin or absent. Based on data in Palmer (1979) for the Middle Jurassic White Limestone Formation (Fig. 4.53). This sequence is broadly comparable to the transition across the present-day Florida Shelf lagoon. Wave action and tidal exchange are greatly reduced over the shelf which results in nearshore areas being effectively tideless. In contrast to (A) and (B) this sequence has formed under a more humid climate and lacks evaporites. Diagram modified from Wright (1984).*

is an example of the so-called schizohaline environments discussed by Folk & Siedlecka (1974).

Storms are important depositional agents in peritidal environments. On the tidal flats of northwest Andros it has been claimed that storms are the exclusive mechanism of deposition (Hardie & Garrett, 1977). Indeed the low tidal energy results in little or no sediment entrainment or transport from the adjacent subtidal area during normal tides. Along the southwest coast of Andros the subtidal zone consists of a platform, 3−5 km wide and less than 1 m deep. During the last 4000 years a series of five, roughly parallel shorelines has developed, progressively seaward on this platform (Gebelein, 1975; Gebelein *et al.*, 1980). Each shoreline is marked by a storm ridge ('hammock'), each of which isolated a linear lagoon behind itself, later filled with muds. However, Bourrouilh-le-Jan (1982) has offered a slightly different explanation for these features, interpreting the hammocks as hurricane trails. If such features can be recognized in the geological record it may be possible to recognize storm-dominated peritidal sequences.

4.3.4d Hydrology

The importance of hydrological processes in peritidal environments has only recently been appreciated. The role of the groundwater system in sabkhas has already been noted in Section 4.3.3g. Such systems are not only important for sabkha chemistry but large systems are necessary for the development and preservation of broad sabkhas (Patterson & Kinsman,

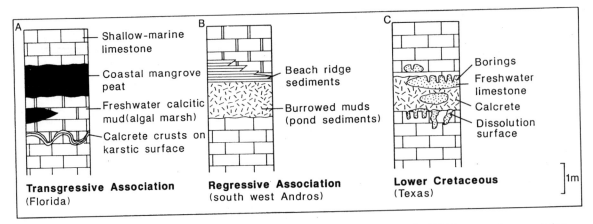

Fig. 4.56 *Peritidal sequences with 'algal marsh' facies. Modified from Wright (1985). (A) Transgressive association based on Florida (see Wanless, 1969; Monty & Hardie, 1976; Enos, 1983). This type of occurrence has a low preservation potential and the peats and algal marsh deposits will be eroded away during the next transgression, except in protected settings. (B) Regressive sequence predicted from southwest Andros Island. From data in Gebelein et al. (1980). (C) Lower Cretaceous of central Texas. The freshwater limestone is petrographically very similar to the algal marsh limestones of Florida. They occur incorporated into a complex calcrete which indicates two periods of exposure. One occurs at the top of the lower marine limestone, the other at the base of the upper marine limestone. The borings were probably made by marine organisms such as pholad molluscs. Based on Halley & Rose (1977).*

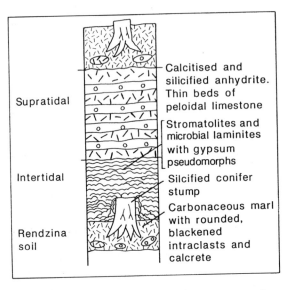

Fig. 4.57 *'Purbeckian cycle' from the Dorset area, southern UK (after West, 1975, 1979). These cycles have been interpreted as a transgressive (soil to intertidal) and regressive (intertidal to supratidal to soil) sequence. However, many criteria for recognizing intertidal facies are absent and the sequence may represent lagoonal or ephemeral lake deposits comparable to those in South Australia (see text). Black-pebbles are common in the paleosols (see Section 4.3.3h).*

1981), to prevent deflation of the sabkhas at their landward ends as progradation occurs.

Groundwaters are important for the diagenesis of peritidal carbonates and their possible role in dolomitization has been discussed by a number of authors (Von der Borch & Lock, 1979; Bourrouilh-le-Jan, 1980; Section 8.7.1).

Groundwater springs emerging on supratidal zones have dramatic effects both physically and chemically on the sediments. Ferguson *et al.* (1982, 1988) have described the processes and products of such springs emerging in the supratidal zone of Fisherman Bay, South Australia, a prograding tidal zone (Burne, 1982; Burne & Colwell, 1982). The products include extensive cementation by aragonite including speleothem-like cements and pisoids, and most characteristically, fracturing related to tepee structures and large polygons. These features are caused by upward-flowing groundwaters and precipitation of cements. The result is a very complex suite of cements and fabrics showing evidence of aragonitic (marine) cements and pisoids, supratidal fabrics with evidence of fracturing, speleothem fabrics (vadose conditions) and common geopetal structures formed in fractures and tepee cavities (Kendall & Warren, 1987). Multiple phases of fracturing occur as the springs shift position and ultimately can result in a very irregular and complex lithosome. Even though such a deposit sounds unusual there are several very well-documented examples in the geological literature, both described before the modern analogue was found. Assereto & Kendall (1977) and Assereto & Folk (1980) have described strikingly similar suites of features from the Triassic Calcare Rosso of north Italy (Fig. 4.58). The spectacular back-reef deposits of the Permian Yates and Tansill Formations of the Guadalupe Mountains of New Mexico (Esteban & Pray, 1983; Warren, 1983) also contain similar features (Fig. 4.45B). In both these sequences fenestral limestones are associated with tepee structures, defining large dish-shaped polygons (up to 10 m across). The polygons provided 'splash-cups' for the formation of pisoids, some of which were pedogenically modified. Evidence of fracturing and early cementation is abundant and replaced giant aragonite cements have been found (Fig. 4.58B, Assereto & Folk, 1980). However, similar features can also form in solar lakes or salt pans, isolated from the sea by barriers while still receiving marine groundwaters (Horodyski & Von der Haar, 1975; Warren, 1982a). These are not strictly peritidal deposits and distinguishing these two types of settings in ancient limestones may be difficult especially after later diagenesis has altered the geochemical signatures of these different groundwaters (Ferguson *et al.*, 1982). Continental groundwater resurgence zones also occur on the sabkhas of Abu Dhabi (Butler *et al.*, 1982) and are marked by large circular to elliptical depressions up to 1 m deep associated with halite cementation and anhydrite diapir structures.

4.3.4e Sediment supply

The importance of sediment supply in controlling the formation of peritidal sequences is generally underestimated. Its role in modern peritidal environments is poorly understood and in ancient sequences it can only be guessed at. Gebelein (1977a) estimated that 94% of the sediment deposited on the Cape Sable tidal flat complex in Florida was derived from outside the complex. The remainder was produced in interior lagoonal areas in the complex. The adjacent subtidal areas supply the sediment enabling most modern tidal flats to prograde. However, this rate of supply must exceed any sea-level rise or recession of the shoreline caused by wave erosion. A decrease in sediment supply along the northwest coast of Andros Island, possibly caused by a change in circulation on the

Bahama Bank, has resulted in an erosive transgression along this coast (Hardie & Ginsburg, 1977). With a greater sediment supply this coastline might prograde as does the adjoining southwest coast.

Sediment supply is probably the main control on sediment thickness in peritidal deposits. If progradation occurs under static sea-level, the thickness of the resulting intertidal unit in the shallowing-up sequence will be equal to the tidal range (Kinsman & Park, 1976). The role of sediment supply in this context is strikingly displayed in the sabkha sequences of Abu Dhabi (Fig. 4.48). The initial transgression along this coast, some 5000 years ago, left only a 0.1 m thick transgressive intertidal unit but the tidal range at this time has been estimated at 1–2 m (Kinsman & Park, 1976). The cause of this anomaly is believed to be the low rate of sediment supply. The effect is also clearly seen in the peritidal cycles in the Calcare Massiccio Formation (Lower Lias, Jurassic) of the central Apennines (Fig. 4.59; Colacicchi *et al.*, 1975). In these cycles the transgressive intertidal member is several times thinner than the regressive intertidal member. This is most likely caused by the differing rates of sediment supply. Indeed, the rate of sediment supply is typically low during transgressions and usually reaches its peak after a lag period (Kendall & Schlager, 1981; Burne, 1982), when water depth has reached a point when carbonate production increases. It could be this uneven rate of sediment supply which creates the typical asymmetrical cycles seen in many ancient peritidal carbonates (Fig. 4.55, Section 2.10.1 and Ginsburg, 1975; James, 1984a; Hardie, 1986a). This asymmetry does not necessarily indicate a rapid transgression as is sometimes suggested. Wave action during this sediment-starved phase results in erosion and reworking with intraclasts of underlying lithologies.

4.3.4f Topography

Topography is an important influence in peritidal environments where depth and tidal influences are small. The narrow shelf of Belize (Central America) contains a number of shelf islands which have developed over a Pleistocene karstic surface during the Holocene sea-level rise (Ebanks, 1975). As a result of the shallow water and the irregular basement, a complex facies mosaic has formed with lagoons, intertidal and supratidal zones around the islands, and shallow intra-island lagoons in the karstic depressions. Topography has also affected the depositional patterns of the Pleistocene of Florida (Perkins, 1977). Pratt &

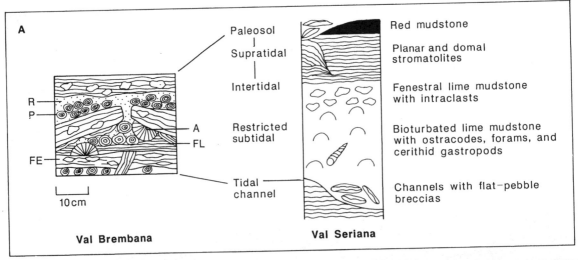

Fig. 4.58 *(A) Variations in peritidal sequences in the Middle Triassic (Ladinian) platform carbonates of Lombardia, north Italy. In the Val Seriana area, northeast of Bergamo, distinctive peritidal cycles occur, 0.3–1.5 m thick, representing asymmetric shallowing-up units capped by paleosols. Some 20 km west in the Val Brembana, these are replaced by the Calcare Rosso. This sequence, up to 45 m thick, consists of highly brecciated, fenestral, pisoid limestones with tepees. A, replaced aragonite cements; FE, fenestral limestone; FL, replaced aragonite flowstone; P, pisoids; R, red mudstone. Based on this author's observations and data in Assereto & Kendall (1977) and Assereto & Folk (1980). (B) Cemented cavities and fractures in tepee limestone, Val Brembana. In these limestones, affected by both marine and freshwater vadose and phreatic conditions, the original tepeed crust constitutes less than 20% of the rock. The spherulites are replaced aragonite cements. Source: Kendall & Warren (1987). Photo courtesy of Christopher Kendall.*

James (1986) have suggested that the St George Group (Ordovician) of Newfoundland was deposited on a large shallow platform dotted with small islands and banks (Fig. 2.23, Section 2.7 and 2.10.3).

4.3.4g Sea-level changes

In the shallow settings that peritidal carbonates form, relatively small changes in sea-level caused by tectonic or eustatic factors will have marked effects on the depositional regime. Sea-level is, however, a dynamic feature in its own right and short-term changes in wind stress, barometric pressure and ocean currents can cause locally significant fluctuations in sea-level (Wanless, 1982). Changes in lithofacies in peritidal carbonates are often interpreted as being caused by sea-level changes or progradation, but the causes may be shifts in such factors as sediment supply, increased wave erosion (e.g. see Strasser & Davaud, 1986) or the development of offshore barriers reducing tidal exchange. With these complex and interactive processes it is not surprising that many ancient peritidal carbonates contain highly complex facies associations. However, many sequences do contain well-ordered, cyclic packages, the possible origins of which are discussed in Sections 2.10 and 4.3.6.

4.3.4h Biological control

Organisms can influence sedimentation in peritidal environments in a number of ways. Microbial mats trap and bind fine sediment and probably encourage aggradation and progradation of tidal flats. Burrowing organisms and rooted vegetation cause mixing of the sediment with the resulting loss of sedimentological detail. Organisms can also influence the whole development of the peritidal lithosome and a striking example has been described from the coasts of southern Australia (Gostin *et al.*, 1984; Belperio *et al.*, 1988).

Around the gulfs and protected embayments of South Australia, a distinctive regressive peritidal sequence has developed over the last 7000 years which reflects the dominance of vegetation-related sedimentary processes. Spencer Gulf is a shallow-marine embayment, having a mainly mesotidal regime with minor wave activity. The climate is dry Mediterranean to semi-arid, and evaporites form locally. The gulf waters are cool–temperate, and the dominant sediment producers are coralline algae, foraminifera, molluscs and bryozoa. The more protected settings have low gradient intertidal flats, colonized by

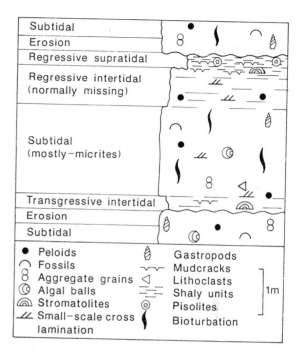

Fig. 4.59 *Peritidal cycle from the Calcare Massiccio Formation (Lower Jurassic) of the central Apennines. Note the thin transgressive intertidal unit compared to the much thicker regressive (progradational) upper intertidal unit. These thickness differences possibly reflect a lower rate of sediment supply during the early transgressive (start-up) stage. Compare with Fig. 4.48. Based on Colacicchi et al. (1975) and Sellwood (1986).*

mangroves, whereas the less protected areas lack mangroves and have extensive bare sand flats and beach-ridge systems. Those areas open to the full influence of the Southern Ocean, such as the Coorong region, have high-energy barrier shorelines.

A distinctive biological zonation has been recognized across the peritidal zone (Fig. 4.60A,B). The subtidal areas down to 10 m are dominated by sea-grass meadows of *Posidonia australis* and *P. sinuosa*. These have formed shoreline-attached platforms or offshore banks. The grass blades support a prolific growth of skeletal forms, mainly foraminifera, coralline algae, bivalves and gastropods. The sediment consists of very poorly-sorted skeletal packstone with abundant fibres of *Posidonia*. The intertidal zone is covered by meadows of the sea-grass *Zostera* (Fig. 4.60A) or is bare, the sea-grass cover being patchy in the lower intertidal zone but denser in the upper zone. The sediment consists of a poorly-sorted mixture of skeletal sand, with a prominent molluscan infauna, and terrigenous mud and sand. The sea-grasses have trapped and bound mud, which provides a soft substrate essential for mangrove colonization. The sand flats merge landward (Fig. 4.60B) into mangrove woodland or into beach ridges, depending on the degree of wave activity (itself dependent on orientation to local winds).

The mangrove zone consists mainly of stands of *Avicennia marina* (Fig. 4.60A), and microbial mats occur within this zone and extend into the upper intertidal zone, which is dominated, in protected areas, by a halophyte samphire marsh with saltbush. The sediments consist of bioturbated muds containing abundant roots, strongly anoxic in the upper few centimetres but oxygenated below. The supratidal zone consists of bare carbonate and gypsum flats with local stranded beach ridges. The sediment is a fenestral mud with gypsum and locally dolomite. Microbial lamination is not preserved in the mangrove to saltbush zone because of bioturbation and it is destroyed by prolonged desiccation in the supratidal zone. Locally small depressions on the supratidal flats act as ephemeral saline lakes (Burne *et al.*, 1980).

The activities of the plant communities are critical factors in the formation of the peritidal deposits, which are aggrading and prograding rapidly. The very high carbonate production rates of the subtidal sea-grass meadows, with accumulation rates in excess of 2 mm yr^{-1} (Belperio *et al.*, 1984a,b), lead to the rapid upgrowth of the banks to sea-level. It is over this subtidal platform that the intertidal zone can rapidly prograde. The sea-grasses trap and bind the sediment with their blades and rhizomes, and similarly the mangrove stands and plants also contribute to the aggradation of the sediment pile. The vegetation zonation also determines the nature and distribution of the carbonate sediment types because each zone has its own distinctive skeletal communities (Cann & Gostin, 1985).

While the biological zonation across the coastal zone is controlled by tidal inundation and the degree of wave exposure, it is the sea-grass-related sedimentation which controls the sedimentary deposit. The sea-grass banks have built locally across embayments along the coast and have isolated saline lakes in their lee. In such cases, the sedimentary record would show the subtidal sea-grass facies sharply overlain by subaqueous evaporites (gypsum) (Fig. 4.60C). Similar 'cut-off' lakes have a marginal facies with microbial bioherms and various structures related to marine

or continental groundwater seepage (Fig. 4.45A; Warren, 1982a,b). Figure 4.60C shows two generalized sections illustrating sequences formed by the progradation of protected tidal flats and where embayments have been isolated to form coastal lakes. However, this sea-grass lithosome is only one of several types of carbonate depositional system found along the southern coast of Australia, controlled by wave regime and climate (Gostin *et al.*, 1988). In areas exposed to major ocean swells the coastline is barred by carbonate sand barriers behind which lagoons with evaporites occur.

An analogy can be drawn with the hypersaline basins of Shark Bay in Western Australia. Here sea-grass banks have formed sills at the mouths of coastal embayments reducing circulation in the embayments resulting in hypersalinity (Section 4.5.8b; Hagan & Logan, 1974b).

4.3.5 Tidal channel model

This model, developed by Shinn *et al.* (1969) working on the northwest coast of Andros Island (Sections 3.2.2b and 4.3.3b), explains the formation of some erosive-based, shallowing-upward cycles as the result of channel migration. It is an attractive model and has found many followers (Loucks & Anderson, 1980; Mitchell, 1985; Ramsay, 1987). However, there are several features which suggest that this model should only be applied with the utmost care (Wright, 1984).

On siliciclastic tidal flats channel migration sequences make up a small percentage of the lithosome, in contrast to fluvial deposits where genetically similar point-bar sequences may be volumetrically very important (Barwis, 1978). This is largely because the interchannel tidal areas receive sediment much more frequently than fluvial overbanks and have higher accumulation rates. This means that aggradation and progradation can be much more rapid than channel reworking. Data from carbonate tidal flats suggest that progradation is a rapid process, while the little evidence available suggests that channel migration rates are very low. For example, Hardie & Garrett (1977) noted little significant migration of the Andros channels over a 25 year period, and Shinn (1986) noted lateral migration of less than 50 m in the last 3000–5000 years in the same area. This lack of migration may reflect the fact that the channels are migrating landward due to local onlap. Shinn has speculated that tidal channel deposits only contribute significantly to a lithosome during transgressive phases while during prograding phases the channels rapidly

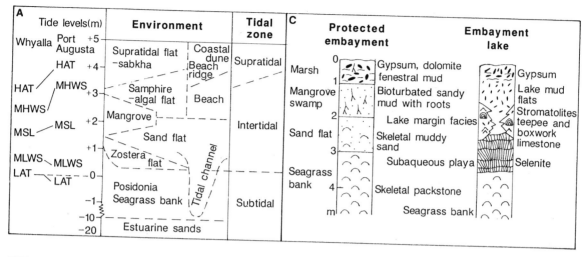

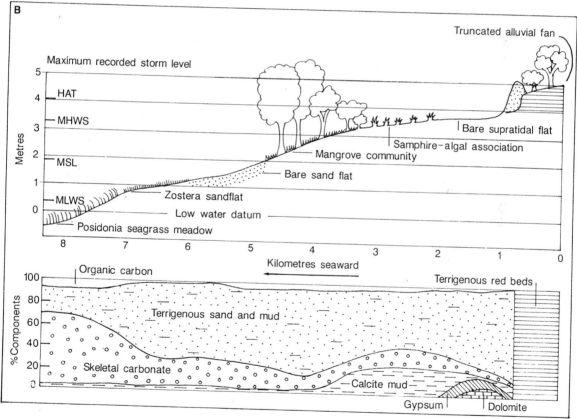

Fig. 4.60 *Sedimentary environments and zonation across the peritidal areas of northern Spencer Gulf, South Australia. (A) Relationship of sedimentary environments to tide levels (based on Port Augusta, north of Redcliff). LAT, lowest astronomical tide; MLWS, mean low water spring tides; MSL, mean sea level; MHWS, mean high water spring tides; HAT, highest astronomical tide. The diagram illustrates both the protected and exposed environment types. (B) Generalized zonation of intertidal environments at Redcliff. Modified from Gostin* et al. *(1984). (C, top right) Generalized lithofacies sequences developed in protected embayments compared to embayments isolated by bank growth, such as Marion Lake. Modified from Belperio* et al. *(1988) and Warren (1982a,b).*

choke with sediment. However, the preservation potential of channel deposits in transgressive sequences will be low as the whole lithosome may be reworked by shoreline erosion.

Along the prograding southwest coast of Andros, the channels have been abandoned and filled, and are not volumetrically important in the lithosome (Gebelein et al., 1980). Another example of channel filling occurs along the prograding coast of northeast Qatar in the Arabian Gulf (Shinn, 1973). The development of these channels is strongly influenced by a buried, cemented layer and the channels are restricted in depth by this layer which forms on the lower intertidal zone in front of the accreting channel belt. Trenches cut across buried channels show a sequence (Fig. 4.61) which consists of a channel lag of winnowed skeletal debris, overlain by a muddy lag deposit with some skeletal grains. These grade up into light grey muds which are disconformably overlain by sandy stromatolitic sediment (Fig. 4.62) and are finally buried by a cross-bedded sand. Shinn interpreted this sequence, by comparison with thalwegs up existing channels, as representing the progressive landward isolation of a tidal channel which became buried by aeolian sediments. These channel-fill units produce sheet-like lag sands, controlled by the underlying cemented layer. An interesting result of this is that the channel bases do not show channel forms and could be mistaken for the transgressive surf-zone layers typically seen in tidal flat sequences. The sequence shown in Fig. 4.61 results from the abandonment of the channels at their landward ends because of the progradation of the tidal flat–channel belt and because of the construction of beach ridges isolating parts of the tidal flats. An obvious result of this is that the tidal flat deposit will be volumetrically (architecturally) dominated by prograding interchannel sequences.

Other factors need to be taken into account such as the nature of the channels themselves. On northwest Andros the channels are very wide, anomalously so for an area with a low tidal range. This has led Bourrouilh-le-Jan (1982) to suggest that they may be maintained by hurricane overflow drainage. In semiarid and arid areas cementation of the channel walls, as described from Shark Bay by Davies (1970a), may considerably reduce the rates of bank erosion.

In summary the tidal channel model has been used rather uncritically in the rock record. In transgressive situations tidal channels may be well developed but their deposits will have a low preservation potential. In prograding sequences the channels will become abandoned and filled. This may result in the preservation of channel forms but evidence suggests that channel deposits will be volumetrically minor in the lithosome. Mitchell (1985) has provided a detailed comparative study of a possible channellized tidal flat sequence from the Ordovician St Paul Group of the Appalachians. Even though channel deposits were recognized they only constituted an estimated 2–3% of the sequence (Mitchell, 1986, pers. comm.). It is reasonable to conclude that the bulk of ancient peritidal deposits will consist of simple tidal flat deposits and not tidal channel fills.

4.3.6 Multiple sequences

Peritidal limestones in the geological record commonly occur in cyclic or rhythmic packages containing from a few to hundreds of individual shallowing-up or more rarely deepening-up units. For example, the Lower Proterozoic Rocknest Formation of Canada contains 140–160 units (Grotzinger, 1986a). The Middle Triassic Latemar carbonate buildup of north Italy contains 500 cycles (Goldhammer et al., 1987). The cause or causes of these packages is still debated but a variety of explanations have been offered (Section 2.10.3). A detailed review of this topic has been given by Hardie (1986b). Most commonly these packages contain shallowing-up units and it is common for these, probably progradational, units to occur within a transgressive sequence. Anderson & Goodwin (1978) have named stacked shallowing-up deposits 'punctuated aggradational cycles' (or PACs). They are similar to the progradational cycles in Holocene sequences showing a basal transgressive, open marine lag formed under conditions of low sediment supply relative to sea-level rise, followed by a prograding sequence.

While most multicyclic packages contain shallowing-up sequences there are also multicyclic, deepening-up units, of which the Late Triassic Lofer cycles are the best known. The Lofer limestones, 1000–1500 m thick, represent a backreef facies developed over what is now the northern limestone Alps of Germany and Austria. The Lofer cycles, some 300 in most sequences, consist of three members (Fischer, 1964, 1975). Member A is developed on the irregular, dissolutional top of the underlying member C. This karstic surface is overlain, and penetrated by fissures containing red and green clays. These horizons have been interpreted as modified soil horizons and may contain rounded clasts of the underlying lithology. The overlying member B (which averages 0.5 m in thickness) consists of laminated and

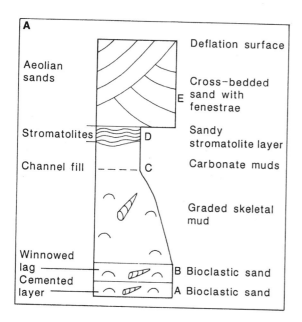

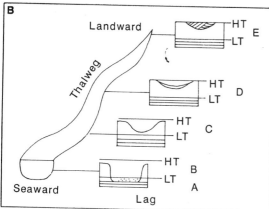

Fig. 4.62 *Abandoned head of a tidal channel on Qatar being filled by stromatolitic sediment. Photo courtesy of David Kitson.*

Fig. 4.61 *Channel sequence developed by tidal channels in northeast Qatar, Arabian Gulf (based on Wright, 1984). Seaward progradation caused the landward isolation and abandonment of the channel. The landward parts of the channel are covered by microbial mats and are eventually filled by aeolian sand. The depth of the channel is controlled by an underlying cemented layer (intertidal beachrock). Based on data in Shinn (1973).*

brecciated dolomitic, fenestral and stromatolitic laminites (intertidal–supratidal). These spectacular laminites have been called *loferites* (Fischer, 1964). Some features also occur in this unit resembling soil textures. Member C, which constitutes over 90% of the sequence, consists of massive bioclastic limestones from 1 to 20 m thick, containing a rich biota of megalodontid and other bivalves, corals, bryozoans, brachiopods, foraminifera, calcareous algae and oncoids. Each cycle represents a rise in sea-level from subaerial (A) to intertidal–supratidal (B) to subtidal (C), and constitutes a transgressive sequence which is abruptly terminated by another subaerial surface. Each cycle (rhythm) represents, on average, about 50 000 years (Fischer, 1964) and they have been interpreted as being eustatically controlled, perhaps by Milankovitch climatic cycles (Fischer, 1975; Kendall & Schlager, 1981). This view has, however, been questioned by Haas (1982) who regards the Lofer cycles of the Bakony Mountains of Hungary as having been controlled by local climatic changes. Even though the Löfer cycles are often regarded as rather unique, because they consist of deepening-up units, regressive sequences with members A-B-C-B have been recorded from the Hungarian Lofer (Haas, 1982; Haas & Dobosi, 1982). Bechstädt *et al.* (1987) have described asymmetric shallowing-upward and deepening-upwards (Lofer-like) cycles from the Triassic Wetterstein Limestone of the eastern Alps. They regard the asymmetry of the latter type as the result of subaerial erosion of the upper part of the cycles, which they believe originally had a shallowing-upward sequence capping each cycle. Cisne (1986) has offered a tectonic model suggesting that the cycles are controlled by earthquakes (see Section 2.10).

In a recent benchmark paper Goldhammer *et al.* (1987) have studied a sequence of 500 Middle Triassic peritidal cycles from north Italy. These metre-thick couplets consist of subtidal units overlain by vadose caps; intertidal—supratidal lithofacies are absent. Within the sequence they have recognized clusters of four, five and six couplets, with five as the dominant type. These *pentacycles*, as they call them, are interpreted as representing 20 000 year Milankovitch cycles, arranged in clusters approximating to 100 000 year cycles. They offered computer simulation sequences to test this hypothesis. This paper, along with those of Grotzinger (1986a) Read *et al.* (1986) and Walkden & Walkden (1990) represent recent studies indicating that modelling and semi-quantification of cyclic peritidal sequences are essential techniques for interpretation. See Section 2.10.3 for further discussion.

4.3.7 Future research

Knowledge of the dynamics of peritidal carbonate environments is still very incomplete and there is little understanding of the real controls on sediment accumulation. The widely used tidal channel model has yet to be tested rigorously to determine if channel migration deposits can contribute to the buildup of prograding tidal flat lithosomes. Many peritidal environments such as mangals and algal marshes have a very poor geological record. Above all there is a need for a more rigorous process-based approach to interpret peritidal sequences, rather than a comparative approach of fitting them to modern analogues. The recent developments in computer modelling and quantification of stacked peritidal cycles is most exciting (Sections 2.10 and 4.3.6) and will provide new insights into the controls on cyclicity. The importance of Milankovitch cyclicity is being widely recognized (Goodwin & Anderson, 1985; Goldhammer *et al.*, 1987; Strasser, 1988) and in the future more evidence for these controls will be found. However, not all peritidal deposits occur on flat-topped platforms or shelves, and recognizing such orbitally-forced cyclicity will be more difficult in sedimentologically more complex settings such as ramps.

4.4 LACUSTRINE CARBONATES

4.4.1 Introduction

Lacustrine carbonates have received relatively little attention from sedimentologists when compared to marine carbonates. This is not surprising because carbonate lake deposits are, volumetrically, of minor importance in the geological record. However, there is increasing interest in these carbonates for not only can they provide very detailed information on palaeo-environments, but they also act as repositories for non-metallic minerals and can contain major hydrocarbon source rocks.

Lakes are dynamic systems, and are particularly sensitive to subtle changes in climate such as fluctuations in precipitation (run-off). This is in contrast to most marine environments which are buffered both physically and chemically from such minor changes. Lacustrine environments, therefore, are much less stable and as a result lacustrine facies patterns are more complex vertically and laterally than those of most marine deposits. Lacustrine carbonates need to be studied on a much finer scale than is usual in marine limestones.

Compared to marine environments, lakes have a much less diverse biota, but biological and chemical processes are much more intimately linked than in the marine realm. Much of the carbonate precipitated in lakes is biogenically induced and there is no direct analogue for many of these processes in marine environments. Perhaps it is in lakes, more than in any other carbonate environment, that the integration of studies of biofacies, lithofacies, petrography, chemistry and diagenesis are essential in order to correctly interpret the environmental record. In addition, there are physical hydrological factors in lakes such as stratification, which again make them more complex sedimentary systems than most marine environments. These all make the study of lacustrine carbonates a particularly challenging field of research.

4.4.2 General settings

Carbonate lakes occur in a variety of tectonic settings and exhibit a great variety of shapes and sizes. Descriptions of present-day carbonate lakes range from very small systems such as Green Lake (Fayetteville, New York) which covers only some 4.3 km^2, to large lakes such as the Great Salt Lake in Utah, which in historical times has covered over 6200 km^2. The several lake basins whose deposits constitute the Eocene Green River Formation of the western USA covered at least 100 000 km^2.

While lakes have a wide variety of origins, large lake systems are mainly tectonic in origin. Extensional tectonic regimes such as rift systems frequently contain thick lacustrine deposits. The lake systems of present-day East Africa are such an example but major lacustrine deposits also occur in the Mesozoic rift systems associated with the opening of the South Atlantic (Brice *et al.*, 1980; Bertani & Carozzi, 1985).

Rapid subsidence rates are common during these passive margin rift phases and very thick sequences may develop. Lakes are also developed along major strike-slip zones such as the present-day Jordan Valley. The thick lacustrine sequences of the Pliocene Ridge Basin of California also developed in a similar setting (Link & Osborne, 1978). Lake basins such as that around Lake Chad in Africa and Lake Eyre in Australia have developed in slowly subsiding zones in cratonic areas. Lake carbonates also occur in foreland basins such as the Early Tertiary basins of the north-central USA.

4.4.3 Hydrology

Lakes may be classified as to whether they are hydrologically open or closed. Open lakes have permanent outlets and are characteristic of exorheic regions where rainfall returns to the sea through the river system. Inflow from the surrounding drainage basin and precipitation are balanced by evaporation and outflow. The result is a lake with both a *relatively* stable shoreline and a stable lake chemistry, with no tendency towards increased salinity or alkalinity. The shorelines may still fluctuate, sometimes considerably as in the case of Lake Malawi (Beadle, 1974).

Hydrologically closed lake systems have no regular outlet and lake levels and chemistries are controlled by the balance between inflow, precipitation and evaporation. Closed-system lakes occur in endorheic and arheic regions. In the case of the former the drainage, if permanent, enters a terminal lake; in the case of the latter no permanent surface drainage occurs (Bayly & Williams, 1974). Two main types of closed lake are recognized: perennial and ephemeral lakes. Perennial lakes contain bodies of water which may last for years or even thousands of years. Such lakes typically show marked fluctuations in lake level. Lake Chilwa in Malawi is a shallow (less than 3 m deep) perennial lake covering 2000 km^2 (Lancaster, 1979). It undergoes seasonal variations in lake level of up to 1 m a year, with a periodic variation of 1−2 m over a 6 year cycle, and is desiccated on average every 68 years. Such fluctuations result from slight changes in run-off in the lake catchment area caused by changes in evapotranspiration, and in the intensity and duration of rainfall. In areas with complex catchments considerable variability can occur in the run-off regime. Lake Bonneville in Utah has completely desiccated 28 times in the last 800 000 years (Eardley et al., 1973).

With changes in climate, open systems may become closed, passing into ephemeral lakes, and likewise closed systems may become open. Any one region may contain exorheic and endorheic or even arheic systems, as occurs today in Victoria, Australia (Bayly & Williams, 1974). During the Early Tertiary in the western USA, large freshwater lakes occurred in the Uinta Basin, in what is now northeast Utah, while in southern Wyoming ephemeral−perennial, alkaline lakes occurred in the Bridger and Washakie Basins (see Section 4.4.7b). Changes in drainage patterns caused by tectonic movements can radically affect the hydrology of an area, changing both drainage patterns and the local climate.

4.4.3a Stratification

One of the most fundamental properties of lakes is their tendency to become stratified. If lake temperature is measured against depth in temperate, tropical and subtropical regions a curve is usually obtained as shown in Fig. 4.63. The major source of heat is from solar radiation and its warming effect decreases with depth. The density of water is mainly a function of its temperature and is greatest at 4°C. As a result the warmer surface waters are lighter and overlie denser, cooler water. In such a thermally-stratified lake the warmer near-surface zone is called the *epilimnion* and the cooler bottom zone the *hypolimnion*. They are separated by the *metalimnion* which is a zone where the rate of change of temperature with depth is rapid. The *thermocline* is the plane of maximum rate of temperature change. The lighter, surface waters are easily mixed by the wind and commonly undergo daily and seasonal circulations. This zone also undergoes

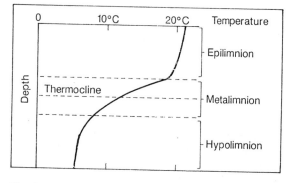

Fig. 4.63 *Temperature distribution and zones in a thermally-stratified lake. Modified from Bayly & Williams (1974).*

free exchange with the atmosphere. However, the hypolimnion does not mix with the surface layer and may become depleted in oxygen. The stability of the stratification depends on the size and shape of the lake. However, cooling of the surface waters, especially in climates with a cool season, can lead to overturning and mixing. If complete mixing occurs the lake is said to be holomictic. Oligomictic lakes undergo mixing at rare intervals. They are typically small, but very deep tropical lakes are typically oligomictic and warm at all depths, with a small temperature difference. Polymictic lakes either never stratify, or display impersistent thermal stratification, often only diurnal. They occur in a variety of settings and are usually exposed to winds.

Density also depends on the amount of dissolved salts and on the amount of suspended sediment in the lake waters. Once thermal stratification is established lakes may also become chemically stratified, especially in saline lakes where a denser, bottom layer brine may occur. In such lakes the stratification is very stable and very little or no mixing occurs. The less saline surface layer, which circulates freely, is called the mixolimnion and is separated from the lower, denser, more saline zone, the monimolimnion, by the *chemocline* (plane of maximum rate of change in salinity). The stability of such systems is often enhanced by dilution of the surface layers by fresh runoff or precipitation. This serves to increase the density gradient reducing the amount of turbulent exchange with the mixolimnion. Such a process, known as ectogenic meromixis, has been offered as a mechanism operating during the deposition of some Eocene lake deposits (see Section 4.4.7b). The remarkable stability of this system is illustrated by the fact that little heat is able to escape upward from the monimolimnion resulting in bottom water temperatures as high as 56°C, and such lakes are called solar lakes.

The important sedimentological effect of this prolonged stratification is that the bottom waters are stagnant and will become anoxic. This results in an absence of a benthos and infauna, and no bioturbation takes place. In addition, the anoxia results in a reduction in the rate of decay of organic matter.

Further details of lake hydrology are to be found in Beadle (1974) and Wetzel (1975).

4.4.4 Environments

Lacustrine environments are largely defined on biological criteria and, as such, they are often difficult to recognize in sedimentary rocks. The classification which follows relates to open and perennial lakes and four main zones can be recognized (Fig. 4.64).

Littoral zone. This is generally taken as the zone of rooted macrophytes and it may extend to depths of 12 m or more, therefore below wave-base. The lower littoral zone is typically colonized by submerged plants such as charophytes, the middle littoral by floating-leafed plants, and the upper littoral by emergent macrophytes such as reeds. The eulittoral zone is the area between the highest and lowest water levels and above this is the supralittoral zone, which is rarely submerged.

The recognition of some of these zones in ancient lacustrine deposits is difficult, if not impossible.

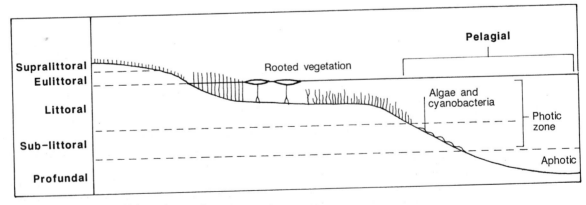

Fig. 4.64 *Subdivision of lake environments.*

Sublittoral zone. This is also called the littoriprofundal zone. It is a zone with fewer green plants, but is still within the photic zone, and may be colonized by lower plants, especially algae and cyanobacteria.

Profundal zone. This is the aphotic zone and typically, but not always, corresponds to the zone below the thermocline.

Pelagial zone. This is also called the limnetic zone and is the open water, planktic zone.

4.4.5 Sedimentary and biological processes

Physical, chemical and biological processes all influence sedimentation in carbonate lakes. The nature and roles of these processes are quite different from those in marine environments.

4.4.5a Physical processes

Wind is the most important physical process in lakes. Water movement is predominantly wind induced, with tidal processes being unimportant in even the largest lakes. Surface, wind-generated waves both effect sediment movement and also lead to turbulence in the epilimnion causing mixing. In wind-exposed shallow lakes wave action may affect the lake bottom resulting in polymixis. In shallow waters, wave action can result in sufficient sediment movement to deter rooted plants such as charophytes, which typically occur in deeper, less agitated bottoms along lake margins. The characteristics of wind-induced surface waves depend on a number of factors (Håkanson & Jansson, 1983) and it is possible to use wave-formed sedimentary structures to estimate the depths and sizes of ancient lakes (Allen, 1981).

Wave-built terraces are a common feature of some lakes and have been recognized in ancient carbonate lake deposits (Section 4.4.7; Swirydczuk *et al.*, 1980). Wave processes in lakes can also result in typical shoreline features such as bars, but these have not been well documented from carbonate lakes. Wave action is also important in the generation of various types of coated grains (Section 4.4.6b).

There are a variety of types of currents in lakes, of which wind-driven ones are the most important. Continued wind stress, causing the piling up of waters in downwind areas, produces return currents which may be pulsed to give lake level oscillations or seiches. Currents may also result from the warming of shallow, nearshore waters or by influxes of river water. These influxes of frequently denser, sediment-laden currents do not always mix with the lake waters but may flow as density currents within the lake waters. A variety of types of flow can occur (Fig. 4.65) such as over-flows, underflows (if the inflow is denser than the hypolimnion) or even as interflows (along the thermocline if the flow is denser than the epilimnion but lighter than the hypolimnion).

Near river deltas, clastic material can be added more or less continuously as river plumes. In addition, turbidity currents, i.e. episodic downslope movements of sediment-laden waters, are especially common. They are also important along steep-sloped lake margins and redistribute material deposited on the slopes. These flows can occur on slopes as low as 5° and result in graded laminae, especially in the profundal zone. In the small, meromictic Green Lake in New York State, locally 50% of the sediment in the profundal zone is resedimented littoral carbonate (Ludlam, 1974; Dean, 1981).

Sediment gravity flows are also important in the formation of littoral benches (wave platforms) which are a prominent feature of many lakes (Section 4.7.7a). These benches prograde into shallow lakes as a result of littoral carbonate being transported across the littoral bench and deposited on migrating bench slopes by sediment gravity flows. In the small, temperate lake basins described by Treese & Wilkinson (1982) from Michigan, allochthonous blocks of littoral carbonates were emplaced by slides into deeper parts of the basin and comprise a significant proportion of the profundal sediments.

Further details of the physical processes affecting lakes are to be found in the reviews by Sly (1978) and Håkanson & Jansson (1983).

4.4.5b Chemical processes

A discussion of the chemical processes in carbonate lakes needs to consider both calcium carbonate deposition in hard-water lakes, and also the deposition and evolution of brines in hydrologically closed systems.

Calcium carbonate in lake sediments has four sources (Jones & Bowser, 1978).
1 Detrital carbonate derived from the hinterland by rivers and by shoreline erosion. This will include reworked lacustrine carbonates exposed during falls in lake level.
2 Biogenic carbonate derived from the skeletal remains of various organisms such as molluscs, charophytes and phytoplankton.

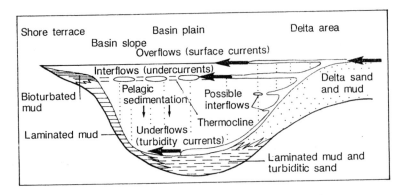

Fig. 4.65 *Sediment dispersal mechanisms and lithofacies for an oligotrophic lake with annual thermal stratification. Based on Sturm & Matter (1978).*

3 Inorganically-precipitated carbonates, much of which is actually biogenically induced.
4 Diagenetic carbonate produced by post-depositional alteration of other carbonate minerals.

The chemistry of calcium carbonate is discussed in Chapter 6, and it has been noted that both temperature and CO_2 pressure are important controls on the precipitation of calcium carbonate. Raising the temperature or lowering the CO_2 pressure will result in precipitation, but the degree of supersaturation resulting from increased temperature is slight so that the removal of CO_2 is the main cause of precipitation in lakes. Temperature-induced precipitation may be more important in the littoral zone of lakes where greater diurnal and seasonal temperature fluctuations occur. Precipitation may also result from supersaturation caused by spring overturning in temperate lakes when the cold hypolimnion water is brought to the surface and warmed rapidly. Temperature-induced precipitation in Fayettville Green Lake in New York, a hardwater temperate lake, is caused by spring and summer warming and results in an annual sedimentation rate of $CaCO_3$ of hundreds of grams per square metre per year (Brunskill, 1969; Ludlam, 1981). Calcium carbonate can be precipitated (or biogenically induced) in the pelagial surface waters to produce 'whitings' (Neev & Emery, 1967; Strong & Eadie, 1978).

Natural degassing to the atmosphere is a slow process and is of minor importance in removing CO_2 from lake waters. The most important process removing CO_2 is photosynthesis, resulting in biologically-induced precipitation of $CaCO_3$. This typically occurs in late spring and summer when photosynthesis is high. Usually precipitation does not occur when supersaturation is reached and precipitation may actually take place when supersaturation is up to 10 times the theoretical saturation level. Proving that precipitation is biologically induced is difficult but it has been invoked by many workers; Mengard (1968) for example, found a linear relationship between the rate at which carbon was fixed by photosynthesis and the rate of depletion of Ca and alkalinity in the lake water, presumably reflecting the precipitation of $CaCO_3$.

Biological induction is not only related to the removal of CO_2 but some organisms utilize bicarbonate directly for photosynthesis, for example charophytes (which typically are more heavily encrusted by $CaCO_3$ than other plants) and some phytoplankton also utilize bicarbonate. This is often necessary because in some lakes CO_2 is depleted more rapidly than it can be replaced from the atmosphere (Wetzel, 1975).

Degassing may also be caused by agitation in wave-influenced areas resulting in mobile sediment grains being coated by $CaCO_3$. Springs emerging in the littoral zone may precipitate carbonate as their waters are warmed or CO_2 pressure is lowered by degassing. A variety of types of encrustation may occur (Risacher & Eugster, 1979), and beachrock may form (Binkley *et al.*, 1980).

Calcium carbonate may also be precipitated where Ca-rich waters flow into CO_3-rich lake waters, or vice versa. This may occur either through springs or by river inflow; the latter may form chemical deltas or plumes. A variety of features result such as tufa or travertine pinnacles (Fig. 4.66; Cloud & Lajoie, 1980; Hillaire-Marcel & Casanova, 1987) or coated grains (Eugster, 1980; Popp & Wilkinson, 1983).

The crystal size and form of the precipitates are influenced by the degree of saturation. In a study of the carbonate precipitates in Lake Zurich, Kelts and Hsü (1978) found that larger, and less well-formed, crystal precipitates formed during the phases of lower supersaturation, and that the smallest crystals were precipitated during the phases of highest supersaturation.

As the precipitates settle out they may undergo dissolution in the undersaturated hypolimnion. Crystal rounding and etching may occur (Brunskill, 1969; Kelts & Hsü, 1978), which may even prevent some precipitates ever reaching the lake floor. However, organic coatings on some crystals may retard dissolution (Otsuki & Wetzel, 1974; Wright et al., 1980). Further dissolution may also occur in the sediment (Kelts & Hsü, 1978).

In open-system lakes with low salinities, the most common mineral precipitate is low-Mg calcite (Chapter 6). The occurrence of other calcium carbonate minerals depends on the Mg/Ca ratio (Müller et al., 1972; Last, 1982). High-Mg calcite is precipitated in waters with a Mg/Ca ratio of 2–12 with the Mg content of the calcite controlled by the Mg/Ca ratio of the waters (Müller & Wagner, 1978). Dolomite is precipitated when the Mg/Ca ratio is between 7 and 12, but may form under lower Mg/Ca ratio conditions if the salinity is low, so that other ions may not interfere with the lattice ordering. Some of the dolomite may be an early replacement of high-Mg calcite. Aragonite precipitates when the Mg/Ca ratio is greater than 12.

Detailed reviews of the geochemistry of calcium carbonate in lakes are provided by Dean (1981) and Kelts & Hsü (1978).

Brine evolution and products. In hydrologically-closed settings, the lake waters become progressively enriched in dissolved ions, and the salinity increases. The composition of the brine, and its precipitates, ultimately depends on the nature of the bedrock around the lake basin and the type of weathering.

Whatever the bedrock, as the brines reach saturation the first precipitates are the alkaline earth carbonates, calcite and aragonite. The nature of the precipitate depends on the Mg/Ca ratio. The precipitation of Ca, Mg and carbonate affects the subsequent evolution of the brine (Fig. 4.67). If the lake waters are enriched with HCO_3^-, compared with Ca and Mg, at the point of precipitation, the brine will follow path I (Fig. 4.67). Na is the most abundant cation in saline lakes, and with the depletion of Ca and Mg, Na carbonates will precipitate next. Minerals such as trona, nahcolite and natronite will be formed and the brine is said to be 'alkaline'. Such sodium minerals are unique to non-marine settings. If the initial waters have Ca and Mg $\gg HCO_3^-$ after the initial precipitates, the brines become enriched in the alkaline earths but depleted in CO_3 and HCO_3^-. If the ratio of

Fig. 4.66 *Travertine spring mound, Mono Lake, California.*

HCO_3^-/Ca and Mg is low, little carbonate may be precipitated, and the brine evolution will follow path II (Fig. 4.67), resulting in the precipitation of sulphates (gypsum). If the ratio of HCO_3^-/Ca and Mg is nearer unity (path III, Fig. 4.67), carbonate precipitation may be extensive and at first Ca will be removed leading to a progressive increase in the Mg/Ca ratio until high-Mg calcite, dolomite or even magnesite will precipitate. Such a trend has been described from the Great Salt Lake deposits of Utah by Spencer et al. (1981). A change from low-Mg calcite to high-Mg calcite to aragonite was detected in a 0.1 m cored interval representing less than 1000 years (Fig. 4.68).

Further critical points are reached in the brine evolution and the final mineral phases to be precipitated are usually Na and Mg salts. However, other salts may form by reactions between the late-stage brines and earlier-formed minerals (Fig. 4.67). Detailed descriptions of brine evolution have been given

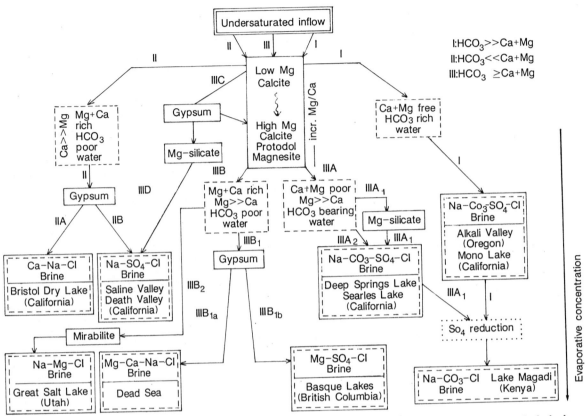

Fig. 4.67 *Flow diagram of brine evolution. The solid rectangles represent critical precipitates and those with dashed borders are typical water compositions. The dash−dot rectangles enclose the final brine types and examples of salt lakes. After Eugster & Hardie (1978) and Allen & Collinson (1986).*

by Eugster & Hardie (1978) and Eugster (1980).

The brine precipitates may form in four main settings: (1) in perennial saline lakes where the precipitates will settle and accumulate on the lake bed, (2) in ephemeral salt pans, (3) as efflorescent crusts on the margins of saline bodies, and (4) as displacive growth within saline mud flats.(Section 4.4.7b). However, not all these products have a high preservation potential. Efflorescent crusts form by updraw from saline groundwater but are easily dissolved with the next storm flooding or rain shower. Displacive precipitates form in the near-surface layers in a similar way to evaporites in sabkhas (Section 4.3.3g; Handford, 1982), and even though they may be replaced during burial, they frequently leave distinctive fabrics indicating their displacive origin. Ephemeral salt pan deposits will have a low preservation poten-

tial also and may be dissolved during the next flooding cycle. Lowenstein & Hardie (1985) have reviewed criteria for their recognition.

The most soluble precipitates in perennial lakes may undergo dissolution as they descend through an undersaturated water zone, and can only accumulate when the whole brine body is at saturation for that mineral. Early diagenetic changes can also occur, such as in the Dead Sea where the surface waters precipitate aragonite and gypsum but the deeper sediments only contain aragonite and some calcite. This has been interpreted as reflecting the reduction of gypsum by sulphate-reducing bacteria to produce H_2S in the bottom waters where the calcium, so released, forms calcite (Neev & Emery, 1967; Begin *et al.*, 1974).

Iron carbonates are commonly found in ancient

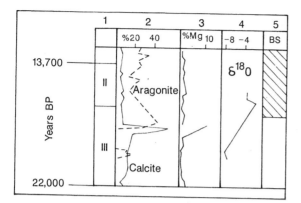

Fig. 4.68 *Geochemical and biological changes in Great Salt Lake, Utah reflecting a drop in sea-level and increase in salinity (closed system) after 15 000 years* BP. *Column 1, sedimentary units; column 2, weight per cent of carbonate; column 3, composition of calcite, mole % MgCO₃; column 4, oxygen isotopic composition of the carbonate; column 5, appearance of brine shrimp (BS). Calcite predominates in the lower part of the sediment interval when lake waters were fresh. The increase in the Mg content of the calcite heralds increased salinities followed by the appearance of the brine shrimp. The δ¹⁸O value reflects increased evaporation. Based on data in Eugster & Kelts (1983) and Spencer* et al. *(1984).*

lacustrine deposits but are apparently less common in present-day lakes. The main minerals are siderite and ankerite, and are particularly common in the Eocene Green River Formation sediments of the western USA (Desborough, 1978). Mg siderite and Mn carbonates have also been recorded from lake deposits. Siderite is a sensitive indication of the chemical milieu for it requires both low sulphide and low Ca^{2+} concentrations, otherwise the Fe^{2+} is taken up with pyrite and the CO_3^{2-} by Ca^{2+}. It requires a low redox potential (Eh) and forms under reducing conditions; it is often found as nodules which have formed at shallow depth in the sediment in the zone of fermentation (Oertel & Curtis, 1972). Wheat grain-like crystals of siderite have been recorded from the Late Neogene deposits of the Black Sea (Hsü & Kelts, 1978) and have been interpreted as lake precipitates and not as diagenetic products. Further details of other carbonate minerals may be found in Kelts & Hsü (1978) and Dean & Fouch (1983).

4.4.5c Biological processes

The influence of the biota on lake sedimentation is even more important than it is in marine carbonate

environments. Plants are particularly important in this respect and produce carbonate by biologically-controlled calcification and by inducing precipitation. In addition, they contribute organic matter to the sediment, often in sufficient quantities to produce potential hydrocarbon source rocks.

The main sediment contributors by direct precipitation are the charophytes, such as the extant forms *Nitella* and *Chara*, which are small aquatic algae. Charophyte remains are ubiquitous in many ancient lake deposits and occur as both calcified reproductive structures and as plant stem encrustations. The female reproductive structure, the oogonium, develops a calcareous outer cover called the gyrogonite (Fig. 4.69). This is usually composed of low-Mg calcite but in saline lakes it may be composed of high-Mg calcite (Burne *et al.*, 1980). Charophytes prefer muddy substrates where they may form extensive meadows but are less common on wave-agitated, coarser-grained substrates. Depending on energy levels and substrate, charophytes inhabit depths usually down to 10–15 m (Stross, 1979). Rather like sea-grasses, charophytes also baffle and trap fine mud as well as produce considerable quantities of carbonate. The stems become encrusted with carbonate and the charophytes are particularly prone to this encrustation because they are capable of using bicarbonate ions for photosynthesis as well as CO_2. They are often more heavily encrusted than other plants because they are more efficient users of bicarbonate as a source of CO_2 (Wetzel, 1975). This gives them an advantage in hard-water lakes where little free CO_2 may be available. Precipitation on the plants is also induced by epiphytic micro-organisms (Allanson, 1973). As a result of differing degrees of external encrustation, and with some internal calcification of the stems, a variety of different carbonate morphologies can result (Schneider *et al.*, 1983). The gyrogonites are relatively resistant to compaction but may disintegrate into small rhomb-like segments (Freytet & Plaziat, 1982); the stems readily disintegrate to form micritic 'marls' which are a very common sediment type in carbonate lakes (e.g. Terlecky, 1974). Charophytes may contribute several hundred grams per square metre per year of fine-grained carbonate (Dean, 1981). While this encrustation is usually low-Mg calcite, aragonite has also been recorded (Müller, 1971). Since these stem encrustations disintegrate easily they do not survive significant transport and if found, well preserved and in quantity in sedimentary rocks, should provide a reasonable indication of shallow water (under 15 m). The gyrogonites, however, are easily transported,

Fig. 4.69 *(A)* Chara, *living plant showing stem, node and gyrogonite. (B)* Porochara, *gyrogonite; fossil form. Photos courtesy of Andrew Leitch.*

especially if desiccated, after which they can float (A. Leitch, pers. comm., 1985).

Charophyte remains in sedimentary rocks were traditionally regarded as good evidence for freshwater environments, but some living forms are able to survive in saline environments with salinities up to 70‰ (Burne *et al.*, 1980). Similar, salinity-tolerant, forms may have occurred in brackish environments in the Middle Jurassic (Feist & Grambast-Fessard, 1984). Racki (1982) has reviewed the occurrence of charophytes in the geological record and regards their present, mainly freshwater, distribution as probably not being representative of their past distributions.

Some microscopic algae also directly calcify such as the coccolith *Hymenomonas* (Hutchinson, 1957) and the chlorophytes *Phacotus* and *Coccomonas* (Kelts & Hsü, 1978; Müller & Oti, 1981). The latter two forms leave calcareous remains of the loricae, a hard protective shell. *Phacotus* has a geological record back to the Upper Miocene and may have been able

to tolerate saline conditions (Müller & Oti, 1981). The significance of these planktonic algae to the sediment budget of carbonate lakes has yet to be fully evaluated but they may be of only minor importance to profundal sediments. This is in marked contrast to the marine realm where biogenic material is the major source of carbonate pelagic deposits (Chapter 5).

The microflora play a crucial role in the biogenic precipitation of calcium carbonate as has already been discussed, but plant and organic matter may also accumulate to form thick peats around lakes or as organic-rich sediments (or gyttia). Sapropel-rich layers may develop in the profundal zone with the organic matter mainly derived from planktonic algae. Such organic-rich sediments are the precursors of source rocks and there are many examples of lacustrine source rocks in the geological record. Many such deposits have developed in deep, stratified lakes such as those which develop in rift systems (Brice *et al.*, 1980; Demaison & Moore, 1980), but others represent shallow, saline–alkaline lakes such as the Eocene Wilkins Peak Member of the Green River Formation of Wyoming (Eugster & Hardie, 1975), the Cambrian Observatory Hill Beds of the Officer Basin of Australia (White & Youngs, 1980) or the Cabacos of the Upper Jurassic of Portugal (Section 4.4.7b; Wright & Wilson, 1985). In shallow, saline–alkaline lakes planktonic and benthic microbial masses are very common and frequently have very high productivities (Bauld, 1981; Kelts, 1982) and are the potential source rock formers.

The flora of lakes (including cyanobacteria) are also important in the formation of bioherms (tufa mounds, stromatolites, oncoids), which are a very common feature of lakes (see Fig. 4.70). They form extensive carbonate deposits in both temperate hard-water lakes and saline–alkaline systems. There are many examples in present-day to subrecent lakes (Dean & Eggleston, 1975; Halley, 1976; Osborne *et al.*, 1982; Cohen & Thouin, 1987; Hillaire-Marcel & Casanova, 1987), and from the geological record (Donovan, 1975; Surdam & Wray, 1976; Riding, 1979; Buchbinder, 1981; Abell *et al.*, 1982; Freytet & Plaziat, 1982; Dean & Fouch, 1983). These bioherms may range from a few centimetres to many metres thick, covering large areas; they may form thick coatings on bedrock outcrops, or may form ledges which build out over the lake along steep lake margins (Eggleston & Dean, 1976). Such bioherms commonly occur in the littoral and sublittoral zone and may extend to depths of 10 m or more. Active stromatolite growth has been recorded at 60 m depth (Gow, 1981). Often the bioherms develop in slightly deeper waters not colonized by higher plants, but their position is dependent on many factors and they are often best developed on shorelines more exposed to sunlight. Far less is known about the influences on their form and distribution as compared to marine stromatolites (Cohen & Thouin, 1987; Hillaire-Marcel & Casanova, 1987), but they may be equally useful for palaeo-environmental interpretation as they have proved in the Eocene Green River Formation (Surdam & Wolfbauer, 1975).

The main organisms responsible for their formation are cyanobacteria (such as present-day genera *Rivularia*, *Schizothrix* and *Scytonema*) and green algae such as *Chaetophora* and *Cladophorites* (Pentecost, 1978; Osborne *et al.*, 1982; Schneider *et al.*, 1983). The actual constructive process firstly involves the trapping and binding of carbonate, and many micro-organisms are able to trap and bind sediment because of their filamentous growth and by virtue of possessing mucous sheaths. Other organisms become encrusted by carbonate, presumably as a result of photosynthetic removal of CO_2, and additional cementation may occur in the local microenvironment (Eggleston & Dean, 1976; Halley, 1976; Riding, 1979; Schneider *et al.*, 1983). The resulting fabrics are typically highly porous but this is dependent on the nature of the calcification. Some forms are composed of dense growths of micritic calcified tubes only a few microns in diameter, representing the encrusted or calcified sheaths of the micro-organisms. The resulting fabric is referred to as a porostromate fabric (Monty, 1981). Bioherms in saline lakes commonly have a less regular, clotted texture which consists of radial clusters of aragonite needles, and may result from the activities of bacteria or unicellular (coccoid) cyanobacteria (Monty, 1976; Buchbinder, 1981). Porostromate fabrics are common in marine stromatolites in pre-Tertiary sequences but appear to be limited to non-marine, or rarely brackish, environments since that time (Richter *et al.*, 1979).

Plants and various micro-organisms are not the only organisms involved in bioherm formation. In the Tertiary lacustrine carbonates of Languedoc (France) mounds up to 10 m high and 0.5–1 m in diameter were formed by algal encrustation on phrygan larval tubes composed of agglutinated hydrobid gastropod shells (Bertrand-Sarfati *et al.*, 1966).

One of the best documented examples of lacustrine bioherms comes from the Upper Miocene Ries Crater of southern Germany (Riding, 1979). The deposits occur in a shallow, circular depression 20–25 km in diameter, which represents an astrobleme. The crater has an inner crater 10 km in diameter which is filled by over 300 m of sediments, including laminated clays and marls which are bituminous and dolomitic. Around the inner basin is a marginal zone some 7 km wide with less than 50 m of lake sediments.

Thick tufa bioherms only occur in this marginal zone, which reach heights of 7 m, and are up to 15 m across. They were built by the green alga *Clado-phorites*, which was externally encrusted by micrite during growth. The coalescence of the encrusted tufts, incorporating detrital material such as peloids, led to the formation of nodules and cones of tufa. Compound cones, which reach heights of 2 m, are grouped to form the large bioherms (Fig. 4.70). During its development the lake basin passed initially from a closed-system to an open-system lake, during which time the bioherms developed in fresh or slightly brackish conditions. Fluctuations in lake level led to phases of emergence and submergence, and during the former, sinter crusts developed and vadose cementation occurred.

Fig. 4.70 *(A)* Cladophorites *cones from lacustrine bioherms, Upper Miocene, Ries Crater, southern Germany. These are compound cones formed by the coalescence of smaller tufts of calcite-encrusted* Cladophorites *(a green alga). These bioherms have undergone a complex diagenetic history including dolomitization. View is 1.4 m wide. (B) SEM photomicrograph of broken* Cladophorites *tubes. The calcitic encrustation occurred during the life of the plant, probably caused by the photosynthetic uptake of* CO_2. *Scale bar is 200 μm. From Riding (1979). Photos courtesy of Robert Riding, University College, Cardiff.*

Oncoids are a particular type of biogenic encrustation and these are discussed in Section 4.4.6b.

The sedimentological role of the lake fauna is also important. Molluscan and ostracode remains are important sediment types in lakes. Such remains may also provide important information of palaeo-environments but studies are still few in number. Ostracodes provide very sensitive indicators of lake chemistries, and so may be useful for reconstructing lake palaeohydrochemistry, at least as far back as the mid-Tertiary (Forester, 1983, 1986; Cohen & Nielsen, 1986).

Arthropods are a common component of lake faunas and the brine shrimp *Artemia salina* is especially common in saline lakes, and pelletization of aragonite muds by the shrimp is an important process. Faecal pellets (peloids) are a common component of many ancient lacustrine carbonates, but compaction of the pellets frequently obliterates their structure such that they are commonly only preserved in shelter pores. Ingestion of pelagial carbonate precipitates by zoo-plankton also results in aggregation and leads to faster settling of the carbonate mud (Brunskill, 1969).

One of the most important biological processes in marine and non-marine settings is bioturbation caused by the benthos and plant roots. Such organisms as chironomid larvae are particularly abundant in the profundal zone of lakes and cause extensive bioturbation in the top 80–100 mm of the sediment. Amphipod crustaceans, oligochaete worms and bivalves are also major burrowers. Detailed reviews of the physical and chemical effects of bioturbation on lake sediments are given by Fisher (1982), McCall & Tevesz (1982) and Håkanson & Jansson (1983). Where the bottom conditions are anoxic, the benthos will be absent, although some soft-bodied burrowing organisms can tolerate low oxygen levels. If the sediment is completely anoxic, no infauna will occur.

Bioerosion is an important process, especially in temperate lakes, and it results mainly from the activities of endolithic micro-organisms and abrasion by grazers. A common product of these processes in temperate lakes are furrowed oncoids and stromatolites (Schneider, 1977; Schneider *et al.*, 1983). Biogenic dissolution also occurs in many lakes (Schröder *et al.*, 1983) and results from seasonal and diurnal changes in CO_2 content (and pH) caused by plant activity.

4.4.6 Sedimentary features

Some sedimentary features are particularly common in lake carbonates and warrant further discussion. These are laminations and coated grains.

4.4.6a Lamination

Fine-scale lamination is a very common feature of lacustrine carbonates but its origins are varied and complex. However, an understanding of these is crucial to appreciate the depositional regime of the lake. Many lacustrine laminites occur as couplets, or even triplets, with the regular superposition of, for example, carbonate or siliciclastic or organic laminae

(Trewin, 1986). These couplets are frequently referred to as varves, but are not varves *sensu stricto*, which represent the seasonal deposits of glacial lakes. Many couplets are *not* seasonal in origin. For couplets or triplets to form there need to be variations in the influx of the components, variations which need not necessarily be seasonally controlled.

In interpreting lamination in lake sediments, three factors must be considered: origin, preservation and diagenesis. As regards origin, there are three sets of processes which cause periodic influxes of sediment into lakes. Physical processes, including sediment gravity flows (turbidites and grain flows), transport sediment either from the littoral zone (Ludlam, 1981) or from sediment-laden river currents. Laminae which result from such processes typically show grading, but reverse grading may also occur if deposition is from grain flows (Swirydczuk *et al.*, 1980). Such laminae will not be seasonal or annual in origin but may reflect episodic events. Couplets of carbonates and siliciclastics are common in many lake deposits and the latter may reflect periods of increased run-off into the lake, as in the case of Lake Lissan (Begin *et al.*, 1974). Sequences of couplets which are regular in thickness may well reflect annual – seasonal variations, but where the thicknesses are irregular, a more episodic origin seems likely. Current activity may also produce lamination in shallow lake waters or on mud flats (Smoot, 1983). Such laminae may contain evidence of scouring, particle alignment and ripples (Spencer *et al.*, 1981, 1984).

Chemical processes also produce lamination, such as the seasonal precipitation of carbonates. During the spring and summer warming, or during mixing caused by overturning, precipitation of carbonate may occur (Dickman, 1985). In perennial saline lakes and salt pans various minerals may precipitate out such as the seasonal calcite and aragonite layers of the Dead Sea or the Pleistocene Lake Lissan. In Dead Sea sediments (Fig. 4.71), couplets consist of light and dark laminae. The white laminae consist of stellate clusters of aragonite needles, $5-10$ μm in diameter. They are present in surface waters in summer and form during phases when the surface water temperatures rise to 36°C. Rarer whitings occur every few years during extremely high temperatures, and result in thick white laminae. The darker laminae consist of clay minerals, quartz grains, detrital calcite and dolomite. Gypsum is also precipitated at the lake surface but is depleted in the surface sediment by sulphate-reducing bacteria (Neev & Emery, 1967; Druckman, 1981). These are therefore seasonal couplets but they may be modified by diagenesis and episodic whitings.

Biological processes are also very important in the formation of layering. During times of surface water carbonate saturation, biologically-induced precipitation may occur. The laminated sediments of Lake Zürich (Kelts & Hsü, 1978) occur in triplets (Fig. 4.72), with a basal organic-rich layer consisting of threads of the filamentous cyanobacterium *Oscillatoria*, iron sulphide pigments, clay-sized detritus, and some calcite and diatoms. This layer represents the autumn–winter phase of settling out from the lake waters. The second layer consists of a lacy frame-

Fig. 4.71 *'Varved' sediments covering stromatolites (lower left) from the west coast of the Dead Sea. The white laminae consist of aragonite precipitated during the summer months. The darker layers consist of detrital quartz, clay minerals, calcite and dolomite. See text and Druckman (1981). Photo courtesy of Y. Druckman, Geological Survey of Israel.*

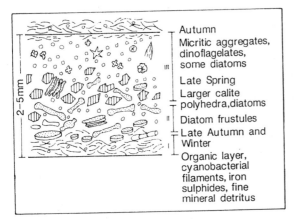

Fig. 4.72 *Schematic diagram showing the composition of a triplet from Lake Zurich, Switzerland. Layer I represents settle out from the lake waters. Much of layer II represents diatom blooms. The decreasing crystal size of calcite in layer III reflects changing saturation levels. Modified from Kelts & Hsü (1978); Allen & Collinson (1986).*

work of diatom frustules, mineral detritus and organic threads. At the top of this layer, large calcite polyhedra (up to 30 μm in diameter) appear mixed with other components. This layer represents diatom blooms. The third layer has a diffuse lower boundary which is less diatomous and contains more calcite. The diatoms represent different species from those in the middle layer and the crystal size of the calcite decreases upwards, giving micrograding which reflects the degree of saturation in the lake waters. This layer represents late spring to summer precipitation following supersaturation.

The preservation potential of lamination is an important consideration. It is most commonly developed in profundal sediments where bioturbation, current activity, slumping and deformation are of minor importance. The delicate lamination will be destroyed by bioturbation by the infauna and by bottom foragers. If the bottom conditions are anoxic (anaerobic) the lamination will be preserved. This is mostly likely to develop in permanently stratified (meromictic) lakes. In holomictic or polymictic lakes, oxygenated bottom conditions may become established long enough for bioturbation to occur. The absence of bioturbation may reflect anaerobic conditions just below the sediment surface, but an epifauna may be present. High salinity conditions also restrict the benthos and laminites develop in very shallow water, such as those of the Dead Sea

(Fig. 4.71), which are deposited in only a few metres of water and are associated with stromatolites.

Currents can destroy fine lamination as well as form it. Lake Urmia in Iran, a large, perennial, hypersaline lake averaging 12 m deep and covering 5000 km² has laminae of aragonite and organic matter. These do not form regular couplets because wind-generated return currents, flowing along the bottom, destroy the couplet. These couplets are, however, well preserved during high evaporation phases when a pycnocline forms, protecting the bottom from currents (Eugster & Kelts, 1983).

Lamination may be deformed in a number of ways. On steeper slopes slumping may occur, and compaction can result in loop bedding and sedimentary boudinage (Dean & Fouch, 1983). Organic-rich 'sludges' will readily creep even on gentle slopes. Seismic activity may also disrupt lamination (Hesselbo & Trewin, 1984). Millimetre-sized gas mound structures also occur in lake sediments and are due to gas bubble generation (Donovan & Collins, 1978; Kelts & Hsü, 1978). Synaeresis cracks are also common in lacustrine deposits (Donovan & Foster, 1972; Plummer & Gostin, 1981), and the displacive growth of evaporites on saline mud flats also disrupts lamination (Hardie *et al.*, 1978; Smoot, 1983). Microbial binding can aid in preserving lamination.

4.4.6b Coated grains

Coated grains are ubiquitous in carbonate lakes and streams and have a variety of origins. Both biogenic and abiogenic forms occur. The former are produced by cyanobacteria and green algae (Fig. 4.73), and have been described from many present-day lakes (Golubic & Fischer, 1975; Jones & Wilkinson, 1978; Schäfer & Stapf, 1978; Murphy & Wilkinson, 1980; Schneider *et al.*, 1983). Such oncoids typically possess a laminated, porous structure, with layers or tufts of micrite tubes representing the encrusted sheaths of cyanobacteria or green algal filaments ('porostromate microstructure'; Monty, 1981). Their microstructure, morphology, and the continuity and form of the laminae can be useful for palaeoenvironmental purposes. As a general rule such oncoids occur in shallow, slightly wave-agitated zones, in more exposed areas than charophytes, which prefer muddier substrates (Schneider *et al.*, 1983). Bioerosion is a common feature of oncoids and leads to furrowing (Jones & Wilkinson, 1978; Schneider *et al.*, 1983). Oncoids have been recorded from a number of ancient lacustrine sequences (Williamson & Pickard, 1974; Link

Fig. 4.73 *Lacustrine oncoids, Mambrillas de Lara Formation, Lower Cretaceous, northern Spain. Note the nucleus is an oncoid fragment. The asymmetry of the oncoid reflects stationary growth. Scale divisions in centimetres. Photo courtesy of Nigel Platt, University of Bern.*

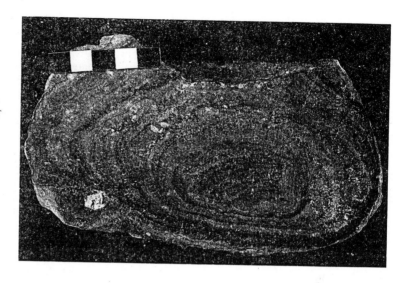

et al., 1978; Anadón & Zamarreño, 1981; Monty & Mas, 1981; Freytet & Plaziat, 1982; Nickel, 1983).

Ooids and pisoids typically form in wave-influenced zones and their mineralogy varies with lake chemistry. Low-Mg calcite ooids occur in lakes with low Mg/Ca ratios but there are relatively few records of such ooids from dilute lakes (Wilkinson *et al.*, 1980). Aragonitic ooids are common in saline lakes, such as the Great Salt Lake, Utah (Kahle, 1974; Sandberg, 1975; Halley, 1977) and Lake Urmia, Iran (Eugster & Kelts, 1983). Bimineraltic ooids, composed of low-Mg calcite and aragonite, have been recorded from Pyramid Lake, Nevada by Popp & Wilkinson (1983). These have formed where Ca-rich spring water mixes with alkaline lake waters. Popp & Wilkinson have suggested that calcite precipitates when the mixing waters are dominated by spring waters (low Mg/Ca ratio), and aragonite when the alkaline lake waters are dominant (higher Mg/Ca ratio).

One of the most detailed descriptions of an ancient lacustrine oolite deposit is the Pliocene Glenns Ferry, Shoofly Oolite of the Snake River Plain, western USA (Section 4.4.7a; Swirydczuk *et al.*, 1979). The oolite has been interpreted as having formed in a low-salinity lake, and the now calcitic ooids show high textural preservation suggesting an original calcitic composition. However, Swirydczuk *et al.* (1979) interpreted the ooids as having been originally aragonite, possibly formed in a low-salinity, high-alkalinity lake comparable to Pyramid Lake in Nevada. Sandberg (1980) has questioned this interpretation, preferring a calcitic precursor for the ooids. It is important to appreciate these arguments if ooid

mineralogy, deduced from fabric preservation, is to be used as an indicator of original lake chemistry in ancient lake deposits.

While ooids may be largely physicochemically precipitated, some of those in the Shoofly Oolite have irregular crenulated outer cortices and their accretion may also have been microbially influenced (Swirydczuk *et al.*, 1979).

4.4.7 Facies models

The study of lacustrine facies models in carbonate lakes is at an early stage in its development with few detailed reviews. Lakes are dynamic systems, especially susceptible to fluctuations in climate, and they are highly variable in their tectonic and sedimentological settings. As a result there is no single set of reliable criteria for recognizing lacustrine deposits. The biota is probably the most reliable feature to use and some minerals are diagnostic of non-marine settings (Section 4.4.5b). Sedimentary structures are less useful and, as already shown, features such as lamination must be interpreted with care. Picard & High have reviewed both the criteria for recognizing lake deposits (Picard & High, 1974) and also their physical stratigraphy (Picard & High, 1981).

Before lake facies are discussed in detail some general features of lacustrine sequences must be stressed. Environmental gradients in lakes are generally marked and this fact, combined with the dynamic nature of lake systems, leads to one of the most characteristic features of lake deposits, that is, *sharp facies changes*. While this is not a feature of any use for recognition of lake deposits, it is worth remembering, and although idealized, transgressive or regressive facies models can be generated for different types of lake settings, the norm in the geological record is for complex and rapid facies changes.

Three main categories of lakes are recognized: hydrologically-open lakes, hydrologically-closed perennial lakes, and hydrologically-closed ephemeral lakes (Section 4.4.3; Eugster & Kelts, 1983). However, as lake systems develop, and as climate and drainage fluctuates, lakes may alter in their hydrological balance. This has certainly been the case in the geological record where there are many examples of this. The Miocene Ries Crater of Germany began as a closed-system lake but developed into a fresh-to-brackish hydrologically open lake (Füchtbauer et al., 1977). The Palaeocene–Eocene Lake Flagstaff of central Utah began as a shallow, freshwater lake (during a pluvial phase), developed into a playa system, and then (during a second pluvial phase) once again became a freshwater lake (Wells, 1983). Hence, when interpreting ancient lake deposits it is often necessary to use several facies models. In addition, lakes are classified on their type of stratification, but this can be difficult to apply to ancient lake deposits. It is often only possible to distinguish between oxic and anoxic bottom conditions which reflect the presence or absence of prolonged thermal or chemical stratification.

4.4.7a Hydrologically open lakes

These are lakes which have an outlet (exorheic), and as a result have relatively stable shorelines. Evaporation and outflow will be balanced by inflow and precipitation, and the lake chemistry will remain dominated by meteoric waters and will be dilute. As a result low-Mg calcite will be the normal precipitate under appropriate conditions. The biota indicates freshwater conditions.

The lake basin (profundal) facies will be controlled by water stratification. If the water column is non-stratified or if anoxia is not permanent, the bottom sediments will show varying degrees of benthic colonization and bioturbation. Sediment gravity deposits may occur.

With reference to the littoral–sublittoral deposits, the recognition of zones based on vegetation type (Section 4.4.4) can be difficult, if not impossible, in ancient lake carbonates. A more pragmatic and sedimentological approach is preferable, using both biological criteria and physical sedimentary structures. In this synthesis, four simple facies models are recognized (Fig. 4.74) in which two categories of lake margin physiography are recognized: steep bench ('shelf') margins, and low gradient ramp margins. Each can be subdivided into low-energy and high-energy (wave-dominated) types.

Bench (steep gradient) margins — low energy (Fig. 4.74A). Carbonate production by plants in the littoral zone can be very high compared to that in the pelagial zone. As a result lake shores can prograde into the lake basin. This produces benches around the lake, whose slopes can be very steep (over 30°) to depths of 25 m or more. Their characteristics depend on energy level (wave action). Around many small temperate lakes with a small fetch and low wave energies, charophyte marls are important deposits (Fig. 4.75A).

Murphy & Wilkinson (1980) have described the facies deposited in and around shoreline benches from Lake Littlefield in Michigan, and this provides a useful model for small carbonate lakes. This lake is a small dimictic lake with a maximum depth of 20 m. The benches can be divided into two zones, the bench platform and the bench slope, each with distinctive lithofacies. The bench slope (Fig. 4.75B), which has a 30° slope, contains two lithofacies. A gastropod micrite dominates its lower part, transitional into the ostracode micrites above, but with a higher carbonate content. It is thinly laminated and contains pyrite. Anaerobic conditions are probably developed just below the sediment surface.

Bivalve and ostracode remains are present but there are no charophyte stems. Evidence of slumping is seen in cores through this zone, which corresponds to the sublittoral zone. Higher up the bench is a sandy micrite lithofacies (Fig. 4.75B) containing sand-sized fragments of encrusted charophyte stems whose concentration decreases with depth. Molluscan debris also occurs. The lower limit of this facies is 4–6 m, representing the lower limit of charophyte growth in this lake. The sediments are structureless.

The bench platform, which lies in 1.5 m of water with a 2° lakeward slope, contains two lithofacies. There is a lower, extensive pisolitic gravel which contains whole and fragmented pisoids, encrusted charophyte fragments and molluscan material, with the content of pisoids decreasing with depth. Charophytes are less numerous than in the upper bench slope sediments, and this probably reflects the coarse substrate. The occurrence of pisoids, in wave-agitated areas in shallower water than charophyte meadows, is a common feature of hard-water lakes (Schneider et

Fig. 4.74 *Schematic facies models for hydrologically open permanently stratified carbonate lakes. (A) Bench margin, low-energy progradation sequence (see Fig. 4.75) based on temperate marl lakes. (B) Strongly wave-influenced, bench margin progradation sequence based on Pliocene Shoofly Oolite (see Fig. 4.76). Skeletal material could replace ooids in other settings and microbial bioherms might also occur. (C) Low-gradient (ramp), low-energy sequence. The low gradients make these systems susceptible to small fluctuations in lake level characterized by alluvial intercalations and extensive pedogenesis. Marsh (paludal) carbonates may also occur (see Fig. 4.77). Such lakes are typically shallow and lack permanent stratification. (D) Low gradient, wave-dominated, barred progradation sequence (see Fig. 4.78). Other variations are possible based on the nature of the shoal belt and the presence of deeper-water microbial bioherms (e.g. Lake Tanganyika; Cohen & Thouin, 1987). T, thermocline. In lakes with prolonged periods of bottom oxygenation the profundal sediments would be bioturbated. These are highly simplified models and other variations should be found, including transitional forms from bench margins to ramps.*

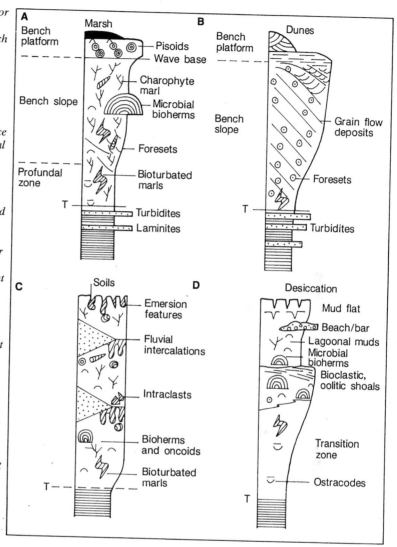

al., 1983). The final lithofacies is a calcareous peat, which consists of pisoid debris with plant and wood fragments overlain by peat. The bench progrades lakeward from the shoreline or from the edges of lakemount islands. The resulting sediment shows a coarsening-up sequence from micritic laminites in the profundal and littoral zone, to sands and gravels in the littoral zone (Fig. 4.75B). Further details of this type of bench shoreline, hard-water temperate lake have been provided by Treese & Wilkinson (1982) from Sucker Lake in Michigan. A similar coarsening-up sequence occurs with bench progradation, although the lithofacies types are more varied. Microbial bio-

herms can comprise a major facies in other low-energy bench margins, in the sublittoral zone (Dean & Fouch, 1983) and reef-sized proportions can be reached (Riding, 1979).

In both Michigan lakes significant resedimentation occurs on the steep slopes, into the profundal areas. This may be by turbidity currents, and in the case of Sucker Lake, large blocks of bench carbonate have slid into deeper waters.

An ancient analogue for a low-energy bench system has been described from the Lower Cretaceous Peterson Limestone of western Wyoming and south-eastern Idaho by Glass & Wilkinson (1980). Much of

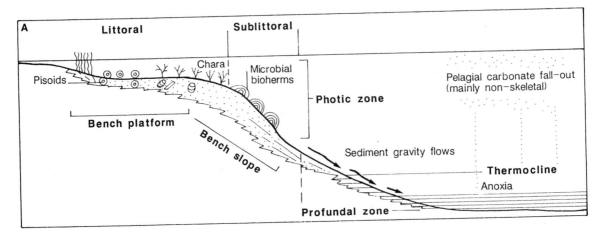

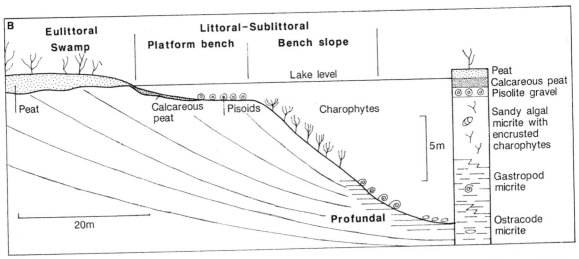

Fig. 4.75 *(A) Simple model of carbonate facies in a low-energy bench margin, hydrologically-open, carbonate-rich lake. The bench slopes may be as high as 39° with vertical heights of up to 25 m or more. In this case the lake is permanently stratified with anoxic conditions in the profundal zone. Charophyte meadows can also occur on the bench slope. (B) Temperate marl bench, Lake Littlefield, Michigan, showing the sequence generated by bench progradation. Modified from Murphy & Wilkinson (1980). Note that in this case no microbial bioherms have developed.*

the lithosome is micritic (low energy) and extensive resedimentation occurred around the bench margins to form graded, silty micrite layers, as well as slumped horizons and diamictites of shallow-water lithologies emplaced by debris flows. Even though the modern analogues for this model are small temperate lakes, the Peterson Limestone covers 20 000 km² and is locally 60 m thick.

Bench margins — strongly wave influenced (Fig. 4.74B). There are very few descriptions of this type in the literature. The best documented example is the Pliocene Shoofly Oolite of the Glenns Ferry Formation, Snake River Plain, northwestern USA (Fig. 4.76; Swirydczuk *et al.*, 1979, 1980). The oolite is up to 35 m thick and can be traced over an outcrop of 45 km. It has been interpreted as a wave-built oolitic bench, with individual bench units up to 12 m in thickness. Three transgressive units occur within

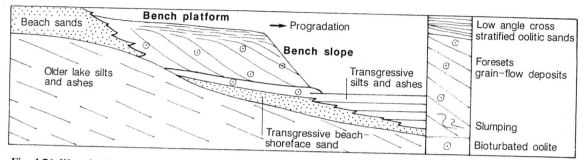

Fig. 4.76 *Wave-built oolitic bench model for the Pliocene Shoofly Oolite, Glenns Ferry Formation. The oolitic benches prograded into the lake following transgressive phases. After Swirydczuk* et al. *(1980).*

the Shoofly Oolite and each consists of two parts, interpreted as bench slope and bench platform deposits. The former unit, which is thicker, consists of lakeward-dipping foresets with an average dip of 26°. The sediments consist of thinly-coated, medium sand-sized ooids. The upper part of the bench slope deposits exhibit reverse grading (suggesting grain flow and avalanching). The lower part is characterized by slumps, dish structures and normally graded laminae. The bench platform exhibits low-angle cross-sets, dipping lakeward at 10° which are transitional into the major foresets. Similar wave-built terraces have been described from Lake Bonneville by Gilbert (1885, 1890).

At the base of each transgressive sequence there is a structureless bioturbated unit up to a metre thick. These have been interpreted as lower-energy deposits, and underwent phosphatization (Swirydczuk *et al.*, 1981).

The simple model offered in Fig. 4.74B is based on the Shoofly example but other wave-dominated sequences could be envisaged, including ones with microbial buildups.

Ramp (low gradient) margins – low energy (Fig. 4.74C). This model envisages very gentle gradients from the eulittoral zone into the lake basin. As a consequence, the littoral zone will be very wide, and even small-scale fluctuations in lake level will result in significant environmental change. Arguably, such fluctuations will occur more readily in hydrologically closed lakes but in this model high salinities are not a common feature. Falls in lake level will expose extensive areas of the lake margins, with desiccation and pedogenic modification. Alluvial intercalations will be prominent in any such sequence. Such lakes may be sufficiently shallow for stratification not to

develop because of wind-generated mixing.

Lacustrine carbonates showing these characteristics have been extensively documented from the Late Cretaceous and Early Tertiary of Europe (Freytet & Plaziat, 1982; Durand *et al.*, 1984; Platt, 1985; Azeredo & Galopim de Carvalho, 1986; Marshall *et al.*, 1988). Similar deposits have been described from the Palaeocene–Eocene of central Utah (Lake Flagstaff) by Wells (1983), and from the Early Cretaceous Draney Limestone of Wyoming and Idaho by Brown & Wilkinson (1981).

Freytet & Plaziat (1982) have beautifully documented one such example from southern France. During the latest Cretaceous and earliest Tertiary, much of southern France (Languedoc) and adjacent areas underwent prolonged periods of continental sedimentation, and locally over 2000 m of continental sediments were deposited. Within these sequences there are a variety of lacustrine and pedogenic carbonates. One interesting feature is the occurrence of pedogenically-modified lacustrine limestones referred to as palustrine limestones by Freytet & Plaziat.

In these lacustrine limestones they recognized three main lithofacies.

1 Micritic limestones (lime mudstone to packstones), which are never highly bioclastic, contain molluscs and charophyte debris (encrusted stems and gyrogonites). These may also be organic rich. The lime mudstones have a clotted appearance resulting from the compaction of pellet muds. The limestones are bioturbated, indicating an absence of permanent stratification and anoxia. The lakes were holomictic or polymictic, although some stratified lake deposits, with finely-laminated sediments containing well-preserved fish, do occur in the post-mid-Eocene lake deposits of eastern Languedoc.

2 The second lithofacies consists of pellet and intraclast grainstones and packstones. The intraclasts are reworked fragments of brecciated lacustrine micrite exposed at the lake margins. The grainstones contain sand- to fine gravel-sized clasts, oncoids and stromatolites built by cyanobacteria, as well as moss tufas.

3 Palustrine limestones consist of pedogenically-modified limestones. The exposure of the lake lime muds resulted in the development of calcareous soils, exhibiting such features as brecciation, root tubules, microkarst, various cavities and internal sediments, iron oxide mottling, terrestrial snails and *Microcodium* (a problematic calcareous pedogenic structure).

These lithologies are found arranged in transgressive—regressive sequences (Fig. 4.77), but more commonly they form incomplete cycles. Some of the exposure-related overprinting, such as root tubes, traverse several sequences. Within the sequences two lithofacies associations are recognized.

1 *Lacustrine association* (Fig. 4.78A,B). These limestones correspond to lithofacies 1. There is generally a low siliciclastic content, which is believed to have resulted from the baffling action of lake margin marshes. However, some deposits do contain mudstones fed from distributaries cutting through the marsh. Exposure features are present, representing falls in lake level.

2 *Palustrine association* (Fig. 4.78C). These are lake margin deposits corresponding to lithofacies 2 and 3.

The Languedoc carbonates represent a predominantly dilute, holomictic—polymictic lake with fluctuating shorelines. Arguably, this latter property is typical of a more closed system, yet even though evaporites did form at times, they were never widespread. The biota and mineralogy indicate freshwater systems which, by virtue of being very shallow with low gradient shorelines, were prone to small-scale fluctuations in lake level.

Ramp (low gradient margins — wave influenced) (Fig. 4.74D). In this model there is a broad shoreline zone, strongly influenced by wave action. The shoreline is characterized by wave-winnowed sands (shell coquinas or ooids), which may be organized into bars. The shoreline does not slope sufficiently for resedimentation to occur or for progradation of a bench.

Cohen & Thouin (1987) have described an approximate modern analogue from Lake Tanganyika, a mildly alkaline lake in the East African rift system (Fig. 4.79). In water less than 2 m deep, shoreline sands give way to bioturbated calcareous silts forming around charophyte meadows, which may extend down

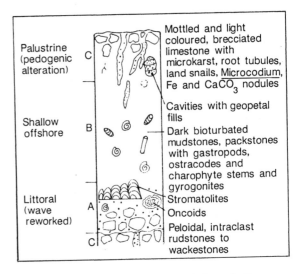

Fig. 4.77 *Transgressive—regressive cycle from the Late Cretaceous—Early Tertiary of Languedoc. (A) Represents the wave reworking of brecciated palustrine limestones at the base of the transgression. (B) A Lower-energy deposit but representing depths of only a few metres. (C) The pedogenically-altered regressive phase. A complete sequence ranges up to 2 m thick. Based on Freytet & Plaziat (1982).*

to 10 m. The baffling action of the plants traps finer sediment. In slightly deeper and more agitated waters, extensive oolitic sands occur, forming belts 1—10 km long and 0.5—1 km wide. These shoals appear to be restricted to coastlines with the lowest gradients. Below fairweather wave-base, about 4 m, the sediment consists of gastropod, ostracode muds or coquinas.

Below 20 m microbial bioherms become more abundant and larger, reaching heights of up to 3 m. These are non-laminated (thrombolites) and are highly porous. Early cementation, probably in the vadose zone as beachrock, is of widespread occurrence. The cements and ooids are high-Mg calcite reflecting the relatively high Mg/Ca ratios.

Ancient examples of wave-influenced ramp margins are known. The Uinta Basin of northeast Utah contains thick sequences of lacustrine deposits formed in fresh to slightly brackish systems, with fluctuating shorelines. The sediments, of Late Cretaceous to mid-Eocene age, occur in the Green River Formation and the Flagstaff Member of the Wasatch Formation. The lake has been interpreted as probably being stratified, with periodically oxygenated bottom waters, and with a depth of 5—30 m (Ryder *et al.*, 1976).

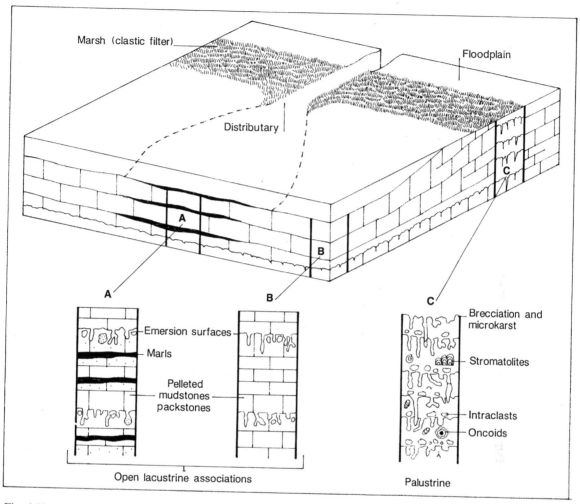

Fig. 4.78 *Lacustrine–palustrine facies associations from the Late Cretaceous–Early Tertiary of Languedoc. Marls are associated with areas near distributary systems (A). Generally the amount of siliciclastic input was low, probably reflecting baffling by shoreline marshes, and areas away from distributaries (B) received little input. Nearshore sequences (C) are comprised of palustrine units (Fig. 4.77) with multiple emergence surfaces. Based on Freytet & Plaziat (1982).*

Two main lacustrine facies association are recognized: open lake and lake margin. Complex intertonguing occurs between the two. The open lake facies, which reach 900 m in thickness in the centre of the basin, consist of dark, organic-rich, mud-supported carbonates and calcareous claystones with thin, horizontal lamination. The carbonate laminae alternate with thin kerogen laminae in couplets, and are interpreted as low-Mg calcite originating from photosynthetically-induced precipitation. The organic laminae have been interpreted as microbial oozes,

and organic matter also occurs as finely-dispersed kerogen in unlaminated beds, and as algal coals with *Botryococcus*.

The lake margin lithosomes vary from 0.3 m to 100 m in thickness, and have traceable widths from several kilometres to more than 50 km. Shallowing-upward sequences can be recognized (Fig. 4.80A). The open lake component consists of kerogenous, mud-supported carbonates with ostracodes. These are overlain by grey, horizontally-bedded, moderately fossiliferous mud-supported limestones. The cycles

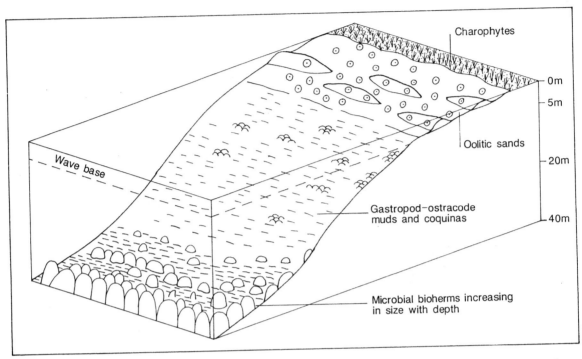

Fig. 4.79 *Schematic diagram of zonation across littoral and sublittoral zones of parts of Lake Tanganyika. Based on Cohen & Thouin (1987).*

are capped by medium-bedded bioclastic, oolitic–pisolitic grainstones with low-angle cross-stratification. Wave and current ripples occur within the upper two units.

Elsewhere in the Green River Formation, Williamson & Picard (1974) have recognized a more complex system with barred shorelines and protected lagoonal environments (Fig. 4.80B).

Another example of a barred shoreline has been documented from the Cretaceous Lagoa Feia Formation, Campos Basin, offshore Brazil (Bertani & Carozzi, 1985). In this case molluscan sands developed as bars during fresher, pluvial phases of the lake. Hydrocarbon reservoirs have been found in leached molluscan limestones.

These simple models are provided only as a rough guide, and much more work is needed to develop usable facies models for interpreting ancient lacustrine carbonates. Other variants to those offered here are possible, and the bench and ramp types are end-members with gradations in between. The nature of the slope will be especially critical in lake basins associated with half-graben structures, with bench margins possibly developed on the footwall sides and ramp margins on the hanging wall side. The differential clastic inputs along these zones may also influence carbonate accumulation.

The openness to wave action will be critical and variations in this can be expected along lake margins. A graphic illustration has been provided by Davis & Wilkinson (1983) from the mid-Tertiary Camp Davis Formation of Wyoming. Along depositional strike, parallel to the lake shoreline, both wave-exposed headlands, along alluvial fan fronts, and protected areas in re-entrants between fans, have been recognized. The former deposits are characterized by oolitic, oncolitic and intraclastic grainstones, while the latter areas were the sites of deposition of root-mottled, charophytic lime mudstones.

4.4.7b Hydrologically closed lakes

The essential property of these lakes is that evaporation exceeds inflow, and no outflow occurs. This results in two features. Firstly, rapid changes in lake

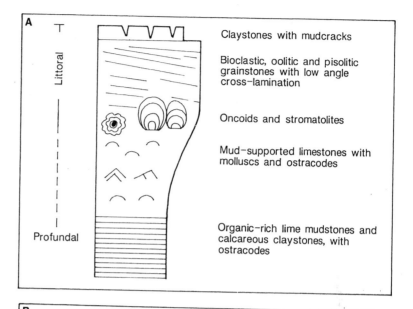

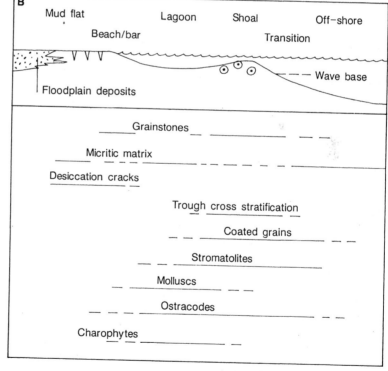

Fig. 4.80 *Green River Formation shoreline deposits. (A) Idealized shallowing-upward profundal–littoral sequence from the Uinta Basin. Thick shoreline deposits contain relatively poorly-ordered packages of these, and other lithologies, reflecting frequent changes in lake level. Wave and current ripples occur in both the upper lithologies. Based on data in Ryder et al. (1976). (B) Simple facies model and selected lithologic properties of shoreline carbonates from the Green River Formation. Based on data in Williamson & Picard (1974).*

level occur-caused by fluctuations in rainfall and run-off, and secondly, the solute content of the water will increase, especially the Mg/Ca ratio, as Ca is depleted by early-stage precipitates. This latter trend is well displayed in the Late Pleistocene–Holocene deposits of

Lake Tanganyika and Lake Kivi in the East African rift system (Stoffers & Hecky, 1978). With time, hydrologically closed perennial lakes may develop into ephemeral lakes and vice versa. Alternatively, tectonic or climatic changes may result in such closed systems becoming hydrologically open. Again, it is stressed that many ancient lake deposits reveal evidence of such hydrological changes.

The broad facies components of open and closed systems will be similar. Stratification may be a consequence of chemical gradients and in highly saline lakes evaporites such as gypsum can precipitate out of surface water and settle to form laminites in the profundal zone. Shoreline deposits will also reflect the degree of salinity and evaporitic mud flats may develop. The biota should be a reliable guide to concentration in ancient sequences, especially ostracodes (Section 4.4.5c). In hypersaline settings, microbial production can be very high, resulting in highly organic-rich sediment (Bauld, 1981). Microbial bioherms are also a significant feature of the littoral and sublittoral zones of such lakes (Halley, 1976).

From a sedimentological viewpoint, the increased concentrations, especially the increased Mg/Ca ratio, will result in the precipitation of high-Mg calcite and aragonite. One peculiarity which can arise in chemically-stratified saline lakes is that a lighter, fresher mixolimnion may contain a freshwater biota while the monimolimnion will be saline. Thus the biota from the lake margin and pelagial zone may indicate dilute water, yet the bottom waters may be hypersaline.

Saline perennial lakes and ephemeral lakes have distinctive shoreline facies. In the latter case evaporites may precipitate out and accumulate on the shallow lake floor, or salt pan (Lowenstein & Hardie, 1985). Around both types of lake there are exposed areas where minerals precipitate out from groundwater, either within or on the sediment. This zone may be extensive, forming a continental sabkha or saline mud flat, while around it is a dry mud flat (or sand flat) with desiccation features and ephemeral salt crusts. The environments and lithofacies of saline lakes have been described in reviews by Eugster & Hardie (1978), Hardie *et al.* (1978) and Eugster & Kelts (1983).

There are several examples of carbonate-rich saline lake deposits in the geological record of which the *Eocene Wilkins Peak Member of the Green River Formation of Wyoming* is perhaps the best documented. The Green River Formation, which outcrops in Colorado, Utah and Wyoming, has received an enormous amount of attention because it not only contains huge reserves of trona, but it is also the largest potential reserve of hydrocarbons (oil shales) in the world. The sequence contains many different types of lake deposit including stratified, open lake

carbonates in the Uinta Basin discussed in Section 4.4.7a. The Wilkins Peak Member is the middle unit of the formation and reaches a maximum thickness of 370 m. It is underlain by the Tipton Member and overlain by the Laney Member. The Wilkins Peak Member represents a relatively arid period of deposition during the evolution of the Lake Gosiute in the Bridger Basin (southwest Wyoming).

A detailed sedimentological study was carried out on this unit by Eugster & Hardie (1975) and seven major lithofacies were recognized (Fig. 4.81A).

1 Flat pebble (intraclast) conglomerates (rudstones) composed of clasts of dolomitic mudstone (Fig. 4.82A), in beds up to 0.2 m thick, which can be traced over large distances. These commonly overlie mudcracked surfaces and the pebbles have been interpreted as rip-up clasts reworked during the transgression of the lake shoreline.

2 Peloidal grainstones. These occur as units from 10 cm to 2 m thick, traceable over large distances. Two lithologies occur with thin-bedded, wave-rippled or wavy-laminated dolomitic peloidal grainstones (Fig. 4.82B), and mudcracked dolomite mudstones. These have been interpreted as the deposits of a fluctuating, wave-influenced, littoral zone, and exposed mud flats respectively. Stromatolites also occur, whose growth forms have been used as indicators of shoreline environments (Surdam & Wolfbauer, 1975). Displacive evaporite minerals such as trona are also present.

3 Dolomitic mudstones. These are thin-bedded mudstones containing both silt-size peloid laminae, which have scoured bases, and graded laminae. Mudcracks are abundant along with other exposure features such as fenestrae. Pseudomorphs of evaporites are common, especially shortite, which may represent a replacement of gaylussite or pirssonite. Some of these mudstones, however, were deposited in deeper perennial lake settings but others represent saline mud flat to dry mud flat deposits (Smoot, 1983).

4 Oil shales. These consist of both organic-rich dolomitic laminites and oil shale breccias. The oil shales contain mudcracks and are associated with beds of trona. The breccias probably resulted from reworking of the mudcracked shales. The oil shales also contain lenses of detrital dolomitic peloidal silt (Fig. 4.82C) interpreted as evidence of wave processes. The organic matter has been interpreted as the product of the settling out of planktic 'algal' blooms or of organic oozes. It has also been suggested that they represent thin, cohesive 'algal' mats (Smoot, 1983), i.e. the remains of benthic microbial mats

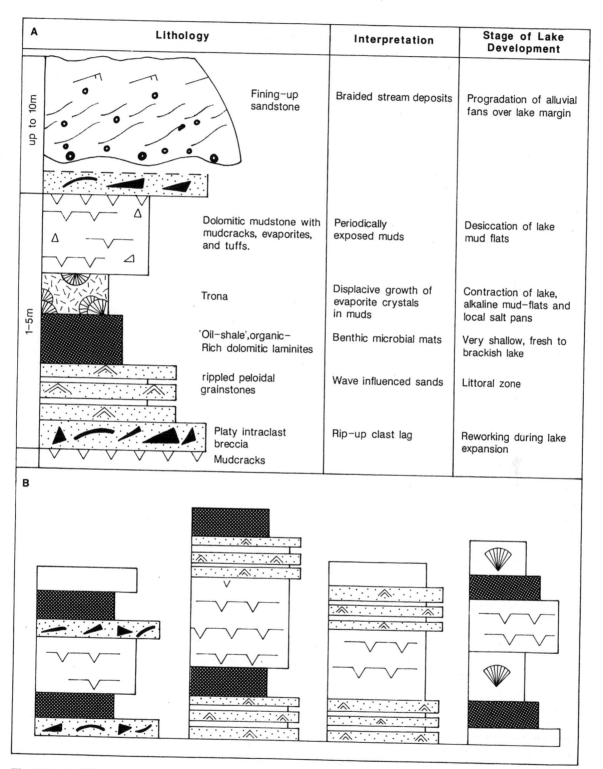

Fig. 4.81 *(A) Wilkins Peak Member, lithofacies and interpretation. (B) Types of facies sequence in the Wilkins Peak Member. Based on Eugster & Hardie (1975) and Smoot (1978).*

Fig. 4.82 *Wilkins Peak Member lithologies, from Green River area, Wyoming. (A) Intraclast lags. (B) Rippled peloidal grainstone. (C) Oil shales with prominent peloidal laminae. Scale divisions in (A) and (C) are 2.4 cm long.*

which are ubiquitous in shallow saline lakes (Bauld, 1981). This lithofacies has been interpreted as the deposit of a perennial, but shallow and occasionally desiccated, saline–alkaline lake. Detailed descriptions of the various laminites in the Wilkins Peak Member, including these oil shales, are given by Smoot (1983).
5 Trona–halite units occur, with the former in beds from 1–11 m in thickness. They are most widespread in the centre of the basin and commonly occur above beds of oil shale. The trona can be compared to deposits of alkaline lakes, such as Lake Magadi in East Africa, having both formed in salt pans and displacively in saline mud flats (Birnbaum & Radlick, 1982). Halite may occur separately or mixed with the trona. The trona deposits are associated with the lower stages of Lake Gosiute and are limited areally to the lowest part of the depocentre.
6 Siliciclastic sandstones. These units occur in tongues thinning into the basin and reach thicknesses of 10 m. These have been interpreted as braided

stream–alluvial fan deposits (Eugster & Hardie, 1975). and sheet flood sandstones also occur (Smoot, 1983).
7 Volcanic ashes are common both as distinctive tuff units and admixed into the mudstones.

These lithofacies occur in units up to 5 m thick, correlated over distances of 24 km. Each cycle represents the expansion of a shallow lake, followed by its progressive shrinkage until a salt pan, saline mud flat and dry mud flat developed. Some 40 major lake level fluctuations occurred during the deposition of the Wilkins Peak Member, as well as many minor transgressions (Smoot, 1983). Periodically, alluvial fan sandstones prograded into the basin during low stands. Eugster & Hardie recognized four basic types of cycle (Fig. 4.80B). This cyclicity (rhythmicity) is better developed in the central parts of the basin, and is less well developed around the basin edge where the units have a more random arrangement of lithofacies (Smoot, 1983). Smoot has also provided detailed lithofacies sections across parts of the basin. Overall, the Wilkins Peak Member represents a dryer phase of Lake Gosiute's evolution, during which time variations in rainfall/evaporation led to phases of complete desiccation. The underlying and overlying members represent less arid phases, which are thought, in part, to reflect basin drainage changes perhaps due to changes in the directions of prevailing storms.

A considerable amount of discussion has taken place over the various origins of the organic-rich sediments of the Green River Formation as a whole (Ryder *et al.*, 1976; Buchheim & Surdam, 1977; Surdam & Stanley, 1979). At times the lakes were probably permanently chemically stratified (ectogenic meromixis) with freshwater overlying denser, saline lake water (Desborough, 1978; Boyer, 1982).

The Cambrian Observatory Hill Beds of the Officer Basin of South Australia are another example of shallow, perennial−ephemeral lake deposits (White & Youngs, 1980). The unit is over 200 m thick and contains cycles 0.5−2 m thick (Fig. 4.83). Organic matter is concentrated in the lower parts of the cycles within mudstones, and has been interpreted as the remains of cyanobacteria which grew during episodic freshwater phases. After this flooding, progressive evaporation occurred and calcite was precipitated, but later replaced by dolomite in a similar way to that observed in the present-day Coorong dolomites of South Australia (Section 8.7.1). Cherts occur which may have formed after magadiite, a sodium silicate which forms in alkaline lakes. With further evaporation, benthic microbial mats built stromatolites, and with complete desiccation, led to desplacive trona (Fig.

4.84) and shortite growth in saline mud flats. Flake breccias also formed.

A more problematic sequence is found in the Upper Oxfordian (Jurassic) Vale Verde of Cabacos Beds of western Portugal (Wright & Wilson, 1985). The sequence consists of thin-bedded limestones and evaporites, and is variable in thickness from 70 m to approximately 200 m. Several lithofacies occur which are locally arranged in cycles (Fig. 4.85). The base of

Fig. 4.84 *Core showing replaced trona crystals from the Observatory Hill Beds (Byilkaoora 1 well). Core is 7 cm wide. Reproduced with permission of the Director General, Department of Mines and Energy of South Australia.*

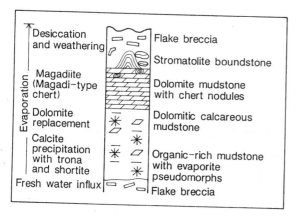

Fig. 4.83 *Idealized facies sequence from the Observatory Hill Beds (Cambrian) of South Australia. These cycles 0.5−2 m thick represent the expansion and subsequent contraction of alkaline lakes. Based on White & Youngs (1980).*

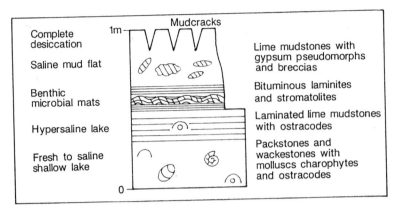

Fig. 4.85 *Lacustrine cycle from the Cabacos Beds, Cabo Mondego, western Portugal. Complete cycles are relatively rare and the whole sequence was deposited in a coast-marginal setting. After Wright & Wilson (1985).*

each sequence is a bioturbated bioclastic limestone representing an expansion of the lake, and contains a biota of charophytes (gyrogonites and stems), molluscs and ostracodes. Finely-laminated, sparsely fossiliferous lime mudstones occur, lacking evidence of an infauna and only containing ostracodes. This is interpreted as a restricted, probably hypersaline, phase of the lake. The unit as a whole contains a high organic content with finely-dispersed kerogen, but the most organic-rich lithofacies are vuggy stromatolitic mudstones with evaporite pseudomorphs. These stromatolitic limestones have been interpreted as the products of benthic microbial mats around the lake margin. Mudstones with evaporite pseudomorphs also occur (saline mud flat deposits) as do numerous mudcracked surfaces, locally with dinosaur footprints (dry mud flat deposits). The overall setting for these shallow, saline lakes was coast marginal and they pass gradationally into lagoonal deposits. Analogies can be drawn with the various coastal lakes of Tunisia, and of the Coorong region of South Australia, whose deposits are also charophyte and evaporite bearing (Burne *et al.*, 1980). Such depositional systems are influenced by marked variations in water salinity caused by strong seasonality and groundwater discharges. Comparisons can also be drawn with the 'Purbeckian facies' of Europe (Section 4.3.4c).

4.4.8 Future developments

Lacustrine environments are more responsive to climatic change than many other depositional systems, and their use in palaeoclimatology is likely to be a major future theme. The role of orbital forcing in this respect is particularly interesting and studies of ancient lacustrine deposits are revealing such controls. The Early Mesozoic Newark Supergroup of eastern North America (Olsen, 1986) and the Middle Devonian Orcadian Basin deposits of Scotland (Astin, 1989) both exhibit evidence of such 'Milankovitch' cyclicity.

The GGLAB project (Global Geological Record of Lake Basins), part of IGCP Project 219 (Comparative Lacustrine Sedimentology in Space and Time) is a forthcoming compilation of data on lakes and lake deposits, being edited by K. Kelts and E. Gierlowski-Kordesch. This will serve as a major source of information for future research and should appear in 1990.

4.5 REEFS

4.5.1 Introduction — classification

Reefs are unique sedimentary systems. The interrelationship of physical, chemical and biological processes makes them especially interesting and, in addition, fossil reefs are major repositories of hydrocarbons. In the past, reef studies were plagued by arguments over terminology which have now largely abated. The controversy over the usage of the term *reef* has been reviewed by Nelson *et al.* (1962), Braithwaite (1973a) and Heckel (1974) and will not be repeated here.

In simple terms two features characterize reefs. Firstly, they are laterally restricted in some way, even though they may cover large areas and/or have significant relief. Secondly, they show evidence of a biological influence during growth, although this is not always clear in some ancient reefs, such as mud mounds (Section 4.5.8). In the past the term *reef* has been used by some workers to describe any discrete carbonate buildup, but Dunham (1970a) suggested a distinction be made between stratigraphic reefs, which are laterally restricted carbonate buildups, perhaps composed of superimposed small reefs, and ecologic reefs which he regarded as rigid, wave resistant, topographically distinct, and biogenically formed. The connotation that to be a 'reef' the structure must have been wave resistant is a requirement of many reef definitions, but proving 'wave resistance' in ancient reefs is difficult. In this chapter the term is used in a general sense for any biologically-influenced carbonate accumulation which was large enough during formation to have possessed some topographic relief. Recent useful definitions have been given by Longman (1981) and informally by James *et al.* (1985). It should be appreciated that a variety of organisms can build reefs (see Fig. 4.88; Longman, 1981), involving many different processes. As a result a wide spectrum of reefs can be formed, but broadly speaking two main types can be recognized: skeletal (frame-built) reefs and reef mounds.

4.5.1a Skeletal (frame-built) reefs

These are reefs built by organisms, usually metazoans, which possessed a rigid, calcareous frame. The recognition of these reefs in the geological record should be based on the presence of *in situ* frame-builders. Such reefs have the potential to be wave resistant, but clearly this will depend on the nature of the skeletal morphology and architecture of the organisms involved. Reefs composed of robust metazoan colonies are capable of forming the major wave-resistant, high-relief, so-called walled-reefs typified by modern barrier reef complexes, and they also form patch reefs, common in shelf lagoons.

4.5.1b Reef mounds

These are carbonate buildups formed biogenically but lacking a prominent *in situ* skeletal framework. They may be rich in bioclastic material or be predominantly mud (Section 4.5.8). Such structures have formed largely by the trapping and binding of sediment by various organisms and by the high local production rate of skeletal material.

Other workers have recognized other divisions within the reef spectrum depending on the roles of the organisms involved (Longman, 1981) and recently James & Macintyre (1985) have offered a classification of reefs recognizing two types of reef mound, namely microbial buildups (formed by algae and cyanobacteria, and including stromatolites) and mud mounds (formed by metazoans and metaphytes). However, some mud mounds, as acknowledged by James & Macintyre, are probably formed by microbial activity

(Section 4.5.8e) and their classification offers no formal criteria to use in classifying ancient reefs.

Since a spectrum exists between frame-built and non-frame-built reefs it might be reasonable to attempt to put a value, for convenience, on the percentage of *in situ* frame present to delimit the two groups. However, many Quaternary reefs have surprisingly low amounts of framework. Cores through such reefs reveal very high percentages (by volume) of cavity and sediment, the former created by the irregular growth of the frame and by alteration processes such as bioerosion. As much as 50% or more of the original frame can be destroyed by physical, biological and diagenetic processes (Longman, 1981). The result is that many ancient reefs apparently lacking an extensive, rigid, interconnected frame, may originally have possessed one during growth. When looking at core material, assessing 'how much coral makes a reef' can be especially difficult, as witnessed by the disagreement over the reefal or non-reefal nature of the Bahamian Bank margins (Beach & Ginsburg, 1982; Mullins *et al.*, 1982).

4.5.1c Reef complexes

Reef complexes are major reefal buildups which are sufficiently large and wave resistant for the development of significant topographic relief so that reef core, fore-reef and back-reef zones are created (James *et al.*, 1985).

4.5.1d Other classifications

Other systems of classifying reefs are based on form, ecological composition and stratigraphic profile. Modern reefs are classified usually on their overall morphology, size and their relationship to nearby coasts (Fig. 4.86). They are also classified on their composition (Rosen, 1975); it is common to describe fossil reefs by reference to the dominant organism, for example rudist reefs and stromatoporoid reefs.

Many major reefal buildups occur along shelf and platform margins and on ramps, and Wilson (1974, 1975) has offered a classification of reefs in these settings (Fig. 4.87), the former two categories, mud mounds and knoll reefs, being types of reef mound with knoll reefs having a higher skeletal component.

4.5.1e Additional terms

A variety of other terms are used to describe biogenic accumulations. The term bioherm is useful to describe

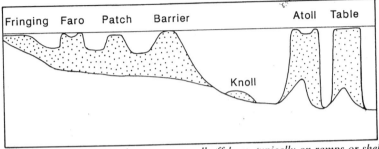

Fig. 4.86 *Shape classification of reefs. Fringing reefs are attached to the coastline; faro reefs are ring-like (atolls) within lagoons, or on atoll margins. Patch reefs are isolated reefs, typically on shelves. Barrier reefs are situated directly out from the coast, separated from it by a lagoon (which may be extensive as in the Great Barrier Reef). Knoll reefs (a term no longer widely used) are isolated reefs situated in deeper water, well offshore, typically on ramps or shelf slopes. An atoll is a ring-like structure with a central lagoon, developed in deeper water. A table reef is a flat-topped reef (platform) situated in deeper water, lacking a central lagoon.*

small, lensoid, biogenic buildups (Cummings, 1932), while the term biostrome relates to more tabular and bedded biogenic deposits. However, these terms have been misused in the literature. In descriptions of recent accumulations the term bank is used to describe any topographic feature on the seafloor.

4.5.2 Reef dynamics

Reefs, more so than other carbonate environments, are complex systems in which biological, physical and chemical factors interplay. There are four main processes which operate to varying degrees in reef formation.

1 Constructive processes. These are the biological processes, such as the direct growth of calcareous organisms, or the effects of organisms such as sediment baffling or binding.

2 Destructive processes. Various processes can damage and destroy the growing reef. These include physical effects such as waves, as well as biological destruction (bioerosion).

3 Cementation. In many reefs, ancient and modern, extensive early cementation has occurred, directly from marine porewaters. Such cementation is an important factor in influencing reef form.

4 Sedimentation. The high degree of biological activity on and around reefs leads to the accumulation of biogenic matter as well as reef-derived detritus.

The great variety of morphologies and internal structures seen in reefs results from the variable roles of each of these processes. Much of our knowledge of them has been derived from detailed studies of Bermudan patch reefs (Schroeder & Zankl, 1974;

Scoffin & Garrett, 1974) and has been extrapolated to other reefs, ancient and modern. Even modern deep-water reef mounds of the Florida Strait (Section 4.5.8d) clearly show the operation of these four processes (Neumann *et al.*, 1977), while the activities of similar processes can be detected in reefs as far back as the Lower Cambrian (James & Kobluk, 1978).

4.5.2a Constructive processes

Reef organisms contribute to reef growth in a number of different ways and can be classified on their sedimentological roles (Fig. 4.88). Heavily calcified forms, especially large individuals or colonies, may act as the building blocks of the reef, constituting frame-builders. Other calcareous forms may encrust these frame units and bind them together. The former are commonly called primary frame-builders (Scoffin & Garrett, 1974) and are represented in modern reefs by scleractinian corals, crustose coralline algae and *Millepora*, a hydrozoan. In ancient reefs this role was taken by scleractinian, rugose and tabulate corals, stromatoporoids, various calcareous algae and stromatolites. The encrusters are referred to as secondary frame-builders, and today consist of crustose coralline algae, serpulids, bryozoans, corals, foraminifers and vermetid gastropods. Many of these secondary frame-builders also occur in cavities within the reef. Other calcareous organisms act as sediment contributors, such as sessile epibenthic organisms like the calcareous alga *Halimeda*. Such sediment contributors are especially important in the formation of reef mounds where the accumulation of locally produced calcareous sediment results in reef growth (Section 4.5.8).

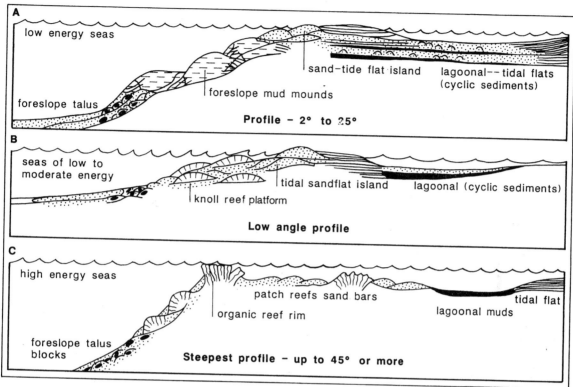

Fig. 4.87 *Three types of reef development on carbonate platform/shelf margins. After Wilson (1974).* **Type 1.** *Downslope mud accumulations. These are linear trends of bioclastic lime mud or belts of mounds located on the foreslope of the margin. Upslope there are sand beaches and islands. Examples include the Capitan Formation of the Permian Reef Complex of West Texas and New Mexico, phylloid algal buildings of Pennsylvanian age and Waulsortian mounds of Europe and North America (which are also associated with ramps). Exact modern analogues are not known.* **Type 2.** *Knoll reef ramps. These reefs consists of linear belts of skeletal (frame-built) knoll reefs, on gentle slope at the outer edge of the shelf margin. Strong wave and current action is absent. Examples include the modern ledge-flat reef of Bermuda, on top of the Bermuda Platform, rudist reefs of shelf margins of the Middle Cretaceous of south Texas and Middle East and Middle and Upper Devonian reefs of western Canada.* **Type 3.** *Frame-built reef rims (walled-reef complexes). These are linear belts of organic reef frame growing up to sea-level and into the zone of turbulence. They may occur as barrier or fringing reefs. They are due mainly to the growth of hexacorals and red algae in post-Triassic reefs. They have steep slopes and extensive talus debris. Examples include modern reefs. e.g. Belize, and Late Devonian Canning Basin reefs of Western Australia.*

However, other organisms are also important, and small or lightly skeletalized epibenthos or non-skeletal organisms can also act as sediment bafflers (trappers) or binders (Fig. 4.88). Sea-grasses are important binders in modern seas and filamentous cyanobacteria have acted similarly since Early Precambrian times.

The contributing roles of the different organisms can be defined even in ancient reefs but various organisms may act in several roles (Fig. 4.89) such as the modern crustose corallines which may constitute primary frame-builders and encrusters.

Major reef complexes only develop when suitable large frame-building organisms are present, but this has only occurred some six or seven times during the geological record (Section 9.2.2 and Fig. 9.4; Longman, 1981; James, 1983). When such forms were absent, smaller reefal buildups or reef mounds occurred, commonly formed by the activities of sediment contributors, binders and precipitators (Section 4.5.8).

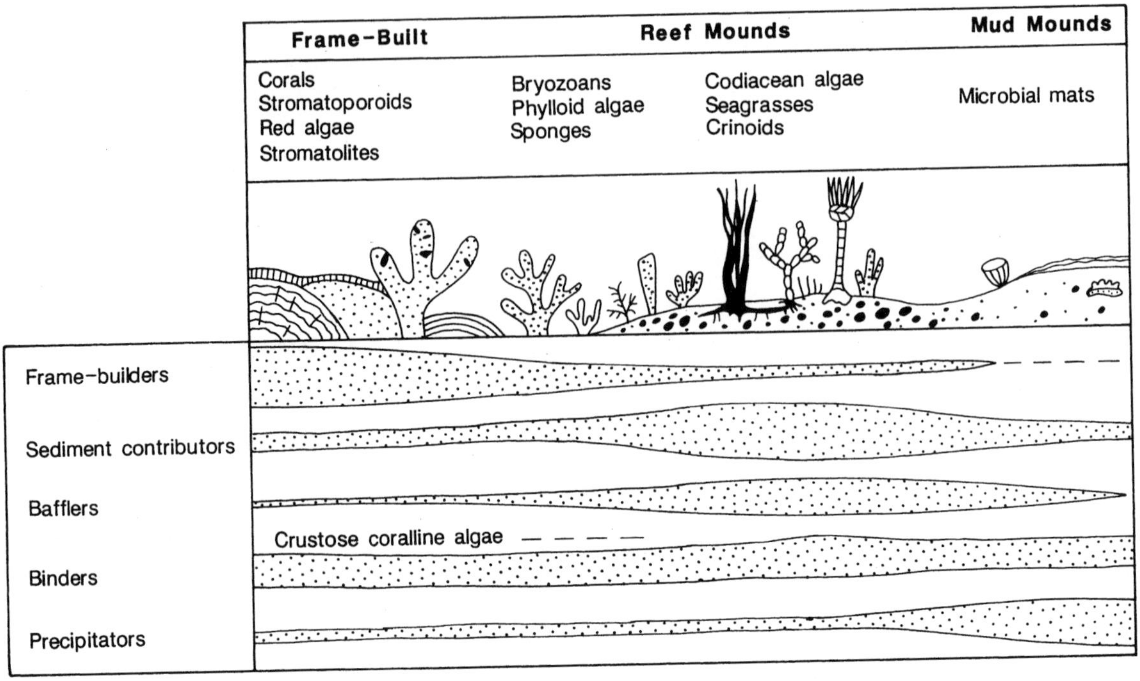

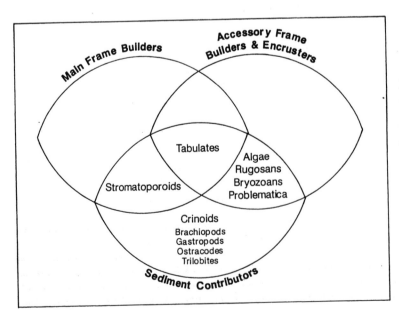

Frame-Built	Reef Mounds		Mud Mounds
Corals Stromatoporoids Red algae Stromatolites	Bryozoans Phylloid algae Sponges	Codiacean algae Seagrasses Crinoids	Microbial mats

Frame-builders

Sediment contributors

Bafflers

Crustose coralline algae — — — —

Binders

Precipitators

Fig. 4.88 *Spectrum of reef types and the roles of the various organisms involved. Stromatolies can act as frame-builders while microbial mats act as bafflers, binders and precipitators.*

Main Frame Builders

Accessory Frame Builders & Encrusters

Tabulates

Algae
Rugosans
Bryozoans
Problematica

Stromatoporoids

Crinoids
Brachiopods
Gastropods
Ostracodes
Trilobites

Sediment Contributors

Fig. 4.89 *Principal sedimentological roles of calcified organisms in Silurian reefs of Europe. Based on Riding (1981). In these reefs many organisms played more than one role and a distinct difference can be made between the main frame-builders (stromatoporoids and tabulate corals) and accessory builders (attached to frames and encrusters). Red algae (solenoporaceans) are abundant in these reefs but, unlike crustose coralline algae in modern-day reefs, these were not able to encrust and acted as accessory frame-builders. The encrusting role was filled by problematica and stromatolites.*

The growth of modern Bermudan patch reefs can be used as a model for understanding ancient reefs. Studies of these reefs have shown that they consist of individual reef 'building blocks' called knobs, 1–5 m in diameter, and up to 3 m high, which grow and coalesce and overgrow one another (Fig. 4.90; Garrett *et al.*, 1971; James, 1983). These consist of corals, algae, hydrozoans and other organisms. Larger structures are, at least initially, constructed of combinations of such knobs. This is the basic mechanism for building framework reefs which can also be identified in ancient reefs (Scoffin, 1972). As a result of the coalescence and overgrowing of frame units, a complex three-dimensional structure develops with numerous irregular growth cavities (Fig. 4.90). It has been estimated that 30–50% of the volume of the Bermudan reefs consists of open cavities and sediment-filled areas (Garrett *et al.*, 1971).

Frameworks built by small skeletal units, especially branching forms, will have even higher cavity-sediment values (60% in Silurian reefs of the Welsh Borderlands, Scoffin, 1972). The numerous cavities are created not only by the irregular foliaceous or branching growth of the frame, but also by bioerosion. These cryptic environments are an important feature of reef growth and contain their own encrusting communities (Jackson & Winston, 1982). Bosence (1984) has named such cryptic frameworks 'secondary frameworks' (Fig. 4.90) and these have been found to constitute up to 7% of Recent coralline algal ridge reefs in St Croix in the Caribbean (Bosence, 1984). Such cryptic frameworks have a long geological history and have been recorded in Early Cambrain reefs (Kobluk, 1981a). During frame growth a characteristic sequence of encrusters develops with, firstly, photophilic forms, such as the crustose coralline algae, but as the structure develops and overgrowth occurs, sciaphilic (shade-loving) forms become more common. The presence or absence of such sequences has been used to calculate the rate of accumulation of reef debris (Scoffin & Hendry, 1984). The growth rates of the frame-builders are important in frame development, for the survival of the frame is a balance between constructive and destructive processes (especially bioerosion). Frame components exposed to bioerosion for long periods will be destroyed, but rapid overgrowth or burial by sediment will isolate them (Scoffin & Garrett, 1974). Large skeletal organisms with a sufficiently continuous organic cover to deter borers will also have higher preservation potentials than vulnerable, small, branched forms with only restricted growth areas. Such taphonomic effects must

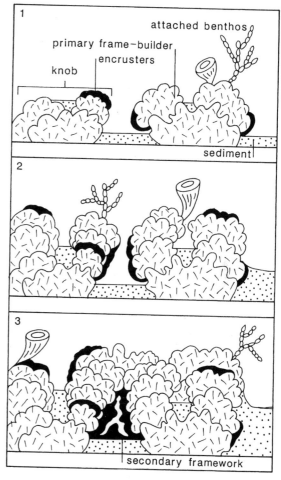

Fig. 4.90 *Schematic diagram of the growth of reef framework. (1) Individual knobs composed of large calcareous metazoans or colonial forms. With growth (2), cavities develop within the framework and are sites of sediment accumulation. Within these cavities secondary frameworks (3) develop composed of cryptic encrusters with sciaphilic replacing photophilic forms.*

be borne in mind when reconstructing the compositions of ancient reef communities.

4.5.2b Destructive processes

Two types of destruction need to be considered: physical destruction and bioerosion.

Physical destruction is a constant process caused by wave and current activity on reefs. Storms and hurricanes are a major influence on reef ecology and sedimentology and it is during such events, though geologically short-lived, that major destruction of the reef communities and sediment transportation occurs (Stoddart, 1969; Maragos *et al.*, 1973; Hernandez-Avila *et al.*, 1977).

The effects of such rare events are difficult to assess in ancient reefs but those of the other category, bioerosion, are ubiquitous. The importance of this group of biological erosive processes in reef sedimentology cannot be overemphasized (Hutchings, 1986). As stated above, any skeletal material exposed in the reef environment will be attacked by a variety of organisms, and will eventually be destroyed unless buried or thickly encrusted (Macintyre, 1984b). The rate of bioerosion almost equals the rate of calcification for modern coral reefs (Hein & Risk, 1975; Scoffin *et al.*, 1980). In the Grand Cayman Islands it has been shown that the rate of bioerosion by clionid sponges is locally *twice* the rate of reef calcification (Acker & Risk, 1985). The estimated rates of bioerosion for various reef taxa have been reviewed by Davies (1983) and Hutchings (1986).

Reef bioeroders can be classified into four categories (Schroeder & Zankl, 1974): borers, raspers, crushers and burrowers.

Borers are particularly important in reef destruction and dead coral on reefs typically exhibits fringes, several centimetres thick, formed by multiple borings (Jones & Pemberton, 1988). The most important borers in reefs include the algae, cyanobacteria, fungi, sponges, sipunculids, polychaetes, molluscs, barnacles and echinoids (Ahr & Standon, 1973; Abbott *et al.*, 1974; MacGeachy & Stearn, 1976; Warme, 1977). The activities of sponges are particularly important (Rutzler, 1975; Hudson, 1977; Moore & Shed, 1977). These range from small cavities formed by clionid sponges to irregular cavities up to 50 mm in diameter created by *Siphonodictyon* (James & Ginsburg, 1979). Boring sponges may date back to the Cambrian (Kobluk, 1981b).

Kobluk & Kozelj (1985) have shown that in the reef framework cavities of the Bonaire reefs (Netherlands Antilles), macroborings show a decrease in their mean size, and in borehole size, with increasing depth. A succession of bioeroders attack dead coral skeletons, initially microborers and later macroborers, with bivalves being 'late-stage' bioeroders (Risk & MacGeachy, 1978). Polychaetes are commonly the first macroborers to colonize a substrate (Hutchings & Bamber, 1985). Both these sets of observations may be useful in palaeoecological studies of reefs.

Raspers are browsing organisms, such as gastropods and echinoids, which scrape the calcareous substrate to remove mainly algal material (Stearn & Scoffin, 1977). Crushers include forms such as the parrotfish (Frydyl & Stearn, 1978). Burrowing is not an important bioerosive process in frame-built reefs, but becomes significant in reef mounds. In living crustose coralline algal reef mounds, such as the maerl facies of temperate and Mediterranean regions, and in the skeletal banks of tropical and subtropical regions, the framework elements which are present are progressively destroyed by burrowing. The result is a skeletal packstone and wackestone matrix (Bosence, 1985). The important lesson here is that even though skeletal frame-builders were present in these reef mounds, the evidence of their presence has been destroyed by bioerosion.

Bioturbation by callianassid shrimps is a significant process in many back-reef settings and influences the resuspension and transport of fine-grained sediment, the grain size distribution and recycling, and can also inhibit the colonization of other benthic organisms (Tudhope & Scoffin, 1984).

Bioerosion has other effects besides the destruction of framework elements (Jones & Pemberton, 1988). The ability of corals to withstand waves depends on the degree of bioerosion (Hein & Risk, 1975; Tunnicliffe, 1982) and extensive boring under overhangs weakens branched forms (Goreau & Hartman, 1963). The various bioeroders produce copious amounts of sediment. Clionid sponge chips make up a major component of the silt and very fine sand component of reef sediments (Fütterer, 1974; Acker & Risk, 1985), and other grain size populations can be attributed to the activities of parrot fish (Gygi, 1975; Ogden, 1977). Conversely, bioerosion also destroys sediment grains and Tudhope & Risk (1985) have estimated that 18–30% of sediment influx into the Davies Reef lagoon in the Great Barrier Reef is lost through bioerosion by micro-organisms, principally endolithic algae.

The most important effect of bioerosion for students of fossil reefs is that it is a major destroyer of primary reefal fabrics. Where bioerosion is greater than carbonate production or sedimentation, near-complete loss of the *in situ* reefal framework may occur (Warme, 1977; Land & Moore, 1980; Scoffin *et al.*, 1980). Multiple generations of boring, sedimentation and cementation can result in the formation of highly complex fabrics (Section 4.5.3). The style of bioerosion seen in modern reefs can be traced back to the Oligocene–Miocene (Pleydell & Jones, 1988), but bioerosion has been a major process since the advent of metazoan reefs.

4.5.2c Sedimentation

Material is supplied to reefs from three main sources: mechanical breakdown of framework material by physical or biological processes (Section 4.5.2b), material contributed by the decomposition of reef dwellers, and material supplied from outside the reef. The most important sediment contributors in modern reefs are *Halimeda*, coralline algae, corals, foraminifera and molluscs (Milliman, 1974). Large benthic forams, a prominent component in many ancient reefs, may produce carbonate at rates comparable to those of corals, coralline algae and calcareous green algae, and are important contributors to reef and near-reef sediments (Hallock, 1981; Scoffin & Tudhope, 1985).

The composition of the resulting reef sediment will depend on a number of biological and taphonomic factors which must be considered when interpretations are made of reef biofacies composition (Section 4.5.7c). Sediment budgets are also important when considering both the development of reef mounds (Section 4.5.8) and frame-built reefs. The role of sediment input in reef accretion is probably underestimated and it can exceed the amount of actual framework production (Hubbard *et al.*, 1986).

Cavities formed on and in the reef by irregular growth and bioerosion are quickly filled by sediments (Schroeder & Zankl, 1974; Scoffin & Garrett, 1974). Fine-grained material can be transported and pumped into the network of cavities near the surface of the reef, and the result is an abundance of geopetal, internal sediments which may show multiple generation of deposition, as well as cross-lamination, grading, and bioturbation (James & Ginsburg, 1979; Watts, 1988). Such sediments tend to be most common in the reef-front and reef-crest zone but they may occur in other parts of the reef. These internal sediments may also become cemented and bored (James & Ginsburg, 1979).

4.5.2d Cementation

Cementation is a major process and is partly responsible for the steep, wave-resistant profiles of many reefs. It is pervasive in the reef-front and reef-crest zone where high water flux occurs as a result of the pumping action by waves (James *et al.*, 1976; James & Ginsburg, 1979; Marshall & Davies, 1981). Reef cements have been the focus of considerable interest and much is now known of the various types and of their distributions. Since many ancient reefs are major

hydrocarbon reservoirs, a thorough understanding of early porosity occlusion by marine cements is essential. Such cements occur abundantly in ancient reefs including Early Cambrian patch reefs (James & Klappa, 1983). Some Devonian reefs in Canada, such as the Golden Spike and Rainbow Reefs, have up to 70% of the volume of cement in the reef-margin zone represented by early marine cements (Walls & Burrowes, 1985).

One interesting feature of cement distribution is its relationship to reef form (Kendall & Schlager, 1981). Reefs with steep profiles, such as modern walled reef complexes, present a prominent surface to wave and current action so that the force of seawater flux is high and extensive cementation occurs. Reefs with low-angle profiles present less of a barrier to the pumping action of waves and currents, and undergo much less cementation. This contrast is well seen by comparing ancient reefs; for example, the steep-fronted Devonian reefs of the Canning Basin show evidence of extensive early cementation (Playford, 1980; Kerans *et al.*, 1986) while the low-relief Devonian, Swan Hills Reef of Alberta shows only minor cementation in the reef front and crest (Wong, 1979; Walls & Burrowes, 1985). Such a relationship can be seen in the Permian Capitan Limestone reefs of New Mexico and west Texas (Babcock, 1977; Yurewicz, 1977a) where the degree of marine cementation increases as the profile of the reef steepens. However, in this case the cement locally reaches extreme amounts, constituting the bulk of the volume of the reef rock (Toomey & Babcock, 1983) and some is probably biogenic in origin.

In present-day reefs a variety of cement types occur with high-Mg calcite generally being much more abundant than aragonite (Section 7.4.1a; Purser & Schroeder, 1986). The former is found in a greater variety of forms including isopachous micritic, bladed and blocky forms (James & Ginsburg, 1979; Friedman, 1985; Pierson & Shinn, 1985). Peloidal high-Mg calcite cements are another common form which consists of peloids, 20–60 μm in diameter with a micritic core, surrounded by 'microspar' crystals (30–40 μm in size) (Section 7.4.1a; Lighty, 1985; Macintyre, 1985).

High-Mg calcite crusts, either micritic or peloidal, and up to several centimetres thick, are a prominent feature of many modern reefs (Marshall, 1983a; Lighty, 1985) including reef fronts (Land & Moore, 1980). They commonly occur as stromatolitic crusts, and even form columnar and branched structures up to 30 mm high (Marshall, 1983a). Similar aragonitic microstromatolites have also been described from the Belize Barrier Reef (James & Ginsburg, 1979). Ancient analogues of these stromatolites have been found in Silurian reefs in Sweden by Watts (1981). Reid (1987) has described abundant peloidal cements and crusts comprising up to 75% of the 'framework' of Upper Triassic reefs from the Yukon Territory, Canada. Surprisingly some high-Mg calcite crusts have preserved porosity in modern reefs by protecting (sheltering) skeletal growth pores from occlusion by sediments (Lighty, 1985).

Aragonite cements are often intraskeletal (James et al., 1976; Marshall, 1983a), and a variety of forms have been described but apparently isopachous cements are rare. Micritic and blocky microspar cements are known, as are fibrous crystals, either having grown epitaxially on coral substrates, or as pore-filling irregular meshworks of crystals. A striking aragonite cement form in modern reefs is the botryoid, up to 50 mm in diameter (Ginsburg & James, 1976; James & Ginsburg, 1979; Marshall, 1983a) and similar forms have also been described from ancient reefs (Mazzullo & Cys, 1979; Aissaoui, 1985; Tucker & Hollingworth, 1986). Such cements are particularly abundant in the Permian reefs of New Mexico and west Texas (Mazzullo & Cys, 1977, 1979; Toomey & Babcock, 1983) (see Fig. 4.115). Locally these botryoidal cements, which are now low-Mg calcite but were originally aragonite (Given & Lohmann, 1985), constitute up to 80% of the reef. They are intimately associated with the problematic alga *Archaeolithoporella*. Aragonite botryoids have been recorded forming beneath various types of living calcareous red algae, and such cements are surrounded by, or contain, organic tissue (Wray, 1977; Walker & Moss, 1984; James et al., 1988). These cements provide a means of attachment for the crustose algal thallus, and some living peyssonnelid algae produce attachment botryoids while the plant tissue is only weakly calcified or even uncalcified. It is possible that extensive botryoidal cement in ancient reefs may represent similar biogenically-produced deposits in which evidence of their makers is lost because of selective diagenesis or lack of calcification (Section 4.5.8f; James et al., 1988).

The nature of the controls on cementation in general is still poorly understood, but as stated above there is a specificity in the distribution of cements in present-day reefs. Cementation is concentrated in areas with high seawater flux. While wave and tidal pumping are usually invoked as the main pumping mechanisms, Marshall (1986) has suggested upwelling currents as an additional mechanism. The rate of reef growth is also a consideration, for in areas of rapid

growth the degree of early cementation is proportionately less than is slow-growing areas (Lighty, 1985). Numerous case studies discussing cementation in present-day and ancient reefs are to be found in Schroeder & Purser (1986; also see Section 7.4.1a and 7.4.4a).

4.5.2e Relative roles of each process

Reef growth must be seen as a dynamic interaction of these four processes. Depending on their relative importance, radically different reef types and internal structures may develop (Fig. 4.91; Schroeder & Zankl, 1974; Scoffin & Garrett, 1974).

The relative roles of these different reef processes are clearly illustrated by comparing present-day high-energy frame-built algal reefs and lower-energy reef mounds (Fig. 4.92; Bosence, 1985). In high-energy settings crustose coralline algal frameworks have a dense wave-resistant structure with a complex fused framework. The high seawater flux encourages cementation and grain-dominated sedimentation. In such settings, the main form of bioerosion is by boring. In contrast, in lower-energy settings, such as the mud mounds of the Florida Keys or the coralline algal 'maerl' facies of temperate and Mediterranean areas, the framework is more open and delicate, with little cementation and with finer-grained, mud-dominated sediments. The substrate is suitable for burrowers, which are the main frame-destroyers, resulting in the breakdown of the original framework components. In this case the differences between a frame-built reef and a reef mound are ultimately related to energy level, which influences frame type, cementation, bioerosion and the sedimentary association, and ultimately the preservation of the reef fabric.

4.5.3 Reef petrography

As a consequence of the complex and variable skeletal frameworks in reefs and because of the multiple phases of boring, sedimentation and cementation reef rock fabrics are often highly complex on the centimetre scale (Fig. 4.93; Ginsburg & Schroeder, 1973; Schroeder & Zankl, 1974; James & Ginsburg, 1979; Land & Moore, 1980). Furthermore, after diagenesis has taken its toll the situation becomes even more complicated. No descriptive system exists to cover such complex fabrics. Embry & Klovan (1971) have proposed a modification to Dunham's classification to describe simple reef fabrics (Fig. 4.94). Their system recognizes an allochthonous component with the terms

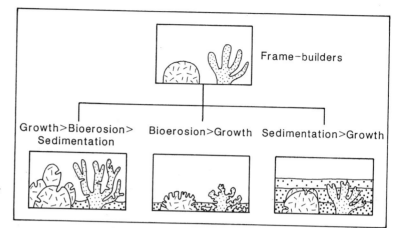

Fig. 4.91 *Schematic diagram to illustrate the roles of growth, bioerosion and sedimentation in the preservation of reef frameworks. Large, massive organisms or colonies have higher preservation potentials than delicate, branched forms.*

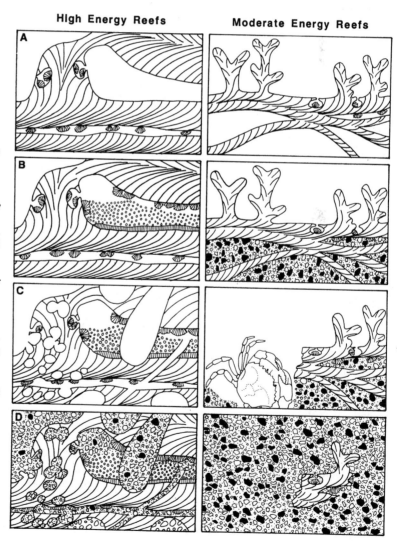

Fig. 4.92 *Contrasting roles of reef processes in high-energy frame-built crustose coralline algal reefs and lower-energy reef mounds. (A) Construction phase: in high-energy reefs a rigid framework is formed by fused thalli. In the moderate-energy reefs the framework is more delicate. In both examples minor intraskeletal cements may occur in conceptacles. (B) Cementation: this is more widespread in the higher-energy reef where a greater seawater flux occurs and where cavities are kept open. In the moderate-energy framework the growth cavities may be filled by fine-grained sediment. (C) Bioerosion: in the rigid high-energy framework borers will be the main agents of bioerosion. In the lower-energy reefs crustacean burrowing destroys the delicate framework. (D) Sedimentation: with further sedimentation the high-energy reef will develop a grainstone matrix with in situ framework, while the moderate-energy framework is destroyed and incorporated into a coralline algal wackestone–packstone. After Bosence (1985).*

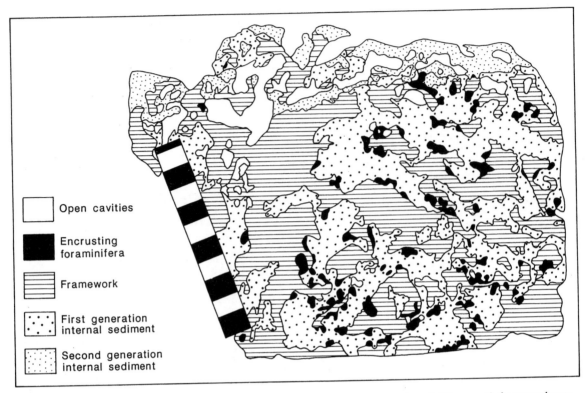

Fig. 4.93 *Tracing of sawed surface of a piece of reef rock from a Bermudan patch reef. The growth framework was formed by crustose coralline algae,* Millepora *and the vermetid gastropod* Dendropoma. *The cavity walls were encrusted by the foraminifera* Homotrema. *The first generation of internal sediment consists of coarse skeletal sand derived from the surface of the reef while the second generation is a lime mud which settles in the smaller, less-agitated pores. Scale divisions are 1 cm. Modified from Ginsburg & Schroeder (1973). This illustrates the complex nature of reef rock fabric formed by the irregular growth of a variety of calcareous organisms, and various phases of internal sedimentation, cementation, new cavity formation (boring) and further sedimentation.*

The legend reads: Open cavities; Encrusting foraminifera; Framework; First generation internal sediment; Second generation internal sediment.

floatstone replacing the term wackestone, and rudstone replacing pack- and grainstone if more than 10% of the particles are larger than 2 mm (Fig. 4.95). The autochthonous component refers specifically to *in situ* reef material with framestone, bindstone and bafflestone. The recognition of these latter two groups is interpretive, and in the case of bafflestone, rather subjective. A modified system was proposed by Tsien (1981) but this has not been widely applied. Riding (1977) proposed a system of classifying reef rock in terms of the relative proportions of three components: *in situ* organism, matrix and cavity. The abundance of the skeletal organisms was further used to define four categories of fabric from dense to sparse types. At first sight a system based on the

recognition of these three elements seems attractive, but recognizing true matrix can be particularly tricky. This is because differentiating micritic cement from mud-grade matrix is difficult even in present-day reef rocks (Friedman, 1985). As a result of multiple phases of boring, internal sedimentation and cementation, reef skeletons are reduced into secondary lime mudstones and packstones (James & Ginsburg, 1979; Land & Moore, 1980). This diagenetic micritization leads to a loss of original frame of between 20 and 70% (Longman, 1981). Additional problems such as distinguishing peloidal cements from a peloidal matrix is another problem in classifying reef rock (Section 4.5.2d).

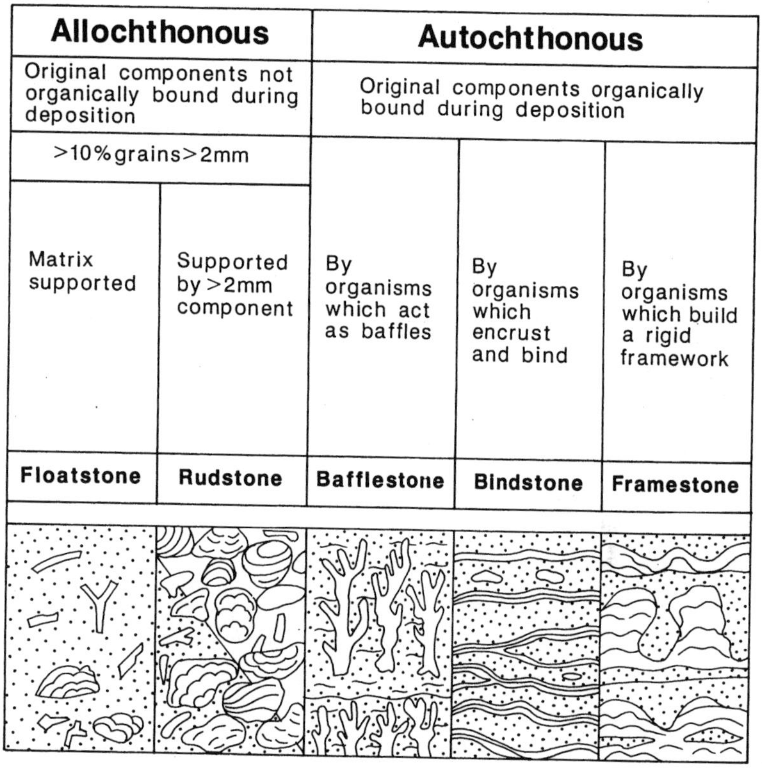

Allochthonous		**Autochthonous**		
Original components not organically bound during deposition		Original components organically bound during deposition		
>10%grains>2mm				
Matrix supported	Supported by >2mm component	By organisms which act as baffles	By organisms which encrust and bind	By organisms which build a rigid framework
Floatstone	**Rudstone**	**Bafflestone**	**Bindstone**	**Framestone**

Fig. 4.94 *Textural classification of reef limestones. Based on Embry & Klovan (1971) and James (1984b).*

4.5.3a Stromatactis

In many fossil reef mounds an important component of the reef fabric is the structure known as stromatactis (Fig. 4.96). Its origin has been, and still is, the subject of much discussion (Wallace, 1987). Stromatactis consists of a centimetre-sized mass of sparry calcite, with a flat to undulose, smooth lower surface, and a digitate upper surface, commonly made up of one or several isopachous crusts of centripetal cement, with or without geopetal sediment fills. It commonly occurs embedded in a lime mudstone or wackestone matrix. This definition (after Bourque & Gignac, 1983) has no connotations of a 'cavity' system or of a replacement origin as some earlier definitions have, but is purely descriptive.

Numerous origins have been proposed for these structures, which have been reviewed by Pratt (1982) and Tsien (1985a). The two main explanations which have been offered are the 'cavity system' theory and the 'recrystallization' theory. In the former case various researchers have suggested that the structures represent cavities formed by the decay of soft-bodied organisms or by the removal/replacement of stromatoporoids, sponges, bryozoans or cephalopods. Bourque & Gignac (1983) noted that the Silurian Stromatactis-bearing reef mounds of Quebec consist, in part, of peloidal networks (concentrations of pellet-like bodies) which resemble the associated sponge *Malumispongium lartnageli*. Stromatactis occurs within these areas and Bourque and Gignac suggested that it represents cavities formed within the sponge framework and also by the decay of uncemented sponge tissue. Playford (1984) and Kerans (1985) have regarded the abundant stromatactis of the Devonian Canning Basin reefs as resulting from bridging by reef organisms, by the decay of soft-bodied organisms (mainly sponges) or by the dissolution of aragonitic shells (mainly of cephalopods). Lees & Miller (1985) interpreted stromatactis occurring in Upper Palaeozoic Waulsortian reef mounds (Section 4.5.8e) as shelter cavities originally formed by fenestellid bryozoans and sponges, but later modified by roof collapse, dissolution and internal erosion.

Various inorganic cavity origins have also been proposed. Bathurst (1980b, 1982) proposed that

Fig. 4.95 *(A) Rudstone made up of branched coral fragments. Oxfordian (Upper Jurassic), Cabo Mondego, Portugal. (B) Framestone of* in situ *corals. Upper Jurassic, Consalacao, Portugal. Scale is 0.12 m.*

stromatactis represents a system of cavities which developed beneath submarine cemented crusts on reef-mound surfaces. Winnowing, he suggested, removed the less cemented material beneath the crusts creating cavities, and similar structures have been noted within the surface crusts of lithified reef mounds (lithoherms) in the Florida Straits (Section 4.5.8d; Neumann *et al.*, 1977). Wallace (1987) has proposed that winnowing takes place in unconsolidated muds s the cavities become progressively more buried sedimentation, current velocities decrease and ntation takes place. Pratt (1982) has also d a winnowing origin for the cavities but of

uncemented layers beneath gelatinous microbially-bound mats. Heckel (1972) proposed that they formed by dewatering, while others have regarded them as formed by slumping and compaction, or even by dissolution during deep burial (Logan & Semeniuk, 1976). Bridges & Chapman (1988) have interpreted cavities in Carboniferous Waulsortian mounds in England as caused by erosion or dissolution. The fluids involved, they believe, were marine phreatic waters released by deep burial compaction.

If the cavities result from purely physical or chemical processes, one is left wondering why they are not common in other types of fine-grained carbonates.

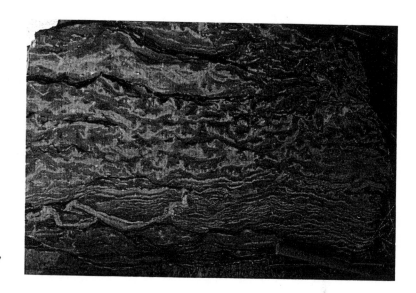

Fig. 4.96 *Various forms of stromatactis from Devonian (Famennian) mud mounds, Croissettes Quarry, Vodécée, Ardennes, Belgium. Photo courtesy of Trevor Burchette.*

Their restriction to mud mounds suggests a genetic link.

A recrystallization origin has been proposed by a number of workers, whereby calcareous algae or stromatoporoids have recrystallized to spar, or where neomorphism of the mud matrix has occurred. Recently Tsien (1985a) has suggested that stromatactis represents the 'recrystallization' of colonies of micro-organisms such as algae, cyanobacteria or bacteria. The isopachous, fibrous calcites which typically occur in the stromatactid structures have usually been regarded as a marine cement but recently Monty (1982) has interpreted some of these as tufa-like and microbial in origin; a view also favoured by Miller (1986). In addition, Van Laer & Monty (1984) have found well-preserved microbial filaments in such calcites. However, these microbial forms may have been inhabiting the cavities (coelobionts), which is a common feature of modern and ancient reefs (Kobluk, 1981a). Perhaps, like so many other geological phenomena stromatactis may turn out to be polygenetic in origin, but it is a striking and very common component of reef mounds.

4.5.4 **Major controls on reef morphology**

In Section 4.5.2 the basic, small-scale reef processes were considered. In this section, the larger-scale factors which affect reef morphology and development are discussed. It is the interaction of these factors, biological growth, topographic effects and sea-level changes, which leads to the great diversity of reef forms.

4.5.4a Biological controls

Reefs are organosedimentary structures and the types which form will ultimately be constrained by the biological potential of reef organisms. Major reef growth at the present time is by scleractinian corals and crustose coralline algae. The hermatypic (reef-building) scleractinians contain symbiotic algae (zooxanthellae), and are limited by plant photosynthesis to well-lit, shallow waters, less than 100 m deep and mainly under 20 m. Temperatures are in the range 16–36°C with an optimum of 25–29°C. Salinity tolerances range from 27–40‰ with an optimum of 36‰. The waters must have low turbidity and good circulation.

Scleractinian zooxanthellate corals are adapted to nutrient-deficient environments. Hallock & Schlager (1986) have argued that nutrient excess can reduce or kill off coral growth. Increased nutrient concentrations (phosphates and nitrates) stimulate plankton growth, which reduces water transparency, and both coral and calcareous algal growth. In addition, high nutrient excess can stimulate the growth of non-calcareous algae and other organisms. Both competition and bio-erosion will increase, reducing reef growth. Nutrient supply can be affected by oceanographic changes, induced by climatic, eustatic and tectonic factors. The role of nutrient excess, and its relationship to such factors as sea-level rise, are important considerations when interpreting ancient reefs.

Under optimal conditions scleractinian growth can be very rapid and 100 m 1000 yr^{-1} is possible (Buddemeier & Kinzie, 1976). However, reef growth is slower because optimal conditions are not reached or maintained, and because of loss by erosion, including bioerosion. Rates of reef growth of 9–15 m 1000 yr^{-1} have been recorded in the Caribbean (Adey, 1978), and 7–8 m 1000 yr^{-1} along the Great Barrier Reef (but with rates as high as 16 m 1000 yr^{-1} for coral frameworks) (Davies & Hopley, 1983; Marshall & Davies, 1984). Different biofacies on reefs also grow at different rates (Davies & Hopley, 1983).

Reef growth is typically greatest on the wave-influenced windward margins of reefs. In the highest-energy zones, such as reef crests, crustose coralline algae dominate, but they are slow frame-builders with mean rates of 3.2 m 1000 yr^{-1} on Caribbean reefs (James & Macintyre, 1985) and 2 m along the Great Barrier Reef (Davies & Hopley, 1983). Along moderate-energy zones acroporoid corals dominate and these have much higher growth rates (James & Macintyre, 1985).

4.5.4b Antecedent topography

Reefs will preferentially develop on topographic highs because shallow-water corals grow faster than deeper forms and sedimentation will be reduced on higher areas (Scoffin et al., 1978). The presence of antecedent topographic features will therefore influence the position and growth of reefs, at least in their early stages. James & Macintyre (1985) have recognized four classes of antecedent topography.

1 *Older reefs.* Many present-day reefs have developed on older reefs as exposed platforms or shelves were flooded during the last sea-level rise. They preferentially grew on the topography afforded by earlier reefs. Examples have been described from the Great Barrier Reef (Marshall, 1983b; Symonds et al., 1983), Belize (Halley et al., 1977), Florida (Shinn et al., 1977), Bermuda (Garrett & Hine, 1979) and other areas. Similar stacking of reefs is commonly seen in the geological record.

2 *Karstic topography.* During Pleistocene sea-level low-stands prolonged subaerial exposure created [karst] surfaces on many platforms and shelves and [af]ter sea-level rises reef growth occurred on [k]arstic highs (Purdy, 1974a,b).

[Eros]ional terraces. Pleistocene low-stands were

associated with wave erosion and terraces along many coastal zones. The terraces were later colonized by reef organisms (Goreau & Land, 1974).

4 *Siliciclastic/volcanic topographic features.* New reef growth has taken place on depositional relief such as the Pleistocene river terraces of Belize (Choi & Ginsburg, 1982). Volcaniclastic deposits are another example (Camoin et al., 1988).

However, as pointed out by James & Macintyre, many reefs at the present time, and in the geological record, have developed where no obvious topographic highs occurred. Antecedent positive topography is not always a requirement; the present-day reef mounds of Florida have developed in areas of negative relief where muds preferentially accumulated and provided suitable substrates for colonization by sea-grasses which helped create the mounds (Section 4.5.8a).

4.5.4c Sea-level changes

Most reefs develop in shallow water and are depth controlled. These two properties make them susceptible to changes in sea-level, which may be caused by eustatic and tectonic effects, and subsidence. Two cases need to be discussed: rising sea-level, and falling sea-level.

Rising sea-level. In this case the relative rates of sea-level rise and reef growth are important (Fig. 4.97). If the rate of sea-level rise is greater than the rate of growth, the reef will be drowned. Such reefs have been referred to as 'give-up reefs' (James & Macintyre, 1985; Neumann & Macintyre, 1985). However, as discussed above, the growth potential of reefal organisms is very high, and it is greater than most tectonic or subsidence rates (Section 2.2.1). The paradox of drowned reefs has been discussed in detail by Schlager (1981), who concluded that particularly rapid sea-level rises or reduction of reef growth by deterioration in the environment are the only reasonable explanations for the drowning of shallow-water reefs. Such deterioration could result from complete killing off of the coral growth by subaerial exposure or possibly by nutrient excess (Hallock & Schlager, 1986). Major sea-level changes may lead to variations in nutrient supply and location of upwelling zones.

A number of drowned Quaternary and older reefs have been documented and are reviewed by Kendall & Schlager (1981). Examples of Quaternary drowned reefs include those of St Croix in the Caribbean

(Adey *et al.*, 1977) and eastern Florida (Lighty *et al.*, 1978). The fate of drowned reefs is to show a progressive decline in their shallow-water biota, and they will subsequently be buried by deeper-water sediments.

If the rate of sea-level rise is not rapid relative to growth, the reef may transgress in suitable geomorphic conditions. Two categories of transgressing reefs have been recognized: retreating reefs and backstepping reefs (Figs 4.97 and 4.98; Playford, 1980). Retreating reefs (perhaps a better term would be onlapping reefs) migrate continuously in a landward direction by progradation of the reef core over its own leeward deposits. This process has been recorded in modern reefs but only during still-stand after reef growth has 'caught up' with sea-level rise (Davies & Marshall, 1980; Macintyre *et al.*, 1981; Marshall & Davies, 1982). Longman (1981) has pointed out that retreating reefs are rare in the geological record. In the well-documented reefal platform margin deposits of the Devonian Canning Basin, such reefs are only weakly developed (Playford, 1980). One possible explanation may be that accumulation rates in back-reef settings are lower than for reef frameworks, and that such areas will deepen more rapidly than the reef flat and so prevent progradation.

Backstepping reefs are more common and are typically associated with platform onlap during transgression (Playford, 1980; James & Mountjoy, 1983). The reef transgresses in stages, moving each time to a shallower, higher position on the platform margin. Well-documented examples have been described from the Devonian of the Canning Basin by Playford (1980; Fig. 4.98), and from the Middle and Upper Devonian of Canada (Bassett & Stout, 1967), where three backstepping reef phases occurred as a major platform transgressed over the Canadian craton (Fig. 4.99).

If the rate of sea-level rise is approximately equal to the growth rate of the reef, continuous upward accretion may occur (Fig. 4.97), giving the so-called 'keep-up reefs' (James & Macintyre, 1985; Neumann & Macintyre, 1985). If, at the start of the rise in sea-level, coral growth rate is in balance with the rate of sea-level rise, then the whole reef will consist of shallow-water facies. However, the typical situation in Holocene reef growth has been for a lag period, after the initial transgression, when the growth rate is low. This lag phase has been recorded from the Great Barrier Reef (Davies & Hopley, 1983), the Alacran Reef of the Yucatan Shelf (Macintyre *et al.*, 1977 and St Croix in the Caribbean (Adey *et al.*, 1977). In these cases the rate of sea-level rise later decreased and reef growth caught up with sea-level. These are the

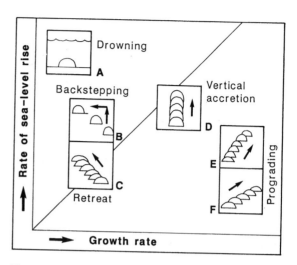

Fig. 4.97 *Schematic diagram showing the responses of reefs to sea-level rise. Growth rate refers to the actual accretionary rate of the reef and not to the biological growth rate. In (A), the rate of sea-level rise greatly exceeds the reef growth rate, and the reef is drowned below the depths of biological growth. In (B), the rate of sea-level rise is pulsed allowing recolonization but in progressively shallower waters (backstepped) (Figs 4.98 and 4.99). In (C), the rate of sea-level rise is nearly balanced by accretion and the reef retreats into shallow water. In (D), accretion and sea-level rise are balanced and vertical accretion occurs. In (E) and (F) the rate of sea-level rise is slow enough for the reef to prograde into deeper water (Fig. 4.100). These geometries relate to reefs on shelf or platform edges but isolated reef complexes will show such responses on all sides, although to varying degrees.*

'catch-up reefs' of James & Macintyre (1985). Such reefs, where lag periods have occurred, will exhibit shallowing-up trends following an initial 'deeper' phase. Local environmental factors can overprint these lag phases and some reefs along the Great Barrier Reef do not exhibit lag phases (Davies & Hopley, 1983).

If the rate of sea-level rise is less than the growth rate, even with a lag period, the reef will quickly grow up to sea-level and begin to expand laterally, in all directions, but usually more so in the windward than leeward directions (Marshall & Davies, 1982, 1984). The reefs will then prograde over their own flank deposits and any reef lagoons will be filled. One of the few long-term estimates of the rates of progradation has been given by Adey & Burke (1977) for the volcanic islands of the Lesser Antilles where 'shelf' width of the fringing reefs has grown at a rate of 0.75 km per million years (0.75 m 1000 yr^{-1}). Evidence

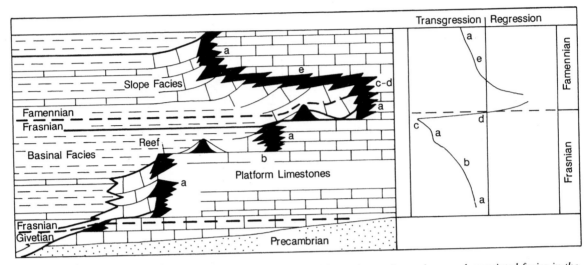

Fig. 4.98 *Schematic cross-section illustrating the development of Devonian reef complexes and associated facies in the Canning Basin of Western Australia, and relative rates of transgressions and regressions during the deposition of the reefs. Black, reef framework and reef flat; a, phase of moderate sea-level rise with vertical reef accretion; b, phase of rapid sea-level rise with backstepping; c, very rapid rise and drowning (not shown); d, phase of sea-level fall and exposure of the reef; e, low rate of sea-level rise with basinward progradation. The pinnacle reefs developed during rapid phases of sea-level rise. Based on data in Playford (1980). See also Fig. 2.14.*

Fig. 4.99 *Backstepping of reefs in Alberta during Middle—Upper Devonian times. Based on Bassett & Stout (1967).*

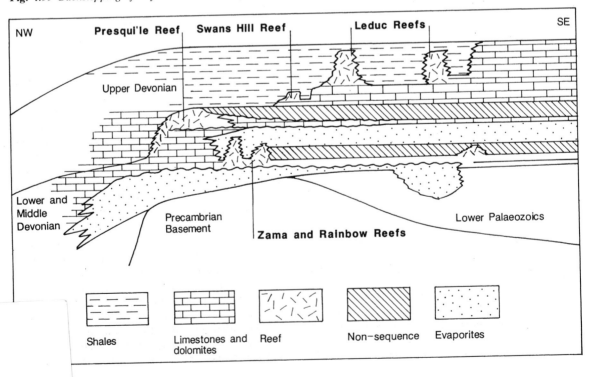

from ancient reefs suggests that such progradation can be a very rapid process. Hubbard *et al.* (1986) have recorded 24 m of reef accretion during the Holocene, from St Croix, with average rates of 0.84–2.55 m 1000 yr^{-1}.

A distinction can be made between progradational situations where the rate of sea-level rise is only slightly less than growth rate and those where the growth rate is much greater (Fig. 4.97). In the case of the former the prograding stacked profile will be relatively steep whereas in the latter the seaward component will be dominant. Examples of both types have been described from the Cretaceous Glen Rose rudist reef complexes of Texas by Bay (1977; Fig. 4.100). Progradational reefal buildups are major components of accretionary platforms and shelf margins (Read, 1982, 1985; James & Mountjoy, 1983) and are typical of passive margin sequences (see Fig. 2.6).

Falling sea-level. Reefs, by virtue of growing in shallow water, are prone to exposure by even minor sea-level drops. If the fall is rapid the reef will be isolated and exposed and coral growth will cease. Subaerial diagenesis will take place and extensive leaching and porosity formation may occur as well as cementation (Section 7.3). Such effects are well documented in Pleistocene reefs (Fig. 4.101; for example Brasier & Donahue, 1985) and in ancient reefs such as the Devonian reefs of Canada (Walls & Burrowes, 1985). Dolomitization, either by marine–freshwater mixing or because of evaporite formation in restricted, hypersaline back-reef areas, is also a common effect (Section 8.7) and is well documented in the Silurian reefs of North America (Sears & Lucia, 1980).

Steep-walled reefs will become isolated regardless of the rate of sea-level fall but during less rapid falls some reefs may prograde seawards giving an offlap relationship of successive reef complexes. Such a situation was widespread during the Messinian (Miocene) sea-level falls in the Mediterranean and is well displayed in the Nijar Basin reefs of southeast Spain (Fig. 4.102; Dabrio *et al.*, 1981). Reef regression can also occur in a stepwise movement with new reefs being developed at a lower level during a still-stand phase after each sea-level fall. Examples of this type of downstepping sequence also developed during the Messinian sea-level falls and have been documented from the Fortuna Basin of southeast Spain by Santisteban & Taberner (1983; Fig. 4.103).

Major reef complexes in the geological record, such as those of the Devonian of the Canning Basin,

Western Australia (Playford, 1980) and of the Devonian of Canada such as the Swan Hills reef complex (Viau, 1983) display evidence of a variety of responses to changes in growth rate, subsidence, sea-level and tectonism (Figs 4.98 and 4.104). A detailed example of the interactions of growth rate, tectonics and eustatic sea-level changes has been given for the Pleistocene reefs of Barbados by Mesolella *et al.* (1970).

4.5.5 **Reef facies: skeletal reef complexes**

A great diversity of fossil and present-day reef types can be recognized and it would be impossible to document all the different varieties of skeletal reefs. Instead two major types will be described in detail: reef complexes (walled reefs of Wilson, 1974) and patch reefs.

The aim of reviewing reef facies is to distill models which can be used to interpret reefs throughout the geological column. However, in using Quaternary examples several problems arise. Firstly, the reefs of today have developed since the last sea-level rise and are therefore 'young' (Longman, 1981), having developed in only the last 6000 years over earlier reef complexes. The striking, steep, lower reef front and fore-reefs of many Caribbean reefs are largely a result of Quaternary sea-level behaviour and may not provide useful analogues for ancient reefs. The organisms on present-day reefs are important in the development of many smaller-scale facies types but many have no direct ancient counterparts (Section 4.5.7).

Any useful model must be derived from an appreciation of the processes specific to reefs and requires the distillation of observations and models from all available descriptions of present-day reefs. Some ancient reefs are sufficiently well exposed so as to allow the broad facies relationships to be recognized at outcrop level, for example the Permian Capitan Reef of New Mexico and Texas or the Devonian reefs of the Canning Basin, Western Australia (Fig. 4.98). However, recognizing reef facies in limited outcrop or core requires careful interpretation of rock types with the integration of palaeontological data.

Two related environmental factors control the sedimentary and biological facies on the reef: depth (light level) and wave energy (stress). The gradients of these two parameters across a reef create sufficient differences to serve as the basis for facies recognition in fossil reefs. Graus *et al.* (1984) Fig. 4.105 have

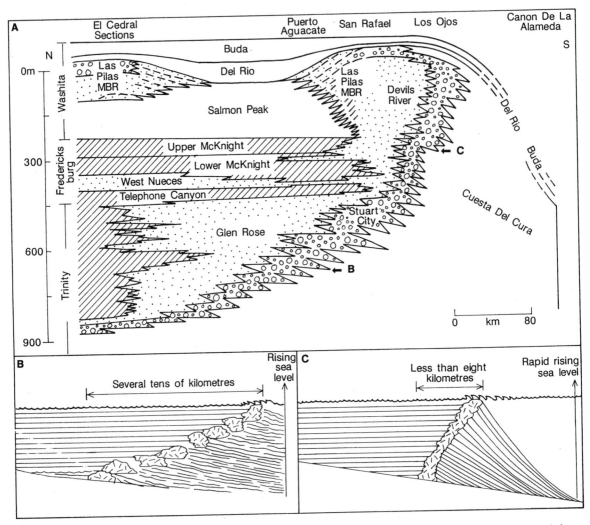

Fig. 4.100 *Reef geometries from the Lower Cretaceous of Texas and New Mexico. (A) Upper Trinity, Fredericksburg and Washita stratigraphic cross-section, northern Coahuila, Mexico. During Upper Trinity times the rate of sea-level rise was relatively low and the reef (rudist reef mounds) grew 400 m vertically over a horizontal distance of 50 km. (B) During Fredericksburg and Washita times the reef, under higher rates of sea-level rise (C), built another 400 m vertically over a horizontal distance of 10 km. Based on Bloxsom (1972) and Bay (1977).*

offered a computer simulation of a Jamaican reef using these two parameters. Figure 4.105 shows the sediments and reef crest–flat zones defined for bottom wave velocity and depth (light). The maximum depth boundaries were plotted using the greatest depths of frame growth heights. The wave velocity boundaries were calculated on such parameters as breaking strength measurements. The model has been used to test the role of hurricane disruption and longer-term processes on reef morphology and zonation.

A large number of subenvironments can be recognized in present-day reefs (Clausaude *et al.*, 1971; Guilcher, 1988) but in ancient reefs it is usually only possible to identify a few simple lithofacies and biofacies. Minor morphological features will have low preservation potentials, and after taphonomic and diagenetic processes have filtered the evidence only broad patterns can be recognized.

Skeletal reef complexes are large buildups represented by reefs developed on the windward margins

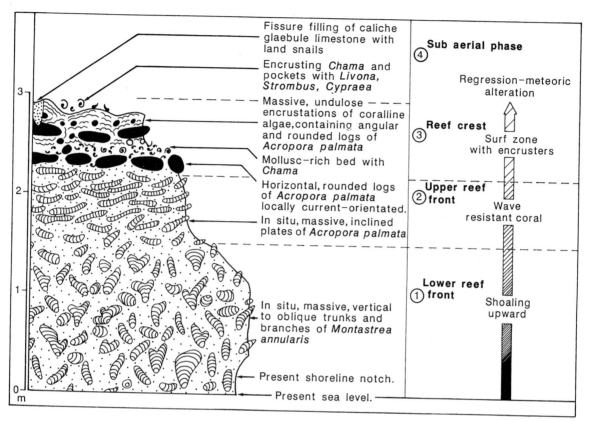

Fig. 4.101 *Schematic section through a regressive reef sequence, the Pleistocene Codrington Formation, Hog Point, Barbuda. Based on Brasier & Donahue (1985).*

of shelves or platforms, and they typically have steep-sided reef fronts (walled reefs). The major zones across such reefs are shown in Fig. 4.106).

4.5.5a Reef front and crest

The reef front can be regarded as that part of the reef extending from the highest point on the reef profile (crest) to a point below which little or no skeletal frame-building occurs. This depth varies according to local conditions but may extend to 70–100 m in modern reefs. In ancient reefs, with different frame-builders, this approximate depth range may have been very different and it should not be used uncritically as a guide.

The highest zone on the reef front, the crest, is the most exposed part of the reef, and is subjected to wave activity. The resulting reef morphology and composition depend on the energy regime (Adey, 1978). In high-energy zones, encrusting organisms

dominate, especially low encrusting growths of coralline algae (Adey, 1975; Adey *et al.*, 1982; Bosence, 1984). Such algal ridges have recently been described in detail from St Croix in the Caribbean by Bosence (1984). The crustose coralline algae predominate in such settings, both because they are physically able to withstand wave action and because they have a high tolerance to the light levels encountered in shallow water (Bosence, 1983c). These crusts typically coat dead coral or coral debris, and other encrusters occur such as vermetid gastropods and foraminifera, as well as non-calcareous algae.

In lower-energy crest zones the hydrozoan *Millepora* may occur, or robust corals such as *Acropora palmata*. The biological gradation of reef communities to decreasing wave action has been described from Caribbean reefs by Geister (1977).

The crest zone also shows the highest levels of skeletal breakage and abrasion with a resulting rubble strewn surface. Levels of bioerosion are also at their

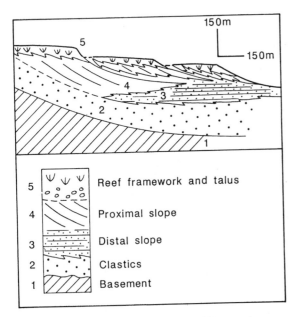

Fig. 4.102 *Messinian (Miocene) reefs of Nijar, southern Spain. During falling sea-level these reefs prograded over their talus slopes. The reef framework consists of vertical coral branches, mainly* Porites. *The proximal slope deposits are sand- and gravel-grade carbonates while the distal slope deposits comprise bioclastic packstones. Units 3–5 are dolomitized. Based on Dabrio* et al. *(1981).*

categorize this zone. Crests, because of their position, are prone to exposure and are more likely to show signs of near-surface or subaerial diagenesis than other reef zones.

Seaward, and in slightly deeper water, coral growth is more extensive, and the hydrozoan *Millepora* is a common form in present-day reefs. This zone typically passes seaward into the zone of spurs and grooves (buttresses and chutes or channels). The spurs are coral- and hydrozoan-covered linear structures, typically oriented parallel to wave direction. They may extend for 150 m seaward to depths of 10 m. They have widths of 5–10 m, with relief of up to 7 m above the adjacent groove (channel) floor. The grooves, typically only a few metres wide, contain skeletal sand and gravel. Spurs and grooves have been described from reef fronts in many regions including Jamaica, Barbados, Bahamas, Belize and Florida (Shinn, 1963; James *et al.*, 1976; Stearn *et al.*, 1977; James & Ginsburg, 1979; Sneh & Friedman, 1980; Shinn *et al.*, 1981, 1982; Rutzler & Macintyre, 1982). The origin of these structures has been debated but evidence points to the spurs being purely constructional features, with wave surge maintaining the grooves (Shinn *et al.*, 1981, 1982).

Similar, but lower-relief structures also occur in more seaward positions on some reefs in water 15–30 m deep (James *et al.*, 1976; Shinn *et al.*, 1981). Some of the deeper-water structures are probably purely erosional features formed during a lower sea-level (Shinn *et al.*, 1981), but others, previously regarded as discrete features, probably connect with their shallow-water counterparts and represent earlier constructional growth, perhaps formed in shallower water during a phase of sea-level rise (Rutzler & Macintyre, 1982).

Spurs and grooves have also been recognized in

highest in this part of the reef (Macintyre, 1984b) and the high seawater flux also results in extensive cementation (Fig. 4.107; Section 7.4.1a).

Reef-crest zones in the geological record are typically recognized on the presence of reef rock dominated by laminar, encrusting forms (bindstones and framestones) with low diversity fossil biotas. Higher levels of early cementation and bioerosion should also

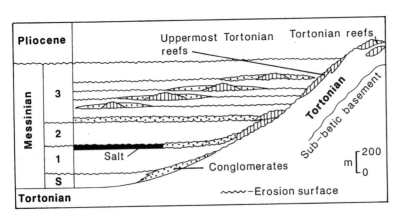

Fig. 4.103 *Downstepping reefs from the Fortuna Basin, southeast Spain. These reefs formed during stages of sea-level fall associated with desiccation during the 'Messinian Salinity Crisis'. 1–3 refer to evaporite phases; S, shallow-water turbidites. The thicknesses of the conglomerates and salts have been exaggerated. Based on Santisteban & Taberner (1983).*

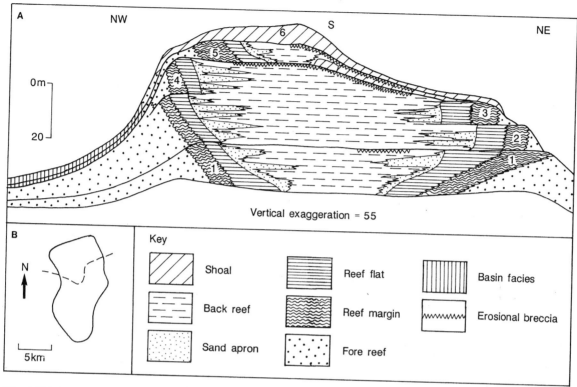

Fig. 4.104 *(A) Cross-section through the Swan Hills reef complex, Upper Devonian (Frasnian), central Alberta, Canada. The atoll-like reef sequence shows six phases of development including phases of progradation, vertical accretion (aggradation) and backstepping. Stage 1 exhibits progradation on both the windward (northeast) and leeward (northwest) sides which was terminated by subaerial exposure. Stages 2 and 3 exhibit vertical growth and only occur on the windward side but marked backstepping occurred after stage 2, caused it is believed by major storms and not sea-level changes. These stages are separated by marine hardgrounds. Similar backstepping occurred after stage 3. Vertical growth occurred during stages 4 and 5, each terminated by subaerial exposure and erosion of the reef on the windward side. After stage 5 the reef had developed a ramp-like profile and stage 6 represents a sand shoal. (B) Outline of reef. Note that these stages are shown so graphically because the vertical scale is exaggerated by a factor of 55. The reef is only 60 m thick but covers an area of 110 km². The approximate direction of the section is shown in the inset. Modified from Viau (1983).*

ancient reefs such as the Miocene of Mallorca (Pomar *et al.*, 1985) and even in the Devonian reef complexes of the Canning Basin, Western Australia (Playford, 1980).

Below the spur and groove zone on some modern reefs there are morphological features which resulted from modifications of the reef front caused by Holocene–Pleistocene sea-level changes. For example, both the Belize (James & Ginsburg, 1979) and Jamaican (Moore *et al.*, 1976) reef fronts have prominent steps (fore-reef escarpment). Such features are unlikely to be present, or recognizable, in ancient reef fronts and, as pointed out by James (1983), most

ancient reef fronts more typically grade into the fore-reef. However, the presence of well-developed, escarpment-type reef fronts should be an indicator of the rate of sea-level changes and original relief in ancient reefs.

Recently Hubbard *et al.* (1986) have documented the styles and rates of accretion of reef-front zones of St Croix.

As water depth increases down the reef front, energy and light levels decrease with subsequent changes in biotas. The reef rock is dominated by framestones with bafflestones and bindstones. Such material will constitute the bulk of the reef 'core' in

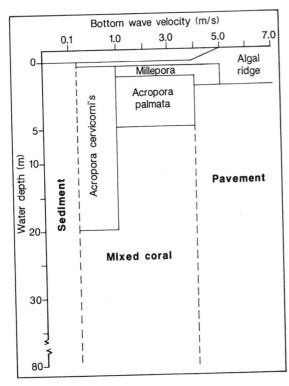

Fig. 4.105 *Coralgal, sediment and pavement zones of a Jamaican reef as defined by bottom wave velocities and depth (i.e. light). Based on Graus* et al. *(1984).*

4.5.5b Fore-reef slope

The fore-reef slope is an area, seaward of the reef front, which grades down into the surrounding basin floor. It is dominated by gravity flow deposits and pelagic—hemipelagic sediments (Section 5.10; Enos & Moore, 1983). Two simple types can be recognized (McIlreath & James, 1984): depositional (or accretionary) reef margins and bypass (escarpment) margins. In the former, the reef front passes downslope continuously into the basin while in the latter a steep submarine escarpment separates the reef front from talus deposits, which themselves grade out into basinal sediments. An analogy can be drawn between the slope and base-of-slope carbonate apron models (Section 5.12).

4.5.5c Reef flat

Behind and partially protected by the reef crest is the reef-flat zone, in which two environments can be recognized: the reef pavement and the sand apron.

The reef pavement is immediately behind the crest and is afforded some shelter by it. On many Pacific atolls a shallow moat zone occurs behind the crest ('algal ridge') which may be occasionally exposed at low tide. The pavement zone is variable in width, from a few metres to over 100 m as on the Belize Barrier Reef (James & Ginsburg, 1979). The depth is typically only a few metres and the zone may be exposed at low tide. The surface is smooth to undulatory with small coral growths, micro-atolls and an algal cover. Commonly, it is rubble strewn with coral and algal debris derived from the reef front. This material can include large boulder-sized material, and much of the debris is encrusted by crustose coralline algae. Bioerosion is extensive with some early cementation (Fig. 4.107; Macintyre, 1984b). Typical reef rock types are bindstones and rudstones, with sparse framestone.

ancient reefs. Longman (1981) has suggested that reef frameworks in ancient reefs are highly reduced remnants of the former reef structure. He stressed that many fossil reefs exhibit only 10—70% framework compared to the 50—100% seen in modern reefs. This loss of material he regarded as both due to bioerosion and early diagenetic processes such as micritization (Land & Moore, 1980; Friedman, 1985).

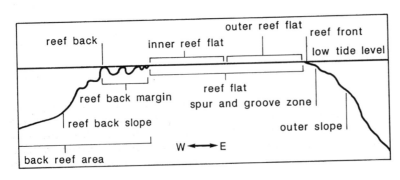

Fig. 4.106 *Geomorphic zones across a ribbon reef from the Great Barrier Reef. Based on Vernon & Hudson (1978). Compare with the profile of Grecian Rocks Reef, Florida (Fig. 3.32).*

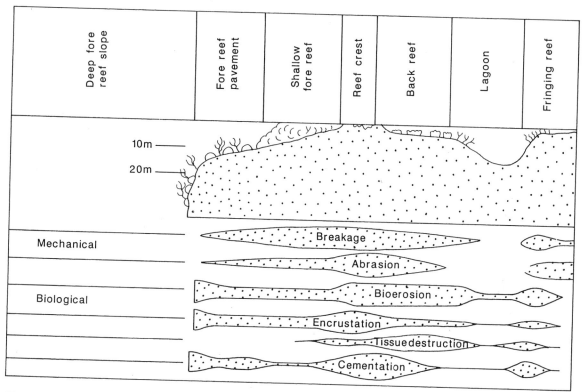

Fig. 4.107 *Schematic diagram showing the relative distributions of processes affecting reefs. The profile is through a barrier reef with lagoon and fringing reef. Based on Macintyre (1984b). Mechanical erosion is most intense in the higher-energy shallow reef front and crest where extensive cementation also occurs because of the wave-and tide-generated flux of seawater.*

The sand apron extends behind the pavement, and water depths range up to 10 m and extend for 160 km parallel to the reef trend. The aprons vary in width from 100 to 200 m wide, as in the case of the Belize Barrier Reef, or form extensive belts such as those behind the Florida Reef Belt or along Bermuda (Section 3.3.1; Enos, 1977a; Garrett & Hine, 1979; Halley *et al.*, 1983). Such sand belts can constitute potential hydrocarbon reservoirs and are commonly more important than the reef front which undergoes extensive early marine cementation (Longman, 1981). Rock types will be grainstones and rudstones which will grade into the back-reef, more muddy lagoonal sediments. Locally and distally the sand apron may be colonized by sea-grasses or algal mats, and will be stabilized, and the resulting sediments will be burrowed muddy sands (packstones–wackestones), reflecting the binding and baffling action of the vegetation.

4.5.5d Back-reef lagoons

Lagoons are protected, lower-energy areas behind the reef crest. Not all reefs have lagoons, for if the reef rim is not continuous, more open circulation will occur and the back-reef will have the aspect of an open shelf or bay. In such settings patch reefs may be separated by inter-reef facies of a more open marine character, rather than protected, lagoonal deposits. The central region of the Great Barrier Reef provides an example of such a setting (Scoffin & Tudhope, 1985).

Lagoons may be of a variety of sizes, from relatively small areas developed within atolls to larger zones behind major barrier reefs. Depth is an important consideration and while many reef lagoons are shallow (under 10 m) some Pacific atoll lagoons are over 70 m deep. Under such circumstances the lagoonal deposits may be of a deeper-water character and could conceivably be misinterpreted for inter-reef sediments. However, Tudhope *et al.* (1985) have described the lagoonal sediments from the deep (80 m deep) Suwarrow atoll in the Northern Cook Islands, which consist of *Halimeda* and foraminiferal sands and gravels. In this case either fine-grained sediment is not produced in abundance, or it is flushed out of the deep lagoon. In many lagoons a talus wedge may border the lagoon on its windward side, prograding from the reef-flat apron.

The essential characteristics of a reef lagoon, which may be recognizable in ancient reef complexes, are that it is protected and has restricted circulation. As a consequence the lower-energy settings may be reflected in finer sediments and a different biota to other reef facies. Lagoons act as traps for sediment and the settling of fine sediment deters many sessile organisms. Many of the corals have growth forms reflecting the higher sedimentation rates (Section 4.5.7a). Calcareous algae such as *Halimeda* and *Penicillus* are a prominent component of lagoons and produce copious amounts of carbonate sand and mud. Sea-grasses and algae may contribute to the formation of carbonate mud banks (reef mounds) in such protected settings. Further sediment may be contributed from terrigenous sources.

Patch reefs are a common feature of reef lagoons and influence sedimentation. In their study of relatively large (1–10 km diameter) patch reefs in the Great Barrier Reef, Scoffin & Tudhope (1985) recognized two lithofacies: proximal-to-framework and distal-to-framework. The proximal lithofacies, which are volumetrically minor, are near to the reef rim or to patch reefs within the lagoon. They consist of coarse, poorly-sorted sediment composed of coral, molluscan and *Halimeda* material, from sand to boulder size. They refer to this lithofacies as the 'talus halo'. The distal lithofacies consists of carbonate sands which are heavily bioturbated by callianassid shrimps (Tudhope & Scoffin, 1984). In the Belize back-reef lagoon a similar broad distinction can be made between a distal shallow, lower-energy 'sand and sea-grass' zone and a proximal sand and rubble zone with local coral growth near lagoonal reefs on the windward side of the lagoon (Rutzler & Macintyre, 1982).

Peritidal environments, including mangrove stands, may also occur within and bordering the lagoon, which may become filled both by a mixture of seaward-derived reef sand apron and talus material, and by peritidal deposits. Kendall & Schlager (1981) have graphically described this filling as the 'bucket principle', where the reefs bordering the platform (or atoll) act as rigid rims, while the interior is filled by lagoonal and tidal deposits.

4.5.6 Reef facies: patch reefs

Present-day patch reefs are well documented and should provide useful models for interpreting the small, isolated patch and knoll reefs so common in the geological record. As these isolated reefs develop, simple facies patterns emerge with a reef core (skeletal reef), reef flank (talus) and inter-reef sediments. However, with time small reefs may develop into larger structures and even relatively small reefs can become differentiated into reef fronts, back-reef zones with lagoons and sand cays.

Recent patch reefs have been described in detail from many areas, including Bermuda (Garrett *et al.*, 1971), Florida (Shinn *et al.*, 1977), the Belize Shelf (Wallace & Schafersman, 1977) and the Great Barrier Reef (Maxwell, 1968; Hopley, 1982; Marshall & Davies, 1984). Developing models from such reefs has to be done bearing in mind the constraints discussed earlier for walled reefs, and that all the present-day patch reefs have developed during the Holocene sea-level rise. Rather than detail all their characteristic it is more fruitful to try to establish general trends in their growth which can be used as a model for ancient patch reefs.

Many present-day patch reefs have grown on Pleistocene bedrock (usually also reefal) whereas others have grown from mud banks (reef mounds). Growth begins as isolated coral knobs which coalesce and the initial coral framework commonly contains large growth cavities. Coalescence results in a more continuous denser cover with a tighter framework (Fig. 4.108; Marshall & Davies, 1984).

As the reef grows upwards into shallower water it will reach wave-base or even grow up into the surf zone. In higher-energy settings this may result in a biological zonation between forms favouring the exposed, windward side, and other forms preferring the more sheltered leeward margins (Section 4.5.7a; Shinn *et al.*, 1977; Wallace & Schafersman, 1977; James, 1983). In more protected settings such differentiation will not develop. The windward margin may become dominated by encrusters such as crustose coralline algae, forming algal crests similar to those of barrier reefs.

Once the reef reaches sea-level, the growth style changes from being a mainly vertical one to lateral accretion in the leeward direction (Scoffin *et al.*, 1978; Stoddart *et al.*, 1978; Davies & Marshall, 1980;

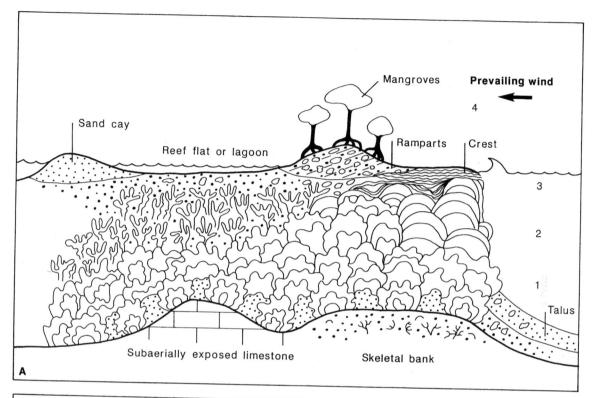

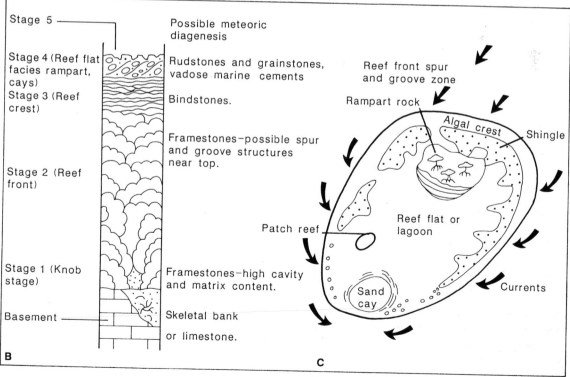

Marshall & Davies, 1982). A reef flat develops behind the windward algal crest zone and results in the refraction of the oncoming waves (Fig. 4.108C). On the Great Barrier Reef shelf, a variety of sediment bodies are deposited on the reef flat including coarse gravel 'shingle ramparts' (bars), near the windward margins, which prograde in a leeward direction. Sediment is dumped at the confluence of the opposing sets of waves refracted around the reef front to form sand cays (Fig. 4.108C). Much of the reef flat is covered by a mobile sand blanket, but it may be locally stabilized by mangroves allowing finer-grained sediments to accumulate. Linear boulder tracts also occur around the leeward margins where local massive coral colonies have been broken and accumulate during storms (Scoffin et al., 1978). As a result of these physical processes the reef becomes aligned with its long axis parallel to the prevailing wind (Fig. 4.108C). Similar wave (wind) influenced reef lithofacies have been recognized in many ancient reefs including Silurian ones of the Great Lakes region (Shaver, 1977). During development, the area of coral growth actually decreases as debris covers much of the original reef area. A similar trend occurs in the development of other reef types (Macintyre, 1984a). Another result of the reef building to sea-level is that exposure will occur. A variety of related diagenetic effects take place including the formation of beachrock (Scoffin & McLean, 1978). Cays may also have limited freshwater lenses, and meteoric cementation and dissolution may result.

Studies of ancient patch reefs commonly show ordered facies and biotic changes which many workers have compared to ecological successions (Section 4.5.7b). Many of the successions compare well with the patch reef model presented here (Fig. 4.108B). Studies of ancient patch reefs should not only seek to examine vertical sequence changes but, if opportunities are available, to study lateral changes which are just as useful for environmental reconstruction. Across the Great Barrier Reef the types of patch reef deposit vary (Maxwell, 1968), with more wave-swept, outer barrier reefs having little sediment accumulations, whereas the inner shelf reefs have well-developed cays and ramparts. Large-scale variations in patch reef morphology and deposits should also be expected across ancient carbonate shelves and platforms. Riding (1981) has described regional variations in patch reef development on the Silurian craton of western Europe, with deeper-water reefs occurring near the deeper platform margin and the higher-energy reefs in the shallower-water areas of the platform interior.

4.5.7 Biofacies

In studying ancient reefs it is essential to integrate both the lithological and palaeontological data. However, there are difficulties in interpreting the latter because direct comparisons between fossil reef-builders and modern forms are not always possible.

Fig. 4.108 (A) Schematic representation of the development of a patch reef. The substrate of many Holocene reefs consists of lithified Pleistocene reef or skeletal banks (reef mound). Stage 1, initial growth as isolated knobs formed by colonial metazoans, with a high percentage of growth cavity. Stage 2, vertical growth, perhaps related to rising sea-level; some lateral growth may occur but the geometry will reflect the relationship of growth rate, sedimentation and sea-level changes. Stage 3, growth into the zone of wave action, followed by differentiation of the reef structure as an energy gradient develops. On the windward side, in higher-energy conditions, a bindstone cap develops at the crest while lower-energy growth forms occur on the leeward side. Stage 4, the energy gradient creates different sedimentary facies on a reef flat with coarse 'rampart' deposits on the windward side, possibly colonized by mangroves, while on the leeward area, converging currents dump sand to form cays. Stage 5 (not shown), sediments accumulated on the reef flat may become lithified by marine vadose processes to give beachrock. After prolonged sedimentation meteoric diagenesis may occur. For further details and sources see text. The complex controls on the evolution of such patch reefs are discussed by Stoddart et al. (1978). (B) Idealized section through the windward zone of an aggrading patch reef. The biota also changes through stages 1 and 3 reflecting higher-energy and shallower conditions. In a prograding situation the base of the sequence would consist of reef talus. (C) Plan view of patch reef based on the inner shelf reefs of the Great Barrier Reef (modified from Scoffin, 1977). Waves are refracted around the reef flanks and impinge at an acute angle to the reef margin. Longshore drift of sand occurs and spits develop, while a low sand cay forms where the opposing wave sets meet. Coarse debris also accumulates behind the reef crest to form ramparts which are large ripple-like structures that prograde leeward and become cemented. Mangroves may colonize the zone behind the ramparts.

4.5.7a Biological zonation on modern reefs

Where strong physical gradients exist across modern reefs, usually due to windward and leeward orientation, a distinctive biological zonation develops. This zonation can simply reflect greater coral growth on the windward side (Shinn *et al.*, 1977), or actual differences in coral morphology, diversity and species composition (Spencer-Davies *et al.*, 1971; Geister, 1977; Wallace & Schafersman, 1977; Done, 1983; James, 1984b). These differences are related to wave energy (wave stress), light levels, the degree of sub-aerial exposure and the sedimentation rate. The resulting growth forms of the corals reflect the relative importance of these factors. (Barnes, 1973; Graus *et al.*, 1977; Chappell, 1980; Graus & Macintyre, 1982). Models of coral (or metazoan) morphology across reefs have been offered by several researchers (Fig. 4.109) but considerable care is needed in applying such generalized models to ancient reefs (Fagerstrom, 1987, pp. 219–223). The shape–environment relationship on Caribbean reefs is largely due to phenotypic plasticity, that is, the growth forms of the reef organisms respond directly to changing environmental factors. However, in Indo-Pacific reefs the changes in growth form reflect species replacement (Barnes & Hughes, 1982). It is essential in interpreting growth forms in ancient reefs to ascertain which of these processes has applied.

Another problem arises in comparing the upper reef front of modern reefs with ancient counterparts. On present-day reefs the coral *Acropora palmata* is the dominant species. It is a robust coral with stout branches (Fig. 3.5) but it is an unusual growth form in such settings, and only appeared in the Pleistocene.

Despite these reservations general trends can be recognized in ancient reefs (Fig. 4.109C; Esteban, 1979; Pomar *et al.*, 1985) and can be integrated into environmental interpretations, although there is always a danger of circular reasoning. Stearn (1982) has provided a critical review of the use of growth form in interpreting ancient reef environments.

4.5.7b Biological succession in reefs

In many ancient patch reefs and mounds reef, sequences of different biofacies and lithofacies occur which have been interpreted as ecological successions (Walker & Alberstadt, 1975; James, 1984b).

Four growth stage have been recognized (Fig. 4.110).

1 Pioneer (or stabilization) stage: this mainly consists of a skeletal accumulation or shoal deposits. In the Palaeozoic, the crinoids, perhaps filling a role comparable to the Cainozoic calcareous green algae, were the major contributors to such accumulations. The typical assemblage consists of sponges, corals, bryozoans and branching red algae. This phase might reasonably be regarded as a reef-mound stage (Section 4.5.8).

2 Colonization stage: this stage marks the appearance of reef-building metazoans, but diversity is commonly still low. Growth forms include branching and encrusting forms, and niche stratification during the stage increases diversity. It is generally a short-lived phase and so is relatively thin.

3 Diversification stage: diversity increases and a greater variety of growth forms occur. This stage forms the bulk of the reef.

4 Domination (or climax) stage: this stage is characterized by lamellar forms (encrusters) giving a typical bindstone or lamellar framestone cap to the reef. Evidence of higher energy is present with rudstone facies, and diversity is low.

Walker & Alberstadt (1975) regarded stages 1–3 as reflecting an autogenic succession, that is, *true* ecological succession. This is an orderly, community-controlled succession in which each stage 'prepares the ground' for the next stage, because each community alters the environment so that it is later replaced by groups of organisms better adapted to the changed environment. Stage 4 is regarded as allogenic (controlled by external factors), perhaps because growth causes the reef to build into shallower, higher-energy conditions, hence the typical reef-crest community. However, in many ancient reefs there is evidence that all the zones developed in shallow water.

The community succession model has subsequently been used in many studies of fossil reefs (Frost, 1977; Klappa & James, 1980; Harland, 1981) and its applicability has been stressed in several review papers (James, 1983, 1984b; James & Macintyre, 1985). However, doubts as to the validity of the model have been raised on both theoretical and practical grounds. The concept of simple ecological succession, an idea developed decades ago, is no longer accepted by ecologists (see review by Crame, 1980). Even in present-day environments it is very difficult to prove categorically that purely intrinsic (autogenic) controls have operated, so there is little chance to do this in fossil deposits.

In a detailed study of exposed Pleistocene reefs along the Kenyan coast, Crame (1980) found that

A	Growth form	Environment	
		Wave energy	Sedimentation
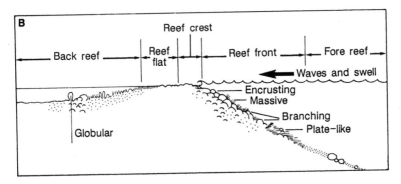	Delicate, branching	Low	High
	Thin, delicate, plate−like	Low	Low
	Globular. bulbous, columnar	Moderate	High
	Robust, dendroid, branching	Mod−high	Moderate
	Hemispherical, domal, irregular, massive	Mod−high	Low
	Encrusting	Intense	Low
	Tabular	Moderate	Low

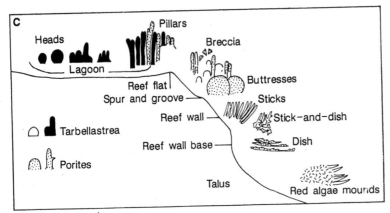

Fig. 4.109 *(A) Growth forms of reef-building metazoans and metaphytes and their relationships to wave energy and sedimentation. (B) Cross-section of a hypothetical reef illustrating the different morphologies of the reef-builders. (A) and (B) based on James (1984b). (C) Morphologies of corals from the Miocene 'reefal unit' of Mallorca, Spain. Based on Pomar et al. (1985).*

shallow patch reefs lacked a clear succession, but possessed a rather random element in their growth. Crame (1981) did recognize some patterns in the early stages of reef growth with pioneer species being replaced by faster-growing branched and platy encrusting types, and there was some evidence that in the longer term the reef would have become dominated by massive domal and encrusting forms. As regards diversity, the evidence was for its control by local environmental factors, whereas in some areas it appeared

Stage	Type of limestone	Species diversity	Shape of reef builders
Domination	Bindstone to framestone	Low to moderate	Laminate encrusting
Diversification	Framestone (bindstone) mudstone to wackestone matrix	High	Domal massive lamellar branching encrusting
Colonization	Bafflestone to floatstone (bindstone) with a mud stone to wackestone matrix	Low	Branching lamellar encrusting
Stabilization	Grainstone to rudstone (packstone to wackestone)	Low	Skeletal debris

Fig. 4.110 *Reef succession within a patch reef core, showing limestone types, relative species diversity and morphology of the reef-builders in each stage. Based on James (1984b).*

that equilibrium assemblages developed *very rapidly* within only 2 m of vertical growth. On a similar theme, Gould (1980) has commented on the scale of the apparent ecological successions described by Walker & Alberstadt. He pointed out that in some of the examples these authors described, the successions reach thicknesses of 70 m, but for true ecological succession to be invoked the conditions must have been unchanging for this period of time. If one uses the rates of growth of modern reefs as a guide such thicknesses reflect periods of 3000 to 10 000 years of stable conditions. Fagerstrom (1987) has reviewed examples of autogenic successions in present-day reefs, most of which develop very rapidly.

Workers on Palaeozoic reefs have also questioned the succession model. Hoffman & Narkiewicz (1977) have recognized a four-stage sequence in Lower to Middle Palaeozoic reefs, but have interpreted it as reflecting changes from bank (reef mound) to reef, as the result of extrinsic, physically-controlled factors, with each stage reflecting different energy levels. Williams (1980) has critically evaluated both these models, in relation to Devonian patch reefs. In this case the study revealed a good fit with the Walker & Alberstadt model, with evidence of the biotic modification of the current regime affecting community development.

Despite the succession model being widely used, surprisingly little effort has been made to study the now considerable database on Quaternary patch reef

successions. Can similar sequences be recognized in modern reefs and do these reflect extrinsic or intrinsic controls? The change from stage 1 to 2, from a reef mound to a frame reef, can be recognized in the reef mounds of the Florida Shelf (Section 4.5.8a), where skeletal accumulations are now overgrown by corals on the windward side of the mounds (Bosence *et al.*, 1985). In this case the succession apparently reflects a change in sea-level.

A comparison can be made between the typical patch reef succession and the Holocene sequence recorded from the Britomart Reef of the central Great Barrier Reef (Johnson *et al.*, 1984). This analogy is made to illustrate the complex interplay of intrinsic and extrinsic factors in a Holocene reef. The Britomart Reef sequence (Fig. 4.111) developed in an exposed shelf environment and, at least on the windward side, shows a sequence from bioclastic rudstones (mound stage) to framestone to bindstone. The coral rudstone unit, which is over 15 m thick, comprises 75% of the reef section. It consists of a poorly-sorted rudstone with a muddy matrix, and minor floatstones. There is little organic binding and the sediment is rich in *Halimeda* plates. It has been interpreted as a reef mound, comparable to *Halimeda* banks found elsewhere in the northern Great Barrier Reef area, which locally have relief of over 20 m (Section 4.5.8c). Its origin may be related to sea-grasses like Tavernier Key and the sea-grass banks of Shark Bay (Section 4.5.8b). Even though the origin of the bank is unclear, it was influenced by rising sea-level, which allowed skeletal material to accumulate *in situ* rather than be removed. As sea-level stabilized, the mound became the site for coral growth and the second unit, the coral framestone developed (Fig. 4.111). The reef grew up into the wave zone, and further developed windward accretion with the formation of the bindstone unit, consisting of algally-bound coral bindstones, with a coarse bioclastic matrix.

The Britomart Reef sequence reflects changes in sea-level, protection from waves, and biogenic processes such as sea-grass growth and later coral colonization. The change from mound to reef is likely to be externally controlled and not simply one of substrate suitability or else such colonization would have occurred earlier. The change from framestone to bindstone, comparable to stages 3 to 4 of Walker & Alberstadt's model, reflects wave exposure and is a common feature of present-day reefs (Adey, 1975; Littler & Doty, 1975; Rosen, 1975) where intense wave action reduces the grazing pressure on the algae, and higher light intensities and wave stress favour crustose coralline algae rather than uncalcified algae or corals.

Britomart Reef does not explain patch reef succession but it does illustrate that reef sequences are the products of complex physical and biological processes.

4.5.7c Reconstruction of fossil reef biotas

Many studies of ancient reefs provide detailed biofacies reconstructions with different communities distributed across the reef profile. Any reconstruction of the original reef biota must be made with the awareness of the varied taphonomic filters which have altered the composition. For example, a well-preserved metazoan biota might be recognized, forming a reef framework, but these are most likely to represent rapidly-growing forms which were not destroyed by bioerosion. Other well-preserved *in situ* fossils may have been buried rapidly or were overgrown and so were isolated from bioerosion. Delicate branched forms will have a lower preservation potential but they may have been volumetrically more important on the living reef than more robust or more rapidly-growing forms (Section 4.5.2b). A striking example of this taphonomic bias has been provided by Hubbard *et al.* (1986) from the shelf-edge reefs of St Croix, US Virgin Islands. On the living reef *Agaricia* represents 54% of the living scleractinian genera, yet it is absent in cores through the reef

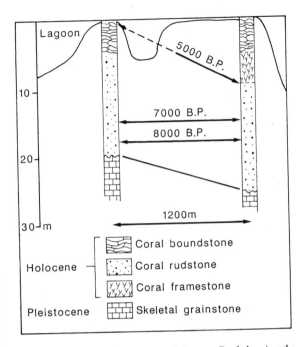

Fig. 4.111 *Sections through the Britomart Reef showing the succession on the windward side. From the bioclastic accumulations (stabilization or pioneer stage), the sequence passes into a framestone unit (diversification stage) and finally to a coral boundstone (domination stage). Such a sequence could be interpreted as due to succession but this ignores several key problems. For example, what is the origin of the lower mound stage and its relationship to rising sea-level (see text)? Why is the skeletal grainstone unit so thick in a relatively exposed shelf environment, unless sheltered by nearby reefs? Why did corals finally colonize the mound, and was this related to a decrease in the rate of sea-level rise? Based on Johnson* et al. *(1984). The labelling of such sequences as 'succession' ignores the experience of studies on Holocene reef sequences where external factors are the major controls.*

framework. This is believed to result mainly by the selective removal of *Agaricia* by bioerosion.

The composition of the skeletal debris in a reef may or may not be a reliable indicator of the original biological composition. It will be related to the composition of the original biota, but it will also reflect the relative production rates of the various organisms. For example, *Halimeda* produces several standing crops a year and so produces far more skeletal debris than other organisms. The preservation potential of such skeletal debris depends on many factors, such as skeletal architecture or size. Some organisms, which are easily broken down during transportation, will be underrepresented in the final deposit. Skeletal architecture also controls bioerosion. Hubbard & Swart (1982) have provided a detailed look at the relationship of environment and preservation for scleractinian corals.

Cryptic habitats, at first sight, may appear relatively good environments for the preservation of biota, being, at least in part, protected from physical erosion. However, taphonomic loss, even of skeletal encrusters, is very high (Rasmussen & Brett, 1985).

In summary, both the composition of the *in situ* biota and the granulometric properties and types of reef detritus are strongly biologically filtered, and what is now seen in a core or exposure is not necessarily a true reflection of the original composition of the reef.

4.5.8 Reef mounds

Much of the terminological and conceptual controversy over the recognition of ancient reefs has centred on whether the fine-grained, matrix-dominated, frame-deficient bioherms in the geological record represent true reefs. Many biologically-influenced buildups lack the remains of prominent frame-building organisms (e.g. Reid & Ginsburg, 1986) and as discussed elsewhere (Section 4.5.2a) large skeletal metazoans capable of building frames were not always present in the past (Longman, 1981; James, 1983, 1984b). These frame-lacking reefs have been termed reef mounds (James, 1983), and were not generally capable of forming reef complexes with wave-resistant reef fronts and crests. However, they show a great variety of forms which reflect the different organisms and processes involved in their formation. There is a spectrum of reef types from rigid frame-built forms to biologically-produced and stabilized mud mounds, and reef mounds represent the part of the spectrum in which *in situ* skeletal structures constitute, at best, only a minor part of the structure.

In general, ancient reef mounds have a lenticular to conical form, commonly with steep slopes. They consist of poorly-sorted bioclastic lime mudstones with some boundstones and typically occur in deeper-water settings than frame-built reefs. As such they correspond to the downslope mud mounds and knoll reef complexes of Wilson's (1974, 1975) classification (Section 4.5.1d).

Despite there being a wide variety of types of reef mound, Wilson (1975) was able to recognize a series of facies associations and defined a complex series of facies belonging to the mound itself and to its environs.

The initial stage of mound growth consists of a basal bioclastic wackestone pile which lacks obvious sediment trappers and binders and may be formed by the mechanical accumulation of skeletal debris. The second stage consists of a lime mudstone core containing organisms presumed to have been capable of trapping fine sediment. This zone is the thickest part of the mound and is commonly associated with such organisms as sponges, bryozoans, phylloid codiacean algae and marine grasses. Stromatactis is also common in this zone (Section 4.5.3a). The third stage represents stabilization of the mound by surface encrusters, represented by a 'crestal' boundstone horizon or mound cap (James, 1983). The fourth stage consists of organic veneers where the mound is protected by a veneer or wall of frame-builders. Later cementation may also occur.

Two additional facies are recognized in the flanks of the reef mound. Flank beds may form if the structure develops near wave-base and will be grainstone dominated. They typically consist of bioclastic material derived from attached reef benthos. Talus deposits represent lithoclastic and bioclastic debris derived from the mound and indicate lithification of the structure and erosion or slumping. Finally, many reef mounds show a capping deposit of grainstones representing a shallower, higher-energy phase. This sequence is rather theoretical and complex, and commonly one or more of the stages or facies is missing. It may serve as a guide to some ancient reef mounds but the sedimentological and biological causes and controls on their formation are not well understood. Reef mounds are also discussed in Section 9.2.4.

Direct analogues for many reef mounds are not known and frequent comparisons are made with the skeletal mud banks of the Florida Shelf and the embayments of Shark Bay, Western Australia. In both these areas shallow mud banks have been studied in some detail (see also Sections 4.3.4h and 3.3.1g).

4.5.8a Florida Shelf

Two types of shallow-marine mud bank can be recognized in this area. In the more restricted areas such as the Florida Bay shelf lagoon (Section 3.3.1g) sinuous linear subtidal banks occur which run for several kilometres and divide the lagoon into 'lakes', 3–4 m deep (Enos & Perkins, 1979). These linear mounds average 3–4 m in height and some are topped by mangrove islands as well as being cut by tidal channels. They consist mainly of burrowed lime mud (Fig. 4.112A) and have steeper windward slopes and gentler sea-grass-covered leeward sides.

Along the inner Florida Shelf, discrete banks occur, up to 3 km long and 1 km wide, orientated roughly parallel to the shelf margin. These mounds are up to 4 m high and are exposed at spring low tides. Two of them have been studied intensively, Rodriguez Bank (Fig. 3.33; Turmel & Swanson, 1976) and Tavernier Key (Fig. 3.34; Bosence *et al.*, 1985). Both developed initially in sheltered depressions where muds accumulated providing suitable substrates for colonization by the sea-grass *Thalassia*. Both mounds exhibit a lower carbonate mud mound stage, presently overlain by a sand–gravel mound stage. The mud mound units are

texturally and compositionally similar to the sediments associated with the *Thalassia* beds in the Florida Bay linear banks such as Cross Bank (Fig. 4.112A). The present surface sediments on the banks contain the coral *Porites* (Fig. 3.33B) and the red alga *Neogoniolithon* in a burrowed muddy gravel to muddy sand. No *in situ* frame is present although extensive bioerosion has destroyed a loose algal framework (Fig. 4.92; Bosence *et al.*, 1985). This coral–algal stage began some 3400 years BP when conditions became less restricted in the area due to a probable rise in sea-level which drowned a seaward protective reef platform (Bosence *et al.*, 1985). The early mud mound stage

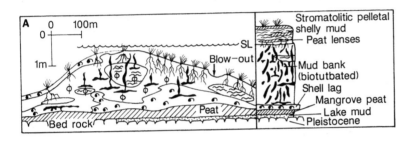

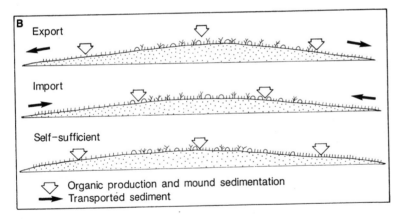

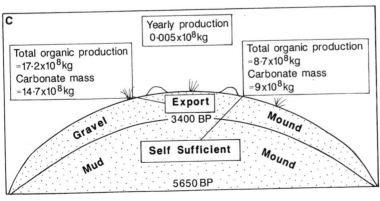

Fig. 4.112 *(A) Generalized cross and vertical sections through a Florida Bay mud mound (Cross Bank). These banks have asymmetric profiles but low slopes (less than 1°). The mound consists of highly bioturbated lime mudstone with zones of bioclastic wackestone, packstone and rudstone. Based on Enos & Perkins (1979) and Sellwood (1986). (B) Models to illustrate the relationship between external and internal sediment supply on carbonate mounds. (C) Sediment production rates and sediment masses for the two stages of growth of Tavernier mound, Florida Keys (B and C based on Bosence* et al., *1985).*

consists of a molluscan mud and was dervied from the breakdown of codiacean algae and *Thalassia* blade epibionts, which produce mud on the banks at the present time.

Although the banks contain a thin sedimentary sequence (Fig. 3.34) they can be used to derive models for understanding ancient reef mounds. Bosence *et al.* (1985) calculated the sediment budget of the Tavernier mound and devised three hypothetical models (Fig. 4.112B). The 'import model' envisages an external source for the carbonate which was produced off the mound and transported on to it and trapped. In the 'self-sufficient model' mound production supplies all the carbonate required for growth, while in the 'export model' production by the mound community exceeds accumulation with material being transported off the mound.

The budget calculations made by Bosence *et al.* (1985) indicated that the mud mound phase represented a self-sufficient situation (Fig. 4.112C). This model also seems to apply to Rodriguez Bank. Calculations show that overproduction is occurring at present during the gravel mound stage for Tavernier Bank and so the export model applies (Fig. 4.112C). Evidence for this comes from the surrounding off-mound sediments which contain up to 24% mound-derived *Porites* and *Neogoniolithon* grains transported during storms. It is during such export stages that flank/talus beds and related intermound sediments would be generated in ancient reef mounds.

Recently, Bosence (1989a) has examined the carbonate budgets for mounds on the Florida Shelf and in Florida Bay. The linear mounds of the inner shelf had previously been regarded as the products of trapping and binding of sediment produced off the banks (import model) (Stockman *et al.*, 1967; Enos and Perkins, 1979). Indeed the inner bay mounds were found to have relatively low production rates and are import mounds whereas those of the outer bay were export mounds. There is an order of magnitude decrease in carbonate production rates from the inner shelf mounds (Tavernier) and outer bay mounds to the inner bay mounds, reflecting the restriction of Florida Bay and the disappearance of faster-producing open marine biotas.

Applying the results of these careful studies to ancient mounds will require both detailed studies of mound and offmound sediments and biotas, and an appreciation of the different production rates associated with fossil biotas. It remains to be seen if this is possible.

4.5.8b Shark Bay

Sea-grass banks occur in the embayments of Shark Bay, Western Australia (see Fig. 4.15), and have been described in detail by Davies (1970b), Hagan & Logan (1974a) and Read (Section 4.1.3e; 1974b). The large sea-grass bank in eastern Shark Bay covers an area of 1036 km^2, with a length of 129 km and an average width of 8 km. The sediment in the bank reaches a maximum thickness of 7.6 m. A comparable fossil bank would be difficult to recognize at outcrop but might be comparable to some of the tabular phylloid algal 'reefs' described from the Pennsylvanian of the Paradox Basin in the USA. In the Edel Province (west Shark Bay) the sea-grass banks reach 6 m in thickness with 2–5 m of relief. They are wedge-shaped or lenticular with slopes of 5° or more. The former occur bordering shorelines while the latter are barrier banks up to 2 km wide, formed across coastal inlets, creating sills with restricted circulation zones in their lee.

The sea-grasses are critical to the growth of the banks, more so than physical factors and they influence sedimentation in three ways.

1 *Sediment contributors.* The epibiota on the grasses, of encrusting foraminifera and red algae, are major sediment contributors, as is the sessile benthos and other organisms.
2 *Sediment bafflers.* The grasses reduce water movement and may locally occur in such dense growths as to act effectively as a barrier.
3 *Sediment binders and stabilizers.* The grasses bind the sediment by means of their dense rhizome system.

The density of sea-grasses is a most important factor. In Edel Province, banks occur in a low-energy setting (Read, 1974b) but consist of skeletal 'grainstones' (mud-free sands), with minor 'packstones' (muddy, grain-supported sands), and are homogeneous, thinly-bedded and cross-bedded. The sea-grass cover is not dense and subsequently little fine-sediment baffling occurs, which is reflected in the matrix-poor sediments. There is a reduced percentage of sea-grass-encrusting epibionts contributing to the sediment, which consists of molluscan foraminiferal and coralline algal material.

In eastern Shark Bay (Wooramel Bank, Davies, 1970b) very dense sea-grass stands occur, which results in the baffling of fine sediment and a high matrix content (up to 30% <62 μm), as well as a higher proportion of encrusting epibionts than in the Edel Province banks. As a result the sediment is more muddy in a relatively high-energy area, a sobering thought for those interpreting ancient sediments. The denser binding of the sediment is also manifested in higher slope angles, especially along tidal channels which dissect the banks.

4.5.8c *Halimeda* mounds

Halimeda mounds, some with sediments up to 52 m thick, have been recorded from the shelf behind the northern Great Barrier Reef (Orme *et al.*, 1978; Davies & Marshall, 1985; Orme, 1985; Phipps *et al.*, 1985) and from an isolated platform, the Kalukalukang Bank, of the eastern Java Sea (Roberts *et al.*, 1987). They have been recorded in the depth range of 20–100 m. These mounds, with relief of over 12 m, are tens to hundreds of square metres in extent, and the larger ones have formed by the coalescence of smaller structures. Some consist almost entirely of *Halimeda* plates whereas others are packstones or wackestones with siliciclastic muds. They have rapid growth rates of 5.9 m 1000 yr^{-1} (Roberts *et al.*, 1987) but also exhibit evidence of major erosion surfaces. In the geological record crinoid banks may be a crude analogue, or codiacean phylloid algal mounds.

4.5.8d Deep-water reef mounds

Corals with symbiotic zooxanthellae require warm, well-lit, shallow waters and form reefs (hermatypic corals). However, corals lacking zooxanthellae (with the misnomer of ahermatypic) *are* capable of constructing reef mounds. Such corals can survive in waters of low temperatures and at depths of over 6000 m, and form reef mounds. These deep-water reef mounds are widespread along the present continental shelves, especially along the eastern Atlantic (Cairns & Stanley, 1981; Newton *et al.*, 1987). Two categories have been described: unlithified mounds and lithified reef mounds or lithoherms.

Lithoherms, a term coined by Neumann *et al.* (1977), are structures found in the Florida Straits, on the lower part of the Little Bahama Bank platform, in water 600–700 m deep. These structures, which form a more or less continuous facies belt of current-oriented mounds (aligned parallel to the platform edge), extend for over 200 km, are 10–15 km wide, and individual mounds are up to 100 m long and 50 m high, with steep slopes of up to 30° (Mullins & Neumann, 1979; Cook & Mullins, 1983). Dense coral growths of *Lophelia* and *Enallopsammia* occur on their upcurrent ends, with no corals on their downcurrent end. The mound biota consists of crinoids,

ahermatypic corals, sponges and alcyonarians.

The surface of the mound is a hard, muddy to sandy sediment, forming a lithified crust, 10–30 cm thick, conforming to the mound shape. The surface crust is cemented by high-Mg calcite, is extensively bored by sponges, and contains geopetal sediments. The flat areas around the mounds are veneered by rippled skeletal muddy sand with planktonic foraminifera and pteropods. The origin of these structures is unclear but Neumann *et al.* have speculated that they resulted from cementation, baffling and the local accumulation of material.

Unlithified deep-water coral buildups have been documented from a number of areas. Mullins *et al.* (1981) have documented coral mounds on the north slope of the Little Bahama Bank. These structures occur in water 1000–1300 m deep and are patchily distributed over a minimum area of 2500 km^2. The mounds are circular to elliptical in form, exhibit 5–40 m of relief above the seafloor and contain a diverse community including ahermatypic corals, alcyonarians, gorgonians, crinoids, sponges, gastropods and polychaetes. They are believed to have formed on seafloor irregularities where strong bottom currents provide oxygen and nutrients. Their growth is due to the local accumulation of skeletal material and to the baffling and trapping of finer sediments. Their development is via a number of stages shown in Fig. 4.113. This model is based on descriptions of analogous exposed Tertiary reefs by Squires (1964). The Bahamian slope facies are described further in Section 5.11.

Intermediate depth reefs (70–100 m) formed by the coral *Oculina* have been described off the east coast of Florida by Macintyre & Milliman (1970) and Reed (1980). Some of these mounds have developed on relict oolitic ridges and have relief of 17–24 m, but they have not been found in deeper waters. Mullins *et al.* (1981) have suggested that the mounds of the type found on the northern slope of the Little Bahama Bank would be preserved as a coral-framework-supported structure. However, Scoffin *et al.* (1980) have described *Lophelia* banks from the margins of the Rockall Bank of the eastern Atlantic which have a contrasting preservation bias. Bioerosion of the frameworks of these thickets and coppices (see Fig. 4.113) is probably more rapid than the rate of burial of the upright colonies. The resulting banks will contain only broken branches, that is a floatstone or possibly rudstone, and not a distinct framework. Such a taphonomic bias should always be considered when interpreting the framework component of any ancient reef or reef mound.

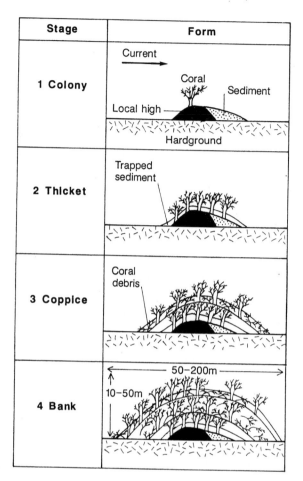

Stage	Form
1 Colony	Current →
	Coral, Sediment, Local high, Hardground
2 Thicket	Trapped sediment
3 Coppice	Coral debris
4 Bank	50–200m, 10–50m

Fig. 4.113 .*Proposed development of deep-water reef mounds. Stage 1, coral colonization of an area of positive relief. Isolated corals modify bottom currents. Sediment accumulates in the lee of the colonies. Stage 2, corals aggregate into thickets. Diversity and niche complexity increase. Baffling and trapping of sediment occurs. Stage 3, addition of skeletal debris from the coral biota. Colonies now aggregated into larger masses (coppices). A diverse benthic fauna occurs. Stage 4, the whole mound accretes and progrades with a circular to elliptical form in plan view. Based on Squires (1964) and Mullins et al. (1981).*

4.5.8e Mud mounds

Mud mounds are a form of reef mound in which the skeletal component is usually very low, although no strict definition exists of how low. James & Macintyre (1985) have suggested a division of reef mounds

into two categories: microbial buildups (such as stromatolites) and mud mounds such as Waulsortian and sponge mounds. However, it is becoming clear that the classic Waulsortian reef mounds, and probably other mud mounds in the geological record, are also microbial in origin. Mud mounds have recently been the subject of much research and a consensus seems to be emerging as to their origins. A review of their characteristics has been provided by Pratt (1982a). These mounds are mainly a Palaeozoic phenomenon but they also occur in the Mesozoic. Essentially, they are large reef mounds, commonly hundreds of metres thick, with evidence of depositional relief of, perhaps, over 200 m. They are typically up to 1 km in diameter, but may occur in extensive complexes covering thousands of square kilometres, and up to a kilometre thick. Internally, they are massive or exhibit a vague bedding with offset growth overlaps, resembling giant cross-sets. The term 'Waulsortian' is used to describe many Palaeozoic mud mounds but, as pointed out by Lees & Miller (1985), the term should be used only after comparisons have been made with the original Waulsortian mounds from the mid-Dinantian of the Waulsort area of Belgium. Waulsortian mounds also occur in mid-Dinantian sequences in England, south Wales, Ireland and New Mexico. Internally, they have been found to contain up to four phases of growth (Lees & Miller, 1985). They have termed these stages A to D (here referred to as 1–4).

Phase 1. This is the basal zone, with a low diversity assemblage of abundant fenestellid bryozoans and crinoid debris. Ostracode, brachiopod and bivalve material also occurs. Stromatactid cavities, filled by sparry calcite, are very common and have apparently formed as shelter cavities beneath large *Fenestella* sheets. This phase represents the 'classic' Waulsortian facies.

Phase 2. Hyalosteliid siliceous sponges characterize this zone and are represented by collapsed and calcitized structures. Fenestellid sheets are less abundant, and consequently so also are the associated spar-filled shelter cavity systems. Moravamminid sponges also occur and the dense micrite muds contain abundant microscopic tubular moulds of filamentous organisms, possibly after cyanobacteria.

Phase 3. Fenestellid bryozoans are absent but plurilocular foraminifera appear. The assemblage is more diverse with aoujgaliid (sponge?) fragments, gastropods, trilobites and echinoids. The tubular moulds occur and porostromate tubes appear for the first time in the buildup, including the form-genera *Girvanella*. Peloids are also common.

Phase 4. Algally-coated grains and micritization, probably by endoliths, are characteristic of this phase. Calcareous algae including dasycladaceans occur as well as radial ooids and fenestellid bryozoans.

Using this division of phases, Lees & Miller were able to classify all the Waulsortian mounds of Europe and New Mexico. Some mounds only contained phases 1–3, other only 2, or 2 and 3, or 2–4, but relatively few contained 1–4, like the Furfooz-type mound of Belgium (Fig. 4.114A). Lees *et al.* (1985) have offered depth values to each phase of the Belgium Waulsortian mounds, based on palaeontological data. Phase 1 represents water deeper than 150 m, probably below 300 m, phase 2 represents water between 120 and 150 m deep, phase 3 between 90 and 120 m, and phase 4 less than 90 m (photic zone) (Fig. 4.114B). The significance of these studies extends beyond the mid-Dinantian and shows how detailed microfacies work can shed light on apparently uniform mud buildups.

As well as growth phases within each Waulsortian mound, there are also lateral variations between mound core, flank and intermound limestones. These have recently been discussed by Bridges & Chapman (1988).

The actual origins of mud mounds is still a matter of debate. Some workers have regarded the mounds as purely hydrodynamic in origin, representing giant mud-grade sand waves. Others have envisaged the mud as having been baffled by organisms, such as the fenestellids and crinoids (Wilson, 1975). Pratt (1982) has suggested that microbial mats may have trapped and bound (and lithified) the muds. All these ideas require an external source of mud, and as pointed out by Lees & Miller (1985) and Miller (1986), the efficiency of this process must have been staggering for such huge buildups to form with 100–200 m of relief. From their study of Waulsortian mounds, Lees & Miller (1985) have noted that the micrite muds in the

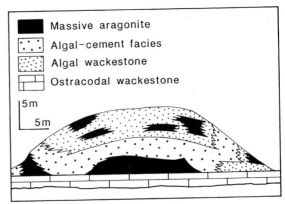

Fig. 4.115 *Schematic cross-section of the Scorpion reef mound of Early Permian age from the northern Sacramento Mountains of New Mexico. The section shows the original reef facies distribution. The structure was later modified by subaerial exposure which created minor relief on the mound surface and was associated with extensive internal brecciation of the reef. The massive aragonite consists of calcite which has replaced radial spherulites of aragonite; the algal cement consists of fibrous cement and skeletal material (bivalves and some phylloid algae); the algal wackestone has been interpreted as a bafflestone formed by codiacean phylloid algae which had leafy calcified thalli. These reefs lack a true frame but were probably quite rigid structures because of the large volume of early cement, much of which was probably formed directly by the phylloid algae (see text). Diagram modified from Mazzullo & Cys (1979) and Toomey & Babcock (1983).*

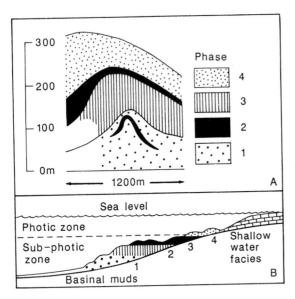

Fig. 4.114 *(A) Schematic cross-section of the Furfooz mound in Belgium, showing the various Waulsortian phases. (B) Depth zonation of possible Waulsortian mounds on a carbonate ramp. The Furfooz mound started growth in subphotic depths (probably greater than 300 m) and as a result of upward growth and sea-level changes, reached the photic zone. Modified from Lees & Miller (1985).*

mounds are texturally and compositionally different from the much thinner muds surrounding the mounds. For this to happen by a trapping and binding mechanism, the organisms must have been capable of selectively excluding argillaceous material, which is more abundant in the intermound sediments. They argue that the muds in the mounds must have been locally produced, were associated with a probable mucilaginous surface cover, and were probably microbially precipitated. Monty *et al.* (1982) and Tsien (1985b) reached similar conclusions for Devonian mud mounds in Belgium. Bridges & Chapman (1988), in a detailed description of Early Carboniferous (Chadian) age mounds from Derbyshire, England invoked a microbial origin for the clotted, micritic reef muds. However, these mounds are elongate and aligned, suggesting that sediment may have been moved and Bridges & Chapman compared the situation to the 'export model' of Bosence *et al.* (1985; Section 4.5.8a).

The microbial explanation may not apply to all mud mounds. Bourque & Gignac (1983) have interpreted the peloidal fabrics of Silurian mud mounds of Quebec as resulting from the decay of calcareous sponges such as *Malumispongium*, which are associated with muds.

4.5.8f Cement reefs

Another rather bizarre type of reef mound consists of large volumes of cement, apparently lacking a frame or abundant skeletal debris. These are best documented from the Upper Palaeozoic, especially the Permian of the southwestern USA (Mazzullo & Cys, 1977, 1979; Toomey & Babcock, 1983) and are associated with phylloid algae or algal-like problematica. In such reefs replaced aragonitic botryoidal cements locally constitute up to 80% of the rock (Fig. 4.115). Their origins remain unclear but the role of peyssonnelid algae is suspected (Section 4.5.2d; James *et al.*, 1988).

Recently, Edwards and Riding (1988) suggested that many 'replaced aragonite' cements represent neomorphosed algal skeletons, drawing an analogy between such 'cements' and recrystallized Silurian solenoporid red algae.

5 Carbonate depositional systems II: deeper-water facies of pelagic and resedimented limestones

5.1 INTRODUCTION TO PELAGIC CARBONATES

Pelagic carbonates are deposits of the open sea, composed chiefly of skeletal debris from planktonic organisms. At the present time, coccoliths, foraminifera and pteropods comprise the calcareous oozes which are widely distributed in a range of deeper-water settings, including the vast areas of the ocean floor, seamounts, mid-ocean ridges and outer continental shelves. Siliceous oozes are composed of radiolarians (zooplankton) and diatoms (phytoplankton). Pelagic oozes are variably admixed with terrigenous material and volcanic ash. In the geological record, pelagic limestones are well represented in the Mesozoic and Cainozoic, but in the Palaeozoic they are less common. There is, of course, a preservation problem with many pelagic carbonates, in that being deposited most extensively on ocean floors, many will be carried down subduction zones and lost to the rock record, or they will be scraped off subducting ocean floor to be preserved as deformed masses in mountain belts. Pelagic deposits have recently been reviewed by Scholle *et al.* (1983), Leggett (1985) and Jenkyns (1986).

5.2 PELAGIC OOZES AND ORGANISMS

Pelagic sediments on the deep seafloors vary considerably in composition depending on the proportions of four major constituents: carbonate skeletons, siliceous skeletons, terrigenous silt and clay and authigenic components, notably Fe−Mn oxides, zeolites and clays. Six major classes of deep sea sediment have been recognized during the DSDP/ODP campaign (see Table 5.1 and initial reports, e.g. Austin *et al.*, 1986) and these include two groups of calcareous sediments: (1) pelagic calcareous biogenic, with 65−100% $CaCO_3$, and (2) transitional calcareous biogenic, with 35−65% $CaCO_3$.

Qualifiers can be added (foram, nannofossil, pteropod) to indicate the dominant type of skeletal grain present. In addition, modern to sub-Recent pelagic carbonates vary in their degree of induration, so that three classes of firmness are recognized: (1) unlithified, ooze if a fine-grained sediment, (2) party lithified (firm or friable), corresponding to the term chalk for finer-grained sediments, and (3) lithified (hard, non-friable, cemented), corresponding to the term limestone (or lithified ooze).

Of the pelagic biogenic calcareous sediments, foraminiferal ooze is the most important type in terms of volume and distribution on the present ocean floors, although the faunal composition does change markedly with latitude in response to water temperatures. More delicate and smaller foraminifera are less resistant to dissolution than larger more robust forams, so that the microfossil composition of an ooze may not reflect the living assemblage of the surface waters. Forams reach 1−2 mm in diameter and may be reworked to produce lime sands, through winnowing out of lime mud. *Globigerina* ooze is a term frequently applied to ooze with more than 30% planktonic foraminifera, although *Globigerina* may not be the most common microfossil. Planktonic foraminifera live within 100 m of the water surface. Benthic forams form less than 1% of the ooze generally.

The yellow−green unicellular algae coccolithophoridae produce spherical tests, 10−100 μm in diameter, called coccospheres. They are made up of platelets 2−20 μm across, termed coccoliths. Coccospheres are present in huge numbers in surface waters, 50 000 to 500 000 per litre being typical. Coccoliths account for much of the fine-grained debris (less than 6 μm) in ooze. However, because of their very small size, coccolith platelets take a long time to settle and are very prone to dissolution during sedimentation. Many coccoliths are probably sedimented as aggregates, produced in faecal pellets. High rates of coccolithophorid productivity occur in regions of upwelling in sub-Arctic and sub-Antarctic waters. However, there are also high rates of diatom productivity in these areas and much of the coccolith material dissolves during sedimentation, so that bottom sediments in high latitudes contain less than 1% coccoliths.

Coccoliths are generally not such important con-

Table 5.1 Lithological classification of deeper-water, fine-grained sediments, as used in the Ocean Drilling Program. Figures in per cent. Based on Austin *et al.* (1986)

Sediment type	Components				
	Terrigenous and volcanic detritus	Calcareous skeletons	Silt and clay	Siliceous skeletons	Authigenic components
Pelagic clay	—	—	—	<30	>12
Pelagic biogenic siliceous	—	<30	<30	>30	—
Transitional biogenic siliceous	—	<30	>30	10–70	—
Pelagic biogenic calcareous	—	>30	<30	<30	—
Transitional biogenic calcareous	—	>30	>30	<30	—
Terrigenous sediments	>30	<30	—	<10	<10

stituents of modern pelagic oozes as the foraminifera, although in some locations, such as the Mediterranean, these nannofossils dominate the carbonate fraction. In the Early–mid-Tertiary, it appears that the coccolith contribution to oceanic sediments was greater than foraminifera. Some coccoliths are important in biostratigraphy (e.g. Bukry, 1981), especially in mid–low latitude oozes, and generally these ones have a greater resistance to destruction and dissolution than the planktonic foraminifera. Discoasters are large coccoliths common in the Tertiary, which are resistant to dissolution and usually have prominent calcite overgrowths.

The pteropods are conical planktonic gastropods reaching a few millimetres in length. Since they are composed of aragonite, the more soluble $CaCO_3$ polymorph, their occurrence in pelagic oozes is restricted to the shallower depths of tropical regions.

Living within deep sea oozes are many deposit-feeding worms of unknown affinities. However, these organisms leave distinctive trace fossils which have been described from numerous deep sea cores. The characteristic assemblage consists of *Chondrites*, *Planolites* and *Zoophycos* (shown in Fig. 5.22); the vertical burrows, *Skolithos* and *Teichichnus*, of suspension-feeding organisms, are rare to moderately common (Ekdale & Bromley, 1984).

5.3 CONTROLS ON PELAGIC CARBONATE SEDIMENTATION

One of the major controls on the sedimentation of pelagic carbonate is the *calcite compensation depth* (CCD). This is the depth, generally of several kilometres, below which $CaCO_3$ does not accumulate since the rate of calcite supply is balanced by the rate of calcite dissolution (Fig. 5.1). The dissolution is a result of the decreased temperature, increased pressure and increased CO_2 content of deeper waters. Although ocean water is undersaturated with respect to $CaCO_3$ below several hundred metres, extensive dissolution of carbonate particles being sedimented is inhibited by organic coatings on grains. Also the sedimentation rate of these fine particles is increased by their occurrence as aggregates in faecal pellets. Towards the CCD, there is a progressive loss of the various skeletal grains, with foraminifera having thick, robust tests persisting to deeper levels than those with more fragile delicate tests. Planktonic foram tests are generally less resistant to dissolution than benthic foraminifera. Below the CCD, radiolarian oozes are deposited in areas of high biological productivity, while red clays occur elsewhere. Resedimentation of pelagic ooze and rapid burial can lead to deposition of carbonate below the CCD. The effect of the CCD is well seen on the flanks of mid-ocean ridges where pelagic carbonates pass laterally into siliceous oozes and red clay as the depth increases away from the ridge axis through thermal subsidence of the volcanic pile (see Fig. 5.7).

At a somewhat shallower depth than the CCD, referred to as the lysocline, there is a pronounced increase in the rate of calcite dissolution (Fig. 5.1). The lysocline is defined by the position of the maximum change in the foraminiferal composition of the ooze due to differential dissolution (Berger & Winterer, 1974), although it is not such an easy level to locate on the seafloor as the CCD. A facies boundary between well-preserved and poorly-preserved foraminiferal assemblages can be located in the floor of

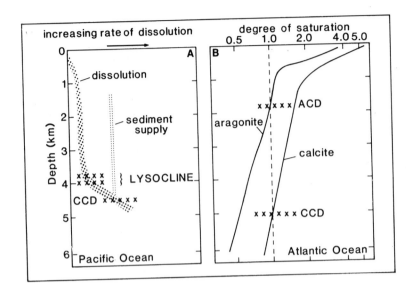

Fig. 5.1 *Carbonate saturation and dissolution in the deep sea. (A) The profile of increasing $CaCO_3$ dissolution with increasing depth for the Pacific Ocean. The lysocline is the depth where the rate of dissolution increases markedly, and the CCD is the depth where the rate of sediment supply is matched by the rate of dissolution and below which therefore the sediments are $CaCO_3$-free. After Jenkyns (1986). (B) The profiles of decreasing degree of saturation for aragonite and for calcite with increasing depth for the Atlantic Ocean. After Scholle et al. (1983).*

the central Atlantic south Pacific Oceans (e.g. Berger, 1970). A compensation depth for aragonite (the ACD) can be recognized by its disappearance from the ooze, at a depth much shallower than the CCD (Fig. 5.1).

The positions of the CCD and ACD are variable in space and time. They are affected by the fertility of surface waters, since this determines the rate of carbonate supply, and by the degree of undersaturation of the deeper waters. These two factors are very much affected by oceanic circulation patterns, temperature and salinity, which are ultimately controlled by plate positions, opening/closing oceans and global sea-level.

In areas of high organic productivity, such as in the Pacific equatorial zone, the CCD is relatively deep at around 4.5–5 km. Away from the equator the CCD rises as productivity decreases towards the central water masses and oceanic gyres, and bottom waters become colder and more undersaturated with respect to $CaCO_3$ (Figs 5.2 and 5.3). In the equatorial Atlantic, the CCD is a little deeper at around 5 km, and it occurs at this depth to 40–50° of latitude, north and south (Fig. 5.3). By way of contrast, the ACD is 2–2.5 km in the Atlantic, but less than 300 m in the Pacific. The dissolution of $CaCO_3$ is also affected by the production rate of CO_2 derived from the breakdown of suspended organic matter. Apart from being related to surface productivity, the CO_2 content of the oceans is also controlled by deep-water circulation and oceanic mixing. 'Older' waters in ocean basins are more aggressive to $CaCO_3$ particles. As a result of more efficient flushing and better circulation, the

north Atlantic has about half the amount of dissolved CO_2 as the Pacific, and this largely accounts for variations in CCD and ACD positions between the two oceans. Close to continental margins, the CCD shallows (Fig. 5.2) due to the high input of organic matter and consequent high CO_2 levels.

High production rates of calcareous plankton presently occur in warm low-latitude surface waters, mostly within 60° of the equator. The siliceous plankton are more common in cooler surface waters, especially where there is an abundant supply of nutrients, as in areas of upwelling. The latter are present off the western side of continents and along oceanic divergences. Siliceous plankton skeletons, composed of opaline silica, are also subject to dissolution during sedimentation and on the seafloor, but the 'opal compensation depth' is much deeper than the CCD. Diatoms are less resistant than radiolarians and sponge spicules.

Studies of cores from the ocean floors (e.g. Berger & Winterer, 1974; Ramsay, 1977; Van Andel et al., 1977) showed that the position of the CCD has varied with time (Fig. 5.4), both within ocean basins and between them. There is a general trend in all ocean basins of relatively shallow CCD during the Eocene, followed by deepening into the Oligocene and Early Miocene. Another shallowing phase into the Late Miocene is evident before a fall to very deep levels in the Pliocene–Quaternary. Differences between the Atlantic and Pacific CCD depths are attributed to oceanographic factors, such as the more vigorous circulation in the Atlantic as a consequence of the

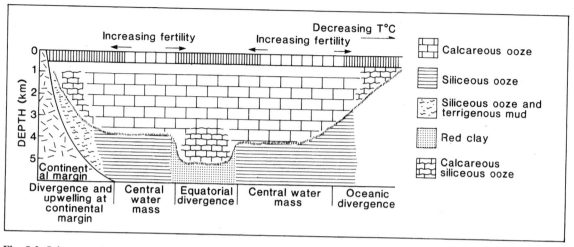

Fig. 5.2 *Schematic diagram of the distribution of deep sea sedimentary facies and position of the CCD relative to major patterns of oceanic circulation and near-surface fertility, based on the Pacific Ocean. After Leggett (1985) after Ramsay (1977).*

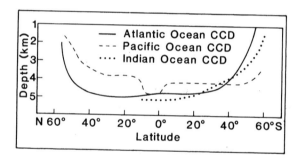

Fig. 5.3 *Position of the CCD within the major oceans for the Pliocene–Quaternary. After Leggett (1985) after Ramsay (1977).*

opening of the Norwegian–Greenland Sea in the Early Tertiary (Ramsay, 1977). The depressed CCD in the Oligocene in all oceans suggests increased ventilation which could reflect the development of high-latitude sea-ice, producing cold oxygenated waters and increased bottom water circulation.

There is apparently a connection between the CCD and global sea-level (Fig. 5.4; Berger & Winterer, 1974; Berger *et al.*, 1981) whereby the CCD rises when sea-level is high through extensive carbonate sedimentation on continental shelves, and the CCD falls during major regressions when the volume of pelagic carbonate sedimentation increases.

Short-term climatic changes may affect pelagic sedimentation. If is now well established from oxygen isotope analyses of Quaternary pelagic sediments (see Section 9.6.2; Fig. 9.20), that there have been subtle variations in climate over the past million years as a result of the Milankovitch cyclicities (see Section 2.10.3). Ancient pelagic facies commonly show Milankovitch-type cyclicities in bed thickness variations, and clay and microfossil contents (see Section 5.7.1). The orbital perturbations are causing fluctuations in insolation, which lead to variations in plankton productivity and/or terrigenous input to the oceans.

Much pelagic sediment is not simply deposited out of suspension, but is affected by deep sea currents and processes of resedimentation. At times of vigorous circulation such as occur during glacial periods, ooze may be eroded or not even deposited, to give hiatuses in the stratigraphic record. Apart from an absence of particular zone fossils, breaks in sedimentation may be marked by hardgrounds or iron–manganese nodules and pavements. Hiatuses are especially common in the Late Eocene–Oligocene, when strong bottom currents flowed northwards from the Antarctic region into the Pacific and Atlantic Oceans. Miocene stratigraphic breaks in the north Atlantic are likely due to south-ward-flowing bottom water from the Arctic. Contour-following geostrophic currents along continental margins and basin slopes rework sediments into lags and contourites and may lead to the development of

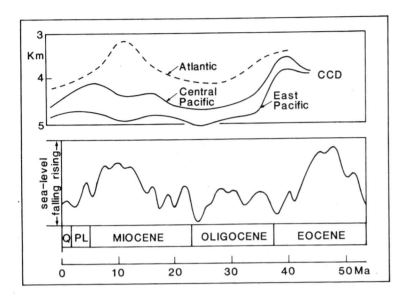

Fig. 5.4 *Variations in depth of the CCD through the Cainozoic (top) compared with global sea-level. After Berger et al. (1981).*

sediment drifts (see Section 5.11); resedimentation by turbidity currents, slides and slumps is, of course, commonplace (Section 5.10).

Resedimentation is common in areas of relief, such as occur along mid-ocean ridges, and thick accumulations of pelagic ooze in depressions, small basins and rift valleys, along these ridges are largely the result of this resedimentation. Turbidites of pelagic materials are most easily recognized where they are deposited below the CCD and so occur within non-carbonate ooze. They generally have a sharp base and a concentration of foraminifera at the base of the turbidite bed, but grading is poorly developed (because of the homogeneous grain size distributions) and cross-lamination generally absent (e.g. Kelts & Arthur, 1981). The upper part of pelagic turbidite beds may be disturbed by burrowing organisms.

5.4 PELAGIC CARBONATE SEDIMENTATION RATES

Rates of pelagic carbonate sedimentation are generally quite low compared with shelf carbonates with the typical range being $0.5-1.5$ g cm^{-2} 1000 yr^{-1} or an average of 13×10^{14} g yr^{-1}. These rates are the order of $10-50$ mm 1000 yr^{-1}, whereas siliceous oozes and red clay accumulate at $1-10$ mm 1000 yr^{-1}. It does appear that there have been strong variations in the pelagic carbonate sedimentation rate through the Cainozoic, with periods of rapid deposition occurring

at $0-6$, $22-30$ and $45-53$ Ma BP (Fig. 5.5; Davies & Worsley, 1981). These patterns are seen in Atlantic, Pacific and Indian Oceans, but there are some differences such as an increase in carbonate sedimentation $12-15$ million years ago in the Atlantic and Indian Oceans, and a corresponding decrease at this time in the Pacific. There is a broad correlation between total pelagic carbonate accumulation and global sea-level, with high rates corresponding to times of maximum exposed continental areas, when the carbonate flux to the seas from rivers was high. Deviations from the pattern are explained by oceanographic factors affecting productivity and dissolution rates, and climatic changes. On the time-scale of 10^4 to 10^5 years, small variations in sedimentation rate can be expected through the effect of the Milankovitch cyclicities on organic productivity (Section 5.3).

5.5 DEPOSITION OF MODERN PELAGIC CARBONATE FACIES

Pelagic carbonates are accumulating in a wide range of oceanic settings and on continental margins at the present time. In fact, some 50% of the seafloor is covered by carbonate ooze (more than 30% $CaCO_3$) and this occurs particularly between latitudes 45° N and S of the equator (Fig. 5.6). Jenkyns (1986) has reviewed in detail the occurrences of modern pelagic sediments and distinguished five depositional locations, based on their tectonic and geomorphologic

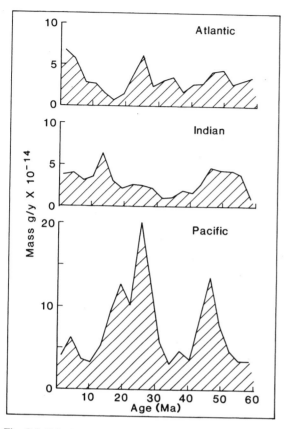

Fig. 5.5 *Pelagic carbonate deposition in the major oceans through the Cainozoic. After Davies & Worsley (1981).*

of low sedimentation and reworking, and foraminiferal sands may form through winnowing. Once the seafloor subsides below the CCD, several hundred kilometres away from the ridge axis, then red clay and siliceous ooze accumulate (Fig. 5.7).

The aseismic volcanic structures category of Jenkyns includes volcanic seamounts, ridges and plateaux, which rise several kilometres above the deep ocean floor. Initial sediments are again iron and manganese-rich muds and altered volcaniclastics, but hydrothermal precipitates are less important. Since many of these structures originally reached close to, or even above, sea-level, shallow-water and reefal carbonates commonly occur on top of the lavas. Subsidence after volcanic activity has commonly led to pelagic facies being deposited over the shallow-water carbonates. Since these aseismic structures occur throughout the oceans and migrate with seafloor spreading, the pelagic deposits upon them will vary according to location with respect to high productivity zones, latitude and depth. Phases of non-deposition, particularly on the flanks of the highs, have resulted in cementation of the carbonates (e.g. Fischer & Garrison, 1967; Milliman, 1971) and precipitation of phosphates (e.g. Marlowe, 1971) and Fe−Mn crusts (e.g. Cronan, 1980). Hiatuses are also produced in these sequences through current action and fluctuations in the position of the CCD.

Deep ocean basins mostly have depths below the CCD so that red clay and siliceous oozes are the typical sediments on the abyssal hills and plains. The red clay consists mostly of illite and montmorillonite, some zeolites and Fe−Mn oxyhydroxides. Much of this material is derived from alteration of volcanic ash, although there is a significant component (10−30%) of wind-blown dust. Ferromanganese nodules (Fig. 5.8) and crusts occur in regions of little or no sedimentation, especially where there are strong bottom currents. Radiolarian oozes are best developed in the equatorial Pacific and Indian Oceans, whereas diatom oozes occur in a global southern high-latitude belt and in the northern Pacific.

Small ocean basins are variable in their pelagic record and may have a substantial terrigenous input. Some, such as the Mediterranean and Red Sea, are sites of calcareous sedimentation; in others, including the Gulf of California and back-arc basins of the northern Pacific, siliceous, particularly diatom oozes are accumulating. In both the Red Sea and Mediterranean,

context: spreading ridges, aseismic volcanic structures, deep ocean bains, small ocean basins and continental margins.

Spreading ridges vary considerably in their topography but generally occur at depths of 2.5−3.0 km, i.e. above the CCD. Dark brown metalliferous deposits typically occur directly upon the basaltic pillow lavas, breccias and sands and many consist of Fe-rich smectites and Fe−Mn oxyhydroxide globules. Sulphides and sulphates are precipitates where heated waters (up to 350°C) are vented on to the seafloor through smokers, with their specialized and often bizarre suite of animals. Crusts of Fe and Mn oxyhydroxides occur on ridge volcanics and may also be of hydrothermal origin. Away from the ridge axis, pelagic carbonates occur particularly in valleys and basins, with much resedimentation taking place off local highs. Cementation of ooze is common in areas

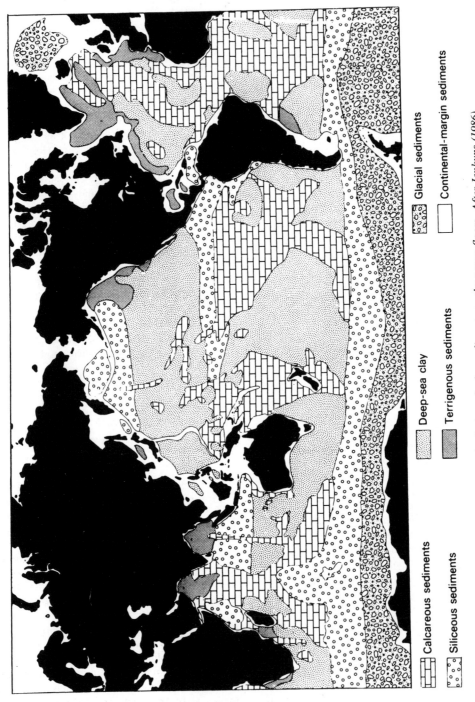

Fig. 5.6 *Global distribution of principal types of pelagic and other sediments on the ocean floors. After Jenkyns (1986).*

Calcareous sediments

Siliceous sediments

Deep-sea clay

Terrigenous sediments

Glacial sediments

Continental-margin sediments

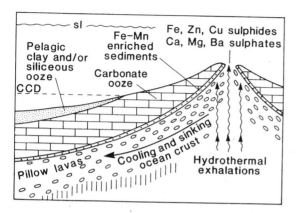

Fig. 5.7 *Schematic distribution of pelagic sediment on a fast-spreading mid-ocean ridge. After Jenkyns (1986) based on several sources.*

organic productivity is relatively low because of the poor circulation (inflowing surface water and out-flowing deep water). In the Mediterranean, micritic high-Mg calcite is being precipitated inorganically and nodules are forming within the carbonate ooze (Milliman & Müller, 1973). In the Red Sea aragonite-cemented pteropod layers and high-Mg calcite lithified nannofossil−foram clasts occur in the Pleistocene (Milliman *et al.*, 1969). In both cases, elevated temperatures and salinities are probably responsible for this seafloor cementation. In eastern Mediterranean Late Cainozoic strata, there are numerous black sapropel layers, generally a few centimetres thick, composed of organic carbon, clays and nannofossils (e.g. Murat & Got, 1987). They formed when the Mediterranean was periodically stratified, with a deep stagnant layer permitting the sedimentation of much organic matter. The sapropels are probably related to the climatic changes of the last million years and the glaciation/deglaciation of northern Europe.

The Gulf of California is a highly fertile small ocean basin because of deep-water inflow, surface water outflow (induced by strong offshore winds) and seasonal upwelling. Muddy diatom oozes are being deposited in the gulf with an organic-rich facies consisting of clay-rich and diatom-rich laminae occurring at depths of 300−1200 m where the oxygen-minimum zone impinges on the basin slope. Of interest in terms of carbonates, is the development of dia-genetic dolomite beds within these anoxic sediments (see Section 8.7.5, and Kelts & McKenzie, 1982). The dolomites, which have distinctive carbon isotopic sig-natures, are related to reduced sulphate levels brought about by organic matter diagenesis.

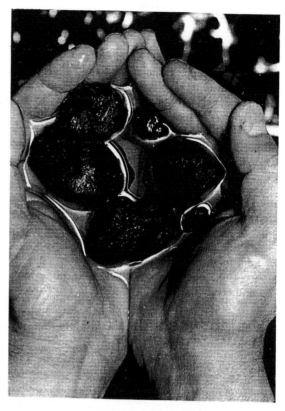

Fig. 5.8 *Manganese nodules from the Blake Plateau, depth 3000 m, off eastern USA. Photo reproduced from* The Smithsonian, *with kind permission of the photographer, Christopher Springmann, Pt Reyes, California.*

Continental margins are the locations of pelagic sedi-mentation where terrigenous debris is excluded, and depths are in excess of, say 50−100 m, so that benthic organisms are restricted and at greater depths, ex-cluded. In many cases, the pelagic oozes are accumu-lating above shallow-water carbonates, deposited at a time of lower sea-level or before major subsidence. The Blake Plateau off the eastern USA is a fragment of continental lithosphere which subsided rapidly in the Early Cretaceous, so that birdseye limestones are overlain by pelagic limestones. Strong currents (the Gulf Stream) over the plateau have resulted in little deposition during the Cainozoic so that ooze is com-monly cemented into hardgrounds and nodules, and phosphatized pelagic limestones, phosphatized lag de-posits, ferromanganese nodules and pavements are common.

Pelagic carbonates occur on the slopes around the Bahamas where they are mixed with bank-derived material to form periplatform ooze (see Section 3.2.1e and Mullins *et al.*, 1984). Nodules and hardgrounds occur in these slope carbonates and resedimentation is important in transporting material to deep water. The Yucatan Peninsula in the Gulf of Mexico is a good example of an open carbonate shelf with inshore reefs and sand bodies that pass seawards into pelagic oozes (Logan *et al.*, 1969). Relict shallow-water carbonates occur on the outer shelf and are being mixed with modern planktonic foraminifera in water depths in excess of 50 m.

These varied occurrences of modern pelagic carbonates all have their equivalents in the geological record, although there are some ancient pelagic limestones such as the very extensive epeiric sea Upper Cretaceous chalks, which have no modern counterparts.

5.6 DIAGENESIS OF MODERN PELAGIC CARBONATES

As a result of deep sea drilling, many studies have considered the diagenesis of pelagic carbonates, and the nature of the pore-fluids. Most pelagic carbonates are homogeneous foram–nannofossil oozes with a simple mineralogy (low-Mg calcite). Generally, they have not been affected by subaerial exposure and meteoric diagenesis and they have simply been gradually buried with marine pore-fluids. Two locations of diagenesis can be distinguished: (1) on and close to the seafloor, where cementation and alteration are most pronounced in areas of low or no sedimentation and high current activity, and (2) during burial where diagenetic processes operate at considerable depths within the sediment.

There are now many records of cementation of oozes on the deep seafloor at depths down to 3500 m, mostly from areas of negligible sedimentation, such as seamounts, banks and plateaux, resulting in the formation of crusts, hardgrounds and nodules (e.g. Fischer & Garrison, 1967). The limestones are commonly bored and may be impregnated with phosphate and Fe–Mn oxides. The cement is generally a micritic to decimicron equant calcite, with a low Mg content reflecting the low temperatures of the deep waters. In shallow-buried periplatform ooze from around the Bahamas, in water depths of 700–2000 m, aragonitic components have been leached, high-Mg calcite grains have lost their Mg and lithification is by low-Mg

calcite (e.g. Schlager & James, 1978; Mullins *et al.*, 1985; Dix & Mullins, 1988). Oxygen isotopic signatures of these deep-water limestones are normally enriched in ^{18}O relative to associated ooze (Fig. 5.9), again reflecting the cold deep waters, which are usually in the range of 3–6°C. Near-surface cementation of pelagic oozes to form chalky layers is occurring in Oligocene-aged sediments of the south Atlantic at a depth of only 13 to 130 m sub-bottom (Wise, 1977). The cement consists of calcite overgrowths on large coccoliths, such as *Braarudosphaera* sp., with the $CaCO_3$ probably derived from dissolution of smaller coccoliths.

Where pelagic carbonate has been deposited more or less continuously, then a progressive increase in lithification is revealed in cores, with ooze passing into chalk and into limestone with increasing depth. The passage from one lithology to another is mostly gradual, although they may be interbedded in the transition zones. The depths of the ooze–chalk and chalk–limestone transitions vary with sedimentation rate and location, but ooze generally occurs to depths of 150–250 m below the seafloor (i.e. in Late Tertiary strata), chalk is typical for Early Tertiary sections (to depths of 500–800 m), and limestone occurs in Early Tertiary–Late Cretaceous parts of the sequence, com-

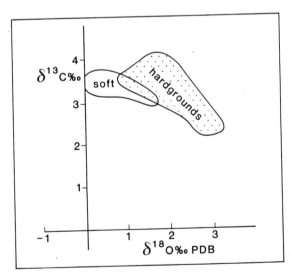

Fig. 5.9 *Carbon and oxygen isotopic cross plot of data from uncemented and cemented pelagic sediment from Tongue of the Ocean, Bahamas. The hardgrounds have a more positive oxygen signature because of cement precipitation from cold, bottom waters. After Schlager & James (1978).*

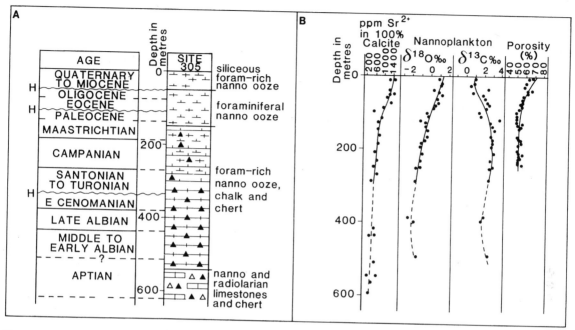

Fig. 5.10 *Lithology and geochemistry of a Pacific Ocean deep sea core. (A) Lithostratigraphy from DSDP Site 305 on the Shatsky Rise, northwest Pacific Ocean, showing increase in lithification with depth. (B) Down-hole changes in porosity, Sr^{2+}, carbon and oxygen isotopes. After Matter* et al. *(1975).*

monly with chert nodules (e.g. Fig. 5.10; Matter, 1974; Garrison, 1981).

Down-core reduction in porosity is very marked and occurs in two stages. In the upper 200 m, porosity decreases from 80% to 60% as a result of rapid dewatering. This takes around 10 Ma. Below 200 m in the deep burial realm, porosity decreases to 40% or less over 1000 m. The process involved here is mainly cementation, and several tens of millions of years are involved.

There are conspicuous changes in fossil preservation down-core. There is a gradual decrease in the abundance of planktonic foraminifera through dissolution, so that the planktonic/benthic foram ratio decreases down-hole. The reason for this is the greater solubility of planktonic foram tests during burial, compared with benthic forams. There is also an increase in the percentage of broken tests down-hole: 20% in ooze, 50% in chalk. Chalks and limestones are produced by compaction and cementation. The latter involves the precipitation of low-Mg calcite within intraskeletal cavities and as overgrowths upon discoasters and large coccoliths. Much of the $CaCO_3$ is derived from the dissolution of micron-sized crystals formed from disaggregation of coccoliths. A comparison of ooze, chalk and limestone is given in Table 5.2.

Schlanger & Douglas (1974) introduced the concept of diagenetic potential to pelagic carbonates to account for the local variations in diagenetic grade observed in cores. They reasoned that the composition and nature of the original sediment, controlled by the conditions of sedimentation, namely water depth, deposition rate, temperature, sediment compaction and grain size, determined the progress of diagenesis, such as the rates of mechanical compaction, grain breakage, grain dissolution and $CaCO_3$ precipitation. The presence of clay in the pelagic carbonate reduces the potential for cementation (Matter, 1974). Argillaceous sediments are lithified by compaction through increasing overburden pressure and expulsion of porewater, contrasting with the dissolution–reprecipitation which dominates lithification of pelagic carbonates. For the latter, gravitational compaction alone is only important at shallow burial depths (50–200 m), where pelagic ooze changes from a soft to stiff consistency through water loss, resulting in a porosity reduction of around 10–20%. Chemical compaction takes over at greater depths as fine particles are lost through dissolution. The presence of clays, biogenic silica and volcanic ash affect

Table 5.2 Characteristic features of pelagic ooze, chalk and limestone

Ooze e.g. Miocene at 100 m subseafloor	Intact, but slightly etched planktonic forams Matrix of well-preserved nannofossils Foram chambers empty Some broken coccoliths Porosity 65%
Chalk e.g. Oligocene at 300 m subseafloor	Corroded forams Discoasters with overgrowths Etched coccoliths Porosity 60%
Limestone e.g. Lower Cretaceous at 1100 m sub- seafloor	Very etched coccoliths Matrix of euhedral−subhedral cement Few forams, no small coccoliths Porosity 35−40%

pelagic carbonate diagenesis in different ways, but these are not completely understood. Clays generally enhance carbonate dissolution, whereas calcite cementation is common near cherts formed from biogenic opaline silica.

Analyses of oxygen isotopes down-core have revealed trends of more negative $\delta^{18}O$ in the pelagic carbonate with increasing age and depth of burial (e.g. Fig. 5.10). Matter *et al.* (1975) interpreted this as reflecting progressive lithification under the increasing temperature of a normal geothermal gradient. More negative excursions from the general trend can result from alteration of volcanic debris in the sediments and the changes which this induces in the pore-fluid composition. The carbon isotope values are generally fairly constant down-hole, indicating little diagenetic alteration and an internal source of carbon for the carbonate in the cements.

Strontium and Mg/Ca ratios in pelagic ooze sequences change in response to dissolution and lithification (Baker *et al.*, 1982; Elderfield *et al.*, 1982). The Sr content of pelagic carbonates generally decreases with burial depth. In modern oozes from the northwest Pacific, the Sr content is around 1300 ppm, and there is a drop to around 700 ppm in mid-Tertiary sediments at a depth of 50−100 m. This coincides with ^{18}O depletion and is where overgrowth cementation is important. The Sr loss arises from the dissolution of high-Sr grains and precipitation of low-Sr cement. Sr continues to be lost until values around 350 ppm in Early Tertiary−Cretaceous strata. Porewaters show an increase in Sr/Ca ratio with depth, through the release of Sr^{2+} from the sediment (Baker *et al.*, 1982).

Dolomite has now been recorded from many pelagic sequences and although in the early days a detrital origin was often advanced, much is now known to be authigenic, commonly forming in organic-rich pelitic sediments (see Section 8.7.5).

5.7 ANCIENT PELAGIC CARBONATES

Like their modern counterparts, ancient pelagic carbonates exposed on land were originally deposited in a range of geotectonic settings. Because the preservation potential of sediments deposited on the ocean floors is not high, the geological record of pelagic carbonates is dominated by those deposited on continental lithosphere in continental margin and epeiric seas. Pelagic carbonates deposited on oceanic lithosphere are mostly preserved in mountain belts associated with ophiolite complexes. The development of pelagic limestones is also time dependent: they are much more common in post-Triassic sequences, through the evolution of nannoplankton and planktonic foraminifera in the Early Mesozoic. Palaeozoic pelagic limestones are thus rather restricted in their occurrences and true pelagic limestones are absent from the Precambrian. The following sections discuss: (1) Tethyan pelagic limestones, well developed in the Mediterranean area and deposited on foundered carbonate platforms and on ocean crust, (2) Devonian Cephalopodenkalk and Griotte of western Europe, mostly deposited on extended continental crust, (3) the Cretaceous−Lower Tertiary chalks of western Europe and USA deposited in an epeiric sea setting, and (4) the Ordovician *Orthoceras* limestones of Scandinavia, deposited on the submerged Baltic craton.

5.7.1 Tethyan pelagic limestones

Pelagic limestones are well known from the Mesozoic of the Mediterranean area where they were mostly deposited on drowned carbonate platforms. During the Triassic the fine-grained, usually red, pelagic limestones were deposited on basinal submarine highs and slopes while shallow-water carbonates were being deposited on platforms; during the Jurassic, the carbonate platforms broke up and subsided and the pelagic facies accumulated upon the shallow-water limestones, as well as in adjacent basins (Fig. 5.11; Bernoulli & Jenkyns, 1974). A variety of rock types comprise the pelagic facies, with stratigraphically condensed limestones occurring upon submarine highs (Jenkyns, 1971) and much thicker sequences of limestones, commonly nodular and with interbedded marls and shales, and resedimented units occurring in more basinal environments.

The fauna of these Triassic—Jurassic pelagic deposits typically consists of ammonites, belemnites, thin-shelled bivalves such as *Bositra*, *Daonella* and *Posidonia*, small gastropods, foraminifera, radiolarians, sponge spicules, ostracods and echinoderm debris (Fig. 5.12A). Locally, there occur microfacies composed of concentrations of one or several of these skeletal components; limestones composed entirely of thin-shelled bivalves, for example, are especially common (Fig. 5.12B). Nannofossils can be seen in these limestones with the electron microscope; some are coccoliths (e.g. Fig. 5.12C), others problematic.

In the condensed facies, there occur Fe—Mn oxyhydroxide pavements, crusts and nodules (e.g. Jenkyns, 1970, 1977) and these are usually associated with hardground surfaces, lithoclasts and borings. Colonies of sessile foraminifera form minute buildups on lithified surfaces (Wendt, 1969). Ammonites are commonly poorly preserved through shell dissolution and corrosion, as they rested on the seafloor (Fig. 5.13A) (e.g. Schlager, 1974). Perhaps surprisingly, oncolites and stromatolites are widespread in the condensed facies, the latter occurring as columnar and domal forms, in some cases upon ammonites (e.g. Massari, 1981). Laminae are usually thick (several millimetres) and poorly defined (Fig. 5.13B).

In many Tethyan pelagic limestones, neptunian

dykes and sills cut the sequence and many extend down into underlying shallow-water carbonates (e.g. Wendt, 1971). They are filled with pelagic sediment, in some cases with a specialized or dwarf fauna, and marine fibrous calcite cement.

The stratigraphically expanded pelagic facies tend to be more nodular with a rhythmic bedding through muddy, marly partings and thin beds (Fig. 5.14). The origin of the nodular texture is thought to be due to local dissolution—reprecipitation of $CaCO_3$ in the marly sediment, with later compaction and pressure dissolution enhancing the nodularity (Jenkyns, 1974). Ammonitico Rosso is a term used for this facies in Italy, although the name is often loosely applied to almost any red pelagic limestone in the Mediterranean area. Hardgrounds and Fe—Mn encrustations are usually absent, although the fauna is similar to the condensed facies, apart from a paucity of nannofossils.

Dark to light grey, cream to white pelagic limestones occur in the Mediterranean area in the later Jurassic—Cretaceous and the local names Maiolica and Biancone are commonly applied to the white varieties which form sequences several hundred metres thick. They are typically even-bedded lime mudstones with planktonic foraminifera (Fig. 5.12D), calpionellids (also protistids), coccoliths, radiolarians, calcispheres and sponge spicules. Ammonites are mostly quite rare, although their aptychi (composed of calcite) are more common. Cherty nodules and layers also occur. In general, the Maiolica—Biancone facies was deposited after breakup of the Triassic—Lower Jurassic carbonate platforms and phase of synsedimentary tectonics, and it blankets the earlier irregular topography.

Slump folds (Fig. 5.15B), breccias (Fig. 5.15C)

Fig. 5.11 *Pelagic limestones of the Alpine Jurassic. Schematic cross-section showing variety and depositional sites of the various pelagic facies upon extended, foundered and drowned Triassic carbonate platform. After Bernoulli & Jenkyns (1974).*

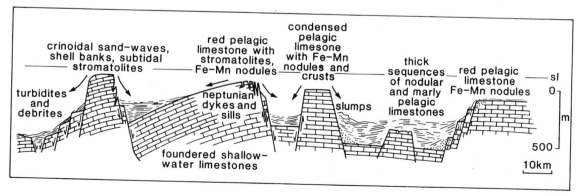

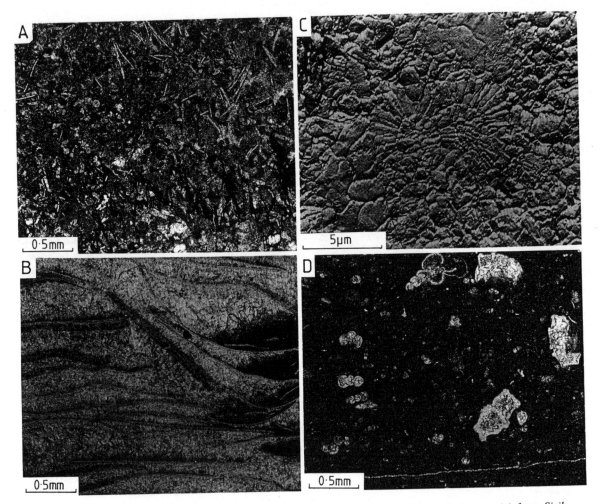

Fig. 5.12 *Pelagic limestone petrography. (A) Photomicrograph of Ammonitico Rosso (mid-Jurassic) from Sicily. Micritic sediment contains abundant thin-shelled bivalve, echinoderm and calcisphere debris. Prominent sutured stylolite has concentrations of haematite along it. (B) Photomicrograph of thin-shelled bivalve limestones (Upper Jurassic) from Alicante, Spain. Some lime mud has filtered down into the sediment to form geopetal structures. Fibrous calcite cement has nucleated upon the* Bositra *shells. (C) Transmission electron micrograph of Jurassic pelagic limestone showing two poorly-preserved coccoliths and much indeterminate micrite. (D) Photomicrograph of Early Cretaceous pelagic limestone (Maiolica) from Sicily consisting of many planktonic forminifera in a dense lime mudstone.*

and slides (Fig. 5.15A) have been recognized within many sections of the expanded facies and attest to mass movement off highs and down slopes. Reworked nodules cemented by calcite spar form the Scheck (Fig. 5.15D), a characteristic microfacies within the Jurassic pelagic sequence of the Salzburg region. Pelagic material also forms graded beds, with sole structures and planar and cross-laminae, deposited from turbidity currents.

It is difficult to be precise about depths of deposition of ancient pelagic facies. The paucity of benthic fossils in the Tethyan pelagic limestones suggests depths in excess of 50–100 m, but then depths could be down to several kilometres, depending on the position of the CCD. The presence of borings and micrite envelopes, and stromatolitic structures, if of algal origin, suggests depths within the photic zone, i.e. less than around 250 m. In the case of the Tethyan

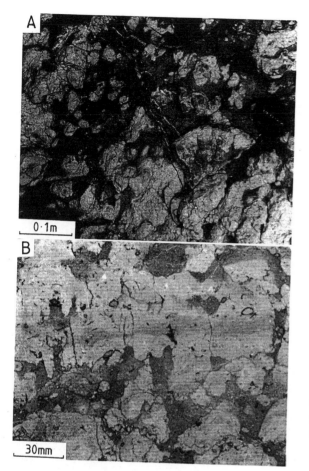

The Jurassic–Cretaceous pelagic rocks of the Alpine region vary in their stratigraphy. In some areas, such as the northern calcareous Alps, red pelagic limestones above platform carbonates pass up into even-bedded cherty limestones and radiolarites. This sequence could indicate continued deepening (Garrison & Fischer, 1969). The breakup of the Triassic–Lower Jurassic carbonate platform took place at different times in different areas (Bosellini & Winterer, 1975; Jenkyns, 1980a) and subsidence rates also varied.

Pelagic facies continue into the Middle–Upper Cretaceous and Tertiary in some parts of the Alpine chain, and in the Venetian Alps of northern Italy, the terms Scaglia Rossa and Scaglia Bianca refer to pink and white thin-bedded micrites with nannofossils and coccoliths similar to the Maiolica and Biancone. The trace fossils *Thalassinoides*, *Chondrites* and *Zoophycos* are common in these rocks (see Fig. 5.21). Thin layers of chert with radiolaria follow the bedding, and there are usually dark shaley partings. Schwarzacher & Fischer (1982) identified a cyclicity in the bed thickness of the Scaglia Bianca and Maiolica, which, assuming depositional rates, is of the order of 100 000 years for bundles of 4 or 5 beds and so 20 000 years for each main bed. Such a pattern is consistent with the orbital perturbations or Milankovitch cycles (see Section 2.10.3), suggesting some climatic influence on sedimentation. Exactly which sedimentary

Fig. 5.13 *(A) Bedding plane of Ammonitico Rosso showing nodular structure, some nodules after burrows and a poorly-preserved, shell-less, ammonite. Mid-Jurassic, Mallorca. (B) Polished surface of Ammonitico Rosso showing columnar stromatolite structures with weak internal lamination. Some lithoclasts with an Fe–Mn oxide coating are visible, as well as stylolites. Verona, Italy.*

pelagic limestones, many could have been deposited from around a hundred to several thousand metres. The seafloor corrosion of ammonite shells (originally aragonite) could indicate a depth between the ACD and CCD, but precisely where these were in the Jurassic is only speculation. Tethyan radiolarites have been interpreted as below CCD deposits, at a depth of several kilometres (Garrison & Fischer, 1969; Hsü, 1976), so that pelagic limestones above or below these cherts could also have been deposited close to this depth.

Fig. 5.14 *Ammonitico Rosso, mid-Jurassic, Mallorca, showing range of textures: thin red marls, nodules in marl, marly limestones, nodular limestones and stylolitic limestones. 3 m of strata seen here.*

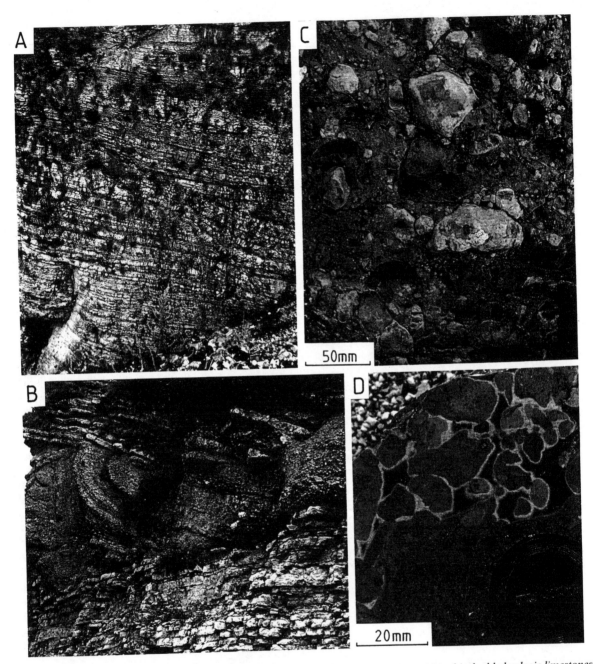

Fig. 5.15 *Resedimentation in Jurassic red pelagic limestones. (A) Major slide plane within thin-bedded pelagic limestones with marly partings. Height of cliff 50 m. Vaiont, Italy. (B) Major slump fold within bedded and marly red pelagic limestones. Height of cliff 10 m. Steinplatte, Austria. (C) Debrite of nodular limestone clasts in marly matrix, Mallorca. (D) The Scheck, reworked nodules and clasts of pelagic limestone (some with Fe–Mn-oxide-coated lithoclasts) cemented by fibrous calcite. A deeper red marly micrite later filtered into the partly-cemented conglomerate. Adnet, Austria.*

parameter is varying in response to the climatic fluctuations is unclear. It could be variations in organic productivity of the plankton or in the amount of seafloor dissolution, the latter especially if depths were close to the CCD. Alternatively, it could be variations in the amount of terrigenous material supplied to the basin. Synsedimentary slides, slumps and breccias are also common in the Scaglia (e.g. Alvarez *et al.*, 1985) and many bedding planes are pressure dissolution seams and stylolites. Some limestone—shale rhythms appear to be more the result of a diagenetic unmixing process, rather than purely of sedimentary origin (Hallam, 1986; Raiswell, 1988).

Organic-rich horizons and oceanic anoxic events.
Within the Jurassic—Cretaceous there are organic-rich horizons, some of which can be correlated over wide areas. One of the best known occurs at the Cenomanian—Turonian boundary, and this has been located in DSDP cores from the Atlantic, Pacific and Indian Oceans, as well as in exposures on land in many parts of the world (Schlanger *et al.*, 1987). Another widespread black bed occurs in the Lower Toarcian *falciferum* zone of the Lower Jurassic (Jenkyns & Clayton, 1986), recorded in the Alpine—Mediterranean region and in northern Europe. In the UK, this horizon is the Jet Rock of Yorkshire. Laterally extensive, thin (<1 m) organic-rich horizons occur in the Palaeozoic too; the Kellwasserkalk of the Upper Devonian, for example, can be located from northern Germany to southern Morocco, and mostly occurs within pelagic limestone (Cephalopodenkalk—Griotte) sequences (next section).

Organic-rich deposits require particular conditions for their formation, notably high organic productivity, poor circulation and low oxygen content of seawater (see Demaison & Moore, 1980). Some have undoubtedly formed through deposition in silled basins (Fig. 5.16), where poor circulation has permitted the development of water stratification and an anoxic lower layer, like the modern Black Sea. The global or regional black beds within more open oceanic settings are ascribed to *oceanic anoxic events* (OAEs): short-lived periods of high organic productivity, coinciding with low levels of oxygen in the ocean (Schlanger & Jenkyns, 1976; Jenkyns, 1980b). Of particular importance is the oxygen-minimum zone, located a few hundred to a few thousand metres below the surface, where maximum oxidation takes place of organic matter being sedimented from near-surface waters. At times of high organic productivity, this zone expands and becomes less oxic, even anoxic, so that

organic matter can be deposited and preserved in areas where the oxygen-minimum zone impinges on the seafloor (Fig. 5.16). OAEs appear to correlate with transgressive events (when shelves are flooded permitting increased levels of organic productivity) and they also tend to occur at times of equable climate, such as the Mesozoic, when there were no ice-caps and oceanic circulation would have been sluggish (Jenkyns, 1980b).

The quite sudden increase in organic matter productivity and burial rate as a result of an OAE leads to a change in the $\delta^{13}C$ signature of seawater. Organic matter is enriched in ^{12}C so that its massive burial leads to a more positive ocean. Carbonate precipitated from seawater during an OAE should thus be enriched in ^{13}C. Such positive excursions have been identified in the Upper Cretaceous (Scholle & Arthur, 1980) and Lower Jurassic (Jenkyns & Clayton, 1986); also see Section 9.6.2.

Also in the Tethyan region there occur *pelagic facies associated with pillow basalts and ultramafics*, which were thus likely deposited on the ocean floor (see Jenkyns, 1986, for a review). These occur in the Jurassic ophiolite belts of the Ligurian Apennines (Decandia & Elter, 1972; Barrett, 1982), in the French Alps (Lemoine, 1972) and in the Vourinos Massif of Greece (Pichon & Lys, 1976). Serpentinites and gabbros close to the sediments are commonly brecciated and veined, with fibrous and sparry calcite and internal sediments filling cavities (Barbieri *et al.*, 1979). These *ophicalcites* have parallels in brecciated rocks recovered from the Mid-Atlantic Ridge. Overlying the volcanics there locally occur a few metres of mineralized, Fe—Mn-rich sediment, before radiolarites (0—200 m thick). These give way to white nannofossil limestones (0—200 m thick) of Maiolica type in the Late Jurassic. The vertical sequence of cherts upon ridge basalts up in to pelagic limestones is the opposite of what is seen in modern oceans (Fig. 5.7). This curious situation could indicate a subsequent depression of the CCD through increased carbonate productivity and abundance of calcareous nannoplankton, or a phase of uplift bringing a deep seafloor above a shallow CCD.

In the Troodos Massif of Cyprus, a piece of Upper Cretaceous seafloor is overlain by umber and radiolarian cherts (Robertson & Hudson, 1973, 1974), occurring in depressions and ponds upon pillow lavas. The umbers of Cyprus are carbonate-free mudstones rich in Fe, Mn and other metals, directly comparable to modern metalliferous spreading-ridge sediments (Robertson & Hudson, 1973). Evidence for black

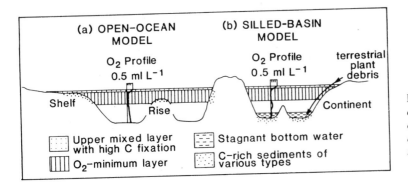

Fig. 5.16 *Model for deposition of organic-rich sediments in (a) an open-ocean model, as during an oceanic anoxic event and (b) a silled-basin model. After Schlanger & Jenkyns (1976).*

smokers in the form of sulphide chimneys has been found recently (Oudin & Constantinou, 1984). Above the radiolarites, pelagic limestones and chalks are developed from the Late Cretaceous into the Tertiary, the whole sequence shallowing-up and capped by Miocene reef and gypsum. Resedimented units are common in the Lower Tertiary part of the sequence and mostly consist of thin, graded calciturbidites with CDE and DE Bouma sequences (see Section 5.10.1). The material in the turbidites is mostly pelagic (foraminifera and nannofossils) with some benthic foraminifera and a significant component of volcanic debris (Robertson, 1976). Chert is widespread in the chalks and consists of nodular and bedded varieties, both clearly formed by replacement. Opal C-T in the form of lepispheres is present (Robertson, 1977). The volcanic–chert–limestone sequence of Troodos could again indicate uplift, rather than the normal long-term thermal subsidence of ocean floor, or variation in the position of the local CCD (Robertson & Hudson, 1974).

Pelagic limestones associated with radiolarian cherts, pillow lavas and basic–ultrabasic rocks also occur in the accreted terrains of western North America, in the Franciscan Complex of California (Alvarez W. *et al.*, 1980), in the California Coast Ranges ophiolite and in the Crescent Formation of the Olympic Peninsula, Washington (Garrison, 1973).

5.7.2 Devonian Cephalopodenkalk and Griotte

Very similar carbonate facies to the Tethyan condensed pelagic limestones and Ammonitico Rosso are widely developed in the Devonian of western Europe. Again, condensed and expanded facies have been recognized, the first termed Schwellen facies from the common development upon submarine highs (swells)

and the second Becken facies, deposited in adjoining basins. Cephalopodenkalk is a term used for the pelagic limestone facies in Germany and Griotte is used in the south of France and Spain. The pelagic limestones are micritic with a dominantly pelagic fauna of cephalopods (goniatites, clymenids and orthocones), thin-shelled bivalves, conodonts, styliolinids and ostracods. Rarely, there are small brachiopods and gastropods, crinoidal debris, calcareous algae, trilobites and solitary corals (e.g. Bandel, 1974; Tucker, 1974). Limestones composed entirely of styliolinids are one distinctive microfacies. Trace fossils are mostly simple burrows and borings into skeletal debris. The origin of the fine-grained matrix in these limestones is unclear, but it could well be derived from the breakdown of carbonate macroskeletons, rather than from microfossils and nannofossils.

In Devonian condensed pelagic limestones, there is abundant evidence of early lithification. Hardgrounds have surfaces encrusted by sessile foraminifera and corrasional and corrosional surfaces can be distinguished (e.g. Tucker, 1974). Lithoclasts are common and they may be coated by Fe–Mn oxyhydroxide crusts. Neptunian dykes and sills cut through the pelagic limestones and in some cases penetrate into underlying shallow-water carbonates. The nodular structure of some Cephalopodenkalk and Griotte types is due to early lithification in the muddy calcitic sediment, with many nodules forming around goniatite shells. Later pressure dissolution and compaction producing stylolites and marly seams (flasers) accentuates the nodular structure, in some cases leading to a *stylobreccia* texture (Fig. 5.17).

The condensed facies is best developed on former submarine highs, which in the Rheinisches Schiefergebirge of West Germany were mostly drowned Frasnian reefs and volcanic seamounts.

Fig. 5.17 *Stylobreccia, a nodular pelagic limestone showing sutured pressure dissolution seams between nodules (or are they clasts?!). Found on a bomb site in west Berlin; probably Jurassic in age.*

Fig. 5.18 *Nodular pelagic limestone interbedded with calcareous shales. Cephalopodenkalk (Upper Devonian), Sauerland, West Germany.*

There are also basement highs, presumably fault-bounded, with cephalopod limestones on top. In the Montagne Noire, southern France, the Griotte was deposited over a drowned shallow-water platform, also from the Frasnian. A Lower Carboniferous Griotte in the Cantabrian Mountains of Spain (Alba Formation) covers several thousand square kilometres and occurs above black shales and cherts (Wendt & Aigner, 1985). In the Upper Devonian of southern Morocco, cephalopod limestones deposited on platforms are apparently interbedded with birdseye limestones containing tepee structures and vadose diagenetic fabrics, suggesting quite shallow depths of deposition for the pelagic facies (Wendt *et al.*, 1984) or rapid sea-level changes. Orthoconic nautiloids are abundant at some horizons and preferred orientations, commonly bimodal, indicate strong wave action in the shallow waters. Sedimentation rates were low for the condensed cephalopod limestones, generally in the range of $1-5$ mm 1000 yr^{-1}.

The cephalopod limestones pass laterally into nodular limestones or Knollenkalk (Fig. 5.18), marly limestones and shales with nodules of the Becken facies, going into deeper, basinal environments. Fossils are less common, although ostracods may be important. A feature of the slope facies is the occurrence of resedimented horizons. Debris flow beds, slide blocks and slumps and turbidites have been recognized in many Upper Devonian pelagic slope

sequences (Szulczewski, 1968; Tucker, 1973; Wendt & Aigner, 1985).

5.7.3 The Chalk

The Upper Cretaceous was a period of high sea-level when substantial water depths (>50 m) existed over the northwest European craton, and the chalk was deposited. A similar pelagic deposit was laid down in the Western Interior Seaway of North America at this time.

In western Europe, the chalk typically has a thickness around 200 m in outcrop, but it thickens up considerably into the North Sea, with more than 1200 m being present in the Central Graben (Fig. 5.19). Chalk is monotonously composed of coccoliths (Fig. 5.20) and rhabdoliths, with many planktonic foraminifera and calcispheres. Important pelagic macrofossils include ammonites, belemnites and crinoids, but in shelf-sea chalks there is a significant benthic fauna of echinoids, bivalves, sponges, brachiopods and bryozoans. Trace fossils (Figs 5.21 and 5.22) are common with *Thalassinoides*, a crustacean dwelling burrow, being particularly widespread in shelf-sea chalks, but conspicuously absent in deeper-water chalks, such as those of the central North Sea. *Chondrites*, *Zoophycos* and *Planolites*, the burrows of deposit-feeding worm-like organisms, occur in both shelf-sea and deep sea chalks, but are generally more

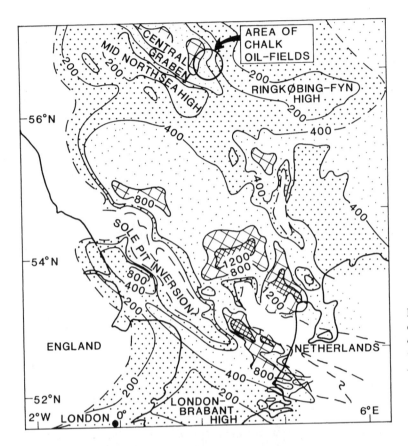

Fig. 5.19 *Isopachyte map of the chalk of the North Sea and surrounding areas. Onshore chalk is generally only a few hundred metres thick, but in parts of the North Sea it reaches over 1000 m. In some areas of the North Sea the chalk is absent as a result of Tertiary inversion. After Hancock & Scholle (1975).*

Fig. 5.20 *Scanning electron micrograph of soft chalk showing well-preserved coccolith debris. Cenomanian Chalk of Sussex, England. Photo courtesy of Peter Ditchfield.*

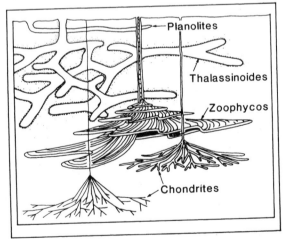

Fig. 5.21 *Tiering of trace fossils in chalk, based on the Upper Cretaceous of Jylland, Denmark. From shallowest to deepest, the trace fossils are* Planolites, Thalassinoides, Zoophycos *and large and small varieties of* Chondrites. *After Bromley & Ekdale (1986).*

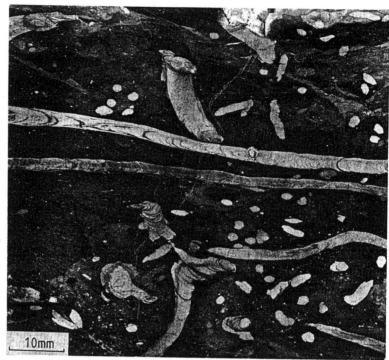

Fig. 5.22 *Trace fossils in shelf-sea chalk, with visibility enhanced by a coating of oil. Burrows of* Chondrites *and* Zoophycos *(showing internal laminae) with pale coloured fills cutting a background fabric indistinctly mottled with* Planolites *and* Thalassinoides. *Maastrichtian marly chalk, Jylland, Denmark. After Ekdale & Bromley (1984). Photo courtesy of Richard Bromley.*

common in the latter (Ekdale & Bromley, 1984). These burrowing organisms live at different depths within the sediment and this tiering of the endobenthos leads to cross-cutting relationships among the trace fossils, the deeper ones cutting the shallower ones (Fig. 5.21; Bromley & Ekdale, 1986). Diagenetic flint nodules are of course a feature of the chalk and many of these have formed by replacing burrow systems (Bromley & Ekdale, 1984).

Much of the European chalk is 100% $CaCO_3$ but clay is important at some horizons, such as the *Plenus* Marl, Middle Hod and Ekofisk Tight Zone of the North Sea Central Graben and Lower Chalk of southern England.

Chalk is very variable in its lithofacies, the result of a wide spectrum of depositional processes (Fig. 5.23). Much shelf-sea chalk, as seen in onshore outcrops, is evenly bedded (Fig. 5.24), with the stratification revealed by slight colour differences, the arrangement of flint nodules, alternation of clay-rich and clay-poor chalks, and variations in bioturbation density. Much of this chalk has an apparent rhythmicity in the stratification, on a decimetre to metre scale, such that the term *periodite facies* has been applied by Kennedy (1987). This small-scale cyclicity

could reflect subtle climatic fluctuations, perhaps due to the Milankovitch effect of orbital perturbations controlling the solar energy input, and this affecting organic productivity, water temperature, clastic input, etc. (Hart, 1987).

At various horizons throughout shelf-sea chalk, there are nodular horizons, omission surfaces and hardground surfaces, which reflect an increasing degree of seafloor cementation (Kennedy & Garrison, 1975). The hardground surfaces are commonly encrusted with bivalves, bryozoa and serpulids, and bored by algae, fungi, sponges, worms and bivalves. Some hardgrounds are mineralized, with glauconite and phosphorite impregnating the sediment, burrow walls and fossils, and occurring as discrete grains. These horizons of seafloor cementation probably formed during periods of reduced sedimentation, perhaps connected with some shallowing of water depth and increased current action. One particular feature of the English chalk is the continuity of beds; individual hardgrounds can be traced over many hundreds or thousands of square kilometres, reflecting the regionally uniform conditions of sedimentation of shelf-sea chalk (Bromley & Gale, 1982). In the chalk of Denmark, bryozoan-rich mounds with several tens of

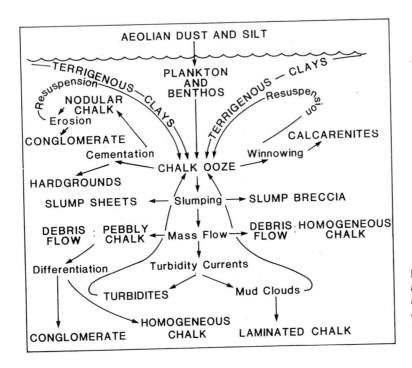

Fig. 5.23 *A flow diagram illustrating the processes of deposition and chalk types produced, based on the features of the Chalk Group of the Central Graben, North Sea. After Kennedy (1987).*

metres of relief are a form of deeper-water organic buildup. In the chalk of the Normandy coast, France (Fig. 5.24), prominent large-scale channels and scours cut into bedded chalk and contain intraclastic conglomerates, echinoderm-rich chalks, hardground surfaces (commonly mineralized), marginal slumped beds and debris flow deposits. These structures are attributed to strong seafloor currents, localized particularly to tectonically-controlled topographic highs and blocks (Quine, 1988).

In the North Sea Central Graben many different lithofacies can be recognized (Fig. 5.23), with redeposited chalk a characteristic feature (e.g. Watts *et al.*, 1980; Nygaard *et al.*, 1983; Kennedy, 1987). Pelagic chalk simply accumulated through settling of coccoliths and coccoliths aggregated into faecal pellets. This facies usually contains a continuously overlapping sequence of trace fossils, from the near-constant sedimentation. Depositional rates are of the order of 150–250 mm 1000 yr^{-1}. Influxes of clay lead to more argillaceous chalks, commonly with rhythmic bedding which may reflect an episodic input. Bioturbation is ubiquitous, leading to disruption of the primary bedding, although at times bottom waters were poorly oxygenated so that burrowers were excluded and the rhythmic bedding preserved. Hardgrounds are less common in deep sea chalks, but

there is a prominent one at the Maastrichtian–Danian boundary (i.e. Cretaceous–Tertiary boundary) in the North Sea, which could represent a short hiatus. Concentrations of foraminifera occur where current winnowing has removed nannofossils. These chalk lags (calcarenites) developed on submarine highs. Sedimentary structures are often difficult to discern in chalk, but staining the cut surface with methylene blue or using an oil (Buchinski's technique) can reveal the structures (Bromley, 1981, see Fig. 5.22). Some laminated units within pelagic chalk may be the deposits of low-density mud clouds and the paucity of trace fossils in these units suggests high sedimentation rates.

The resedimented units vary from thin turbidite beds to huge slumps and slide sheets (Fig. 5.25). In the Tor Formation and also the Ekofisk Formation, debris flow deposits are abundant. They consist of chalk pebbles in a fine matrix, in beds up to several metres thick, rarely up to 30 m. Some pebbles were clearly only semi-lithified and are streaked out. 'Shredded' textures also occur through breakup of clasts. Turbidites are sharp-based graded beds with a foraminiferal packstone lower part passing up into foram–wackestone/mudstone. In the absence of physical sedimentary structures, homogeneous chalk can be divided into debris flow and pelagic types on the

Fig. 5.24 *Chalk cliffs west of Etretat, northern France. Rhythmically-bedded chalk in the upper part, channel structures in the lower part. Cliffs 50 m high. Upper Cretaceous (Coniacian–Santonian). Photo from slide courtesy of Malcolm Hart.*

basis of trace fossils (Nygaard *et al.*, 1983). Debris flow units are burrowed in their upper parts only, mostly by *Chondrites* and *Zoophycos*, save for escape burrows such as *Trichichnus*, contrasting with pelagic chalk, where there is commonly a continuously over-lapping sequence of these trace fossils. Escape burrows like *Trichichnus* are common in all allochthonous beds. Large-scale slide masses, tens of metres thick and probably hundreds to thousands of metres in lateral extent, occur in the Ekofisk Formation and show slump folding of beds. Although often difficult to recognize in cores, they can be picked out on dipmeter logs as packets of bedded chalk with a steeper dip than strata above and below (Fig. 5.26). Porosity and formation density-related logs can also be used to recognize some slumped intervals, debrites and turbidites (Kennedy, 1987). Streaked-out bur-rows, convolutions and a secondary lamination in these slumped units are produced by internal sediment sliding. An allochthonous origin for sections of the

chalk helps to explain some of the stratigraphic com-plications revealed by the micropalaeontology. Inter-bedding of Maastrichtian and Danian Chalk is easier explained by the occurrence of Maastrichtian slide blocks in a Tertiary sequence than Danian microfossils already existing in the Cretaceous (Gartner & Kearn, 1978). Figure 5.27 presents a facies model for the Tor Formation of the Central Graben, Norwegian Sector.

There has been much discussion over the depths of deposition of chalk, but clearly it varied from around a hundred metres to a few hundred metres for shelf-sea chalk, to many hundreds and perhaps thousands of metres or more for deep sea chalk, such as that of the Central Graben of the North Sea.

The diagenesis of the chalk has received a lot of attention, especially since it is an important hydro-carbon reservoir. The diagenesis is variable; there are horizons which are well cemented and have quite low porosities, the nodular chalks and hardgrounds for instance, but much of the chalk is only weakly lithified

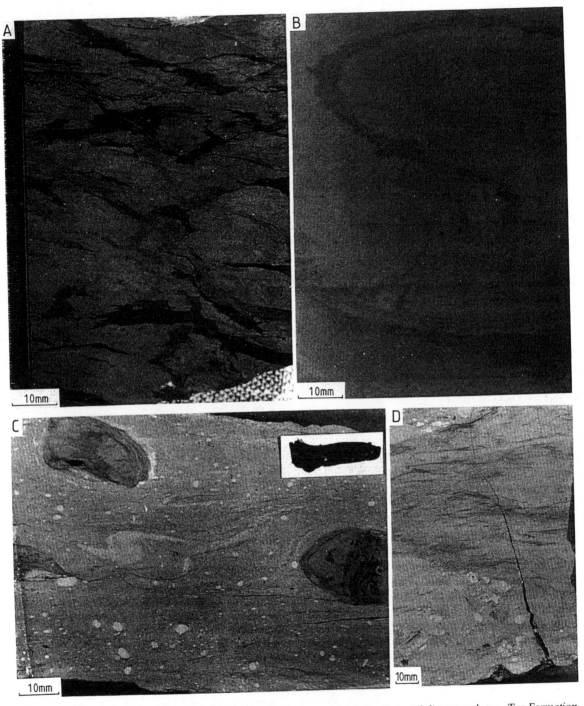

Fig. 5.25 *North Sea Chalk. (A) Plastically deformed burrows in marly chalk. Note stylolite towards top. Tor Formation. (B) Slump fold in Ekofisk Formation. (C) Debris flow deposit with streaked out and folded clasts and rounded dark clasts. Ekofisk Formation. (D) Debris flow deposit in lower part with angular and rounded clasts passing up into periodite facies with streaked-out dark burrows. Ekofisk Formation. After Kennedy (1987), photos courtesy of Jim Kennedy.*

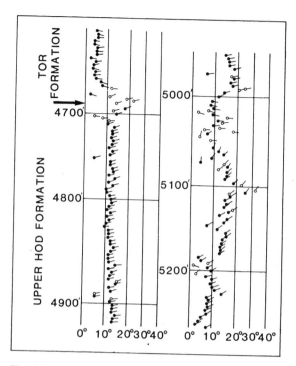

Fig. 5.26 *Dipmeter data from Upper Cretaceous Chalk of Central Graben, North Sea. The background structural dip is given by the uniform low values of 5° E in the lowest part of the Tor Formation. The thick packet of higher dip angle (10−14°) between 4720' and 4970' is a slide block. The more erratic dips below 4970' are slumps and debris flow deposits. Depths in feet!*

and retains a high porosity (35−45%). High porosities are maintained in spite of deep burial. In the North Sea, for example, reservoirs with 25−40% porosity occur at much greater depths than would be expected from typical limestone porosity losses (Fig. 5.28). Slow sedimentation rates, much bioturbation and bottom current activity, all enhance the chances of early lithification and porosity loss. Rapid pelagic sedimentation and resedimentation by debris flows and turbidity currents result in highly porous sediments with little early cementation. The clay content is also important since this generally inhibits early cementation, but it does lead to significant compaction in later diagenesis, with much nannofossil dissolution and flattening of burrow systems. The uncemented nature of much chalk is related to several factors, but especially its mineralogy and pore-fluid composition. The chalk's original mineralogy is overwhelmingly low-Mg calcite, the most stable carbonate mineral, so that there is little aragonite for early dissolution to supply $CaCO_3$ for cementation. With most chalks, and especially those of the North Sea, pore-fluids were of marine origin and there was no early flushing through with meteoric waters, as has happened with many other limestones. The pore-fluids would have been Mg-rich, which would have inhibited the precipitation of calcite. Another factor, important in some onshore chalks, is that they have not been buried very deeply and release of $CaCO_3$ by pressure dissolution may have been limited. The Mg-rich connate porewaters

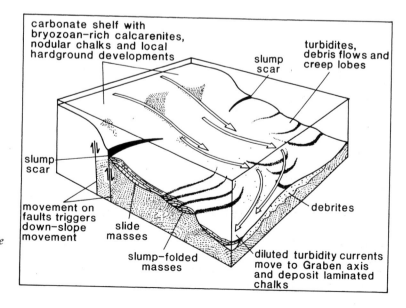

Fig. 5.27 *Depositional model for the Tor Formation in the Norwegian Sector of the North Sea Central Graben. After Kennedy (1987).*

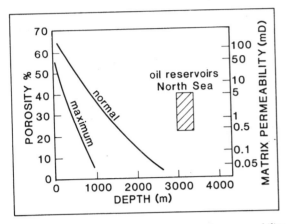

Fig. 5.28 *Typical trends of decreasing porosity—permeability for pelagic carbonates with increasing depth: the normal curve is for pelagic carbonates buried with marine porewaters, whereas the maximum curve is for those which have experienced early freshwater diagenesis. The field of North Sea-Chalk is for the rapidly-deposited, overpressured type which mostly forms the oil reservoirs. After Scholle* et al. *(1983).*

may also have reduced the effects of pressure dissolution (Neugebauer, 1974). The burial diagenesis of chalk is considered to be similar to that described in Section 5.6 for pelagic sediments recovered from the ocean floors: early mechanical compaction and dewatering reduce porosity from 70 to 50%, and then chemical compaction, causing dissolution of fine particles and

reprecipitation of $CaCO_3$ as overgrowths on coccoliths and forams, reduces the porosity to less than 20%. Low porosity, well-cemented chalks consist of a dense mosaic of calcite crystals in which few skeletal fragments can be identified (Fig. 5.29). It does appear as if there is a positive correlation between the preservation of nannofossils and porosity and permeability, measuring the former by the percentage of whole tests to whole plus fragmented tests (Feazel et al., 1985). High porosity intervals contain well-preserved nannofossils, with minimum fragmentation. This feature is likely a primary effect through a strong sedimentational control. Where there is more clay present, then pressure dissolution leads to the formation of wispy clay seams (flasers) (Garrison & Kennedy, 1977). The occurrence of very porous zones (up to 50% porosity) in chalk at great depths (3 km) in the North Sea could be a function of high pore-fluid pressures and/or early oil entry. These high porosities are typical of chalks buried to around 1 km, not 3 km. Where connate pore-fluids are prevented from escaping by permeability barriers, then high pore-fluid pressures can develop. In these overpressured zones, the intergranular stress is reduced so that the drive towards pressure dissolution is also reduced. Overpressuring can thus retard mechanical and chemical compaction and lead to a preservation of high porosity. Oil in a rock greatly reduces the rate of diagenetic processes and in the North Sea Chalk there is some evidence for early oil entry which would have arrested burial diagenesis. Where chalk pores were water-

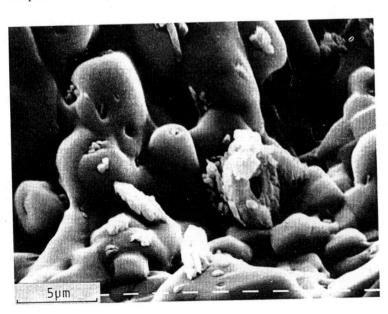

Fig. 5.29 *Scanning electron micrograph of well-cemented chalk with syntaxial overgrowths of calcite on coccolith and foram debris. Cenomanian Chalk of Sussex, England. Photo courtesy of Peter Ditchfield.*

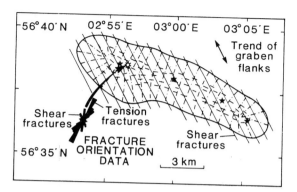

Fig. 5.30 *Schematic diagram illustrating the orientations of tension and shear fractures in Maastrichtian Chalk of the Albuskjell Field, Central Graben, North Sea. The elongate, domal shape of the field is due to Tertiary halokinesis of underlying Zechstein salt. Tension fractures are radial and concentric and were produced by the diapirism, whereas the shear fractures parallel the trend of the Central Graben (north-northwest–south-southeast) and are the result of regional tectonism. After Watts (1983).*

wet, then burial diagenesis would have continued along the normal route. The effects of this are well seen in the Lower Hod Formation, where oil-bearing zones have a porosity in excess of 30%, whereas nearby oil-free zones, which are not overpressured, have much lower porosities (<15%) and are more tightly cemented (Van den Bark & Thomas, 1981; Hardmann, 1982).

The fine-grained nature of chalk means that matrix permeability is low, but two processes have operated to give moderate to high permeabilities. The resedimentation noted above has resulted in some form of sediment reconstitution, which has raised the matrix permeability. More important though is the development of tension fractures, related to salt doming by Zechstein evaporites. These fractures have a radial and concentric arrangement around the domed chalk (Fig. 5.30) contrasting with tight, shear fractures which were produced by the regional stress pattern (northwest–southeast, parallel to the trend of the Central Graben) (Watts, 1983).

Oxygen isotope data from onshore chalks confirm the general trend of burial diagenesis: there is a positive correlation between porosity and $\delta^{18}O$ (Scholle, 1975, 1977). Much porosity loss stems from cementation and as this takes place at higher temperatures during increased burial, the $\delta^{18}O$ of the precipitated calcite becomes more negative. The more porous chalks have a more positive $\delta^{18}O$, indicating little

diagenetic change. Hardgrounds within the chalk generally have much heavier $\delta^{18}O$ since cementation is taking place on the seafloor and water there is colder than near the sea surface where most of the carbonate is formed by microfossils and nannofossils (Scholle, 1977). Oxygen isotope data from the Danish sub-basin and Central Graben of the North Sea region (Jorgensen, 1987) show that outcrop and near-surface chalks have similar $\delta^{18}O$ values (0.0 to -2.5‰) whereas chalks from deep wells show more negative values, down to -7.0‰ (Fig. 5.31A). The near-surface chalks are uncemented and only slightly altered diagenetically so that the $\delta^{18}O$ values are taken to reflect the $\delta^{18}O$ of Cretaceous seawater, with the variation representing some temperature gradients within the sub-basin. With the more deeply-buried chalks, $\delta^{18}O$ correlates negatively with depth within individual wells, but differences (Fig. 5.31A) between wells in the Central Graben are the result of later halokinetic movements. There is a positive correlation of $\delta^{18}O$ with porosity and $\delta^{18}O$ with per cent microspar, and a clear negative correlation of porosity with depth (Fig. 5.31B). These features reflect the increasing effects of burial diagenesis with depth, and the precipitation of microspar at elevated temperatures during burial. The $\delta^{18}O$ data from deep wells in the Danish sub-basin show values (-1.0 to -2.0‰) similar to those from outcropping chalk down to 600 m and then more negative values to -4.0‰ at greater depths (Fig. 5.31A). Above 600 m there is insignificant microspar in the coccolith mud, but below, it increases in amount. These trends are interpreted in terms of the onset of pressure dissolution: down to 600 m, porosity loss is by mechanical compaction whereas below this porosity is reduced by chemical compaction through recrystallization and pressure dissolution and the precipitation of microspar.

The carbon isotopic signatures of North Sea Chalks, ranging from $+0.5$ to 3.0‰, are acceptable as normal marine values and do not suggest any significant alteration during diagenesis (Jorgensen, 1987). This is confirmed by a lack of correlation of $\delta^{13}C$ with porosity, depth and microspar content.

Carbon isotope stratigraphies through the chalk have revealed short-term excursions which can be correlated between sections in Mexico, western Europe and DSDP cores from the Atlantic (Scholle & Arthur, 1980). The excursions are mostly positive and are interpreted as reflecting global eustatic sea-level variations and oceanic anoxic events, when increased rates of organic matter burial led to a depletion of the oceanic $\delta^{13}C_{TDC}$ in ^{12}C (Section 9.6.2).

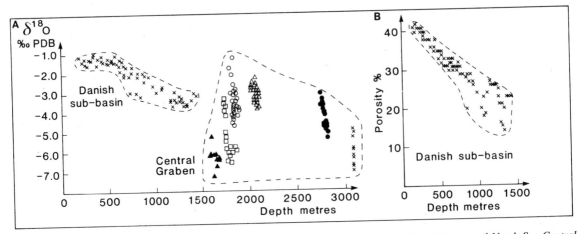

Fig. 5.31 *Oxygen isotope and porosity data from chalk in deep wells from the Danish sub-basin and North Sea Central Graben. (A) Trend of δ¹⁸O with depth of burial. In the less deeply-buried chalks of the Danish sub-basin, there is a marked depletion in ¹⁸O below 600 m, which is taken to reflect the onset of pressure dissolution. In the Central Graben field, the different symbols are for different wells, and a clear depletion in ¹⁸O with increasing depth is revealed for each individual well. The different depths of the chalk sequences from well to well are the result of Tertiary halokinesis. (B) Porosity versus depth for deep wells from the Danish sub-basin. After Jorgensen (1987).*

In the Upper Cretaceous Interior Seaway of the western USA, chalks such as the Greenhorn Limestone, Niobrara Chalk and Austin Chalk, occur in the central, deeper parts of the basin, passing laterally into shelf and shoreline clastics. The chalks generally have a high clay content and are intensely burrowed (e.g. Frey & Bromley, 1985). There is a greater range of benthic organisms compared to European chalks but coccoliths and foraminifera, commonly in pellets, still dominate the sediment. Hardground and nodular horizons formed through early lithification are poorly developed. Periods of low oxygenation of bottom water led to deposition of more organic-rich sediments with little benthos. The Austin Chalk of Texas, buried down to 3 km, shows similar burial diagenetic textures to North Sea Chalk, with evidence of both mechanical and chemical compaction (Czerniakowski *et al.*, 1984). Porosity also decreases with increasing burial. However, oxygen isotope data from micrite, intergranular cements and fracture fills from depths down to 3 km show very little deviation from marine values determined from oysters. This uniformity is interpreted as the result of closed-system burial diagenesis, with the increased temperatures during burial history having little effect on the cements.

5.7.4 Ordovician *Orthoceras* limestones

One other extensive, epeiric sea pelagic deposit occurs in the Ordovician of Scandinavia, where *Orthoceras* limestones were deposited over 500 000 km² of Baltic Shield during a prolonged high stand of sea-level. The micritic limestones are mostly grey-green to brown in colour, and range from thin-bedded to nodular, shaley types. The limestones are a condensed sequence, being only 50 m thick; they accumulated at an estimated sedimentation rate of 1 mm 1000 yr⁻¹. Hardgrounds and discontinuity surfaces are common and are bored and mineralized with glauconite, phosphorite, pyrite and haematite. Seafloor corrosion of cemented surfaces and shells also occurred. Buckled and folded beds and tepees resulted from seafloor precipitation of cement and expansion of the surface layer (Lindstrom, 1963). The sediment itself consists of much skeletal debris, from trilobites, nautiloids, echinoderms, ostracods, brachiopods and gastropods, in a fine-grained matrix of indeterminate origin. There is a significant non-carbonate component, largely derived from wind-blown volcanic ash (Lindstrom, 1974). The more nodular facies consists of regularly-arranged layers of calcite nodules in shale, with burrows crossing between nodule and shale demonstrating a very early origin of the nodules, probably in a diagenetic unmixing mechanism.

5.8 PELAGIC FACIES MODELS

Compared with some facies, the sedimentation of many pelagic carbonates is relatively simple, involving only deposition from suspension and vertical aggradation. Generally, there is little lateral movement of sediment unless there are strong bottom currents or deposition took place upon a slope, leading to resedimentation by turbidity currents and debris flows. Facies models for pelagic sediments can be discussed at two levels: in terms of the pelagic limestone lithofacies itself with the spectrum of its development, and in terms of the stratigraphic sequences in which the pelagic limestone lithofacies occur.

A major control on pelagic carbonate lithofacies is sedimentation rate, determined by organic productivity (controlled by many factors), seafloor currents and circulation patterns, and depth, especially relative to the CCD. These various parameters also control the amount of seafloor cementation and dissolution. Pelagic carbonates typically form a continuum of lithofacies (Fig. 5.32) from: (A) thick sequences of parallel, thin-bedded limestones, perhaps with some lamination and bioturbation, where deposition was near-continuous, out-of-suspension, (B) bioturbated and nodular limestones where a slower sedimentation rate allowed animals to churn up the sediment and destroy any lamination, (C) limestones showing evidence of cementation in the form of intraclasts and hardgrounds with surfaces encrusted and bored, (D) limestones with evidence of seafloor dissolution in the form of corrosional hardgrounds and moulds of aragonitic bioclasts, and (E) pelagic skeletal sands where current winnowing removes lime mud leaving the larger pelagic bioclasts (e.g. foraminifera, styliolinids, thin-shelled bivalves) to comprise the sediment.

Fe–Mn encrustations and nodules will be common in categories (C) and (D) and there may be important stratigraphic breaks and omission surfaces. This lithofacies spectrum can reflect: decreasing sedimentation rate, which could result from increasing depth as the CCD is approached; decreasing depth and more current reworking; no depth changes but decrease in organic productivity and variations in bottom water circulation. The various lithofacies commonly develop in different palaeotopographic situations; (C), (D) and (E) on topographic highs for example and (A) and (B) on deep-water slopes and subsided platforms. Pelagic lithofacies can also be divided on colour into red, brown, grey, green, white, etc., reflecting the iron content (may indicate proximity to land masses) and degree of oxygenation of the seafloor, determined by oceanic circulation and organic matter supply. At the extreme of poor circulation and seafloor anoxia, organic-rich limestones and black shales may accumulate.

There is also a spectrum of pelagic lithofacies with increasing clay content: from pure limestones, through limestones with shaley partings and pressure dissolution seams to nodular limestones and then shales with dispersed calcareous nodules of early diagenetic origin. This limestone to shale spectrum may result from: (1) topography: pelagic limestones preferentially deposited on topographic highs and shales with nodules in topographic lows, (2) increasing clastic input/decreasing carbonate productivity, or (3) increasing depth, with loss of $CaCO_3$ as the CCD is approached.

As is clear from the sections on modern and ancient pelagic carbonates (5.5, 5.7), pelagic limestones are deposited in a range of depositional settings, giving a variety of vertical sequences. Two end-members are an association with (1) shallow-water carbonates and (2) pillow lavas. The first type of sequence most commonly develops in continental margin settings as

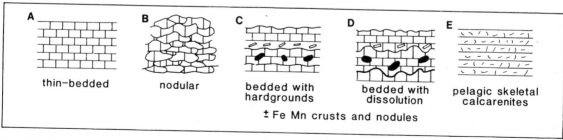

Fig. 5.32 *The spectrum of pelagic lithofacies, resulting largely from variations in depth, current activity and clay content.*

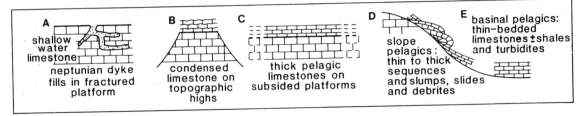

Fig. 5.33 *The range of depositional sites for pelagic carbonates, from dyke fills and seamounts through to slopes and basins.*

a result of drowning of a carbonate platform through eustatic sea-level rise or breakup and subsidence of a platform through extensional tectonics. A range of facies sequences develops depending on rate of subsidence, rate of pelagic sedimentation, depth (especially relative to the photic zone on the one hand and the CCD on the other), seafloor topography, bottom water circulation and degree of anoxia. At one extreme, pelagic sediment occurs in neptunian dykes within the shallow-water carbonates (Fig. 5.33A), or upon the platform limestones (Fig. 5.33B), especially where they form isolated fault-blocks and seamounts; here little pelagic limestone is deposited, hardgrounds and omission surfaces and Fe−Mn crusts and nodules are common. At the other extreme, thick sequences of pelagic limestone may develop upon a subsided carbonate platform if deposition is continuous and there is little winnowing or dissolution of sediment. On slopes (Fig. 5.33D), pelagic facies will commonly contain slumps, slides and debrites, and thin or thick sequences may develop, depending on the amount of downslope resedimentation (see next sections). In basins (Fig. 5.33E), thick pelagic sequences may accumulate, but they will commonly be shaley, and turbidites may be interbedded.

In a vertical sequence, the facies succession will depend on subsidence and sedimentation rate, clastic input, etc., but if subsidence outpaces pelagic carbonate deposition, then radiolarian cherts and siliceous shales can be expected as water depths exceed the CCD.

Where associated with pillow lavas, pelagic carbonates may well include horizons of volcaniclastics, metalliferous sediments derived from hydrothermal activity, and Fe−Mn oxide crusts and nodules, especially if sedimentation rates are low. Since thermal subsidence generally occurs after volcanic activity, pelagic limestones upon pillow lavas can be expected to pass vertically upwards into deeper-water facies,

such as radiolarian cherts. Variations in this rather simple, depth-determined, vertical sequence will arise from changes in the position of the CCD, variations in organic productivity, and the local tectonic context.

5.9 INTRODUCTION TO RESEDIMENTED CARBONATES

In many ancient carbonate sequences, much redeposition of shallow-water carbonate sediment into deep water has taken place. Reef debris beds, carbonate megabreccias and limestone turbidites have been described from many formations and these, usually coarse-grained and porous deposits, are important petroleum reservoirs in such places as Mexico and the Permian Basin of Texas. Modern resedimented carbonates are less easy to study but in recent years much data has been obtained from drilling and work with submersibles, especially off the Bahama Banks (e.g. Mullins & Neumann, 1979; Schlager & Chermak, 1979; Crevello & Schlager, 1980; Mullins *et al.*, 1984), and also off Belize and Jamaica (e.g. Land & Moore, 1977). Bank margins of the Bahama Platform were drilled during the Ocean Drilling Program (Austin *et al.*, 1986).

Deeper-water sandstones have been intensively studied over the last two decades and our understanding of siliciclastic sediment gravity flows, turbidite facies sequences and facies models is at an advanced stage. Redeposited carbonates on the other hand have only just been given the detailed attention they deserve, and there is still a long way to go before the depositional processes are unravelled and the end-member facies models are deduced. Data from deep-water siliciclastic formations cannot simply be applied to their carbonate counterparts. For example, the submarine fan model has proved very useful in explaining many ancient siliciclastic turbidite sequences, but there are few deep-water resedimented carbonate formations which can be accommodated within a submarine fan model. In many cases, it appears that an apron model is more applicable. In the following sections, the depositional processes of resedimentation are discussed first, then the recent work on modern carbonate slopes and platform margins is outlined. Facies models for deeper-water resedimented carbonates are then presented, before ancient examples are described.

5.10 PROCESSES OF CARBONATE RESEDIMENTATION

Submarine mass transport takes place by a variety of processes: rock falls, slides, slumps and sediment gravity flows, with the last including debris flows, turbidity currents and grain flows (Table 5.3). In addition, permanent/semi-permanent deep-marine bottom currents may rework the resedimented carbonates, as well as pelagic sediments. These deep sea processes form a continuum of transport/depositional mechanisms from the carving off of a block of reef rock to the deposition out of suspension of a pelagic foraminifera (Fig. 5.34). Of the numerous papers dealing with deeper-water mass transport, recent reviews include: Lowe (1982); Cook & Mullins (1983); Cook *et al.* (1983); Enos & Moore (1983); Pickering *et al.* (1986); Stow (1985, 1986).

Rockfalls. These generally occur along steep slopes and involve single blocks, which may be large, or many clasts in a range of sizes. Rockfalls feed talus piles at the base of the slope, or blocks may be left on the slope. Little transport takes place, other than the freefall, rolling or sliding of the blocks down the slope, although debris flows could be initiated by rockfalls on to the talus fans. Rockfalls are most common along platform margins which are fault-bounded, or where there is a steep fore-reef slope and active reef growth.

Slides and slumps. These terms are often used interchangeably or are loosely defined. To some workers, a slide conjures up the idea of mass sediment movement, *en bloc* upon a glide plane, with little internal deformation, whereas slump refers to mass movement with significant internal deformation of the sediment. Within one sediment package which has undergone mass movement, there will likely be a range of internal deformation structures. Slides have been divided into translational and rotational types. In translational slides, the shear plane is parallel to the bedding, as a planar or gently undulating surface, whereas in rotational slides (slumps to some people) the shear plane is concave-up and there is commonly a backward rotation of the sediment mass. The internal deformation varies from simple, large-scale folds, usually asymmetric and overturned in the direction of movement, to loss of bedding through homogenization and mixing, to brecciation of competent beds. Slides and slumps may develop into debris flows through break-up of beds and continued flow. On carbonate slopes, gravitational creep along detachment surfaces is common where sediments are unlithified and leads to creep lobes and hummocks on the seafloor.

Sediment gravity flows. These are flows of sediment moving downslope under the action of gravity. Although there is a spectrum of processes operating and one flow type may develop into another during the sediment's transport, five end-member types of sediment gravity flow can be distinguished, based on their rheology (liquid vs. plastic behaviour) and particle support mechanism (Table 5.3; Lowe, 1982). In *turbidity currents*, the sediment is supported by the fluid turbulence and low- and high-density flows are distinguished. In *fluidized flows*, the sediment is supported by upward-moving pore-fluid. *Liquefied flows* are ones where the sediment is not fully supported, but the grains are settling through the pore-fluid, which is displaced upwards. *Grain flows* show plastic behaviour and the sediment is supported by dispersive pressure arising from grain collisions. *Debris flows*, also called *mudflows* or *cohesive flows*, are ones where the sediment is supported by a cohesive matrix. Sediment is deposited from decelerating gravity flows by

Table 5.3 Classification of laminar sediment gravity flows based on flow rheology and particle support mechanisms. After Lowe (1982)

Flow behaviour	Flow type	Sediment support mechanism
Fluid	Turbidity current	Fluid turbulence
	Fluidized flow	Escaping pore-fluid
	Liquefied flow	Escaping pore-fluid
Plastic	Grain flow	Dispersive pressure
	Mudflow/cohesive debris flow	Matrix density and strength

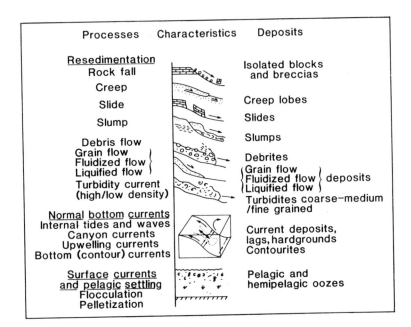

Processes	Characteristics	Deposits

Resedimentation
Rock fall

Creep

Slide

Slump

Debris flow
Grain flow
Fluidized flow
Liquified flow

Turbidity current
(high/low density)

Normal bottom currents
Internal tides and waves
Canyon currents
Upwelling currents
Bottom (contour) currents

Surface currents
and pelagic settling
Flocculation
Pelletization

Isolated blocks
and breccias

Creep lobes

Slides

Slumps

Debrites
Grain flow
Fluidized flow } deposits
Liquified flow
Turbidites coarse-medium
/fine grained

Current deposits,
lags, hardgrounds
Contourites

Pelagic and
hemipelagic oozes

Fig. 5.34 *Schematic representation of the main transport and depositional processes in the deep sea. After Stow (1985).*

two different mechanisms. In fluid flows, the grains are deposited individually, either from the bed load (traction sedimentation) or from suspension, and the bed is deposited from the base upwards. Debris flows on the other hand deposit sediment as the applied shear stress falls below the yield strength of the moving material. The flow thus freezes, either *en masse*, or from the outside inward.

5.10.1 Turbidity currents and limestone turbidites

The deceleration of turbidity currents deposits beds with a range of internal structures and these change within a single bed along the transport path. Low-density and high-density turbidity currents can be distinguished, giving medium- to fine-grained turbidites and coarse-grained turbidites respectively (Fig. 5.35).

Where sediments are finer than medium sand, they can be maintained in suspension by the fluid turbulence to give *low-density turbidity currents*. During deceleration, sediment transport changes from suspension to bed load. The deposits of these currents are well documented in the siliciclastic literature and they are characterized by graded beds showing Bouma T_{bcde} sequences (Fig. 5.35L). The lower horizontally-laminated division (b) and cross-laminated division (c) are deposited by traction sedimentation, the upper horizontally-laminated division (d) is deposited more

directly from suspension, with some traction, and the pelite interval (e) is deposited entirely from suspension.

Turbidites deposited from low-density flows are typically in the range of 0.05–0.3 m thick. Depending on the rate of deposition and grain size distribution, not all of the Bouma divisions may be present and downcurrent, the lower divisions are gradually cut out (Fig. 5.36). Limestone turbidites deposited from low-density turbidity currents can be identical to their siliciclastic counterparts. Commonly, because of the grain size distribution, the internal structures may not be immediately apparent.

High-density turbidity currents can transport much coarse sediment, of coarse sand to pebbles and cobbles. Finer sediment contributes a matrix buoyancy lift, which, with the fluid turbulence and dispersive pressure resulting from grain collisions, helps transport the coarse material. High-density sand-dominated flows are thought to have a basal traction layer, where flat-lamination and cross-stratification develop to give an S_1 division to the turbidite bed (Fig. 5.35G). Shallow scour structures may occur. Deceleration results in the transfer of material from suspension to bed load and a traction carpet may develop giving centimetre-thick, inversely-graded coarse sand to granule horizontal layers (an S_2 division). Coarse traction carpet units in the range 0.05 to 0.15 m thick may occur at the base of medium- to

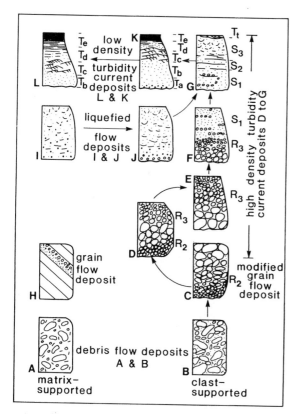

Fig. 5.35 *The deposits of sediment gravity flows. Lines with arrows connect deposits which may be parts of an evolutionary spectrum. No scales given, but bed types A through G are typically a metre or more in thickness, I through L a metre or less, and H is simply cross-bedding with each cross-bed the order of 10 mm thick. After Lowe (1982).*

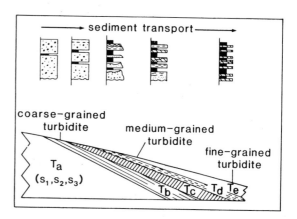

Fig. 5.36 *Proximal–distal changes in turbidites: downcurrent decrease in bed thickness, grain size and sand/mud ratio, and lateral changes in internal structures. After Stow (1986).*

fine-grained turbidite beds and constitute the graded T_a Bouma division. The uppermost division (S_3) of a high-density turbidity current bed is deposited out of suspension and may have a massive appearance or be normally graded; water escape structures', especially dish and pillars structures, may develop (Lowe, 1975). After deposition of the coarse-grained, high-density suspended sediment load, the residual turbidity current with finer sediment is free to deposit a 'normal' turbidite bed, rework the sediment surface or move on to deposit a graded bed in a more distant location.

High-density turbidity currents can transport much gravel, usually in the form of a highly concentrated traction carpet and in suspension just above. Sedimentation of the gravel takes place by freezing, once the flow velocity falls below a critical value. An inversely-

graded basal gravelly layer (R_2) deposited from the traction carpet, overlain by a normally-graded layer (R_3) deposited from suspension, will characterize the turbidite beds deposited from these coarser, high-density flows (Fig. 5.35D; Lowe, 1982). In very proximal areas, traction structures may form to give cross- and flat-bedded conglomerates (R_1). Sand and finer grades of sediment can remain in suspension, forming high-density sand flows and low-density turbidity currents, and be deposited beyond the area of gravel deposition. As they travel along and deposit sediment, turbidity currents will change in character and the features of the turbidite bed will also change: gravelly facies will dominate in the proximal area, passing laterally into sandy traction beds, both deposited from high-density flows, and more classic turbidites with Bouma sequences (Figs 5.36 and 5.37) will be deposited further downflow from low-density turbidity currents.

Limestone turbidites can show all the features described above, and pebble–cobble-grade calcirudites and rudstones are common in basin margin and toe of slope settings close to sites of active reef growth on shelf margins. However, there are hydro-dynamic differences between siliciclastic and bioclastic grains. The latter have a lower specific gravity, much lower if they are porous, as in crinoids for example, and they are commonly irregular in shape. Other carbonate grains, such as ooids and peloids, may show a narrow spread of grain sizes, so that internal beds and laminae are difficult to discern. A different sequence of internal structures is seen in some limestone

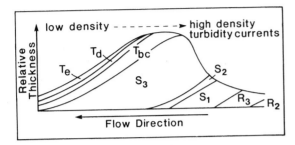

Fig. 5.37 *Schematic diagram of the downslope changes in turbidite bed organization from a high-density to low-density flow. After Lowe (1982).*

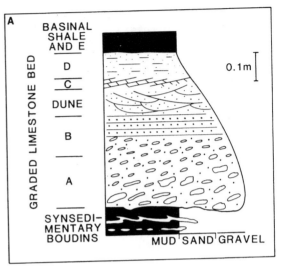

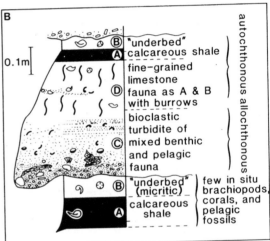

Fig. 5.38 *Limestone turbidites. (A) Turbidite from the Cow Head Breccia, Upper Cambrian, Newfoundland, showing internal divisions and a large-scale cross-stratified unit between B and C. After Hubert* et al. *(1977). (B) Turbidite from the Garbeck Limestone, Upper Devonian, Sauerland, West Germany, showing development of underbed. After Eder (1970).*

turbidites, which might reflect hydrodynamic differences. For example, Devonian crinoidal turbidites from southwest England (Tucker, 1969) show a lower (T_b) and upper (T_d) horizontally-laminated division in beds 0.3 to 0.7 m thick, with a central unit (0.1–0.2 m) of a single set of cross-bedding. A similar cross-stratification is seen in Ordovician calciturbidites from southwest Scotland (Hubert, 1966), but in the lower part of the beds, and in calcarenites associated with the Cow Head Breccia (Hubert *et al.*, 1977) trough cross-beds occur between Bouma B and C divisions (Fig. 5.38A). One other way in which limestone turbidites may differ is in the presence of an underbed, a fine-grained, 0.01–0.1 m thick basal layer to the turbidite, well developed when the limestones are interbedded with shale (Fig. 5.38B). The underbed is diagenetic in origin and is formed by dissolution of carbonate from the turbidite and its reprecipitation immediately below the bed. Immediately above the turbidite, there may be another fine-grained limestone layer with a rich benthic fauna (Fig. 5.38B). Deposition of the turbidite evidently led to a more fertile seafloor and carbonate precipitation there. Limestone turbidites with underbeds are well developed in the Devonian–Carboniferous 'Plattenkalk' of Germany (Eder, 1970, 1982).

Apart from the internal structures and grain size changes through turbidite beds, described above, sole structures are common on the bases of the beds. These commonly take the form of flutes and grooves, but broad scours are common in limestone turbidites.

Turbidity currents, like most flows, are subject to surging, mostly involving an abrupt increase in flow velocity followed by a gradual deceleration. Surges are likely to be most common in the early stages of flow and will give rise to repetitions of grading, amalgamated beds and internal erosion surfaces.

5.10.2 **Other sediment gravity flows**

Liquefied flows and fluidized flows are probably not important depositional processes in the submarine environment. They can form through liquefaction after sediment collapse and slumping, but in many situations they are likely to accelerate downslope and develop into turbidity currents. Where deposition does take place from a liquefied flow, then the bed is usually of fine sand to coarse silt grade in the case of siliciclastics, but, because of hydrodynamic equivalence, carbonate liquefied flow deposits could be coarser. Beds may be massive or graded, with water escape structures (Fig. 5.35I,J). The upper surface of such beds may be reworked into traction structures by currents in the ambient water produced by shear at the upper boundary of the liquefied flow. Liquefied flows are laminar in character, rather than turbulent, so that basal surfaces tend to be flat and unscoured.

Grain flows are produced by the dispersive pressure of colliding grains and it is thought that steep slopes (18–28°) are required to maintain the flow (Lowe, 1976; Middleton & Hampton, 1976). It has been suggested that true grain flow deposits cannot be thicker than 0.05 m, because the dispersive pressure is unable to support a greater thickness of sediment against gravity (Lowe, 1976). Grain flow deposits may show inverse grading, since larger grains are pushed towards the zone of least shear stress near the top of the flow. Avalanching down the lee slope of dunes and sand waves mostly takes place by grain flow (Fig. 5.35H).

In *modified grain flows*, the grain dispersive pressure is increased through buoyant lift of the larger particles by a dense sediment–water matrix, through a shear transmitted down to the flow from an overlying current or from excess pore-fluid pressure. These modified grain flows are cohesionless gravel–sand–mud slurries and could operate on gentle slopes. Quite thick (>0.4 m) beds result, consisting of pebbles and cobbles in a poorly-sorted sand–mud matrix (Fig. 5.35C). Inverse grading of cobbles is common; more pebbly beds tend to be ungraded or show weak inverse grading. Modified grain flows may develop into high-density turbulent turbidity currents, so that the gravelly bed is overlain by a normally-graded bed. Mullins & Van Buren (1979) have described a modern modified carbonate grain flow deposit from off the Bahama Platform (Fig. 5.39). The resedimented bed of shallow-water carbonate, 2.4 m thick, consists of four units: I, a basal part 0.07–0.08 m of mixed platform–hemipelagic sediment; II, a 0.72 m thick, inversely graded unit of medium to very coarse sand with pebbles towards the top; III, a 1.25 m massive non-graded unit of very coarse sand with scattered pebbles; and IV, a 0.35 m coarse to fine sand normally graded unit. Units I–III are thought to have been deposited by a modified grain flow with a muddy matrix providing additional support to the dispersive pressure from sand- to cobble-grade clast collisions.

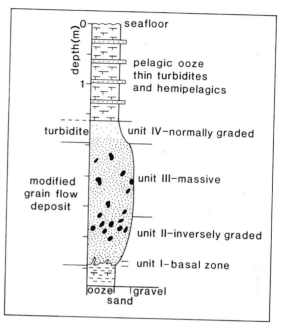

Fig. 5.39 *A modified grain flow deposit capped by a normally-graded turbidite, from a core taken in 4000 m of water, east of Little Bahama Bank. After Mullins & Van Buren (1979).*

Unit IV is interpreted as the deposit of a turbidity current, formed by dilution of the modified grain flow with ambient water and the generation of turbulence. This particular modified grain flow deposit does not appear to be laterally persistent for more than a few kilometres, and its development could be a result of the relatively steep slope along this part of the Bahama Platform margin.

In *debris flows*, clasts are supported by the cohesiveness of a finer sediment–water matrix and there are few collisions to produce a significant dispersive pressure, as in grain flows. Debris flows are familiar transport processes on alluvial fans and in upper submarine fan and submarine canyon environments. The strength of the muddy matrix is sufficient to transport very large blocks (many metres across). Debris flow deposits (Fig. 5.35A,B), called debrites by some people, vary from mud-dominated, with few scattered clasts, to clast-dominated, where a muddy matrix of as low as 5%, is sufficient to reduce the effective weight of the blocks by exerting a buoyant lift. These clasts move mainly by rolling and sliding and so may show a preferred orientation of long axes. Most debris flow deposits do not show any sorting or grading of clasts, although the latter may protrude above the top of the deposit.

With modern carbonate debris flow deposits from off the Bahamas, mud-supported and grain-supported types are distinguished (Mullins *et al.*, 1984). In the former, rounded limestone clasts, mostly of the upper-slope facies, are randomly distributed in a matrix of pelagic lime mud. These deposits reach 5 m in thickness and occur in proximal parts of the lower slope. Clast-supported debris flow deposits, occurring in distal portions of the lower slope, are less than a metre thick, non-graded muddy gravels contrasting with associated turbidites which are matrix-free and graded. Two-layer deposits of a lower debrite part passing up into turbidite also occur.

Ancient carbonate debris flow beds have been referred to as *megabreccias*, where angular clasts larger than 1 m across are conspicuous (Cook *et al.*, 1972; Mountjoy *et al.*, 1972). Some blocks in mega-breccias have been mistaken for patch reefs before now (e.g. Conaghan *et al.*, 1976; Read & Pfeil, 1983). Important additional criteria for the identification of carbonate debris flow deposits include: (1) disorientation of geopetal fabrics and stratification from one clast to another, (2) association of megabreccias with deeper-water lime mudstones and calciturbidites, and (3) juxtaposition of blocks of a range of facies. Mega-breccias occur in sheets and linear, channel forms, but generally they are not persistent for more than a few kilometres.

5.10.3 Bottom currents and contourites

Deep-marine bottom currents are well documented from modern oceans and arise from thermohaline circulation, internal waves and tides. Such currents may lead to breaks in the stratigraphic record on the deep seafloor, and cementation and mineralization of sediments (Section 5.3). More commonly, reworking by the bottom currents results in sorted pelagic fossil silts and fine sands, and reworked tops to sediment gravity flow deposits. The latter are likely to occur within or close to deep-water channels where bottom currents are more common. Contour-following (geostrophic) currents deposit muddy and sandy contourites which may show sharp or gradational bases or bases with lags, some cross- and planar-lamination, starved ripples and bioturbation (Pickering *et al.*, 1986). They may be difficult to distinguish from turbidites in the rock record, although palaeocurrents should be parallel to the strike of the palaeoslope. Sediment drifts in deep water off the northwestern

corners of the Great and Little Bahama Banks are the result of converging geostrophic currents (Section 5.11).

5.10.4 Classification of deep-water facies

In the deep-water siliciclastics—submarine fan literature, several schemes have been put forward for classifying the facies. One frequently used is that of Mutti & Ricci-Lucci (1978 and others) involving an alphanumeric code; it provides a convenient way of quickly describing the gamut of deep-water facies in the field, and facies associations and sequences can be more easily identified. A recent review by Pickering *et al.* (1986) has modified the Mutti & Ricci-Lucci scheme into seven facies classes (based on grain size), divided into 15 facies groups (based on internal organization of structures and textures), which in turn are divided further to make 41 individual facies on the basis of internal structures, bed thickness and composition. In most instances, each of the facies results from a particular depositional mechanism and since carbonates are resedimented by the same general processes as siliciclastics, this facies scheme can be applied directly to deep-water carbonates too. Some modification may be needed at the individual facies level where resedimented carbonates show deviations from the siliciclastic norms or where beds have the same field appearance but different microfacies. Care has to be exercised when using lithofacies codes of this type, that beds are not pigeon-holed into existing categories for the sake of proper description.

5.11 MODERN RESEDIMENTED CARBONATES: THE BAHAMA SLOPES AND BASINS

In recent years, much information has been obtained on the deeper-water carbonate facies around the Bahama Platform, particularly the northern part, and this has contributed greatly to our understanding of ancient resedimented carbonate slope and basin deposits. The setting of the Bahamas has been outlined in Section 3.2 and the general arrangement of banks and the major channels (deep-water basins) is shown in Fig. 5.40. In most areas there is a marginal escarpment, just off the platform towards the top of the slope in quite shallow water (20–40 m to 150–180 m). The slope here is commonly in excess of 45° and a similar marginal escarpment occurs off other carbonate banks and shelves in the Caribbean, such as Jamaica (Land & Moore, 1977) and off Belize

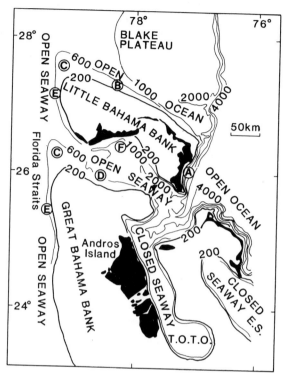

Fig. 5.40 *Types of platform margin in the Bahamas.*
(A) Escarpment margin. (B) Base-of-slope apron.
(C) Sediment-drift prism. (D) Diagenetic ramp. (E) Lower-
slope bioherms. (F) Gravity flow margin. After Mullins &
Neumann (1979) and Mullins (1983).

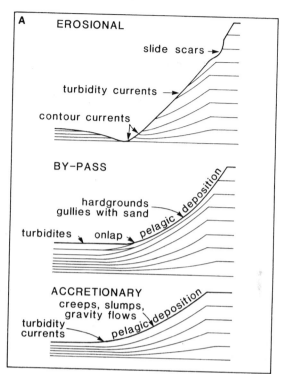

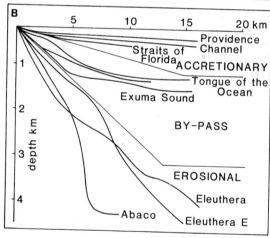

Fig. 5.41 *Division of Bahamian carbonate slopes into*
erosional, bypass and accretionary types. (A) Schematic
illustration of slope profiles and processes. (B) Slope profiles
of the Bahamas, vertical scale exaggerated four times over
horizontal. After Schlager & Ginsburg (1981).

(James & Ginsburg, 1979). Beyond the marginal escarpment there is much variation in the nature of the slope. In some areas, such as off the northwestern parts of the Little and Great Bahama Banks, the slope is very gentle, at an angle of only a few degrees or less into the basin. Elsewhere there are very steep zones, with 30–60° slopes, and even local vertical cliffs, such as on the eastern sides of the Little and Great Bahama Banks. The basins themselves are at depths from 800 m, as in the Florida Straits, to over 4000 m, at the eastern end of Providence Channel. Schlager & Ginsburg (1981) recognized three types of platform margin: erosional, bypass and accretionary, and the type appears to depend largely on the slope profile (Fig. 5.41).

Erosional slopes are mostly steep, with sediment gravity flows, contour currents and rockfalls all contributing to the erosion. The rocks exposed range from Tertiary to Pleistocene. *Bypass margins* are sites

of pelagic deposition and hardground development during pauses in sedimentation. Sediment gravity flows bypass the slopes, and their deposits accumulate in the adjoining basin and onlap the slope facies.

Accretionary slopes are generally less steep and sediment gravity flow deposits are common, especially in the lower slope. The Bahamian slopes can also be classified on their oceanographic setting into those facing: (1) the open ocean, where there is unlimited fetch, (2) open seaways, such as the Florida Straits and northwest Providence Channel, and (3) closed seaways, namely Tongue of the Ocean and Exuma Sound (Fig. 5.40, Mullins & Neumann, 1979). Also important is whether the slopes are in a leeward or windward location. As explained in Section 4.2.1, along windward platform margins, much coarse carbonate is moved on to the platform, whereas sediment is commonly moved off-platform, on to the adjoining slope and into the neighbouring basin, along leeward platform margins.

Depositional slopes in the Bahamas, below the marginal escarpment, can commonly be divided into: (1) an upper gullied slope and (2) a non-gullied lower slope or basin-margin rise, which may pass laterally to a flat basin floor. On the northern slope off the Little Bahama Bank, the upper slope is moderately steep (4°) and occurs in water depths of 200–900 m (Mullins *et al.*, 1984). Many distinct submarine canyons or gullies, 1–3 km across and 50–150 m deep, cut into the upper slope (Fig. 5.42). They grade from V-shaped at their heads to more U-shaped at the mouths. Most canyons terminate at the bottom of the upper slope, with only a few persisting across the lower slope as broad shallow channels. The lower slope has a quite smooth surface, although there are numerous small hummocks, 1–5 m high, of uncertain origin, and larger, 5–40 m high, ahermatypic coral mounds (Mullins *et al.*, 1981). In the Tongue of the Ocean, the gullied upper slope (6–9° dip) has smaller gullies, a few tens to hundreds of metres wide, and 20–100 m deep, commonly flanked by erosional cliffs of chalk and limestone (Schlager & Chermak, 1979).

The submarine canyons and gullies are considered by Mullins *et al.* (1984) to be formed by slope failure, giving slides and slumps, and then erosion and enlargement by sediment gravity flows. Hooke & Schlager (1980) suggested that the canyons developed as a consequence of the change from an accretional to an erosional slope as the adjacent platform built up.

5.11.1 Bahamian slope sedimentary facies

The *gullied upper-slope sediments north of the Little Bahama Bank* are largely periplatform oozes (Fig. 5.43). They consist of aragonite (~50%), high-Mg calcite (~30%) and low-Mg calcite (~20%), with some of this material being bank-derived lime mud transported in suspension by storm and tidal currents, and some pelagic skeletal debris (coccoliths and planktonic foraminifera). Bank-derived carbonate sand is absent since the northern slope of the Little Bahama Bank is in a windward setting and coarse material is transported on-platform. Along the western leeward slope, lime sands and gravels make up a substantial part of the upper-slope facies (Mullins & Neumann, 1979). Submarine cementation is common in the shallower parts of the upper slope, with hardgrounds being locally developed in several hundred metres of water (200–400 m), passing downslope into deeper-water (400–600 m) oozes containing carbonate nodules. A similar downslope decrease in the amount of cementation occurs in the upper-slope facies north of the Great Bahama Bank (Mullins *et al.*, 1980). The degree of lithification in these slope deposits is largely a function of the sediment's composition (higher contents of metastable, platform-derived aragonite and high-Mg calcite lead to an increased potential for cementation) and ocean bottom current activity which promotes cementation.

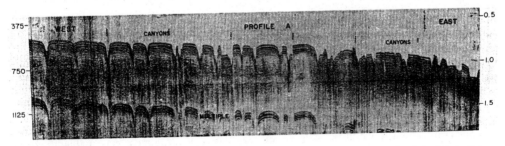

Fig. 5.42 *Air-gun seismic reflection profile parallel to the upper slope to the north of the Little Bahama Bank showing prominent canyons and gullies. From Mullins* et al. *(1984).*

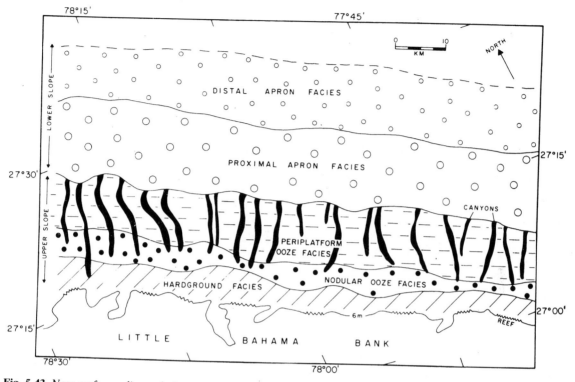

Fig. 5.43 *Near-surface sediment facies map of the slope to the north of Little Bahama Bank. From Mullins* et al. *(1984).*

In deeper water on the upper slope, periplatform oozes are uncemented and bioturbated. Cyclic variations in the mineralogy of the periplatform ooze have been detected in cores from the upper slope of the Tongue of the Ocean (Kier & Pilkey, 1971; Droxler *et al.*, 1983) and from north of the Little Bahama Bank (Mullins *et al.*, 1984). Regular fluctuations in the percentage of aragonite plus high-Mg calcite are interpreted as reflections of Quaternary sea-level oscillations by Kier & Pilkey (1971), with aragonite plus high-Mg calcite-rich layers corresponding to high sea-level stands when platforms were submerged and much shallow-water lime mud was transported offbank. During low sea-level stands, platforms were exposed and calcite-dominated oozes (from pelagic microfossils) were deposited on the upper slope. In contrast, Droxler *et al.* (1983) suggested that the cycles were produced by periodic dissolution of the aragonite, as a result of fluctuations in the aragonite compensation depth. They also believed that a large part of the high-Mg calcite in the oozes was precipitated close to the seafloor and was not bank-derived.

The channels of the upper slope are floored with coarse sand and boulders, and exposed rocks in the sides are bored, encrusted and coated with Fe—Mn oxides.

The lower-slope sediments north of the Little Bahama Bank consist of around 60% sediment gravity flow deposits, interbedded with periplatform ooze. The lower slope is an apron of resedimented carbonates and periplatform oozes (Fig. 5.43), which can be divided into proximal and distal apron facies. In the proximal facies, thick (up to 5.5 m) mud-supported debris flow deposits and coarse, gravelly turbidites (T_a) up to 2.5 m thick, are common, passing to thinner, coarse-grained turbidites (<1 m), thin, finer-grained turbidites (0.2−0.3 m), thin (<1 m) grain-supported debris flow deposits, and two-layer deposits, in the distal apron facies (Fig. 5.44). The clasts in the coarse turbidites and debris flow deposits are all planktonic foraminifera—pteropod biomicrites cemented by peloidal high-Mg calcite. In fact, most of the material in the sediment gravity flow deposits is resedimented pelagic and periplatform ooze and limestone from the upper slope.

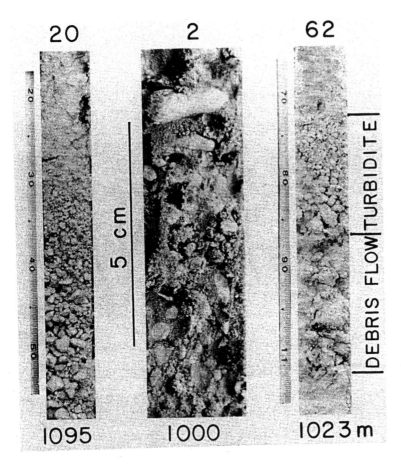

Fig. 5.44 *Piston cores from proximal apron facies, slope north of Little Bahama Bank. Core 20 is a typical proximal, normally-graded (T_a) turbidite. Core 2 is a typical debris flow deposit with chalk lithoclasts in a mud matrix. Core 62 is a two-layer sediment gravity flow deposit with a lower debrite unit overlain by a normally-graded turbidite. Water depths also given. After Mullins et al. (1984).*

Seismic reflection profiles of the Little Bahama Bank north slope have revealed numerous minor detachment surfaces in Pliocene to Recent sediments of the lower slope, resulting from gravitational creep and large-scale rotational movements (Fig. 5.45). Creep lobes are formed, with one of the main controls being the degree of lithification. Large-scale slump masses were deposited on the lower slope during the Middle Miocene, possibly the result of a regional tectonic event, and a system of channels and levees was developed upon the lower slope from the Upper Miocene to the Pliocene (Harwood & Towers, 1988).

Resedimented facies also occur along the *lower-slope aprons of Tongue of the Ocean and Exuma Sound* (Schlager & Chermak, 1979; Crevello & Schlager, 1980). In the Exuma case, a much higher percentage (50–70%) of shallow-water material occurs in the clastic carbonates. Three sediment types were identified: (1) clean, graded sand and gravel turbidites, (2) massive poorly-sorted sand and gravel

deposited from modified grain flows and high-density turbidity currents, and (3) muddy gravelly debris flow deposits (Crevello & Schlager, 1980). Two turbidites (layers I and II) were mapped out from cores and shown to be lobe-shaped beds (Fig. 5.46A), that thin and fine away from the source, where thickness exceeds 0.4 m. Bouma sequences revealed in cores are simple T_{a-e} graded layers of T_{bcde} and T_{cde} sequences. The turbidity currents were apparently generated at a point source and travelled 15–25 km, depositing sediment on the lower slope and flat basin floor. Layer I is a peloidal, skeletal lime sand largely of platform origin, whereas layer II is a fine peloidal, globigerinid–pteropod and *Halimeda* sand, i.e. a mixture of shallow- and deep-water material A debris flow–turbidite sheet (layer III) mapped out by Crevello & Schlager covered a much greater area: a large part of the lower slope, where it attained a thickness of 3 m, and much of the basin floor of Exuma Sound where it was 1–2 m thick in the axial

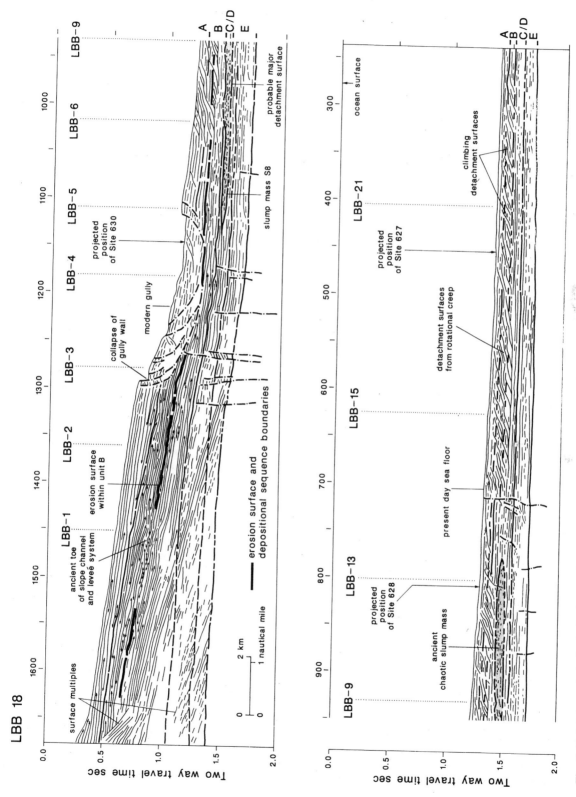

Fig. 5.45 *Interpretation of seismic section across the lower slope north of the Little Bahama Bank showing area of gully collapse, seismic facies units A through E, erosion and detachment surfaces and slump masses in the shallow subsurface. After Harwood & Towers (1988).*

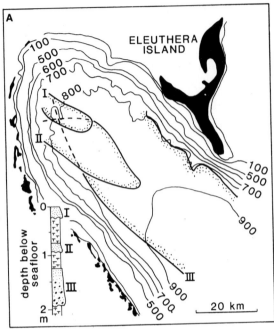

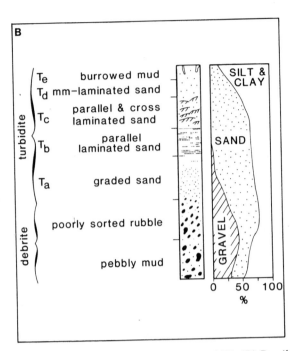

Fig. 5.46 *Debris flow and turbidity current deposits in Exuma Sound. (A) Distribution of layers I, II and III. (B) Detail of layer III (0.7 m thick) showing lower debrite and upper turbidite units. After Crevello & Schlager (1980).*

part, thinning to a few centimetres at the basin margins (Fig. 5.46A). The debris sheet consists of pebbly mud at the base (only present in the proximal region), overlain by muddy gravel (occurring in the proximal region and along the basin axis), capped by clean, graded gravelly sand and muddy silt, with Bouma a to e divisions (Fig. 5.46B), recorded over the whole basin floor and on topographic highs. This complex unit was deposited from a debris flow and overlying turbidity current, the whole flow having an estimated thickness of 130–160 m. It appears that the source was a 20–30 km sector of the upper slope in the northwest part of Exuma Sound which collapsed and the slump developed into the sediment gravity flow that travelled several hundred kilometres southeast down the basin axis. Layer III is a mixture of platform-derived sediment (40%), lithoclasts of platform material (10%) and lithoclasts of deep-water chalk. The frequency of sediment gravity flows into north-west Exuma Sound is one every 10 000–13 000 years, but overall, the frequency of flows into the whole basin will be much higher (1 in 500 to 1 in 1000 years) since the source area is the very long linear extent of the platform-margin upper slope around the Sound.

It has been suggested that carbonate resedimentation was more common during high sea-level stands (Mullins, 1983). At such times, overproduction of shallow-water carbonate on the Bahama Platform would have led to a higher sedimentation rate on adjoining slopes (4 to 6 times the background pelagic sedimentation rate). This in turn would have resulted in overloading and steepening of the slope and instability, leading to failure and the generation of sediment gravity flows. During lower sea-level stands, the Bahama Platform would have been subaerially exposed and mainly pelagic sediments would have been deposited on the slopes.

The origin of the hummocky surface which occurs on the lower slope off the Little Bahama Bank, and elsewhere is unclear. The hummocks could be creep lobes, masses of debris or even single blocks; they could also be buried *coral mounds*. The latter are a common feature of the lower slope to the north of the Little Bahama Bank in water depths of 1000–1300 m (Mullins *et al.*, 1981) and comparable, but lithified coral mounds (*lithoherms*) occur in the Straits of Florida at depths of 600–700 m (Neumann *et al.*, 1977). The unlithified coral mounds on the northern slope of Little Bahama Bank vary from 5 to 40 m in relief and are covered with a great variety of branched and solitary ahermatypic corals, as well as many other invertebrates. It is thought that the mounds develop on seafloor highs where there are strong bottom currents providing oxygen and nutrients to the fauna. The mounds probably develop from local coral colonies to thickets to coppices to banks, with sediment trapped between the coral branches and a gradual increase in ecological complexity of the associated fauna (Fig. 4.113; Mullins *et al.*, 1981).

The deep water lithoherms of the Florida Straits occur in areas of strong bottom currents (up to 0.5–0.6 m s^{-1}). They have a relief up to 40 m and are up to 100 m long, oriented parallel to the current. The lithoherms appear to be accreting into the current (southwards) while the downcurrent ends are undergoing bioerosion. Thickets of the branching ahermatypic coral *Lophelia* and *Enallopsammia* colonize the upcurrent zone and trap much sediment. Submarine cementation is widespread, producing hardgrounds and cemented surfaces encrusted and bored by a variety of organisms (see also Section 4.5.8d).

Carbonate contourite deposits are well developed in regions of strong bottom currents, such as in the Florida Straits off the northwest corners of the Great and Little Bahama Banks (Fig. 5.40; Mullins, 1983). Lobes of carbonate sand derived from the adjacent shallow-water platform and from pelagic skeletons have accumulated where northward- and westward-flowing contour currents converge; much winnowing of the sand takes place, and ripples are common. Submarine cementation is extensive, producing hardgrounds and nodular horizons, and shallowly-dipping (to the north) planes are revealed in seismic reflection profiles.

5.11.2 Facies models from the Bahama slopes

A consideration of the occurrences of the various facies described in the preceding sections enabled Mullins & Neumann (1979) and Mullins (1983) to identify eight main facies models for the slopes of the Bahama Platform (Fig. 5.47), four of which are very relevant to ancient deep-water resedimented carbon-

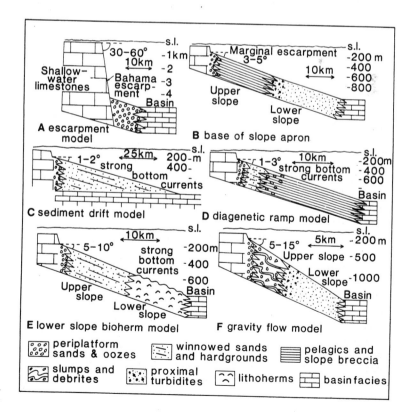

Fig. 5.47 *Facies models for Bahamian slopes. After Mullins & Neumann (1979) and Mullins (1983).*

ates. For the open-ocean margins, the escarpment model and base-of-slope apron model are both important. In the *escarpment model* (Fig. 5.47A), a steep (30–60°), high-relief (4–5 km) escarpment is mainly a site of resedimentation, supplying a narrow (<10 km) base-of-escarpment talus prism. Rockfalls, slides, debris flows, modified grain flows and high-density turbidity currents can be expected to operate in this setting. Sediment is derived both from the shallow-water platform edge and from the slope itself (periplatform ooze and limestone). The narrow talus facies belt passes oceanward into hemipelagic muds and basinal turbidites (T_{bcde} and thin graded beds). This facies model applies particularly to the eastern side of Little Bahama Bank (off Abaco Island), and should be appropriate for ancient platform margins where a very steep slope, perhaps fault-bounded, existed. The *base-of-slope apron model* (Fig. 5.47B) is derived from the facies distribution on the northern margin of the Little Bahama Bank, discussed earlier (see Fig. 5.43), and described in detail by Mullins *et al.* (1984). A *gullied upper slope* is a site of periplatform ooze sedimentation and resedimentation down submarine canyons and gullies and in slumps and slides to the *lower slope* where sediment gravity flow deposits accumulate. Facies of the lower slope can be divided into proximal and distal parts, with thicker, muddy debris flow deposits and coarse turbidites in the proximal region, and thinner, gravelly debris flow deposits and finer turbidites in the distal facies area, both with interbedded periplatform ooze. Intra-sediment creep is a major process within unlithified lower-slope sediments. On the upper slope, periplatform facies vary from limestones with hardground surfaces, through nodular horizons to uncemented ooze as depth increases down the slope and bottom currents become weaker. The degree of lithification is a major factor in slope processes, affecting the formation of gullies by wall collapse on the upper slope and permitting gravitational creep on the lower slope. The basinal facies consists of hemipelagic ooze with thin turbidites.

Along open seaways in the Bahamas, namely the Florida Straits and Providence Channels, Mullins & Neumann (1979) recognized four slope facies models: (1) a sediment drift model, (2) a diagenetic ramp model, (3) a lower-slope bioherm model, and (4) a gravity flow model, of which (1) and (2) are important to facies analysis of ancient deep-water carbonates. The *sediment drift model* (Fig. 5.47C) applies to locations where strong, convergent bottom currents rework and deposit sediment to form a lobe of sand of both shallow-water and pelagic origin, as off the northwest corners of the Little and Great Bahama Banks. An absence or paucity of sediment gravity flow deposits, the occurrence of bedded packstones and grainstones (contourites), perhaps with a detectable original dip and much evidence of seafloor cementation, will characterize the facies of this sediment drift model. Slides and slumps, and sediment gravity flows could occur off these contourite lobes into the adjoining basin.

The *diagenetic ramp model* (Fig. 5.47D) accounts for slopes which are quite gentle (1–2°) and where there is little mass sediment transport (slope north of Great Bahama Bank for example). Periplatform facies then show the downslope passage from hardgrounds to nodular oozes to unlithified ooze. Resedimentation only involves the periplatform deposits; being a windward margin, there is a paucity of platform-derived sands.

The *lower-slope bioherm model* (Fig. 5.47E) simply has a zone of deep-water coral mounds which grade upslope into periplatform sands and oozes, with hardgrounds, and downslope and basinwards into hemipelagic oozes and distal turbidites. The *gravity flow model* (Fig. 5.47F), based on the slope south of the Little Bahama Bank, is poorly known, and probably little different from the open-ocean, apron model. Slopes are steeper, with many gullies. The lower-slope apron is a narrow wedge of sediment gravity flow deposits. Being a leeward platform margin, a greater contribution of shallow-water sediment can be expected in the apron.

For the closed seaways of the Bahamas, Mullins & Neumann (1979) identified the concentric facies belt model and the debris sheet model. The *concentric facies belt model*, derived from the Tongue of the Ocean and Schlager & Chermak (1979), is basically the same as the base-of-slope apron model, with a different megascale geometry for the apron (concentric as opposed to linear). There are small, local differences: the marginal slopes are steeper and facies belts narrower in the Tongue of the Ocean, and sediment is supplied from windward, leeward and tide-dominated platform margins. Also, a basinal facies of pelagic ooze is not developed because of the quite small, closed nature of the seaway and sediment supply from three sides. The *debris sheet model* is based on the occurrence of the layer III debris flow–turbidite bed in Exuma Sound. In terms of generalized facies models, all that needs to be noted is that enormous slides can occur on carbonate slopes and develop into major debris flows and turbidity currents which can deposit substantial thicknesses of shallow-water and upper-slope material over a vast area of the lower slope and in the basin. This facies model is similar to the basin-fill turbidite facies model, originally put forward for deep-water siliciclastics, and discussed in section 5.12.

5.11.3 Other modern carbonate slopes

Studies of the submarine slopes off Jamaica (Goreau & Land, 1974; Land & Moore, 1977), Belize (James & Ginsburg, 1979) and Grand Cayman (Rigby & Roberts, 1976) show similar morphologies and facies patterns to the escarpment and apron facies models derived from the Bahamas. A very steep (30° to vertical to overhanging) slope or *wall* from around 40 to 160 m depth is skirted by a talus of reef-derived blocks (up to 10 m diameter) which passes into a slope of sediment gravity flow deposits and periplatform ooze. As in the Bahamas, gullies and canyons occur on the upper slope, feeding small fans and lobes on the lower slope. Much cementation and micritization of upper-slope sediments is occurring off Jamaica, especially at depths of 200–300 m in the region of the thermocline (Land & Moore, 1980). Some dissolution of aragonite bioclasts is also taking place. Reefs with spurs and grooves along the platform margins pass into a sand-covered fore-reef slope, steepening to a brow or drop-off at the top of the steep slope or wall.

Off the Florida Reef Tract there is a slope reaching 10°, down to a terrace at a depth of 30–50 m, before the main slope into the Florida Straits. Much reef debris is transported onshelf, since this is a windward setting, but some is deposited on the slope and in an apron along the toe of the slope on the terrace (Enos, 1977a).

5.12 FACIES MODELS FOR RESEDIMENTED CARBONATES: APRONS AND FANS

From the work on modern slope and basin resedimented carbonates described above, the *base-of-slope apron facies model* (Figs 5.47B, 5.48B) is the most useful for studies of ancient deep-water carbonates. In the base-of-slope apron model, shallow-water debris is shed into the basin from a line source, the whole platform margin (rather than a point source as in the submarine fan model), and much of this material bypasses the upper slope or is only deposited there temporarily, with pelagic sediment, before re-sedimentation by sediment gravity flows, slumps and

slides to the lower-slope apron. The submarine apron is a wedge of shallow-water and pelagic sediment and clasts with little vertical organization of facies. However, lobes of talus, perhaps showing beds thinning and fining upwards, may be recognizable, reflecting the development, operation and abandonment of submarine canyons and gullies as conduits for sediment transport from the upper to lower slope. The escarpment model (Fig. 5.47A) will have less general application, since there were few ancient carbonate platforms which stood up many hundreds or thousands of metres above the basin floor. The sediment drift model (Fig. 5.47C) will only be relevant to very specific oceanographic and geomorphologic situations, where strong and convergent bottom currents reworked sediment into lobes. The diagenetic ramp model (Fig. 5.47D) is more applicable to slopes of net sedimentation, where mostly pelagic deposits accumulate and little resedimentation takes place. When comparing modern and ancient carbonate slopes, it must be borne in mind that many of the upper slopes off modern reefs in the Caribbean–Bahamas–Florida region are really relic features from the Pleistocene and formed through subaerial erosion and karstic dissolution at times of low sea-level stand. The upper-slope deposits, as well as the reefs themselves, are commonly just veneers over Pleistocene surfaces.

Most attention has focused on the slopes around the Bahamas which descend into very deep water (800–4000 m), and the slopes off Jamaica, Belize and Grand Cayman are similar. By way of contrast, many ancient carbonate slope facies have been deposited in water depths of a few hundred or few tens of metres, and in many cases there was a gentle slope into the basin from the platform margin, without any escarpment. The slope deposits may interfinger with the shallow-water deposits. Thus, as discussed by Mullins & Cook (1986), there is a need for a *slope apron facies model*, where the sediment gravity flow deposits extend up to the adjacent shelf/slope break without an upper-slope bypass zone (Fig. 5.48A). In this model, sediment is supplied from the whole length of the platform margin and the deposits consist of broad sheets of debris, with only local channelling. Organized vertical sequences of facies are unlikely to develop. Modern slope aprons of this type have not been described, although some parts of the shelf slope off the Florida Reef Tract may be an analogue, since here reef debris is being deposited on a terrace, some tens of metres below the shelf-break.

For the majority of ancient resedimented carbon-

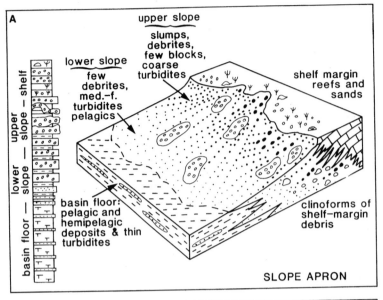

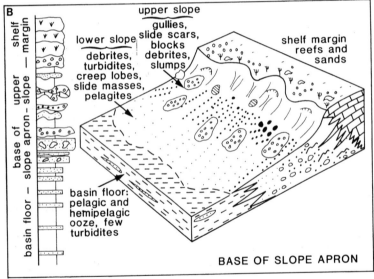

Fig. 5.48 *Facies models and sequences for carbonate slope aprons. (A) Slope apron, where slope debris beds pass up into platform/shelf-margin facies. (B) Base-of-slope apron, where the upper slope is largely bypassed by shelf-margin sediment gravity flows and debris mostly accumulates at the toe of the slope. After Mullins (1983).*

ates, the slope apron and base-of-slope apron facies models would seem to be most appropriate, with the former developing along rather gentle (<4°) and the latter along steeper platform to basin slopes. The *submarine fan model*, exhaustively applied to deep-water siliciclastics (e.g. Bouma *et al.*, 1986), has only been applied to a few carbonate sequences (e.g. Ruiz-Ortiz, 1983; Wright & Wilson, 1984) and there do not appear to be any modern carbonate submarine fans.

A submarine fan occurs in a lower slope/basin rise setting and is generally fed by one major channel (Fig. 5.49). The latter has levees and usually branches and becomes shallower across the fan. Channels may be active or abandoned and they may braid or meander. At the ends of the channels, lobes of sediment may occur (suprafan lobes). Channels have mostly lost their identity on the outer fan, which has a smooth surface passing gradually into the basin plain. Inner fan facies are typically debris flow and high-density turbidity current deposits, commonly channellized. Thin, fine-grained turbidites are usually interbedded and represent levee and interchannel

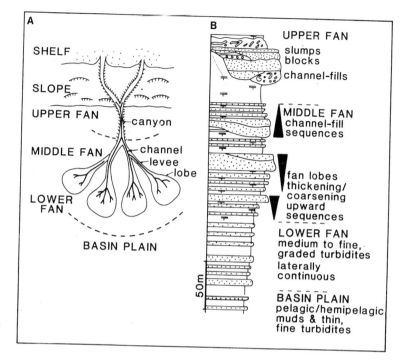

Fig. 5.49 *Submarine fan model. (A) Classic model of fan with suprafan lobes, some active, some inactive. (B) Classic sequence produced through submarine fan development and progradation. After Stow (1986) based on other sources.*

deposits. In the mid-fan area, thinning- and fining-upward packets of turbidites may occur, representing the filling of broad distributary channels. In the lower part of the mid fan, the suprafan lobes consist of thickening- and coarsening-up turbidites. Beds are laterally impersistent and palaeocurrents become more divergent down-fan. The outer fan facies consists of laterally persistent turbidites, interbedded with hemipelagic or pelagic background sediment. The adjacent basin plain facies will be largely fine-grained mudstones with thin distal turbidites.

Progradation of a submarine fan typically generates a coarsening-and thickening-up pile of sediments (Fig. 5.49) with basinal mudrocks passing up into thin turbidites (outer fan), to packets of thicker turbidites (mid fan), to channellized coarse turbidites and debris flow deposits (inner fan). Such a sequence could be generated by a single fan, which is then overlain by shallow-water sediments, or it could be a lobe of a much larger fan, which is eventually abandoned as a new lobe develops on another part of the fan.

The few ancient carbonate submarine fan formations appear to show identical facies and facies sequences to their siliciclastic analogues (Section 5.13; Ruiz-Ortiz, 1983; Wright & Wilson, 1984), so that the siliciclastic fan facies models can be used for predicting facies distributions on carbonate fans. It may well be

that with more detailed work and the recognition of other carbonate submarine fans, some differences will emerge so that the existing fan facies models will need to be modified for carbonates. The apparent absence of modern and paucity of ancient carbonate fans would appear to be due to the fact that sediment is supplied to slopes from the whole length of a platform margin (so that a talus apron dominates along the lower slope) rather than from several point sources spaced at substantial distances along the margin, which could supply material to discrete submarine fans. The predominance of aprons could be a result of the steep marginal escarpment which characterizes most platform margins in the Bahamas–Caribbean area and the absence of major canyons cutting right back to the shallow-water platform itself.

For siliciclastic turbidites, there is also a *basin-fill turbidite facies model* where there is little deposition on a submarine fan or apron at the lower slope/basin rise but instead, sediment gravity flows transport most of their sediment directly into the basin. Basin floors or plains vary considerably in size, from a few tens of square kilometres in small fault-bounded basins to many hundreds of thousands of square kilometres in the case of ocean floors and abyssal plains. They also vary considerably in depth, from a few hundred metres to several kilometres in the case of ocean basins and submarine trenches. Basin plains generally have little relief but there may be topographic highs or banks, plateaux and seamounts. Major controls on sedimentation on basin plains are basin size, sediment supply and provenance, and tectonic activity. Two end-member basin plains are the undersupplied and oversupplied types (Stow *et al.*, 1984). Undersupplied basin plains are generally large basins with minimal sediment supply and stable tectonic setting. Sediments tend to be fine grained, with a large pelagic–hemipelagic component. Oversupplied basin plains are generally smaller in extent, in more tectonically active areas and the sites of deposition of much coarse sediment derived from surrounding shallow-water platforms.

In basin-fill turbidite sequences, the beds in any vertical section through the turbidite pile will generally show similar features (thickness, grain size and internal structures), and proximal facies can be distinguished from medial and distal facies. Although there may be a large-scale upward coarsening and thickening, or fining and thinning of beds through a basin-fill turbidite sequence, there is generally no rapid interbedding of thick and thin beds (channel and overbank), or packets of coarsening and thickening beds (suprafan lobes) or fining and thinning beds (distributary channel fills), as is typical for submarine fan sequences. The pattern may be complicated, however, if sediment is being supplied from several source areas. Proximal facies of basin-fill turbidites are generally coarse-grained, thick T_{a-e} sequences, perhaps with traction carpet deposits, abundant scour, flute and groove sole structures, and amalgamated beds. Debris flow deposits may also be present, and some channelling. Distal facies are much thinner turbidites, T_{bcde}, T_{cde} or simple graded beds. Sole structures will be mostly tool marks and there will be much pelagic or hemipelagic material between the redeposited shallow-water beds. The basin-fill turbidite facies model should be applicable to many resedimented carbonate sequences where most deposition took place on the basin floor and slopes were bypassed so that fans and aprons were not developed.

5.13 ANCIENT RESEDIMENTED CARBONATES

Shallow-water carbonates in deep-water settings are common throughout the geological record. Many are closely associated with platform margins and appear to be of the slope apron type. Submarine fan carbonates are rare, but those described are very similar to their siliciclastic counterparts. Basin plain carbonate turbidites are well known from many deep-water sequences, although commonly it is difficult to locate the source of the shallow-water material.

5.13.1 Ancient apron carbonates

Good examples of resedimented limestones deposited in slope aprons adjacent to carbonate platforms occur in the Cambro-Ordovician of western North America (California, Nevada, western Canada) and eastern USA (Appalachians). Devonian reefs in western Canada and Western Australia have thick talus aprons and in the Alpine Triassic, platforms were surrounded by beds of shallow-water debris. Carbonate aprons are also well developed in the Permian of Texas and New Mexico and in the Cretaceous of Mexico and elsewhere, adjacent to rudist banks and buildups. Calciturbidites and megabreccias filled Lower Jurassic rift basins along the Tethyan continental margin in Switzerland and France.

Extensive platform carbonate sedimentation during the Cambrian and Ordovician on the North American craton led to much resedimentation of shallow-water carbonate along the continental margins of the time, located from eastern California to Alberta and from Virginia to Newfoundland. In *eastern Nevada, the platform Whipple Cave Formation and Lower House Limestone* (600 m thick) consist of typical shallow-water facies including stromatolites, bioherms, fenestral limestones, edge-wise conglomerates and grainstones (Cook & Taylor, 1977). An upward shallowing trend is taken to indicate seaward (westward) progradation of the platform margin. The equivalent *Hales Limestone* (150 m thick) in *central Nevada* (see Fig. 2.15 for general stratigraphy) consists of fine-grained, thin-bedded limestones with interbedded coarse sediment gravity flow deposits and slumped units. The thin-bedded limestones have a millimetre-scale lamination, a lack of bioturbation and traction structures, and a fauna of sponge spicules and trilobites. The depositional environment was evidently relatively deep. Overfolded, contorted and brecciated horizons in this deeper-water lime mudstone–wackestone facies indicate downslope sediment sliding and slumping. Sheet-like and channellized conglomerates of the deeper-water facies were deposited by debris flows which were initiated on the upper slope. Other conglomerates contain shallow-water clasts, showing that material was also being shed off the platform margin. Thin grainstone beds (<20 mm) of shallow-water skeletal debris with cross-lamination and ripples are interpreted as contourites, formed by deposition and reworking of sediment by slope-parallel bottom currents. Cook & Taylor (1977) proposed a slope origin for the Hales Limestone in view of the abundance of slump and debris flow deposits and paucity of turbidites. Later, Cook & Mullins (1983) presented evidence for local carbonate submarine fans in the Hales Limestone, with inner fan feeder channel breccias associated with submarine slides, mid-fan upward-thinning braided channel-fill turbidites and breccias, outer fan lobe upward thickening and coarsening turbidite packets, and fan fringe and basin plain turbidites and mudrocks. The fan and slope sequence, some 500 m thick, evidently prograded westwards, over the basin plain facies, which is around 1000 m thick. The 'fans' appear to form more of a talus wedge, however, and are not discrete entities like siliciclastic fans, fed by one major canyon. In eastern California and western Nevada (Kepper, 1981) faulting interrupted a Middle Cambrian ramp, and slope apron breccias deposited against this fault were derived from the newly-created shelf margin and from older ramp limestones exposed in the footwall (Fig. 5.50).

Further north, *in the Middle Cambrian of the southern Canadian Rocky Mountains* (McIlreath, 1977), a bypass, escarpment shelf margin has been recognized. Deposition of *Epiphyton* algal boundstone (now dolomitized) formed the 'thick' *Cathedral Formation* reef (Fig. 5.51) with a near-vertical cliff 120–300 m high facing into the basin, against which reef debris accumulated. Individual reef blocks up to 30 m in diameter in this 'thin' Cathedral Formation were generated by rockfalls, and blocks reaching 20 m across occur in debris flow units. The shelf basin transition was evidently a near-vertical wall for at least 16 km, with no suggestion of canyons or channels cutting into it. Vertical upbuilding of the *Epiphyton* reef must have been very rapid to produce the cliff. Sea-level may well have been rising during this time and seafloor cementation could have contributed towards the stability of the cliff, so reducing the amount of reef debris entering the basin. After Cathedral Formation times, some mud was deposited in the basin and then a wedge of limestone, the Boundary Limestone, was deposited against the escarpment of the former reef. This limestone is composed of shallow-water lime sands, shed off the shelf produced by the earlier Cathedral reef, and deeper-water lime mud. It forms a wedge or apron at the foot of the escarpment, and the lack of evidence for sediment lobes suggests that submarine fans were

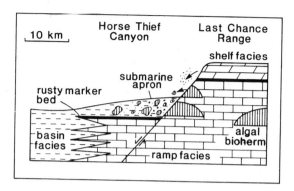

Fig. 5.50 *Schematic cross-section of Upper Cambrian shelf with base-of-slope debris apron, Nevada, USA. The shelf margin was formed by faulting of an earlier carbonate ramp. After Kepper (1981).*

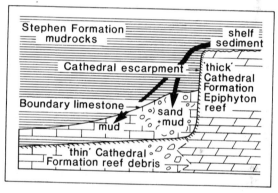

Fig. 5.51 *Schematic cross-section of Middle Cambrian shelf margin with base-of-slope debris apron, Rocky Mountains, British Columbia. The Cathedral escarpment reached 300 m in height. After McIlreath (1977).*

not present. Intraformational truncation surfaces in the Boundary Limestone wedge are interpreted as submarine slide surfaces. Within a short distance (1–2 km), the mixed shallow-water/deep-water sediment passes into a thin basinal lime mudstone. The escarpment model derived from the open-ocean, eastern side of the Little Bahama Bank (Fig. 5.47A) would be applicable to this Middle Cambrian situation in the Rocky Mountains, although the escarpment was much lower (maximum height 300 m) than the modern ones. This Middle Cambrian shelf–basin transition is an example of a bypass margin (Section 2.5.3). Also in the Cambrian of western Canada, Krause & Oldershaw (1979) described a variety of

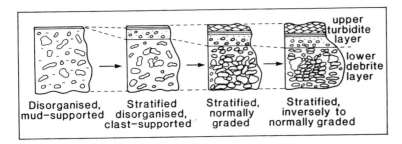

Fig. 5.52 *Carbonate breccia spectrum derived from study of resedimented slope limestones in the Lower Cambrian of North West Territories. Downcurrent, a mud-supported debrite passes laterally into a clast-supported debrite and then stratified breccias with an upper turbidite layer, as a result of deposition from a two-component sediment gravity flow. Beds typically a metre or more in thickness. After Krause & Oldershaw (1979).*

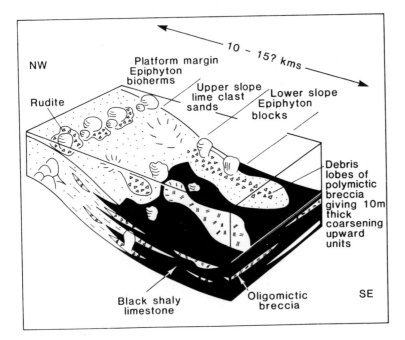

Fig. 5.53 *Facies model for Cambrian shelf margin with slope apron in the Shady Dolomite of southwest Virginia, Appalachians, USA. After Pfeil & Read (1980).*

carbonate breccias, which formed an evolutionary sequence developed during downslope transport (Fig. 5.52). Of particular note is the occurrence of breccias capped by graded calcarenites. These beds resulted from a two-component sediment gravity flow, of a debris flow lower part and turbidity current upper part (see Figs 5.44 and 5.46 for modern examples).

Along the eastern side of North America, the *Early Palaeozoic Cow Head Breccia of Newfoundland* (Hubert *et al.*, 1977; Hiscott & James, 1985) contains many carbonate debris flow units (megabreccias), with blocks up to 200 m across. Hubert *et al.* (1977) believed that the debris flows were shed off narrow elongate carbonate platforms oriented northeast–southwest, but Hiscott & James (1985) demonstrated

that transport was largely to the southeast off a carbonate platform developed along the trailing edge of the Cambro-Ordovician North American continent. Slump folds are common, but they have a complex geometry. The breccias are composed of clasts of a variety of shallow-water facies and plates and chips of thin-bedded limestone, mostly identical to the inter-breccia facies. The lower surfaces of breccia beds are generally non-erosional and smooth, except for linear scour structures, a few decimetres wide and deep. Hiscott & James (1985) distinguished five types of conglomerate: a facies A of graded–stratified conglomerates (<1 m thick) which have cross-stratified and ripple-laminated upper parts; a facies B (<2 m) of poorly-sorted limestone-plate conglomerates and a facies C of limestone-chip conglomerates (also <2 m) with floating exotic boulders; facies D are like C, but thicker (up to 5 m) with more exotics which may be concentrated at the bed base, dispersed through the bed or even protrude above the bed's upper surface; facies E breccias are up to 50–100 m thick with very large blocks. Hiscott & James (1985) interpreted facies B through E as the deposits of viscous debris flows, whereas type A were the products of turbidity currents. Hubert *et al.* (1977) described graded calcarenites interbedded with the breccias as showing a modified turbidite-type Bouma sequence (Fig. 5.38A) with a unit of trough cross-bedding between the flat-bedded B and cross-laminated C divisions.

It is of interest to note that the Cow Head Breccia accumulated over some 70 Ma but is only 300–500 m thick. It is thus a starved sequence, with most time taken up by the thin-bedded shale and limestone (hemipelagic and weak turbidity current sedimentation); the breccias represent very rare events, on the order of one every 300 000 years (Hubert *et al.*, 1977).

In the Appalachians, Lower Palaeozoic off-platform carbonates have been described by Keith & Friedman (1977), Reinhardt (1977) and Pfeil & Read (1980). In *the Cambrian Shady Dolomite of Virginia*, *Epiphyton* boundstone reefs and grainstones occur at the platform margin and pass seaward into thick slope sequences of bedded limeclast packstone–grainstone, breccia beds and thin-bedded black shaley limestones (Fig. 5.53; Pfeil & Read, 1980; Read & Pfeil, 1983). The last facies are non-graded to graded, massive or laminated (T_a, T_{a-b}, T_b) beds with some rippled upper surfaces, interpreted as distal turbidites. Of note is the occurrence of pull-apart structures, which range from brittle fractures, giving polygonal, mudcrack-like patterns, to soft sediment boudins. They are probably related to downslope sediment creep and are a feature of thinly interbedded limestone–shale slope deposits. The breccias are debris flow beds, with some occurring in crude coarsening-upward packets (<10 m thick) of black shaley limestone passing up into graded and laminated grainstones and then pebbly to disorganized polymictic breccias, with clasts up to 1 m across. Such packets are thought to represent prograding lobes of rudite on the lower slope (Fig. 5.53). In the lower-slope facies there occur blocks of *Epiphyton* boundstone, emplaced by sliding (Read & Pfeil, 1983). The lower-slope basinal facies are black shaley limestones with oligomictic breccias consisting of clasts of the former. The Bahamian base-of-slope apron facies model would appear to be appropriate for many of the Cambro-Ordovician slope

carbonates of North America. Escarpments do not appear to have been present, except in the Cathedral case, and discrete submarine fan systems were either absent or poorly developed. Reinhardt (1977) and Keith & Friedman (1977) have suggested the presence of fan facies in the Cambro-Ordovician of Maryland, New York and Vermont, but exposures are poor and details are not provided.

During the Devonian, carbonate platforms were widely developed and resedimented deep-water carbonates are associated with many of these. In western Canada, foreslope breccias occur around many of the carbonate buildups (e.g. Cook *et al.*, 1972) and in some instances, the clasts consist predominantly of cemented shelf-margin grainstone (Hopkins, 1977). The slopes from the buildups into the basins were mostly of the order of 5–10° and it appears that the basins were being rapidly filled, so that there was not a great difference in height between shelf and basin. Commonly the foreslope breccias have been dolomitized, as in the Miette Buildup (Section 8.7.4).

In the *Devonian of western Europe*, off-platform debris beds have been referred to as allodapic limestones (Meischner, 1964) and many of these are graded and laminated turbidite beds (e.g. Tucker, 1969; Eder, 1970, 1982; Burchette, 1981). Nodular, pelagic limestones, deposited on slopes and topographic highs in and around Devonian basins were commonly slumped and resedimented into breccias. In the Canning Basin of Western Australia, spectacular debris beds (Fig. 5.54) occur adjacent to the Givetian–Frasnian reefs (for stratigraphy see Fig. 4.98; Playford, 1980). Breccias and megabreccias, skeletal rudstones and grainstones pass downslope into thinner graded limestones interbedded with basinal shales. Features of particular interest include geopetal structures up to an angle of 10° to the bedding, indicating the original gradient, oncolites which rolled downslope and then resumed upward growth so that they also record the original incline of the beds (Fig. 5.54C), and columnar stromatolites which grew at a depth of some 45 m down on the fore-reef slope. In addition, enormous blocks of reef rock occur in basinal shales, a kilometre or two from the shelf edge. Similar allochthonous clasts up to 1 km across are common in the Lower Devonian Nubrigyn algal reefs of eastern Australia (Conaghan *et al.*, 1976).

The classic *Permian Capitan Reef of Texas and New Mexico* which developed along the margin of the Delaware Basin, has a well-developed slope apron of resedimented carbonates (for details see Newell *et al.*, 1953; Wilson, 1975). Of particular interest is the

Fig. 5.54 *Slope facies in the Upper Devonian of the Canning Basin, Western Australia. (A) View of fore-reef slope showing debris beds with original dip (back-reef limestones in distance are horizontal) and several large blocks of reef rock. (B) Reef debris beds with original dip and large block of reef rock towards top which is part of a debrite. (C) Slope limestones (original dip) with oncolites, the topmost ones with a vertical conical cap from renewed growth while upon the slope. (D) Debrite of shallow-water, reefal limestone clasts in marly matrix. From slides courtesy of Colin Scrutton.*

progradation of the reef for many kilometres over its debris beds. The reef talus occurs in large-scale foresets or *clinoforms* which have a maximum dip of 30°. The slope area was up to 10 km wide and evidently the Delaware Basin floor was about 700 m below the shelf-margin reefs. Clinoforms have been recognized in the subsurface of the Permian Basin from seismic data (Mazzullo, 1982). In the Lower Clear Fork and Wichita Groups, rapid progradation of shelf margins extended the platforms around the Midland Basin and was coincident with periods of low sea-level stand. Sarg (1988) has interpreted the Delaware Basin Permian strata in terms of sequence stratigraphy and depositional systems tracts.

Prograding platform margins are spectacularly developed in *Triassic strata of the Dolomites in northern Italy* (see Fig. 2.13; Bosellini, 1989). During the Ladinian and Carnian, buildups and lime sands developed around the margins of an indented shallow-water carbonate platform and around isolated platforms and banks (Leonardi, 1967; Bosellini & Rossi, 1974; Gaetani *et al.*, 1981). Much of the strata is heavily dolomitized but the reefs appear to have been formed by binding and encrusting blue–green algae, Porifera and *Tubiphytes*, with much early cementation. Undolomitized blocks of this reef rock (the so-called Cipit boulders) occur in the basinal San Cassian Formation and are thought to have been derived from the platform margin (Biddle, 1981). Carbonate platforms prograded by the rapid redeposition of shallow-water material on to the slopes into the basin, by debris flows, turbidity currents, rockfalls and slides. Megabreccias, breccias, graded beds, isolated blocks (the Cipit boulders) and slumped and folded strata are all present. Progradation was not always continuous, but often episodic, as shown by the interbedding of basinal, pelagic sediments with the breccias on the slope. The maximum slope angles varied from 20 to 40° and heights of the platform edge above the basin ranged from 200 to 800 m. Laterally extensive talus aprons developed rather than large discrete fans, as material was shed off the whole length of the platform and bank margins. However, thin coarsening-upward sequences can be recognized and probably represent sediment lobes and small fans.

Progradation generated large-scale inclined beds, i.e. the clinoforms (see Fig. 2.13C), which show a range of relationships with basinal and platform sediments (Fig. 5.55; Bosellini, 1984). The different types of progradation are controlled by: (1) the rate of basinal sedimentation, (2) the rate of platform subsidence, (3) the platform width, (4) the basin depth, and (5) sea-level changes. At the top of platforms (Fig. 5.55A, B and C), *offlap* relationships developed when lateral progradation kept pace with vertical aggradation, so that the platform-margin buildups grew up and over the inclined talus deposits. Offlap is typical of sedimentation during a relative rise in sea-level, and this relationship is well seen in the Lower Ladinian, when it is thought that tectonic subsidence was the cause. *Toplap* is a simple lateral progradation without any vertical upbuilding, and is produced during a relative still-stand of sea-level. Many of the Carnian platforms show toplap. When there is a fall in sea-level, platforms can be exposed so that clinoforms show *erosional truncation*. This is usually a disconformity with karstic relief. Peripheral Ladinian platforms, such as the Civetta and Sciliar, show this relationship as a result of emergence for several Ma. At the toes of clinoforms, three relationships are seen: horizontal, climbing and descending lower boundaries (Fig. 5.55D, E and F; Bosellini, 1984). In the first case, the basinal sedimentation rate is very low, so that the platform margin progrades *horizontally* to produce a tabular body of inclined strata. The Catinaccio Platform prograded in this fashion for about 9 km into a starved basin. The thickness of tabular talus bodies in the Dolomites ranges from 400 to 900 m. Where the basin floor is aggrading, then *climbing* progradation takes place

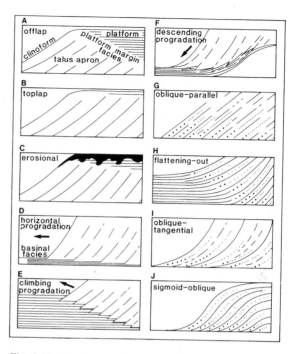

Fig. 5.55 *Clinoform configurations from the Triassic of the Dolomites, Italy. A, B and C upper boundary relationships. D, E and F lower boundary relationships. G, H, I and J geometry of clinoforms. After Bosellini (1984).*

and the lower boundary of the talus packet rises stratigraphically. If there is little change in sea-level, then the talus unit thins up, as the basin floor rises through sedimentation. Climbing progradation is common in the Carnian platforms since the basin was being supplied with much detritus from erosion of Ladinian volcanics. The Sella Platform talus climbed at least 500 m, while prograding laterally 1500 m. *Descending* progradation occurs where the platform margin prograades into a pre-existing depression. The talus sequence thickens outward and clinoforms become longer and thinner. Descending relationships are common around the Carnian platforms which grew next to or on top of the older Ladinian platforms.

The clinoforms themselves show a range of shapes (Fig. 5.55G, H, I and J), described in similar terms to those used in seismic reflection studies (Mitchum *et al.*, 1977). Oblique–parallel clinoforms are most common where the talus is a megabreccia, and then the largest blocks occur at the toe of the slope. Oblique–tangential and flattening-out clinoforms occur where the sediment is mud, sand and gravel.

Bosellini (1984) recognized two types of basinal sequence which developed adjacent to the prograding platforms: deepening-upward and shallowing-upward. In the first case, relative sea-level was rising during sedimentation and the basinal sequence shows pelagic and hemipelagic Plattenkalk (flat, thin-bedded limestone) and Knollenkalk (nodular limestone) passing up into limestone turbidites and breccias derived from the advancing platform. Shallowing-upward basinal sequences developed when sedimentation exceeded subsidence/sea-level rise, so that the basin was gradually filled and the advancing talus sequence thins and climbs laterally.

Termination of Ladinian and Carnian platform-margin sedimentation in the Dolomites was brought about by drowning or subaerial exposure as a result of relative sea-level changes and, at the end of the Ladinian, by filling of the basin with a vast thickness of volcanic rocks (see Fig. 2.13).

In the *Cretaceous*, well-developed talus aprons are associated with some platform-margin rudist buildups. These are seen in Provence and the Spanish Pyrenees (e.g. Simo, 1986), and especially in Mexico (Carrasco, 1977; Enos, 1977b). The hydrocarbon-bearing *Tamabra Limestone* in the subsurface of Mexico is a basin-margin deposit of debris flow breccias and rudist fragment grainstones to wackestones which interdigitate basinwards with pelagic microfossil wackestones (Tamaulipas Formation). The shallow-water material is resedimented from the El Abra Limestone of reef and grainstone facies which rimmed the Golden Lane Atoll. An escarpment existed, up to 1000 m high, against which the debris beds accumulated in a wedge 200–400 m thick. Porosity in the Tamabra Limestone is largely the result of dissolution and fracture of rudist skeletons (Enos, 1988). The escarpment model, with a basal talus apron, derived from the eastern side of the Bahama Platform (Fig. 5.47A), would be applicable to this Middle Cretaceous platform margin in Mexico.

In a quite different tectonic setting from the examples described above, resedimented carbonates were deposited in base-of-slope aprons in half-grabens developed during Jurassic extension along the *Tethyan continental margin in Switzerland and France* (Fig. 5.56; Eberli, 1987). Lithofacies recognized include megabreccias and conglomerates deposited from debris flows, unsorted breccias resulting from rockfalls and avalanches, a variety of calciturbidites with complete to part Bouma sequences deposited from high- to low-density turbidity currents, calcarenites from modified grain flows, and marl and pelagic limestone representing the background sediment. Shallow-water carbonates forming on the footwall tilt block

margins were sedimented into the rift basins, along with clasts of older limestone from the footwall block itself. Eberli (1987) identified three types of megabreccia: (1) chaotic, (2) bimodal where a basal conglomerate passes up into chaotic breccia, and (3) graded where a conglomerate occurs at the top of the chaotic part and is commonly overlain by a turbidite. These megabreccia types are thought to represent a proximal to distal trend. The resedimented carbonates form a wedge of debris along the faults bounding the half-grabens and in a vertical sequence they show a thinning- and fining-upwards megacycle from the rockfall breccias at the base, through the megabreccias, thick-bedded turbidites to thin-bedded turbidites (Fig. 5.56B). A smaller-scale thinning-fining-upwards cyclicity on the scale of 10−20 m occurs within the thick-bedded calciturbidites and could reflect the filling of broad, shallow depressions and channels on the talus apron. The non-cyclic thin-bedded turbidite facies association extends over most of the basin, indicating that much of the original relief had been buried. Away from the half-graben boundary fault, distal turbidites and pelagic−hemipelagic deposits accumulated on the hanging-wall tilt block.

5.13.2 Ancient carbonate submarine fan sequences

Although there have been frequent references to resedimented limestones being of submarine fan origin (e.g. Reinhardt, 1977; Bosellini *et al.*, 1981; Cook & Mullins, 1983), there are very few detailed descriptions available. The best documented are from the *Jurassic of Spain* (Ruiz-Ortiz, 1983) and *Jurassic of Portugal* (Wright & Wilson, 1984). In the *Loma Del Toril Formation in the Betic Cordillera*, 50 m of clast-supported and matrix-supported limestone conglomerates in amalgamated beds pass laterally into pelagic limestones. The conglomerates are interpreted as submarine fan valley deposits. They are overlain by 100 m of mostly limestone turbidites with some conglomerates, that are organized into distinct thinning-and fining-upward cycles. The turbidites show a range of internal structures, but some beds consist entirely of cross-bedding. Thick, commonly channellized limestones are interbedded with thinner turbidites. These calcarenites are interpreted as distributary channel fills and overbank deposits of a mid-fan region. Succeeding thin turbidites, commonly T_c or T_{cde}, interbedded with pelagic limestones are outer fan and basin plain facies. Poorly-developed thickening- and coarsening-upward cycles may represent depositional

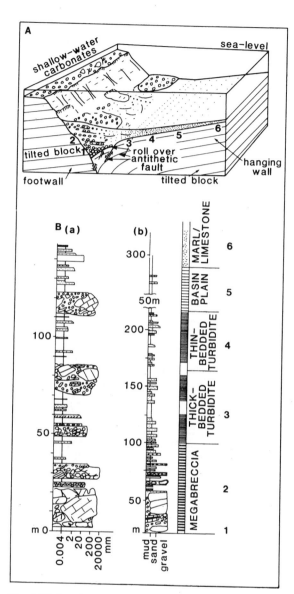

Fig. 5.56 *Resedimented carbonates in Liassic rift basins of the western Alps. (A) Facies model for rift-basin deposition: carbonates are derived from the shallow-water shelf on the footwall block and from the slope. Six facies associations are recognized: (1) base-of-fault scarp association: breccias, (2) megabreccia association: megabreccias intercalated with calciturbidites, marls and limestones, with slumps, (3) thick-bedded turbidite association: conglomerates and turbidites in thinning- and fining-upward cycles, (4) thin-bedded turbidite association, (5) basin plain association of thin turbidites and much hemipelagite, (6) marl/limestone. (B) Two logs through rift-basin sequences in the Swiss Alps; (a) base-of-slope and megabreccia facies association and (b) complete sequence. After Eberli (1987).*

lobes on the outer fan. Contemporary fault movements were responsible for the development of the basin in which this fan was deposited and probably they affected the fan facies geometry and sequence too. The sediments themselves are a mixture of pelagic and shallow-marine skeletal grains, with some oolitic lithoclasts from an older formation.

In the *Toarcian of the Peniche Peninsula, Portugal* (Wright & Wilson, 1984), a 300 m sequence from restricted basin through outer fan to mid fan is preserved (units 1 to 7, Fig. 5.57A). The lower units consist of hemipelagic mudstone with *Zoophycos*, ammonites, belemnites and thin-shelled bivalves (*Bositra*). In unit 3, limestone turbidites are thin (average 0.17 m) and persistent, with common grooves and much bioturbation of thinner beds and the tops of thicker beds. The outer fan turbidites of units 4 and 5 show thickening- and coarsening-upward packets (Fig. 5.57B), which may have resulted from lobe progradation. Thicker and coarser turbidites, with broad channels and amalgamated beds (Fig. 5.57C) occur in units 6 and 7, along with thin turbidite interbeds. These units were deposited in distributary or feeder channels on the mid fan and the complex pattern of the channels is taken to suggest that a braided system existed. The thinner beds are overbank deposits. In unit 7, some beds have medium-scale, planar cross-bedding (0.5 m high sets), with bipolar orientations. These could arise from tidal currents reworking the sand. Units 3 to 7 of the Peniche sequence record the progradation of a carbonate submarine fan into the area. However, it should be noted that the outcrops only give a two-dimensional picture, so that the fan may not have had the classic geometry of modern siliciclastic fans. In addition, the palaeocurrents do not show a radiating pattern, although regional information indicates that the fan was quite localized. The limestones are mostly composed of ooids and micritized grains derived from a sand shoal complex to the west.

There is always a problem in correctly identifying the sedimentary environment of resedimented carbonates from well cuttings, when the presence of shallow-water grains would normally suggest a shallow-marine facies. In the Peniche region, this problem has arisen in an exploration well, but the clue is in the nature of the associated mudrock cuttings which contain the pelagic bivalve *Bositra*. This indicates that basinal facies are present so that the oolitic cuttings were from resedimented limestones and not a platform carbonate (Wright & Wilson, 1984).

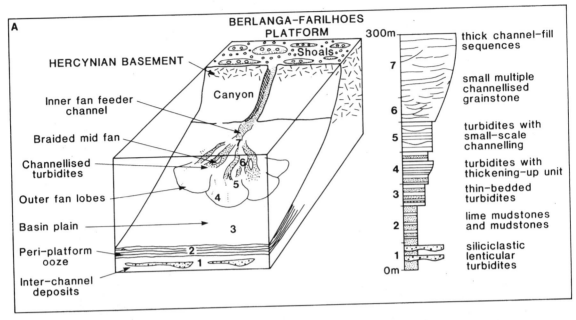

Fig. 5.57 *Carbonate submarine fan sequence from the Upper Jurassic at Peniche, Portugal. (A) Facies model and vertical sequence. (B) Thickening-up and coarsening-up unit of oolitic turbidites from suprafan lobe deposition. (C) Cross-bedded, oolite from channelled upper mid-fan area. After Wright & Wilson (1984).*

5.13.3 Basin plain limestone turbidite sequences

Calciturbidite sequences not related to distinct aprons and fans occur in basins which were directly receiving shallow-water debris, with little being deposited on basin slopes. The material may have been funnelled into the basin along major canyons, which were not preserved, bypassing the slope.

The basinal *Mississippian Rancheria Formation of New Mexico and Texas* (Yurewicz, 1977b) consists of lime–silt peloidal grainstones interbedded with lime mudstone and spiculitic wackestone. The former are structureless to finely-laminated beds 0.04–0.3 m thick, with some cross-lamination and normal grading. T_{bcd} beds are present. Deposition from turbidity currents is most likely although reworking by bottom

Fig. 5.58 *Crinoidal turbidites from the Upper Devonian of Cornwall, England. (A) Field view of turbidite sequence, which is structurally inverted, showing bed continuity, large scours on the bases of beds (seen on upper surfaces in this view) and sharp tops to beds. Cliff is 50 m high. (B) One turbidite bed with coarse basal layer, passing up into parallel-laminated division and then cross-lamination with a convolution at the top.*

50mm

currents is possible. Crinoidal grainstones towards the basin margin in the Rancheria Formation were deposited from higher-density turbidity currents. Intraformational disconformity surfaces within the Rancheria are interpreted as the result of erosion by deep sea currents. Some surfaces truncate the bedding and probably represent broad channels; others are planar with relief of less than a metre. Similar age limestones in Utah (Great Blue Formation) form an 800–1400 m sequence of dark thin-bedded calci-turbidites in basinal mudrock (Bissel & Barker, 1977). These bathyal sediments filled a depression which was the forerunner of the Oquirrh Basin, that persisted until the end of the Permian.

In the *Devonian and Carboniferous of southwest England*, basin plain carbonate turbidites are well developed in the Marble Cliff and Westleigh Formations. In both cases, the locations of the shallow-water platforms which supplied the sediment are unknown. The Marble Cliff crinoidal turbidites (Fig. 5.58) commonly have three internal units of T_{bcd}, where division c is a single cross-bed set, up to 0.3 m thick (Tucker, 1969). Some beds have very coarse skeletal bases which are traction carpet deposits (Fig. 5.58B).

6 Carbonate mineralogy and chemistry

6.1 INTRODUCTION

The raw material that carbonate rocks are built from, by definition, must consist of more than 50% carbonate minerals. The most important carbonate minerals volumetrically are calcite ($CaCO_3$), dolomite ($Ca \cdot Mg(CO_3)_2$) and aragonite ($CaCO_3$). To investigate processes such as growth and dissolution of these minerals it is first necessary to understand their crystal structure. Carbonate mineralogy is here considered first using ideal, stoichiometric structures, then inorganic solid solutions and finally, complex biogenic precipitates.

Trace element and isotope analyses have been applied to a range of carbonate-related problems over the past two decades. A general understanding of these techniques is a prerequisite to some later sections in this book. Unlike structure, where consideration of the solid alone is appropriate, a knowledge of the behaviour and composition of the mother liquor during carbonate precipitation is vital to understanding both trace element and isotopic composition of the solid. Equally important to any quantitative consideration of trace element or isotopic composition is the distinction between precipitation that occurs in an equilibrium situation and precipitation where kinetic factors modify or control partitioning between the liquid and solid.

6.2 CRYSTAL CHEMISTRY OF COMMON CARBONATES

The basic structural unit common to all carbonate minerals is the CO_3 group. This is constructed of three oxygen atoms forming the corners of an imaginary equilateral triangle which is centred by a carbon atom. The C−O bonds are short, resulting in a close approach between oxygen atoms which, in turn, produces a very strong structural unit. The C−O bond is stronger than the metal−O (M^{2+}−O) bond present in any carbonate.

Aragonite is orthorhombic and structurally different from rhombohedral calcite and dolomite. The simpler rhombohedral structures are considered first. Three different unit cells have been used for rhombohedral carbonates (Fig. 6.1A). The hexagonal unit cell emphasizes the alternating layers of metal atoms and CO_3 groups along the c-axis (Fig. 6.1).

6.2.1 Calcite and dolomite structures

Calcite and dolomite are chemically integral to the $CaCO_3$−$MgCO_3$ solid solution series but, despite strong similarities, they are structurally different. A perspective view of a small portion of the calcite structure, Fig. 6.2A, emphasizes the alternating layers of Ca atoms and CO_3 groups. Each CO_3 group within one layer has a common orientation which is 180° reversed in each adjacent layer (Fig. 6.1B,C). The Ca atom is co-ordinated with six oxygens from different CO_3 groups to form a slightly distorted octahedron (Fig. 6.2C). Each oxygen in the CO_3 group is bonded to one carbon and two Ca atoms from adjacent cation layers (Fig. 6.2B). In calcite, the C−O bonds are coincident with the three a-axes.

The structure of dolomite is most simply pictured by substituting Mg atoms into alternate Ca layers of the calcite structure. The substitution, however, introduces a change in bond strengths. This causes a displacement of atoms and results in dolomite possessing a lower degree of symmetry than calcite. The difference in bond length between the Ca−O and Mg−O bonds (Fig. 6.3B) causes the oxygen (in 3-fold co-ordination) to be displaced towards the Mg. This displacement is accommodated by all the CO_3 groups being uniformly rotated around the 3-fold axis relative to their position in calcite. The sense of rotation in successive layers is the same, so the oxygen position lies off the diad axis where it is in calcite and successive CO_3 groups are no longer related to a c-glide as in calcite.

The magnitude of distortion of the $M^{2+}O_6$ octahedra is related to mineral stability. The CaO_6 octahedra in dolomite are smaller and less distorted than CaO_6 in calcite (Figs 6.2C and 6.3C). The MgO_6 octahedra in dolomite are smaller and less distorted than

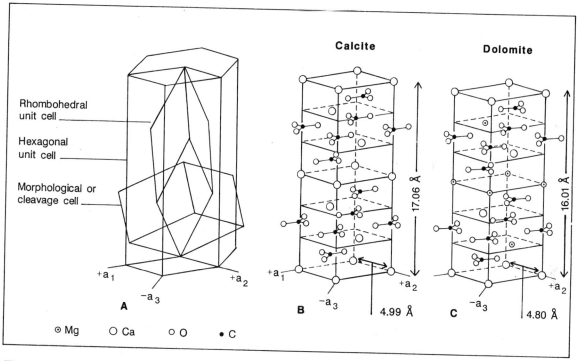

Fig. 6.1 *(A) Diagrammatic relationship between rhombohedral, hexagonal and morphological unit cells and the hexagonal crystallographic axes. (B) Hexagonal unit cell of calcite, apparent rectilinear shape due to perspective; a_1, a_2 and $a_3 = 4.99$ Å; $c = 17.06$ Å. (C) Hexagonal unit cell of dolomite; a_1, a_2 and $a_3 = 4.80$ Å; $c = 16.01$ Å.*

in magnesite (Reeder, 1983b). The metastable behaviour of magnesian calcites is perhaps related to their $M^{2+}O_6$ octahedra being considerably more distorted than the proportional average of magnesite and calcite end-members.

6.2.2 Aragonite structure

The CO_3 group in aragonite is virtually identical to that in the rhombohedral carbonates. In dolomite the C atom is displaced from the plane of the three oxygen atoms towards the layer containing the smaller Mg (compared to Ca) atoms; in aragonite the C atom is displaced (0.06Å; Speer, 1983) towards the nearest Ca layer. The calcium atoms are arranged in pseudo-hexagonal layers parallel to (001) with an ABAB layer sequence along c (Fig. 6.4). The calcium layers are separated by two distinct layers of CO_3 groups. The Ca atoms are bonded to nine oxygens, involving two oxygens at the edge of three CO_3 groups and one oxygen at the apex of three CO_3 groups (Fig. 6.4).

6.2.3 Solid solutions

The Ca position in calcite can be substituted by a variety of elements. Eight divalent cations are known to exist as calcite isotypes: Ca (calcite), Cd (otavite), Mn (rhodochrosite), Fe (siderite), Co, Zn (smithsonite), Mg (magnesite), and Ni: the mineral names are in parentheses. Some structural properties of these isotypes, such as M—O bond length and cell volumes, vary almost linearly with cationic radius. The axial ratio c/a and $M^{2+}O_6$ octahedral distortion of these isotypes, however, show no clear relationship to ionic radius. The lack of correlation between M—O bond length and c/a ratios is probably due mainly to the electronic configuration rather than ionic size (Reeder, 1983b).

The presence of isomorphous solid solutions between some of the end-member calcite-type carbonates can be explained, despite incomplete correspondence between cationic radius and crystal parameters, on the basis of cationic size. The magnitude of the difference between end-member cationic

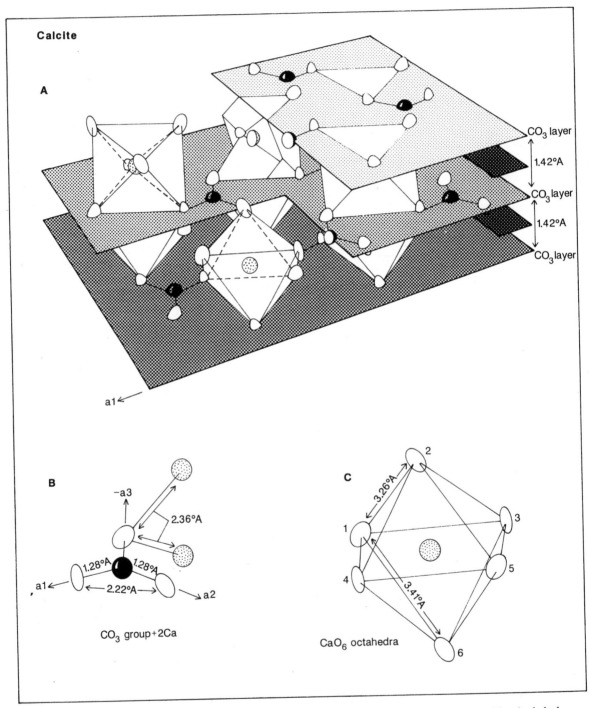

Fig. 6.2 *(A) Perspective view of portion of calcite lattice in which the layered nature is emphasized by shaded planes: black spheres = carbon atoms stippled spheres = calcium atoms and unshaded ellipsoids = oxygen atoms. CO_3 groups intersected by three full planes. Position of Ca planes shown by partial planes at right margin. CaO_6 octahedra between CO_3 layers unshaded. (B) CO_3 group with one oxygen showing two bonded Ca ions. (C) CaO_6 octahedron showing bond lengths.*

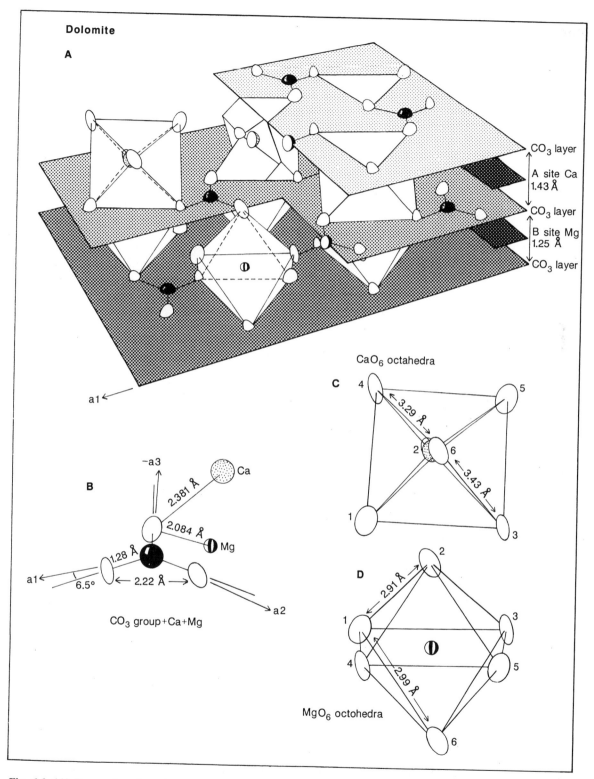

Fig. 6.3 *(A) Perspective view of portion of dolomite lattice, symbols as for Fig. 6.2 with vertically striped spheres = magnesium ions. Note rotation of structural units relative to a_1 direction and different layer thicknesses compared to Fig. 6.2. (B) CO_3 group. Note 6.5° rotation of C−O bonds relative to a_1 axis. (C) CaO_6 and (D) MgO_6 octahedral bonding.*

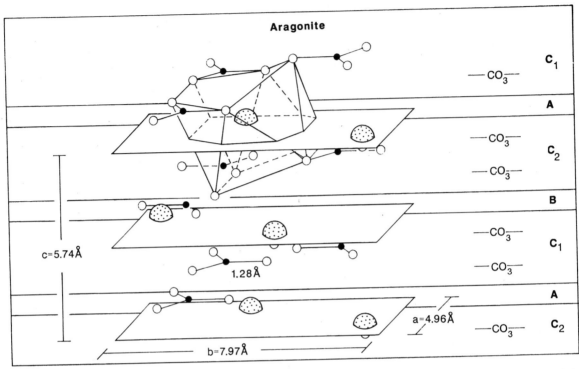

Fig. 6.4 *Projection emphasizing the layered nature of the aragonite structure against a (100) plane. Three unshaded horizontal planes intersecting stippled calcium atoms (ABAB layering). Black spheres = carbon atoms, unshaded spheres = oxygen atoms. The CaO_9 co-ordination polyhedron characteristic of the orthorhombic carbonates outlined. Unit cell dimensions shown.*

radii allows two categories of solid solution series to be established: in the first, where this difference is small, complete miscibility exists, and in the second, where this difference is large, only limited miscibility exists (Fig. 6.5). The second category includes the most important natural solid solution series $CaCO_3-$$Ca\cdot Mg(CO_3)_2-MgCO_3$. Five ordered double carbonates are known of which only dolomite and ankerite occur commonly in nature. Ankerite or ferroan dolomite, $Ca\cdot(Mg, Fe)(CO_3)_2$ may be regarded as dolomite with limited Fe substitution (up to approximately 70 mole % $Ca\cdot Fe(CO_3)_2$ in dolomite; Goldsmith, 1983). The A site (Fig. 6.3A) is probably completely filled by Ca, leaving other cations and any excess Ca to occupy the smaller B site. The relationship between cationic size and stability of the double carbonates is not straightforward. The difference in size between Ca and Mg is greater than between Ca and Fe (Fig. 6.5) yet dolomite $Ca\cdot Mg(CO_3)_2$ is stable and $Ca\cdot Fe(CO_3)_2$ does not exist. Distortion of the MgO_6 octahedra in dolomite is similar to FeO_6 dis-

tortion in ankerite but the linked distortion of the CaO_6 octahedra is considerably greater in ankerite than in dolomite. Rosenberg & Foit (1979) have suggested that increasing Fe substitution in dolomite increases octahedral distortion which eventually destabilizes the structure.

Substitution for Ca in the aragonite structure is exemplified by end-member isomorphous strontianite ($SrCO_3$), cerussite ($PbCO_3$) and witherite ($BaCO_3$): synthetic rare earth element isomorphs are also known. Alkaline earth double carbonates (alstonite and paralstonite, $Ca\cdot Ba(CO_3)_2$) occur naturally and are structurally related to aragonite. At sedimentary temperatures and pressures, little miscibility exists between these minerals. Recorded compositions of natural members of the $CaCO_3-SrCO_3-BaCO_3-$$PbCO_3$ system show some substitution. Of this group, only aragonite is of any significance to sedimentary carbonate rocks. Aragonite may contain up to 14 mole % $SrCO_3$ and 2.5 mole % $PbCO_3$ (Speer, 1983).

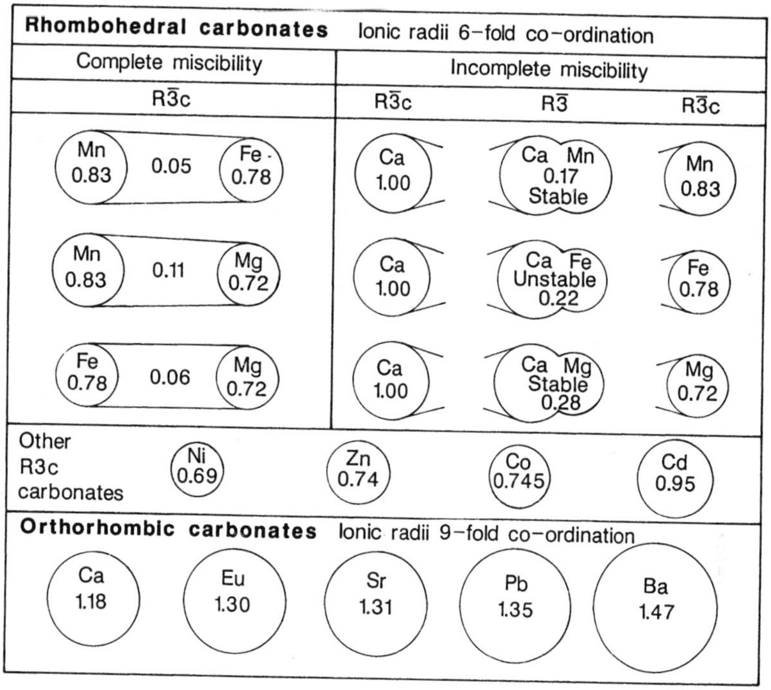

Rhombohedral carbonates		Ionic radii 6–fold co-ordination		
Complete miscibility		Incomplete miscibility		
$R\bar{3}c$		$R\bar{3}c$	$R\bar{3}$	$R\bar{3}c$
Mn 0.83 — 0.05 — Fe 0.78		Ca 1.00	Ca Mn 0.17 Stable	Mn 0.83
Mn 0.83 — 0.11 — Mg 0.72		Ca 1.00	Ca Fe Unstable 0.22	Fe 0.78
Fe 0.78 — 0.06 — Mg 0.72		Ca 1.00	Ca Mg Stable 0.28	Mg 0.72
Other R3c carbonates	Ni 0.69	Zn 0.74	Co 0.745	Cd 0.95
Orthorhombic carbonates		Ionic radii 9–fold co-ordination		
Ca 1.18	Eu 1.30	Sr 1.31	Pb 1.35	Ba 1.47

Fig. 6.5 *Cationic radii and carbonate space groups. The diameters of 'cation' circles are proportional to their radii. Linked cations show the extent of miscibility for some rhombohedral carbonates. Central figures in these solid solution series are differences (in Å) between end-member radii.*

6.2.4 Phase relationships

CaCO₃. Aragonite and calcite are the only common, natural $CaCO_3$ minerals: a third $CaCO_3$ polymorph, vaterite, is known to occur naturally (see Table 6.1) and is frequently encountered in experimental work. Vaterite is metastable, of low density and its hexagonal structure has some similarities with the aragonite structure but differs appreciably from calcite (Easton & Claugher, 1986).

At elevated temperatures and/or pressures five other $CaCO_3$ polymorphs are known from experimental work. The phase relations of $CaCO_3$ and ikaite ($CaCO_3 \cdot 6H_2O$) are shown in Fig. 6.6. Ikaite forms at subzero temperatures and is preserved as calcite pseudomorphs (Shearman & Smith, 1985; Jansen *et al.*, 1987). Monohydrocalcite, $CaCO_3 \cdot H_2O$, another hydrated calcium carbonate mineral is, like ikaite, metastable at the Earth's surface.

The best documented phase boundary in the $CaCO_3$ system is between calcite and aragonite (from 100°C to 600°C). The addition of strontium drops this boundary to lower pressures but unnaturally large amounts (greater than 50 mole % $SrCO_3$) are required to depress the boundary to 1 kbar (Carlson, 1980). The outcome of this is that at normal Earth surface temperatures and pressures calcite is the only stable $CaCO_3$ phase.

CaCO₃–MgCO₃. The phase behaviour of the $CaCO_3–MgCO_3$ series has been investigated above 400°C (Fig. 6.7; Goldsmith, 1983). Extrapolation of phase boundaries to sedimentary temperatures indicates that only magnesite, dolomite and calcite with a few mole % $MgCO_3$ are expected to be stable under Earth surface conditions.

6.2.5 Magnesian calcites

The crystal structure and phase relationships described above apply to the simplest, ideal carbonates. Marine pelagic oozes and some lacustrine carbonates are composed predominantly of the predicted, thermodynamically stable calcite. Shallow-marine, tropical carbonates, however, are composed predominantly of aragonite and calcite with high Mg concentrations, that is, metastable phases (see Table 6.1). Nonstoichiometric, disordered metastable dolomite is the commonest type of Recent dolomite. Dolomites traditionally are treated separately from other carbonates; see Chapter 8.

Calcites with more than a few mole % $MgCO_3$ are

Table 6.1 Listing of carbonate biominerals. After Lowenstam & Weiner (1983)

		C A V M D A
Monera		C A M
Protoctista	Dinoflagellata	C
	Haptophyta	C ?
	Phraeophyta	A
	Rhodophyta	C A V
	Chlorophyta	A
	Siphonophyta	C A
	Charophyta	C
	Foraminifera	C A
	Mixomycota	C X
Plantae	Bryophyta	C
	Tracheophyta	C A V
Animalia	Porifera	C A
	Cnidaria	C A X
	Platyhelminthes	C X
	Bryozoa	C A
	Brachiopoda	C
	Annelida	C A V
	Mollusca	C A V M
	Arthropoda	C A V X
	Sipuncula	C A
	Echinodermata	C D
	Chordata	C A V M X

C = calcite including Mg calcite; A = aragonite; V = vaterite; M = monohydrocalcite; D = dolomite; X = amorphous carbonate; ? = not fully identified.

termed magnesian calcites (dividing line often taken at 4 mole %). In marine waters many calcareous skeletons, cements and sometimes ooids are composed of magnesian calcite (Table 6.1; Chapter 1). Magnesian calcites also from some freshwater tufas and cements.

The distribution of modern marine magnesian calcite skeletons shows a clear relationship between mole % $MgCO_3$ and latitude (Fig. 6.8). The Mg content of these magnesian calcites has been positively correlated with temperature, for in cements the Mg content also diminishes with increasing depth and cooler waters in the ocean. However, Mg content and temperature also correlate with saturation state (CO_3^{2-} concentration). Biological calcification may be related not only to external factors such as temperature and saturation state but also to internal factors related to the biota's physiology. Clearly, great caution or further understanding is required to interpret the Mg content of magnesian calcites (further discussion Sections 6.3.3 and 7.4.3).

The quantitative determination of mole % $MgCO_3$ in calcites has been attempted through X-ray diffraction. Synthetic magnesian calcites produced at equilibrium above the calcite−dolomite solvus (Fig. 6.7) have been used. A clear relationship has been found between $MgCO_3$ content and unit cell parameters (Bischoff *et al.*, 1983) despite variations in the methods used to produce the high temperature magnesian calcites. Cell volume and axial ratio (c/a) of calcites with up to 23 mole % $MgCO_3$ deviate from a straight-line

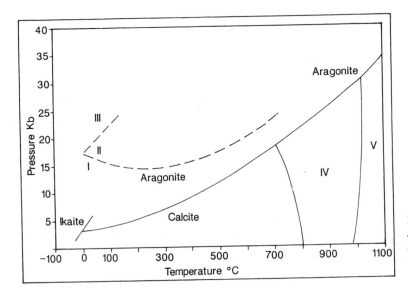

Fig. 6.6 *Phase relations of the CaCO$_3$ system and CaCO$_3$·6H$_2$O. 1−V, calcite polymorphs; 1−III are metastable. After Marland (1975) and Carlson (1980).*

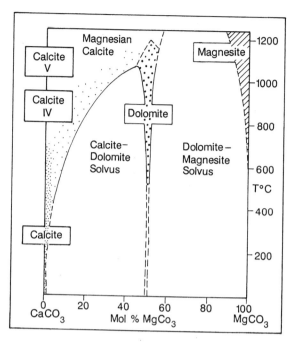

Fig. 6.7 *Phase relations of the CaCO₃−MgCO₃ system. High temperature margin to dolomite field conjectural. After Goldsmith & Heard (1961a).*

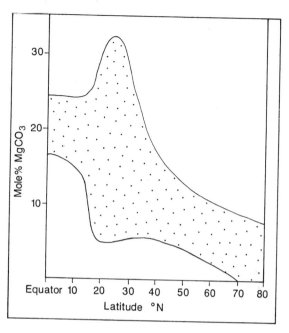

Fig. 6.8 *Range of mole % MgCO₃ in skeletal magnesian calcites plotted against latitude. Data (from Chave, 1954) include red algae and a variety of invertebrate phyla.*

relationship between the calcite and magnesite end-members. However, low temperature synthetic material and natural skeletal calcites of known Mg content correlate poorly against unit cell parameters derived from the high temperature calcites. The unit cell volumes (in the range 4−16 mole % MgCO₃) and axial ratios (c/a) of skeletal material are greater than those of the high temperature phases of similar composition. These differences would result in inaccuracies of over 5 mole % MgCO₃ in skeletal material if estimated from diffraction data (Mackenzie *et al.*, 1983). The determination of Mg content is better done by direct methods such as electron microprobe or inductively coupled plasma analysis.

Examination of high temperature synthetic and skeletal magnesian calcites by Raman spectroscopy (Bischoff *et al.*, 1985) produces a somewhat similar response to X-ray diffraction. Raman spectral shifts and band halfwidths increase regularly with increasing magnesium content of synthetic calcites but show much more scatter for skeletal calcites. The increase in spectral shifts and band halfwidths is partly caused by the effects of Mg on neighbouring CO₃ groups but is largely due to tilting of the CO₃ group out of the

basal plane (containing the a-axes). This rotation of the CO₃ groups towards the c-axis allows shorter Mg−O bonding. This effect reaches a maximum in calcite with 25 mole % MgCO₃ and in turn this may explain the deviation of diffraction data for magnesian calcites from a straight-line interpolation between calcite and magnesite.

Raman spectroscopy reveals that some skeletal magnesian calcites contain HCO₃, whilst infrared spectroscopy shows the presence of O−H bonds in skeletal calcites. The O−H bond may be due to the presence of bound water which would have considerable effect on mineral properties (Mackenzie *et al.*, 1983).

The substitution of Mg into the calcite lattice causes positional disorder in all magnesian calcites. The regular X-ray diffraction and Raman response of high temperature magnesian calcites to mole % MgCO₃ compared to the irregular behaviour of skeletal magnesian calcites indicates that the irregular behaviour could be caused by: (1) irregular or poorly-crystallized material, (2) inhomogeneities in chemical composition (high- and low-Mg domains), (3) substitution by other cations (Mn, Fe, Sr, Ba, Na, K, etc.), and/or (4) substitution by other anions (HCO₃, SO₄ etc.).

6.2.6 **Biomineralization**

Carbonate sedimentary particles are largely biologically produced or induced. The most obvious biominerals are those that form skeletons to provide support or protection but biominerals may also act as waste disposal sites, storage deposits, eye lenses, gravity receptors and so on. Carbonate biominerals have been recorded by Lowenstam & Weiner (1983; see Table 6.1). The degree of biological involvement in the production of biominerals varies widely. Lowenstam (1981) has proposed two categories of biomineralization.

6.2.6a Biologically-induced mineralization

Here biotic influence is weakest and mineralization results from reaction between metabolic byproducts and ions in the external environment. The crystal habit of these precipitates is similar to that produced inorganically in the external environment, and their orientation is random. A reaction that is particularly important in affecting carbonate equilibria is the fixation of carbon dioxide by bacteria and algae:

$$CO_2 + 2H_2Z \rightarrow (CH_2O) + H_2O + 2Z \qquad (1)$$

Algae affect this reaction during photosynthesis when H_2Z is water. If the reduction of CO_2 occurs in waters containing calcium bicarbonate, precipitation of $CaCO_3$ occurs:

$$Ca(HCO_3)_2 \rightarrow CaCO_3 \downarrow + H_2 + CO_2 \qquad (2)$$

However, CO_2 may be fixed both in the absence of oxygen, when H_2Z is an inorganic sulphur or organic compound, and in the absence of light, by nitrifying or iron-oxidizing bacteria. The gas produced by bacterial metabolism may diffuse some distance before reacting with metal ions in the external fluid to produce mineral precipitates. In these circumstances the connection between these extracellular precipitates and biological induction becomes tenuous. Many carbonate concretions are formed through bacteriological reactions, yet concretions are seldom regarded as biominerals.

6.2.6b Organic matrix-mediated mineralization

Here every mineral unit is enveloped and commonly permeated by organic material. The mineral species, its crystallographic orientation and its construction are under direct biological control. The chemistry of this organic matrix is best known in molluscs, vertebrates and some protoctistids. In molluscs, controlled nucleation and growth of carbonate take place on the surfaces of organic sheets that separate each mineral lamella. In some molluscs (e.g. *Nautilus*; Weiner & Traub, 1984), fabric elements of the organic matrix and crystalline phase have a common orientation (implying epitaxy). The organic matrix possesses a regular structural repeat pattern that matches the crystalline lattice. However, tissues that are mineralized with amorphous minerals suggest that epitaxy is not essential for biomineralization. Weiner & Traub (1984) have separated peptides unique to the calcite and aragonite layers of the *Nautilus* shell suggesting that specific proteins act as templates for mineralization. The arrangement of components in mineralizing mollusc tissue is shown in Fig. 6.9.

Considerable attention has been paid to biomineralization of the coccolithophorids, which are unicellular marine algae. The coccoliths of *Emiliania huxleyi*, an abundant component of the marine phytoplankton, are synthesized intracellularly in a vesicle derived from the Golgi apparatus. When calcification is complete the coccoliths are extruded on to the surface of the cell. The coccoliths consist of calcite and organic material (polysaccharide). It is not known whether the organic material coats the surface or is distributed throughout the calcite. Each coccolith is made of a number of units, each one consisting of a basal single crystal sheet with an upper hammer-shaped projection consisting of a mosaic of small domains (30–50 nm diameter) with no strong orientation (Westbroek *et al.*, 1986). The acidic polysaccharide present in the coccolith vesicle is thought to regulate calcification, for it is able to bind Ca^{2+} and strongly influences $CaCO_3$ growth (De Jong *et al.*, 1986). The form of the coccoliths is controlled by a series of morphological changes in the membranes of the coccolith vesicle. Different types of biomineralization occur within *Emiliania huxleyi* and other coccolithophorids (Westbroek *et al.*, 1986).

Molluscs and coccolith biomineralization are used as examples here because considerable research effort has been devoted to them. The diversity of process and product in biomineralization is enormous. One organism may produce both biologically-induced and matrix-mediated biominerals. The distinction between these two types of biomineralization is often blurred and unclear. Matrix proteins are claimed by some to be the common thread to many types of biomineralization but their role is controversial. Much still needs to be discovered about the biochemistry of mineralization.

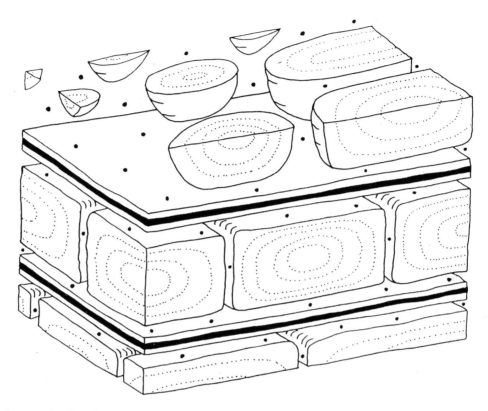

Fig. 6.9 *Diagrammatic example of molluscan biomineralization. The central layer and lower half layer are completely mineralized by aragonite single crystal tablets (shown with growth rings). The upper half layer is undergoing mineralization, the aragonite crystals being seeded in a layer composed of acidic amino acids (large dots). The black layer (B-chitin) and unshaded layers (silk, fibroin-like proteins) separate the mineralizing layers. After Weiner & Traub (1984).*

It is clear that many biologically-produced crystals are quite unlike inorganically-precipitated crystals. Matrix-mediated skeletal units such as echinoderm ossicles are exquisitely sculptured, microporous crystals. The organisms which produce these must be able to distort normal lattice growth; the so-called single crystals are mosaics of crystallites or domains of similar orientation separated by organic matter.

Calcareous biominerals are likely to be the most important source of lime sediments. These exceptional particles may be crystalline or amorphous; they are usually chemically heterogeneous and impure. Their margins are characterized by extremely complex shapes possessing enormous surface area with high surface free energy values. Such particles, on release from their organic tissue on death of the parent, are unlikely to act like grains of inorganic carbonate.

6.3 TRACE ELEMENTS

Minor and trace constituents are found in natural calcite, dolomite and aragonite. Perhaps the simplest method of incorporating these constituents is where a guest ion substitutes for a host or stoichiometric ion of similar charge and radius in the lattice. However, constituents may occur interstitially between lattice planes, be absorbed, occupy lattice defects, become occluded as solid or liquid inclusions, and occur interstitially along crystal boundaries. When a compatible trace component substitutes for a lattice calcium in calcite (a guest/host substitution) the concentration of that trace element can be predicted by means of its distribution coefficient.

When the volume of solid precipitated (containing the guest constituent) is insignificant relative to the volume of the mother solution, then:

$$\left(\frac{^{m}\text{Tr}}{^{m}\text{Ca}}\right)_{s} = k \left(\frac{^{m}\text{Tr}}{^{m}\text{Ca}}\right)_{1} \qquad (3)$$

where m is the molar concentration; Tr is the guest or trace constituent; Ca is the host constituent or Ca^{2+} in the case of calcite; s is solid and 1 is liquid; k is the partition coefficient. The mother solution is effectively an infinite reservoir, and therefore the composition of the precipitate also remains constant. A similar situation arises during homogeneous precipitation when the trace component in the precipitate constantly read-justs to changes in the trace concentration in the liquid. The entire crystal will possess a uniform composition, at equilibrium with the trace to calcium ratio in the final liquid. Hence:

$$D = \left(\frac{^{m}\text{Tr}_{s}}{^{m}\text{Ca}_{s}}\right)_{f} \bigg/ \left(\frac{^{m}\text{Tr}_{1}}{^{m}\text{Ca}_{1}}\right)_{f} \qquad (4)$$

where f is the final concentration and D is the homogeneous distribution coefficient.

In a closed system (see later) the precipitation of the solid can affect the trace to host ratio of the liquid (heterogeneous precipitation). The concentration of the component in the liquid relative to the host changes as precipitation proceeds, so the concentration of that constituent in the calcite will vary; no readjustment takes place in the solid and zoned crystals result:

$$\log \left(\frac{^{m}\text{Tr}_{i}}{^{m}\text{Tr}_{f}}\right) = \lambda \log \left(\frac{^{m}\text{Ca}_{i}}{^{m}\text{Ca}_{f}}\right) \qquad (5)$$

where i is the initial concentration, and λ is the heterogeneous distribution coefficient.

The simplest equation (3) requiring equilibrium to be established is now used to illustrate partitioning in calcite. When the value of k is more than 1, the trace will be partitioned preferentially into the solid. The greater the numerical value of k, the greater will be partitioning such that the guest element will be more effectively scavenged from the fluid into the precipitate. With k less than 1, the ratio of trace to calcium being incorporated into calcite is less than that in the fluid. If $k = 1$ then no partitioning occurs, and the trace to calcium ratio will be the same in the fluid and the precipitate.

The concentrations of five cations for three natural fluids are shown in Fig. 6.10. In these three liquids Mg^{2+} and Sr^{2+} have higher concentrations than Fe^{2+} and Mn^{2+}; $k Mg_{c}^{2+}$ and $k Sr_{c}^{2+}$ are less than 1, and $k Fe_{c}^{2+}$ and $k Mn_{c}^{2+}$ are greater than 1. According to equation (3), the relative amounts of these cations

will be reduced or reversed in calcites precipitated from these fluids. Most notable is the case of Fe^{2+} being next to the least important cation in both meteoric and formation waters but the most important (excluding Ca^{2+} which acts as host) in their precipitated calcites.

The element concentrations for calcite shown in Fig. 6.10 should not be accepted without further consideration. For instance, both Fe and Mn are assumed to be in their divalent state but this is highly unlikely for open marine waters and many meteoric waters. Here, the Fe and Mn are oxidized (Fe^{3+} and Mn^{3+} or Mn^{4+}) and the much larger radii of these ions would make them unlikely guests in the calcite lattice. Magnesium and strontium are unaffected by changes in redox potential.

Natural calcium carbonates are mainly biogenic in origin. Skeletal precipitates mediated by an organic matrix are precipitated not from the ambient waters but from body fluids. This is illustrated by planktonic foraminifera and coccolithophorida which produce calcite skeletons with Mg concentrations in the hundreds of parts per million level. The Mg concentration in calcite precipitated from marine water would be predicted (Fig. 6.10) to be in the 10 000 ppm range. Similarly, aragonitic molluscs produce strontium-poor shells with 2000 ppm rather than 10 000 ppm as would be predicted from equation (3). However, many Recent calcareous skeletons do possess Mg and Sr concentrations close to that predicted by partitioning theory.

Marine water is a relatively well-mixed reservoir compared with meteoric and formation waters which have highly variable chemistries. Accepting a uniformitarian approach it should be possible to predict the composition of ancient marine, inorganic precipitates. The principal components of seawater are believed to have remained constant throughout the Phanerozoic (Holland, 1978), but caution is necessary; Graham et al. (1981), for example, documented possible Sr/Ca variations of 15% to 25% during the Cenozoic. See Section 9.3 for further discussion.

The inorganic origin of marine and most superficial calcareous precipitates is difficult to prove, but it is generally accepted that many calcites precipitated from subsurface meteoric and formation waters are inorganic. When considering such diagenetic products the size of the fluid reservoir becomes important, for the composition of the fluid in a small reservoir will change on precipitation of solids unless the k values are equal to unity. This is most simply illustrated by a system where a finite amount of the reactants and

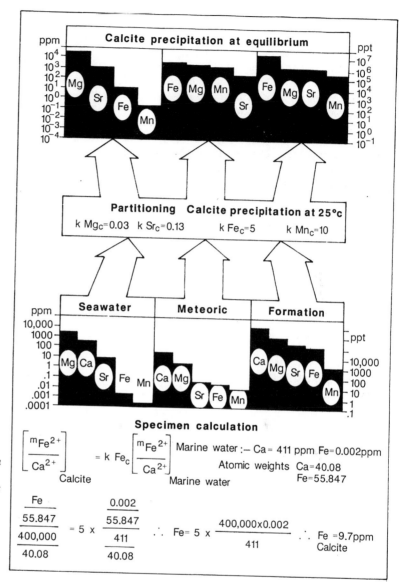

Fig. 6.10 *Diagram to illustrate the calculation of trace element abundances in calcite precipitated from waters of known composition using partition coefficients. The cation concentrations in seawater are well documented, those used for meteoric water approximate river water and those used for formation water approximate an oil field brine. The partition coefficients used are arbitrarily chosen and it is assumed precipitation occurs at equilibrium.*

their product(s) are mutually in contact in a sealed volume, i.e. a closed system.

6.3.1 **Closed system**

A convenient way to picture a closed system is a box containing aragonite and distilled water. The aragonite is being transformed to calcite due to the solubility difference between the two minerals. When the liquid and aragonite come into contact aragonite will dis-

solve; it will continue to dissolve until calcite saturation is exceeded, for aragonite is more soluble than calcite. At this stage calcite will precipitate as the liquid is supersaturated with respect to calcite, and aragonite will continue to dissolve for the same liquid is under-saturated with respect to aragonite. What will be the partitioning behaviour of Sr and Ca as this reaction proceeds?

During the first stage of aragonite dissolution the $^mSr/^mCa$ ratio in the liquid will be the same as that in the aragonite. If the Sr content of the aragonite is 10 000 ppm, the $^mSr/^mCa$ ratio of the aragonite and the liquid is 0.011 (the Ca abundance in $CaCO_3$ is 400 000 ppm). The first increment of calcite to precipitate will have a $^mSr/^mCa$ ratio one tenth of that in the liquid, assuming $kSr_c = 0.1$. The excess Sr not incorporated in this first calcite remains in the fluid and is added to the Sr released on dissolution of further aragonite to produce the second increment of calcite. The $^mSr/^mCa$ ratio in the liquid has now increased so the second increment of calcite will contain more Sr than the first. With each unit of aragonite dissolved and calcite precipitated the excess Sr remains in the liquid until the time when the calcite being precipitated has the same $^mSr/^mCa$ ratio as the dissolving aragonite. At this time the $^mSr/^mCa$ ratio of the liquid is ten times that of the precipitate and a

steady state exists until aragonite dissolution ceases. Figure 6.11 illustrates such a system.

The behaviour of Mg is similar to that of Sr in the closed system just described, for both have k values below unity. In Fig. 6.11 the concentration of Mg in aragonite is taken as 1000 ppm and $kMg_c = 0.01$. It can be seen that steady state is not achieved between liquid and solid for $^mMg/^mCa$ as soon as for $^mSr/^mCa$. In a closed system, steady state is reached most rapidly for those elements whose k values are both below and closest to unity.

Iron and manganese behave differently from Sr and Mg as their k values exceed unity. If $kFe_c = 5$, then the calcite precipitated would attempt to draw an $^mFe/^mCa$ ratio five times greater than that provided by the dissolving aragonite. In a closed system there is nowhere else to derive Fe so every increment of

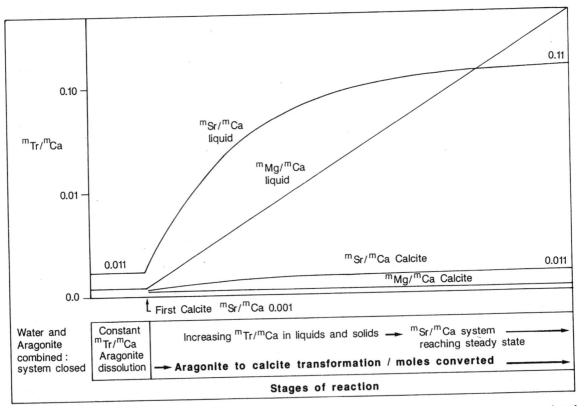

Fig. 6.11 *Stages of reaction (or moles converted) during the aragonite to calcite transformation in a closed system plotted against the $^mSr/^mCa$ ratios in the liquid. Once precipitation of calcite commences, partitioning affects the $^mSr/^mCa$ ratio until a steady state is reached when the $^mSr/^mCa$ ratio in the dissolving phase is identical to that in the precipitating phase. The $^mSr/^mCa$ ratio is shown for the liquid before and at the time of first calcite precipitation (0.011); the first precipitate (0.001); the final liquid (0.11) and the final precipitate (0.011) therefore $kSr_c = 0.1$ and $kMg_c = 0.01$. See text.*

calcite would contain the same Fe/Ca ratio as that released by aragonite dissolution.

The notion that the aragonite to calcite transformation proceeds as a series of steps or increments is a device to picture the process. The course of the reaction, plotted in Fig. 6.11 as continuous curves, shows that these steps are infinitesimally small. The positions of the curves in Fig. 6.11 may be changed by varying the composition of the dissolving solid(s), the composition of the fluid, temperature (which would affect k values), and so on.

6.3.2 Open system

The closed system considered above involved reaction in a closed box. If two opposite ends of the box are opened allowing throughput of liquid, an open system results. To make this more realistic, a tilted layer of aragonite is considered in Fig. 6.12A. Infiltration of meteoric water (containing negligible Sr and Ca) occurs at the updip end and no exchange is allowed outside the layer as the water moves unidirectionally downdip. Conceptually, it is convenient to consider the reaction as a number of steps, each step being the passage of one unit of water through the layer effecting the aragonite to calcite transformation repeatedly. The first unit of water to pass through the layer shows the same reaction stages as the closed system but the reaction and its products are spread out, from where the liquid first precipitates calcite, in a downdip direction (Fig. 6.12B).

Dissolution of carbonate where water first penetrates the aragonite layer causes the scarp to retreat (Fig. 6.12C). Concomitant with retreat of the exposed margin, all reaction stages move downflow along the reaction path. Retreat of the dissolution stage means that not only is aragonite dissolved but also calcite derived from previous transformations. As this calcite has a lower $^mSr/^mCa$ ratio (0.001) than the aragonite (0.011), the $^mSr/^mCa$ ratio of the fluid derived from their dissolution is a combination of the two (less than 0.011). The first calcite precipitated from this fluid consequently has a lower Sr concentration (Fig. 6.12C) than the first calcite precipitated from the first passage of water through the aragonite layer (Fig. 6.12B). As the percentage of calcite to aragonite dissolved in this first stage increases, so the Sr content of the first calcite diminishes. If only calcite had dissolved during this first stage, that calcite would be a collection of first calcites precipitated from many episodes of infiltration, each consecutive episode producing a first

calcite with diminishing Sr content. Consequently, if it were possible to precipitate from such an all-calcite dissolution stage, the precipitated calcite would have a maximum $^mSr/^mCa = 0.0001$ (Sr = 100 ppm). The involvement of calcite in the dissolution/precipitation process during open-system transformation clearly distinguishes it from a closed-system reaction and allows a much lower Sr concentration in calcite to be achieved.

The open system depicted in Fig. 6.12 may be modified. The starting layer could contain a mixture of minerals. The boundaries between reaction stages may become more complex if the transmissivity of the unit becomes anisotropic. The system can be made open in all directions allowing cross-formational flow from the confining beds above and below the aragonite layer. Fluid could be introduced at any place along the reaction path, and so on.

In an open system the Tr/Ca ratio may be changed in the fluid by losing Ca or gaining the trace element from a non-carbonate phase. If $k\mathrm{Fe_c} = 5$, dissolving aragonite provides a particular $^mFe/^mCa$ ratio which is 1/5 that required by the k value for the precipitating calcite. An episode of calcite precipitation occurs from a reservoir derived by aragonite dissolution. Precipitation begins with an $^mFe/^mCa$ ratio 5 times that of the fluid. Iron will be rapidly exhausted in the fluid and in the precipitated calcite. However, if immediately after precipitation starts the fluid is removed from the system, calcite is preserved with an $^mFe/^mCa$ ratio 5 times that of the aragonite. This process may be repeated; the bulk composition of the calcite depends on how rapidly the fluid is removed from the precipitation site once precipitation has begun or on the 'openness' of the system. This reaction involves an episode of dissolution, the build up of a reservoir, and precipitation which is curtailed by loss of the reservoir from the system. This may be thought of as modifying the Tr/Ca ratio by loss of Ca.

An increase in the $^mFe/^mCa$ ratio of the liquid may also be achieved by adding Fe^{2+} from an external source. This could occur through some non-carbonate mineral reaction which releases Fe but not Ca, such as a reaction that strips Fe^{2+} from clays.

The open system has been pictured as a bed of aragonite subjected to rapid throughput of meteoric water. The water develops Tr/Ca gradients which increase for components whose k values are less than one and vice versa. The alteration of marine carbonate ooze as it is buried beneath the deep ocean floor may also exemplify an open-system reaction, but here the transport mechanism is diffusion rather than

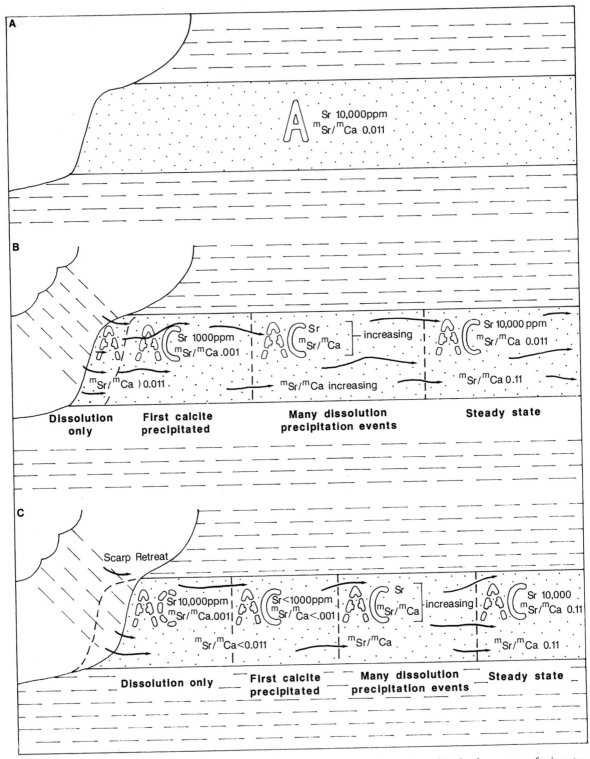

Fig. 6.12 *Evolution of $^mSr/^mCa$ ratio in a titled layer of aragonite being transformed to calcite by the passage of rainwater through the layer: an open system with $kSr_c = 0.1$. (A) Initial layer before water penetrates. (B) The effects of the passage of the first unit of water through the layer. (C) The effects of the second unit of water to pass through the layer. The width of reaction zones arbitrary. See text.*

advection, and the gradients developed are much move severc (Baker *et al.*, 1982).

The use of equation (3) in calculating the trace element concentration of a precipitate, knowing the Tr/Ca ratio of the liquid, depends on several factors: the trace component must substitute for a stoichiometric ion, precipitation must be at equilibrium, and the value of k must be known. Ions of the appropriate size and valence which substitute in lattice sites are welcome guests and the only ones which it is useful to consider here. While Sr^{2+} will substitute for lattice Ca^{2+} in aragonite and calcite it can also reside in non-lattice sites! Analyses of carbonate minerals show that they also include 'unwelcome guests': ions of the wrong size and valence for Ca^{2+} substitution, such as K^+ in calcite.

A unique k value exists for each equilibrium partitioning reaction at a specific temperature and pressure. During attempts to determine k values, it has been found that kinetic factors, such as precipitation rate, influence partitioning. When this occurs, application of the thermodynamically-derived k value for carbonate precipitation is inappropriate.

Attempts to determine partition coefficients for carbonates have either used experimental methods or looked at natural systems. In the latter case the Tr/Ca ratios of both the solid and the liquid are determined. The reaction must be extant, but in complex natural systems it is difficult to identify all controlling factors. A variety of experimental methods have been used to precipitate calcite and aragonite, but not dolomite (see Section 8.5). The most common methods involve either the so-called free-drift or chemo-stat methods. In the free-drift method, the compositions of the fluid and solid are allowed to change during the course of the reaction. It is assumed the solid does not recrystallize in response to the changing composition of the fluid. In the chemo-stat method the precipitate is of uniform composition which is achieved by adding to the solution what is lost due to precipitation. Spontaneous or homogeneous nucleation has been used, but offers little control over crystal form. The technique offering best control is the chemo-stat method involving growth on introduced seeds.

The distribution coefficient determined in these natural and experimental systems will be referred to as k_e, the effective distribution coefficient. The value of k_e may only be strictly applicable to the conditions of reaction in which it was determined.

6.3.3 k_e for alkaline earth elements Sr and Mg

Initially, it was believed that $k_e Sr_c = 0.14$ at 25°C (Holland *et al.*, 1964a) and that only temperature significantly altered that value (at 100°C $k_e Sr_c = 0.08$; Holland *et al.*, 1984b). Subsequent work has shown that temperature has only a slight effect, and that $k_e Sr_c$ has a range of values. The $^m Sr/^m Ca$ ratio of the liquid has an important effect as does precipitation rate and the presence of cations both larger and smaller than Ca^{2+} (such as Ba^{2+} and Mn^{2+} respectively; see Fig. 6.13). The variability of $k_e Sr_c$ is due to either variation in the co-ordination environment at the site of Sr incorporation on to the growing crystal surface or to Sr being incorporated into lattice and defect sites (Pingitore & Eastman, 1986). Whatever the underlying cause, differing values of $k_e Sr_c$ can be assigned to precipitation which occurs at differing rates in solutions of variable $^m Sr/^m Ca$ ratios (Fig. 6.13B,C).

A much higher $k_e Sr_c$ value has been proposed for magnesian calcite, with exact values varying proportionally to the $MgCO_3$ content of the precipitate from 0.195 to 0.344 as $MgCO_3$ varies from 2.7 to 11.2 mole % (Mucci & Morse, 1983). This was suggested to be due to the small Mg^{2+} ion distorting the lattice, creating sites more appropriate to the large Sr^{2+} ion rather than Ca^{2+}. Mucci & Morse (1983) also found $SrCO_3$ content to be independent of precipitation rate (but see comments below on $k_e Mg_c$ value in magnesian calcite).

The distribution coefficient for Sr in aragonite as determined by free-drift experiments is generally just above unity (Kinsman & Holland, 1969; Kitano *et al.*, 1971).

The $k_e Mg_c$ value like $k_e Sr_c$ is below unity but is strongly influenced by temperature unlike $k_e Sr_c$. Some (Lahann & Siebert, 1982; Given & Wilkinson, 1985) believed saturation or precipitation rate to be of over-riding importance for $k_e Mg_c$, whilst others (Morse, 1985; Burton & Walter, 1987; Howson *et al.*, 1987) believed it of no importance at low saturations (1.2 to 17.0; where 1 is saturated, less than 1 is undersaturated, and more than 1 is oversaturated).

The relationship between $k_e Mg_c$ and temperature is shown in Fig. 6.14 by lines connecting three data sets. These lines show similar slopes although their position in $^m Mg/^m Ca$ to $k_e Mg_c$ space is variable. The value of $k_e Mg_c$ is thought to be influenced by the $^m Mg/^m Ca$ ratio of the liquid, but at low ratios (see Fig. 6.14) a wide range is found, even though the two chemo-stat experiments were comparable (Mucci &

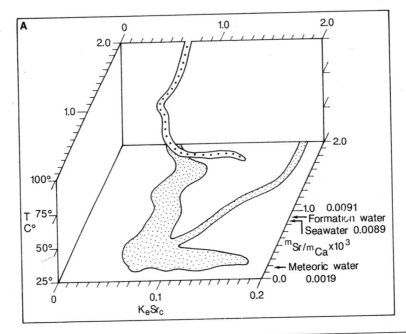

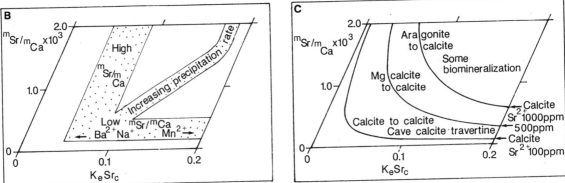

Fig. 6.13 (A) Perspective plot of k_eSr_c (x-axis) against $^mSr/^mCa \times 10^3$ of liquid (y-axis) against temperature °C (z-axis). Data plotted at 25°C from Pingitore & Eastman (1986); at 98°C from Katz et al., (1972). Notice data at 98°C and 25°C overlie, implying little temperature effect on k_eSr_c. (B) Causes for the variation in k_eSr_c at 25°C (Ichikuni, 1973; Pingitore & Eastman, 1986). (C) Calcites of constant Sr concentration, and various carbonate reactions and precipitates shown for 25°C on an $^mSr/^mCa$ versus k_eSr_c plot.

Morse, 1983; Howson et al., 1987 give k_eMg_c values of 0.027 and 0.03 respectively). However, Howson et al. (1987) found no change in k_eMg_c value with $^mMg/^mCa$ ratio of the liquid between 0.15 and 1.04, the range Mucci & Morse predicted should show increasing k_eMg_c with decreasing $^mMg/^mCa$. Howson et al. (1987) suggested that when $^mMg/^mCa$ is less than 1, k_eMc will be 0.03, but when $^mMg/^mCa$ is more than 5, k_eMg_c will be 0.01. The intermediate region of rising

$^mMg/^mCa$ and falling k_eMg_c is one of increasing Mg surface adsorption.

Modern marine magnesian calcite precipitated at 25°C contains 12 mole % $MgCO_3$ but the prediction from experimental work (using $k_eMg_c = 0.123$; Mucci & Morse, 1983) is 8 ± 1 mole % $MgCO_3$. Mucci et al. (1985) believed this to be due to the presence of dissolved organic matter in seawater, a substance omitted from their experimental work. To test this

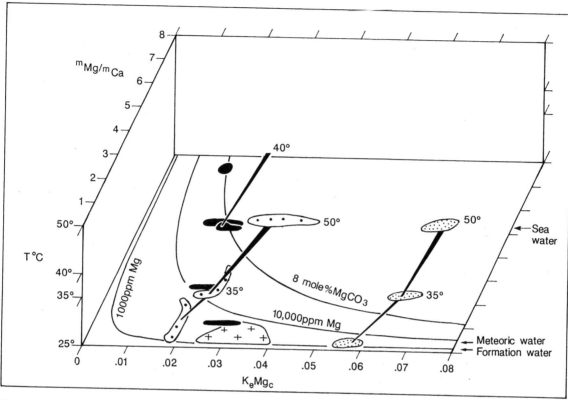

Fig. 6.14 *Perspective plot of k_eMg_c (x-axis) against $^mMg/^mCa$ of liquid (y-axis) against temperature °C (z-axis). Calcites of constant composition (1000 ppm Mg; 10 000 ppm Mg which approximately = 3.5 mole % MgCO3; and 8 mole % MgCO3) plotted on 25°C surface. Data from Mucci & Morse (1983) in black, Katz (1973) fine stipple, Oomori et al. (1987) coarse stipple and Howson et al. (1987) crosses. Data at different temperatures connected by lines; point at 40°C from Mucci (1987).*

hypothesis they placed seeds in natural environments containing organic matter; overgrowths were produced with 11–14 mole% MgCO3. The growth rate on the seeds was much slower than that predicted from their experimental study at seawater saturation; again organic matter was proposed to inhibit growth.

Magnesian calcite and calcite should be distinguished when considering partitioning of trace elements. Mg is a major constituent of marine magnesian calcite rather than a trace element. Magnesian calcites are metastable precipitates which possess complex microstructures with high dislocation densities and compositional heterogeneities at the atomic scale (Blake & Peacor, 1981). Such a mineral does not lend itself to analysis by partition theory.

For aragonite, k_eMg_A = 0.00016 (Oomori *et al.*, 1987); temperature seems to have a negligible effect between 10 and 50°C.

6.3.4 k_e for transition metals

The transition metals have not been as thoroughly investigated experimentally as the alkaline earths (some data exist for Mn, Fe, Co, Zn, and Cu; Veizer, 1983b). The transition metals behave differently from the alkaline earths during calcite precipitation, for the former have distribution coefficients above one and the latter below one. The transition metals are enriched in calcite, whereas the alkaline earths are depleted relative to the Tr/Ca ratio in the parent solution. Increasing precipitation rate causes their distribution coefficients to converge towards unity; the concentrations of transition metals in calcite decrease, whereas those of alkaline earths increases with increasing precipitation rate.

The k_eMn_c values calculated from experimental and natural data are plotted in Fig. 6.15, which shows

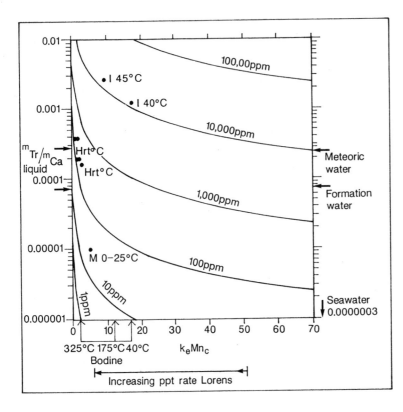

Fig. 6.15 *Plot of $k_e Mn_c$ (x-axis) against $^m Mn/^m Ca$ of liquid (y-axis). Curves are constant Mn composition in ppm for calcite. Data for calcites from Ichikuni (1973) = 1, Heijnen (1986) = H, rt°C = room temperature and Michard (1968) = M. Bodine et al. (1965) and Lorens (1981) gave ranges of $k_e Mn_c$ but no solution data. The $^m Mn/^m Ca$ ratios of meteoric water (river), formation water (oil field brine) and seawater are given.*

the relationship between the distribution coefficient, the Tr/Ca ratio of the liquid and the Tr concentration in the solid; clearly changes in the distribution coefficient are most sensitive to trace element concentrations in the solid at low values.

For aragonite, $k_e Mn_A = 0.86$ (Raiswell & Brimblecombe, 1977) and is independent of temperature (30–60°C), Mn concentration (up to Mn/Ca = 2.1–3), and ionic strength (up to 5 M NaCl).

6.3.5 k_e for alkali metals

The amount of alkali metals Na^+, Li^+, K^+ and Rb^+ incorporated in aragonite follows the proximity of their ionic radii to that of Ca^{2+}; Na^+ is the closest to Ca^{2+} and thus occurs most abundantly. The amount of these metals incorporated into calcite decreases with increasing ionic radius irrespective of proximity to Ca^{2+} ($Li^+ > Na^+ > K^+ > Rb^+$; Okumura & Kitano, 1986). It appears that alkali metals substitute for calcium in the aragonite structure, whereas in calcite they occupy interstitial positions. Two alkali

metal ions substitute for one Ca^{2+} to maintain charge balance. The amount of alkali metal in calcite can be correlated with crystal growth rate (saturation state), and by implication the number of crystal defects (Busenburg & Plummer, 1985). Experimentally-determined $k_e Na_c$ values vary by an order of magnitude. Sodium concentration has been used as a palaeosalinity indicator but is, in fact, a better measure of crystal growth rate than of the Na^+ concentration of parent waters. The presence of magnesium enhances alkali metal incorporation into calcite but inhibits incorporation in aragonite. The presence of Na^+ decreases Li^+, K^+ and Rb^+ incorporation into aragonites.

6.3.6 k_e for anions

Sulphate in calcite has been examined experimentally by Busenburg & Plummer (1985). Substitution of the larger SO_4^{2-} ion for the smaller CO_3^{2-} ion in the calcite structure increases proportionately with the crystal growth rate.

6.3.7 **Discussion**

The mode of incorporation of ions into the carbonate lattice determines trace element concentration. The range of constituents within carbonate crystals indicates that more than one mode of incorporation takes place. Ions ideally suited to replace Ca^{2+} in lattice positions may also be incorporated in non-lattice sites. Lattice defects become increasingly available with greater precipitation rates. The state of the trace ion, whether it is paired or complexed with other ions in the aqueous phase also influences incorporation. Most ions are hydrated, and as the ionic strength of a solution increases more ions form pairs and/or complexes which affect the availability of the ions to participate in precipitation. The nature of the absorbed layer next to the growing crystal surface is also of vital importance. The need for greater understanding at the atomic level is demonstrated by the occurrence of *sector zoning* in carbonates. Reeder & Prosky (1986) attributed differences in trace element concentrations in different sectors of a synchronous growth layer to different co-ordination environments for the attachment of ions on different crystal faces. The effect of crystallographic form on trace element partitioning in carbonates has yet to be investigated experimentally.

In the early 1970s relatively little attention was paid to kinetic effects on trace element partitioning. If carbonates precipitated at equilibrium the quantitative interpretation of trace element concentrations through k values would be simple. It is interesting to note that Plummer & Busenberg (1987) have calculated kSr_A to be 0.095 at 25°C. This value is more than an order of magnitude lower than k_eSr_A (1.1) and is also the other side of unity. It is k_eSr_A that closely predicts the Sr concentrations of inorganic marine aragonites. Ranges of k_e values are known for some elements which are appropriate for particular reactions and waters. The coverage is patchy; further work is required. Knowledge of the mechanism of trace element incorporation at the atomic scale is rudimentary and there is little understanding of partitioning in solutions with very high ionic strength which may be most realistic for diagenesis at a depth of several kilometres in sedimentary basins.

6.4 **ISOTOPES**

The 92 naturally occurring elements can be divided into about 300 stable and 1200 unstable or radioactive isotopes. The nucleus of an element contains a set number of protons but isotopes of that element have different numbers of neutrons. Hence isotopes of the same element have slight differences in mass and energy which cause differences in physical and chemical properties. These differences are generally greatest for elements of low atomic number. In a molecule with two isotopes, the isotope of lighter mass possesses weaker bonds and is more reactive than the heavier isotope. A change in the ratio of the two isotopes during a reaction or process is called a *fractionation*. Each isotope reaction obeys the Laws of Mass Action and is described by a temperature-dependent equilibrium constant or fractionation factor α:

$$\alpha_{A-B} = \frac{R_A}{R_B} \qquad (6)$$

where R_A is the heavy to light ratio in phase A; R_B is the heavy to light ratio in phase B.

The fractionation factor is correlated to absolute temperature by:

$$\ln\alpha = AT^{-2} + BT^{-1} + C \qquad (7)$$

where A, B and C are coefficients determined experimentally and T is the absolute temperature in kelvin.

Quantitative treatment of isotopte data is only possible if isotopic equilibrium is established, if isotopic fractionation is insignificant or if the kinetics of reaction are fully understood, in much the same way as was found with trace element partitioning. Aspects of stable isotope fractionation applied to water, the carbon cycle and carbonates are considered next. The isotopic variability of reservoirs, reservoir mixing and mineral–water reactions are dealt with later where appropriate; further information on isotope geology can be found in Hoefs (1988), Arthur *et al.* (1983) and Faure (1986).

6.4.1 **Stable hydrogen and oxygen isotopes in water**

Hydrogen has two stable isotopes: ^1H, abundance 99.985% and ^2H or D (deuterium), abundance 0.015%, which because of their proportionally large mass difference show the largest fractionation of any element. Oxygen has three isotopes: ^{16}O (99.763%), ^{17}O (0.0375%) and ^{18}O (0.1995%). Combinations of these five isotopes give nine different possible arrangements for a water molecule.

Considering oxygen, the range of ^{18}O concentration in oxygen of terrestrial materials is 1900–2100 ppm. The absolute concentration of ^{18}O is difficult to measure, but its ratio against ^{16}O can be determined more easily and accurately. Usually, the $^{18}O/^{16}O$ ratio is measured by comparison with a standard having a known $^{18}O/^{16}O$ ratio. This relative difference or δ value is given by:

$$\delta^{18}O = \frac{R_x - R_{std}}{R_{std}} \tag{8}$$

$R_x = {}^{18}O/^{16}O$ in sample; $R_{std} = {}^{18}O/^{16}O$ in standard.

The standard used for water is SMOW (standard mean ocean water). The δ value is generally expressed in per mil (‰):

$$\delta^{18}O = \left[\frac{({}^{18}O/^{16}O)_x - ({}^{18}O/^{16}O)_{SMOW}}{({}^{18}O/^{16}O)_{SMOW}} \right] \times 1000. \tag{9}$$

So a $\delta^{18}O$ of −5‰ would mean that the sample (x) has an $^{18}O/^{16}O$ ratio 5‰ (or ½%) lower than the standard. Conversely, a positive value would indicate heavy isotope enrichment of the sample relative to the standard.

The $\delta^{18}O$ fractionation which occurs when water vapour condenses to liquid can be expressed by equation (6) and the δ notation:

$$\alpha_{l-v} = \frac{\delta^{18}O_l + 1000}{\delta^{18}O_v + 1000} \tag{10}$$

where 1 is liquid; v is vapour.

Most values of α are close to one, so that the isotopic enrichment (ε) for a reaction pair can be conveniently expressed as:

$$\epsilon_{A-B} = (\alpha_{A-B} - 1) \times 1000 \tag{11}$$

which reports the isotopic separation in per mil caused by the reaction.

6.4.1a Condensation of water

The circulation of water in the hydrosphere and atmosphere has been compared to a global distillation process. Surface water evaporates to form water vapour chiefly at low latitudes. The water vapour condenses to form water droplets (clouds). Clouds cool as they move to either high altitudes or high latitudes causing coalescence of the water droplets, which when reaching a critical mass, fall as rain.

The vapour pressure of water is proportional to its mass, so $^1H_2{}^{16}O$ has a higher pressure than $^2H_2{}^{18}O$. Consequently, during condensation, the liquid is enriched in ^2H and ^{18}O whereas the remaining water vapour is enriched in the lighter isotopes. Several

experimental determinations of the equilibrium fractionation factors between vapour and liquid have been made for δ^2H and $\delta^{18}O$. Those reported by Majoube (1971) are:

$$\delta^2H: \ln\alpha = \frac{24\cdot844}{T^2} \times 10^3 - \frac{76\cdot248}{T} + 52.612 \times 10^{-3} \tag{12}$$

$$\delta^{18}O: \ln\alpha = \frac{1\cdot137}{T^2} \times 10^3 - \frac{0\cdot4156}{T} - 2.0667 \times 10^{-3} \tag{13}$$

where T is the absolute temperature.

If the loss of rain from a cloud is a process which occurs at equilibrium then the isotopic composition of the precipitate and its diminishing cloud reservoir can be modelled by the Rayleigh distillation equation:

$$R_t = R_o f^{(\alpha-1)} \tag{14}$$

where R_t is the isotopic ratio at time t; R_o is the original isotopic ratio; f is the fraction of vapour remaining in cloud at time t; and α is the fractionation factor.

Considering oxygen and the δ notation equation 10:

$$\frac{R_t}{R_o} = \frac{\delta^{18}O + 1000}{(\delta^{18}O)_o + 1000} = f^{(\alpha-1)}$$

$$\delta^{18}O_v = [(\delta^{18}O)_o + 1000]f^{(\alpha-1)} - 1000 \tag{15}$$

The changes in $\delta^{18}O$ during condensation of a cloud are graphed in Fig. 6.16. The closed system is impossible in nature as it would require rain condensed from the cloud to stay in the cloud and continually re-equilibrate with subsequent rain until all vapour had condensed. The open system requires rain, once condensed, to leave the cloud without further reaction. The paths for vapour and liquid towards increasingly negative values with a diminishing reservoir have parallels with the trace element modelling discussed above. The process of 'rainout' from clouds leads to increasingly negative values for rain along a cloud's flight path and there should be a correlation between δ^2H and $\delta^{18}O$ values. A poleward gradient towards increasingly light $\delta^{18}O$ values is shown in Fig. 6.17. The principal reason for cloud condensation is lowering temperatures, and the pattern shown in Fig. 6.17 can be correlated with temperature and secondarily with latitude and altitude. The correlation with temperature is not linear and continental areas are more depleted in ^{18}O than oceanic islands at the same temperature. Craig (1961) pointed out the linear corre-

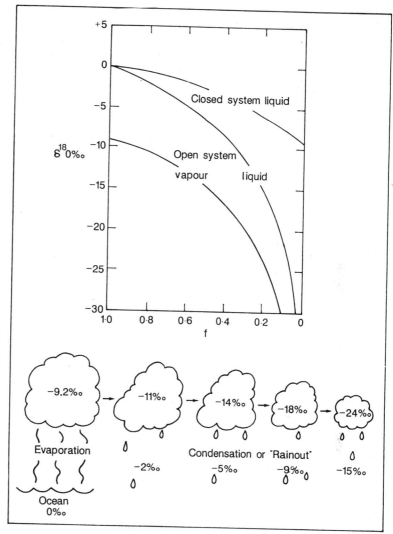

Fig. 6.16 *Top: $\delta^{18}O_{SMOW}$ plot shows Rayleigh distillation in a closed and open system. In the closed system the product accumulates and is well mixed at all times. In the open system the heavier liquid is immediately removed on condensing from the vapour which causes the vapour to become more negative. Temperature constant. The lower diagram depicts the open system as applied to rain. The $\delta^{18}O$ values for cloud vapour and rain can be read from the graph at f values of approximately 1, on formation of the cloud and when 80%, 60%, 40% and 20% of the cloud vapour remains.*

lation between δ^2H and $\delta^{18}O$ for meteoric water on a global scale which has a slope of approximately 8:

$$\delta^2H = 8\delta^{18}O + 10 \qquad (16)$$

A $\delta^2H/\delta^{18}O$ plot is shown in Fig. 6.18. The linear correlation between δ^2H and $\delta^{18}O$ for meteoric water on a global scale is known as the meteoric water line (MWL).

6.4.1b Evaporation

The process of evaporation may be visualized as a series of steps (Craig, 1965). Initially, vapour released at the liquid surface is at equilibrium with the liquid.

The fractionation involved is the reverse of condensation (equations (12 and 13)) and results in the vapour being depleted in ^2H and ^{18}O relative to the liquid. The vapour then diffuses away from the liquid and this involves further depletion in ^2H and ^{18}O due to the differing diffusivities in air of the various hydrogen and oxygen isotopes. Finally, the vapour escapes to the atmosphere without further fractionation, and atmospheric vapour penetrates the diffusion layer to condense at the liquid surface. Evaporation therefore is different from condensation because of the additionation fractionation due to diffusion. Hence the net fractionation is a combination of equilibrium fractionation and a kinetic enrichment factor $\Delta\epsilon$. The

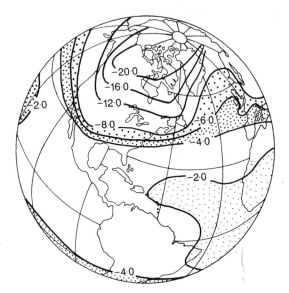

Fig. 6.17 *Isodelta lines centred on America for mean annual* $\delta^{18}O$ *precipitates. After Yurtsever (1976).*

magnitude of $\Delta\epsilon$ can be approximated (Gonfiantini, 1986) by:

$$\Delta\epsilon^2H\text{‰} = 12.5^{(1-h)} \qquad (17)$$
$$\Delta\epsilon^{18}O\text{‰} = 14.2^{(1-h)} \qquad (18)$$

where h is the relative humidity.

The course of isotopic change during evaporation may be plotted on a $\delta^2H/\delta^{18}O$ diagram. Evaporation which occurs in a dry atmosphere departs most from the meteoric water line; as humidity rises the effect of $\Delta\epsilon$ diminishes until at 100% humidity, evaporation of meteoric water would plot along the MWL (Fig. 6.18).

The course of isotopic evolution in a closed water body during evaporation to dryness is shown for various humidities in Fig. 6.19. At 95% relative humidity the expected enrichment in heavy isotopes rapidly levels off to a steady state producing a convex-upwards curve. This steady state is controlled by a balance between the isotopic composition of the evaporating water and that of the back-condensing atmospheric vapour. The short evaporation path for 95% humidity shown in Fig. 6.18 is due to this steady state, for once reached, further change does not occur. In contrast, evaporation which occurs in dry air describes a concave-upwards curve (Fig. 6.19) showing a dramatic enrichment in heavy isotopes which continues until theoretically the last water shows infinite enrichment

(no light isotopes remain). The course of this dry air Rayleigh-type process, plotted in Fig. 6.18, has a slope of 3.7.

Three additional factors affect evaporation of saline waters. Firstly, increasing salt concentration decreases the thermodynamic activity of water and the evaporation rate. On evaporation of seawater, as salinity rises the rate of change towards heavier isotopes slows and eventually reverses producing a hooked plot on the $\delta^2H/\delta^{18}O$ diagram. Secondly, water entering the hydration sphere of some salts may be isotopically different from free water and that difference varies with the salt involved. Finally, some salts which precipitate during evaporation, such as gypsum, contain water of crystallization which usually is isotopically different from that in the remaining liquid. These effects are only important in saline lakes, sabkhas and the final stages of lake evaporation.

The processes of condensation and evaporation are basic to understanding isotopic differentiation in the hydrological cycle. However, seldom do natural systems behave as simply as described above. Clouds are not only generated at the equator, nor do they move in straight lines from the equator to the poles shedding their condensed water as they go. The isotopic composition of rain is often modified by evaporation after it has left its parent cloud. Hail may rise and fall several times within the cloud acquiring ice of differing isotopic composition before deposition. Lakes seldom have no inflow or outflow or evaporate to complete dryness in an atmosphere of uniform humidity. The natural system is complex and thus departs from the picture of the hydrological cycle being a global distillation column.

6.4.2 Carbon

Carbon has two stable isotopes: ^{12}C (98.89%) and ^{13}C (1.11%). Carbon isotopes are usually analysed as CO_2 gas, and the $^{13}C/^{12}C$ ratio of the specimen is compared against values for CO_2 released from the standard (PDB), a belemnite from the Cretaceous Pee Dee Formation, South Carolina, on digestion in phosphoric acid. Global carbon can be divided into two reservoirs: the oxidized reservoir, principally as CO_2, HCO_3^{2-} and carbonate minerals, and the reduced reservoir as organic compounds, fossil fuels and the native element. There is a constant exchange between the oxidized and reduced carbon reservoirs which maintains a balance. The processes which effect this cycling of carbon compounds also involve isotopic fractionation. Atmospheric CO_2 is a well-mixed reser-

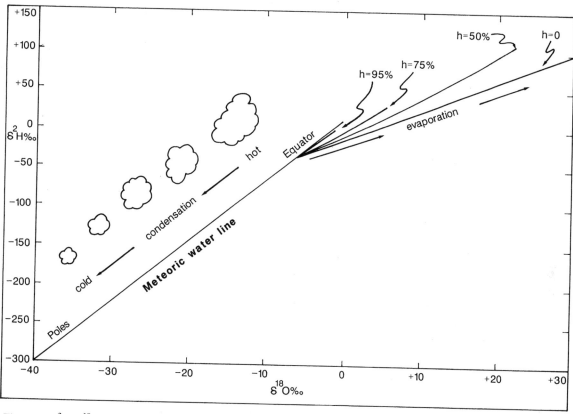

Fig. 6.18 $\delta^2H/\delta^{18}O$ plot. The meteoric water line (MWL) defined by Craig (1961) as $\delta^2H = 8\delta^{18}O + 10‰$ corresponds to mean δ^2H and $\delta^{18}O$ values of global precipitation samples. Rayleigh-type cloud distillation figuratively shown. Evaporation curves are shown for a water on the MWL of $\delta^2H = -38‰$ and $\delta^{18}O = -6‰$ which evaporated into an atmospheric vapour (when present) of $\delta^2H = -86‰$ and $\delta^{18}O = -12‰$, reasonable for a low- to mid-latitude coastal lake. After Gonfiantini (1986). h = relative humidity.

voir that forms the link between relatively ^{12}C-enriched organic compounds through photosynthesis and carbonates via exchange reactions with aqueous HCO_3^{2-} causing relative enrichment in ^{13}C.

6.4.2a Photosynthesis

Three different pathways exist whereby land plants fix atmospheric CO_2 as organic compounds during photosynthesis.

The majority of land plants, including most terrestrial higher plants, use the Calvin pathway, also known as the C_3 cycle because the first sugars to be synthesized are 3-carbon sugars. In this cycle, atmospheric CO_2 is dissolved in the plant's cytoplasm with a slight fractionation ($\alpha = 1.001$; Vogel *et al.*, 1970). The aqueous CO_2 is directly fixed to 3-phosphoglyceric

acid with the aid of an enzyme RUBISCO (ribulose diphosphate carboxylase) located inside the cell's chloroplasts, which effects another fractionation. Plants with C_3 metabolism have $\delta^{13}C$ values of -25 to $-28‰$.

Most tropical grasses use a second process, the Hatch–Slack or Kranz cycle, also known as the C_4 cycle as the first sugars synthesized are 4-carbon sugars. Here, CO_2 gas is dissolved as HCO_3^{2-} with considerable fractionation ($\alpha = 1.0075$; Vogel *et al.*, 1970). Atmospheric CO_2 has a $\delta^{13}C$ of $-7‰$, and therefore the carbon source for C_4 metabolism is approximately 0‰ from $\alpha = 1.0075$. The HCO_3^{2-} is fixed as oxaloacetic acid by the enzyme PEPC (phosphoenolpyruvate carboxylase) which is located outside the chloroplast. Plants with C_4 metabolism have $\delta^{13}C = -12$ to $-14‰$.

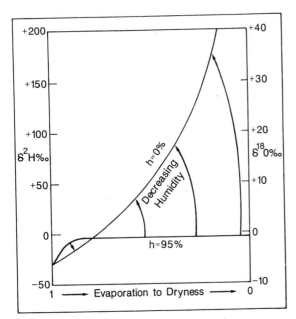

Fig. 6.19 *Evolution of $\delta^2 H$ and $\delta^{18}O$ during evaporation of a closed water body (initial water $\delta^2 H = -38‰$ and $\delta^{18}O = -6‰$) into a uniform atmosphere at $\delta^2 H = -86‰$ and $\delta^{18}O = -12‰$. h = relative humidity. A family of curves for intermediate humidities exist (not shown) between the lower h = 95% curve and the upper h = 0% curve. The arrows indicate deflection of these curves with decreasing humidity. $\delta^2 H$ and $\delta^{18}O$ curves rationalized into one. After Gonfiantini (1986).*

The differentiation of organic compounds involves further fractionation; for instance, sugars have heavier $\delta^{13}C$ values than lipids within the range of $\delta^{13}C$ values for both C_3 and C_4 plants. Environmental factors may also affect the $\delta^{13}C$ content of photosynthetically-fixed carbon (see Deines, 1980, for further details).

A third photosynthetic process is found in plants of the Crassulaceae family (CAM, Crassulacean acid metabolism), which are water storage plants such as cacti, epiphytes and orchids. Their $\delta^{13}C$ values reflect both C_3 and C_4 pathways; they possess both RUBISCO and PEP carboxylases and are the only plants which can fix CO_2 in the dark (using the C_4 pathway) and in light (using the C_3 pathway).

The $\delta^{13}C$ values of animal organic compounds reflect the $\delta^{13}C$ composition of the food consumed. All organic matter (plant or animal) that decays or is buried with sediment falls within a range of $\delta^{13}C$ values from approximately -5 to $-35‰$ (Stahl, 1979).

6.4.2b Organic matter degradation during shallow burial

Organic matter is 'burnt' rapidly in oxic environments by free oxygen in respiration. The breakdown of organic tissues in anoxic environments is much more complicated, involving hydrolysis and eventually a variety of bacteriologically-mediated oxidation and fermentation reactions linked with multivalent inorganic species of O, Fe, Mn, N, S, and C which act as electron acceptors.

The oxidation of organic matter to CO_2, whether by aerobic or anaerobic bacteria, involves little fractionation, so the CO_2 produced will retain the $\delta^{13}C$ value of its source material. However, when methane formation is involved, severe fractionation occurs. CH_4 is produced in Recent anoxic sediments by bacterial methanogens by either CO_2 reduction or acetate fermentation:

$$CO_2 + 8(H) \rightarrow CH_4 + 2H_2O \tag{19}$$

$$CH_3COOH \rightarrow CH_4 + CO_2. \tag{20}$$

The two metabolic pathways operate in both saline and freshwater, although reaction (19) is dominant in saline and (20) in freshwater. The range of $\delta^{13}C$ values for CH_4 and coexisting CO_2 produced by these reactions is shown in Fig. 6.20. The CO_2 reduction data (Whiticar *et al.*, 1986) were collected over a temperature range of $-1.3°C$ to $58°C$ which overlie the equilibrium fractionation $\alpha = 1.07$ for $20°C$. $\delta^{13}C$ fractionation during CO_2 reduction is probably kinetically controlled but this is more obvious for the acetate fermentation data, whose temperature range of $5°C$ to $19°C$ is clearly below that for any reasonable value.

An important process for carbonates occurs when CH_4, produced in anoxic sediment, becomes bacterially oxidized to CO_2 as it escapes upwards through overlying oxic sediment and water. The process is well understood where aerobic methanotrophs utilize molecular oxygen:

$$CH_4 + 2O_2 \rightarrow CO_2 + 2H_2O \tag{21}$$

The process involved is kinetically controlled, enriching ^{13}C in the residual CH_4, and ^{12}C in the respired CO_2 (the opposite effect to CH_4 generation). The fractionation involved ranges from 1.002 to 1.014 (Whiticar & Faber, 1986). There is, however, indirect evidence that rising CH_4 is commonly largely consumed before reacting with surficial oxic sediments; it is oxidized at the base of the sulphate reduction zone

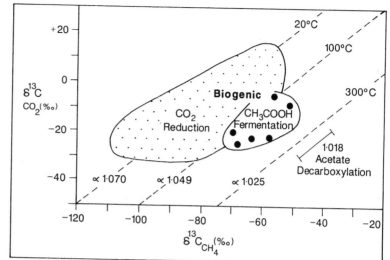

Fig. 6.20 *Plot of $\delta^{13}C_{CO_2}$‰ against $\delta^{13}C_{CH_4}$‰. Equilibrium fractionation CO_2-CH_4 $\alpha = 1.070, 1.049$ and 1.025 for $20°C, 100°C$ and $300°C$ respectively from Bottinga (1969). Biogenic reduction and fermentation areas for natural coexisting CO_2-CH_4 pairs from Whiticar et al. (1986). Solid line represents experimental CO_2-CH_4 products of an acetate decarboxylation at $300°C$ from Kharaka et al. (1983).*

(Devol *et al.*, 1984). The microbiology of this reaction (22) is poorly understood and the CH_4-CO_2 fractionation involved is unknown.

$$CH_4 + SO_4^{2-} \rightarrow HS^- + HCO_3^- + H_2O \qquad (22)$$

6.4.2c Organic matter degradation during deep burial

Continued burial of organic matter causes thermal alteration producing coal, liquid hydrocarbons, CO_2 and CH_4. The $\delta^{13}C$ of coal and oil does not change during maturation, the variations found reflecting the type of original source matter (for reviews see Deines, 1980 and Schoell, 1984). The thermogenic gases are formed by cracking petroleum and/or kerogen at two stages of burial. The first is concurrent with oil formation, producing CH_4 mixed with C_2-C_5 gases (wet gases); the second, on deeper burial, affects both oil and kerogen, and produces principally CH_4 (dry gas). The characterization of natural gases is reviewed by Schoell (1983).

Thermal cracking to produce CH_4 involves kinetic fractionation, the breaking of the weaker $^{12}C-^{12}C$ bond from the parent organic matter being favoured over the $^{12}C-^{13}C$ bond which, in turn, is favoured over the $^{13}C-^{13}C$ bond. Experimental thermal decarboxylation of acetic acid (reaction (20)) at $300°C$ produced both CH_4 and CO_2 (Kharaka *et al.*, 1983). The fractionation involved was 1.018 (Fig. 6.20).

The differing CH_4-CO_2 fractionations (Fig. 6.20) involved in CH_4 production can be compared by a Rayleigh-type distillation process (equation (14)). The

strong preference for ^{12}C fractionation into biogenic CH_4 causes the source CO_2 pool to become enriched in ^{13}C. By the time half the CO_2 reservoir is converted, the remaining CO_2 has $\delta^{13}C = +29‰$; when 20% of the original CO_2 reservoir remains, $\delta^{13}C = +97‰$ (Fig. 6.21). Thermogenic CH_4, by contrast, forms with $\alpha = 1.025$. When half the source CO_2 remains the reservoir still has a negative $\delta^{13}C(-3‰)$. At 20% remaining $\delta^{13}C = +20‰$ (Fig. 6.21). Strictly α, the thermodynamic fractionation factor, should not be used for CH_4-CO_2 which is kinetically controlled.

The $\delta^{13}C$ of hydrocarbons buried progressively more deeply becomes heavier as crude oil is thermally destroyed. The $\delta^{13}C$ change has been attributed to loss of isotopically light gases during thermal cracking, but sulphate reduction may also be involved. In sediments bearing anhydrite, the SO_4^{2-} reacts with H_2S to produce elemental sulphur which, in turn, oxidizes the crude oil (Sassen, 1988). Elemental sulphur may also oxidize CH_4:

$$4S° + CH_4 + 2H_2O \rightarrow 4H_2S + CO_2 \qquad (23)$$

The fractionation involved is not yet known.

6.4.2d Atmospheric carbon

There are several carbon-bearing gases in the atmosphere (CO_2, 320 ppm and in decreasing abundance CH_4, CO, CCl_4 and freon group compounds). Their concentration is low but their effect profound on environmental conditions, which is leading to concern over anthropogenic additions. Atmospheric CO_2 has

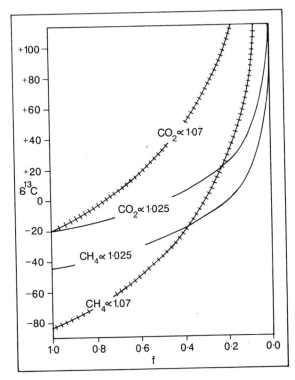

Fig. 6.21 *Rayleigh model for CO_2-CH_4 fractionation (equation (14)). Comparison using $\alpha = 1.07$, appropriate for CH_4 production by biogenic CO_2 reduction at ~20°C and $\alpha = 1.025$, appropriate for thermochemical CH_4 production at ~300°C. See text.*

an average $\delta^{13}C = -7‰$, but variations do exist. Lighter values and increased concentrations occur over marine waters, forest and grasslands at night when plant respiration adds CO_2 with $\delta^{13}C$ $-21‰$ to $-26‰$.

Atmospheric CO_2 is in isotopic equilibrium with dissolved carbon in marine surface waters, hence changes in atmospheric CO_2 levels are likely to affect carbonate precipitation (see Section 9.3).

6.4.3 Carbonates

The oxygen stable isotope composition of carbonates may be reported against either the general oxygen standard SMOW or the PDB scale (both mentioned above). Oxygen is normally measured at the same time as carbon against PDB on CO_2 gas produced by phosphoric acid digestion of carbonate; for details of fractionations involved see Friedman & O'Neil

(1977). Conversion from one scale to the other is straightforward.

$$\delta^{18}O \text{ (calcite SMOW)} = 1.03086\, \delta^{18}O \text{ (calcite PDB)} + 30.86 \tag{24}$$

$$\delta^{18}O \text{ PDB} = 0.97006\, \delta^{18}O \text{ SMOW} - 29.94 \tag{25}$$

6.4.3a Equilibrium fractionation

The oxygen isotopic fractionation in the calcium carbonate–water–bicarbonate system was the first to be suggested for palaeotemperature reconstruction of ancient oceans (Urey, 1947). The fractionation was established using calcareous skeletons by Epstein *et al.* (1953), whose expression was modified by Craig (1965) to:

$$t(°C) = 16.9 - 4.2\,(\delta c - \delta w) + 0.13\,(\delta c - \delta w)^2 \tag{26}$$

where δc is the $\delta^{18}O$ of CO_2 produced by reaction of $CaCO_3$ in phosphoric acid at 25°C, and δw is the $\delta^{18}O$ of CO_2 in equilibrium with water at 25°C, both PDB.

This expression (26) compares closely with that determined by O'Neil *et al.* (1969) from inorganically-precipitated calcite:

$$10^3 \ln\alpha = 2.78\,(10^6\, T^{-2}) - 2.89 \tag{27}$$

and by Bottinga (1968) from theoretical calculations.

The palaeotemperature expression (26) is apposite when considering the $\delta^{18}O$ composition of ancient calcites. There are three unknowns in (26) only one of which can be measured directly: the $\delta^{18}O$ of the calcite. When estimating the temperature and the $\delta^{18}O$ of the water in such a situation it is useful to consider the range of equilibrium conditions that are feasible. This is readily done using a plot of these three variables as in Fig. 6.22. For instance, if it were possible to precipitate marine calcite at equilibrium in ocean water ($\delta^{18}O = 0‰$ SMOW) with a temperature range from 0–30°C, then the $\delta^{18}O$ of the calcite would vary from +3‰ to −3‰ PDB respectively.

The temperature dependence of the $H_2O-CaCO_3$ aragonite fractionation has been determined by Grossman & Ku (1986) for a number of aragonitic marine skeletons. A combination of all their data gives:

$$t\,(°C) = 20.6 - 4.34\,(\delta^{18}O \text{ Ar} - \delta w). \tag{28}$$

which yields a similar slope to equation (26).

Theoretically, fractionation should occur between calcite and aragonite due to differences in internal vibrational frequencies of the carbonate ions. Ar-

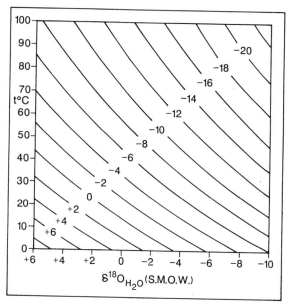

Fig. 6.22 *Equilibrium relationship between $\delta^{18}O$ of calcite, temperature and $\delta^{18}O$ of water; x-axis represents $\delta^{18}O$ value of water (SMOW); y-axis represents temperature between 0 and 100°C. The curved lines represent constant $\delta^{18}O$ values (PDB) for calcite calculated from expression (26).*

agonitic benthic foraminifera are enriched by 0.6‰ relative to calcitic benthic foraminifera, which is the same ^{18}O enrichment as is obtained for inorganic aragonite–calcite precipitates (Tarutani *et al.*, 1969), and may be regarded as an equilibrium fractionation.

The $\epsilon\,^{18}O$‰ values for pairs relevant to carbonate precipitation are plotted in Fig. 6.23. It can be seen how the water–aragonite and water–calcite pairs are strongly temperature dependent but the aragonite–calcite pair is practically temperature independent.

Tarutani *et al.* (1969) reported that Mg calcite precipitated inorganically at 25°C is enriched in ^{18}O relative to calcite by 0.06‰ per mole % $MgCO_3$.

$\delta^{18}O$ fractionation for water–dolomite is a problem because at low temperatures dolomite has not been precipitated experimentally. Extrapolation of high temperature data yields values which indicate sedimentary dolomite should be enriched in ^{18}O relative to calcite by 3–6‰. However, analyses of co-existing (but not necessarily coeval) natural calcite–dolomite occurrences sometimes show no fractionation at all (Land, 1980). It is likely that sedimentary dolomites which possess a range of compositions and structures also have a range of isotopic behaviour, but

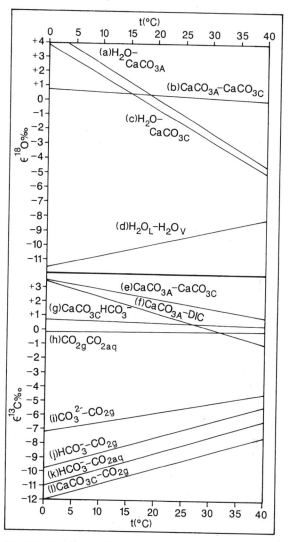

Fig. 6.23 *The ^{18}O and ^{13}C equilibrium separation factor ϵ for various pairs important in carbonate precipitation. A = aragonite; C = calcite; g = gas; aq = aqueous; v = vapour; l = liquid and DIC = dissolved inorganic carbon which equates with HCO_3^- + 0.3‰ (Grossman & Ku, 1986). Source of expressions and data: a, b, e and f (Grossman & Ku, 1986); d (Majoube, 1971); c, g, h, j and k (Salomons & Mook, 1986) and i and l (Turi, 1986).*

at present an $\epsilon^{18}O = +3$‰ to $+4$‰ is commonly assumed for the calcite–dolomite pair (see Section 8.6).

The $\epsilon^{13}C$ isotope separations for species important to carbonate precipitation are shown in Fig. 6.23. It can be seen that the $\epsilon^{13}C$ of calcite at 25°C is separated from CO_2 g by 10‰ which means calcite precipitated at equilibrium with atmospheric CO_2 ($\delta^{13}C = -7$‰) in equilibrium with ocean water, has a $\delta^{13}C$ value of $+3$‰. The aragonite–calcite line in Fig. 6.23 shows a moderate temperature dependence; a palaeotemperature scale based on $\epsilon^{13}C_{A-C}$ would have a great advantage in that it is independent of any knowledge of the isotopic composition of water. However, the preservation of ancient aragonite is rare and carbon is prone to non-equilibrium effects.

6.4.3b Non-equilibrium fractionation

The palaeotemperature equation (26) was established on molluscan skeletal carbonate. Much work on palaeoclimatology of the Tertiary is based on isotopic analysis of foraminifera. In both cases it was essential to establish that the carbonate precipitated at isotopic equilibrium; this is not always the case, however, for skeletal precipiates (see Fig. 7.6). Echinoderms clearly exert physiological control on their endoskeleton for not only do they precipitate calcite several per mil out of equilibrium, but particular elements of the skeleton are consistently different isotopically from other elements of the same test (Weber, 1968). Mixing of dissolved inorganic bicarbonate with respiratory CO_2 may be responsible. The photosynthetic symbiotic algae in hermatypic corals are blamed for their divergence from equilibrium (Weber & Woodhead, 1970), and so on.

Non-equilibrium precipitation also applies to inorganic carbonates where precipitation rate has an effect as noted by Turner (1982) in experimental work, and by Turi (1986) in natural travertines.

6.4.4 Radiogenic isotopes: strontium

Strontium has four naturally occurring stable isotopes: ^{88}Sr (82.53%), ^{87}Sr (7.04%), ^{86}Sr (9.87%) and ^{84}Sr (0.56%). Strontium-88, 86 and 84 are not part of any decay series; their abundance is constant. Strontium-87 occurs naturally and is generated by radioactive decay of rubidium-87. The precise Sr isotopic composition of a Rb-bearing rock or mineral depends on the Rb/Sr ratio and age, which is the basis for the Rb–Sr method of radiometric dating. However, this technique cannot be used on carbonates because they exclude Rb from their structure, while acting as repositories for Sr.

Measured $^{87}Sr/^{86}Sr$ ratios are normalized to $^{86}Sr/^{88}Sr = 0.1194$ to correct for variable mass frac-

tionation in the mass spectrometer. $^{87}Sr/^{86}Sr$ is not significantly fractionated by natural processes so may be used to study the provenance of strontium.

6.4.4a $^{87}Sr/^{86}Sr$ and natural waters

The concentration of Sr in carbonates, as discussed above, is related to the Sr/Ca ratio of the water from which it precipitated through k_e values. On the other hand, the $^{87}Sr/^{86}Sr$ ratio of carbonates is the same as that in the water because no significant fractionation occurs on precipitation. The $^{87}Sr/^{86}Sr$ ratio of the water, in turn, is related to the rocks and minerals which have donated Sr to the water. River waters from a single drainage basin may have widely differing $^{87}Sr/^{86}Sr$ values. One stream may drain and interact with granitic and metamorphic rocks (water $^{87}Sr/^{86}Sr = 0.7189$; Sr concentration 0.06 mg 1^{-1}) whilst another drains an area of marine limestones (water $^{87}Sr/^{86}Sr = 0.7142$; Sr concentration 0.4 mg 1^{-1}). If the streams carry equal discharge, at their confluence the resulting $^{87}Sr/^{86}Sr$ ratio will be 0.7097, a value much closer to the second stream's $^{87}Sr/^{86}Sr$ ratio because of its greater Sr concentration which dominates the mixture (Fisher & Stueber, 1976).

The $^{87}Sr/^{86}Sr$ provenance in a drainage basin may vary with time either by the deposition of new sediments, such as volcanic ash ($^{87}Sr/^{86}Sr \sim 0.7037$) or by erosion to expose older rocks, such as feldspathic basement ($^{87}Sr/^{86}Sr \sim 0.7198$). Also, the $^{87}Sr/^{86}Sr$ ratio being released to the water will increase with time if the source contains Rb. Any carbonates precipitated from the waters will record the $^{87}Sr/^{86}Sr$ value of that water during precipitation and this will be preserved (barring diagenesis) in Rb-free carbonates.

Groundwaters (Collerson et al., 1988) and subsurface brines (Stueber et al., 1984) also show great variability in $^{87}Sr/^{86}Sr$ values due to water-rock interaction.

Strontium has a residence time in the oceans of 4–5 million years and it takes approximately 1000 years to mix the oceans; consequently, Sr concentration in the oceans is uniform (8 ppm). As the half-life of Rb is very long (4.89×10^{10} years) and its concentration in ocean water is low (120 ppb), insignificant amounts of radiogenic ^{87}Sr are produced *in situ* in ocean water in 5 million years. Consequently, the $^{87}Sr/^{86}Sr$ values of the oceans at any one time is constant. However, it is well established (e.g. Burke et al., 1982; Koepnick et al., 1985) that the $^{87}Sr/^{86}Sr$ ratio of the oceans has changed with time (Fig. 9.21)

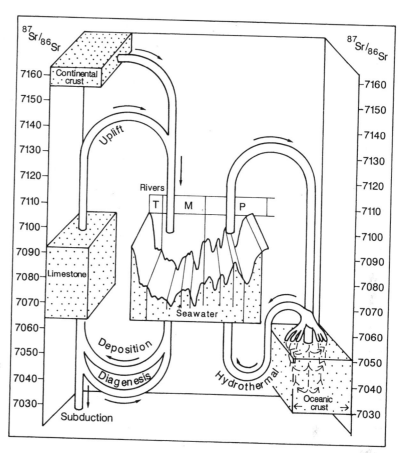

Fig. 6.24 *The Phanerozoic marine strontium cycle. Seawater curve from Burke et al. (1982). P = Palaeozoic, M = Mesozoic and T = Tertiary. 'Piping' filled with aqueous fluid except subduction limb which removes carbonates from the immediate cycle.*

due to variations in supply from sources with different $^{87}Sr/^{86}Sr$ values (Fig. 6.24).

The Earth formed from the solar nebula with a $^{86}Sr/^{86}Sr$ ratio of 0.699. In Archean carbonates, which are rare, $^{87}Sr/^{86}Sr$ ratios are similar to the contemporary mantle (-0.700 to 0.701). During the Proterozoic $^{87}Sr/^{86}Sr$ rose irregularly until the Cambrian when values were slightly above those of seawater today. The two primary sources of Sr to the ocean are the continental crust (by subaerial weathering and river transport) and the oceanic crust (by hydrothermal activity at the mid-ocean ridges and submarine alteration of basalts). By the beginning of the Phanerozoic era these two sources had evolved distinctly different $^{87}Sr/^{86}Sr$ values. The initial differentiation of crust from mantle which began early in Earth history, led to Rb being preferentially partitioned into the crust. Consequently, the $^{87}Sr/^{86}Sr$ ratio of the crust has and will

evolve higher $^{87}Sr/^{86}Sr$ values more rapidly than the mantle. The fluctuations in $^{87}Sr/^{86}Sr$ ratio of Phanerozoic ocean water never reached the values of the two primary sources (Fig. 6.24). This was partly due to the buffering effect of carbonates, whic reintroduce older marine Sr into seawater by diagenetic reactions in marine carbonates within advective or diffusive connection with seawater, and by dissolution of uplifted marine carbonates by meteoric waters. In addition, the two primary sources are mixed together in seawater.

It is perhaps ironic that marine carbonates, which exclude Rb from their lattice and consequently rule out Rb−Sr dating techniques, provide a tool through knowledge of oceanic $^{87}Sr/^{86}Sr$ variation for high-resolution stratigraphic dating (DePaolo, 1986; Elderfield, 1986).

7 Diagenetic processes, products and environments

7.1 INTRODUCTION

The diagenesis of carbonate sediments encompasses all the processes which affect the sediments after deposition until the realms of incipient metamorphism at elevated temperatures and pressures. Diagenesis includes obvious processes such as cementation to produce limestones and dissolution to form cave systems but it also includes more subtle processes such as the development of microporosity and changes in trace element and isotopic signatures. Diagenetic changes can begin on the seafloor, as the grains are still being washed around or as a reef is still growing, or it may hold off until burial when overburden pressure has increased or pore-fluid chemistry has changed so that reactions are then induced within the sediments.

Most modern and many ancient marine carbonate sediments originally consist of a mixture of aragonite, high-Mg calcite and low-Mg calcite, and in the Recent the first two minerals are commonly precipitated as cements from marine pore-fluids during early diagenesis. However, aragonite and high-Mg calcite are metastable and in most instances they are converted to calcite, in one way or another, sooner or later, in a limestone's history. Low-Mg calcite is the stable form of $CaCO_3$ and is the cement most commonly precipitated from meteoric waters in the near-surface and from basinal fluids in the deeper subsurface. Unless dolomitized, the majority of ancient limestones are composed entirely of low-Mg calcite. Studies of limestone diagenesis thus often involve trying to identify the original mineralogy of the various cements, their significance in terms of pore-fluid chemistry and diagenetic environment and the relative timing of precipitation and any alteration. The manner in which the grains themselves are preserved is also most important. Such studies involve routine petrography, CL (cathodoluminescence), scanning electron microscope (SEM) examination and geochemical analysis for isotopic and trace element signatures (see Tucker, 1988 for information on these techniques).

Diagenesis includes six major processes: cementation, microbial micritization, neomorphism, dissolution, compaction (including pressure dissolution) and dolomitization. The last is a process affecting some 30–40% of limestones and is dealt with separately in Chapter 8. The patterns of diagenesis may vary greatly from one limestone formation to another, and there are frequently variations both laterally and vertically within one limestone sequence. Major controls on the diagenesis are the composition and mineralogy of the sediment, the pore-fluid chemistry and flow rates, geological history of the sediment in terms of burial/uplift/sea-level changes, influx of different pore-fluids and prevailing climate.

During diagenesis carbonate sediments may gain or lose porosity. With increasing depth of burial, there is generally a decrease in porosity (see Section 7.6.1 and Fig. 7.31C; Schmoker & Halley, 1982), but there are late processes of dissolution and fracture which can restore higher porosity values. The understanding of porosity formation and occlusion is another major aim of diagenetic studies.

From the data available on limestone diagenesis broad diagenetic models can be developed to predict the patterns. These models, discussed in Section 7.7, can be of use in predicting reservoir quality in limestones and dolomites. However, there are still many features of carbonate diagenesis which appear to be quite random in their distribution and development, so that specific diagenetic models need to be erected for many limestones.

Carbonate diagenesis operates in three principal environments: the marine, near-surface meteoric and burial environments, and there are features of the cement fabrics and other textures which are diagnostic of a particular diagenetic environment. These environments are discussed in later sections of this chapter, but diagenesis in the pelagic, deep-marine realm was presented in Section 5.6. Finally, as with other aspects of carbonate sedimentology, the principle of uniformitarianism does not apply one hundred per cent throughout the rock record. There are differences in the mineralogy of marine precipitates during the Phanerozoic for example, so that modern marine cements are quite different from those in the mid-Palaeozoic and Middle–Late Mesozoic. Although ancient marine cements are disscussed in this chapter, more consideration of the causes of temporal variations in diagenesis and limestone geochemistry are given in Chapter 9, the Geological Record of Carbonate Rock.

Carbonate diagenesis is discussed at length in the book of Bathurst (1975) and recent compilations of useful papers include Roehl & Choquette (1985), Schneidermann & Harris (1985), Schroeder & Purser (1986), Choquette & James (1988) and Tucker & Bathurst (1989). Reviews include Moore (1979), Longman (1980), James & Choquette (1983, 1984) and Choquette & James (1987).

7.2 DIAGENETIC PROCESSES: A BRIEF INTRODUCTION

Cementation. The precipitation of cements in carbonate sediments is a major diagenetic process and takes place when pore-fluids are supersaturated with respect to the cement phase and there are no kinetic factors inhibiting the precipitation. Petrographic and geochemical studies of these cements enable deductions to be made of the environment and conditions of cementation. Organic geochemical influences are important in some instances. Aragonite, high-Mg calcite, low-Mg calcite and dolomite are the common carbonate cements in limestones and they comprise a range of morphologies. Less commonly, ankerite, siderite, kaolinite, quartz, anhydrite, gypsum and halite are the cements. The identification of cement is mostly straightforward; it is precipitated into cavities of many types and between grains, and many cement crystals show particular fabrics indicative of void-filling. The actual mineralogy and fabric depend chiefly on the composition of the pore-fluids (especially on the Mg/Ca ratio and whether the fluids were marine, meteoric or basinal), the carbonate supply rate and on the rate of precipitation. Some carbonate cements are less easy to identify. With micritic limestones, for example, it is often possible to demonstrate early lithification if intraclasts and hardground surfaces are present, but the cement itself is usually impossible to resolve from the sediment particles. In many reef rocks, there are peloids which could be precipitates rather than altered grains or faecal pellets. A similar type of cement can be precipitated in different diagenetic environments. Equant, drusy calcite spar, the major pore-filling cement in limestones, can be precipitated in near-surface meteoric environments or in the deep burial

realm. In the Ordovician and Jurassic, there is evidence to suggest it even precipitated on the shallow seafloor. With calcite spar, then, careful attention to the textures and geochemistry (trace elements and isotopes) is necessary to deduce the precipitational environment.

Cementation of limestones requires an enormous input of $CaCO_3$ and an efficient fluid flow mechanism for complete lithification. The source of $CaCO_3$ varies with the different diagenetic environments: in the marine realm it is seawater; in the meteoric and burial environments it is mostly dissolution of the sediment itself.

Microbial micritization. This is a process whereby bioclasts are altered while on the seafloor or just below by endolithic algae, fungi and bacteria. The skeletal grains are bored around the margins and the holes filled with fine-grained sediment or cement. Micritic envelopes (Section 1.3 and Fig. 1.11) are produced in this way and if the activity of the endolithic microbes is intense, completely micritized grains are the result. The original skeletal nature of such grains is often difficult to determine. The usually rather irregular shape of these micritized grains distinguishes them from the micritic faecal pellets (see Fig. 3.16).

Neomorphism. This term was introduced by Folk (1965) to cover processes of replacement and recrystallization where there may have been a change of mineralogy. Recrystallization, strictly, refers to changes in crystal size without any change of mineralogy. Since many carbonate sediments originally consist of a mixture of calcite and aragonite, the term recrystallization cannot properly be applied to replacement textures and neomorphism is used instead. Neomorphic processes take place in the presence of water through dissolution–reprecipitation; that is, they are wet processes. Dry, solid-state processes, such as the *inversion* of aragonite to calcite or *recrystallization* (*sensu stricto*) of calcite to calcite, are most unlikely to take place in limestones, since diagenetic environments are always wet (see Bathurst, 1975 for extended discussion). Most neomorphism in limestones is of the aggrading type, that is leading to a general increase in crystal size, and this occurs chiefly in fine-grained limestones, resulting in microsparitic patches, lenses, laminae and beds. The opposite, degrading neomorphism, is not common, but has been recorded from crinoidal limestones as a result of burial diagenesis and perhaps incipient metamorphism. Although microbial micritization of skeletal grains leads to a fine-grained texture, it is not a neomorphic process.

One other common neomorphic process is *calcitization*, whereby aragonitic grains and cements are replaced by calcite. Dolomite can also be calcitized, as well as evaporite minerals in limestones. The important point about calcitization is that it is a *replacement* process; it involves gradual dissolution of the original mineral and precipitation of calcite, so that usually some minute relics of the original shell or cement are retained in the neomorphic calcite. The calcitization may involve a thin film, whereby there is dissolution of aragonite on one side and precipitation of calcite on the other. A range of replacement fabrics are produced, depending on the saturation state of the pore-fluid. Calcitization results in fabrics which contrast with those produced through wholesale dissolution of aragonite and subsequent filling of the void by calcite, which then is actually cement.

Dissolution. Carbonate sediments and cements and previously lithified limestones may undergo dissolution on a small or large scale when pore-fluids are undersaturated with respect to the carbonate mineralogy. Individual grains may be dissolved out, especially if they are of a metastable mineralogy like aragonite. Aragonite is less stable than calcite and the solubility of calcite increases with increasing Mg^{2+} content; calcite with 12.5 mole% $MgCO_3$ has a similar solubility to aragonite in distilled water (Walter, 1985). Dissolution is particularly important in near-surface meteoric environments, but in the marine realm, seawater is undersaturated *re* aragonite below a depth of several hundred metres in the equatorial Pacific so that aragonite dissolution can take place there too. With limestones, vugs, potholes, caverns and cave systems may develop as a result of dissolution through karstification. This may happen soon after deposition, or much later when limestone is uplifted. Paleokarsts are commonly developed below unconformities (see Section 7.5).

Compaction. When carbonate sediments are buried under an increasing overburden, then if they are not already cemented, grain fracture takes place and porosity is lowered by a closer packing. This is *mechanical compaction*. Eventually grains begin to dissolve at point contacts to produce sutured and concavo-convex contacts. This *chemical compaction* also takes place in previously lithified limestones to generate stylolites and dissolution seams, the latter sometimes referred to as flasers. Several hundreds to thousands of metres or more of overburden are generally neces-sary to produce these compactional structures. Tectonic stresses may also produce such pressure dissolution effects, and extensional and compressional fractures too, which may be filled by calcite to give veins.

7.3 DIAGENETIC ENVIRONMENTS

Three major diagenetic environments are distinguished: the marine, near-surface meteoric and burial environments (Fig. 7.1). Marine diagenesis takes place on the seafloor and just below, and on tidal flats and beaches. In the open marine environment, the processes operating depend very much on water depth and latitude; along the shoreline climate is a major factor. Meteoric diagenesis can affect a sediment soon after deposition, if lime sand is thrown up on to a supratidal flat by a storm for example, or there is shoreline progradation/slight relative sea-level fall so that rainwater now falls on the carbonates. In additon, limestones may be uplifted and exposed at the Earth's surface many millions of years after deposition and then subjected to meteoric diagenesis. This mostly involves limestone dissolution and karstification, and the effects of soils. With meteoric diagenesis, and to a much lesser extent with marine diagenesis, an important distinction is made between the vadose zone (above the water-table) and the phreatic zone (below). Meteoric diagenesis does not just operate in continental areas, but also along shelf margins or upon platforms where islands have developed and on atolls and isolated platforms where sediments rise above sea-level. Meteoric lenses develop in these situations (Fig. 7.2) and, depending on climate, they may extend down considerable distances (many hundreds of metres). In addition, meteoric diagenesis is not confined to the near-surface. Where there is an aquifer with strong hydraulic head and recharge descending into the deep subsurface, then meteoric diagenesis can take place at these greater depths. There are locations, for example, where meteoric water emerges on to the seafloor, in both shallow and deep water. The burial environment is perhaps the least well known and understood and is from below the zone affected by surface processes, from tens to hundreds of metres depth, down to several thousands of metres or more, where the zone of metamorphic dehydration reactions is reached. Choquette & Pray (1970) introduced the terms eogenetic, mesogenetic and telogenetic for early near-surface, burial and uplift/unconformity-related diagenetic processes respectively, but they are not widely used. Early diagenetic, referring to near-surface, and

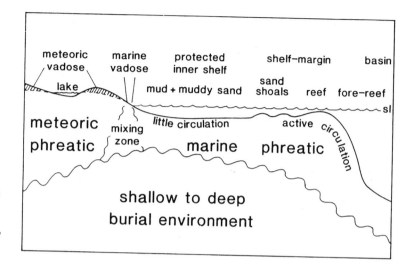

Fig. 7.1 *Carbonate diagenetic environments, schematically drawn for a rimmed shelf with unconfined aquifers. Where there are confined aquifers, through the presence of impermeable stratal layers, then it is possible, for example, for meteoric water to penetrate deep beneath the marine shelf, and even emerge on to the seafloor.*

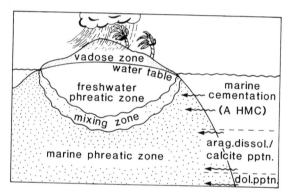

Fig. 7.2 *Diagenetic environments for an isolated platform, atoll or shelf margin, where an island is present with a freshwater lens. With increasing depth, seawater will become less saturated with respect to CaCO₃ so that aragonite may be dissolved and calcite precipitated, and eventually dolomite may be precipitated. This part of the model could apply to any platform or shelf margin. Depths will depend on the saturation state of seawater at the time.*

late diagenetic, referring to burial, are frequently employed.

In the shallow subsurface, where the marine and meteoric waters interface, there is a fourth diagenetic environment: *the mixing zone.* This environment features prominently in studies of dolomitization (see next chaper). The geometry of the mixing zone varies along a shoreline, depending on the hydrostatic head, rock porosity–permeability and presence of confined/unconfined aquifers. It may also vary through time, on an annual time-frame if there are marked wet and dry seasons, or on a longer time-scale if there are relative sea-level changes or climatic fluctuations.

The diagenetic environments pass vertically and laterally one into the other. A carbonate sediment may also pass from one environment to another with time, deposition and burial, sea-level changes and/or vertical tectonic movements. The sequence of diagenetic events and cement types can be predicted. For example, in the case of a fall in sea-level, the diagenetic environment of a sediment would change from marine phreatic to mixing-zone to meteoric phreatic and then meteoric vadose. There need not be evidence for all these environments preserved but thinking of the geological history, especially the burial history, of a limestone helps in knowing what cement fabrics and textures to expect (see Section 7.7).

There have been numerous studies of modern marine and meteoric diagenesis and the data collected have helped considerably in our understanding of ancient limestones. However, the Recent does not provide all the answers for two main reasons. Firstly, there is a growing body of evidence for subtle variations in seawater chemistry through the Phanerozoic, and in the Precambrian too, so that marine precipitates were of a different mineralogy and fabric at certain times in the past (see Section 7.4.2 and 9.3). Secondly, as a result of the drastic sea-level changes over the last one million years or so, most modern carbonate shelf sequences have been exposed to meteoric waters and karstification, and Recent shallow-marine sediments are a thin veneer (a few metres thick), mainly less than 5000 years old. The slow rate of some diagenetic processes means that the products are poorly represented in Quaternary carbonates. In addition, it is not possible to study Quaternary shallow-water carbonate sediments which have been buried to any extent in marine pore-fluids. Many ancient carbonates were formed and then gradually buried in seawater without any meteoric influences. It is only in the pelagic realm that one can study the effects of burial on a continuously subsiding Quaternary to Tertiary sedimentary sequence (see Section 5.6), but then most pelagic sediments have a dominantly low-Mg calcite mineralogy, whereas tropical shallow-water carbonates have high contents of metastable aragonite and high-Mg calcite.

7.4 SHALLOW-MARINE DIAGENESIS

7.4.1 The Recent

In low-latitude, shallow-marine environments, seafloor diagenesis mainly involves the precipitation of cements and the alteration of grains by microbial micritization and borings by other organisms. Cementation is most widespread in areas of high current activity, such as along shorelines and shelf margins, where seawater is pumped through the sediments, but it also occurs in areas of evaporation, as on tidal flats and beaches. Micritization of grains by endolithic algae, fungi and bacteria takes place almost everywhere but it is most prevalent in quieter-water locations where there is little sediment movement. Thus in the shallow-marine realm, it is possible to distinguish three areas: (1) the *active marine phreatic* where porewaters are constantly being replenished and cementation is common (e.g. reefs, sand shoals), (2) the *stagnant marine phreatic*, where there is little sediment or pore-fluid movement, microbial micritization of grains is ubiquitous and cementation is limited (e.g. onshelf lagoons), and (3) the *marine vadose*, where cementation chiefly occurs through evaporation of seawater and there may also be microbial effects (e.g. beaches, tidal flats). Modern marine diagenesis is discussed here under the subheadings reefs, carbonate sands and intertidal–supratidal zones.

In mid–high latitudes, shallow-marine carbonates are rarely cemented. Shallow seawater becomes undersaturated *re* $CaCO_3$ away from the subtropics, and carbonate grains are thus more liable to suffer dissolution (e.g. Alexandersson, 1978). Cement has been found within rhodoliths in the Skagerrak, where seawater is undersaturated, suggesting a biochemical

influence on precipitation (Alexandersson, 1974). Endolithic borings are common in skeletal debris of temperate latitudes, but most borings are empty, rather than filled with micrite.

7.4.1a Marine diagenesis in modern reefs

As discussed in Chapters 3 and 4, reefs are complex environments of construction by framework and encrusting skeletal organisms, destruction by physical and biological processes, sedimentation of debris, and cementation. In reefs, both depositional and diagenetic processes operate together: reefs are growing, being cemented, being wave attacked and bioeroded, all at the same time. Biogenic alteration and disintegration of framework and skeletal grains is ubiquitous, through the boring activities of, especially, microbial organisms, clionid sponges and lithophagid bivalves (e.g. Bromley, 1978). Distinctive borings are left on a micron to centimetre scale, and in the case of sponges, substantial quantities of angular carbonate chips (tens of microns across) are produced. Internal sedimentation of fine-grained detritus into primary and secondary cavities is another important very early diagenetic process in reefs. Some organisms (coelobites) such as serpulids and foraminifera live in these cavities and contribute to the sediment. By far the most important, and certainly the most-studied, diagenetic process in reefs is cementation. The precipitation of cements is a feature of most modern and many ancient reefs, and it does of course contribute to the stability of the reef structures. In some ancient reefs, marine cements are so common that the term cement-stone has been used. Notable occurrences of cements in modern reefs include those in Jamaica (Macintyre *et al.*, 1968; Land & Goreau, 1970), Bermuda (Ginsburg *et al.*, 1971; Schroeder, 1972a), Belize (James *et al.*, 1976; James & Ginsburg, 1979), the Great Barrier Reef, Australia (Marshall & Davies, 1981; Marshall, 1983a, 1986) and French Polynesia (Aissaoui & Purser, 1985; Aissaoui *et al.*, 1986).

Although a wide variety of cement morphologies are present in modern reefs (Fig. 7.3), they are essentially of only two mineralogies: aragonite and high-Mg calcite. The common types of aragonite cement are acicular crystals occurring as isopachous fringes, needle meshworks and botryoids, and micron-sized, equant crystals (micrite). High-Mg calcite cement occurs as acicular-bladed isopachous fringes, equant crystals, micrite and peloids.

Acicular aragonite (Fig. 7.4A,B) mostly consists of needle-like crystals typically 10 μm across and 100

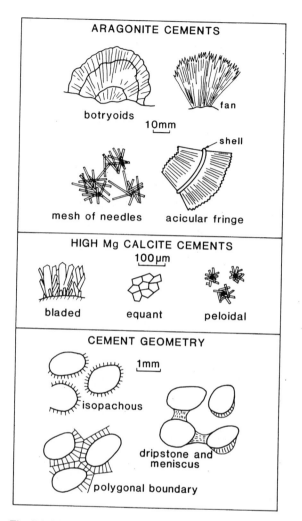

Fig. 7.3 *Modern marine cements and their geometries.*

some cases, the crystals are in optical continuity with crystals of the substrate and then the fringe normally consists of densely-packed, regularly-arranged, parallel needles. This is particularly the case where the fringe has developed on aragonitic molluscan fragments. Aragonite cement layers on coral skeletons are generally less regular (reflecting the more complex coral ultrastructure). Although there is commonly this substrate control of epitaxial overgrowth, aragonite cements are also developed on calcitic substrates. In some reef cavities, the aragonite needles have no preferred orientation but are simply a mesh of fine crystals partially to completely filling the pore. Micritic aragonite cements do occur in reefs but it is usually difficult to rule out a lime mud origin. The needle meshworks and micritic aragonite cements are common within silt-sized internal sediments in intraskeletal and interskeletal cavities.

The most conspicuous aragonite cement is the botryoidal form. The structures may reach 100 mm in diameter and are isolated or coalescent mamelons (Fig. 7.4C). They consist of fans of elongate euhedral fibres, commonly twinned to give a pseudohexagonal shape. Concentric growth zones are usually seen in thin sections and some have been bored by endolithic organisms. This cement form is a feature of the Belize reefs (James & Ginsburg, 1979), occurring in various cavity types and not showing any particular growth direction. In examples described by Aissaoui (1985) from the Pleistocene and Miocene of the Red Sea, the botryoids are late, forming after a phase of dissolution in one instance and after dolomitization and karstification in the other.

High-Mg calcite bladed cements are common in many reefs, although notably rare in some others compared to aragonite (e.g. Florida Reef Tract). The crystals are in the range of 20 to 100 μm long and less than 10 μm wide; generally they show a gradual increase in width along their length and then have an obtuse pyramid termination. Isopachous fringes are most common, in some cases with several generations of cement; the crystals commonly form tight clusters

μm long, but in some instances reaching 500 μm. They are elongate parallel to the crystals's c-axis and have straight extinction. Terminations are pointed or chisel-shaped and twinning is common. Where the acicular crystals form a fringe, it is generally isopachous. In

Fig. 7.4 *Modern reef cements. (A) Acicular aragonite cement fringes of various thicknesses in a coral skeleton; also present is a small aragonite botryoid (centre), a collection of high-Mg calcite peloids and micritic internal sediment. Florida, crossed polars. (B) SEM view of acicular aragonite cement crystals within a coral. Florida. (C) Botryoid of aragonite within coral skeletal cavity. Belize, crossed polars. (D) Reef debris (mostly calcareous algae and foraminifera) cemented by an isopachous fringe of bladed high-Mg calcite. Belize. (E) Reef debris cemented by high-Mg calcite fringe, with peloids occurring within the cavity centre. Belize, crossed polars. (F) SEM view of micritic high-Mg calcite cement. Florida. Photos courtesy of Ian Goldsmith.*

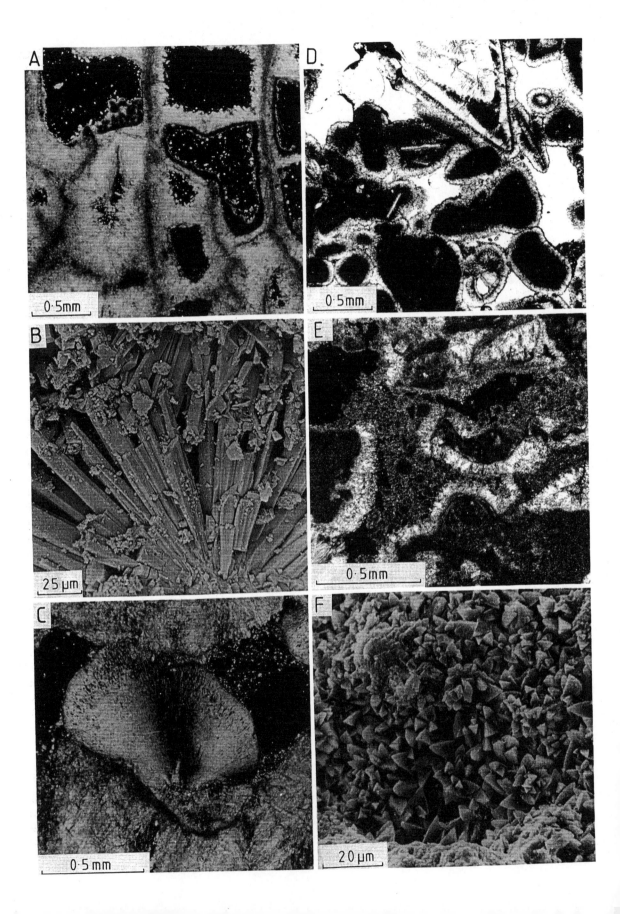

or bundles in these fringes (Fig. 7.4D,E). Some have more of a palisade structure with all crystals parallel. They may be closely associated with internal sediments and occur both before and after aragonite cement. These high-Mg calcite fringes are particularly well developed in Mururoa Atoll (Aissaoui *et al.*, 1986; Aissaoui, 1988) and Belize (James & Ginsburg, 1979). Schroeder (1972b) has described this bladed cement forming palisades around algal filaments from the Bermuda reefs.

Micritic high-Mg calcite cement is common in the Belize and Bermuda reefs and in the Great Barrier Reef, mostly as 2–8 µm sized rhombs with curved faces (Fig. 7.4F). They form coatings up to 20 µm or more thick around grains and lining interskeletal and intraskeletal cavities. Such cement fringes should be distinguished from micrite envelopes which are generally the microbially micritized outer part of bioclasts and not a coating. Where micritic cements are precipitated in abundance, there soon becomes a problem of distinguishing them from mechanically-deposited lime mud. Lithified internal sediments in reefs often do not have obvious fibrous cements, and they are probably cemented by micritic high-Mg calcite. More 'muddy' microfacies can be generated by such precipitation: a grainstone could become a packstone.

A more blocky, equant, coarser type of high-Mg calcite, reminiscent of sparry calcite (low Mg) typical of meteoric and burial environments (later sections) has been described from the reefs of Bermuda (Schroeder, 1972a) and the Pleistocene Hogsty Reef of the Bahamas (Pierson & Shinn, 1985); but it is exceedingly rare. The Bermuda example is a mosaic of equant (up to 60 µm diameter) crystals (17 mole% $MgCO_3$), lacking a drusy fabric, occurring in serpulid tubes. In the Hogsty case, the magnesian calcite (12 to 3 mole% $MgCO_3$) occurs as a last void-filling marine cement after acicular aragonite.

A most common feature of modern reef rocks is the presence of abundant peloids composed of high-Mg calcite (Fig. 7.4A,E). They are spherical to subspherical in shape and average around 40 µm in diameter, with a range mostly of 20–60 µm. They apparently result from two phases of precipitation: early, probably rapid growth of micron–submicron-sized anhedral crystals, associated with organic matter (the inner part of the peloids) and a later, apparently slower growth of larger (4–30 µm) euhedral, rhombic crystals (the outer part). The peloids form a range of textures reflecting packing density, from a very open 'grainstone' texture to a packstone, where they can only be distinguished from a featureless lime

mudstone by the presence of faint spherical structures. In the latter case, loss of peloid definition is probably due to further micritic calcite precipitation between peloids. The peloids occur in interskeletal and intraskeletal cavities, but they also form surficial crusts on corals of the southeastern Florida Reef (Lighty, 1985). The origin of the peloids has been much discussed (e.g. Lighty, 1985; Macintyre, 1985; Chafetz, 1986) with five likely explanations: (1) an algal origin, (2) a replacement texture, (3) detrital sediment, (4) the product of pelletizing organisms, or (5) an *in situ* precipitate. Some peloidal structures are cross-sections of calcified algal filaments (cf. Schroeder, 1972b), but the restricted size range, consistent texture, uniform crystal size and euhedral crystal outer part, and monomineralogy suggest that they are *in situ* precipitates. From study of the nuclei of the peloids with SEM, there is an indication of a microbial involvement with precipitation occurring within and around clumps of bacteria (Chafetz, 1986).

Where there have been detailed studies of the distribution of cements in reefs, then it does appear as if there is more cementation along the windward margins, where there is more active circulation of seawater through the reef framework (e.g. James *et al.*, 1976; Aissaoui & Purser, 1985; Lighty, 1985). In many reef rocks, on a somewhat smaller scale, the cement distribution is quite varied: both aragonite and high-Mg calcite may be present, some cavities may have both, others just one cement type, and still others may be empty. There may or may not be internal sediments. In the southeastern Florida Reef (Lighty, 1985), aragonite appears to precipitate first, close to the growing coral, whereas high-Mg calcite cements are apparently a little later. The peloids may relate to the presence of bacteria and the breakdown of organic tissue. Variations in the degree of cementation are strongly controlled by rates of fluid flow through the reef, and these are determined by local permeabilities. Computer modelling of coral skeletons has shown the importance of micropermeability contrasts in the amount of cementation taking place (Goldsmith & King, 1987). Fluid flow rates may also be a factor in the mineralogy and fabric (see Section 7.4.3 and Given & Wilkinson, 1985).

7.4.1b Marine diagenesis in carbonate sands

The most conspicuous early diagenetic process affecting shallow-marine carbonate sands is cementation. It tends to occur in areas of strong currents and waves, especially where sedimentation rates are low. Cements are initially precipitated within bioclasts in intra-skeletal pores, shells and tests. Gastropod and foraminiferal grains, for example, commonly have marine cements within them. The precipitation of intergranular cements leads to the formation of hard-grounds and crusts. The hardgrounds probably develop first just below the surface, where grains are not being moved very frequently, but seawater is continuously being pumped through. During storms, the hardgrounds may be exposed on the seafloor, and then they can be encrusted and bored. Abrasion from sand moving across a hardground may produce a flat surface, cutting the borings and encrusters. Modern hardgrounds are well documented from the Arabian Gulf, off Qatar (Shinn, 1969) and from the Bahama Platform, in the region of Eleuthera Bank (Dravis, 1979). In the Qatar case, the lithification has resulted in an expansion of the cemented layer so that a polygonal fracture system is developed, with buckling of the crust and the formation of pseudoanticlinal structures (tepees). The seafloor limestone pavement is bored by lithophagid bivalves and sponges, and these truncate the grains and cements. At least four cemented hardgrounds occur off Qatar, each 0.05–0.1 m thick, separated by uncemented sand. The hardgrounds in the Bahamas are forming on the sediment surface and the degree of lithification decreases downward. Buried crusts also occur in the oolitic sands. Acicular aragonite is the dominant cement, but algal filaments, which are commonly calcified, are also involved and bind and cement the ooids.

Disruption of seafloor surficial crusts by storms, waves and burrowing organisms leads to intraclasts, which may be reworked. The cements in these hard-grounds are mostly acicular aragonite fringes (Fig. 7.5A) and micritic high-Mg calcite. Where well developed around grains, the aragonite fringe gives rise to a polygonal pattern where the crystals from the different grains meet (Fig. 7.3). This is a feature seen in ancient cemented grainstones (e.g. see Fig. 7.14)

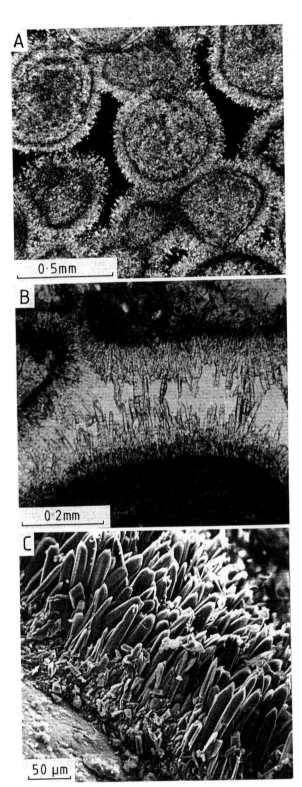

Fig. 7.5 *Cemented lime sands. (A) Ooids cemented by isopachous fringe of acicular aragonite. Shallow subtidal hardground. Bahamas, crossed polars. (B) Beachrock from Great Barrier Reef area, showing coral (upper) and calcareous algal grains cemented by thin, dark layer of micritic high-Mg calcite and then acicular aragonite. (C) SEM of beachrock showing aragonite needles growing on micrite upon the grain (bottom left).*

and generally indicates the presence of a marine cement.

Incipient cementation takes place in more protected areas on the Bahama Platform to give aggregates, collections of grains cemented by micritic aragonite often called grapestones (see Figs 1.8 and 3.15). Initially, the grains are bound together by algal filaments and encrusting tubular foraminifera. Some alteration of magnesian calcite to aragonite also takes place as the grapestones are cemented (Winland & Matthews, 1974). Formation of these aggregates appears to occur in relict lime sands in areas of slow sedimentation where there is occasional turbulence. The major process affecting grains in the quieter-water areas is micritization by endolithic algae, fungi and bacteria (Bathurst, 1966; Golubic *et al.*, 1975; Kobluk & Risk, 1977a,b). The boring algae are most common in the shallow-water areas (<50 m depth) giving micrite envelopes around bioclasts through to completely micritized grains (see Fig. 1.11). The bores from the algae are 5–15 μm in diameter and should only occur in grains within the photic zone, unless transport into deeper water has taken place. The microboring takes place via chemical dissolution of the $CaCO_3$ and Tudhope & Risk (1985) have shown that this is a major factor in shallow-marine carbonate budgets. Algal filaments also occur upon grains and these may be calcified to form a micritic coating (Kobluk & Risk, 1977b). Below the photic zone (100–200 m depth), grains are still being bored but mainly by fungi, which produce bores of diameter 1–2 μm (Zeff & Perkins, 1979). The bores produced by these various microscopic organisms are filled by micritic high-Mg calcite predominantly, also aragonite, but it is difficult to be certain whether the micrite is a cement or sediment. Micritization also weakens grains considerably, making them more susceptible to comminution by physical processes.

In these shallow-water carbonate sands and muddy sands, the seafloor diagenetic process operating depends basically on the energy level. In high-energy areas, such as shelf margins, lime sands are produced in abundance and seawater is pumped through them, increasing the chances of cementation. The porous nature of the sediment also promotes cementation and hardground development. In more protected areas, more muddy sands accumulate, there is less water movement through the sediments, porewaters are more stagnant, and cementation is only on a local scale, producing grapestones and intragranular cements. Microbial micritization is intense in these areas.

7.4.1c Marine diagenesis in modern intertidal–supratidal zones

Along tropical beaches, precipitation of cement may lead to the formation of beachrock (reviewed in Scoffin & Stoddart, 1983). The lithification mostly takes place below the beach surface, but beachrock is often exposed through storm erosion. It is then liable to be coated by algae, encrusted, bored and grazed by intertidal organisms. Beachrocks are commonly jointed, normal and parallel to the shoreline, and they may be eroded into intraclasts to form gravels and recemented to form intertidal conglomerates.

The most common cement is acicular aragonite (Fig. 7.5B,C), which in the lower intertidal zone forms isopachous fringes, since here pores are fluid-filled for most of the time. In the mid–high intertidal zone, asymmetric aragonite fringes may occur in dripstone and meniscus forms (Taylor & Illing, 1969). Micritic high-Mg calcite is a common grain coating (Fig. 7.5B,C) and it may also show vadose textures (e.g. Meyers, 1987). Micritization of grains and even cements in beachrock is common, and calcified microbial filaments occur within the micritic cements. If there is a high meteoric groundwater table in the backshore area, then it is possible for low-Mg calcite cement to be precipitated in the upper intertidal part of beachrock. Beachrocks in temperate parts of the world (e.g. southwest UK) are mostly cemented by low-Mg calcite precipitated from meteoric waters at the back of the beach. With beach progradation or slight sea-level fall, beachrocks may be subjected to meteoric diagenesis: dissolution and pedogenesis (e.g. Beier, 1985).

The origin of most breachrock cements is evaporation and CO_2-degassing of seawater. The constant pumping of seawater through the sand as the tide rises and falls is important too. Fluctuations of pH and P_{CO_2} due to photosynthesis of algae and the effects of bacteria living in and on the sediment may also contribute to the precipitation, particularly of micritic cements.

Cementation of carbonate sediments also takes place on tidal flats and in supratidal zones frequently flooded by seawater. Dolomitization may also take place in these situations, to produce surficial crusts (see Fig. 3.25) and dolomitic layers beneath the surface (see Section 8.7.1). Precipitation of $CaCO_3$ cements on tidal flats leads to the development of cemented pavements and these are commonly polygonally cracked, buckled, folded and thrust into tepee structures, not unlike those noted in the preceding section.

Tepees are well developed on the upper intertidal–low supratidal flats of the Trucial Coast and around shallow, coastal lakes of South Australia and also Lake Macleod of Western Australia fed by seawater seepage through barriers (Kendall & Warren, 1987). Continental groundwaters may recharge on to the supratidal flats and contribute to crust and tepee formation (e.g. Ferguson *et al.*, 1982). One characteristic feature of groundwater recharge and emergence on to the surface is the development of vertically-laminated sediments and cements within the cracks between polygons. Cements in the crusts are again typically acicular aragonite and micritic, high-Mg calcite, but they may be of low-Mg calcite if there is a strong meteoric groundwater influence. Underneath the cemented slabs dripstone cements, aragonite botryoids, pisoids and sheets of floe carbonate may develop (e.g. Ferguson *et al.*, 1982; Handford *et al.*, 1984). *Floe calcite* (or aragonite) is an extremely thin sheet of carbonate precipitated on the surface of a pool of water by rapid degassing of CO_2 and evaporation. The sheets are quickly broken up by movement of the water and the intraclasts are then deposited of the floor of the pool, often in an imbricated fashion. Floe calcite (or aragonite) is commonly associated with tepee structures, and can form a substrate for further cement precipitation. Pisoids forming in these environments are analogous to cave pearls; they mostly consist of alternations of micrite (sediment and/or cement) and fibrous cements (aragonite or calcite) and may show evidence of asymmetric, stationary growth.

7.4.2 Geochemistry of modern marine cements

There are few trace element–stable isotope analyses of modern marine cements, mainly because of the difficulties of extraction, although the aragonite botryoids are easy to analyse (e.g. James & Ginsburg, 1979; Aissaoui, 1985). It appears that aragonite cements contain 8000–10 000 ppm strontium, a value similar to that of other inorganic precipitates like ooids. Na will be around 2000 ppm and Mg 1000 ppm or less. Magnesian calcite cements generally have between 12 and 19 mole % $MgCO_3$, but the strontium value is quite low, around 1000 ppm.

The oxygen isotopic signatures of shallow-marine cements and sediments (Fig. 7.6) depend largely on the seawater $\delta^{18}O$ composition and the temperature. With marine cements, the values are commonly those predicted for precipitation in equilibrium with sea-

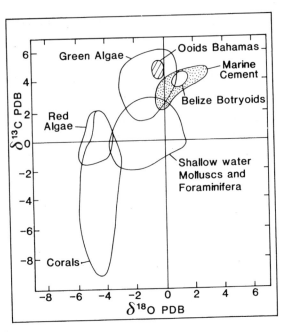

Fig. 7.6 *Stable isotope signatures of Recent skeletal grains, ooids and marine cements. After various sources including Land & Goreau (1970), Milliman (1974), James & Ginsburg (1979), and Anderson & Arthur (1983).*

water or they are slightly heavier than would be expected (Gonzales & Lohmann, 1985). The $\delta^{18}O$ range is typically −0.5 to +3‰, and high-Mg calcite cements may show a slight enrichment in ^{18}O over aragonite, because of a fractionation effect. Biogenic carbonate sediments show quite a range of $\delta^{18}O$ (Fig. 7.6) but many do precipitate carbonate with a $\delta^{18}O$ value suggesting isotopic equilibrium with the waters in which the organisms were living. This is particularly the case for molluscs, brachiopods and planktonic foraminifera. Marine organisms with $\delta^{18}O$ deviating from the predicted isotopic equilibrium values include echinoderms, corals and red algae, and generally the values are lighter (^{16}O-enriched) than expected (see review of Anderson & Arthur, 1983). Inorganic precipitates such as ooids have $\delta^{18}O$ close to equilibrium values (0‰).

Carbon isotopic signatures of marine cements and sediments (Fig. 7.6) again depend on seawater $\delta^{13}C$, but also on any organic mediation or involvement in precipitation. Cements and ooids from tropical environments like the Bahamas and Belize have $\delta^{13}C$ values between +2 and +5, typically around +4‰. In some instances high-Mg calcite peloids may be a little depleted relative to aragonite cements in the same sample. This could indicate some minor input of organically-derived carbon for the peloids, or it could be a reflection of a faster growth rate. There are some marine cements which have quite extreme $\delta^{13}C$ values; high-Mg calcite in nodules from the Fraser Delta, British Columbia ($\delta^{13}C = -7$ to $-59‰$, Nelson & Lawrence, 1984) for example, and botryoidal aragonite from North Sea pockmarks ($\delta^{13}C = -56‰$, Hovland et al., 1987). However, these are rather special cases, and the very negative $\delta^{13}C$ is reflecting an organic involvement, mostly anaerobic bacterial fermentation, the production of methane and its oxidation. Biogenic grains have $\delta^{13}C$ values ranging from near equilibrium to substantially less (Fig. 7.6). Corals for example may have $\delta^{13}C$ down to -8, where equilibrium values are around 0 to $+2$. The deviations from equilibrium, these 'vital' effects exerted by organisms, are mainly due to the mixing and isotopic exchange between seawater CO_2 and respiratory CO_2 (usually depleted in ^{18}O and ^{13}C) at or near the site of skeletal precipitation. Although shallow-marine carbonate sediments and cements do exhibit a range of isotopic values, this range is usually distinct in a $\delta^{13}C-\delta^{18}O$ cross-plot from cements and other diagenetic products of meteoric and burial diagenetic environments.

7.4.3 Modern marine cementation: discussion

There are a number of prerequisites for marine cementation to occur.

1   There is a suitable stable substrate. From a thermodynamic point of view, homogeneous nucleation, that is nucleation direct from solution without a substrate, is energetically unfavourable. A substrate is necessary for heterogeneous nucleation to occur.

2   There is a lack of mechanical abrasion. Delicate, early crystals are easily destroyed by mechanical abrasion, so that the most common areas for cementation are within intraskeletal cavities and within framework porosity in a reef. Lime sands may be cemented to form hardgrounds (Section 7.4.1b), but it mostly begins below the sediment–water interface where there is rarely any grain movement or there is some algal binding initially.

3   Seawater is supersaturated with respect to $CaCO_3$. This is the case for low-latitude, shallow seas.

4   Water exchange rates are high. Calculation of the molarity of supersaturated seawater re Ca^{2+} and

HCO_3^- indicates that many thousands of pore volumes of seawater must pass through a pore in order to fill it with cement. As only $CaCO_3$ above the saturation level will be removed from the solution, and this is likely to be an inefficient process, the figure is probably the order of 100 000. Thus cementation (in general) will take place preferentially where there is an active pumping mechanism to force large quantities of water through a sediment body. Shelf margins and high-energy shorelines where there is intense wave, storm and tidal activity are favoured sites, and so reefs and lime sands in these areas are frequently being cemented.

5   Porewaters should be oxygenated. The precipitation of cements, especially aragonite, is favoured under such conditions.

6   Time is needed. Cementation probably requires substantial amounts of time in the shallow subtidal environment, less so along shorelines and on tidal flats where evaporation promotes precipitation. Cements will be better developed in areas of slow sediment accumulation or slow reef growth.

As noted above, seawater is supersaturated re $CaCO_3$ and so it might seem surprising that carbonate precipitates are not more abundant. One factor is nucleation energy (see Berner, 1980) but this is negligible where there is an abundance of stable nuclei upon which heterogeneous nucleation can take place. It has been suggested that the nuclei (mostly skeletal debris) are covered in some sort of mucilaginous organic material or membrane which prevents the precipitation of $CaCO_3$ (Berner et al., 1978). Also important are the inhibitory effects of Mg^{2+}, phosphate and sulphate ions on carbonate precipitation (see Berner, 1975; Reeder, 1983; Walter, 1986). Magnesium ions have a marked retarding effect on the precipitation of calcite out of seawater, but not on aragonite. The reason is that Mg^{2+} ions fit more easily into the calcite lattice, but since Mg^{2+} ions are smaller than Ca^{2+}, they have a higher surface charge and this attracts a larger hydration sphere. The inhibitory effect relates to the high energy required to dehydrate the Mg^{2+} ions as they are incorporated into the calcite crystals. Dissolved phosphate inhibits both calcite and aragonite precipitation (the former more so), by adsorption on to the crystal surface and blocking of nucleation and growth sites. Sulphate in seawater apparently inhibits calcite precipitation considerably more than aragonite.

7   With regard to the mineralogy of modern marine precipitates, it is far from clear as to what controls whether aragonite or magnesian calcite is precipitated.

The kinetic inhibition of calcite precipitation by Mg^{2+} and SO_4^{2-} in seawater favours aragonite precipitation. In low Mg^{2+} waters, such as meteoric fluids, low-Mg calcite is the dominant cement mineralogy and aragonite is exceedingly rare. There is a temperature effect too: with increasing temperature (over the range 5 to 37°C) there is a greater increase in the precipitation rate of aragonite over that of calcite (Burton & Walter, 1987) and there is an increase in the amount of Mg^{2+} in calcite too. This can explain the decrease in aragonite cement abundance and decrease in Mg^{2+} content of magnesian calcite with increasing depth in tropical seas. A temperature effect can also account for the predominance of calcitic-secreting organisms in temperate seas and aragonitic-secreting organisms in tropical seas. Thermodynamically, aragonite is equivalent to a calcite with approximately 12 mole% $MgCO_3$, so that a magnesian calcite with more than 12 mole% will be less stable than aragonite (Walter & Morse, 1984), and the growth rate of calcite will become progressively more inhibited as the porewater Mg^{2+} content and Mg/Ca ratio increase, leading to a kinetic advantage for aragonite precipitation. In terms of dissolution of carbonate skeletons, important during diagenesis, the microstructural complexity is also a factor; aragonite grains with more complex microstructures can dissolve more rapidly than the supposedly less stable, higher-Mg calcites (Walter, 1985). This shows that the effect of reactive surface area is greater than thermodynamic stability.

An organic geochemical control on cement mineralogy is also a possibility, as occurs in biomineralization. Some proteinaceous molecules capable of chelating Ca^{2+} and CO_3^{2-} ions do so in a specific stereochemical geometry so that a particular cement mineralogy is preferred. Ions held in a 9-fold co-ordination geometry will nucleate aragonite, and ions in a 6-fold co-ordination will nucleate calcite (Degens, 1976). Organic compounds in nuclei could exert a control on cements precipitating upon them, with further growth taking place purely inorganically. It has often been noticed that cements of a particular mineralogy are preferentially precipitated on to a surface of the same mineralogy. A close similarity of substrate and cement at the atomic structural level will lower the nucleation energy and allow crystal growth in optical and lattice continuity. However, although there is certainly a control here, there are many cases of aragonite crystals growing on a calcitic substrate, and vice versa, and of aragonite and high-Mg calcite cements being interlaminated.

One possible further physical control determining mineralogy is the rate of supply of CO_3^{2-} ions (Given & Wilkinson, 1985). It is suggested that high rates of carbonate supply favour aragonite precipitation over high-Mg calcite (see Fig. 7.7). Within cavities, fluid flow rates will be high initially and so aragonite should precipitate. As pore throats become progressively more restricted by the precipitation of aragonite cement, fluid flow rates should decrease and once some threshold is passed high-Mg calcite should precipitate. In addition to temperature and pore-fluid Mg/Ca ratio, the Mg^{2+} content of calcite also appears to relate to crystal growth rate: the slower the growth rate, the more Mg^{2+} is expelled from surface layers and the lower Mg^{2+} content of the crystal.

The morphology of marine cements is basically either acicular or micritic; the larger, more equant calcite typical of meteoric and burial environments is exceedingly rare in shallow-marine environments. Several explanations have been put forward to account for this. Folk (1974) suggested that the Mg^{2+} ions in

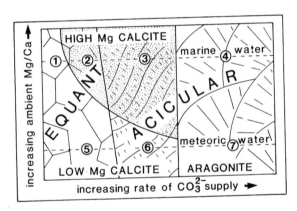

Fig. 7.7 *Schematic illustration of relation between fluid Mg/Ca ratio, rate of carbonate ion supply, crystal morphology (equant or acicular) and mineralogy of inorganic precipitates (low-Mg calcite, high-Mg calcite or aragonite). Mg/Ca ratio of the meteoric line is 0.3 and of the marine line 5.2. Low-Mg calcite is up to 9 mole % $MgCO_3$ and high-Mg calcite is above. The positions of modern, naturally occurring precipitates shown are (1) equant calcite spar cement in cold, deeper-water, low-latitude sediments and shallow-water temperate sediments, (2) equant high-Mg calcite cement in reefs (rare), (3) acicular high-Mg calcite cement in reefs and lime sands, (4) acicular aragonite cements in reefs and lime sands, (5) equant low-Mg calcite spar cements in meteoric environments, (6) acicular low-Mg calcite in speleothems and travertines, and (7) acicular aragonite in speleothems (rare). After Given & Wilkinson (1985).*

seawater poison the calcite lattice and inhibit growth on all faces except that normal to the c-axis, which is the fastest growth direction anyway. The incorporation of a small Mg^{2+} ion is thought to cause distortion of the lattice. As a result of the calcite structure, alternating layers of Ca^{2+} and CO_3^{2-} ions normal to the c-axis (Section 6.2.1), an Mg^{2+} ion incorporated on to the side of a growing crystal causes more distortion than if it landed in the middle of a continuous sheet of Ca^{2+} ions. Another factor is the effect of surface charge (Lahann, 1978). Through the excess of cations to anions for the $CaCO_3$ system in seawater, the surfaces of calcite crystals have a net positive charge. The highest charge density is on the c-axis faces and so these attract the largest number of anions resulting in fastest growth in the c-axis direction. In fact, the availability of CO_3^{2-} ions is the rate-limiting step in crystal growth (Given & Wilkinson, 1985). This can account for the acicular nature of most aragonite cements and some high-Mg calcite cements (the bladed type) in shallow-marine sediments (see Fig. 7.7). The precipitation is mostly driven by CO_2-degassing from agitated seawater, and this generates CO_3^{2-} ions from the dissociation of HCO_3^-. Where carbonate supply rate is much lower, then with much slower crystal growth, more equant forms are the result (Fig. 7.7). This is typically the case with cements of meteoric as well as burial environments (Sections 7.5 and 7.6.2). For the microcrystalline morphology, the important point is relative rates of nucleation and crystal growth. With acicular cements, crystal growth occurs on an individual or a few nucleation centres, so that the rate of growth is rapid compared to the rate of nucleation. With the micritic cements, nucleation rates are very high compared to the rate of crystal growth. Thus many nuclei are formed, but they do not have the chance to grow to any substantial size. Rapid crystal growth also accounts for the high Mg^{2+} content of the micritic magnesian calcites.

7.4.4 Marine diagenesis in the geological record

Many shallow-water limestones in the geological record show abundant evidence of marine diagenesis. Many limestones have a first generation of cement which can be interpreted as marine. However, it should be stressed at the outset, that many ancient marine cements do not have an exact modern equivalent or they are very poorly represented in Cainozoic limestones. There is quite a variety of ancient marine carbonate cements; in terms of mineralogy, aragonite,

high-Mg calcite and low-Mg calcite were all precipitated on the seafloor at one time or another and their morphology ranged from acicular to micritic, columnar to equant. There has been a problem of identification of original mineralogy and fabric of ancient marine cements because of the lack of some modern analogues. Too rigid adherence to the 'present is the key to the past' maxim has led to difficulties in the interpretation of ancient cements. Ancient aragonite and high-Mg calcite in limestones are normally low-Mg calcite in limestones now, so that careful petrographic and geochemical study is necessary to identify original mineralogy. It is also now known that at certain times in the Phanerozoic (middle-Palaeozoic and Middle–Late Mesozoic) calcite was the dominant marine precipitate, while aragonite and calcite (maybe magnesian) were common cements in the Late Precambrian, Late Palaeozoic–Early Mesozoic and Cainozoic (Sandberg, 1983, 1985; Wilkinson et al., 1985; see Section 9.3).

General features of marine cements are: (1) they are the first cement generation, (2) they are commonly fibrous (acicular or columnar), (3) they mostly form isopachous fringes around grains, (4) they may be cut by borings or include microfossils, (5) they may be associated with internal sediments, (6) they are succeeded by sparry calcite, (7) the crystals are non-ferroan and non-luminescent, and (8) they may occur in intraclasts. More specific features are discussed below.

7.4.4a Marine diagenesis in ancient reefs

Most ancient reefs have well-developed marine cements, in many cases very obvious even in the field. Several generations may be present with small differences in fabric between them, reflecting some evolution of the marine porewaters. Many reef cements are isopachous fringes of fibrous calcite; this type is prominent in mid-Palaeozoic, especially Devonian, reefs and mud mounds. Cements are also very conspicuous in Permian and Triassic reefs, such as those of the Permian Basin in Texas and New Mexico and those in the Alps of western Europe (e.g. Fig. 7.8). Where the distribution of marine cements in an ancient shelf-margin reef complex has been well documented, then it often appears that they are most common along the seaward margin of the reef and most porosity is occluded there. A number of Devonian reefs such as those of the Canning Basin, Western Australia (Kerans et al., 1986) and the Golden Spike, Leduc Reef of western Canada (Walls, 1983; Walls & Burrowes, 1985) show this preferential

Fig. 7.8 *Marine cement in a Triassic reef, Hafelekar, Austria. Isopachous layers of fibrous calcite cementing fore-reef debris and known locally as 'Grossoolith'. Lens cap 6 cm diameter.*

zone of marine cementation in framework and shelter porosity along the seaward reef margin. Fore-reef debris is commonly well cemented too in reef complexes, and cements may occur in clasts reworked into the talus, demonstrating synsedimentary, marine precipitation (e.g. from the Devonian of Germany, see Krebs, 1969).

Internal sediments are commonly closely associated with marine cements, and these may contain microfossils and fine skeletal debris. These sediments are different from vadose silts, which are internal sediments consisting of cement fragments, thought to be derived from internal erosion in reef and other cavities during exposure and meteoric diagenesis (Dunham, 1969a). *Neptunian dykes* are also a feature of ancient reefs, mostly forming through differential settling and compactional fracture of the lithified rock mass. The fractures generally run parallel to the shelf margin and may lead to rotational failure and sliding of huge blocks of reef rock. The fissures are usually filled by marine sediments and cements, unless they formed during subaerial exposure, and many phases of fill are common. The internal sediments commonly

form geopetal structures, enabling depositional dips to be measured and the distinction of allochthonous reef blocks from *in situ* patch reefs and mounds in reef slope deposits. Some neptunian dykes contain fossils of organisms which inhabited these submarine fissures (e.g. *Renalcis* encrusting dyke walls in Devonian reefs) or used them for nesting or mating (e.g. fissures in Permian reefs of northeast England full of nautiloids and dykes in Jurassic reefs of Sicily with dwarf ammonites). Spherical 'spar balls' of fibrous calcite, like cave pearls, form in submarine fissures.

In some reefs, cementation is really a major factor in the formation of the reef itself, creating a wave-resistant structure. This important role is well illustrated in the Canning Basin reefs, where Late Givetian/Early Frasnian reefs have little marine cement and were only low-relief banks, but the succeeding Frasnian and Famennian reefs have abundant cement, aiding the formation of precipitous reef scarps elevated high above the adjacent basin floor. Very steep slopes were maintained in the upper fore-reef area and blocks of cemented reef debris are common near the base of the marginal slope (Kerans *et al.*, 1986).

7.4.4b Marine diagenesis in ancient lime sands and muds: hardgrounds

In non-reefal sediments, field evidence for marine diagenesis may be difficult to find. The main feature of seafloor cementation of shelf sands and muds is the presence of hardgrounds. These generally form just below the seafloor, but they are most easily identified where they were exposed and exhumed on the seafloor, and then encrusted and bored. Hardground surfaces exposed in this way vary from smooth bedding planes, where there has been a degree of corrosion by lime sand moving over the surface, to more irregular surfaces where nodular-bioturbated, patchily-cemented layers were exposed after more gentle erosion, and perhaps were subjected to some corrosion and dissolution or mineralization. Hardground surfaces may be encrusted by such organisms as oysters, serpulids, crinoids, sponges, calcareous algae and corals. They are bored by polychaete worms (such as *Trypanites* and *Polydora*), sponges, lithophagid bivalves and endolithic algae. With some hardground surfaces, repeated erosion of the surface can be identified by truncated borings, particularly of the lithophagid bivalves which have such a distinctive Cod's bottle (flask) shape. Associated with hardgrounds there are commonly intraclasts, which may themselves be encrusted and bored.

Hardgrounds are best developed in areas of slow sedimentation and high current activity (see Section 7.4.1b). Thus they may become impregnated with minerals such as iron hydroxides (goethite–limonite), phosphorite and glauconite. This is the case with some hardgrounds in the deeper-water chalk, described in Section 5.7.3. As hardgrounds develop from loose sediments through firm grounds to lithified layers, there is a gradual change in the fauna. Burrowing may have been intense before the lithification, as sedimentation rate slowed down. The ecological succession with hardground development is well documented for Jurassic examples (e.g. Fursich, 1979; Gruszczynski, 1986).

Where lime sands were cemented to form hardgrounds then the marine cements are usually obvious in thin section. They are typically isopachous fibrous calcite fringes around grains (Fig. 7.9), possibly with a polygonal compromise boundary where fringes meet (as in Fig. 7.14). A marine, synsedimentary origin is best confirmed where borings cut grains and their cement coating (Fig. 7.9). This has been well illustrated from the Jurassic of the Paris Basin by Purser (1969). In some grainstone hardgrounds, it appears that equant sparry calcite was the marine precipitate (Wilkinson *et al.*, 1982; Wilkinson *et al.*, 1985) and the significance of this is considered in Section 9.3. Where hardgrounds have formed in more muddy sediments, then it is usually impossible to recognize the cements, which are also fine grained.

7.4.4c Marine diagenesis in ancient intertidal–supratidal facies

There are very few records of ancient beachrock; this might seem surprising in view of its fairly widespread occurrence along modern tropical shorelines. Ancient beach limestones are known but they are not as common as shallow subtidal carbonates. Beach facies have characteristic sedimentary structures (low-angle, planar cross-stratification, truncation surfaces, keystone vugs (etc.), see Section 4.1.3a and Fig. 4.7B) and any first generation fibrous or micritic calcite occurring in such a facies may be a beachrock cement.

Evidence of a cemented surface, in the form of boring and encrusting organisms, would indicate exposure of the beachrock by wave erosion. The presence of dripstone and meniscus morphologies in the fibrous and micritic cement would further support a beachrock interpretation and upper intertidal marine cementation. Intraclast breccias could be associated. Beachrock has been described from the Cretaceous of

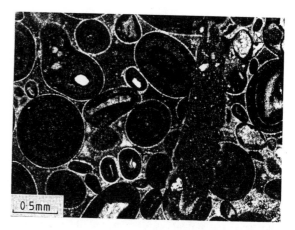

Fig. 7.9 *Oolitic sand cemented in the submarine environment to form a hardground which was then bored. Photomicrograph shows thin isopachous fringe of fibrous calcite around ooids and then micrite partly filling the intergranular porosity. The micrite may be an internal sediment or in part a cement. The ooids, micrite and fibrous calcite are cut by a boring which is itself filled with micrite. The early cementation prevented any compaction of this oolite which has grains in point contact. Middle Jurassic, Cotswolds, UK.*

Texas (Inden & Moore, 1983) and the Proterozoic of North West Territories, Canada (Donaldson & Ricketts, 1979).

Intertidal–supratidal carbonate facies are widely represented in the geological record and there are many descriptions of the more diagenetic features of these rocks resulting largely from cementation from marine waters. Polygonal, tepee and pseudoanticlinal structures and associated cements and pisoids are well developed in the Permian Reef Complex of New Mexico and Texas (Smith, 1974; Dunham, 1969b; Assereto & Kendall, 1977; Esteban & Pray, 1983) and in the Triassic Calcare Rosso of Italy (Assereto & Folk, 1980). With many tidal flat facies, the early marine diagenetic story is complicated by the precipitation of dolomite and evaporite minerals, and also the development of soil horizons, laminated crusts and palaeokarsts as a result of meteoric effects. Ancient peritidal facies are considered in detail in Section 4.3.

7.4.5 **Ancient marine aragonite cements**

Aragonite is not normally preserved in the geological record and so petrographic and geochemical evidence is required to identify cements in a limestone which originally had an aragonite composition. The original marine aragonite is mostly replaced by calcite either in a near-surface meteoric or burial diagenetic environment. It may also be dolomitized. The gross morphology of the cements is only an indication of original mineralogy where botryoids are present. Calcite does not appear to have formed such structures in ancient marine environments, although speleothemic calcite may be botryoidal. Isopachous marine cement fringes could have been aragonitic or calcitic originally, so this morphology only indicates phreatic precipitation.

Many former aragonite cements (and aragonite bioclasts and ooids) were calcitized in a thin-film/replacement-front mechanism so that there is some degree of retention of original texture (Fig. 7.10). The calcite crystals are generally large, irregular to equant in shape, without a drusy fabric, and they cross-cut the original aragonite crystal texture of the acicular fringe or botryoid (e.g. Mazzullo & Cys, 1979; Sandberg, 1985). The original aragonite fabric may be retained to a greater or lesser extent through the presence of minute relics of the aragonite, best seen with the SEM as 1–10 μm inclusions with a preferred orientation in the replacement calcite (Fig. 7.10C; e.g. Sandberg, 1985; Tucker & Hollingworth, 1986). Organic matter which was present between aragonite fibres may be retained in the replacement calcite too. One particular feature which has been used to identify replaced aragonite cements is the presence of 'square-ended terminations' on the cement fringe (Folk & Assereto, 1976; Loucks & Folk, 1976). These may be the replaced terminations of large, single aragonite crystals or bundles of near-parallel acicular aragonite crystals.

Neomorphic calcite after aragonite cements (and bioclasts) is commonly pseudopleochroic in shades of brown (Hudson, 1962). This is thought to be the result of occluded organic matter. Where, during aragonite stabilization, the rate of dissolution of aragonite increased, large voids would have developed. Patches of coarse, equant, clear calcite within irregular neomorphic mosaics are the result of filling of such dissolution cavities (Fig. 7.10D). Where complete dissolution of aragonite has taken place, then no relict structures will be present in calcite, which is a cement, filling the void, and positive identification of the former presence of aragonite is very difficult. In some cases, a more fibrous elongate calcite ('ray crystals') does replace the original aragonite, with replacement crystal elongation following that of the host (e.g. Mazzullo & Cys, 1979; Tucker &

Hollingworth, 1986). Calcite after aragonite may retain a geochemical signature: a high strontium content (>1000 ppm) in neomorphic calcite could reflect replacement of aragonite with its 8000–10000 ppm Sr (e.g. Davies, 1977; Scherer, 1977; Tucker, 1985b).

Ancient aragonite cements do not have a uniform distribution through time, but are most common from mid-Carboniferous through to Early Jurassic and in Cainozoic limestones (Sandberg, 1985). They are especially common in Upper Palaeozoic and Triassic reefs (e.g. Loucks & Folk, 1976; Davies, 1977; Scherer, 1977; Mazzullo & Cys, 1979; Given & Lohmann, 1985; Tucker & Hollingworth, 1986).

The isotopic signatures of ancient aragonite cements have been altered to varying extents as a result of the calcitization. In one detailed study from the Permian Reef Complex of Texas and New Mexico (Given & Lohmann, 1985), isotopic analyses of former aragonite botryoids give a linear covariant trend (Fig. 7.11), reflecting random proportions of two types of calcite, one luminescent, one not, replacing the aragonite. The non-luminescent calcite is the most positive ($\delta^{13}C = +5.3‰$; $\delta^{18}O = -2.8‰$), and it is suggested that this calcite was precipitated early in a low water/rock ratio, 'closed' system and that the isotopic composition is derived from the original marine cement. The luminescent calcite has a variable, more negative isotopic composition (average $\delta^{13}C = -3‰$; $\delta^{18}O = -8.2‰$) and was precipitated at the same time as pore-filling equant spar. These calcites are interpreted as meteoric in origin and reflect the establishment of a meteoric phreatic system within the reef complex (Given & Lohmann, 1986).

7.4.6 Ancient marine calcite cements

The majority of marine calcite cements in limestones are of the fibrous type, that is crystals have a significant length elongation, mostly parallel to the c-axis, making the crystals lengthen fast. In addition, some early equant sparry calcite cements in limestones could be marine, but most are undoubtedly meteoric or burial, and some syntaxial echinoderm overgrowths could be marine. Micritic calcite and peloidal calcite cements in limestones also occur but are not common.

Fibrous calcite is the major cement type in most ancient reefs of the Phanerozoic and it also occurs in grainstone hardgrounds and in many cavities within tidal flat facies. Two broad varieties of fibrous calcite are the *columnar* form, where the length to width ratio is more than 6:1 and crystals are more than 10 μm wide, and the *acicular* form, being needle-like and

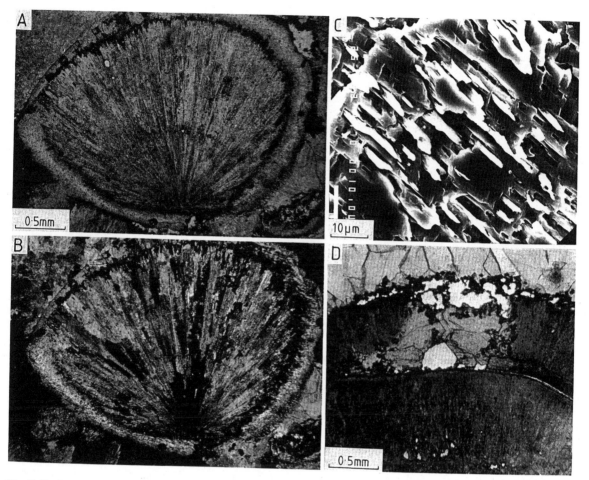

Fig. 7.10 *Former aragonite cements from the Upper Permian reef of northeast England. (A) and (B) Calcitized aragonite botryoid occurring within a brachiopod shell. (A) Plane polarized light. (B) Cross polars. (C) SEM of calcitized aragonite botryoid showing relics of aragonite and many small holes in coarse calcite. (D) Fringe of former aragonite cement around a bivalve shell showing calcitized fabric with relic acicular structure and area of complete aragonite dissolution, now filled with clear calcite spar cement. The white areas are vugs formed by much later dissolution.*

less than 10 μm wide. Fibrous calcite mostly forms isopachous layers and it may be closely associated with internal sediments. In some large reef framework or mud mound stromatactis cavities, numerous generations of fibrous calcite are present. The crystals of the more columnar varieties of fibrous calcite, which reach several millimetres in length, show a range of fabrics with radiaxial (most common) and fascicular-optic (less common) as end-members (see Figs 7.12 and 7.13). Many fibrous calcite crystals

are unit extinguishing with straight twin planes and slightly irregular intercrystalline boundaries. They may be capped by terminations or syntaxially overgrown by later equant, drusy calcite spar. These fibrous crystals may form a palisade fringe (all crystals parallel) or they may comprise fans giving a radial-fibrous texture (Fig. 7.12), whereby each crystal has unit extinction but it is part of a larger structure of swinging extinction. In both fascicular-optic fibrous calcite (FOFC) and radiaxial fibrous calcite (RFC)

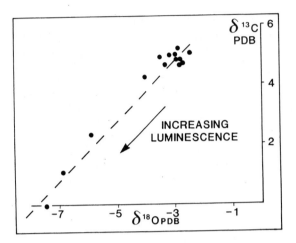

Fig. 7.11 *Stable isotope analyses of calcitized botryoidal aragonite from the Permian Capitan Reef, Texas. The covariant trend reflects a varying degree of alteration of the original marine isotopic signature. After Given & Lohmann (1985).*

each crystal has undulose extinction, as a result of divergent fast vibration directions (optic axes) in the first case and convergent in the second. Thus to distinguish between these two fibrous calcite types, it is necessary to observe the direction of extinction swing in each crystal; in RFC crystals it is the same as the direction of turning of the microscope stage, whereas in FOFC crystals it is in the opposite direction. In addition, twin planes are curved: concave away from the substrate (FOFC). Subcrystal are often seen in these fibrous crystals. One feature of fibrous calcite is that the crystals are turbid with inclusions (Fig. 7.13); these may be fluid or mineral inclusions.

The origin of the rather complicated fabrics of fibrous calcite cement has been much discussed in the literature, with replacement of an acicular precursor being popular (e.g. Kendall & Tucker, 1973). The lack of modern radiaxial fibrous calcite was an influence in this interpretation. However, it has been argued recently (Kendall, 1985) that most of the features of fibrous calcite are primary, with the characteristic fabric of convergent fast vibration directions in the more common radiaxial fibrous calcite being produced by a process of asymmetric growth as the calcite crystals were undergoing split-growth. Support for a primary origin comes from the discovery of radiaxial calcite in Miocene limestones from Enewetak Atoll in the Pacific (Saller, 1986) and from the Pleistocene of Japan (Sandberg, 1985). In Enewetak, the radiaxial calcite occurs in Miocene strata at a depth of 375–850 m. At these depths, marine dissolution of aragonitic bioclasts has taken place. The fibrous calcite has a variable Mg^{2+} content (1.6 to 11.1 mole% $MgCO_3$), and $\delta^{18}O$ values (−1.8 to +0.4‰) are consistent with precipitation from seawater. Dolomite a little lower in the sequence may also have been precipitated from seawater (see Section 8.7.5). The virtual absence of radiaxial fibrous calcite from Quaternary limestones and its abundance in the past, especially in Palaeozoic reefs and mud mounds, does suggest some change in seawater chemistry (see Section 9.3). From the Enewetak occurrence, it appears that a seawater undersaturated with respect to aragonite, active circulation of seawater during shallow burial, and perhaps fluctuations in degree of calcite supersaturation and crystal growth rate could be the factors controlling radiaxial calcite precipitation (Saller, 1986).

The more acicular variety of fibrous calcite also

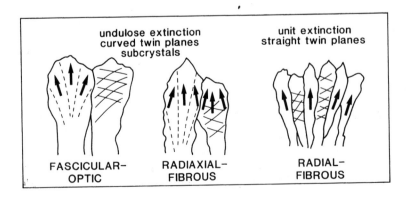

Fig. 7.12 *Fibrous calcite: the fabris of radiaxial fibrous calcite, fascicular-optic fibrous calcite and radial-fibrous calcite. The arrows show the fast vibration directions in each case, and the dashed lines in RFC and FOFC represent subcrystal boundaries. After Kendall (1985).*

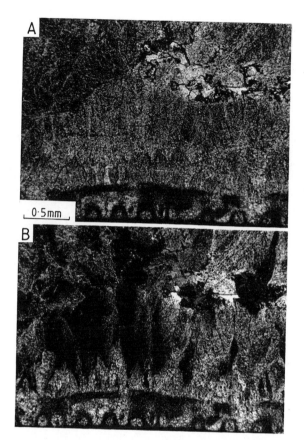

0·5mm

Fig. 7.13 *Radiaxial fibrous calcite upon a large foraminifera. Permian, Capitan Reef, Texas. (A) RFC in plane polarized light showing numerous minute, dark inclusions (perhaps of microdolomite) giving the crystals a cloudy appearance which contrasts with the much clearer calcite spar in the cavity centre. (B) RFC under crossed polars showing the undulose extinction which moves across each crystal in the same direction as the microscope stage is being turned.*

occurs in reef cavities, but it is particularly common in grainstones. It usually forms fringes up to a few hundred microns thick and shows polygonal compromise boundaries between fringes (Figs 7.14 and 7.15). The absence of any dissolution effects or replacement textures in this cement type suggests an original calcite mineralogy.

One important consideration with these marine calcite cements is whether they were originally low-Mg (1–4 mole% $MgCO_3$) or high-Mg (11–19 mole % $MgCO_3$) calcite. Since all magnesian calcites lose Mg^{2+} during diagenesis, there is a problem. However, there are two indicators of an original high Mg content:

(1) microdolomite inclusions, and (2) a magnesium memory. The process whereby magnesian calcite is converted to low-Mg calcite has been referred to as *incongruent dissolution* (Land, 1967; Bathurst, 1975) as it appears that $MgCO_3$ is lost into solution without any disruption of the calcite lattice structure, at least at the microscopic level. However, in some cases, it appears that microscopic dolomite crystals are precipitated within the calcite crystal during this process. These so-called *microdolomites* occur in many fibrous calcite cements (e.g. Lohmann & Meyers, 1977) as well as in originally high-Mg calcite skeletons such as echinoderm grains (Leutloff & Meyers, 1984). Microdolomites are best seen with the SEM (e.g. Fig. 7.15B). The formation of these crystals reflects the degree of openness of the diagenetic system: if there is a high water/rock ratio, then they will be rare or absent. In a more closed system (low water/rock ratio), microdolomites should be abundant.

Even though magnesian calcites lose Mg^{2+} during diagenesis , they may still retain sufficient to indicate an original high level. Microprobe traverses across fibrous calcite fringes commonly reveal higher Mg^{2+} contents (up to several mole % $MgCO_3$) than adjacent grains or later, clear spar (mostly less than 0.5%) (e.g. Marshall & Ashton, 1980; Prezbindowski, 1985). One other suggestion for the identification of

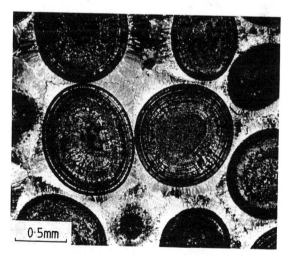

0·5mm

Fig. 7.14 *Fibrous calcite marine cement (primary) around ooids (also primary calcite) with prominent polygonal compromise boundary between fringes. These more acicular crystals are in optical continuity with radial-fibrous crystallites of the ooids. Jurassic Smackover Formation, subsurface Arkansas. Crossed polars.*

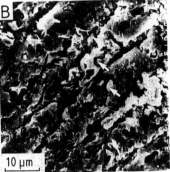

Fig. 7.15 *Fibrous calcite and microdolomite, Upper Permian reef, northeast England. (A) Acicular calcite upon a brachiopod shell and bryozoan (crossed polars) which under the SEM (B) is seen to contain minute crystals of dolomite (arrowed), which could indicate an original high-Mg calcite mineralogy for the fibrous calcite.*

high-Mg calcite is that it may preferentially take up iron, if this is available in the pore-fluids at the time of stabilization (Richter & Fuchtbauer, 1978). Some ferroan fibrous calcites therefore may originally have been magnesian calcites, or ferroan zones within fibrous calcite may originally have been zones of higher Mg^{2+} (the stability of magnesian calcites decreases with increasing Mg^{2+} content).

Although it is now generally accepted that most fibrous calcite cements were primary marine precipitates they may still have been affected by diagenetic processes and neomorphosed to some extent. Evidence of alteration of fibrous calcite is best seen with the luminoscope. Unaltered marine calcite should be non-luminescent, since seawater contains negligible amounts of manganese and this is the activator of luminescence. Where neomorphism has occurred, then bright luminescent patches will be present. They commonly form along intercrystalline boundaries of the original fibrous crystals. Quaternary, marine fibrous magnesian calcite cements from Mururoa Atoll have elongate, intercrystalline cavities and voids along cement layers as a result of meteoric dissolution (Aissaoui, 1988). Some alteration of the fibrous calcite is very likely in the more magnesian varieties, when the $MgCO_3$ is lost and microdolomites form. It is during this stabilization that isotopic exchange may occur.

Fibrous calcites will have had very low Fe and Mn contents originally, since they are marine precipitates and seawater has negligible amounts of these trace elements. Na and Sr would have been of the order of a few thousand and 1200–1500 ppm respectively, like modern marine magnesian calcite. During diagenesis, the fibrous calcites may well have picked up Fe and Mn, if these were available and pore-fluids were re-

ducing, and especially if there was a significant Mg^{2+} content to begin with. Na and Sr contents would be lowered somewhat during diagenesis (because of the below 1 value of the distribution coefficients) to many tens or a few hundred ppm, in the same way that Mg^{2+} is lost.

If there is little neomorphism or burial alteration of fibrous calcite, then the carbon and oxygen isotopic signatures will reflect the seawater isotopic composition, as will values from calcitic fossils and sediments. Such a picture is well illustrated in analyses of Jurassic hardgrounds by Marshall & Ashton (1980) where marine isotopic values are given by whole-rock samples from the hardground surface, and values become more negative down from the surface as the proportions of marine cements decrease and depleted burial cements increase (Fig. 7.16A,B). Cretaceous radiaxial fibrous calcites from the Stuart City Trend, Texas, have $\delta^{18}O$ values averaging $-2.5‰$ and $\delta^{13}C$ values of $+2.0‰$ (Prezbindowski, 1985). These values are compatible with precipitation from seawater of $\delta^{18}O = 0$ to -1 and indicate little neomorphic lateration. Where there has been a variable degree of neomorphism of the fibrous calcite, then the isotopic signatures may well form a covariant linear pattern, with the least altered, most marine values being the more positive ones. The reason for this is that most later diagenetic calcite is more depleted in ^{18}O than marine calcite because of the higher temperature of precipitation in the burial environment and the fractionation effect, and/or the generally more negative $\delta^{18}O$ of meteoric/burial waters relative to seawater (see Fig. 7.16B,C). A covariant $\delta^{18}O-\delta^{13}C$ pattern was obtained by Given & Lohmann (1985) from former magnesian calcite in the Capitan Reef, similar to their trend for the replaced aragonite cements (Fig. 7.11).

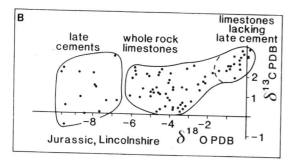

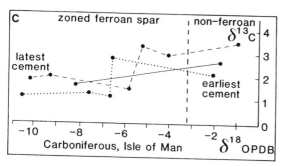

Some fibrous calcites do not show any covariant trend, but are generally more depleted in ^{18}O than would be expected. Such is the case with radiaxial fibrous calcites from the Devonian Golden Spike Reef, Alberta (Walls *et al.*, 1979) with $\delta^{18}O$ values of -5.8 to $-7.7‰$ (Fig. 7.17). Sediments and fossils have similar $\delta^{18}O$ compositions and mid and late burial sparry calcites are even more negative (-9 and $-13‰$). Apparently unaltered Devonian brachiopods from North America (Popp *et al.*, 1986) have most positive $\delta^{18}O$ values around $-3‰$ and this is taken to be a marine value. (Thus Devonian seawater was 2 to 3‰ lighter than modern seawater.) The depletion in ^{18}O of the Golden Spike marine grains and fibrous cements is attributed to isotopic re-equilibration at elevated burial temperatures. The $\delta^{13}C$ values of the Golden Spike marine components are $+1.6$ to $+3.8‰$; these are typical marine values and do not suggest any diagenetic alteration of the carbon isotopes. In fact, it is quite usual to find that original $\delta^{13}C$ signatures are retained during diagenesis, while $\delta^{18}O$ values become lighter. This is because the carbon reservoir of pore-fluids is generally very small compared to the carbon reservoir of the host rock, and there is little fractionation of the carbon isotopes with the increasing temperature that many limestones are subjected to during burial.

Equant sparry calcite is the typical cement of meteoric and burial environments (Sections 7.5 and 7.6.2) but in some Ordovician and Jurassic hardgrounds, this type of cement appears to be the seafloor

Fig. 7.16 *Stable isotopes, marine and burial cementation. (A) Diagram illustrating the changes in cement type and stable isotope signatures from a hardground surface with borings and abundant marine cements down to a grainstone with only burial calcite spar, 20 cm below. Middle Jurassic, Lincolnshire Limestone, England. (B) Isotopic data from the Jurassic Lincolnshire Limestone showing trend to more negative $\delta^{18}O$ values with increasing content of late spar. (A) and (B) after Marshall & Ashton (1980). (C) Isotopic signatures of early to late calcite cement phases in Carboniferous limestone from Isle of Man, UK. Later cements are more depleted in ^{18}O and more ferroan than early cements. After Dickson & Coleman (1980).*

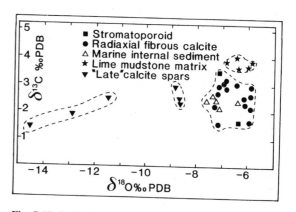

Fig. 7.17 *Stable isotopic data from the Devonian Golden Spike Reef, Alberta, illustrating the differences between sediments, marine cements and fossils and later burial calcite spars. After Walls et al. (1979).*

cement (Wilkinson *et al.*, 1982; Wilkinson *et al.*, 1985). In addition, syntaxial calcite overgrowths on echinoderm debris also occur in these hardgrounds. These cements are closely associated with all the characteristic features of hardgrounds, including truncation by endolithic borings and lithoclast boundaries. It would seem then that coarse sparry calcite has been a marine precipitate, even though at the present time such cement is exceedingly rare in shallow-marine settings. Echinoderm overgrowths do occur in modern deeper-water slope deposits (and in the shallow-water Miocene limestones beneath Enewetak), but they are a very common feature of Palaeozoic and Mesozoic grainstones. Traditionally, they have also been regarded as meteoric or burial cements, but it could be that some of these were also precipitated in the shallow-marine environment. Echinoderm overgrowths are usually compositionally zoned in a complex fashion best observed with cathodoluminescence. In some cases, the early zones are cloudy and inclusion-rich from the presence of microdolomites. This could indicate a magnesian calcite overgrowth initially, which would suggest precipitation from more of a marine pore-fluid. Meyers & Lohmann (1978) have argued for marine and marine–meteoric mixing-zone origins for such microdolomite-rich syntaxial cements. The early diagenetic origin of such cloudy overgrowths can be demonstrated where there was next a phase of compaction before the precipitation of further, now clear, overgrowths (Hird & Tucker, 1988).

Micritic cements in limestones are very difficult to distinguish from lime mud sediment. They were clearly precipitated early in many fine-grained limestones, since evidence of compaction and fossil fracture is rather rare (although see discussion on experimental compaction in Section 7.6.3). Marine precipitation of micrite has clearly taken place in hardgrounds developed in fine-grained sediments, but it is difficult to prove an original aragonitic or magnesian calcite origin for the sediment, let alone the cement (see Section 7.6.4 on diagenesis of micrites and Lasemi & Sandberg, 1984). Micritic cements can be recognized in grainstones where they occur as an isopachous fringe or have a meniscus geometry between grains (see Meyers, 1978 for an example). However, where they are pore filling, then it is difficult to prove a cement origin.

Micritic peloidal cements occur in ancient reefs and have to be distinguished from peloids of faecal or micritized bioclast origin with care. They are abundant in some Triassic reefs (e.g. Reid, 1987), and show many of the features described earlier from Recent examples (Section 7.4.1a). Peloidal cement, coatings and matrices in Upper Jurassic of southeast England have isotopic signatures suggesting a microbial origin (Sun & Wright, 1989).

7.4.7 Marine dissolution

For the most part, modern sedimentary grains of aragonite, high-Mg calcite and low-Mg calcite are quite stable in marine waters. However, the saturation state of seawater decreases with depth (see Fig. 5.1, Section 5.3) and a point is reached where seawater becomes undersaturated with respect to aragonite, a depth of around 300 m in the Pacific and 2000 m in the Atlantic. Magnesian calcites will be susceptible to alteration too at similar depths. Carbonate sediments on continental slopes (periplatform ooze) are subject to a degree of dissolution, depending on the mineralogy. This is well documented from Exuma Sound, Bahamas, by Dix & Mullins (1988). More importantly, shallow-water platform-margin carbonates buried a few hundred metres through continuous sedimentation and subsidence are also candidates for some dissolution. They will be particularly vulnerable to dissolution where seawater undersaturated *re* aragonite is actively pumped into a platform margin by oceanic and tidal currents (see Fig. 7.2). Seawater circulation appears to be quite vigorous within atolls, and aragonite dissolution is taking place at a few hundred metres depth in Miocene limestones beneath Enewetak Atoll in the Pacific (Salter, 1986).

In the geological record, it is likely that at times the depth to aragonite dissolution was shallower than at present in the Pacific, and, indeed, there is evidence for aragonite dissolution on shallow seafloors in the Jurassic and Ordovician (Palmer *et al.*, 1988). These were times when seawater was saturated just with respect to calcite, and not aragonite as well (see Section 9.3). Traditionally, aragonite dissolution has been regarded as largely a near-surface meteoric diagenetic process.

7.5 METEORIC DIAGENESIS

7.5.1 Introduction

The meteoric diagenetic realm is one of critical importance for understanding limestone formation. It is the zone where rainfall-derived groundwater is in contact with sediment or rock. Unlike the burial diagenetic realm, it is relatively accessible at outcrop and through coring. The first investigations of meteorically-altered Pleistocene carbonates showed them to be both cemented by calcite and stabilized into low-Mg calcite (Land *et al.*, 1967; Gavish & Friedman, 1969). These are the two fundamental properties of ancient limestones, and many workers believed that meteoric diagenesis was the main cause of the carbonate sediment to limestone transformation.

However, many limestones in the geological record did not undergo meteoric diagenesis, while others had only the weakest of meteoric overprints. The meteoric diagenetic realm is very important but it is not the only setting where major diagenetic change takes place.

A review of meteoric diagenesis has been provided by James & Choquette (1984), and a summary of subaerial exposure criteria is to be found in Esteban and Klappa (1983). Longman (1980, 1982) has also provided reviews of meteoric diagenesis.

7.5.2 **Meteoric environments**

Within the meteoric diagenetic realm several zones can be recognized (Fig. 7.18). These each have distinctive suites of processes and products (Steinen, 1974; Allan & Matthews, 1982) which enable similar zones to be recognized even in very ancient (Precambrian) limestones (Beeunas & Knauth, 1985). These environments are dynamic and small-scale fluctuations in sea-level can create complex diagenetic stratigraphies.

The major zones are shown in Fig. 7.18. The *water-table* is the surface where atmospheric and hydrostatic pressures are equal. It separates the vadose and phreatic zones. If the aquifer is open to the atmosphere, via the vadose zone, it is said to be unconfined, but if it is contained by impermeable units (aquicludes), it is said to be confined.

When an unconsolidated carbonate sediment, for example an oolite shoal, becomes subaerially exposed, water movement will be through a relatively homogeneous network of pores. This is referred to as '*diffuse flow*' and in such systems the water-table is well defined and relatively regular in extent. However, as lithification occurs, joints and fractures develop and water movement is by *conduit flow* (or *free flow*). Under such circumstances caves develop and larger-scale karstic features evolve. The water-table is then highly irregular in geometry.

In the vadose zone the pores periodically contain water or air or both. Water drains under gravity and

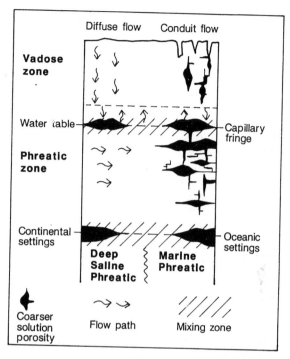

Fig. 7.18 *Groundwater zones. Flow may be through simple pore networks (diffuse flow) or through fractures (conduit flow). Extensive dissolution can occur in mixing zones, as a result of mixing corrosion, in the vadose–phreatic capillary zone, and especially in the lower phreatic zones. Scales are highly variable.*

this can be very rapid in fractured rocks. Two subzones can be identified: the upper zone of infiltration and the lower zone of percolation. The two most important processes are gravitational drainage and desiccation by evaporation and evapotranspiration, which interact to create distinctive cement geometries (Section 7.5.4b). Water passing through this zone contains both atmospheric CO_2 (and other 'acidic' gases), soil-derived CO_2 and organic acids. Not only do these increase the aggressiveness of the fluid, but also add organic carbon to the groundwater.

The water-table is a critical interface in diagenesis. It separates the zone of intermittent saturation and drying (vadose zone) from the phreatic zone with permanent saturation. Transitional between the vadose zone and the phreatic is the capillary rise zone (capillary fringe zone) where water is drawn up by capillary action. There are such marked differences in processes and products above and below the water-table that its position can even be recognized in ancient subaerially exposed sequences (Wright, 1982, 1986b). In diffuse systems its surface is a subdued representation of the topographic relief and is generally very flat in modern carbonate platform and shelf settings.

The phreatic zone is the zone in which all the pore space is filled by fluid. In continental settings this zone passes down into deeper, usually more saline fluids. Most subaerially exposed carbonate sediments will be relatively close to marine waters, either as isolated islands (cays), reefs, platforms or ramps and shelves. In such cases marine waters may underlie or border the phreatic-waters (phreatic lens). On low relief, subaerially exposed platform areas, such as Andros Island, the meteoric phreatic lens may be very thin or brackish. If the hydrostatic head is sufficient, the meteoric phreatic–marine phreatic interface may be displaced seawards so that freshwater may emerge at the seafloor kilometres or even tens of kilometres offshore (Johnson, 1983; Chafetz *et al.*, 1988).

One special case is where the phreatic lens occurs on an isolated cay or island (Fig. 7.19). The meteoric water floats on the denser, marine phreatic and, in unconfined reservoirs, the phreatic lens extends below sea-level approximately 40 times the height of the water-table above sea-level. This is known as the Ghyben–Herzberg relationship (Davis & DeWiest, 1966; Todd, 1980). This is a very significant 'effect' by which even small drops in sea-level, causing subaerial exposure, can lead to the development of relatively deep meteoric lenses. The relationship assumes no mixing between the meteoric and marine waters, which does, in fact, occur. However, the relationship does apply to lenses bounded by the 50% isochlor surface (Vacher, 1978), but in strongly heterogeneous aquifers, such as reefs, the lens has a very complex geometry (Buddemeier & Oberdorfer, 1986). The rate of fluid movement (flux) is a critical consideration regarding all diagenetic environments, and is a major control on the rate of meteoric diagenesis. Vertical and lateral flow will be greatest in the upper phreatic lens and is much lower in the deeper parts where stagnation can occur. In exposed carbonate provinces, a variety of processes, often marine related, can cause movement of the phreatic zone, e.g. wave and tidal action, or refluxing of dense platform brines or Kohout convection (Section 8.7.5).

Between the meteoric phreatic and the underlying marine waters is a zone of diffusion and mixing known as the transition or mixing zone (Back *et al.*, 1979, 1984). This is an important site for both dolomitization (Section 8.7.3) and dissolution (Section

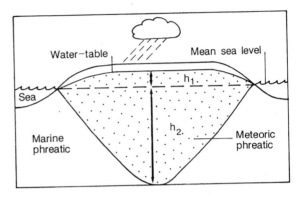

Fig. 7.19 *Ghyben–Herzberg model. A freshwater lens 'floats' on seawater on a homogeneous porous island aquifer. The depth of the freshwater lens below mean sea-level* (h_2) *is approximately 40 times the height of the water-table above mean sea-level* (h_1). $h_2 = h_1 (d_{fw}/d_{sw} - 1)$, *where* d_{fw} *and* d_{sw} *are the densities of freshwater and seawater respectively. This model assumes no mixing of the freshwater and seawater.*

7.5.3a). Its thickness ranges from a few metres to 10 m, as on Andros Island (Smart *et al.*, 1988), but its thickness varies with changes in permeability and flow rate. It typically thickens at its seaward margin because of the higher flux of discharging meteoric water.

7.5.3 Meteoric processes

Three main processes operate during meteoric diagenesis: dissolution, precipitation (cementation) and mineralogical transformation.

7.5.3a Dissolution

Undersaturated meteoric water will dissolve carbonate sediments and limestones. The meteoric waters are acidified by atmospheric and soil CO_2, as well as by soil acids. Aragonite, being more soluble than calcite, will undergo more rapid dissolution. The solution chemistries are complex and are discussed by Thrailkill (1968), Plummer *et al.* (1979), Bogli (1980), Drever (1982), Trudgill (1985), and Lohmann (1988). Mixing zones are sites of major dissolution caused by 'mixing corrosion', the chemistry of which has been succinctly reviewed by James & Choquette (1984). Basically this occurs when two solutions, saturated with respect to a given mineral phase, are mixed and undersaturation or supersaturation can occur (Figs 7.20 and 7.21). This will result when waters with different P_{CO_2}, temperatures, salinities, calcite saturations and pHs mix (Bogli, 1964; Runnels, 1969). Such mixing corrosion may take place when vadose and phreatic waters mix, or at the base of the phreatic lens with more saline or marine phreatic waters. Recently, Smart *et al.* (1988) have suggested that microbial oxidation of organic matter is an important process in mixing zones, producing CO_2.

Mixing-zone waters can be especially aggressive and quite spectacular degrees of corrosion can take place, often resulting in highly fretted 'Swiss-cheese' style of dissolution (Back *et al.*, 1986; Smart *et al.*, 1988). Extensive cave development also takes place in this zone (Vernon, 1969; Smart *et al.*, 1988) and subsequent collapse can even control coastline shapes (Hanshaw & Back, 1980; Back *et al.*, 1984). The mixing zone has the potential for extensive dolomitization and dissolution, and highly porous dolomites of this type are known (Ward & Halley, 1985).

7.5.3b Cementation

The meteoric diagenetic realm is not only the site of extensive dissolution, but is also one of cementation. Aragonite and high-Mg calcite are more soluble than low-Mg calcite in meteoric waters. Their dissolution leads to supersaturation with respect to calcium carbonate and, in the meteoric waters with a low Mg/Ca ratio, low-Mg calcite precipitates. By this means carbonate sediments are able to undergo dissolution while they are being cemented. The mechanisms of precipitation in the phreatic zone are poorly under-

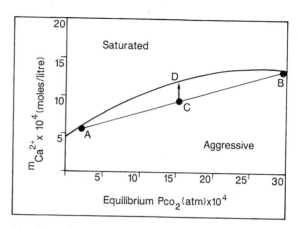

Fig. 7.20 *Mixing corrosion effect. The solid curve shows the solubility of $CaCO_3$ with respect to the total CO_2 in solution. If two liquids, A and B are mixed, an undersaturated solution results, C. It evolves by dissolution of calcite to equilibrium at D. From various sources.*

stood, but in the vadose zone degassing of CO_2 (including CO_2 removal by plants), evaporation and evapotranspiration are the main causes.

7.5.3c Processes of transformation

During meteoric diagenesis both aragonite and high-Mg calcite are 'replaced' by low-Mg calcite (Fig.

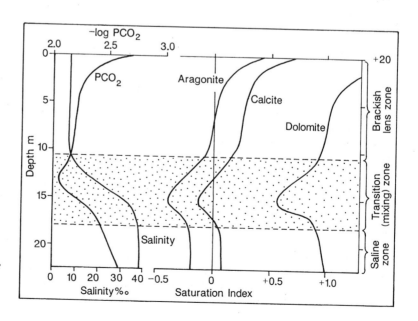

Fig. 7.21 *Vertical variation of salinity, P_{CO_2}, and the saturation indices of aragonite, calcite and dolomite through the meteoric– marine mixing zones within a 'blue hole', Andros Island. After Smart et al. (1988).*

7.22). As a result of these mineralogical changes, geochemical changes also occur (Fig. 7.23; Gavish & Friedman, 1969) involving losses of cations present in the original marine carbonates and gains of other trace or minor elements from the meteoric 'fluid' (Section 6.4). Strontium (from aragonite) and magnesium (from high-Mg calcite) are depleted (Benson, 1974), while Mn and Fe (typically present in greater amounts in groundwater than seawater) may be incorporated into the diagenetic low-Mg calcite. Changes in the isotopic composition of the carbonate also occur (Fig. 7.23; Section 7.5.4c) Physical transformations also occur. Aragonite is either dissolved out or is replaced via a very fine-scale process to leave relic fabrics (Sections 1.3 and 7.1). The replacement of high-Mg calcite is less clearly understood. At the light-microscope level of examination, replaced magnesian calcite grains appear little altered, but under the electron microscope considerable alteration of the original ultrastructure is usually seen to have taken place. The process operates by microscale dissolution and precipitation (Manze & Richter, 1979). The solubility of magnesian calcites depends on the Mg content and on the microstructure of the particle (Walter, 1985). The transformation appears to take place in stages (Richter, 1979), probably with a little loss of original structure occurring at each stage. Hence calcites with a very high Mg content will probably undergo more extensive structural change than those with a lower Mg content.

7.5.3d Controls on meteoric processes

While the style of meteoric diagenesis that a carbonate deposit undergoes will in part be controlled by whether the material is unconsolidated (diffuse flow) or lithified, bedded and fractured (conduit flow), the main controls are mineralogy and climate. A carbonate body with a high content of aragonite (and high-Mg calcite) will have a much greater diagenetic potential (Section 9.5) than one composed of more stable low-Mg calcite. Geochemical changes will be more marked as the former stabilizes to calcite. Thus the effects of meteoric diagenesis will be more marked during periods when aragonite was the main inorganic precipitate in the seas, and when aragonitic organisms predominated in shallow-marine environments (Sections 9.3–9.5). This does not imply that during calcitic phases meteoric diagenesis is not significant, but rather that its effects may be more 'muted'.

Climate is a critical control, for not only does it influence the availability of meteoric water, and its

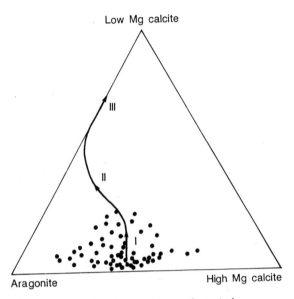

Fig. 7.22 *Triangular diagram showing the typical mineralogical evolution of Quaternary marine sediments during prolonged meteoric diagenesis. Stippled area represents range of marine sediments. I, initial cementation by low-Mg calcite. II, loss of Mg from high-Mg calcites. III, dissolution of aragonites and cementation by low-Mg calcite. Modified from Tucker (1981).*

flux, but also temperature, and related factors such as soil cover and vegetation. Under more arid conditions, low rainfall will result in relatively slow rates of meteoric alteration and little water will infiltrate into the aquifer directly through the vadose zone. Carbonates will accumulate slowly in the upper part of the vadose zone and the water-table will be deep. Low rates of groundwater movement result in little phreatic cementation. In more humid conditions net dissolution may occur, with little carbonate accumulating in the soil or vadose zone, but also, as a consequence of high rates of dissolution and a high flux, extensive cementation may take place. These are two extremes and most subtropical carbonates undergo meteoric diagenesis in an intermediate setting. The importance of climate is well illustrated by comparing subaerially-exposed carbonates during different Quaternary climatic phases. Ward (1978b) and McKee & Ward (1983) have detailed such differences from the Pleistocene and Holocene aeolianites of northeast Yucatan, Mexico. In this area the Upper Pleistocene carbonate aeolianites were subaerially exposed during what Ward (1978b) interpreted as an

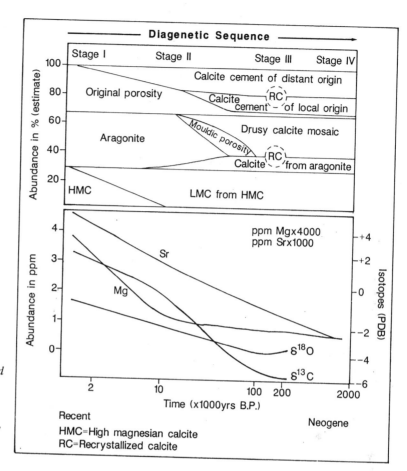

Fig. 7.23 *Textural, mineralogical and geochemical changes with age in Quaternary carbonates from the Mediterranean coast of Israel (modified from Gavish & Friedman, 1969). Note the early disappearance of high-Mg calcites.*

arid phase. The unconsolidated carbonate sands underwent little cementation and exhibit features such as finely-crystalline cements, rhizocretions, needle-fibre calcite and calcrete crusts. In the younger Holocene carbonate aeolianites, coarse calcite cements are common, including even some of the youngest deposits locally being totally cemented. However, rhizocretions, needle-fibre calcite and calcrete crusts are absent. The present-day climate in the area is humid. During the Late Pleistocene the more arid climate resulted in a lower flux of meteoric waters, with less dissolution and less cementation. Similar climatic controls on diagenetic style and porosity evolution have been documented by Calvet (1979) and Calvet *et al.* (1980) from the Pleistocene of Mallorca, Spain, and by Harrison (1975) from Barbados. Similar climatic controls have been invoked for different styles of diagenesis in Early Carboniferous oolites and peritidal limestones in south Wales (Fig. 7.24; Hird

& Tucker, 1988; Wright, 1988). Within this thick ramp sequence (Section 2.6.3; Figs 2.20 and 4.24) there are a number of oolitic sand bodies developed during progradational and regressive phases and capped by subaerial exposure surfaces (Wright, 1986a). Some oolites, for example the Brofiscin and Gilwern Oolites (Figs 7.24 and 9.13), underwent extensive early meteoric cementation and the top of the Gilwern Oolite exhibits a well-developed 'humid'-type karst horizon (Wright, 1982). However, the Gully Oolite and the peritidal Llanelly Formation (which directly overlies the Gilwern Oolite) exhibit very minor pre-burial cementation but do possess minor karstic surfaces with calcrete crusts, rhizocretions, needle-fibre calcite and, in the case of the Llanelly Formation, well-developed semi-arid paleosols and minor evaporites. This lack of early cementation later resulted in extensive mechanical and chemical compaction of both these units with burial (see Fig. 9.13).

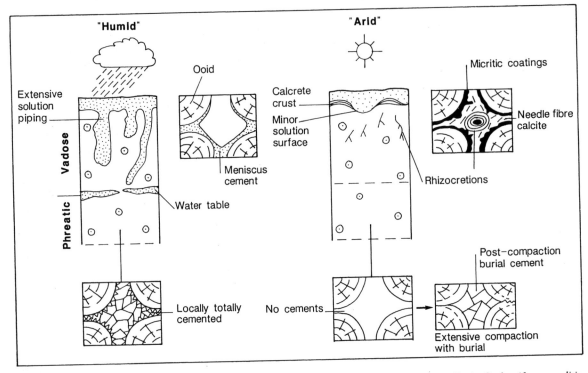

Fig. 7.24 *Contrasting styles of karstification, meteoric cementation and porosity evolution in Early Carboniferous oolitic limestones from south Wales. Based on Hird & Tucker (1988) and Wright (1988).*

7.5.3e Rates of meteoric diagenesis

A number of studies on the rates of meteoric diagenetic alteration have been carried out (Land *et al.*, 1967; Gavish & Friedman, 1969; Calvet & Esteban, 1977; Halley & Harris, 1979; Reeckman & Gill, 1981). The changes take place in a relatively ordered way (Fig. 7.23) but the rates vary depending primarily on climate (Harrison, 1975) and local hydrological conditions. It would be wrong to offer any average figures when so much variability exists between different areas. For example, complete mineral stabilization may have occurred in under 100 000 years on the Israel coast (Gavish & Friedman, 1969; Buchbinder & Friedman, 1980). In South Australia, the same effect requires exposure of 400 000–700 000 years (Reeckman & Gill, 1981). On Barbados, in arid areas, aragonite and high-Mg calcite persist in limestones up to 300 000 years old, while in more humid areas aragonite is absent in reef limestones older than 200 000 years old (Harrison, 1975). Halley & Harris (1979), from their study of Joulter's Cay in the Bahamas which has been exposed for 1000 years,

estimated complete stabilization in only several tens of thousands of years.

7.5.4 Products of meteoric diagenesis

The products of meteoric diagenesis can be grouped into dissolution features, cementation (including calcretes) and features related to mineralogical stabilization such as petrographic and geochemical changes.

7.5.4a Dissolution features

Dissolution on a microscale creates porosity but when significant dissolution occurs the term karst is applied. Karstic features have been extensively documented (e.g. Bogli, 1980; Trudgill, 1985), and the recognition of palaeokarst in ancient carbonate sequences has become commonplace, and often has economic implications. Many aspects of this topic are discussed in the recent volume edited by James & Choquette (1988).

In simple terms two sorts of karst can be defined, referred to as surface and subsurface karst. Surface karst develops at the air–limestone, or soil–limestone interface. A remarkable variety of forms of small-scale features develop, known as karren (Bogli, 1980; Allen, 1982; James & Choquette 1984). Similar forms have been recognized in ancient limestones (Walkden, 1974; Read & Grover, 1977; Wright, 1982; Kobluk, 1984; Desrochers & James, 1988), but recognizing such surfaces can be difficult. Some dissolution surfaces develop in the *subsurface* at lithological discontinuities, referred to as interstratal karst, and both irregular submarine hardgrounds and stylolite surfaces can be mistaken for palaeokarstic surfaces. Read & Grover (1977) and Wright (1982) have discussed criteria for differentiating surface palaeokarst from other similar features.

Subsurface karst is represented by fissures and caves, and Ford (1988) has recently provided a review of cave formations. Palaeokarst equivalents have also been documented but perhaps one of the most striking and common examples are solution-affected neptunian dykes. These are sediment bodies of younger material filling fissures in limestones. 'Neptunian' refers to the fact that they contain marine-derived sediments. They have been documented widely from ancient carbonate platform sequences. Modern examples of such huge fracture systems have been documented from the platform margins of the Bahama Banks (Smart *et al.*, 1988) and many such ancient and modern fissure systems have been affected by karstic processes. Vera *et al.* (1988) have provided a carefully documented example of such fractures from a Jurassic platform margin of the Subbetic Zone of Spain. These karsted fracture systems not only represent major porosity zones but also influence the hydrology of platform margins (Smart *et al.*, 1988).

One common feature associated with karstic surfaces is red-coloured argillaceous material, commonly called *terra rossa* (Esteban & Klappa, 1983). The origins of such material have been the subject of much debate, and while some is residual in origin, much is aeolian derived. The use of the term has been discussed by Wright & Wilson (1987).

Palaeokarstic intervals are of major significance for hydrocarbon exploration and recently a number of studies have been published showing the diversity of palaeokarst reservoir styles. Of special note are those by Craig (1988), Kerans (1988) and Wigley *et al.* (1988). The three palaeokarsts contrast strikingly and serve to illustrate major aspects of 'palaeokarst facies'. Craig (1988), in an elegant study of the Permian Yates Field of west Texas, provided an example of a relatively short-lived, small-scale palaeokarst developed around a series of low-relief

shelf-edge islands. In another west Texas reservoir, although of Ordovician age, Kerans (1988) described areally extensive and thick breccias formed by large-scale cave collapse within the Ellenburger Group. In this case a very uniform dissolution zone developed, at shallow depth, within a large subaerially exposed. Early Ordovician platform. Wigley *et al.* (1988) described a more complex reservoir system from the Amposta Marino Field of offshore northeast Spain. Here, a mature palaeokarst, developed in Cretaceous limestones, was overprinted by a Miocene marine–meteoric mixing zone, enhancing the connectedness of the reservoir interval. The 'Yates' study illustrates the integration of subsurface data and simple hydrogeological models to understand dissolution porosity distribution. The 'Ellenburger' study reveals a palaeokarstic origin for thick fractured and brecciated zones and its implications for re-evaluating other fracture reservoirs are clear. The 'Amposta Marino' example shows how, during the evolution of any palaeokarst, multiple events can complicate the geometry of karstic intervals, for most major palaeokarsts are polyphase.

One peculiar product of undersaturated meteoric vadose waters is dissolution at grain to grain contacts (Knox, 1977). This process, referred to as 'vadose compaction', has been invoked by a number of authors including Dunham (1969b), Becher & Moore (1976), Clark (1979, 1980), Sellwood *et al.* (1985) and Hird & Tucker (1988) to explain overcompacted limestones. Sarkar *et al.* (1982) have also illustrated similar features from Precambrian oolites from India, but assumed they resulted from simple mechanical plastic deformation. In many cases the early compaction is seen in leached grainstones where micrite envelopes are deformed. The author has noted the same phenomenon in leached molluscan grainstones in aeolianites in eastern central Oman. In such cases it is possible that the organic-rich micrite envelopes deform plastically at shallow depth. Plastic deformation does occur in undersaturated solutions even at shallow depths (Meyers 1980). Whatever its origin, overcompaction in vadose-altered limestones, related to dissolution, appears to be common and may provide a criterion for recognizing meteoric alteration zones.

7.5.4b Carbonate cementation

Cementation is a common feature in both the vadose and phreatic zones but the style differs markedly between them.

One distinctive type of meteoric cementation consists of zones, in some cases many metres thick, of fine-grained secondary carbonate, termed *calcrete* or *caliche* (Goudie, 1983). Typically, these are micritic or microspar grade cements which fill pore spaces but can also be replacive and displacive in origin. Calcretes occur both in the vadose zone, usually as distinctive soil horizons, or within, or just below, the capillary rise zone (Seminiuk & Meagher, 1981; Arakel & McConchie, 1982; Carlisle, 1983). In the latter situation they are termed phreatic or groundwater calcretes.

Some workers have restricted the term to prominent carbonate horizons developed only in soils, where they constitute calcic or petrocalcic horizons, the latter term being used if complete induration has taken place. However, it is very difficult to differentiate between pedogenic, capillary fringe and phreatic forms and the term calcrete should be used for the whole family of near-surface secondary carbonate accumulations, and prefixes such as pedogenic, vadose or phreatic should be applied when appropriate.

A considerable literature has developed on the morphologies and microstructures of calcretes, both in continental and marginal marine settings. The recognition of such horizons in ancient sequences provides evidence of subaerial exposure, and criteria, especially petrographic, for recognizing them are well documented. Esteban & Klappa (1983) provided an illustrated guide to recognizing calcretes. Regardless of whether such calcretes occur in siliciclastic or marginal marine limestones, a simple genetic classification can be made based on their microstructure (Fig. 7.25). Two end-member types are recognized (Wright, in press). Alpha calcretes exhibit little or no evidence of biological activity and consist of dense crystalline carbonate (micrite or microspar), which commonly has a rhombic habit. The crystal mosaics are commonly irregular, resembling neomorphic microspars, but detailed examination under cathodoluminescence often reveals that the irregulary shaped crystals show multiple phases of precipitation and dissolution, creating the irregular crystal outlines. Multiple zoning is a surprisingly common feature in many calcrete

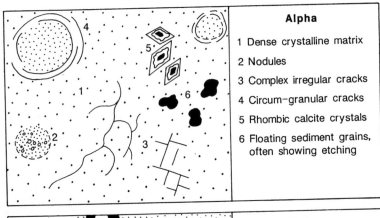

Alpha

1 Dense crystalline matrix

2 Nodules

3 Complex irregular cracks

4 Circum−granular cracks

5 Rhombic calcite crystals

6 Floating sediment grains, often showing etching

Beta

1 Microbial micritic coatings

2 Needle fibre calcite

3 Calcified microbial tubules

4 Rhizocretions

5 Alveolar septal fabric

6 Calcified faecal pellets

Fig. 7.25 *'End-member' types of calcrete fabric. In beta calcretes a wide variety of rhizocretions can form, with laminated or micritic walls. Alveolar septal fabric consists of narrow curved septa of parallel-oriented needles of calcite. The septa are the sites of mycelial bundles in which the needles have precipitated. They occur in open pore spaces between grains or within rhizocretions.*

Fig. 7.26 *Development of pedogenic calcrete profiles with time. On non-calcareous substrates (A) calcrete develops in a typical sequence from nodular to massive to brecciated. By stage 4 the horizon (petrocalcic horizon) is usually impermeable and subsequent precipitation occurs at the top of the horizon from ponded waters to give a laminar horizon. With time these horizons become brecciated and karsted. Multiple coated pisoids will form if a slope is present and movement downslope takes place. In exposed carbonate sediments (B) the sequence is similar but nodules do not commonly develop. On carbonate sands or other porous carbonates, much of the accumulation of secondary carbonate takes place by cementation and not by displacive crystallization and nodule growth. The light-coloured carbonates are*

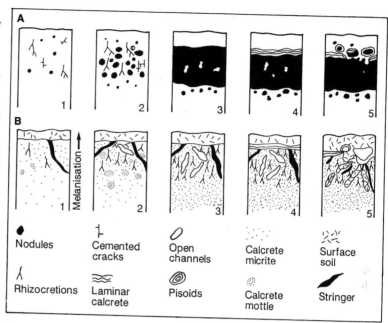

initially darkened (melanized) by organic staining. The cementation is prevasive but is locally patchy. Rhizocretions are especially abundant and dark micritic stringers probably result from larger roots. Complete induration can occur (stage 4) followed by laminar crust growth. Typically, brecciation is assoicated with the introduction of fine siliciclastic material (terra rossa). (A) is based on classification of Machette (1985) which recognizes two divisions corresponding to 4 above, with thin incipient laminar crust (stage 4 of Machette) and a thick laminar crust (stage 5). (B) is based partly on Arakel (1982). In both cases the laminar calcretes may represent calcified root mats (Wright et al., 1988).

calcites and must reflect phases of suboxic conditions during calcite precipitation, perhaps microbially induced.

Alpha fabrics also exhibit a variety of features related to replacive and displacive growth, such as embayed grains and floating-grain textures. Displacive growth and desiccation also result in numerous fine fractures in the matrix, commonly partially or wholly filled by more coarsely crystalline calcite. The fine crystal size of alpha calcretes is generally believed to be a function of rapid precipitation and small pore size. Alpha fabrics occur in all forms of calcrete.

Beta calcretes (Fig. 7.25) show high degrees of biological activity, especially by fungi. Rhizocretions (root encrustations, moulds or petrifications) are common as are needle-fibre calcites (fungally-precipitated needles of low-Mg calcite: Wright, 1986c; Phillips & Self, 1987). Beta calcretes are most common on carbonate substrates within the soil or deeper, related to phreatophytes. The controls on the distribution of

these two types are unclear but climate–vegetation factors and the composition of the parent material appear to be most important.

Calcretes are typically composed of low-Mg calcite but dolomitic forms (dolocretes) are surprisingly common. Aragonite and high-Mg calcite are relatively uncommon in calcretes. The mechanisms for precipitation are very poorly understood, but the loss of H_2O by evaporation and evapotranspiration, and CO_2 loss (degassing) into the atmosphere, appear to be the main causes.

There are three main sources for the carbonate. Material can be added to the profiles from dust, dissolved and reprecipitated lower down at zones of high moisture tension. Other carbonate can be sourced from groundwater, but in limestones much is supplied by the local dissolution of the substrate.

Calcretes develop in a series of stages, with variations dependent on the nature of the substrate (Fig. 7.26). Complete induration and plugging of the horizons can lead to the development of accretionary laminae or to root mat calcification on the upper surface (in pedogenic calcretes). Brecciation, usually caused by root activity, may follow, breaking up the laminar, and underlying horizons. Coated grains develop on slopes where soil creep moves the fragments (Fig. 1.5). Detailed accounts of profile forms have been given by Netterberg (1980) and Arakel (1982). The time required to form such sequences has only really been studied in detail in the, mainly, alpha calcretes of the southwest USA (Machette, 1985) and little work has been done on the lengths of time needed to form beta calcretes. However, any rate estimates would depend on many local variables such as climate and vegetation, and could only be used in a relative sense as a guide to exposure time in ancient sequences.

In the vadose zone, as a consequence of the percolation of meteoric waters, some pores will be alternately filled with water and air. Desiccation results in the localized precipitation of calcite in distinctive geometries (Fig. 7.27). In general, vadose cements are patchily distributed, being more prevalent in finer-grained layers. The cements are typically finely crystalline, a consequence of the rate of precipitation, and crystal terminations are poorly developed. Syntaxial overgrowths also develop and many crystals may exhibit 'bioerosion' caused by various micro-organisms (Jones, 1987). Water is held near grain contacts by capillary forces, and as a result cements form in these zones with similar forms (meniscus cements) (Figs 7.27 and 7.28). However, their preservation potential is low for they become overprinted by later cements, but cathodoluminescence can often reveal their presence. As a result of the preferential cementation of pore throats, the pore itself becomes more rounded in outline (Figs 7.27 and 7.28). Water will collect on the undersides of grains (Fig. 7.27) as pendant droplets. After multiple phases of drainage and precipitation, a visible cement will develop with a similar pendant (gravitational or microstalactitic) form.

Calcite crystals, commonly rhombic, formed in the upper part of the vadose zone can be transported in suspension, and deposited as *crystal silts*, which may exhibit grading or ever. cross-lamination.

Many of the fabric types can also form in vadose marine settings but the precipitates are typically high-Mg calcite or aragonite (Meyers, 1987). Further details of vadose cementation styles have been discussed by many authors (Ward, 1975, 1978a; Badiozamani *et al.*, 1977; Halley & Harris, 1979; Binkley *et al.*, 1980; Longman, 1980).

In the phreatic zone the pores are permanently filled with water. The cementation style is more even with crystal growth being more continuous and slow, resulting in larger crystal sizes than in the vadose zone. The phreatic lens can extend to considerable depths, for example, to 600 m depth below the Florida Peninsula (Schmoker & Halley, 1982).

Distinguishing meteoric phreatic cements from burial cements is difficult, if not impossible without 'thermometry' from fluid inclusions or oxygen isotopes (Sections 7.6.2a,b). Phreatic cements are typically clear, blocky calcites, low in iron. Many authors have referred to the cathodoluminescence behaviour of meteoric calcites (Section 7.6.2c; Choquette & James, 1988; Mussman *et al.*, 1988; Neimann & Read, 1988). Such cements are typically non-luminescent with minor bright zones, reflecting the apparently oxidized nature of the shallow meteoric waters. Down dip, reducing conditions develop resulting in calcites capable of luminescence. These petrographic changes reflect the well-documented chemical gradients in carbonate aquifers (Edmunds & Walton, 1983) which are low in nitrates (for suboxic reactions) and have limited Mn and Fe, and rapidly pass from oxic to anoxic conditions. The oxic phreatic zone typically corresponds to the unconfined aquifer, with free gas exchange.

The palaeohydrogeology of ancient aquifers has been reconstructed using both petrography, including cathodoluminescence, and geochemistry (Meyers, 1974, 1978; Grover & Read, 1983; Dorobek, 1987; Mussman *et al.*, 1988; Niemann & Read, 1988). Such studies have revealed the enormous areal extent of ancient meteoric phreatic aquifers (Meyers, 1974, 1978) and their depths (Mussman *et al.*, 1988). However, the remarkable extent and geochemical uniformity of some of these ancient aquifers is very surprising and it seems likely that more complex systems were the norm, especially where regional groundwaters were sourced from varied, non-carbonate terrains.

7.5.4c Mineralogical stabilization and geochemical changes

During meteoric diagenesis the conversion of high-Mg calcite and aragonite to low-Mg calcite, and the addition of low-Mg calcite cement, results in significant chemical and textural changes.

The types of textural change during mineral stabilization have already been discussed for aragonite (Section 1.3) and high-Mg calcite (Section 7.5.3c).

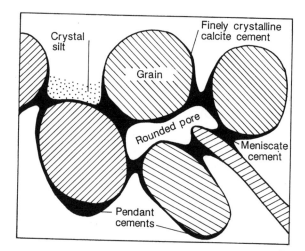

Fig. 7.27 *Common cement geometries in vadose zones (see text).*

However, it has been suggested that the nature of the aragonite transformation during meteoric diagenesis varies depending on whether replacement takes place in the vadose or phreatic zone. Pingitore (1976) has shown that corals altered in the vadose zone exhibit fabric-selective calcite mosaics with ghost-fabrics resembling calcitized bivalves (Section 1.3). The delicate replacement mechanism was presumably via a very thin film of fluid. However, corals altered in the phreatic zone exhibit poor preservation with coarser, cross-cutting calcite mosaics, whereby replacement takes place via a chalky zone several millimetres wide. Calcitized aragonitic grains are not a common feature of vadose zones and they should not be used as a criterion for recognizing that zone.

As a consequence of the mineralogical changes during meteoric diagenesis, chemical changes also occur. Mg and Sr are lost from high-Mg calcite and aragonite respectively and other cations substitute into low-Mg calcite as a function of calcite structure and cation availability.

During mineralogical changes and cementation the stable isotopic compositions of the unstable grains change from a marine to a meteoric signature. These changes have been documented in several studies of Quaternary limestones (Gross, 1964; Allan & Matthews, 1977, 1982; Humphrey *et al.*, 1985; Beier, 1987). Typical trends are shown in Fig. 7.29. The main features are a more negative $\delta^{13}C$ caused by the addition of light carbon (^{12}C) from the soil, and a more negative $\delta^{18}O$ due to the addition of lighter, meteoric-derived ^{16}O. In addition $\delta^{13}C$ values should be lightest in the upper part of the profile where more organic 'soil' carbon (enriched in ^{12}C) is available. It has been suggested that $\delta^{18}O$ values become heavier in the vadose zone, reflecting the loss of ^{16}O by evaporation (James & Choquette, 1984). However, as pointed out by Cerling (1984), from a study of pedogenic carbonates, evaporation is probably of little

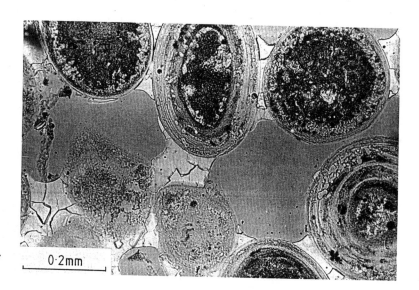

Fig. 7.28 *Meniscus cement, vadose zone, Joulter's Cay. Bahamas.*

0·2mm

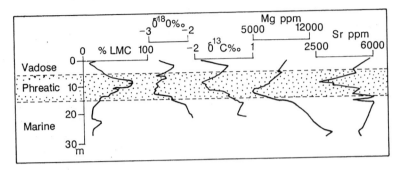

Fig. 7.29 *Mineralogical and chemical profile from a borehole in Barbados, West Indies. Much of the aragonite and high-Mg calcite has been converted to low-Mg calcite (LMC), hence the lower Mg and Sr values. Modified from Humphrey* et al. *(1985).*

importance in precipitation in the soil zone compared to evapotranspiration, which is not believed to result in fractionation.

A number of authors have used these isotope trends to detect palaeoexposure surfaces (Videtich & Matthews, 1980; Allan & Matthews, 1982; Wagner & Matthews, 1982; Beeunas & Knauth, 1985). The use of this technique must be seriously reviewed in the light of the points raised in the next section.

7.5.5 Summary

Most carbonate sediments, by virtue of having been deposited in very shallow waters, are prone to being affected by subaerial diagenesis as a consequence of even relatively small-scale changes in sea-level. It is therefore hardly surprising that meteoric diagenetic products are so commonly encountered in ancient limestones. There have been many studies where simple diagenetic sequences have been recognized, reflecting the evolution of the sediment from marine to meteoric, to burial settings. However, such simple diagenetic sequences may be deceptive and one has to ask: are diagenetic features in limestones as unrepresentative of 'reality' as are, for example, fossil assemblages of their original communicaties? A bench-mark study, shedding light on this question, is that of Matthews & Frohlich (1987). They generated, by computer, a model of platform or shelf-margin diagenesis based on the impact of high-frequency glacio-eustatic sea-level fluctuations. The sequences generated are such that it would be practically impossible to unravel the actual events from the limestones using traditional techniques, a consequence of the complex diagenetic patterns produced. However, simple diagenetic sequences do occur in the rock record. This could reflect simpler sea-level histories or a preservation or observation bias on the data. Such

studies as Matthews & Frohlich's should not deter attempts to unravel diagenetic sequences, but offer a new and important perspective on the real meaning of such sequences in ancient limestones. Marked variability in the successions and distributions of near-surface diagenetic phases should be the norm as studies of some Cenozoic limestones have shown (Jones *et al.*, 1984).

Finally it needs to be asked: how important is meteoric diagenesis in limestone formation? Even a cursory examination of long-exposed Pleistocene limestones will reveal the presence of the two features which distinguish limestones from carbonate sediments: mineral stabilization and cementation. No one would doubt the 'ability' of the meteoric diagenetic realm to stabilize sediments, but while some workers (Friedman, 1975; Longman, 1980) have stressed that meteoric diagenesis can also bring about major cementation, others have questioned this view (Halley & Schmoker, 1983). Several studies on meteorically-stabilized limestones have shown that little porosity loss occurs during meteoric diagenesis (Halley & Beach, 1979; Halley & Schmoker, 1983). From studies of the Pleistocene Miami Limestone of the Florida Keys, Schmoker & Hester (1986) have suggested that after complete mineral stabilization, the porosity loss would be only 15–20%. To paraphrase Schmoker & Hester: 'Meteoric diagenesis may set the stage for subsequent porosity evolution, but the most *significant* porosity loss still occurs during burial diagenesis'.

7.6 DIAGENESIS IN THE BURIAL ENVIRONMENT

7.6.1 Introduction

Of the major environments of carbonate diagenesis, that in the burial realm is the least understood, but it is the one through which most ancient limestones have passed, and many have spent tens to hundreds of millions of years there. Burial processes, particularly cementation, compaction and pressure dissolution, operate over a considerable range of depth, pressure and temperature and in pore-fluids of varied salinity, chemistry and origin. Burial diagenesis is generally taken to begin below the depth where sediments are affected by near-surface processes of the marine and meteoric environments. The effects are progressive, subtle at first, obvious later, so that it is usually not possible to pinpoint the onset of burial diagenetic effects and in ancient limestones it is often very difficult to relate the products of burial diagenesis to precise depths. Shallow-marine carbonates enter the burial diagenetic realm from one of two standpoints: they may be buried directly from the marine environment, with marine pore-fluids, or they may be affected by meteoric diagenetic processes first, so that sediment mineralogy is stabilized and pore-fluids are fresh or mixed marine–meteoric. The early diagenetic history of a carbonate sediment has important considerations during later, burial diagenesis.

The composition of subsurface (or formation) waters in many sedimentary basins is now well documented, but there is still much discussion over the origin of the fluids (e.g. Collins, 1975; Hanor, 1983; Land, 1987). Subsurface waters are generally brines, with salinities reaching several 100‰ and in many basins there is an increase in salinity with depth (Fig. 7.30A). Various mechanisms have been put forward to account for the great increase in salinity relative to seawater or meteoric water, including dissolution of evaporite horizons (but these are absent in some sedimentary basins), infiltration of brines formed by surface evaporation, and membrane filtration or reverse osmosis, whereby a residual brine is produced by preferentially forcing water molecules upward through semipermeable shale horizons. The chemical composition of subsurface brines is also rather different from surface waters. Expressing the composition relative to seawater (see Fig. 7.30B for the Illinois Basin), it is usually found that subsurface brines are strongly depleted in K^+, Mg^{2+}, and SO_4^{2-}, slightly depleted in Na^+ and considerably enriched in Ca^{2+}. HCO_3^- is depleted in highly saline brines but enriched in moderately saline brines, and the relative proportions of both Ca^{2+} and K^+ increase with increasing salinity. There is also a higher molar Ca/Mg ratio (average 3.5) in subsurface water compared to seawater (0.19).

Subsurface brines are derived from mixtures of meteoric water, seawater, and juvenile and magmatic waters. In many sedimentary basins, there does appear to have been a greater input of meteoric water compared with the other sources (e.g. in Illinois, Michigan and Alberta Basins). In the much studied Gulf Coast Basin of the southern USA, there are regional variations in brine chemistry with dissolution of Jurassic halite giving highly saline NaCl brines and low-salinity waters arising from dilution by waters released during clay mineral transformations (Morton & Land, 1987).

With subsidence and increasing overburden, sediments and pore-fluids are subject to increasing temperature and pressure. The temperature increase depends on the geothermal gradient (see Fig. 7.31A), and it is generally in the region of 25°C per km. One consequence of increasing temperature is that some chemical reactions are likely to proceed more quickly. This is particularly the case with dolomite precipitation where some kinetic inhibiting factors are less influential at the higher temperatures. In addition, the solubility of calcite decreases with rising temperature and so should also precipitate more easily at depth. There are of course many other depth–temperature-related mineral–chemical processes, involving the clay minerals, gypsum–anhydrite and organic matter. Lithostatic pressure results from the mass of sediment overburden and increases with depth in a near linear fashion (Fig. 7.31B). Pore-fluid pressure increases with depth as the force of the overlying mass of pore-fluid increases; in porous sediments this is the hydrostatic pressure. In low permeability sediments, however, the pore-fluid pressure is often higher (i.e. the rocks are overpressured), and the lithostatic pressure is the maximum, when the pore-fluid is actually supporting the rock mass itself. The development of high pore-fluid pressures is important in burial diagenesis since it can prevent or retard mechanical and chemical compaction, and cementation. It may also lead to fracturing. In carbonate rocks, overpressuring is likely to develop where grainstones and other porous facies are confined between compacting lime mudstones or shales. The formation of CO_2, CH_4 and other natural gases from organic matter maturation also contributes to high pore-fluid pressures.

Pore-fluid flow rates in the subsurface are very low indeed compared to the near-surface. Although difficult to measure, rates will be of the order of less than $1-10$ m yr^{-1}; this contrasts with seawater being pumped through shelf-margin reefs at rates up to 2000 m day^{-1} (Buddemeier & Oberdorfer, 1986). It follows that rates of cement precipitation, dependent

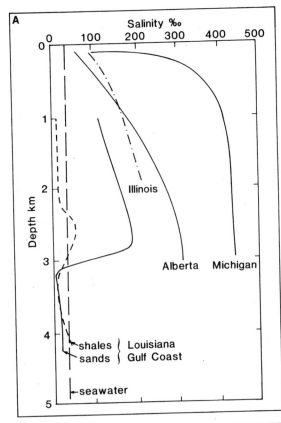

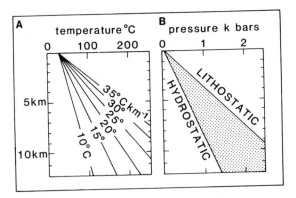

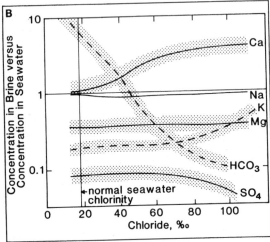

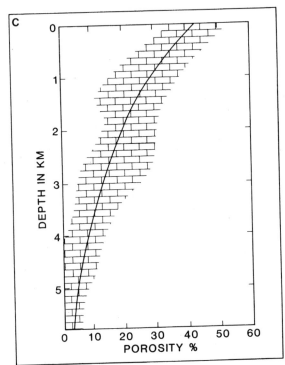

Fig. 7.30 *(A) Generalized salinity variations with depth for formation waters from the Alberta, Illinois, Louisiana Gulf Coast and Michigan Basins (B) Generalized variations in chemical composition of subsurface brines in the Illinois Basin with increasing chlorinity. Values are normalized with respect to seawater of the same chloride content. From Hanor (1983).*

Fig. 7.31 *(A) Temperature changes with depth for various geothermal gradients. (B) Pressure changes with depth. The position of the hydrostatic pressure line is dependent on the salinity of the pore-fluids. Most pore-fluid pressures are somewhat higher than the hydrostatic line. (A) and (B) after Choquette & James (1987). (C) Porosity versus depth for carbonate rocks of south Florida. The curved line is the average and the ornament shows the range. After Halley (1987).*

on fluid flow rates, will be much lower in the burial environment, where pore-fluids are generally also supersaturated *re* $CaCO_3$. Undersaturated porewater does occur, notably where there is much organic matter degradation and high CO_2 pressure, and then carbonate dissolution is possible (see Section 7.6.5). The flow of subsurface waters is controlled by the hydraulic head, a measure of the relative elevation and the absolute fluid pressure of the porewater, and the osmotic head, a measure of the thermodynamic activity of the H_2O molecules in the pore-fluid. Water flows from areas of high to low total head, and two major influences here in a sedimentary basin are gravity-driven downward freshwater recharge into an aquifer, and upward formation water movement through mudrock compaction and dewatering (see hydrology texts for further information, e.g. Freeze & Cherry, 1979).

With increasing burial, there is a progressive decrease in porosity. This is discussed in Section 5.6 for pelagic sediments and the same trend, although with much scatter, is apparent for shallow-water carbonates (see Fig. 7.31C; Schmoker & Halley, 1982). With limestones from Florida, initially porosity is in the range of 30 to 60%, depending on the original texture of the sediment (Enos & Sawatsky, 1981) and on the amount of meteoric dissolution. By around 2 km of burial, porosity has been reduced by half, but then further loss is at a slower rate and by 6 km the average is 5%. Rates of porosity loss are generally lower in carbonate sands and higher in lime muds (see Section 7.6.3).

Burial diagenesis has been discussed at length in the literature and recent reviews include Bathurst (1980a, 1986), Wanless (1983), Scholle & Halley (1985), Choquette & James (1987) and Halley (1987).

7.6.2 Burial cementation

Cements precipitated in the burial environment are mostly some form of clear, coarse calcite spar. There are four mosaic types (Fig. 7.32): (1) drusy, equant calcite, the most common and present in the majority of limestones, (2) poikilotopic calcite, (3) equant−equicrystalline mosaics of calcite spar, and (4) syntaxial calcite spar. Some prismatic calcite may be of burial origin and baroque dolomite is a common late precipitate too (Section 8.4.3). Calcite spar is also the common cement of near-surface meteoric environments and it occurs there in the same mosaic types. Hence, there is often a problem of ascertaining the origin of the calcite spar and careful petrographic and geochemical arguments are usually required.

Drusy calcite spar is a characteristic pore-filling cement with an increasing crystal size towards the cavity centre. This fabric originates from the preferred growth direction of calcite along the c-axis and competitive growth. When crystallites have nucleated upon a substrate those with their c-axes at a higher angle to the substrate will grow faster and overtake others with c-axes at a lower angle (Fig. 7.32B). As a result of this growth competition, crystals in a drusy mosaic have a preferred orientation of optic axes normal to the substrate. Drusy calcite spar also typically has plane intercrystalline boundaries, referred to as compromise boundaries, generated by two crystals growing alongside each other at a similar rate (Fig. 7.32B). The origins of the fabrics of drusy calcite spar are fully discussed by Bathurst (1975) who also used the enfacial junction (a triple junction of intercrystalline boundaries where one is 180°) as a criterion for a cement origin. Recently, graphical modelling of cements by Dickson (1983) has shown the importance of the calcite rhombohedral crystallographic form in the type of crystal mosaic generated and the complex shapes of crystals in a two-dimensional section. In addition, although acute-closed crystallographic forms give rise to length-fast crystals in a mosaic (and this is the most common situation), obtuse forms result in length-slow crystals. One further feature of drusy calcite spar is the presence of growth zones. These may be revealed with staining (Dickson, 1966) or with cathodoluminescence (Fig. 7.33). Generally, calcite spar has a dull luminescence, but it often shows bright zones. There may be a pattern of non-luminescence to bright to dull with progressive burial and evolution of porewaters (see later section).

Poikilotopic calcite spar consists of large crystals including several grains (see Fig. 7.35). They may reach several millimetres or more in diameter. They have probably resulted from a very low nucleation rate of calcite crystals and slow growth. They may precipitate from pore-fluids only just supersaturated *re* $CaCO_3$.

Equant−equicrystalline mosaics of calcite spar are not common, and many cases of this texture may be a function of the plane of the section — a cut through calcite spar parallel to the substrate. Alternatively, it may be a texture produced by neomorphism of a pre-existing cement, although some relics would be expected.

Syntaxial calcite spar is a feature of grainstones containing echinoderm fragments (e.g. Fig. 7.34) and

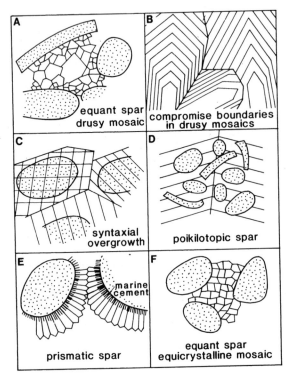

Fig. 7.32 *Calcite spar: sketches illustrating the common types.*

these may also be very large crystals which include other grains. Echinoderm overgrowths are also commonly zoned (e.g. Evamy & Shearman, 1965; Meyers, 1974), in the same way as drusy calcite spar, although the early zone(s) may have precipitated in a near-surface marine, meteoric or mixing-zone environment. Burial syntaxial overgrowths are generally clear, relative to inclusion-rich and cloudy earlier overgrowths (see Section 7.4.6).

Prismatic calcite consists of coarse, elongate crystals which line cavities or occur upon marine fibrous cement and are then succeeded by calcite spar. In some cases, the prismatic calcite is an overgrowth on the marine cement or on bioclasts. Such calcite could be an early burial cement (Choquette & James, 1987), although the precise origin is unclear, whether it is a mixing-zone or subsurface meteoric zone precipitate.

The determination of a burial origin for calcite spar can be made on several textural criteria. It is necessary to evaluate the timing of spar precipitation relative to

compactional features; if there is clear evidence for the spar being precipitated after mechanical or chemical compaction, then a burial origin is confirmed. Where there is no evidence for this, then a burial origin is not ruled out, but it would have to be established on geochemical grounds (such as negative $\delta^{18}O$, see next section). Textural criteria indicating a burial origin for spar include: broken and collapsed micrite envelopes present within calcite spar, fractured grains and ooids included within spar and sutured or concave-convex contacts between grains before spar precipitation (see Figs 7.35, 7.38 and 9.13B).

7.6.2a Calcite spar geochemistry

Compared with marine cements, burial spar is typically depleted in Mg, Sr and Na, but it may have significant amounts of Fe and Mn. Many burial cements are ferroan, as shown by staining, and this is usually the result of precipitation from negative Eh porewaters and Fe^{2+} in solution derived from clay minerals and shale beds. Manganese reaches many hundreds of ppm typically. Burial spar contrasts with near-surface meteoric spar in the presence of significant Fe^{2+} and Mn^{2+}, since most meteoric groundwater is well oxygenated and therefore does not contain these elements in solution.

Burial calcite spar is generally more depleted in ^{18}O than marine or earlier meteoric cements (e.g. Figs 7.16 and 7.17). The main reason for this is the higher temperature of precipitation in the burial environment and the fractionation effect. Taking the difference in oxygen isotopic composition between calcite spar and marine cements or marine components in the same rock and using the isotopic fractionation equation for calcite (see Section 6.4 and Fig. 6.22) gives an indication of the burial temperature of precipitation. However, one complication here is the isotopic composition of the subsurface fluids; they are very likely to be different from seawater. If there is a strong meteoric water input to the formation fluids then the $\delta^{18}O$ composition may well be lighter than marine water. On the other hand, more saline waters are commonly more enriched in ^{18}O. The carbon isotopic signature of burial spar is usually similar to a marine value or a little depleted. The reasons for this are that most of the carbon for the CO_3^{2-} is derived from the limestone itself (by dissolution somewhere in the system) and there is little fractionation of $^{13}C/^{12}C$ with increasing temperature. Slight depletion is due to the incorporation of some light CO_2 from organic matter degradation. In organic-rich limestones,

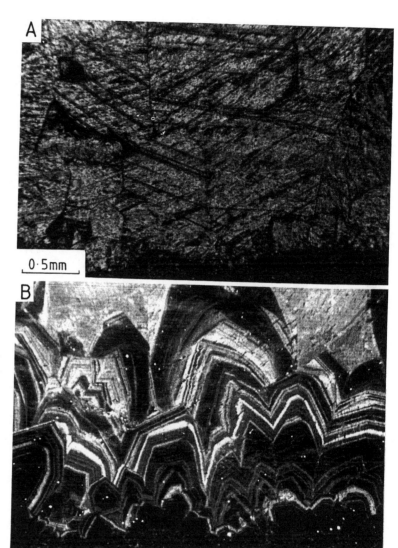

0·5mm

Fig. 7.33 *Equant calcite spar under plane polarized light (A) and same view under cathodoluminescence (B). The drusy fabric of increasing crystal size towards the cavity centre is visible in (A), and in (B) the growth zones the beautifully seen, reflecting subtle changes in porewater chemistry. The cathodo-luminescence pattern here is of a thin, non-luminescent early part with two bright subzones and then a much thicker zone of bright luminescence (pink) with thin yellow subzones. The central part of this particular cavity is filled with a more dull luminescing spar without subzones. Upper Triassic (Keuper), south Wales.*

strongly negative $\delta^{13}C$ signatures are possible for burial spar.

There have been numerous studies of calcite spar geochemistry (see papers in Schneidermann & Harris, 1985 for example) but one illustrating the trend of increasingly negative $\delta^{18}O$ with later precipitates is Dickson & Coleman (1980). This case presented analyses of successive growth zones from the same, large spar crystals in a Lower Carboniferous grainstone and revealed a trend from $\delta^{18}O = -2.5\permil$ to $-12.4\permil$ in early to late zones (see Fig. 7.16C). Zoned poikilotopic spar crystals in the Jurassic Smackover

Formation show a similar trend and are interpreted as burial cements (Moore, 1985). Marshall & Ashton (1980) analysed trace elements and isotopes from a Jurassic marine-cemented hardground down into a grainstone without marine cements and only burial spar. $\delta^{18}O$ and to a lesser extent $\delta^{13}C$ become more negative away from the hardground in whole-rock analyses, reflecting the increase in isotopically-depleted burial spar (Fig. 7.16A), and extracted late cements have quite negative $\delta^{18}O$ of $-8.5\permil$ average (Fig. 7.16B). There is also an increase in Fe^{2+} and decrease in Mg^{2+} down from the hardground.

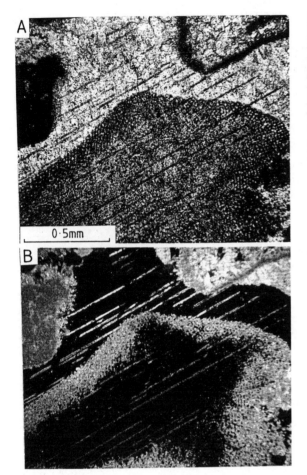

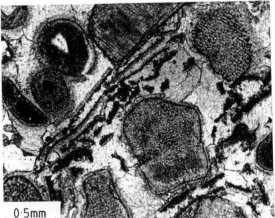

Fig. 7.35 *Mechanical compaction in a skeletal, oolitic grainstone resulting in collapse of micrite envelopes (indicating earlier aragonite dissolution) and spalling of a marine fringing cement off a micritized ooid (upper left). The calcite spar, which is of the poikilotopic type, is clearly a post-compaction precipitate. Upper Jurassic, Yorkshire, UK.*

Fig. 7.34 *Syntaxial calcite spar overgrowth on an echinoderm fragment. (A) Plane polarized light. (B) Crossed polars. Originally the grain was very porous and the minute intraskeletal holes around the outside of the grain were filled by micrite. Part of this may be the result of seafloor micritization by endolithic algae. The porosity of the central part of the grain was filled later by calcite spar. The overgrowth spar is in optical continuity with the grain as clearly shown by the common extinction and twin planes continuous from grain to overgrowth. The dusty appearance of the grain and spar could be due to the presence of microdolomite inclusions, although they could also be fluid-filled inclusions or empty micropores. Mid-Cretaceous (Urgonian), Provence, France.*

7.6.2b Fluid inclusions in calcite

Crystals of calcite spar (and also dolomite) commonly have fluid inclusions up to 10 μm or more in diameter and their study may give useful information on temperature of precipitation and nature of the diagenetic fluids. Fluid inclusion studies have featured prominently in research into ore geology and mineralization, but have only recently been used in carbonate diagenetic projects. The inclusions may be formed as a crystal is growing or they may form subsequently through fracturing of a crystal, fluid injection and fracture healing. When formed, many of the minute cavities are completely filled by a homogeneous saline solution but at Earth surface temperature and pressure, these inclusions are two-phase; a vapour bubble is present in addition to the solution. Re-homogenization of the two phases is accomplished by heating on the microscope stage to give a homogenization temperature which will only be the same as the trapping temperature if the effects of pressure are zero. To determine the actual trapping temperature, a correction is applied which takes into account the fluid composition. An assessment of this is made by measuring the depression of the freezing point and then the initial melting temperature. The techniques of fluid inclusion study are discussed in Barker & Halley (1986). There are many pitfalls in interpreting fluid inclusion data; particular attention has to be given to the possibility of physical re-equilibration of the calcite due to the increased overburden pressure during burial; then the inclusions simply reflect bottom-hole temperatures. Two-phase inclusions are common in burial calcite cements in the Smackover Formation of the Gulf Rim subsurface and homogenization temperatures are in the range 60 to 160°C. From the freezing point measurements (-15 to -30°C), the fluids are likely to have salinities from 5 to 9 times that of seawater (see Moore, 1985). This is in the range of present Smackover pore-fluids. It is possible to extract the fluid from inclusions and analyse it directly with ICP or some other technique.

Some inclusions are filled with liquid hydrocarbons and these are recognized by their fluorescence under UV light. The colour of the fluorescence emission varies with the °API gravity of the oil. The timing of hydrocarbon migration can be estimated relative to diagenetic events by studying the hydrocarbon-filled inclusions in different generations of calcite spar (e.g. Burruss et al., 1985).

7.6.2c Cement stratigraphy and cathodoluminescence (CL)

One aim of carbonate diagenesis research is to produce a well-documented burial history of the formation, relating cement phases to evolving porewaters and temperatures, palaeohydrology, compaction, organic matter diagenesis and hydrocarbon formation. In many cases, the broad pattern of burial and uplift of a limestone can be deduced from an assessment of the regional geology, and then the diagenetic events can also be considered in terms of a time-frame. One technique which has been employed in these studies is cement stratigraphy (Meyers, 1974). Drusy calcite spar and echinoderm syntaxial overgrowths filling cavities commonly show compositional zoning on a microscale as a result of subtle variations in trace elements. The zoning records crystal growth histories and changes in porewater chemistry during crystal growth. The cement zones can be studied up through a sequence and laterally across a region to compile a stratigraphy, and features such as compactive fracture, dolomite and sulphide precipitation, dissolution events, influx of hydrocarbons and cements in intraclasts at unconformities, are used to fit the zones into a time-frame of geological and other diagenetic events.

Zonation in cement crystals is best revealed by using a luminoscope and observing the luminescence patterns (e.g. Figs 7.33 and 7.36), but it can also be seen to an extent by staining with Alizarin Red S and potassium ferricyanide (e.g. Dickson, 1966). With CL, the luminescence is mostly a function of Fe^{2+} and Mn^{2+} contents, with Mn^{2+} being the activator and Fe^{2+} the inhibitor, but other elements are involved such as Pb^{2+} and rare earths (activators) and Ni^{2+} and Co^{2+} (quenchers) (Machel, 1985). The concentrations of these elements necessary for luminescence do appear to vary, but generally more than several tens of ppm Mn^{2+} are required, and once Fe^{2+} is in excess of around 2 wt % $FeCO_3$, crystals do not luminesce (see Pierson, 1981; Fairchild, 1983; Machel, 1985; Ten Have & Heijnen, 1985, and Hemming et al., 1989). The luminescence colour and intensity in calcite and dolomite probably relate to the ratio of Fe and Mn rather than to the actual concentrations. The zonation of the carbonate crystals is a reflection of fluctuations in the chemistry of the pore-fluids, but also of changes in the rate of crystal growth (Ten Have & Heijnen, 1985).

In many limestones, there is a similar pattern of luminescent zones in calcite spar cement: the spar commonly shows a non-luminescent–bright–dull zonation with increasing burial (Fig. 7.36; e.g. Frank et al., 1982; Grover & Read, 1983; Frykman, 1986). This sequence is thought to relate to the increasingly reducing nature of pore-fluids during burial. The early non-luminescent zone is precipitated from oxidizing porewaters, where Mn^{2+} and Fe^{2+} are not present in the positive Eh solution. This zone may show thin luminescent subzones, reflecting relatively short periods of porewater stagnation and the precipitation of Mn^{2+}-bearing calcite. The bright zone records the onset of permanently reducing conditions when Mn^{2+} ions (if available) are present in the waters and iron is being taken out as the sulphide through the reduction of SO_4^{2-}. The reducing conditions are brought about by the stagnation of the oxidizing waters and/or the thermal decomposition of organic matter in the sediments. The succeeding dull zone is a ferroan (and maybe manganoan) calcite precipitated from more negative Eh waters, below the stability field of FeS_2. This zone is precipitated during deeper burial and commonly follows a phase of fracturing. Any synsedimentary marine cements are generally non-luminescent (because of the absence of Mn^{2+} and positive Eh of seawater). CL study may also reveal phases of dissolution between zones, seen as disconformities cutting across earlier zones or as patches of the later zone within the earlier. Internal sediment of crystal fragments can also be distinguished from cement by its CL appearance.

Within a sedimentary basin the cement stratigraphy

A

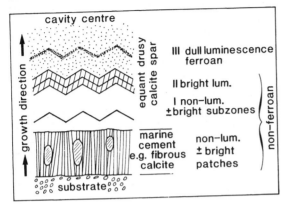

Fig. 7.36 *The common pattern of luminescence seen in calcite cements (equant drusy calcite spar and syntaxial overgrowth spar) of non-luminescent—bright—dull, which reflects precipitation in oxic (marine and near-surface meteoric to shallow burial), through suboxic to anoxic diagenetic environments with increasing burial. (A) Diagrammatic cavity fill. (B) and (C) An example from the Silurian of Sweden of CL zonation in a syntaxial overgrowth on a crinoid fragment. Four cement zones are recognized: zone 1 is cloudy in plane polarized light (B) and dull with bright specks in CL (C). This zone is interpreted as a marine precipitate and the inclusions/bright specks are microdolomites, suggesting that the marine precipitate was originally high-Mg calcite. Zone 2 is clear in plane polarized light and non-luminescent under CL and is separated by a dust line from zone 3, which is brightly luminescent with fine subzones. Zone 4 is dull and subzoned and occludes remaining pore space. Zones 2—4 were precipitated during shallow burial, probably from meteoric water. For further information see Frykman (1986). Photos courtesy of Peter Frykman.*

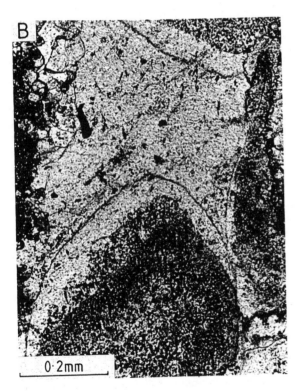

may show little variation or there may be regional differences. Individual cement zones may be synchronous across a basin (as in the Mississippian Lake Valley Formation of New Mexico, Meyers, 1978), but in other cases they are clearly diachronous (e.g. Middle Ordovician of Virginia, Grover & Read, 1983). In the first instance, cement precipitation relates to aquifers developed beneath a regional unconformity on the top of a platform carbonate sequence. In the second case, zoned cements occur preferentially in one part of the basin, where aquifers were subjected to meteoric recharge from upland areas. Deeper burial cements were precipitated from warm, reducing brines expelled from compacting basinal shales. In the Helderberg Group of the Appalachians (Dorobek, 1987), regional CL zonation indicates shallow burial spar cementation from meteoric groundwaters which become progressively more reducing as they flowed downdip. Thus non-luminescent cement with bright subzones passes downdip into dull cement with bright subzones and then basinward into non-zoned dull cement. The gravity-driven meteoric groundwaters flowed laterally into the Helderberg Basin along confined aquifers in interbedded sandstones for at least 150 km. Connate marine pore-fluids were expelled and it is thought that eventually the meteoric waters discharged on to the seafloor, as occurs at the present time on the continental shelf off the eastern USA. In many of the studies of cement stratigraphy, trace element, isotope and fluid inclusion data provide further constraints on the nature of the pore-fluids (e.g. Meyers & Lohmann, 1985; Dorobek, 1987; Kaufman *et al.*, 1988).

7.6.2d Sources of CaCO₃ burial for cementation

There has been much discussion of the origin of the $CaCO_3$ for burial calcite spar cementation in limestones, but little agreement (see Bathurst, 1975). Some process of dissolution is probably involved but it is unclear where this takes place. Meteoric groundwaters entering a carbonate formation are likely to be under-saturated at first and should dissolve grains, those of aragonite and high-Mg calcite mineralogy more readily than those of low-Mg calcite, and those composed of a finer crystalline texture more rapidly than those of a coarser nature. Once a supersaturated state is reached, then calcite can be precipitated as the cement (the generally low magnesium content of meteoric water accounting for the low-Mg calcite mineralogy). Thus meteoric fluids moving down a confined aquifer into a sedimentary basin will, in time, cement the sediment with $CaCO_3$ derived from updip. In the deeper sub-surface, dissolution of $CaCO_3$ particles in shale beds takes place as they undergo compaction, especially if there is much organic matter being decomposed. This input of $CaCO_3$ promotes supersaturation of pore-waters in adjacent limestones and leads to cementation. Marine porewaters in carbonate sediments will already be supersaturated *re* $CaCO_3$ and during burial

calcite will be precipitated if the Mg/Ca ratio is reduced, as would occur through Mg^{2+} loss to clays for example. Increasing temperature will also promote calcite precipitation. Pressure dissolution is commonly invoked as the major source of $CaCO_3$ and in many limestones lacking early cements, grain–grain dissolution can be clearly observed to have taken place before calcite spar precipitation (e.g. Hird & Tucker, 1988, see Fig. 9.13). In some limestones, low porosity zones occur in close proximity to stylolites, suggesting that $CaCO_3$ from the dissolution has been reprecipitated close by (e.g. Wong & Oldershaw, 1981). Although pressure dissolution is clearly ubiquitous in the burial realm (see next section), it is unlikely to be able to supply $CaCO_3$ for cementation at shallow to moderate burial depths unless there is considerable updip movement of pore-fluids. In many limestones, calcite spar precipitation *begins* a long time before pressure dissolution *in that rock*.

7.6.3 Compaction

Increasing overburden stresses on carbonate sediment result in the formation of a great range of compactional textures and fabrics, some visible in the field, others more subtle. The importance of compaction is probably greatly underestimated in limestones, but it can account for much thickness reduction in carbonate sequences. Compactional processes and products are generally split into two categories: *mechanical* (or physical) and *chemical*. Mechanical compaction begins soon after deposition whereas chemical compaction mostly requires several hundred metres of burial. The original nature of the sediment, particularly the amount of lime mud and clay present, is an important control on the types of compactional structure formed, and the early diagenetic history, especially the degree of cementation, is also significant. Laboratory experiments on compaction (e.g. Shinn *et al.*, 1977; Shinn & Robbin, 1983; Bhattacharyya & Friedman, 1984) have provided useful insights and constraints on the real but largely inaccessible world of burial compaction.

Mechanical compaction begins as soon as there is overlying sediment with simple dewatering, re-arrangement of particles and closer grain packing. The effects are most marked in more muddy sediments where there is significant porosity loss and considerable reduction in sediment thickness in the first few metres to several tens of metres of burial. In grainy sediments, the closer packing leads to a change from a more cubic to a more rhombohedral arrangement of grains (Fig. 7.37). In these early stages of compaction, a preferred orientation of elongate bioclasts may develop parallel to the stratification/seafloor, normal to the principal stress direction. The next stage in mechanical compaction actually involves the fracture of grains and early cement fringes and the ductile deformation of grains and sediment. The burial depths at which these processes begin to operate are difficult to pinpoint. In cores through Quaternary−Tertiary shallow-water limestones, from Pacific atolls, grain fracture is common at several hundred metres depth (e.g. Schlanger, 1964; Saller, 1984). Experiments with ooids at pressures from 824 to 1565 kg cm^{-2}, equivalent to burial depths of 3.5 to 6.5 km, produced many fractured grains, as well as pressure dissolution between grains (Bhattacharyya & Friedman, 1984). Porosity changes here were from an initial value of 40% down to around 20%. With lime muds, experiments on cores of modern sediments show that sediment thickness can be reduced by 50% or more under pressures equivalent to less than 300 m of burial (Shinn & Robbin, 1983). Accompanying porosity reduction is from initial values of 65−75% down to 35−45%. Significant further reductions in sediment thickness did not take place even with pressures equivalent to more than 3000 m of burial.

The textures produced by mechanical compaction are readily recognized and occur in most limestones. Only where there has been considerable early cementation is mechanical compaction prevented. In grainstones and packstones, grains may be broken and the fracture surfaces can be seen to be controlled by the fabric of the particle. Thus ooids with a strong radial fabric (typically calcite originally) will fracture across the grain, following the fibrous crystal boundaries, whereas an ooid with a concentric fabric (typically aragonite originally) will break and crack along the lamellae so that cortical layers may spall off (Fig. 7.38). Early cement fringes around grains may also become detached and fragment. Micrite envelopes will collapse and break if the host bioclast has been dissolved earlier (Fig. 7.35); this is a common feature of originally aragonitic shells subjected to early meteoric dissolution. Of the grains, plastic deformation mostly affects faecal pellets, with their homogeneous, micritic nature. The grains may be flattened into spindle shapes and show planar or curviplanar contacts with more resistant grains. Where pellets are protected from the overburden pressure, as beneath shells in umbrella cavities, then the original shape and grain packing are maintained.

In lime mudstones and wackestones, the effects of

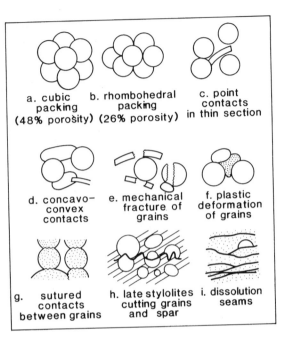

Fig. 7.37 *Grain packing, grain contacts and mechanical and chemical compaction textures in limestones.*

Fig. 7.38 *Concavo-convex and microstylolitic contacts in Smackover Oolite, Jurassic, Louisiana subsurface. Notice also the spalling and fracture of thin oolitic coating off the central grain. Intergranular porosity later filled by very coarse, poikilotopic calcite spar (white).*

mechanical compaction are well seen where burrows have been compressed. Commonly, the compaction is most obvious where there is a significant clay component, and then burrows in the more clay-rich layers are normally compacted to a greater extent than those in more lime mud-rich beds (e.g. Byers & Stasko, 1978; Gaillard & Jautée, 1987). In the experiments of Shinn & Robbin (1983), circular burrows were compacted to ellipsoidal structures and open burrows lost completely, under pressures equivalent to less than 300 m of burial. Intertidal fenestrae also may be obliterated. Early lithified thin layers and nodules and bioclasts in muddy sediments may be fractured. Tabular intraclasts and shells may be rotated more into the plane of the bedding. The so-called *molar-tooth structures*, common in Precambrian limestones, are syneresis crack fills subjected to compaction and ptygmatically folded. Desiccation crack fills may also go the same way.

Laminae and bedding in more muddy sediments are commonly draped over rigid objects, such as fossils and nodules on a small scale and reefs and mud mounds on a larger scale. In organic-rich sediments, wispy laminae and black, wavy stringers of carbonaceous matter may be present. In many micritic limestones, fossils are preserved full-bodied and in vertical orientations with no signs of compaction, and birdseyes and peloids retain their original shape. This has led to the suggestion that many lime muds are lithified early so that mechanical compaction is prevented (e.g. Zankl, 1969; Bathurst, 1970). However, although early cementation has clearly taken place in many instances, as witnessed by hardgrounds and intraclasts, in the majority of micrites such features are absent and cement cannot be discerned from sediment so that proof of early lithification relies on the absence of mechanical compactional structures. It should be borne in mind, however, that in the laboratory experiments of Shinn *et al.* (1977) and Shinn & Robbin (1983), shells are generally only fractured when they come into contact with other shells. Where floating in lime mud, shells are rarely broken, even where original sediment thickness is reduced by half.

One important effect of the experimental compaction of skeletal wackestones is the generation of skeletal packstones as a result of the closer packing of particles. In nature, the presence of flattened burrows and fractured grains should enable packstones produced by mechanical compaction to be distinguished from original packstones.

Chemical compaction and pressure dissolution are very important burial processes. Apart from producing a range of dissolution textures, they also result in the dissolution of grains and sediment, and this may be a significant source of $CaCO_3$ for burial cementation. Pressure dissolution arises from the increased solu-

bility of material at grain contacts and along sediment interfaces as a result of applied stress. A thin film of water exists at the site of pressure dissolution and ions going into solution may join the main pore-fluid reservoir of the limestone. A chemical potential gradient is set up with ions moving away from the pressure dissolution site by diffusion or solution transfer. The $CaCO_3$ in solution may be reprecipitated as a cement in the immediate vicinity, where the stress field is lower, or carried away in an active pore-fluid system to a more distant site of precipitation. The stresses arise from increasing overburden but also from tectonic stress. Much cleavage in limestones for example involves pressure dissolution. The dissolution of limestone leads to the accumulation of insoluble residue which mostly consists of clay minerals, iron oxides/hydroxides and organic matter. For discussions of the pressure dissolution process see Weyl (1959), Bathurst (1975), De Boer (1977), Robin (1978), Guzzetta (1984) and Tada *et al.* (1987).

There have been numerous descriptions and classifications of pressure dissolution structures (recent examples include Logan & Semeniuk, 1976; Wanless, 1979; Buxton & Sibley, 1981; Bathurst, 1987), and here the three categories of the last two papers are used: (1) fitted fabrics, (2) dissolution seams, and (3) stylolites.

A *fitted fabric* is a pervasive framework of interpenetrant grains, and the surfaces between the grains are generally slightly sutured to curved to planar. The term microstylolite has been used for these surfaces which do not extend beyond the two grains in contact. Fitted fabrics are usually the result of grain–grain dissolution operating pervasively throughout a zone. They occur on a microscale between skeletal grains or ooids (Fig. 7.38), and the pressure dissolution has normally taken place before any significant cementation. Early, near-surface cementation may in fact prevent this fabric from developing (Bathurst, 1987; Hird & Tucker, 1988). Fitted fabrics may also occur on a macroscale, where intraclasts, burrow fills, or even early diagenetic nodules are involved. The *stylobreccia* fabric of Logan & Semeniuk (1976) is a type of fitted fabric and an example is illustrated in Fig. 5.17.

Dissolution seams are smooth, undulose seams of insoluble residue (Fig. 7.39) which lack the distinctive sutures of stylolites (see below). They are the non-sutured seams of Wanless (1979). They mostly pass around and between grains, rather than cutting through them, and they are usually anastomosing. They occur in swarms and are common in nodular

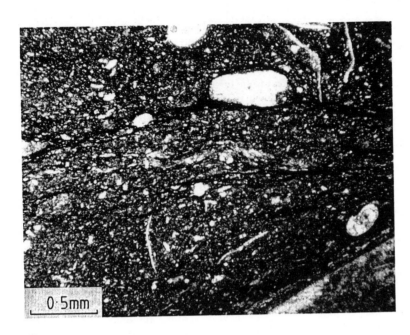

Fig. 7.39 *Dissolution seams with insoluble residue (clay) concentrated along them. Upper Devonian, Griotte (pelagic limestone), southern France.*

limestones, where they wrap over nodules and lead to the development of *flaser* structure (e.g. Garrison & Kennedy, 1977; also see Section 5.7.3). Dissolution seams tend to be common in more argillaceous limestones, and develop preferentially along thin clay layers or at the junctions of clay-rich and clay-poor limestones. Dolomite forms along these dissolution seams in some instances (e.g. Wanless, 1979).

Stylolites are a prominent feature of many limestones and are a serrated interface between two rock masses, with a sutured appearance in cross-section. The amplitude of the suture is much greater than the diameters of the sediment grains. Stylolites transect the rock fabric, cutting across grains, cement and matrix indiscriminately. They are normally absent in limestones with more than 5–10% clay.

Pressure dissolution is a major process in enhancing and producing bedding planes. In many shelf and platform limestones, the 'bedding planes' are not primary depositional surfaces but they have been produced by pressure dissolution during burial diagenesis. This is most clearly demonstrated where, for example, a 'bedding plane' occurs within an event bed such as a storm layer, or where stylolitically-defined 'bedding plane' cross-cuts a primary bedding surface (Simpson, 1985). These pseudobedding surfaces produced by pressure dissolution account for much layering in both shallow- and deep-water limestones. In many instances there is a regular alternation

of 'hard' limestones showing few pressure dissolution effects and 'fissile' limestones with abundant dissolution seams in wackestones and fitted fabrics in grainstones (Bathurst, 1987). Bedding planes are preferentially developed in the fissile limestones (Fig. 7.40) and show no relation to primary depositional planes, which mostly were destroyed by bioturbation. It appears that the hard layers were selectively cemented early, and then mechanical and chemical compaction affected the less cemented layers to produce the fissile limestones and the bedding-planes. The significance of this is that (1) cementation of sediment was taking place periodically during shallow burial beneath the seafloor and (2) this was controlling the pattern of later diagenesis. Point (1) is particularly important and has not been appreciated before in diagenetic studies.

7.6.4 Lime mud diagenesis and neomorphism

Modern lime muds have a variable mineralogy depending on the source of the material, in particular the proportion coming from breakdown of green algae (aragonitic), red algae (calcitic, high Mg) and coccoliths–planktonic foraminifera (calcitic, low Mg), and the amount of abiogenic aragonite and high-Mg calcite precipitated. With ancient fine-grained limestones, it is usually very difficult to determine the

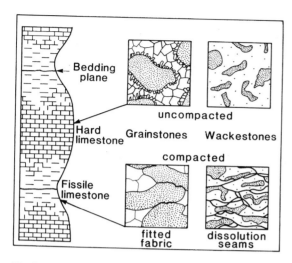

Fig. 7.40 *Uncompacted, hard limestones alternating with fissile limestones showing fitted fabrics and dissolution seams. This is the result of episodic subseafloor cementation. After Bathurst (1987).*

origin of the particles and under the SEM many are seen to be composed of equant micrite crystals, with diameters averaging 2−3 μm. Aggrading neomorphism has commonly taken place in fine-grained limestones to give mosaics of microspar crystals (generally in the range of 5−10 μm) and coarser pseudospar crystals (>30 μm). This neomorphism may take place during early meteoric diagenesis or during burial diagenesis. It may also occur on weathering. In pedogenic limestones, coarse displacement−replacement crystal mosaics are common.

In thin section, *neomorphic spar* can be recognized by: (1) irregular, embayed to curved intercrystalline boundaries, contrasting with the commonly planar intercrystalline boundaries of spar cement, (2) an irregular crystal size distribution and patchy development, (3) gradational and irregular boundaries to the areas of neomorphic spar, and (4) the presence of skeletal and other grains floating in coarse spar (see Fig. 7.41). Under CL, neomorphic spar crystals may each show concentric growth zones, rather than be part of a larger-scale zonation, as is seen in cavity-filling sparry cements. This aggrading neomorphism probably takes place by the dissolution of some crystals and the syntaxial overgrowth of others. The factors promoting this are unclear but clay content and/or Mg^{2+} ions may well be involved. Neomorphic effects are most prominent in pure micrites; where there is a significant clay content, then aggrading

neomorphism appears to be inhibited. In one case study (Bausch, 1968), neomorphic spar was only present in Jurassic limestones with less than 2% clay and in another, the Cretaceous Edwards Formation of Texas (Longman & Mench, 1978), microspar only occurs in the updip part where porewaters are fresh; it is absent downdip where formation water is saline. Folk (1974) and Longman (1977) have suggested that the presence of Mg^{2+} ions prevents neomorphism by forming a 'cage' around the micrite crystals. If this is removed by meteoric water flushing or adsorption of the Mg^{2+} ions on to clay minerals, then microspar crystal growth takes place.

Studies of Pleistocene microcrystalline limestones formed from aragonite-dominated lime mud have questioned the importance of aggrading neomorphism. With material from Barbados and the Bahamas, Steinen (1978, 1982) showed that there is a good deal of cementation taking place in primary voids to give the coarser crystal patches. The same is the case in Florida samples where Lasemi & Sandberg (1984) found abundant aragonite relics in micrite and microspar crystals. This suggests a one-step calcitization process leading to both fine and coarse crystals, rather than microspar developing from the aggrading neomorphism of micrite, which itself replaced the aragonite needles. SEM studies of ancient micrites by Lasemi & Sandberg have shown that aragonite-dominated precursor muds (ADPs) can be distinguished from calcite-dominated precursor muds (CDPs) on the basis of aragonite relics and pits, with ADPs occurring at least back to the Ordovician. Wiggins (1986), studying Pennsylvanian microspar in biomicrites, was able to demonstrate the importance of aragonite in the original lime mud from high Sr/Mg ratios, the presence of celestite and 'needle-mud' pseudomorphs. This microspar is also interpreted as the result of aragonite dissolution−calcite precipitation, rather than micrite neomorphism, and isotope evidence suggests this took place in relatively fresh, but highly exchanged and reducing fluids.

7.6.5 **Burial dissolution**

Although it is generally the case that limestone porosity decreases with increasing depth as a result of cementation and compaction, porosity can be created at depth through dissolution. It may also be created through fracturing of course. Burial dissolution is well documented from sandstones where carbonate cements, lithic grains and feldspars may be partly to completely dissolved. In limestones such dissolution is

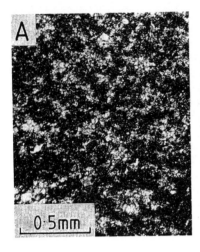

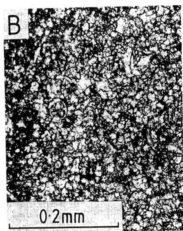

Fig. 7.41 *Neomorphic spar. (A) Patches of microspar in a micritic pelagic limestone resulting from aggrading neomorphism. Upper Devonian, West Germany. (B) Coarse microspar mosaic of equant crystals with floating skeletal debris. Pseudobreccia, Lower Carboniferous, northwest England.*

not widely recognized, but it may be more important than generally thought. Secondary porosity of this type is recorded from the subsurface Smackover Formation (Druckman & Moore, 1985) where it occurs as fabric-selective dissolved grains and solution-enlarged intergranular voids. In the Madison Limestone reservoir of the Williston Basin, secondary porosity also occurs (Elliott, 1982).

The dissolution of $CaCO_3$ in the deep burial realm is generally attributed to the development of pore-waters with high P_{CO_2} formed during the thermal decarboxylation of organic matter. Strongly acidic formation waters may also be produced by sulphate reduction. Such corrosive fluids are most likely to be formed during compaction and thermal maturation of organic-rich shales, so that limestones in the vicinity of such deposits may be expected to develop this dissolution porosity. Hydrothermal fluids may also cause dissolution of limestone or dolomite of course and Pb−Zn mineralization may be expected in vugs and cavities formed in this way. Burial dissolution of sulphate evaporites may lead to Ca^{2+}-rich fluids capable of dissolving dolomite and/or dedolomitization. Dolomite collapse breccias may form in the subsurface updip from dissolving evaporites.

7.7 DIAGENETIC SEQUENCES AND MODELS

Although the patterns of diagenesis vary considerably in limestone formations, there are common sequences of diagenetic fabrics which can be recognized in many limestones. These can be used to erect end-member diagenetic models, analogous to facies models, which provide a summary of information and can be used as a predictive tool when studying the diagenesis of a new formation. The sequence of diagenetic events in a carbonate sediment depends on many factors. To begin with there is the sediment itself, its grain size and texture on the one hand and its mineralogy on the other. These features determine the sediment's diagenetic potential (see Section 9.5) and also its permeability. The latter is most important for rates of dissolution and cementation since it is a major control on the rate of fluid flow through the sediment. Diagenetic processes operate at much higher rates in active zones where pore-fluids move relatively rapidly, compared to stagnant zones, where pore-fluid movement is low or zero. Then there is the nature of the pore-fluids themselves, whether they are meteoric, marine, mixed or hypersaline, their degree of saturation *re* $CaCO_3$, their Mg/Ca ratio, SO_4^{2-} content, etc. Climate is important for early diagenesis both in the marine and meteoric environments. More arid climates promote seafloor cementation and supratidal dolomitization, while more humid climates are likely to favour meteoric dissolution and cementation and mixing-zone dolomitization (see Section 9.5). Pedogenesis and karstification will also reflect the prevailing climate. The style of early diagenesis can have a profound effect on the pattern of later diagenesis, particularly in terms of both physical and chemical compaction. Significant early lithification either by marine or meteoric cements can prevent compaction during burial. Later diagenesis of carbonates is also affected by the burial history and amount of overburden, degree of fracturing, and dewatering of basinal muds. The influx of hydrocarbons into a sequence generally leads to a cessation of carbonate diagenetic processes, since these mostly operate through the medium of porewater.

Four end-member early diagenetic models are presented in Fig. 7.42. In models A and B, early diagenesis takes place in marine pore-fluids, with no seafloor cementation, but grain micritization in model A and seafloor/subseafloor cementation in model B. In models C and D, the early diagenesis takes place in the meteoric environment, with much dissolution in model C and much cementation in model D. In both C and D, high-Mg calcite grains will be converted to calcite (low Mg), but in C aragonite is likely to be dissolved out and the voids left empty, whereas in D, the aragonite grains are more likely to be calcitized. In later diagenesis, there are two end-member models (E and F), one with much compaction, both mechanical and chemical, through a lack of early cementation, and the other with little compaction. There will be similarities between models A and C and models B and D in their response to burial in that the first two will suffer much compactive fracture of grains and pressure dissolution between grains before burial sparite cement precipitation (model E), whereas B and D will suffer little compaction in view of the early cementation generating a framework resistant to overburden stresses (model F). Late through-going stylolites may occur in all models. In both models A and B, if there is no subsequent near-surface meteoric diagenesis then aragonite will be preserved into the burial environment where it is likely to be calcitized, perhaps after some compactive fracture of grains in A.

The reason for the differences in early diagenesis between models A and B could simply be a function of fluid movement rate: a more stagnant pore-fluid in model A and a more active pore-fluid in model B. Model A would be typical of an onshelf, lagoonal muddy sand whereas B would be more typical of a shelf-margin sand body or reef in a high-energy environment with much water flowing through the sediment. The mineralogy and fabric of the marine cements in model B would be determined by seawater composition and flow rates (see Sections 7.4.3 and 9.3). Models A and B would pass into the burial environment with marine pore-fluids if subsidence rates were sufficiently high or sea-level was rising in a transgressive mode so that subaerial exposure did not take place. If subaerial exposure were to occur then the diagenetic path would be to model C or D. The two scenarios C and D could reflect different positions within one meteoric environment: C being in an area

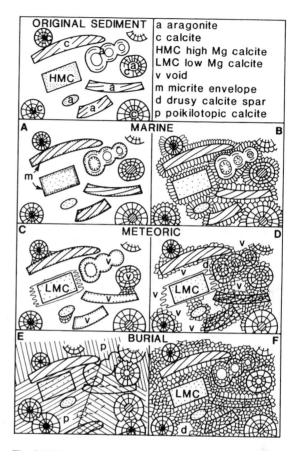

Fig. 7.42 *End-member diagenetic models for marine, meteoric and burial environments. Model A is for marine diagenesis in a poorly agitated or stagnant location where grain micritization is dominant. Model B is for marine diagenesis in a high-energy location where there is much cementation through seawater pumping. Model C is for meteoric diagenesis in an active zone of much dissolution, whereas model D applies to a zone of more CaCO₃-saturated meteoric water where cementation and grain replacement are features. Model E applies to the burial diagenesis of sediments which have little near-surface cement and so compaction is extensive, whereas model F is appropriate for sediments which have been well lithified before any significant compaction. See text for further discussion.*

of considerable throughput of undersaturated freshwater, as in an upper vadose zone or very active phreatic zone, and D being from a site where meteoric waters had become supersaturated *re* CaCO₃ through leaching from strata above or updip, as in the lower part of a vadose zone or in a moderately active phreatic zone. Alternatively, diagenetic model C could reflect a more humid climate compared with D. Evidence of the emergence surface could well occur above limestones with types C and D diagenesis, seen in the form of a paleokarst or paleosol for example.

In all scenarios, one could expect to see the non-luminescent—bright—dull pattern of CL response in the calcite spar as a result of the oxidizing to gradually reducing nature of pore-fluids with increasing burial (see Section 7.6.2). The early meteoric spar in D would be non-luminescent, with bright subzones if the pore-fluid became stagnant at times.

The models A through F have been presented simply showing the precipitation, dissolution and alteration of $CaCO_3$. Dolomite is a common precipitate too of course during diagenesis (Chapter 8). It could be precipitated in both A and B, from seawater less saturated in $CaCO_3$ pumped through the sediments in B, or from sulphate-reduced seawater in A. With diagenetic models C and D, dolomite could be precipitated if the sediments passed into a meteoric—marine mixing zone (see Section 8.7.3), and there was active fluid flow there for a sufficiently long period of time. Dolomite could also be precipitated during burial in both E and F, especially if there were compacting basinal muds close by expelling Mg^{2+}-rich fluids. Under a more arid climate, dolomite could be precipitated within supratidal sediments, and if evaporites were also being precipitated, seepage reflux of dolomitizing brines would be possible.

Closely associated limestones with different styles of diagenesis have been described from a number of formations. In Upper Carboniferous cycles, Heckel (1983) demonstrated that carbonate grainstones deposited during the transgression had a diagenetic style like model A with overpacked grains, neomorphosed aragonitic bioclasts and late ferroan calcite cement, whereas regressive shoal-water grainstones were much more variable, with diagenetic styles like B, C and D. The transgressive grainstone had marine pore-fluids into burial diagenesis whereas the regressive grainstones went through a phase of meteoric diagenesis before burial. Some Lower Carboniferous oolites in south Wales have contrasting diagenetic histories which reflect the prevailing climate (see Section 9.5 and Hird & Tucker, 1988).

Taking modelling of diagenesis a stage further, Matthews & Frohlich (1987) have used a computer program to work out the pattern of diagenesis in bank-margin carbonates subject to high-frequency glacio-eustatic sea-level fluctuations. The outcome is a very complicated picture as the diagenetic environments, meteoric, mixing zone, marine, move up and down through the sediments with the changing position of sea-level. Fortunately, most ancient platform carbonate sequences were not deposited during rapid sea-level changes of amplitude 100 m, as occurred during the last million years.

8 Dolomites and dolomitization models

8.1 INTRODUCTION

The geology of dolomite and dolomitic rocks is a major field of research. There are many unsolved problems and gaps in our knowledge with this mineral and its occurrence in the geological record. Dolomite the mineral, $CaMg(CO_3)_2$, has a complicated crystal structure, should precipitate from seawater but does not, and has an uneven distribution through time. Dolomite the rock is commonly porous and hence a target in petroleum exploration. It is also a host for sulphide mineralization.

Dolomite is a problem mineral since it is difficult to study its formation in laboratory experiments. Being a highly ordered mineral and requiring time to precipitate, it has not yet proved possible to synthesize dolomite at sedimentary temperatures using natural waters. The chemical controls on dolomite precipitation have thus been difficult to elucidate and have mostly been extrapolated from high temperature experiments. Another point which requires explanation is that seawater is supersaturated with respect to dolomite, but because of kinetic factors, it is not being precipitated very widely in the normal marine environment. In addition most of the modern occurrences are in intertidal–supratidal, often evaporative, environments, and this has clouded the interpretation of ancient dolomites. In fact, modern dolomites were really only discovered in the late 1950s and early 1960s.

It was realized early on that dolomites were far more common than limestones in the Precambrian and it was thought that their abundance decreased from the Palaeozoic through the Mesozoic–Cenozoic (e.g. Chilingar, 1956; Ronov, 1964). The lack of good modern analogues for most ancient dolomites and a firm belief in uniformitarianism led to the notion of the 'dolomite problem' (e.g. Zenger, 1972). However, it is now known that the present is not the key to the past for many aspects of sedimentology, and the dolomite distribution through time is not a simple decrease in abundance to the present (Given & Wilkinson, 1987), but another facet of secular variations and the

geotectonic control on sedimentary and diagenetic environments (Chapter 9).

In the past there has been much discussion over the primary (i.e. direct precipitation) versus replacement origin of many dolomites, but particularly the fine-grained dolomites of peritidal facies. The current view is that primary dolomite is very rare, only forming in certain lakes and lagoons, such as in the Coorong and saline lakes of southern Australia (Section 8.7.1), and that most of the dolomite in the geological record is of replacement origin. However, dolomite cements are common and are precipitated directly from pore-fluids during early and late diagenesis (Section 8.4.2).

Two major considerations in the formation of dolomite are the source of the Mg^{2+} ions and the process whereby the dolomitizing fluid is pumped through the carbonate sediments. Seawater is Mg^{2+}-rich and thus the obvious source, but because of the kinetic obstacles to dolomite precipitation in the marine realm, in most dolomitization models seawater is chemically modified to a greater or lesser extent. Five broad categories of dolomitization model are currently available for the interpretation of ancient dolomites: evaporative, seepage–reflux, mixing-zone, burial and seawater models (some shown in Fig. 8.1; Section 8.7). The popularization of modern evaporative and sabkha dolomite in the 1960s led to the interpretation of many ancient dolomites as supratidal in origin. Extending this *evaporative model* (Fig. 8.1A), Mg^{2+}-rich, high Mg/Ca ratio, hypersaline fluids descending into the subsurface were the basis of the *seepage–reflux model* (Fig. 8.1B) and evaporative drawdown model (Fig. 8.1C), devised to account for dolomitized intertidal–subtidal facies, usually beneath evaporites. In the early–mid 1970s, dolomites with no evaporite association were commonly interpreted as the result of the *meteoric–marine mixing-zone model* (Figs 8.1D,E), whereby dilution of seawater was thought to overcome the kinetic hindrances to dolomite precipitation. Evaporation of recharging Mg^{2+}-rich meteoric groundwaters is the basis of the *Coorong model* (Fig. 8.1F). *Burial dolomitization* (Fig. 8.1H) of platform-margin carbonates by Mg^{2+}-rich fluids

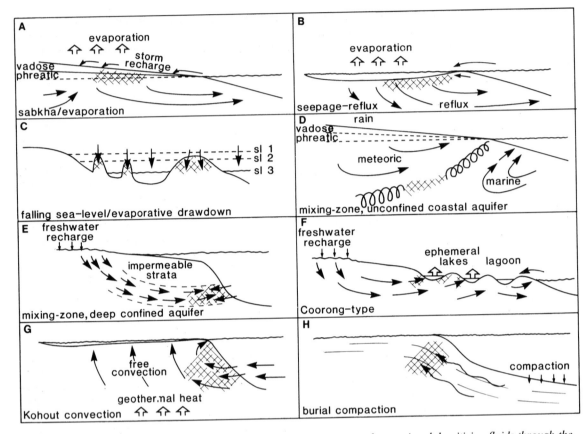

Fig. 8.1 *Models of dolomitization, illustrating the variety of mechanisms for moving dolomitizing fluids through the sediments. In part after Land (1985). Also see Fig. 8.31 for seawater dolomitization models.*

expelled from basinal mudrocks became popular in the late 1970s although the source of the Mg^{2+} and efficiency of the mechanism have been questioned. More recently (1980s), dolomites have been reported forming from normal or only slightly modified seawater in a variety of settings from supratidal to reef to pelagic/hemipelagic and from at to far below the sediment–water interface. These occurrences are the basis for a group of *seawater dolomitization models* (see Fig. 8.31), which include the Kohout convection model (Fig. 8.1G).

8.2 THE DOLOMITIZATION REACTION

The conditions of dolomite formation are difficult to

determine in the laboratory since no-one has been able to precipitate it at low temperatures from natural waters. For direct precipitation one can write:

$$Ca^{2+} + Mg^{2+} + 2(CO_3^{2-}) \rightleftharpoons CaMg(CO_3)_2 \quad (1)$$

and an equilibrium constant K is derived from

$$K = \frac{[Ca^{2+}] \, [Mg^{2+}] \, [CO_3^{2-}]^2}{[CaMg(CO_3)_2]} \quad (2)$$

where the brackets indicate activities of dissolved and crystalline species. The value of K at low temperatures is not precisely known, but from several lines of reasoning (see Hsü, 1967), it would appear to be of the order of 1×10^{-17}. Knowing the approximate activities of Ca^{2+}, Mg^{2+} and CO_3^{2-} in seawater, calculation of the ion activity product ($10^{-15.01}$) shows that seawater is supersaturated with respect to dolomite by nearly two orders of magnitude (Lippmann, 1973). However, dolomite rarely precipitates out of normal seawater, probably for the kinetic reasons outlined below.

Most dolomite forms by replacing $CaCO_3$ and this is often expressed by:

$$2CaCO_3 + Mg^{2+} \rightleftharpoons CaMg(CO_3)_2 + Ca^{2+} \qquad (3)$$

Calculation of the free energy change of this reaction ΔG_r° for aragonite to dolomite gives -1.83 kcal mole^{-1}, indicating that the thermodynamic drive is towards dolomitization. The equilibrium constant for equation (2):

$$K_{cd} = \frac{[Mg^{2+}] [CaCO_3]^2}{[Ca^{2+}] [CaMg(CO_3)_2]} \qquad (4)$$

can be reduced to:

$$K_{cd} = \frac{[Mg^{2+}]}{[Ca^{2+}]} \qquad (5)$$

taking the solid phases as unity (Carpenter, 1980). Hsü (1967) gave a figure of 0.67 for K_{cd}, meaning that the reaction will go to the right and $CaCO_3$ will be dolomitized in solutions where $[Mg^{2+}]/[Ca^{2+}]$ is greater than 0.67. Now seawater has an Mg/Ca molar ratio of around 5.2, clearly showing that $CaCO_3$ should be dolomitized in seawater. Again, the reasons for this not happening on a large scale are kinetic.

Equation (3) above can be criticized in that only Mg^{2+} is brought into the system and Ca^{2+} has to be taken out, otherwise the Mg/Ca ratio of the solution will decrease and the drive towards dolomitization will also decrease. An alternative equation (Lippmann, 1973) is:

$$CaCO_3 + Mg^{2+} + CO_3^{2-} \rightleftharpoons CaMg(CO_3)_2 \qquad (6)$$

wherein both Mg^{2+} and CO_3^{2-} ions are supplied by the dolomitizing fluid and all Ca^{2+} ions of the precursor $CaCO_3$ are incorporated into the dolomite. The free energy of this reaction is -13.24 kcal mole^{-1}, implying an even greater drive towards dolomitization.

To dolomitize completely a limestone formation a vast amount of Mg^{2+} must be pumped through the sequence. Apart from an appropriate chemical environment, an Mg^{2+} source and efficient transport system are needed. Potential fluid sources are seawater, and subsurface fluids of marine and/or meteoric origin, and in addition Mg^{2+} could be released from high-Mg calcite and smectite clays.

Land (1985) has calculated that to dolomitize a cubic metre of typical Recent marine carbonate sediment (40% porosity, 6.3 mole % $MgCO_3$) would require around 650 pore volumes of seawater. If seawater is diluted with meteoric water, then the volume needed increases dramatically (6500 pore volumes for seawater diluted 10 times). Much less fluid is needed if seawater is evaporated; at halite saturation only 30 pore volumes are required. Without an external source of Mg^{2+}, the amount of dolomite that can form is very limited, being dependent on the amount of high-Mg calcite present. A sediment solely composed of high-Mg calcite (19 mole % $MgCO_3$) would only give around 30% dolomite.

The only abundant source of Mg^{2+} ions for early diagenetic surface and near-surface dolomitization is seawater. It contains 1290 ppm Mg (0.052 moles l^{-1}) and 411 ppm Ca (0.01 moles l^{-1}), i.e. an Mg/Ca weight ratio of 3.14 or molar ratio of 5.2. By way of contrast, meteoric water, although variable, has much lower values of both; for example, average river water has 15 ppm Ca and 4 ppm Mg giving a molar Mg/Ca ratio of 0.44. Thus seawater is generally considered the source of the Mg^{2+} ions for most early dolomitization, but this water is usually modified in the various dolomitization models in vogue. Dolomitization in the burial environment can involve modified marine waters buried along with the sediment or formation waters originally of meteoric origin which have acquired Mg^{2+}. Formation waters vary considerably in composition (see Section 7.6.1), but generally they have a much lower Mg/Ca ratio (1.8 to 0.04) than seawater (5.2), as a result of Mg loss to clay minerals and early dolomite formation, and Ca release from limestones undergoing mineral stabilization and pressure dissolution. The clay mineral transformation smectite to illite during deep burial does release Mg^{2+}, but there has been much discussion over whether there is sufficient for more than local dolomitization of limestone adjacent to shale (see Section 8.7.4 on burial dolomites).

That dolomite is not widely precipitating out of seawater and $CaCO_3$ is rarely being dolomitized on the seafloor are attributed to the kinetic problems of forming the dolomite mineral. The most important of these are (1) the high ionic strength of seawater and fast carbonate precipitation rates, (2) hydration of the Mg^{2+} ion, and (3) the low activity of CO_3^{2-}. The main problem with precipitating dolomite is that it is a highly ordered mineral (Section 8.3), and in a saline solution like seawater, which is also supersaturated with respect to $CaCO_3$, the other carbonate minerals, aragonite and high-Mg calcite, with their simpler structures, are precipitated preferentially. Fast rates of carbonate precipitation also hinder dolomite precipitation, again favouring the more easily formed $CaCO_3$ minerals, and where dolomite does precipitate from very saline solutions, as in evaporitic settings (Section 8.7.1), then it is poorly ordered and Ca-rich, with ions out of their correct lattice position trapped by the fast-growing crystals.

Magnesium ions are more strongly hydrated in seawater than Ca^{2+} ions and they are also involved in ion pairing. This means that there are fewer Mg^{2+} ions available for reaction and that Ca^{2+} ions more easily enter growing Ca–Mg carbonate crystals. Once incorporated on to a growing crystal surface, the hydrated Mg^{2+} ions then provide a barrier for the CO_3^{2-} anions. The hydration of Mg^{2+} is more easily overcome through crystal growth at higher temperature, so that this is a factor often cited in favour of burial dolomitization.

Dolomite apparently forms from the carbonate anion (CO_3^{2-}) rather than the bicarbonate anion (HCO_3^-) since the former has the ability to dislodge water dipoles from Mg^{2+} ions on the growing crystal surface (Lippmann, 1973). This suggests that dolomitization is favoured from solutions with high alkalinity and high pH, where the CO_3^{2-} ion dominates over the HCO_3^-. In seawater (pH 7.8–8.2) the alkalinity is mostly in the form of HCO_3^- and unionized H_2CO_3, rather than CO_3^{2-}. The low activity of CO_3^{2-} in seawater means that there are insufficient CO_3^{2-} ions to break through the hydration barrier on the Mg^{2+} ions.

The kinetic hindrances to dolomite precipitation from seawater can be overcome by evaporating seawater, diluting it, raising the temperature or lowering the SO_4^{2-} content.

With the evaporation of seawater, gypsum–anhydrite and aragonite are precipitated and this raises the Mg/Ca ratio of the pore-fluids. In addition, with increasing salinity and consequent decreasing activity of water in solution, the proportion of less strongly hydrated Mg^{2+} ions should increase markedly. Eventually dolomite is precipitated in the abundant supply of Mg^{2+} ions. Dolomite formation in this style is well documented from the Arabian Gulf sabkhas, where Mg/Ca ratios range from 2.5 to 27 and according to Patterson & Kinsman (1982) dolomite is being precipitated where the Mg/Ca ratio is in excess of 6. Evaporative dolomite itself is unlikely to form very thick dolomite sequences but the dolomitizing fluids developed in such a setting are employed in the seepage–reflux model (Section 8.7.2).

Diluting seawater can overcome the problem of high ionic strength, and of the interfering effect this has on dolomite precipitation, and the problem of rapid rates of precipitation. If seawater is diluted with meteoric water, then the Mg/Ca ratio of 5 is maintained and it should be easier to precipitate dolomite (Fig. 8.2; Folk & Land, 1975). Unfortunately, it has not been possible to confirm this experimentally, because dolomite precipitation at low temperature is such a slow process, but there are geochemical arguments from trace elements and isotopes that many ancient dolomites have formed from dilute waters. Dilution of seawater is the principle of meteoric–marine mixing-zone dolomitization. Seawater provides the Mg^{2+} and the meteoric water permits the dolomite precipitation by overcoming the kinetic inhibitions (Land, 1973a,b; Folk & Land, 1975). An important feature of this mixing model is that whereas dolomite is supersaturated in solutions with more than 5% seawater, calcite is undersaturated with up to 50% seawater. The basis for this situation is that the solubility curves of calcite and dolomite are not linear. Runnels (1969) showed that if the solubility curve of a mineral in the presence of added electrolyte (such as NaCl) was non-linear, then it would be possible to produce an undersaturated solution by mixing two saturated solutions, or to get supersaturation from two undersaturated solutions. Badiozamani (1973)

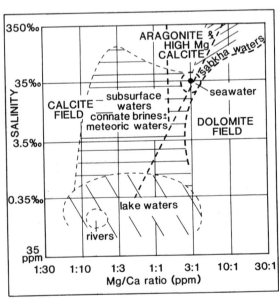

Fig. 8.2 *The precipitational fields of calcite, dolomite and aragonite plus high-Mg calcite in terms of salinity and Mg/Ca ratio of natural waters. After Folk & Land (1975).*

determined the effect of mixing Yucatan groundwater with seawater on the degree of saturation of calcite and dolomite. This showed that the mixture was less saturated in calcite than the groundwater, until 50% seawater, but that the dolomite saturation increased continuously as seawater was added (Fig. 8.3A). Thus in the interval 5–50% seawater, the solution should be capable of dolomitization by replacement of calcite. Although this dilution dolomitization mechanism overcomes several kinetic problems, it has been suggested by Carpenter (1976) that diluting seawater decreases the thermodynamic drive towards dolomitization by reducing the all important Mg/Ca ratio. Taking mixing a stage further, Wigley & Plummer (1976) noted the importance of the P_{CO_2} of the waters, since this has an important effect on carbonate saturation. They were able to show that with high P_{CO_2} dolomitization of calcite could take place in mixtures of groundwater and 35–75% seawater.

Lately, there have been more strong criticisms of meteoric–marine mixing (e.g. Machel & Mountjoy, 1986; Hardie, 1987). Hardie, for example, showed that if solubility constants for the more soluble disordered dolomites are used in Badiozamani's model (Fig. 8.3A), rather than values for ordered dolomite, then the range of fluid compositions for dolomitization is considerably reduced (Fig. 8.3B). In addition, it is true to say that significant large-scale replacement of calcite by dolomite does not appear to be taking place in modern coastal mixing zones, although this may be a question of time. Modern mixing zones have only been established for a few thousand years.

The presence of SO_4^{2-} anions in waters and certain organic compounds may be kinetic inhibitors of dolomite precipitation. Experiments by Baker & Kastner (1981) on the dolomitization of calcite and aragonite revealed that the SO_4^{2-} content of the fluids was an important control. For calcite, a little SO_4^{2-} (less than 5% of its seawater value) is sufficient to inhibit strongly its dolomitization, whereas aragonite can still be dolomitized at somewhat higher SO_4^{2-} contents. It was suggested that this is a prime reason for the absence of dolomite precipitation out of normal seawater. In situations where the SO_4^{2-} content is lowered, then dolomitization can be expected. One effective mechanism for this is microbial reduction of sulphate, taking place in organic-rich sediments. Dolomite forming in such anoxic marine pelagic sediments is now well documented from DSDP cores (e.g. Gulf of California, Kelts & McKenzie, 1982), and the $\delta^{13}C$ values support the involvement of CO_2 from organic matter diagenesis (Section 8.6). Microbial

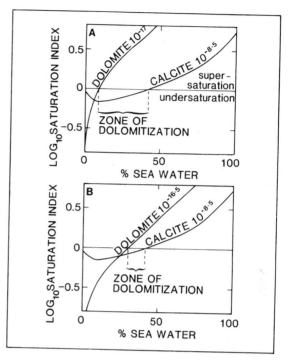

Fig. 8.3 *Mixing-zone dolomitization. Solubility curves for dolomite and calcite in meteoric water with increasing content of seawater. Dolomitization is considered to take place in waters supersaturated with respect to dolomite but undersaturated with respect to calcite. (A) For dolomite with solubility product $K = 10^{-17}$, as used by Badiozamani (1973). (B) For disordered dolomite with $K = 10^{-16.5}$, as used by Hardie (1987). Note that in (B) the zone of dolomitization is much reduced. After Hardie (1987).*

sulphate reduction will also take place in intertidal–supratidal sediments where cyanobacterial mats are being decomposed. A byproduct of sulphate reduction is increased alkalinity through production of HCO_3^- and CO_3^{2-} from the organic matter, and, as noted earlier, this also favours dolomitization. The lowering of pore-fluid SO_4^{2-} values can also occur through the precipitation of $CaSO_4$ minerals gypsum and anhydrite in evaporitic settings. Lower SO_4^{2-} could thus explain sabkha dolomite rather than the high Mg/Ca ratios. Because of the low SO_4^{2-} content of most meteoric waters (mean river water 1.2 ppm), dilution of seawater (2650 ppm SO_4^{2-}) also lowers the sulphate content. Although the significance of sulphate in dolomite precipitation is not everywhere accepted (e.g. Morrow & Ricketts, 1986; Hardie, 1987), it is another factor to consider, even though the actual mechanism of inhibition is not known.

Experiments with organic acids, such as aspartic acid, which occur in marine carbonates, have revealed that these also inhibit dolomitization (Gaines, 1980). When the organic material was oxidized, then dolomitization of aragonite (ground ooids) proceeded more rapidly. From this, one could suggest that dolomitization of $CaCO_3$ is more likely to take place where organic matter in the carbonates has already been degraded.

8.3 DOLOMITE CRYSTAL STRUCTURE, STOICHIOMETRY AND ORDERING

Dolomite is a rhombohedral carbonate consisting of alternating layers of carbonate anions and the cations. In 'ideal' dolomite, there is an equal number of Ca and Mg ions, and these are arranged in separate sheets, with CO_3^{2-} planes between (see Section 6.2.1). The carbonate anions are triangular in form and have similar orientations in each layer, but these are reversed in successive planes. The layered structure is oriented normal to the c-crystallographic axis of the dolomite crystal, which belongs to the $\bar{3}$ class of the trigonal system (trigonal subsystem of the hexagonal system in North American terms). The smallest unit cell is an acute primitive rhombohedron of one $CaMg(CO_3)_2$ with its long axis oriented parallel to the c-axis. Formerly, dolomite crystals were described using rhombohedral axes, but now they are more commonly described by the hexagonal axial system, with the rhombohedron arranged in an obverse position. the height of the hexagonal unit cell is the same as for the acute rhombohedral unit cell, although there are three horizontal axes at 120° and it contains three $CaMg(CO_3)_2$. Further details of dolomite structure are given in Section 6.2.1 and by Reeder (1983).

Many, if not most, natural dolomites are not stoichiometric, and so do not have the ideal molar ratio of $CaCO_3/MgCO_3$ of 50:50 ($Ca_{0.5}Mg_{0.5}CO_3$). Instead, there is commonly an excess of Mg, up Ca/Mg=58:42, and less commonly an excess of Ca^{2+}, up to $Ca_{48}Mg_{52}$. In addition, other ions, such as Fe, Mn, Na and Sr can substitute for the cations. The effect of Ca substitution for Mg is to increase the lattice spacing since Ca^{2+} is a larger ion than Mg^{2+}. This can be detected by X-ray diffraction (XRD) since the lattice reflections occur at lower 2θ angles as the Ca content increases. XRD is thus commonly used to determine the Ca/Mg ratio of a dolomite, by measuring the displacement of the d_{104} peak relative to a standard such as halite or fluorite. Lumsden

(1979) produced an equation relating mole per cent $CaCO_3$ (N_{CaCO_3}) to the $d_{[104]}$ spacing measured in angstrom units (d):

$$N_{CaCO_3} = Md + B$$

where M is 333.333 and B is -911.99. The d_{104} spacing for 50.0% $CaCO_3$ is taken as 2.886 Å and for 55.0% $CaCO_3$ as 2.901 Å, based on Goldsmith & Graf (1958a).

Iron can substitute for the cations in dolomite to give ferroan dolomite (>2 mole% $FeCO_3$) with the term ankerite used for high iron Ca–Mg carbonates which approximate to $CaMg_{0.5}Fe_{0.5}(CO_3)_2$. Most of the Fe^{2+} replaces Mg^{2q}, as is shown by the common positive correlation between mole % Ca and mole% Fe. In view of the slightly larger size of Fe^{2+} (ionic radius 0.86 Å) relative to Mg^{2+} (0.80 Å), with more than a few mole% $FeCO_3$ there is noticeable increase in the lattice spacing of d_{104} (Goldsmith & Graf, 1958b; Runnels, 1970; Al Hashimi & Hemingway, 1974). In addition, the intensities of the XRD reflections are weaker in ferroan as opposed to non-ferroan dolomites (Goldsmith & Graf, 1958b). See Hardy & Tucker (1988) for an account of the XRD technique for dolomites.

The segregation of the cations into separate sheets in the dolomite lattice results in a set of superstructure reflections with X-ray diffraction (peaks 021, 015 and 101), which are not present in the structurally similar calcite. The sharpness and relative intensities of these ordering peaks can be used to give a measure of the degree of ordering of the dolomite crystal. The greater the ratio of the heights of ordering peak 015 to diffraction peak 110, the higher the degree of order.

Dolomites which are non-stoichiometric are generally less well ordered than 'ideal' dolomite, through the occurrence of some Ca ions in the Mg sheet (or vice versa). It is theoretically possible for a 50:50 Ca/Mg carbonate to have no ordering reflections, if the cation sheets in the lattice are equal mixtures of Ca and Mg. In practice, all naturally occurring dolomites are ordered to an extent (otherwise, strictly, the mineral is not dolomite) with most modern dolomites showing poor ordering reflections, compared with many ancient dolomites. At high temperatures (over 100°C), the amount of disorder increases, and above 1200°C, complete disorder may exist (Goldsmith & Heard, 1961b).

The term *protodolomite* was introduced by Goldsmith & Graf (1958a) for poorly-ordered dolomite manufactured in the laboratory and this term has been employed for some modern dolomites.

Discussion of the word protodolomite (Gaines, 1977, 1978; Deelman, 1978b; Gidman, 1978; Land, 1980) has concluded that the term should be kept for synthetic dolomites, and if a naturally occurring Ca−Mg carbonate has the ordering reflections, no matter how weak, then it is a dolomite. Many dolomites with an excess of Ca are simply referred to as calcian dolomite.

An effect of poor ordering and non-stoichiometry is that the dolomite crystals are metastable and more soluble, compared to 'ideal' dolomite. Stabilization of calcian dolomite to a more ideal type will mostly take place by dissolution reprecipitation, since solid-state diffusional processes only operate at the submicron scale and are slow. More stoichiometric dolomites are apparently developing from less ordered and Ca-rich dolomites in the shallow subsurface of the Abu Dhabi sabkha (McKenzie, 1981) and on the tidal flats of Sugarloaf Key, Florida (Carballo et al., 1987).

Apart from providing a characterization of ancient dolomite, the stoichiometry and ordering can be used to distinguish between different types of dolomite. Lumsden & Chimahusky (1980) and Morrow (1978, 1982a) identified three broad groups of dolomite, based on stoichiometry, texture and whether associated with evaporites or not: (1) coarsely-crystalline, sucrosic dolomites which are generally near-stoichiometric (mode 50.0−51.0% $CaCO_3$), (2) finely-crystalline dolomites not associated with evaporites which are generally Ca-rich (54−56% $CaCO_3$), and (3) fine-grained dolomites associated with evaporites which are also nearly stoichiometric (mode 51.0−52.0% $CaCO_3$). Types 2 and 3 are usually early diagenetic, near-surface in origin. The underlying cause of these associations is thought to be the salinity and Mg/Ca ratio of dolomitizing solutions, with a climatic control important for groups 2 and 3. Where there is an evaporite association (group 3), indicating an arid climate, then pore-fluids are likely to have had a high Mg/Ca ratio from precipitation of gypsum−anhydrite and aragonite. It is contended that the abundance of Mg^{2+} ions in the fluids would result in near-stoichiometric dolomite. The calcian dolomites of group 2 are thought to have formed from solutions with lower Mg/Ca ratios, such as occur in mixing zones, which are more active during humid climatic times (Section 8.7.3). Group 1 dolomites are generally of late diagenetic burial origin and the near stoichiometry could reflect slow growth from dilute solutions, possibly aided by elevated temperatures. Stoichiometry and ordering data from Dinantian dolomites of south Wales (Hird, 1986) show that tidal flat dolomicrites are less well ordered than mixing-zone and burial

types (Fig. 8.4).

Using new and published XRD data, Sperber et al. (1984) obtained two pronounced modes at 51 and 55 mole % $CaCO_3$ in Phanerozoic dolomites which ranged from 48 to 57 mole% $CaCO_3$ (Fig. 8.5A). They also found a bimodal distribution in the percentage of dolomite in carbonate rocks (Fig. 8.5B): a mode at 97% dolomite (dolostones) and at 20% (dolomitic limestones), indicating that carbonate rocks are either partially or completely dolomitized. Sperber et al. (1984) suggested that the dolomitic limestones, which generally contain rhombs of calcian dolomite, originated in diagenetically closed systems during high-Mg calcite dissolution−low-Mg calcite and dolomite precipitation, so that in these rocks there was an internal supply of Mg^{2+}. The dolostones, on the other hand, consist of more stoichiometric dolomite and are thus considered to have originated in more open diagenetic systems. A trend towards more stoichiometric dolomite in older dolostones is evident from the data of Lumsden & Chimahusky (1980) and Sperber et al. (1984) and there is also a broad correlation of increasing stoichiometry with increasing crystal size. Both these features are consistent with dolomites undergoing dissolution reprecipitation− during diagenesis with the formation of more stoichiometric, better-ordered dolomite from a less

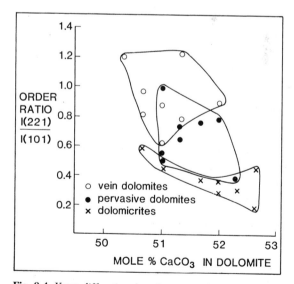

Fig. 8.4 X-ray diffraction data from some Lower Carboniferous dolomites from south Wales. The three types of dolomite, peritidal dolomicrites, pervasive mixing-zone dolomitized subtidal carbonates and burial, 'vein' dolomites, commonly baroque and developed along joints, plot in different fields with some overlap. After Hird (1986).

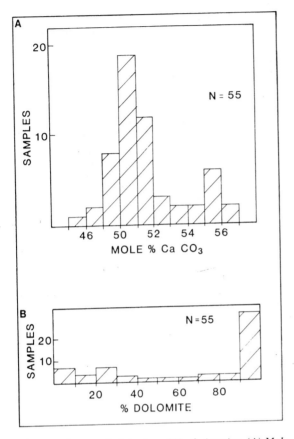

Fig. 8.5 *Phanerozoic dolomites of North America. (A) Mole % CaCO₃ in dolomite of 55 samples showing two groupings, one stoichiometric, the other calcian. (B) Per cent dolomite in the samples showing two groupings, one of 90–100% dolomite (dolostones), the other of 10–30% dolomite (dolomitic limestones), with few between. After Sperber* et al. *(1984).*

stoichiometric, poorly-ordered original precipitate.

As a result of the complex structure of dolomite crystals, defects in the lattice are common. These microstructures are on a scale of 1 to a few hundred angstroms and are studied with the transmission electron microscope (TEM), a relatively new technique as applied to carbonates (Wenk *et al.*, 1983). Lattice defects are generally classified on their geometry: *point defects* arise from vacant atom spaces, foreign atoms or atoms between lattice planes; *line defects* or *dislocations* are due to displacement of the lattice along a line; and *planar defects* include twin boundaries, stacking faults and antiphase and interphase boundaries, where a whole plane of atoms is displaced. Microstructures are produced during crystal

growth, deformation and phase transformation, so that their study can throw light on the origin (e.g. cement versus replacement) and history of the dolomite crystals. The effect of stress on crystals is to produce dislocations as a result of movement on crystallographically-defined slip planes. Transformation microstructures occur where there is a lattice mismatch across domain boundaries in a crystal. Domains are commonly produced when a crystal attains a more ordered structural or chemical state. Antiphase boundaries (APBs) occur where two domains can be brought to coincidence by a lattice translation; twin boundaries occur where a rotation is required and inversion boundaries where an inversion of the lattice is needed. Lattice defects commonly produced during crystal growth include dislocations and twin planes (can be similar to those produced during deformation and transformation), growth bands and stacking faults. A common feature of calcian dolomites is a modulated structure (Reeder, 1981), with a lamellar spacing of 100–200 Å and an orientation parallel to one of the cleavage rhomb faces. 'Tweed structures' arise where there are two modulations but the second one is usually coarser and more irregular and is interpreted as growth banding. Modulated structures appear to be coarser in older rocks, but they are apparently absent in stoichiometric dolomites which are generally homogeneous in microstructure. Modulated structures have been attributed to cation disordering through the excess Ca, but they could also reflect some disorder in the anion sublattice (Reeder & Wenk, 1979; Reeder, 1981). Baroque dolomites (Section 8.4.3) contain complex arrays of fault-like growth defects, as well as the modulated structures. Recent dolomites have a range of microstructures from irregular heterogeneous with many growth defects (dislocations, faults and subgrain boundaries) to modulated structures. It is likely that the high defect density of modern dolomite decreases its stability relative to more ordered stoichiometric dolomite.

8.4 DOLOMITE PETROGRAPHY

Dolomite occurs in a wide range of crystal forms, fabrics and mosaics, just like calcite, although there are some specific types of dolomite with no calcite equivalent. Much dolomite of course is a replacement of pre-existing CaCO₃ and there is a wide range of replacement fabrics developed, but dolomite is commonly a cement too, filling voids in the same way as calcite spar (Section 7.6.2).

8.4.1 Replacement dolomite

Replacement of $CaCO_3$ by dolomite ranges from fabric destructive to retentive and from fabric selective to pervasive. The important factors here are grain original mineralogy and crystal size, timing of dolomitization and nature of dolomitizing fluids. The shapes of dolomite crystals in replacement mosaics vary from anhedral to euhedral rhombs, with the terms xenotopic and idiotopic referring to the mosaics (Fig. 8.6). Sibley & Gregg (1987) placed emphasis on the nature of crystal boundary shapes, recognizing non-planar, planar-e (euhedral) and planar-s (subhedral) types (Fig. 8.6).

Dolomite crystals can also faithfully pseudomorph the crystal forms of the $CaCO_3$ precursor. This happens when the dolomite nucleates in optical continuity with the $CaCO_3$ precursor. The important factor for pseudomorphic replacement is the number of nucleation sites (Sibley, 1982). For most fossils and ooids, composed of numerous microcrystals, many nucleation sites are required to retain the crystallographic orientations of the precursor crystals, otherwise, where there are few nucleation sites the dolomite crystals will grow and replace many microcrystals of the grains, producing non-pseudomorphic crystals, with optical orientations unrelated to the precursor microcrystals. In general, fine, precursor $CaCO_3$ mosaics are more likely to be replaced pseudomorphically, since nucleation sites are likely to be abundant, although large calcite crystals, such as comprise echinoderm grains, can be pseudomorphically replaced, since few nucleation sites are needed. Two controls on dolomite replacement are precursor mineralogy and grain/crystal size. Many authors have noted that high-Mg calcite and aragonite grains are preferentially dolomitized relative to low-Mg calcite, and commonly the high-Mg calcite grains are dolomitized with good fabric retention (Fig. 8.7). Well-preserved but dolomitized coralline red algal, foraminiferal and echinoid grains, all originally of high-Mg calcite, have been observed by Murray (1964), Land & Epstein (1970) and Sibley (1980, 1982). Coralline algal grains have a cryptocrystalline structure which is commonly replaced by microcrystalline dolomite (<10 μm; Sibley, 1982). Intraparticle voids are usually filled by the dolomite too. Foraminifera, such as *Amphistegina*, with a radial extinction pattern may be faithfully replaced by dolomite but in the Pliocene Seroe Domi Formation, many more are partially to completely dissolved out. Where echinoids and forams had been converted to

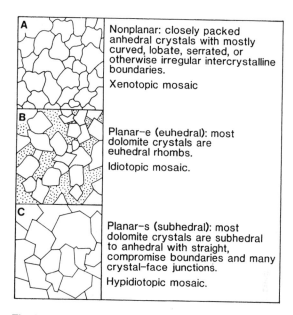

Fig. 8.6 *Three common dolomite textures. (A) Non-planar crystals in a xenotopic mosaic. (B) Planar-e crystals (e for euhedral) in an idiotopic mosaic. (C) Planar-s crystals (s for subhedral) in a hypidiotopic mosaic. After Sibley & Gregg (1987).*

low-Mg calcite before dolomitization, then they were resistant and only suffered minor non-pseudomorphic replacement. Aragonite grains, such as mollusc and coral, are either replaced by anhedral dolomite mosaics or dolomite cement fills the skeletal moulds (Fig. 8.7), indicating a phase of aragonite dissolution prior to dolomite precipitation (e.g. Land, 1973a, in the Hope Gate). Low-Mg calcite grains tend to resist dolomitization, as is seen in many ancient dolomites where brachiopods are still calcite in a matrix of dolomite. Alternatively here, it could be that there were more nucleation sites in the matrix compared to the brachiopods, because of crystal size differences.

An experimental study of skeletal dolomitization by Bullen & Sibley (1984) confirmed some of the above observations. They found that echinoids, coralline algae and foraminifera are dolomitized with fabric retention, although the dolomite crystals are somewhat coarser than original high-Mg calcite crystallites, whereas bivalves, gastropods and corals (aragonite) were dolomitized with fabric destruction, and less quickly. Interestingly, mimic replacement of the high-Mg calcite grains was still observed when they were converted to low-Mg calcite before dolomitization, suggesting a crystal size control rather

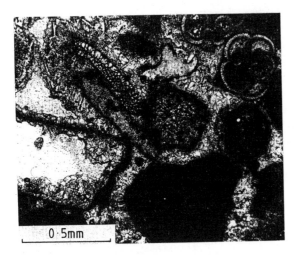

Fig. 8.7 *Dolomitized Pleistocene limestone, Aruba, Netherlands Antilles. Calcareous algae and foraminifera have been replaced with no fabric destruction, contrasting with the molluscan grain on the left which was originally composed of aragonite, now dissolved out (micrite envelope retained shape) and the void is being filled by a dolomite cement. Dolomite between grains is a replacement of marine cement and is a cement itself. Sample courtesy of Duncan Sibley. Photomicrograph plane polarized light.*

than a mineralogical control. High-Mg calcite grains of course do contain up to 19 mole% $MgCO_3$ and it has frequently been suggested that this aids dolomitization (e.g. Blake *et al.*, 1982).

Lime muds are commonly dolomitized preferentially (e.g. Murray & Lucia, 1967) and the many nucleation sites result in a fine-grained dolomite replacement and therefore retention of the sedimentary structures and gross textures (e.g. Fig. 8.8). The latter is well seen in modern supratidal dolomites (e.g. Shinn, 1983a) where micron-sized dolomite has largely replaced aragonite needles. Some of this micritic dolomite could be a direct precipitate. Although difficult to prove in carbonate tidal flats, such an origin is likely for dolomite within a siliciclastic tidal flat in Baja California (Pierre *et al.*, 1984).

Xenotopic to idiotopic mosaics of anhedral to euhedral dolomite crystals (Fig. 8.6) comprise pervasively dolomitized limestones, and commonly they do not preserve much of the original sedimentary fabric (Fig. 8.9). The term sucrosic is often applied to a porous mosaic of euhedral rhombs. Dolomite crystals in these mosaics usually have cloudy centres and clear rims (e.g. Murray, 1964; Sibley, 1980, 1982). The cloudiness arises from the presence of inclusions, mineral relics of the $CaCO_3$ precursor and empty or fluid-filled microcavities. The mineral relics suggest that during the early stages of dolomitization, the

Fig. 8.8 *Tidal flat dolomicrites with good textural preservation. (A) Dolomite crust from the Bahamas showing peloidal-micrite, algal filaments and irregular fenestrae. (B) Dolomitized stromatolitic laminites with evidence of desiccation, thin laminoid fenestrae beneath laminae, intraclasts, and coarse grainstone layers of probable storm origin. Caswell Bay Mudstone, Lower Carboniferous, south Wales.*

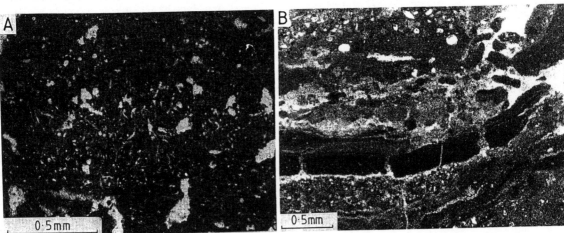

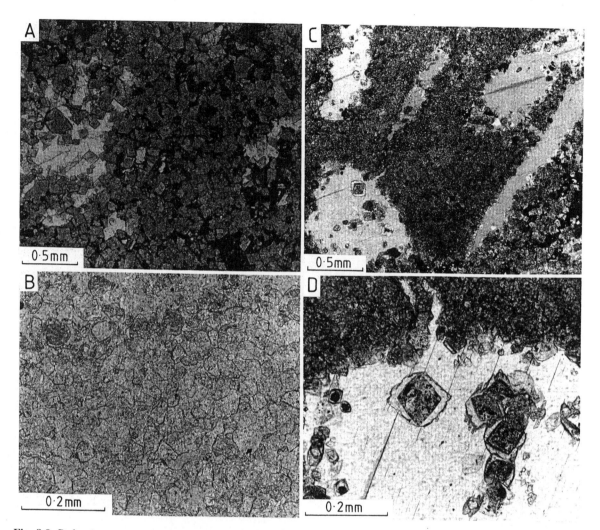

Fig. 8.9 *Dolomite textures. (A) Idiotopic mosaic of euhedral dolomite rhombs. After dolomitization, there was patchy poikilotopic anhydrite precipitation (white crystals) and then oil entered the rock. Smackover Formation, subsurface Arkansas. (B) Xenotopic mosaic of anhedral dolomite crystals. (C) Dolomitized nummulitic packstone with fine xenotopic mosaic and nummulite moulds now filled with poikilotopic gypsum. (D) Detail of (C) showing cloudy rhombs and clear outer zone, slightly etched, within nummulite mould. Eocene, Tunisia.*

fluids were unable to dissolve the $CaCO_3$ completely. There are usually chemical differences between the inner and outer zones of dolomite rhombs; CL and back-scattered electron microscopy (BSEM) commonly reveal zonations due to Fe and Mn contents (Figs 8.10 and 8.11; Choquette & Steinen, 1980; Fairchild, 1980b; Gawthorpe, 1987). Trace element zonations across rhombs have been taken to indicate precipitation of the clear rims from more dilute solutions (Land *et al.*, 1975) and the latter would also

explain the loss of $CaCO_3$ relics if the solutions had also become undersaturated with respect to calcite (Sibley, 1982).

With many pervasive dolomites, it is likely that precipitation did not take place in one event, but over a much longer period of time, from shallow into deeper burial. Commonly, evidence for this is seen in the form of fracturing of early dolomite rhombs and overgrowth by later dolomite or in the occurrence of dissolution surfaces within rhombs (Figs 8.10 and

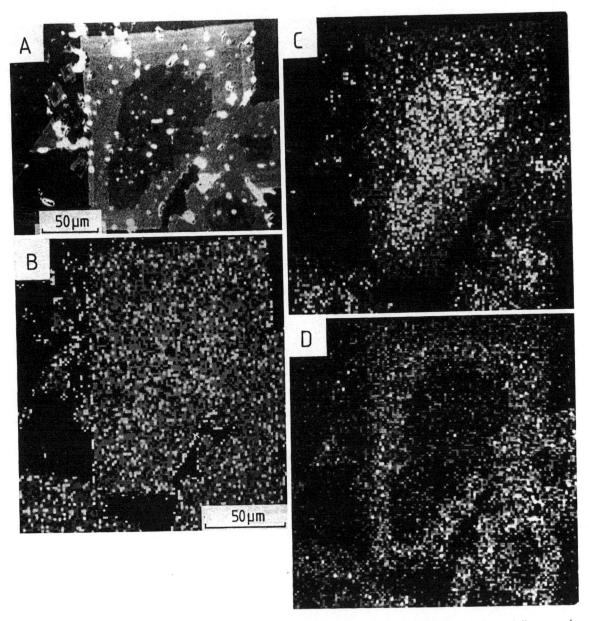

Fig. 8.10 *Zonation in a dolomite rhomb. (A) Back-scattered electron micrograph (BSEM) showing a dull core and lighter rim (the irregular boundary indicating some etching before precipitation of the rim). The lighter zones suggest the presence of elements with higher atomic numbers compared to the dull core. (B) DIGIMAP view of rhomb in (A) for Ca^{2+} showing uniform distribution. (C) DIGIMAP view for Mg^{2+} showing depletion in crystal rim relative to core. (D) DIGIMAP view for Mn (Fe is similar) showing low content of core (dark) compared to rim. Pendleside Limestone, Lower Carboniferous, northwest England. Photos courtesy of Rob Gawthorpe.*

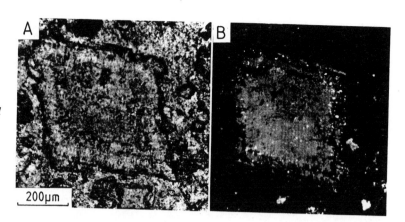

Fig. 8.11 *Dolomite rhomb with cloudy centre and clear outer part, and from CL evidence of etching and partial dissolution of early precipitated dolomite. (A) Plane polarized light. (B) Cathodoluminescence. Pendleside Limestone, Lower Carboniferous, northwest England. Photos courtesy of Rob Gawthorpe.*

8.11). Pre- and post-compactional rhombs may be evident from inclusion patterns (e.g. Figs 8.12 and 8.13) in partly dolomitized limestones.

Reviewing the occurrence of xenotopic and idiotopic dolomite mosaics, with new experimental work, Gregg & Sibley (1984) suggested that the temperature at which the crystals were growing was a major factor. They noted that xenotopic texture is typical of pre-Cainozoic dolomites whereas idiotopic texture is common in dolomites of all ages. From crystal growth theory, it appears that at low temperatures a smooth crystal surface is energetically favoured, so that crystal mosaics consist of euhedral—subhedral crystals, whereas above a 'critical roughening temperature' (CRT), a rough surface is favoured, leading to a mosaic of anhedral crystals. Gregg & Sibley (1984) suggested that the CRT for dolomite was between 50 and 100°C, so that xenotopic textures should result from burial dolomitization of limestone or burial recrystallization of earlier-formed near-surface dolomite. Some idiotopic textures can be expected to form at higher temperatures where crystals grow into cavities or are affected by impurities such as clay and organic matter. Since calcite has a CRT around 25°C, idiotopic textures are rarely developed, and most neomorphic limestones have xenotopic mosaics.

In addition to temperature affecting the dolomite texture, the saturation state of the dolomitizing fluid is also important. Where this is high, then dolomitization is likely to be pervasive and all components, whatever their mineralogy and crystal size, will be replaced. Where the saturation state is lower, then only the more susceptible components (aragonite plus high-Mg calcite, and finely crystalline low-Mg calcite) will be dolomitized. Unreplaced $CaCO_3$ allochems

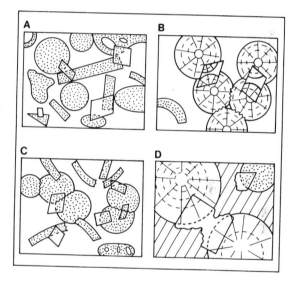

Fig. 8.12 *Common dolomite rhomb relationships in limestones. (A) Pre-compaction rhombs, grains in point contact. (B) Compaction after rhomb precipitation so that ghost textures in rhombs are displaced relative to grains. (C) Post-compaction rhombs including fractured grains and stylolites. (D) Pre-sparite rhombs etched and replaced.*

may dissolve out later to give moulds. Thus, to describe dolomite textures completely, attention should be focused on the matrix and voids and on the preservation state of the allochems, in terms of whether they are unreplaced, partially to completely replaced or moulds, and whether the replacement is mimic or not. Sibley & Gregg (1987) have provided such a scheme (Table 8.1), which includes the crystal boundary shape (planar to non-planar) and crystal size distribution (unimodal to polymodal).

Table 8.1 Terms for describing dolomite textures. After Sibley & Gregg (1987)

Dolomite crystal size	Unimodal or polymodal
Dolomite crystal shape	Planar or non-planar
CaCO$_3$ allochems ⎫	Unreplaced or replaced or moulds
CaCO$_3$ matrix ⎬	If replaced: partial or complete,
CaCO$_3$ cement ⎭	mimic or non-mimic
Void-filling dolomite	Limpid, rhombic, drusy, baroque

Fig. 8.13 *Pre-compaction dolomite rhomb. Notice that the edge of the ooid is displaced within the rhomb. Gully Oolite, Lower Carboniferous, south Wales.*

It is usually very difficult to demonstrate recrystallization of earlier-formed dolomite, although there is much circumstantial evidence that it has taken place. For example, much dolomite of modern evaporitic settings is very fine grained with crystals 0.5–5 µm across (Section 8.7.1), whereas ancient supratidal dolomites are generally coarser (5–20 µm). Recrystallization of early dolomite during burial can be expected from the CRT approach and also since the early dolomites are usually non-stoichiometric and poorly ordered, and the crystals are very small and thus more reactive. Burial recrystallization could also account for depletion in trace elements Sr and Na, and the picking up of Fe and Mn, since the process would have to be wet dissolution–reprecipitation. Isotopic signatures, especially $\delta^{18}O$, can also be expected to change. Coarser dolomite crystals, such as are developed in the mixing-zone and seepage–reflux models, are likely to be less susceptible to recrystallization since they are generally more nearly stoichiometric and better ordered. These dolomites can still be expected to recrystallize to xenotopic textures, however, if the CRT is exceeded during later diagenesis.

8.4.2 Dolomite cements

Although dolomite is largely a replacement, cements of this mineral are common. In Plio-Pleistocene dolomites of the Caribbean–Bahamas, dolomite cement

occurs between dolomitized grains and in the voids left by skeletal dissolution (e.g. Fig. 8.7; Land, 1973a; Supko, 1977; Sibley, 1980, 1982; Kaldi & Gidman, 1982). The dolomite may occur as a clear drusy spar, directly comparable to calcite spar (Section 7.6.2), although commonly the cement is simply a single isopachous layer of large rhombs lining the cavity. These crystals are frequently described as *limpid* because of their clear, glassy appearance. Emphasis has been put on these, suggesting they are a typical product of mixing-zone dolomitization (Folk & Land, 1975). In some instances dolomite cement contains zones of syntaxial calcite (Land, 1973a; Ward & Halley, 1985; Humphrey, 1988), indicating fluctuations in porewater chemistry. In the Pleistocene mixing-zone dolomites of Yucatan (Ward & Halley, 1985), dolomite cements have zones defined by variations in Mg/Ca ratio of the dolomite or of alternations of calcian dolomite with high-Mg or low-Mg calcite. Higher Ca dolomite zones are commonly leached out. A cement sequence from stable limpid calcian dolomite to less stable, more calcian dolomite and calcite is interpreted as the result of an increase in the amount of freshwater in the mixing zone.

In older dolomites, cements are also common and where a good drusy fabric is present, then delicate zones may be revealed with cathodoluminescence, including a similar broad pattern of non-luminescent–bright–dull zones (e.g. Tucker, 1983a). Internal discontinuities may also be revealed by CL, produced by a phase of dissolution interrupting cement precipitation (e.g. Fairchild, 1980b, 1982; Tucker, 1983a).

8.4.3 Baroque dolomite

One particular type of dolomite which may be a cement or a replacement is baroque dolomite, also called 'saddle' or 'white sparry' dolomite and known to mineral collectors as pearl spar. It is characterized by a warped crystal lattice, so that it has curved crystal faces, curved cleavage planes and markedly undulose extinction (Fig. 8.14; Radke & Mathis, 1980). Baroque crystals larger than a millimetre are usually a composite of subcrystals, giving the crystal a stepped surface. Fluid inclusions (usually of hypersaline brine), and calcite relics if the dolomite is a replacement, give the crystals a cloudy appearance in thin section and the pearly lustre in hand specimens. Gypsum−anhydrite inclusions also occur (e.g. Beales & Hardy, 1980). Baroque dolomite is commonly Ca-rich and it may have a substantial iron content (e.g. up to 15 mole % $FeCO_3$ in Woronick & Land, 1985). Variations in composition of individual crystals are commonly pronounced, occurring between different growth layers giving a distinct zonation and within the zones themselves (e.g. Radke & Mathis, 1980).

Baroque dolomite occurs as both a cavity fill and a replacement. In its cavity-fill mode, it usually has a xenotopic texture of irregular crystal boundaries within the mosaic but curved, scimitar-like terminations into the cavity. Baroque dolomite may show an astropetal fabric, i.e. only occurring on the floor of cavities, as a result of precipitation in a water−oil-filled cavity and only the low part being water wet. As a replacement, it commonly forms coarse xenotopic mosaics (Fig. 8.14A,B), usually with only a vague indication of the pre-existing limestone fabric. Fluorescence microscopy may reveal information on the original microfacies (e.g. Dravis & Yurewicz, 1985, figs 1 and 2). Alternatively, the baroque dolomite rhombs may be scattered through a limestone, or preferentially replace certain grains such as ooids. It also occurs in veins and fractures, and is commonly associated with epigenetic sulphide mineralization. Baroque dolomite is also commonly associated with hydrocarbons (e.g. Fig. 8.14D) and this has been used to suggest that it forms within the oil window, at temperatures of 60−150°C (Radke & Mathis, 1980). Sulphate-reducing processes in the presence of organic matter can occur at these temperatures and could be instrumental in the precipitation of the dolomite, by lowering the SO_4^{2-} content and increasing the alkalinity (Section 8.2). The curious lattice distortion, which is a feature of this dolomite type, is attributed to variations in the concentration of calcium ions adsorbed on to the growing crystal surface (Radke & Mathis, 1980).

8.5 TRACE ELEMENT GEOCHEMISTRY OF DOLOMITE

Trace elements in dolomites can be used in a similar

manner to the major cations Ca and Mg (Section 8.3) to give information on the nature of the dolomitizing fluids. The trace elements usually studied are Sr, Na, Fe and Mn, and their concentrations are determined by the same factors discussed in Section 6.3 for the trace elements in aragonite and calcite: (1) the concentrations of the minor elements in the pore-fluids, (2) the degree of openness (or water/rock ratio or solute index) of the diagenetic system, and (3) the effective distribution coefficient k_e of the trace element (Me) between dolomite and water:

$$(^mMe/^mCa)_{dol} = k_{e_{dol}} \times (^mMe/^mCa)_{water}$$

There has been much discussion over the actual values of trace element distribution coefficients in carbonates (e.g. Land, 1980; Kretz, 1982; Veizer, 1983a,b) and of their use in interpreting carbonate chemistry (see Section 6.3). One of the main problems is that theoretical distribution coefficients only apply to trace elements in lattice sites, but there are several other modes in which trace elements can be incorporated into crystals, hence the concept of effective k's which will be different from theoretical k's. With dolomites, the experimental values of k_{dol} are difficult to obtain, since dolomite cannot be precipitated in the laboratory at realistic diagenetic temperatures from natural waters. Effective distribution coefficients appear to be quite variable for each element, being particularly dependent on temperature, rate of precipitation and other kinetic factors.

As emphasized in Chapter 6, trace element data can help in the interpretation of carbonate diagenesis, but should not be used in isolation. Consideration of the sedimentologic and petrologic context of the dolomites is essential for realistic interpretation of their chemistry.

8.5.1 Strontium in dolomite

It has been suggested that the distribution coefficient for Sr in dolomite should be around half that for calcite on the grounds that the Sr largely substitutes for Ca rather than Mg ions in the dolomite lattice (Behrens & Land, 1972; Kretz, 1982). However, as noted in Section 6.3.3, there is much disagreement over the theoretical $k_{calcite}^{Sr}$ value, with 0.14 being proposed by Kinsman & Holland (1969) from the experimental precipitation of calcite from seawater and 0.07 by Katz et al. (1972) for calcite replacement of aragonite. Experimental determinations of k_{dol}^{Sr} at high temperatures (250−300°C) by Katz & Matthews (1977) gave 0.025 and by Jacobson & Usdowski (1976)

Fig. 8.14 *Baroque (or saddle) dolomite. (A) Plane polarized light of xenotopic texture. (B) Crossed polars showing undulose extinction. Gully Oolite, Lower Carboniferous, south Wales. (C) Scanning electron micrograph showing curved crystal faces. Pendleside Limestone, Lower Carboniferous, northwest England. Photo courtesy of Rob Gawthorpe. (D) Dolomitized catagraph (calcified algal grains) grainstone with fenestral vug occluded by baroque dolomite crystal, precipitated after influx of bitumen. Notice fragments of bitumen forming geopetal sediment within the dolomite. Dolomie Inférieur, Late Precambrian, Anti-Atlas, Morocco.*

gave 0.07. As with Mg/Ca ratio, the Sr content has been used to place constraints on the composition of the dolomitizing fluids and on the dolomitization model (e.g. Veizer *et al.*, 1978; Veizer, 1983a). Since seawater has such a high ionic strength compared with meteoric water, mixtures with more than 5% seawater will have an Sr/Ca molar ratio similar to seawater, and dolomites precipitated from a mixture with more than 20% seawater could have a typical marine Sr content

(whatever that is, considering the uncertainties of k_{dol}^{Sr} and the effects of local conditions of precipitation). Taking Jacobson & Usdowski's k, this would be around 550 ppm. Dolomites with much more Sr than this should be the precipitates of hypersaline fluids where strong evaporation and gypsum precipitation have elevated the $^mSr/^mCa$ ratio.

In situations where the water/rock ratio is low (a relatively closed system), then the mineralogy of the precursor carbonate is important. Where aragonite is being dolomitized, the dolomite may have a marine-type Sr content (500–600 ppm) since the $^mSr/^mCa$ ratio of aragonite is similar to seawater. Where calcite, with its lower $^mSr/^mCa$ ratio, is dolomitized, then low-Sr dolomite is formed. Marine high- and low-Mg calcite with 1000–2000 ppm Sr will be replaced by dolomite with several hundred ppm Sr, but diagenetic calcite with its much lower Sr, will be replaced by very Sr-depleted dolomite. Thus the *timing* of dolomitization is a most important factor in determining the dolomite Sr content: early dolomitization of marine carbonates (A, HMC and LMC) will result in Sr-rich dolomite, relative to later dolomitization of *stabilized* marine carbonates, composed of diagenetic LMC, by Sr-poor dolomite.

Now turning attention to actual dolomites, modern supratidal dolomite from the Arabian Gulf, Bahamas and Florida Keys has around 600 ppm Sr (Behrens & Land, 1972; Land & Hoops, 1973). Note that this is what one would predict with the Jacobson & Usdowski k^{Sr}_{dol} and it is about half of the Sr content of marine calcite. Modern dolomites occurring within pelagic sediments of the Guaymas Basin, Gulf of California and precipitated from pore-fluids originally of marine origin, have 300–700 ppm Sr. Knowing the Sr/Ca ratio of the pore-fluids, Baker & Burns (1985) calculated an average k^{Sr}_{dol} of 0.06.

Ancient dolomites have variable Sr contents, a few tens to hundreds of ppm is the typical range. As a generalization, early diagenetic dolomites, many dolomitized tidal flat micrites for example, have higher Sr contents than later diagenetic coarsely crystalline dolomites.

The very extensive Plio-Pleistocene dolomites of the Caribbean–Bahamas region, interpreted as mixing-zone dolomites, mostly have 70–250 ppm Sr (e.g. Land, 1973a; Supko, 1977) although the 120 Kyr Falmouth dolomites of Jamaica (Land, 1973b) have 3000 ppm Sr (!). The latter is attributed to a nearly closed system (low water/rock ratio) with dissolution of coral and algal aragonite supplying the Sr. Dolomites beneath Pacific atolls, also attributed to mixing-zone processes, have 150–230 ppm Sr (e.g. Rodgers *et al.*, 1982).

Whatever values of k^{Sr}_{dol} are believed, it is a low figure, below unity, and this means that in any recrystallization of dolomite, Sr will be lost. This is an important consideration, since many ancient dolomites have low Sr contents and interpretations have been made on this basis for dilute water dolomitization (mixing zone). However, if the dolomite had undergone recrystallization, then initially quite high Sr values could be reduced significantly. Evidence that this has occurred in some formations is seen in the relationship between Sr content and crystal size, where Sr decreases with increasing crystal size, the latter reflecting the degree of recrystallization (Dunham & Olson, 1980; M'Rabet, 1981). As noted earlier, poorly-ordered calcian dolomites, typical of modern evaporitic environments, are metastable, and should recrystallize with concomitant loss of Sr.

The most useful way to use Sr data is to compare different types of dolomite within one carbonate formation or several related formations. Two examples can illustrate the point. In the Lower Palaeozoic of Arctic Canada, high Sr contents occur in early diagenetic peritidal dolomites in the Cape Storm (average 216 ppm), Somerset Island (161 ppm) and Peel Sound (180 ppm) Formations, relative to medium to coarsely crystalline dolomitized open marine limestones of the Lang River (72 ppm) and Allen Bay (66 ppm) Formations (Veizer *et al.*, 1978). The first group were interpreted as early replacements of aragonitic muds, some of hypersaline environments, whereas the second group were interpreted as hyposaline replacement of carbonate sediments, which may have undergone stabilization to diagenetic LMC before the dolomitization. In the Lower Cretaceous of Tunisia, four types of dolomite were identified by M'Rabet (1981): burial, karst-related, evaporitic sabkha and lacustrine. The evaporitic dolomites have the highest Sr (average 115 ppm), relative to the burial (40 ppm), karst (30 ppm) and lacustrine (100 ppm) dolomites, and these are taken to reflect variations in the chemistry of the dolomitizing fluids.

8.5.2 Sodium in dolomite

As with Na in calcite (Section 6.4), there are uncertainties of where the Na is located in the dolomite crystal: substituting for cations or in fluid inclusions. Although the k^{Na}_{dol} is though to be very low indeed, like that of calcite (2×10^{-5}), the high $^mNa/^mCa$ ratio of seawater means that dolomite precipitated from marine-derived fluids can have a high Na content. Modern dolomite from Florida, Bahamas, Arabian Gulf and Baffin Bay has 1000–3000 ppm Na (Land & Hoops, 1973). The Plio-Pleistocene mixing-zone dolomites of the Caribbean–Bahamas have several hundred ppm Na (Land, 1973b; Sibley, 1980), although the Pacific atoll dolomites have 500–800 ppm (Rodgers *et al.*, 1982). Many ancient dolomites have only a few hundred ppm Na.

As with Na in limestones, there are many problems in interpreting Na data and great care must be exercised in comparing Na contents of dolomites from widely different locations and ages. Like Sr data, Na values can be used to distinguish dolomite types within a formation. For example, in the Lower Palaeozoic dolomites of Arctic Canada cited above (Veizer *et al.*, 1978), the early diagenetic higher-Sr dolomites also have higher Na than the later diagenetic dolomites. Veizer *et al.* (1978) showed that their dolomitized hypersaline facies generally contained more than 230 ppm Na, whereas the open marine facies had less.

8.5.3 Iron and manganese in dolomite

These two elements are ones which tend to be picked up by carbonate precipitates during diagenesis, rather than lost as is generally the case with Sr and Na. The reason for this is two-fold: (1) iron and manganese are present in very low concentrations in seawater, but they can be present in significant amounts in diagenetic pore-fluids, and (2) the k_{dol}^{Fe} and k_{dol}^{Mn} are high values, greater than unity, so that Fe and Mn are preferentially taken into the dolomite lattice relative to Ca. Apart from cation availability, the redox potential of the pore-fluids is important for these two elements, with reducing conditions favouring their occurrence in pore-fluids. Thus early near-surface dolomites tend to have low Fe and Mn contents, since most near-surface fluids are oxidizing, contrasting with later, burial dolomites, which may have high Fe and Mn through precipitation from negative Eh pore-fluids in which Fe^{2+} and Mn^{2+} are in solution.

Ferroan dolomite has more than 2 mole % $FeCO_3$ and is a common late diagenetic precipitate. For example, McHargue & Price (1982) described ferroan dolomite as scattered rhombs in argillaceous lime mudstones and post-compaction sparry cements in limestones of Pennsylvanian age. The close association of the ferroan dolomite with clays led these authors to suggest that the Fe, Mg and Ca were released during the conversion of smectite to illite at elevated temperatures during burial. Wong & Oldershaw (1981) also described late-stage ferroan dolomite, baroque in character, from the Devonian Kaybob Reef of Alberta. The predominance of this cement in reef-slope and reef-margin facies was interpreted as suggesting that dolomitizing fluids had come from basinal mudrock dewatering (also see Section 8.7.4). In the mid-Ordovician Trenton Formation of the Michigan Basin (Taylor & Sibley, 1986), three

dolomite types are recognized on the basis of petrography and geochemistry, with one type ('cap' dolomite, occurring directly below the Utica Shale) having very high Fe (3−13 mole% $FeCO_3$). Early diagenetic fine-grained regional dolomite and very late, fracture-related dolomite (baroque) have low Fe (0.06−0.34 and 0.03−1.93 mole % respectively). The ferroan cap dolomite is thought to have formed by interaction of the Utica Shale the Trenton Limestone, with the main control being availability of Fe, itself controlled by S^{2-} availability. The ferroan cap dolomite is poorly developed in the basin centre, where deposition was continuous, since abundant organic matter was used by SO_4^{2-}-reducing bacteria and most of the iron was precipitated out as sulphide. On the southern margin of the basin, where the cap dolomite is well developed, much organic matter was oxidized during subaerial exposure so that SO_4^{2-}-reducing bacteria were limited, leaving the iron available for incorporation into dolomite. Again, the source of the iron is most likely to have been the closely associated organic-rich shales; iron oxide coatings are common on clays, iron is commonly bound with organic matter, and iron occurs within clay mineral structures.

8.6 STABLE ISOTOPE GEOCHEMISTRY OF DOLOMITE

General principles of stable isotope geochemistry as applied to carbonates have been outlined in Section 6.4. Stable isotope data are being increasingly used in the interpretation of dolomites and when combined with trace elements and petrography they can help greatly to elucidate their origin. There are, however, several problems with regard to the interpretation of the isotopic composition of dolomites. The main problem is that the relationship between temperature, $\delta^{18}O_{water}$ and $\delta^{18}O_{dolomite}$ is imprecisely known. Experimental work, mainly at high temperatures (given in parentheses) has resulted in four fractionation equations ($T = °K$):

$$10^3 \ln\alpha_{dolomite-water} = 3.2 \times 10^6 T^{-2} - 1.50$$
(300−510°C, Northrop & Clayton, 1966)

$$10^3 \ln\alpha_{dolomite-water} = 3.34 \times 10^6 T^{-2} - 3.34$$
(350−400°C, O'Neil & Epstein, 1966)

$$10^3 \ln\alpha_{dolomite-water} = 3.23 \times 10^6 T^{-2} - 3.29$$
(100−650°C, Sheppard & Schwarcz, 1970)

$$10^3 \ln\alpha_{dolomite-water} = 2.78 \times 10^6 T^{-2} + 0.11$$
(25−79°C, Fritz & Smith, 1970)

Curves from these equations are plotted in Fig. 8.15 for water of $\delta^{18}O = 0‰$, to show the relationship between temperature and $\delta^{18}O$ of precipitated dolomite. The curves show that the relative uncertainty for temperature is around 15°C for a given dolomite $\delta^{18}O$. Working back from a dolomite $\delta^{18}O$ and an assumed temperature, a range of 4‰ for the water is obtained. Figure 8.16 shows temperature against $\delta^{18}O_{dolomite}$ for waters of various isotopic compositions from Land (1983), and Fig. 8.17 shows temperature against $\delta^{18}O_{water}$ to show the isotopic composition of dolomite from Woronick & Land (1985), both based on a fractionation equation of $10^3 \ln\alpha_{dolomite-water} = 3.2 \times 10^6 T^{-2} - 3.3$.

Also plotted in Fig. 8.15 is the calcite–water fractionation curve, $10^3 \ln\alpha_{calcite-water} = 2.78 \times 10^6 T^{-2} - 2.89$, which is well constrained. This shows that dolomite will be about 3 to 6‰ heavier than coprecipitated calcite. There has been much discussion over the $\Delta^{18}O_{dol-cal}$ (e.g. Degens & Epstein, 1964; Fritz & Smith, 1970; Katz & Matthews, 1977) with no firm conclusions having been reached (Land, 1980), other than that the equilibrium value of $\Delta^{18}O$ at 25°C is somewhere between 1 and 7‰ and possibly is not constant. The main problem here is that it is difficult to prove coexistence of calcite and dolomite; the minerals commonly occur together, but that is not to say they were precipitated from the same water. Much dolomite of the geological record has replaced $CaCO_3$, but even if this is very early, there could well have been some isotopic evolution of the dolomitizing fluids, from those out of which the $CaCO_3$ was precipitated. Data from modern tidal flats generally show an enrichment of the dolomite in ^{18}O over host $CaCO_3$ sediment, but the difference is variable. For example, in the Arabian Gulf sabkhas, an equilibrium $\Delta^{18}O$ of 3.2‰ was proposed by McKenzie (1981) for coexisting calcite and dolomite which are thought to be in isotopic equilibrium with their environment at a temperature within the sediment of 35°C. This dolomite–calcite fractionation figure of +3.2‰ agrees well with experimental values obtained by Fritz & Smith (1970) from protodolomite–calcite and Matthews & Katz (1977) from dolomitization of $CaCO_3$. However, the $\Delta^{18}O$ values of McKenzie's sediment samples range from −2.2 to +3.2‰, and much of this is due to isotopic exchange with porewaters through changing temperatures as the sediment is buried and dolomite continues to form.

From the study of coexisting metamorphic dolomite and calcite, Sheppard & Schwarcz (1970) predicted a +2.4‰ enrichment of dolomite over calcite

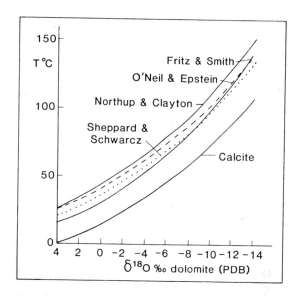

Fig. 8.15 *Dolomite–water fractionation equations as a function of temperature plotted for a $\delta^{18}O_{water}$ of 0‰ SMOW. The calcite–water equation is also shown. After Land (1983).*

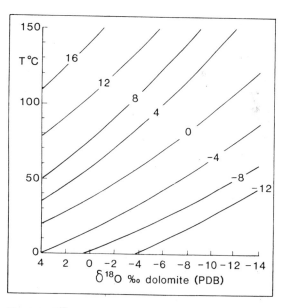

Fig. 8.16 *$\delta^{18}O_{dolomite}$ versus temperature for various $\delta^{18}O_{water}$ values derived from the fractionation equation: $10^3 \ln\alpha_{dolomite-water} = 3.2 \times 10^6 T^{-2} - 3.3$. After Land (1983).*

in ^{13}C at 20°C. However, from coexisting dolomite and calcite in the Abu Dhabi sabkha, little difference in $\delta^{13}C$ is recorded (McKenzie, 1981). Tan & Hudson

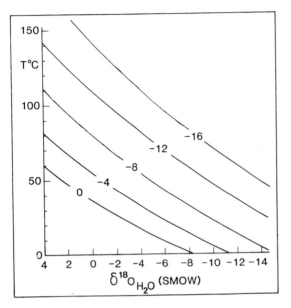

Fig. 8.17 *Curves for* $\delta^{18}O_{dolomite}$ *(PDB) derived from same equation as in Fig. 8.16. After Woronick & Land (1985).*

(1971) obtained a small but position $\Delta^{13}C_{dol-cal}$ from some Jurassic carbonates.

The oxygen isotopic composition of dolomite reflects the temperature of precipitation and the isotopic composition of the dolomitizing fluids (Figs 8.16 and 8.17). The latter can be influenced by the isotopic composition of the $CaCO_3$ minerals being replaced, but since pore-fluids have abundant oxygen, generally precursor minerals only have an effect in low water/rock ratio or closed diagenetic systems. In the majority of cases, since dolomitization and recrystallization take place in the presence of water (dry, solid-state diagenetic processes are rare), the dolomite precipitated has a $\delta^{18}O$ value determined by the pore-fluid composition and temperature. By way of contrast, the $\delta^{13}C$ value of dolomite is generally strongly influenced by that of the precursor $CaCO_3$. Pore-fluids mostly have low carbon contents initially so that the $\delta^{13}C$ value of $CaCO_3$ being dolomitized is commonly retained by the dolomite. In addition, there is little isotopic fractionation of $^{13}C/^{12}C$ with temperature. As discussed for limestones in Section 6.4, the $\delta^{13}C$ value gives information on the source of the carbon in the carbonate. Values between 0 and +4‰ $\delta^{13}C$ are typical marine signatures. However, where there has been much organic matter diagenesis, then more extreme values of $\delta^{13}C$ are likely. Very negative values, down to $-20‰$, indicate that C in the CO_3^{2-} is derived

to a greater or lesser extent from the organic matter, which generally has a $\delta^{13}C$ in the range -22 to $-30‰$. Very positive values, up to $+15‰$, can result from the fermentation of organic matter, where methanogenesis results in very ^{13}C-enriched CO_2 (Irwin *et al.*, 1977).

Oxygen and carbon isotopes of some Cainozoic dolomites are plotted in Fig. 8.18. Modern dolomites from the Arabian Gulf have a $\delta^{18}O$ range from 0.0 to $+3.9‰$, heavier relative to the field of the marine $CaCO_3$ sediments there, as noted earlier. Although much of the enrichment can be attributed to the dolomite–calcite fractionation, some could be due to evaporative concentration of the seawater which leads to a ^{16}O-depleted fluid and isotopically heavier precipitates therefrom. The $\delta^{13}C$ is +2 to +4‰, well within the field of marine $CaCO_3$, suggesting that the $\delta^{13}C$ has been inherited from the $CaCO_3$ the dolomite replaced, or is derived from dolomitizing fluids (modified seawater) which had a similar $\delta^{13}C$ value to the seawater from which the $CaCO_3$ grains were precipitated. Baffin Bay dolomites are also positive with $\delta^{18}O = +4$ to $+5‰$, but the $\delta^{13}C$ is much more variable (-3 to $+4‰$). This presumably reflects the variable incorporation of organically-derived light CO_2 into the CO_3^{2-}.

Plio-Pleistocene supposed mixing-zone dolomites of the Caribbean–Bahamas region show a wide range of isotope values (Fig. 8.18), but average $\delta^{18}O$ is $+3.0‰$ and $+2.1‰$ for southeast Bahamas (Supko, 1977), $+2.2‰$ for the Hope Gate of Jamaica (Land, 1973a), $-1‰$ for the Falmouth of Jamaica (Land, 1973b) and $+3.25‰$ for Barbados (Humphrey, 1988). Although there is overlap, the oxygen data are not quite so heavy as the Baffin Bay–Abu Dhabi evaporitic dolomites. With $\delta^{13}C$, averages are $+0.9‰$ and $+1.8‰$ for southeast Bahamas, $+1.2‰$ for Hope Gate, $-8.4‰$ for Falmouth and $-14.4‰$ for Barbados. The interesting point that emerges here is the great variation; some of these proposed mixing-zone dolomites are very depleted in ^{12}C and this clearly indicates an important contribution from organically-derived CO_2, probably derived from soil gases or sulphate reduction. With some dolomites regarded as mixing zone in origin, such as the Hope Gate and Yucatan cases (Fig. 8.18), there is a covariant trend of $\delta^{13}C$ and $\delta^{18}O$. This can be said to support such an origin, with the trend reflecting the mixing of two waters of differing end-member compositions. Some ancient dolomites also show a positive correlation of $\delta^{13}C$ and $\delta^{18}O$ (e.g. Zhao & Fairchild, 1987; Zempolich *et al.*, 1988).

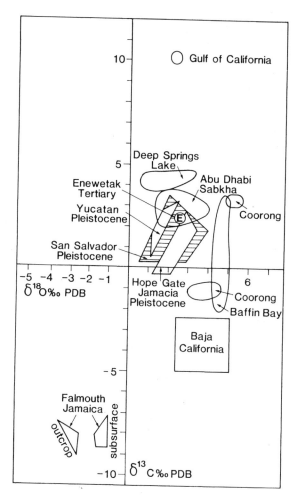

Fig. 8.18 *Carbon–oxygen isotopic compositions of some Recent and Pleistocene dolomites. From sources given in text.*

Bellanca *et al.*, 1986). In this case dolomite precipitation is interpreted as having taken place in the sulphate-reducing zone of the pelagic sediments. Sedimentation rate is the main control on the depth to the various organic diagenetic zones and location of dolomite precipitation. Fast sedimentation rates in the Guaymas Basin quickly conveyed the diatomaceous muds into the zone of methanogenesis, whereas slower rates of the Tripoli permitted the sediments to stay longer in the SO_4^{2-} reduction zone (Kelts & McKenzie, 1984).

Older dolomites show a wide range of oxygen isotope values (e.g. Fig. 8.19) and these are frequently interpreted by comparison with modern and Plio-Pleistocene dolomites. There are great dangers in just comparing numbers, and the dolomite's petrography and trace element content have to be considered to place constraints on the isotope interpretations. One particular problem is that there have been secular fluctuations in seawater $\delta^{18}O$ and $\delta^{13}C$ (Veizer & Hoefs, 1976; Section 9.6.2), so that there will be variations through time of the isotope signatures of dolomite where seawater is involved.

Many ancient platform dolomites show evidence of early, near-surface dolomitization, but they commonly have quite negative $\delta^{18}O$ values. Although some of this can be attributed to the effects of more depleted seawater, another possibility is that it reflects recrystallization of early dolomite or precipitation of further generations of dolomite, at higher temperatures during burial and/or from ^{18}O-depleted pore-waters. Where burial dolomite has been identified and analysed, then usually this does have moderately negative $\delta^{18}O$ (−5 to −10‰). For example, in the Devonian Miette Buildup of Alberta, late diagenetic, post-lithification mosaic dolomite has −5‰, and later sparry dolomite associated with fractures and brecciation has −9‰ (Fig. 8.19A, Mattes & Mountjoy, 1980). Likewise, fracture-related dolomites of the Trenton Formation have the most negative $\delta^{18}O$ of −9‰, whereas the regional early diagenetic near-surface dolomite, probably of mixing-zone origin, has −6.8‰ (Taylor & Sibley, 1986). The more negative burial dolomites are normally interpreted as having formed at higher temperatures (~50–100°C) at depth, but this does involve assumptions of the $\delta^{18}O$ of the water. Plotting the isotope data from different types of dolomite within one formation can be instructive. Figure 8.19B distinguishes between the tidal flat dolomicrites, mixing-zone stratal dolomites and burial 'vein' dolomites of the Dinantian of south Wales (Hird *et al.*, 1987). The first group have marine-

The anoxic dolomites from the Guaymas Basin have oxygen from −4 to +4‰, but the $\delta^{13}C$ is very distinctive, with extremely positive values, up to +13.8‰, average +9.8‰ (Kelts & McKenzie, 1982, 1984). These figures are interpreted as showing that dolomite precipitation has taken place in the zone of active methanogenesis, where heavy CO_2 is produced (Irwin *et al.*, 1977). Ancient anoxic dolomites equivalent to the Guaymas Basin occurrences are characterized by extreme $\delta^{13}C$ signatures due to the strong organic diagenesis taking place as the dolomite is precipitated. The Miocene Tripoli Formation of Sicily, an euxinic basin deposit, has dolomite with $\delta^{13}C$ ranging from 0 to −22‰ (Kelts & McKenzie, 1984;

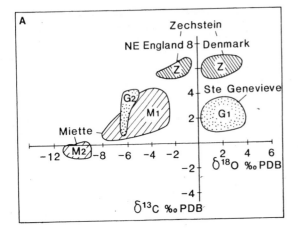

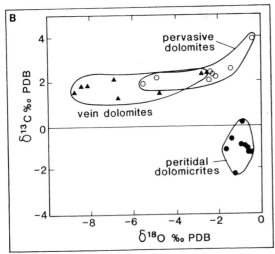

Fig. 8.19 *Carbon−oxygen isotopic compositions of ancient dolomites. (A) The Upper Permian (Zechstein) dolomites of western Europe (Z) of probable seepage−reflux origin, the Lower Carboniferous (Mississippian) dolomitized Ste Genevieve limestones of Illinois of probable mixing-zone origin (G1) with later ferroan dolomite cements having much lighter $\delta^{18}O$ (G2), and dolomitized Devonian limestones of the Miette Buildup, western Canada, of probable burial/ mudrock-dewatering origin (M1), with later, white sparry dolomites (M2) having more negative $\delta^{18}O$ values. Data from Clark (1980), Choquette & Steinen (1980) and Mattes & Mountjoy (1980). (B) Peritidal dolomicrites of penecontemporaneous origin, prevasive dolomites of probable mixing-zone origin and burial, vein (baroque) dolomites, from the Lower Carboniferous of south Wales. After Hird* et al. *(1987).*

type $\delta^{18}O$ values but $\delta^{13}C$ is negative, suggesting incorporation of CO_3^{2-} from organic matter diagenesis. The other dolomite types have typical marine values of $\delta^{13}C$, inherited from the precursor limestone. The near-surface mixing-zone dolomites have a range of $\delta^{18}O$, which could reflect precipitation from waters with a range of compositions from marine to fresh, and the burial dolomites are moderately to very depleted in ^{18}O, signifying precipitation at a range of temperatures and/or from waters of varying isotopic composition.

8.7 MODELS OF DOLOMITIZATION, MODERN AND ANCIENT EXAMPLES

Models for the dolomitization of carbonate sediments can be divided into five categories: evaporitic (sabkha), seepage−reflux, meteoric−marine mixing, burial, and seawater models (Fig. 8.1). Each involves a different type of dolomitizing fluid, mode of flow and geological setting(s), but there is overlap between the models, several could apply in one setting or to one formation, and the product of a particular model may not be very distinctive petrographically or geochemically. Before one of the models can be applied confidently, the palaeogeography and facies, context and distribution, petrography and geochemistry of the dolomite all need to be known.

8.7.1 Evaporative dolomitization

Most dolomites forming at the present time are in evaporitic environments. Dolomite crusts occur on the tidal flats of Andros Island, Bahamas (Shinn *et al.*, 1965), Sugarloaf Key, Florida (Shinn, 1968a) and Bonaire, Netherland Antilles (Deffeyes *et al.*, 1965). Dolomite is being precipitated within high intertidal−supratidal sediments of the Trucial Coast sabkhas (Illing *et al.*, 1965; McKenzie, 1981; Patterson & Kinsman, 1982), within coastal evaporitic lakes of the Coorong, South Australia (Von der Borch, 1976; Von der Borch & Lock, 1979), in saline lakes of Victoria, Australia (Deckker & Last, 1988) and in hypersaline lagoons of Baffin Bay (Behrens & Land, 1972) and Kuwait (Gunatilaka *et al.*, 1984).

In the *Arabian Gulf sabkhas*, briefly described in Section 3.4.4, water is mainly supplied by flood recharge during the winter and spring when extra high tides and storms carry water on to the supratidal flats, particularly along old channel ways (McKenzie *et al.*, 1980; Patterson & Kinsman, 1982). Flood recharge leads to a relatively short-lived downward movement of water through the sediments to join a net seaward flow of groundwater. For much of the year, intense heat over the sabkhas results in evaporation from the capillary zone above the water-table. This induces an upward flow of groundwater to the capillary zone (a process called evaporative pumping by Hsü & Siegenthaler, 1969), until the water-table falls below a level where capillary evaporation can operate. Changes in porewater chemistry across the sabkha (Fig. 8.20) result from evaporation–mineral precipitation and mixing of the marine-derived waters with continental brines (Butler, 1969; Hardie, 1987). Dolomite occurs in areas with seasonal to rare flood recharge, where capillary evaporation and evaporative pumping are important. Areas of more intense dolomitization are closely related to the positions of former tidal channels, and the precipitation is taking place between 0.15 and 0.55 m below the surface, in intertidal facies. McKenzie *et al.* (1980) found that the Mg/Ca molar ratios in areas of dolomite are between 2.5 and 7.0, lower than where flood recharge is more frequent (Mg/Ca = 7 to 27) and aragonite and gypsum are being precipitated, rather than dolomite. On the other hand, Patterson & Kinsman (1982) determined a Mg/Ca molar ratio >6, pH 6.3 to 6.9 and a lower SO_4^{2-} content than seawater in areas of dolomitization. The last results were from microbial reduction and gypsum precipitation. These authors also concluded that the dolomite mostly forms by replacing aragonite according to:

$$Mg^{2+} + 2CaCO_3 \rightleftharpoons CaMg(CO_3)_2 + Ca^{2+}$$

with the released Ca^{2+} being taken up by further gypsum precipitation. McKenzie (1981) found some evidence to suggest that the dolomitization of aragonite takes place via an intermediate high-Mg calcite phase as documented by Katz & Matthews (1977) in an experimental study. Recently, Hardie (1987) has argued for direct precipitation of dolomite within the sabkha rather than dolomitization of aragonite linked to raising of porewater Mg/Ca ratios by gypsum precipitation.

The sabkha dolomite itself is calcium-rich (52–54.6 mole% $CaCO_3$) and poorly to moderately ordered (McKenzie, 1981). Of interest is that the degree of order increases with distance from the shoreline (0.22 at 1 km and 0.58 at 7 km), as a result of progressive recrystallization through an 'ageing' of the crystals. In addition, dolomite crystal shape becomes more prefect and crystal size increases inland across the sabkha, from 1–2 µm rhombs 1 km from the lagoon, to 2–5 µm rhombs 4 km inland and 20 µm

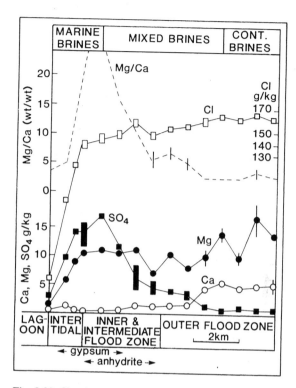

Fig. 8.20 *Chemistry of pore-fluids across the intertidal flat and sabkha of Abu Dhabi, Arabian Gulf. Data from Butler (1969). After Hardie (1987).*

aggregates at 6 km. Stable isotope data indicate precipitation from brines at temperatures between 34 and 49°C (McKenzie, 1981).

On the west side of *Andros Island*, in the Bahamas, a dolomitic crust occurs above high tide level on the flanks of palm hammocks, former beach ridges (see Fig. 3.25; Shinn *et al.*, 1965; Hardie, 1977; Shinn, 1983a). It can also be traced laterally beneath intertidal muddy sediments which have buried the supratidal crust as a result of the relative sea-level rise over the past few thousand years (Fig. 8.21). The crust is mostly broken into polygonal slabs which can be moved and transported by storms. The dolomite has a Ca excess (55 mole% $CaCO_3$) and ordering peaks are weak.

Dolomitic supratidal crusts are usually ascribed to precipitation from evaporated seawater, with the early formation of aragonite and gypsum leading to a higher Mg/Ca ratio of porewaters to facilitate the dolomitization of the surficial sediment layer (Shinn, 1983a). As in the Trucial Coast sabkhas, it is generally considered that seawater is supplied by storm flooding of the supratidal flats and evaporation-

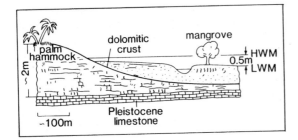

Fig. 8.21 *Schematic cross-section of a palm hammock on the tidal flats of Andros Island, Bahamas, showing dolomitic crust, now forming in the upper intertidal/low supratidal zone, with earlier crust occurring beneath the low intertidal–shallow subtidal sediments. After Shinn* et al. *(1965).*

induced capillary action on sediment porewaters. However, almost identical dolomite crusts on Sugarloaf Key, Florida (Shinn, 1968b), have recently been reinterpreted as precipitates from only slightly modified Florida Bay water pumped through the sediments by tidal action (Carballo *et al.*, 1987; see

Section 8.7.5), so that a similar seawater origin could apply to other supratidal dolomite crusts, without the need to invoke intense evaporation.

The *Coorong of South Australia* is an area of ephemeral alkaline lakes behind a modern beach barrier, which are fed by the sea and continental groundwater discharging from an unconfined, seaward-flowing aquifer (Figs 8.1F and 8.22; Von der Borch, 1976). The pH of lake waters ranges from 8 to 10, Mg/Ca ratios vary from 1 to 20, and for several months of the year some lakes dry out completely. When lakes are full, the aquatic grass *Ruppia* is abundant and algal mats cover the sediment surface. Microcrystalline magnesite, hydromagnesite, aragonite, high- and low-Mg calcite and dolomite are being precipitated in the lagoons and ephemeral lakes. The dolomite mostly occurs in the more distal (landward) lakes, which annually pass through a desiccation phase. Dolomitic carbonates underlie the adjacent coastal plain, formed as successive beach–lagoon complexes were stranded during depositional regression. Much of the Coorong dolomite occurs as 0.5 μm spherular

Fig. 8.22 *Dolomites of the Coorong region, South Australia. (A) The coastal plain of Quaternary barriers and interdune flats where ephemeral lakes occur and dolomite is being precipitated. (B) Stratigraphic log of a core from an ephemeral Coorong lake. After Von der Borch (1976) and Von der Borch & Lock (1979).*

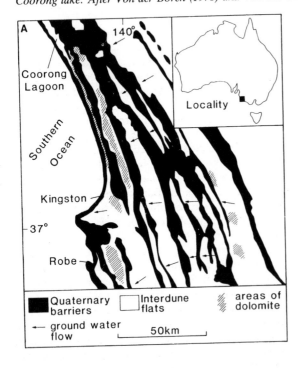

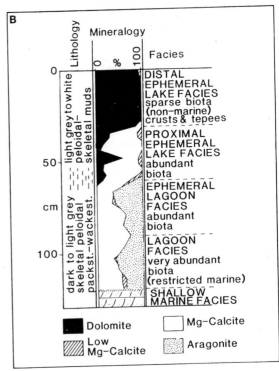

aggregates which may have formed from a Ca−Mg carbonate gel (Von der Borch & Jones, 1976). Much of the dolomite has been called protodolomite, since it only has the approximate chemical composition of dolomite (up to 8 mole% enriched in Ca) and the ordering XRD superstructure reflections are absent. The dolomite occurs as a yoghurt-like pelletal mud which forms indurated crusts and flake breccias, algal laminites and polygonal desiccation cracks around lake margins. As a result of the annual lake desiccation−flooding cycle, groundwater resurgence produces tepees and mud extrusions along polygon boundaries (Von der Borch & Lock, 1979). Isotope data and SEM observations (Botz & Von der Borch, 1984) show that there are two types of dolomite, a finer-grained type precipitated from evaporitically-modified continental groundwater and a coarser type probably formed by the replacement of aragonite. These two dolomite types are thought to form at different times in the annual lake cycle of flooding and desiccation. In spite of the strong evaporation during the dry season, no sulphate or chloride minerals are precipitated. This is probably due to reflux of brines out of the lakes to join the seaward-flowing groundwater.

The mechanism of dolomite precipitation in the Coorong is not fully understood. It could relate to mixing of marine and continental water but evaporation of the continental water is probably the main forcing mechanism. Much of the Mg^{2+} is thought to come from the groundwater and it could be derived from leaching of Quaternary basic volcanics in the recharge areas marginal to the coastal plain. High alkalinity in lake waters is a factor favouring dolomite precipitation (Section 8.2); the removal of SO_4^{2-} from sediment porewaters during diagenesis of grass and algal debris could also be significant.

Coroong-type deposition results in a distinctive sedimentary package (Fig. 8.22B), several metres thick, of shallow-marine subtidal facies (deposited seaward of a barrier), passing up into lagoonal aragonitic peloidal muds (deposited behind a newly-formed barrier), overlain by ephemeral lake deposits, with the dolomite, as the strandplain prograded further seawards and the lake became more distant from marine influences (Von der Borch, 1976). Evidence for groundwater resurgence would occur towards the top of the sequence (tepees and mud injection/extrusion structures) and the unit would be capped by a paleosol, probably of calcrete type.

Although the Coorong dolomites have been known since the 1920s, the model has not had much application to the geological record. An exception is the Yalco Formation of the mid-Proterozoic McArthur Group of northern Australia (Muir *et al.*, 1980). The Yalco mainly consists of fine-grained dolomite with little evidence for the former presence of evaporites, but with many sedimentary structures identical to those of the Coorong, including evidence for movements of the contemporaneous water-table. The Coorong model can result in extensive dolomites during depositional regression where there is a substantial seaward flow of continental groundwaters, derived from a more pluvial recharge area, which feed coastal lakes and lagoons in a more arid, evaporitic setting.

In the continental *saline lakes of Victoria*, the stoichiometric but poorly-ordered dolomite appears to be a direct precipitate from the high alkalinity, high Mg/Ca ratio waters (Deckker & Last, 1988). There is a moderate amount of sulphate present in the lake waters, but this does not appear to inhibit dolomite precipitation. The high Mg/Ca ratio probably reflects weathering of basaltic rocks in the area.

Modern dolomites precipitating in hypersaline marine lagoons are recorded from *Baffin Bay, Texas* (Behrens & Land, 1972) and *Kuwait* (Gunatilaka *et al.*, 1984). In the Kuwaiti lagoon, depth 0.5 m, salinity is mostly 42−44‰, occasionally up to 52‰, but periodically the lagoon is flushed out during heavy rain storms to reduce salinity considerably (16‰). The dolomite averages $Ca_{51.8}Mg_{48.2}$ and is poorly ordered. It occurs in the upper 0.05 to 0.1 m of cores, making up to 4% of the mostly aragonitic pellet muds. The dolomite forms rhombs which appear to be replacing aragonitic pellets and needles. The sediments are organic-rich and pore-fluid sulphate content is low compared with overlying seawater. Gunatilaka *et al.* (1984) attributed the dolomite precipitation largely to the reduced SO_4^{2-}, with the possibility of some significance in the periodic dilution of lagoon water, in a variant of the marine−meteoric water dolomitization scenario (Section 8.7.3). Dolomite is also being precipitated in sabkhas around this Kuwaiti lagoon.

Baffin Bay also has a variable salinity (brackish to hypersaline) and a typical marine Mg/Ca ratio. The dolomite consists of 1−5 μm equant grains and occurs in organic-rich sediments. Carbon isotopes range from −4 to +4‰, indicating a variable contribution from organically-derived CO_2. Oxygen values, however, are enriched in ^{18}O with +4 to +5.2‰, which could indicate precipitation from evaporated, hypersaline pore-fluids.

As is clear from the above, there is a lot of information available on these modern evaporitic dolomites and consequently many ancient dolomites have been interpreted using these as analogues. Evaporitic dolomites will be associated with all the indicators for intertidal–supratidal facies, such as fenestrae, microbial laminae and evaporite pseudomorphs, described in Section 4.3. There need not be any evidence for evaporites, since although these dolomites are forming in hypersaline environments, evaporites are not always precipitated or preserved, as in the Coorong and Bahamas cases. Geochemically, these evaporitic dolomites can be expected to show relatively high Sr and Na values (Section 8.5), and the $\delta^{18}O$ may be extra positive if much evaporation has taken place. Ancient evaporative dolomites are commonly more nearly stoichiometric than other dolomite types (Section 8.3; Lumsden & Chimahusky, 1980). Petrographically, evaporative dolomite is mostly fine grained, although modern examples generally consist of 1 to 5 μm rhombs and ancient ones have crystals more in the range 5 to 20 μm. This suggests that recrystallization has taken place during later diagenesis which would allow the crystals to become better ordered.

There are numerous examples of ancient tidal flat evaporative dolomites including the Lower Carboniferous Aghagrania Formation of Ireland (West et al., 1968) and Caswell Bay Mudstone of south Wales (see Fig. 8.8B; Hird et al., 1987). They also occur in many Upper Cambrian through Devonian carbonates of eastern USA.

8.7.2 Seepage–reflux dolomitization

This model involves the generation of dolomitizing fluids through evaporation of lagoon water or tidal flat porewaters and then the descent of these fluids into underlying carbonate sediments. The model was applied by Adams & Rhodes (1960) to account for the dolomitization of the Permian reef complex of west Texas, where back-reef, shelf and lagoonal carbonates such as the Carlsbad Limestone are dolomitized, but the shelf-edge Capitan Reef is not. They envisaged the precipitation of gypsum in the evaporitic shelf lagoon behind the reef to raise the Mg/Ca ratio, and then the descent of the dense and hot, highly alkaline Mg^{2+}-rich hypersaline brines through permeable lagoonal back-reef carbonates, displacing less dense marine porewater. They imagined that dolomitization of more porous zones could take place to depths of several hundred metres, and they cited the occurrence

of halite cements in the dolomite as testament to the passage of hypersaline brines.

Seepage–reflux in a modern setting was proposed by Deffeyes et al. (1965) from studies of a supratidal, gypsum-precipitating lake (Pekelmeer) on Bonaire, Netherlands Antilles. The lake is separated from the sea by a beach-barrier ridge of coral–algal debris, and some dolomite occurs in and around the lake. It was suggested that seawater fed the lake through the barrier (seepage) and that the dense Mg-rich lake waters descended through underlying carbonates (reflux) to dolomitize them (Fig. 8.1B). Deffeyes et al. cited dolomitized Plio-Pleistocene reef limestones elsewhere on the island as the product of earlier seepage–reflux from a supratidal zone, subsequently removed by erosion. Unfortunately, later work (Lucia, 1968; Murray, 1969) showed that there were no dolomites beneath the Pekelmeer and that for most of the year, seawater entered the lake through springs and fractures in the underlying Pleistocene limestone. If reflux was taking place, it was for only a short time in the summer, when the springs were inactive. In addition, the Plio-Pleistocene dolomites which are widely developed in the Caribbean–Bahama region, are now interpreted as the product of the mixing-zone model (Sibley, 1980).

Two minor occurrences of seepage–reflux dolomite have been described by Müller & Tietz (1971) and Kocurko (1979), from the Canary Islands and San Andres, Columbia, respectively. In both cases dolomitization is taking place where brines, formed by the evaporation of sea-spray in the supratidal zone, descend into underlying limestones.

The seepage–reflux model relies on Mg^{2+}-rich fluids being produced by evaporation and gypsum precipitation descending into the subsurface through density differences with porewaters in underlying sediments. Although there are no good modern analogues, the model has been applied frequently to ancient dolomite sequences. It is a popular model for dolomites associated with evaporites, and one could expect the dolomite to be relatively enriched in trace elements and ^{18}O, as a result of precipitation from evaporated seawater. Where there are sabkha evaporites, then dolomitized intertidal–subtidal facies beneath them could be the product of downward movement of Mg^{2+}-rich fluids. For example, in the Mississippian Red River Group of the Williston Basin, carbonates beneath nodular gypsum–anhydrite are dolomitized (Asquith, 1979). Intertidal facies are tight microcrystalline dolomitic mudstones whereas underlying subtidal facies are porous, sucrosic dolomites, which are oil reservoirs in the Big Muddy Creek Field of Montana.

The *Zechstein (Upper Permian) dolomites of northwest Europe* have been interpreted as seepage–reflux in origin by Smith (1981) and Clark (1980). Dolomitized algal–bryozoan reefs, oolites and intertidal to shallow subtidal skeletal grainstones to mudstones occur around the margins of the Zechstein Basin, and are overlain and laterally adjacent to gypsum–anhydrite and halite, which formed during subsequent partial to complete isolation of the basin from the world's ocean (Fig. 8.23A). The dolomites are thought to have formed by reflux of brines during precipitation of overlying gypsum–anhydrite, or during evaporative drawdown of the basin, when sulphate and especially halite were precipitated in the basin (Fig. 8.23B). The isotopic signatures of some

Zechstein dolomites are shown in Fig. 8.19A.

The seepage–reflux model has also been applied to the *Lower Cretaceous Edwards Formation in Texas* (Fisher & Rodda, 1969). Rudist reefs were developed along the platform margin (the Stuart City and central Texas reef trends, Fig. 8.24), and two on-shelf evaporitic lagoons, the Kirschberg and McKnight, were sites of gypsum–anhydrite precipitation. Carbonate grainstones were deposited behind and around the shelf-edge reefs and these are preferentially dolomitized where closely associated with the onshelf evaporites. The rudist bioherms with a dense lime mudstone matrix have mostly escaped dolomitization. Mg-rich brines are thought to have migrated downwards and outwards from the lagoons into the permeable grainstones. Tidal flat dolomites, with all the usual indicators of exposure,

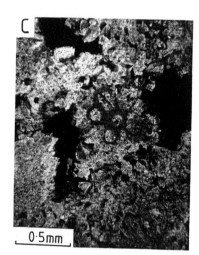

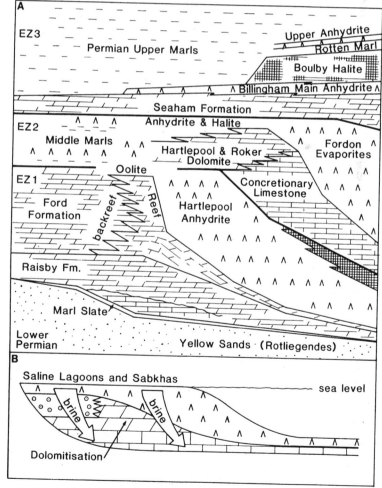

Fig. 8.23 *Dolomites of the Upper Permian (Zechstein) of northeast England. (A) Stratigraphy–lithology showing three carbonate–evaporite cycles. The lower part of the Raisby Formation is still limestone; the Concretionary Limestone is mostly a dedolomite. (B) Seepage–reflux model for Zechstein dolomitization. After Smith (1981) and Clark (1980). (C) Coarse dolomite, bryozoan in centre and vuggy porosity. Crossed polars.*

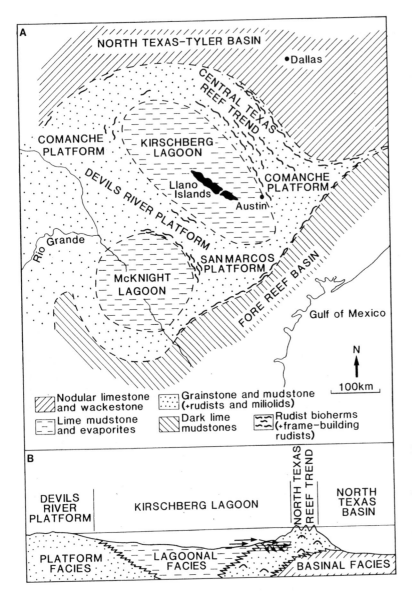

Nodular limestone and wackestone

Lime mudstone and evaporites

Grainstone and mudstone (+rudists and miliolids)

Dark lime mudstones

Rudist bioherms (+frame-building rudists)

Fig. 8.24 *Dolomites of the Lower Cretaceous (Edwards Formation) of Texas. (A) Palaeogeographic map showing shelf-margin reef trends and two onshelf lagoons. (B) Schematic north—south cross-section and seepage—reflux model for dolomitization of the shelf-margin grainstones. After Fisher & Rodda (1969).*

also occur in the Edwards Formation, located particularly on the southwestern side of the Kirschberg Lagoon, on the Devils River Platform.

8.7.3 Mixing-zone dolomitization

The logic behind this model is that it is easier to precipitate dolomite from a dilute solution, so that if seawater with its Mg/Ca molar ratio of 5.2 is mixed with freshwater, the high Mg/Ca ratio is maintained but some of the kinetic obstacles due to the high ionic strength of seawater are removed (Section 8.2; Folk & Land, 1975). The model stems from the work of Hanshaw *et al.* (1971) on the groundwaters in Tertiary limestones of the Floridan Aquifer. They showed that the brackish waters beneath the potable waters were potential dolomitizing fluids, with a Mg/Ca ratio approaching unity. Two key features of the model are that seawater supplies the Mg^{2+} and that active groundwater movement pumps the dolomitizing solution through the limestones. Although the mixing-zone model is attractive, it has to be said that it has received much criticism lately (see for example Land, 1985; Machel & Mountjoy, 1986, 1987; Hardie, 1987). Doubts have been raised over the mechanism itself (see Section 8.2) and dolomite is not being precipitated in modern mixing zones to the extent envisaged, although of course there is the time element. Modern coastal mixing zones have only been established for a few thousand years. Plio-Pleistocene dolomites interpreted as mixing zone in origin are widely developed in the Caribbean–Bahamas region and are common beneath Pacific atolls.

Studies of the partly dolomitized, now Pliocene, *Hope Gate Formation of Jamaica* led Land (1973a) to invoke meteoric dilution of seawater as the mechanism. The dolomite replaces lime mud and coralline algal grains (originally high-Mg calcite) with good textural retention, but many corals and molluscs, originally aragonite, are voids filled with coarse dolomite spar crystals (a cement). The dolomite has a low Sr (200 ppm) and Na (350 ppm) content compared with the original $CaCO_3$ grains and this, together with the occurrence of large dolospar crystals, led Land to implicate dolomitization by dilute waters, rather than hypersaline brines which was the fashion at the time. The $\delta^{18}O$ of the Hope Gate dolomite is low positive (+2.2‰ average), less heavy than sabkha dolomite (Section 8.7.1, also see Fig. 8.18). $\delta^{13}C$ is also low positive (1.2‰ average); this is close to Recent Jamaican reef sediments (variable, but +2.1‰ average), suggesting derivation of most C in the CO_3^{2-} from the replaced $CaCO_3$. Diagenetic calcite in the Hope Gate, mostly precipitated after the dolomite, is much lighter in both $\delta^{18}O$ (−2.4‰) and $\delta^{13}C$ (−5.5‰), consistent with a meteoric origin. Thus Land (1973a) proposed that the dolomitization took place in the meteoric–marine mixing zone (Fig. 8.25) and continued meteoric flushing led to the calcite precipitation. Calculations (Land, 1973b) showed that adding 3–4% seawater to Jamaican meteoric water would give a solution oversaturated with respect to dolomite with a Mg/Ca activity molar ratio greater than 0.8 (that required for dolomite precipitation). Land emphasized the high Pco_2 value of Jamaican meteoric water and the role this could have had in inducing precipitation through degassing and mixing with seawater (Land, 1973a,b).

The Hope Gate is erosionally overlain by the *Upper Pleistocene Falmouth Formation*, showing that the Hope Gate dolomitization took place before Falmouth deposition. It is possible that the Falmouth limestones are being dolomitized at the present time below the groundwater table where mixing with

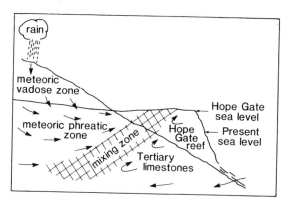

Fig. 8.25 *Mixing-zone model for dolomitization of the Hope Gate Formation of Jamaica. After Land (1973a).*

seawater occurs (Land, 1973b). The Falmouth dolomite is similar to that in the Hope Gate, although the $\delta^{13}C$ is very negative (−8.4‰), indicating strong organic matter diagenesis during dolomitization, and the $\delta^{18}O$ is low negative (−1‰), consistent with the influence of meteoric water (see Fig. 8.18). Of note is the very high Sr (3000 ppm), derived from skeletal aragonite and indicating closed-system diagenesis with little or no meteoric flushing.

On the *tidal flats of southwest Andros Island*, dolomite does occur in the shallow subsurface (Gebelein *et al.*, 1980), and a mixing-zone interpretation has been applied (Gebelein, 1977b). As described in Sections 3.2.2b and 4.3, the tidal flats of southwest Andros are wider, have fewer channels, and the offshore waters are more restricted, compared with the better known tidal flats of northwest Andros, described by Shinn *et al.* (1969) and Hardie (1977). The distinctive features of the southwest tidal flats are hammocks, former beach ridges rising 1 to 1.5 m above the surrounding algal marshes, channels, ponds, intertidal flats and lagoons. As noted in Section 8.7.1, lithified dolomitic crusts are present on the slopes of the hammocks. Freshwater lenses occur beneath the hammocks, whereas more normal seawater, locally evaporated, occurs on and beneath the surrounding lower areas. Calcium-rich 'dolomite', 38–44 mole % $MgCO_3$ but without ordering peaks, is present within the tidal flat sediments of southwest Andros and preliminary studies (Gebelein, 1977b) suggested that it had formed in the mixing zone associated with the freshwater lenses beneath the hammocks. More detailed work, however (Gebelein *et al.*, 1980), showed that there

was not a good correlation between dolomite occurrences and position of the mixed or fresh groundwaters. The 'dolomite' mostly occurs as well-formed
1 μm rhombs growing around and engulfing aragonite
needles, and contrasting with less well-formed rhombs
of the lithified dolomitic crusts. Some aragonite
and high-Mg calcite dissolution and low-Mg calcite
replacement are occurring beneath the hammocks in
the freshwater lenses. Although dolomite is not
forming to the extent envisaged on southwest Andros,
abandoned beach ridges with their freshwater lenses
on an extensive tidal flat do provide a means for the
dilution of seawater. This variant on the mixing-zone
model could be an alternative to the evaporative/
sabkha explanation for dolomites in tidal flat facies.

The meteoric–marine mixing-zone model has been
popular for cases where there are no evaporites associated with the dolomites, subtidal facies are dolomitized, and the dolomitization event was relatively
early (i.e. near-surface, before any compactive fracture
of grains). It is also able to account for low trace
element (Sr, Na) contents, light $\delta^{18}O$, and a positive
correlation of $\delta^{18}O$ with $\delta^{13}C$ from the mixing of two
waters of differing isotopic composition. Mixing-zone
dolomites can be expected to develop extensively
during major regressive periods, when platform carbonates are being deposited. Seaward progradation of
a carbonate shoreline will be accompanied by a progradation of the meteoric–marine mixing zone.
Active circulation and pumping of water through the
carbonate sediments are important in this model (as
in the others) and these will be determined to a large
extent by the climate. Groundwater circulation will
be more active under a humid climate with strong
seasonal rainfall, than under a more arid climate.

Since the mixing-zone model depends on the
effective movement of fluids, there will be a strong
palaeogeographic control on the distribution of
mixing-zone dolomites. They should be located in the
more landward parts of a carbonate platform
sequence, and they may relate to more porous facies,
which acted as conduits for the groundwater.
Illustrating these points, *Early–Middle Palaeozoic
subtidal carbonates in Nevada* consist of dolomites in
the east (to landward) and limestones in the west,
basinwards (Fig. 8.26, Dunham & Olson, 1978, 1980).
The palaeogeographic control and early shallow
subsurface origin of the dolomite, together with an
absence of evaporites, point to the mixing-zone model.
It is envisaged that meteoric waters entered the
carbonate package during subaerial exposure of the
carbonate platform at a time of regional regression, so
that the dolomite–limestone boundary broadly reflects the subsurface penetration of freshwater lenses.
These Palaeozoic dolomites have low Sr contents
(average 50 ppm) and negative $\delta^{18}O$ (−0.8 to −6.7‰),
consistent with dolomitization from dilute waters
(Sections 8.5 and 8.6), although bear in mind that
$\delta^{18}O_{seawater}$ was more negative in the Palaeozoic (see
Fig. 9.19).

In the *Mississippian Ste Genevieve Limestone
of the Illinois Basin*, subtidal lime mudstones–
wackestones are dolomitized in lens-shaped bodies up
to 12 m thick, 0.5–2.5 km across and 1–5 km long,
with an east–west to northeast–southwest orientation. The dolomite bodies, which have porosities up
to 40% and are oil-productive, occur beneath elongate
lenses of ooid grainstone which have a similar orientation (see Fig. 4.22). Choquette & Steinen (1980)
suggested that dolomitization began at the base of

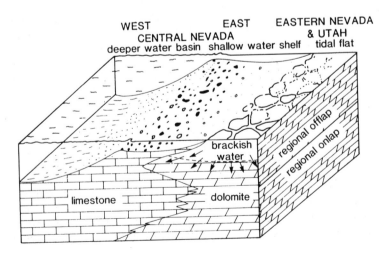

WEST   EAST   EASTERN NEVADA
CENTRAL NEVADA   & UTAH
deeper water basin   shallow water shelf   tidal flat

brackish water

regional offlap

regional onlap

limestone

dolomite

Fig. 8.26 *Mixing-zone model for
dolomitization of Lower Palaeozoic
carbonates of the Western Cordillera
in Nevada and Utah. Emergent areas
along the eastern part of the platform
were sites of freshwater recharge, and
lateral migration of recharge areas
with regional onlap and offlap
resulted in migration of the
limestone–dolomite boundary with
time. After Dunham & Olson (1978).*

freshwater lenses in the oolites, established after some exposure, and continued into later diagenesis with the oolites acting as conduits for groundwater coming from recharge areas to the northeast. It is thought that the mixing-zone dolomitization could have been taking place within the sediments when buried a few tens to hundreds of metres beneath the seafloor. Freshwater aquifers beneath the seafloor are well known off the Florida coast and are parts of dynamic seaward-flowing hydrologic systems (Kohout, 1967). The Ste Genevieve dolomites have low Sr (166 ppm) but high Na (810 ppm), and quite heavy $\delta^{13}C$ (+2.4‰) and $\delta^{18}O$ (+2.0‰) (Fig. 8.19A). The $\delta^{13}C$ suggests C mostly from the original marine carbonate and the $\delta^{18}O$ is similar to the Pleistocene Hope Gate mixing-zone dolomites (although, again, remember there have been secular variations in $\delta^{18}O_{seawater}$).

8.7.4 Burial dolomitization

Evidence for dolomite precipitation during burial can be found in many dolomitic rocks, but whether whole carbonate formations can be dolomitized at depth is still a matter of debate. The principal mechanism advocated in this model is the compactional dewatering of basinal mudrocks and the expulsion of Mg^{2+}-rich fluids into adjacent shelf-edge and platform carbonates (Fig. 8.1H). The Mg^{2+} source invoked is the porewater (generally modified seawater) and clay mineral changes. The composition of subsurface fluids is not well known but in fact many formation waters have lower Mg/Ca ratios (1.8–0.04) than seawater (5.2) (Collins, 1975). This loss is mainly due to the formation of chlorite and to precipitation of dolomite during shallow burial. Formation waters derived from evaporite sequences, however, are likely to have higher Mg/Ca ratios than seawater and a higher Mg^{2+} content.

The transformation of clay minerals with increased burial and rising temperature is well documented and it is frequently suggested that Mg^{2+}, along with Fe^{2+}, Ca^{2+}, Si^{4+} and Na^+ are released on the conversion of smectite to illite (e.g. Boles & Franks, 1979; Mattes & Mountjoy, 1980; McHargue & Price, 1982) with Ca^{2+} and Si^{4+} being released early (to precipitate as calcite and quartz cements) and the Fe^{2+} and Mg^{2+} later. Basinal shales are commonly organic rich, and the diagenesis of this organic matter would contribute CO_3^{2-}. However, dolomite is not a major cement in thick sandstone–shale sequences, such as the Tertiary of USA Gulf Coast (Boles & Franks, 1979; Milliken *et al.*, 1981); rather calcite and ankerite

$(CaFe_{0.5}Mg_{0.5}(CO_3)_2)$ are the carbonates precipitated.

One major problem with the burial–compaction model is the long distance transport of the pore-fluids into adjacent shallow-water carbonates. Much porewater moves upwards rather than laterally during compaction (Magara, 1978), although lateral movement may be significant where shales are interbedded with porous calcarenites and sandstones. In addition, there is no continuous source of fluid in basinal muds, and to supply sufficient Mg^{2+}, Morrow (1982b) has calculated that compaction of several hundred cubic centimetres of shale would be required to dolomitize 1 cm^3 of laterally adjacent limestone. In a similar vein, mass balance calculations by Given & Wilkinson (1987) have shown that Mg^{2+} from clays could only account for some 10% of the dolomite within the Michigan Basin, indicating that there is insufficient Mg^{2+} within mudrock sequences for massive dolomitization.

On the plus side, dolomitization should proceed more easily at depth, where higher temperatures mean that some of the kinetic hindrances to dolomite precipitation are diminished (Section 8.2). The proportion of hydrated Mg^{2+} ions should decrease, and reaction rates of dolomite precipitation should increase. There is also more time available in the burial environment.

Dolomite is common in thin limestones close to clay beds (McHargue & Price, 1982) and it does occur along pressure dissolution seams in limestones (Logan & Semeniuk, 1976; Wanless, 1979). These occurrences are generally on a local and small scale, however. Wanless (1979) suggested the Mg^{2+} was derived from high-Mg calcite grains which had not stabilized nearer to the surface. This could only happen if there had been no early meteoric diagenesis.

The derivation of Mg^{2+} for dolomitization from high-Mg calcite has been termed *solution cannibalization* by Goodell & Garman (1969). They invoked this mechanism for cyclic Plio–Pleistocene carbonates beneath Andros Island, where dolomite appears to have formed after a phase of uplift and erosion. They imagined that dissolution of metastable high-Mg calcite enriched pore-fluids in Mg^{2+}, and that these dense brines then percolated downwards to dolomitize underlying strata. Although such a mechanism could operate on a local scale, there is insufficient Mg^{2+} in high-Mg calcite to generate stratal dolomites (Given & Wilkinson (1987) calculated that 10% was the maximum amount of dolomite that could form through the release of Mg from high-Mg calcite). From studies of lacustrine carbonates, Müller *et al.* (1972) proposed that high-Mg calcite could act as a nucleating point for dolomite, in view of their crystallographic similarity. High-Mg calcite as a half-way stage in the dolomitization of aragonite has been detected by Katz & Matthews (1977) in laboratory experiments and it could be performing the same role in the Abu Dhabi sabkhas (McKenzie, 1981). Thus, it appears that high-Mg calcite could not be the only source of Mg^{2+} for dolomitization, but the presence of this $CaCO_3$ phase could promote the formation of dolomite.

Burial dolomites can be expected to have negative $\delta^{18}O$ values from precipitation at the higher temperatures at depth. Trace elements Sr and Na would mostly be low since generally the original carbonate sediments can be expected to have stabilized to low-Mg calcite and the dolomitization would lower these elements further. Baroque dolomite (Section 8.4.3) is one particular type of dolomite which appears to be largely of burial origin.

One fully documented ancient dolomite interpreted as of burial origin is the *Upper Devonian Miette Buildup of Alberta* (Mattes & Mountjoy, 1980). The dolomite preferentially occurs along a kilometre-wide zone at the margin of the reef complex, and was precipitated during burial after calcite cementation. The dolomitizing fluids are thought to have come mainly from dewatering of basinal mudrocks adjacent to and below the Miette Buildup. The $\delta^{18}O$ of the main replacement dolomite type is quite negative (-5‰), whereas the $\delta^{13}C$ ($+3$‰) has a more normal marine value (see Fig. 8.19A). The last dolomite phase, a baroque dolomite occurring in fracture fills, has a $\delta^{18}O$ of -9.3‰, consistent with precipitation at a higher temperature. Trace elements are low in Miette dolomites (Sr: $30-100$ ppm, Na: $300-400$ ppm).

Burial dolomitization is also invoked for the regionally extensive *Cambrian Bonneterre Dolomite in southeast Missouri*, a less than 6 m thick package at the base of a shallow-water carbonate sequence (Gregg, 1985). The dolomite is a coarse, non-ferroan xenotopic mosaic of crystals with undulose extinction, and vugs and fractures are lined by dolomite cement. $\delta^{18}O$ for the dolomites is variable but averages -7.3‰, whereas host limestones are close to -7.8‰. Dolomitization is attributed to warm, basin-derived waters circulating through underlying sandstone and the basal carbonates. These waters also precipitated Mississippi-Valley type lead–zinc sulphides in a stromatolite reef above the dolomite, and the dolomites associated with these ores have a similar CL microstratigraphy and isotopic signature to the dolomite cements in the Bonneterre.

In the dolomitized *Lost Burro Formation (Devonian, California)*, Zenger (1983) described many features indicating burial dolomitization: (1) coarse crystals (up to several millimetres in diameter), (2) a zebroid texture consisting of bands of matrix dolomite alternating with white dolomite spar in elongate cavities, (3) healed microfractures, (4) dolomite 'tongues' and dolomite fronts cutting across bedding in limestone, (5) relict stylolites within the dolomite mosaics, (6) net textures where seams of dolomite enclose millimetre to decimetre patches of limestone, through dolomitization along stylolites, and (7) stable isotope data where $\delta^{18}O$ averages -7.3‰, which can be explained as the result of precipitation from hot subsurface fluids.

A strong case for burial dolomitization was presented by Gawthorpe (1987) for the *Pendleside Limestone (Dinantian, Bowland Basin, northern England)* deposited in a carbonate slope environment with much interbedded mudrock. The wide range of dolomite textures were all developed after calcite cementation and neomorphism, and some were post-stylolitization. the dolomite varies from near-stoichiometric to ankeritic with 20 mole% $FeCO_3$. $\delta^{18}O$ of the dolomites is more negative than the host limestones, although $\delta^{13}C$ is similar. Dolomitization is ascribed to Mg^{2+}-rich fluids expelled from basinal mudrocks (the Bowland Shales) into interbedded, porous, coarse-grained carbonates, with the most intense alteration taking place at the margins of the limestone beds (Fig. 8.27). Sandstones in the sequence are also dolomitized close to the shales. A vuggy, secondary porosity developed in partly dolomitized beds through a phase of leaching, and some of these vugs were later filled with saddle dolomite (Fig. 8.14C).

8.7.5 Seawater dolomitization

In most of the models discussed in previous sections, the source of Mg^{2+} is seawater and the kinetic problems in precipitating dolomite from normal seawater are overcome by either diluting it or evaporating it. The models then provide a mechanism for driving the dolomitizing fluid through–the carbonate sediments. Land (1985) has suggested that with little modification, seawater itself may be able to dolomitize if there is an efficient mechanism for pumping it through carbonate sediments. Kastner (1984) has been advocating dolomitization from seawater if only the SO_4^{2-} content is lowered (see Section 8.2). Discoveries of apparently

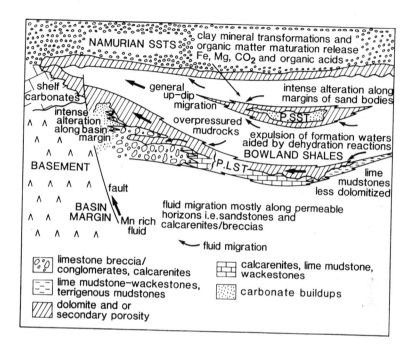

Fig. 8.27 *Burial dolomitization model through mudrock dewatering for Lower Carboniferous sediments, the Pendleside Limestone (P.LST) and Pendleside Sandstone (P.SST), of the Bowland Basin, northwest England. After Gawthorpe (1987).*

marine dolomite are now being made and new models are being proposed which implicate seawater alone as a dolomitizing fluid.

Cores from *Enewetak Atoll* in the Pacific contain dolomitized Eocene coralline algal–rotaline foram grainstone, at depths between 1250 and 1400 m below sea-level (Saller, 1984). The dolomite was precipitated after brittle fracture of grains. The latter is first observed at 610 m depth and suggests that dolomitization has taken place since Early Miocene time, when the Eocene strata would have been buried to that depth. Strontium isotope ratios ($^{87}Sr/^{86}Sr$) on the dolomites confirm this, by indicating precipitation between latest Early Miocene and the present. It appears that ocean water is circulating freely, deep within the Enewetak Atoll so that the only fluid available for the dolomitization would have been seawater. Freshwater or hypersaline water could not have penetrated sufficiently far down for mixing-zone or seepage–reflux models to be viable. The $\delta^{18}O$ values of the dolomite (+2.5‰) are relatively heavy and can be interpreted as reflecting precipitation from normal seawater at low temperatures (15–18°C). Seawater at the depth of the dolomite is 10–20°C, much cooler than surface Pacific water (28°C). The $\delta^{13}C$ of +2.3‰ suggests a marine source of CO_3^{2-}. The calcite saturation depth in this region of the Pacific is around 1000 m, and Saller (1984) reasoned

that seawater would still be supersaturated with respect to dolomite at and below this depth. The driving mechanism for seawater circulation through the atoll is likely to be oceanic tides, which appear not to see the atoll, and thermal convection due to the higher heat flow through the volcanic basement of the atoll. The Enewetak dolomites are thus being formed where seawater is pumped through the shallow-water limestones at a depth where the seawater is undersaturated with respect to calcite. The model deriving from this (Fig. 8.28) could very well be applicable to ancient platform-margin carbonates, where dolomites preferentially occur in fore-reef and reef facies, but are absent in platform-interior facies. Although in the Enewetak case, dolomitization is taking place in very deep water, it is quite likely that at times in the past the depth to the level of calcite undersaturation was shallower.

In a similar manner to Enewetak, it does appear that ocean water is being pumped through the Bahama Platform too, where ocean currents such as the Gulf Stream impinge on the Bahama escarpment. It also transpires that dolomitization of Pleistocene carbonates is taking place beneath the mixing zone on the Bahamas, within these circulating saline waters. The latter are undersaturated or only just saturated with respect to $CaCO_3$, but well saturated with respect to dolomite (see Fig. 7.21; Smart *et al.*, 1988b). This

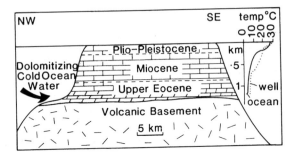

Fig. 8.28 *Model for dolomitization of Eocene limestones of Enewetak Atoll by cold ocean water. The temperature profile of water within the atoll from a deep well is closely parallel to that of the adjacent open ocean, suggesting that ocean water freely circulates through the atoll. There is an increase in temperature near the bottom of the well suggesting that circulating water is removing geothermal heat from Enewetak's volcanic basement. After Saller (1984).*

ocean current pumping process could be a major dolomitization mechanism since it provides a means of transporting huge volumes of seawater through a carbonate platfrom.

Large-scale circulation of seawater into carbonate platform margins was proposed by Kohout (1967), and Simms (1984) has considered the implications of *Kohout convection* for dolomitization. Mullins *et al.* (1985) invoked this process to account for diagenetic changes in periplatform oozes off the Little Bahama Bank where high-Mg calcite and aragonite are lost and calcite (low Mg) gained by the sediment. Kohout convection occurs in response to horizontal density gradients between cold marine waters adjacent to a carbonate platform and geothermally-heated groundwaters within the platform (Fig. 8.1G). The ocean waters are drawn into the platform margin, displacing the warm, less dense porewaters within the platform which emerge as springs on the platform or platform edge (e.g. Kohout, 1967; Fanning *et al.*, 1981). The open convection cell will operate where ocean water depths are in the range of 1 to 3 km (Simms, 1984), and it provides a mechanism whereby cold seawater undersaturated with respect to high-Mg calcite and aragonite but saturated with respect to low-Mg calcite, or undersaturated with respect to all three $CaCO_3$ minerals but saturated with respect to dolomite, can be pumped through platform-margin carbonates. Kohout convection will be a long-lived process operating on the time-scale of the platform itself, so that potentially it is a very important mechanism for dolomitization. In addition, the

Kohout convection cell may interact with the regional, meteoric groundwater system, as occurs in the Floridan Aquifer (Fig. 8.29). Mullins *et al.* (1985) and others have shown that Mg is leached from high-Mg calcite grains in periplatform ooze and this could well be taken up by inflowing ocean water to raise its potential to dolomitize platform-interior carbonates. Although much of this is hypothetical, dolomite has been reported in periplatform oozes (Mullins *et al.*, 1984, 1985) and the oxygen isotopic compositions (+2.9 to +5.1‰) support precipitation and equilibration with deep, cold bottom waters. In addtion, the $\delta^{13}C$ signature of around −2.6‰ is a typical marine carbonate value. It does not suggest a necessary link with organic matter diagenesis and reducing conditions, as is envisaged for the Gulf of California dolomites which have very enriched to depleted carbon isotope values (Section 8.6; Kelts & McKenzie, 1984). In the Florida subsurface, Eocene limestones are dolomitized and several studies have suggested a mixing-zone origin (e.g. Randazzo & Cook, 1987). However, warm (40°C) chemically-altered seawater discharging on the gulf side of Florida, has lost Mg^{2+} and gained Ca^{2+}, relative to normal seawater. Fanning *et al.* (1981) interpreted this as the result of dolomitization of the porous, aquifer limestones, at a depth of 500−1000 m, by the seawater alone.

Another mechanism for pushing seawater through a carbonate platform suggested by Simms (1984) is that of *reflux*. Shallow water on carbonate platforms frequently becomes a little more saline than ocean water through the areal extent of the platform, summer evaporation, and restriction and weak circulation caused by platform-margin reefs, sand bodies and islands. On the Bahama Platform (Section 3.2), salinities up to 42−45‰ are regularly developed over large areas, and in very protected parts, they reach 80‰. Simms (1984) was able to show by experiment that, if the porewaters within the platform were of normal marine water (35‰), downward reflux of only slightly hypersaline seawater (42‰) would occur. The process would be much more vigorous if the salinity contrast was greater. Reflux is thus another mechanism for pushing seawater through a platform carbonate sequence and this could result in subsurface dolomitization, especially if there were slight modifications to the seawater chemistry to reduce the kinetic hindrances to dolomite precipitation, such as a higher Mg/Ca ratio, lower SO_4^{2-} or higher CO_3^{2-} (see Section 8.2).

Tidal flat dolomites precipitated from little-altered marine waters are forming on Sugarloaf Key, Florida and Ambergris Cay, Belize, and the mechanism for

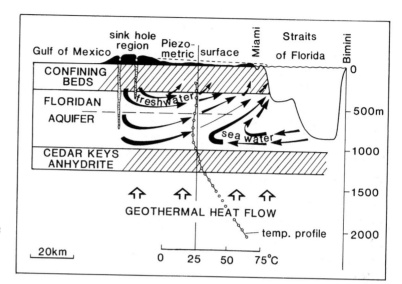

Fig. 8.29 *Groundwater flow system in the Florida subsurface, showing the effect of thermal convection. After Kohout (1967).*

the seawater moving through the sediment is *tidal pumping*. In both cases, dolomitic crusts are forming on high intertidal–low supratidal muds which overlie porous Pleistocene oolites. Although formerly ascribed to evaporative processes (Shinn, 1968b), the Sugarloaf dolomite is apparently being precipitated from Florida Bay water only slightly modified by evaporation and/or sulphate reduction (Carballo *et al.*, 1987). The water is pumped through the sediments during spring tides and most dolomite occurs where the Holocene muds are thin. The initial precipitates are 0.1 to 0.3 μm subrounded crystallites which develop into 1–5 μm euhedral Ca-rich, poorly-ordered dolomite rhombs. At Ambergris Cay, Belize, dolomite replaces high-Mg calcite grains and mud and is a cement too (Mazzullo *et al.*, 1987). Although there are seasonal influxes of meteoric water which etch the dolomite and leach aragonitic grains, the dolomite isotope data ($\delta^{18}O = +1.2$ to 2.2‰) are taken to indicate precipitation from near-normal marine water, rather than hypersaline or brackish water.

Minute quantities of dolomite ($Ca_{1.22}Mg_{0.78}$ have recently been found within a submarine hardground associated with a fringing reef off Jamaica (Mitchell *et al.*, 1987). The dolomite is a syntaxial cement upon high-Mg calcite grains (echinoderms and red algae) and cements. The origin of the dolomite is unclear. Mixing with meteoric water after heavy rains is a possible influence, and some sulphate reduction may be involved too, since the $\delta^{13}C$ value of 1.6‰ is slightly less positive than the associated high-Mg

calcite cements. The $\delta^{18}O$ signature of +2.00‰ is consistent with precipitation from seawater, and the Sr (600 ppm) and Na (1300 ppm) contents are similar to, or slightly less than, other marine dolomites.

In the *hemipelagic environment*, dolomites are now known to be forming during burial in organic-rich muds from porewater of marine origin (Baker & Burns, 1985). In the Gulf of California (Kelts & McKenzie, 1982, 1984), at water depths of hundreds of metres, dolomite is first encountered in ooze at 60 m sub-bottom. Firm through to lithified and impermeable beds of micritic dolomite occur at deeper levels and it appears that most of the dolomite is a cement in these generally $CaCO_3$-poor muds (Fig. 8.30). Some pelagic microfossils are replaced whereas others show evidence of dissolution. The dolomites are usually Ca-rich to stoichiometric (42–51 mole% $MgCO_3$) with the latter tending to occur in more lithified layers. The dolomite is being precipitated in the zone of methanogenesis (below the zone of sulphate reduction) where fermentation of organic matter produces very heavy CO_2 and extremely ^{13}C-depleted methane. The dolomite thus has very positive $\delta^{13}C$ values with an average of +9.8‰ (maximum +13.8‰). Oxygen values are variable since they reflect the depth and temperature of precipitation and dolomite is being precipitated over several hundred metres of burial. With individual beds of dolomite, the central parts have positive $\delta^{18}O$ (around +3‰) and there is a trend to negative values (to −7‰) towards the outer parts of the bed, clearly showing that the bed has

Fig. 8.30 *Euhedral dolomite crystals engulfing diatom debris. Guaymas Basin, Gulf of California. From Kelts & McKenzie (1982), with permission of Ocean Drilling Program, Texas A & M University.*

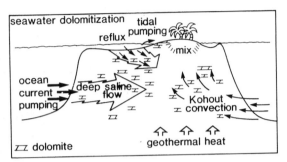

Fig. 8.31 *Seawater dolomitization models for a carbonate platform: ocean current pumping, reflux of slightly hypersaline lagoon waters, tidal pumping along shorelines and Kohout convection.*

accreted during burial and rising temperature. Pore-fluids are strongly depleted in SO_4^{2-} and alkalinity reaches high values.

On the Pacific side of Baja California, dolomite is forming close to the sediment–water interface in hemipelagic slope oozes (depths 200–1500 m) which are within the oxygen-minimum zone (Shimmield & Price, 1984). Dolomite is absent from adjacent shallower-water shelf and deeper-water oceanic muds. Precipitation is taking place in the sediment's reduction zone and is attributed to the high organic content of the mud and low sedimentation rates. Mg^{2+} is thought to be supplied from overlying seawater, with bioturbation aiding mixing. Two other factors of significance here are again low SO_4^{2-} and high alkalinity.

Dolomite forming in muds in fore-arc basins off Peru (Kulm *et al.*, 1984) shows a much greater range of $\delta^{13}C$ (+20 to −16‰) than the Gulf of California dolomite. The very positive values come from organic-rich muds of the Lima Basin and reflect precipitation in the zone of bacterial fermentation. The more negative values come from less organic-rich muds of the Trujillo Basin, deposited at a lower sedimentation rate, where diagenesis of the organic

matter was largely by microbial sulphate reduction, resulting in light CO_2.

These 'organic' dolomites in hemipelagic sediments generally occur in areas of high organic productivity, often associated with upwelling and oxygen-minimum zones, and although the actual mechanism for the dolomite precipitation is unclear, it could well be related to the lowered SO_4^{2-} and increased alkalinity of the marine porewaters arising from organic matter diagenesis (Baker & Burns, 1985). Micritic dolomites in ancient deeper-water sequences could well be analogous to these modern occurrences (e.g. some Monterey Formation dolomites, California, Garrison *et al.*, 1984; Burns & Baker, 1987, and the Tripoli Formation, Messinian, Sicily, Kelts & McKenzie, 1984; Bellanca *et al.*, 1986), but the point to make, again, for the interpretation of ancient platform dolomites, is that the mineral can be precipitated out of normal seawater, in this case only modified by diagenetic reactions with organic matter.

From this section, it should be apparent that in certain situations dolomite can be precipitated from seawater, although slight modifications to its composition may be necessary. One of the main requirements in many of the cases is the efficient pumping of seawater through the sediments, and several different mechanisms to do this have been described here. They are summarized in Fig. 8.31. The concluding message from this chapter is that an open mind should be kept when studying ancient dolomites (also see Section 9.7) and that dolomitization from just plain old seawater should not be discounted.

9 The geological record of carbonate rocks

9.1 INTRODUCTION

Since James Hutton (1727–1797) and Charles Lyell (1800–1860) advanced the doctrine of uniformitarianism, it has been tacitly assumed that the present is the key to the past for all aspects of sedimentary geology. However, in recent years, it has been increasingly appreciated by carbonate sedimentologists that this tenet can only be applied to carbonate rocks in very general terms, and that there are many instances where this principle does not hold. There have been variations in the abundance of some facies through the Phanerozoic and there are certain facies which have only developed at particular times. Some of these fluctuations in facies occurrence are a response to global sea-level changes whereas others reflect more the changing fortunes of organisms, the evolution of new groups and the demise of others.

There is now much evidence for temporal fluctuations in the mineralogy of marine carbonate precipitates through the Phanerozoic, with aragonite and high-Mg calcite predominating at certain times and low-Mg calcite at others. This is taken to reflect subtle changes in the chemistry of seawater. Also, on a broader scale, it has long been known that dolomites are more common in the Early Palaeozoic and Precambrian than in more recent times.

Thus, with carbonate rocks it is very dangerous to assume the principle of uniformitarianism. At the present time, sea-level is relatively low, there are no epeiric seas and sea-level has fluctuated drastically over the last few million years in response to the 'Great Ice Age'. The facies, mineralogy and early diagenesis of Holocene carbonate sediments reflect these recent changes in sea-level so that they can be used to interpret Pleistocene carbonate formations. However, care has to be exercised in using the present to understand pre-Pleistocene rocks. For limestones, the principle of uniformitarianism is better rewritten as 'the present is the key to perhaps the Pleistocene' (Neumann & Land, 1975 quoting Lowenstam), with the past a guide to the future. This chapter outlines some of the changes in facies and mineralogy of carbonates which can be perceived in the rock record.

9.2 CARBONATE FACIES THROUGH TIME

The major controls on carbonate facies are geotectonics and climate, as discussed in Chapter 2. These two factors determine the relative positions of sea-level, and geotectonics control the type of carbonate platform. Compilations of sediment lithologies through the Phanerozoic show that there is a broad correlation between carbonate rock occurrence and global sea-level. More shallow-marine carbonates were deposited at times of higher sea-level stand/continental submergence, than at times of continental emergence (Fig. 9.1, from Kazmierczak et al., 1985, but note the curious low the Middle–Upper Cambrian).

Carbonate platforms covered by shallow epeiric seas on a craton or continental scale only develop when global sea-level is high. Thus the Lower to Mid-Palaeozoic was a time of extensive shallow-water platform carbonate sedimentation over much of the North American and Russian cratons. Larger areas of the cratons were flooded during this time than in the major Mesozoic transgression (Fig. 9.1B; Hallam, 1977), when carbonates were also widely deposited. During the Upper Cretaceous high sea-level stand, deeper-water chalks (Section 5.6) were deposited over much of western Europe, from Ireland to the Russian platform, and within the North American mid-continental seaway. Epeiric platforms on a more regional scale are much more dependent on the local tectonic regime, rather than global sea-level, with the location of faults and subsidence rates affecting facies variations and sediment thickness. Such platforms are well developed in the British Lower Carboniferous and in the Tethyan region during the Triassic–Jurassic. Rimmed shelves and ramps are also less controlled by global sea-level and more by local and regional tectonics, so that the deposits of both settings can be found throughout the geological record.

9.2.1 Carbonate sands

Carbonate sands have been deposited in a similar

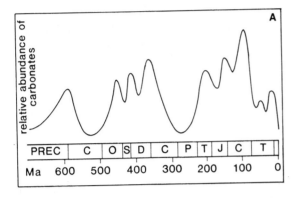

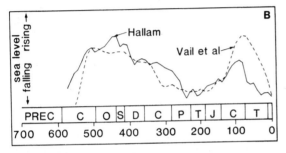

Fig. 9.1 *Carbonate abundance and global sea-level. (A) Relative abundance of marine carbonate rocks during the Phanerozoic. After Kazmierczak et al. (1985). (B) Pattern of global sea-level change during the Phanerozoic from work of Hallam (1977) and Vail et al. (1977). After Given & Wilkinson (1987).*

range of higher-energy environment throughout the geological record, so that lithofacies, sedimentary structures and sequences are comparable to modern occurrences (see Sections 4.1 and 4.2). However, in detail, there are differences in the composition of the carbonate grains. Obviously, there will be variations in the composition of bioclastic sands, since these reflect the types of carbonate-secreting organisms living at the time (see Fig. 9.2). Palaeozoic skeletal grainstones are usually dominated by brachiopod and crinoidal debris, with contributions from bryozoans and corals, whereas bivalve material is more common in Mesozoic grainstones. Cainozoic bioclastic lime sands are mostly mollusc rich, with brachiopod and echinoderm material insignificant. Some Precambrian grainstones are composed of calcified algal clusters.

Non-skeletal grainstones are peloidal and oolitic, but there do appear to have been peaks of ooid precipitation during the Phanerozoic. Wilkinson *et al.* (1985) found peak abundances of oolite during the Late Cambrian, Late Mississippian, Late Jurassic and Holocene (Fig. 9.3). Comparison with the global sea-level curve shows that these periods coincide with the first order time of global transgression and global regression. They do not coincide with times of low global sea-level, but this is not surprising since carbonates are generally less abundant then, when the areas of shallow-marine deposition are more restricted. Curiously, the peaks do not coincide with periods of

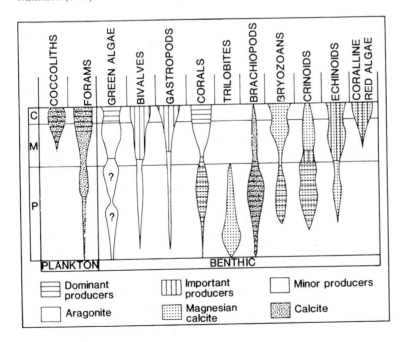

Fig. 9.2 *Approximate diversity, abundance and composition of principal groups of calcareous marine organisms. After Wilkinson (1979).*

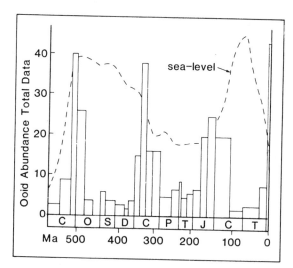

Fig. 9.3 *Abundance of oolite in Phanerozoic limestone sequences. Note that oolite commonly formed during transgressions and regressions but was rare during continental emergence and submergence. After Wilkinson* et al. *(1985).*

high sea-level either, when significantly larger volumes of carbonate were deposited. Oolites are conspicuously reduced in abundance in mid-Ordovician to Lower Mississippian and Upper Cretaceous to mid-Tertiary limestones. Two possible explanations for this conundrum are (1) that during continental submergence water depths were too great for the development of the wave-agitated environments needed for ooid precipitation, or (2) that some chemical factors prevented ooid precipitation, in spite of there being suitable physical conditions (Wilkinson *et al.*, 1985). The latter is considered more likely since carbonate production rates are sufficiently high to keep up with rates of transgression and many shallow-water carbonate facies, but with few oolites, were deposited during these submergent times. Wilkinson *et al.* (1985) suggested that the $P\text{co}_2$ levels, which are invoked as an explanation of the dominantly calcitic mineralogy of marine precipitates during submergent modes (Section 9.3), may have reached sufficiently high values to inhibit physicochemical carbonate precipitation by lowering the degree of carbonate saturation.

9.2.2 Reef complexes and patch reefs

A more obvious variation in carbonate facies through time is seen in the development of reef complexes.

Many modern, shelf-margin, reefs consist of a framework built by scleractinian corals, encrusted and bound by crustose coralline algae, and a well-defined facies model has been deduced with reef-flat, reef-crest and reef-shope zones (see Section 4.5). This facies model can be applied broadly to many ancient shelf-margin reef complexes, but in detail there are often important differences. This is especially the case where the reef core is concerned, since in many instances, the organisms involved did not have the ability to produce a rigid framework and/or encrusting and binding organisms were not present. Ancient carbonate buildups vary considerably in internal structure and texture depending on the roles played by the organisms. Some reefs did have frameworks produced by rigid, massive and branching skeletal organisms such as corals and stromatoporoids, whereas others only consisted of many *in situ* skeletons which trapped and baffled sediment. Buildups have also been produced by non-skeletal organisms, such as grass and algae, through trapping and binding of sediment.

The geological record of organisms capable of producing large, rigid, branching, tabular or massive skeletons has not been continuous. There have been seven major phases (Fig. 9.4; Longman, 1981; James, 1983): (1) Middle-Upper Ordovician bryozoan—stromatoporoid—tabulate coral reefs, (2) Silurian and Devonian stromatoporoid—coral reefs, (3) Upper Permian sponge—calcareous algal reefs, (4) Upper Triassic and (5) Upper Jurassic scleractinian coral—stromatoporoid reefs, (6) Upper Cretaceous rudistid bivalve reefs and (7) Tertiary—Quaternary scleractinian coral—red algal reefs. On a broad scale, all these reefs can show similar facies patterns to modern coralgal reefs (phase 7, see Section 4.5), with the best development along shelf margins, a seaward fore-reef talus apron, and a shoreward back-reef skeletal sand facies passing into a shelf lagoon. In detail, there will still be many differences, principally in how the organisms behaved and interacted in the reef-core facies, but also in such features as the degree of synsedimentary cementation and the extent of bioerosion, especially by boring organisms.

Patch reefs, by way of contrast, have a more or less continuous record through the Phanerozoic (Fig. 9.4) although there is a conspicuous absence of such reefs in Early Triassic, Late Devonian and Late Ordovician times. These discrete buildups, bioherms, reef mounds, banks and knolls, are also variable in the types of organism(s) involved and the roles they played in patch reef construction. They have been

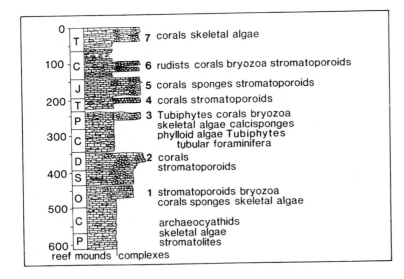

Fig. 9.4 *Generalized occurrence of major reef complexes (1 through 7) and patch reefs/reef mounds through the Phanerozoic with the principal skeletal elements involved. After James (1983).*

formed by most organisms at one time or another and vary considerably in size, shape, faunal diversity and associated facies. Apart from the obvious colonial skeletal organisms, such as corals, stromatoporoids, bryozoans and sponges, at times patch reefs have been constructed by organisms not normally thought of as reef buildings, such as oysters, brachiopods, gastropods, foraminifera and serpulid worms. Many are simply accumulations of skeletal debris rather than true reefs with a framework.

Thus for both shelf-margin reef complexes and patch reefs, there are major changes in the faunal compositions, mode of construction and facies through the Phanerozoic, all basically reflecting the evolution and demise of carbonate-secreting and sediment-trapping organisms. Looking at reef-building taxa through the Phanerozoic, James (1983) has suggested that there have been two major cycles of reef development (Fig. 9.5), a shorter earlier one of around 240 Ma from the Cambrian through the Devonian, and a longer, later one of around 340 Ma from the Carboniferous to Recent. In both cycles, there was an early period with few reef-building organisms and these were mainly small branching and encrusting forms, giving rise to patch reefs, and a later period when a variety of large skeletal metazoans gave rise to complex reef structures. In cycle I, the calcified algae *Epiphyton*, *Girvanella* and *Renalcis* are important components of Cambrian reef mounds and they occur in reef complexes formed by sponges, corals and stromatoporoids until the Devonian. In cycle II, phylloid and other algae, *Tubiphytes* and foraminifera

form early reef mounds and then occur as accessory components to later reef complexes also formed by sponges, corals and stromatoporoids. The end-Cretaceous extinction event allowed the scleractinian corals to dominate in the Cenozoic. Archaeocyathids and rudist bivalves are two distinct, short-lived groups of reef-building organisms in the Lower Cambrian and Upper Cretaceous respectively.

9.2.3 Peritidal facies and stromatolites

Peritidal facies have distinctive characteristics (Chapter 4) which have been recognized in limestones of most ages throughout the geological record. However, one common component of peritidal sequences, stromatolites, does show some variation in abundance and diversity through time: they are a major feature of Precambrian carbonates and become less important into the Phanerozoic. *In the Precambrian*, stromatolites grew in basinal to shallow subtidal to supratidal environments (see Hoffman, 1976, for an example), but it is in the formation of reefs, or buildups, that they achieved their greatest diversity. They take on a wide range of growth forms, from branching to columnar to domal types (e.g. Fig. 9.6), and these may occur on a range of scales within one stromatolite buildup. Bioherms surrounded by oolites, biostromes laterally persistent for many kilometres and shelf-edge reefs cut through by tidal channels, have all been described. Planar and domal stromatolites, commonly desiccated, are a characteristic feature of Precambrian tidal flat facies and intraclasts derived

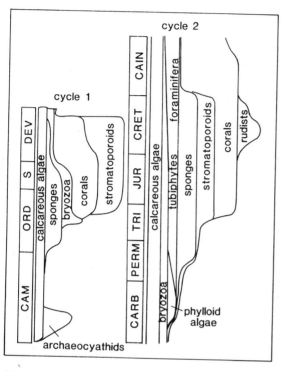

Fig. 9.5 *Distribution of major skeletal organisms involved in reef construction through the Phanerozoic, illustrating the apparent presence of two cycles. The widths of the various taxa on the figure reflect their relative abundance and importance. After James (1983).*

from these laminated sediments can form thick flake-stone units, with channels, imbrication and cross-bedding (see Tucker, 1977 for an example). *In Phanerozoic peritidal limestones, planar stromatolites (cryptalgal laminites) are of course very common, but the domal and columnar forms do appear to be less important.*

In the early days of stromatolite studies, it was usually assumed that they only formed in intertidal environments, by comparison with the majority of modern microbial mats then known from tidal flat (e.g. Andros Island, Black, 1933, and Monty, 1967, 1972) and foreshore environments (e.g. Shark Bay, Logan, 1961). Later, more critical consideration of ancient stromatolites (e.g. Playford *et al.*, 1976; Tucker, 1977) showed that many had to be of shallow to deep subtidal origin, and a better knowledge of modern environments revealed that there are stromatolites growing in the subtidal too (e.g. in Shark Bay, Playford *et al.*, 1976; and in the Bahamas, Gebelein, 1969; Dill *et al.*, 1986, see Fig. 3.12).

Precambrian stromatolites are frequently used for stratigraphic purposes and individuals, particularly the domal and columnar types, are given generic and species names in a similar manner to trace fossils. The number of columnar stromatolite forms recognized in the Precambrian rises to a maximum in the Late Riphean and then drops sharply into the Vendian with very few in the Cambrian (Fig. 9.7A). The macro and microstructures of stromatolites can be diagnostic of a particular stratigraphic horizon (e.g. papers in Walter, 1976). It might seem surprising that stromatolites can be used in this way, since most of the algal species involved had evolved by the Early Proterozoic, and stromatolite morphology is very much influenced by environmental factors. However, it appears that the stromatolite structure does vary in response to changes in the composition of the microbial mat community which may involve several algal species and many types of bacteria (Gebelein, 1974b). Microbial mat communities may be destroyed and reformed with a slightly different make-up by major regressive—transgressive events, so that a particular stromatolite microstructure may be diagnostic of a certain time interval.

Several authors have explained the apparent decline in stromatolite abundance and diversity from the Precambrian into the Phanerozoic (Fig. 9.7B) as a consequence of metazoan evolution. One popular explanation involves the evolution of the herbivorous gastropods which would have grazed upon the stromatolite-forming microbial mats (Garrett, 1970). At the present time, microbial mats on tidal flats are often best developed when or where gastropods are excluded, as a result of very high or very low salinities for example. The occurrence of columnar stromatolites in Shark Bay has been attributed to the hypersalinities lowering the gastropod grazing pressure. The development of burrowing organisms in the Cambrian, which could disrupt stromatolite fabrics that did form, has been suggested as a cause of decreasing preservation potential of stromatolites (Garrett, 1970). The development of encrusting skeletal organisms was thought by Monty (1973) to have limited blue—green algal colonization of substrates, particularly in reef settings. Monty also proposed that the rise of red and green calcareous algae in shallow-marine environments left only marginal marine (intertidal—supratidal) and freshwater settings for the cyanobacteria.

More recently, Pratt (1982b) has collected data to show that there are many Phanerozoic reefs which do have stromatolites as an important component. The

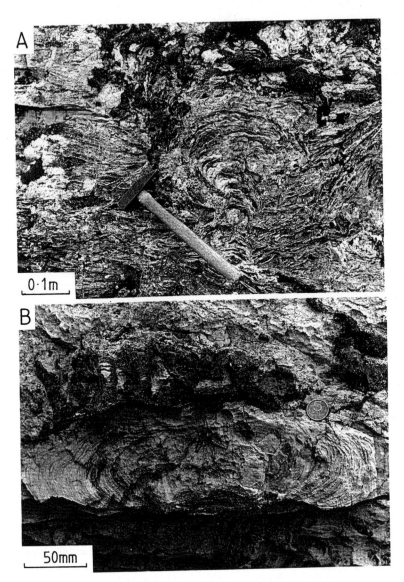

Fig. 9.6 *Precambrian stromatolites from the Porsanger Dolomite, Upper Riphean, Finnmark, Norway. (A) Cabbage-shaped stromatolite, part of small biohermal structure occurring within cross-bedded grainstones. (B) Cushion-shaped stromatolite with small digital, columnar stromatolites on top; basal part of a biostrome, 10 m thick, which extends for several kilometres. For details see Tucker (1977).*

Devonian Canning Basin reefs (Playford *et al.*, 1976), the Permian of west Texas and northeast England, Triassic reef mounds of Catalonia, and the Miocene reefs of Spain (Dabrio *et al.*, 1981) are examples. In modern reefs, branching and domal stromatolite-like structures are being found in deeper-water fore-reef environments (e.g. James & Ginsburg, 1979; Land & Moore, 1980), although the exact nature of the organisms involved is unclear. Pratt (1982b) considered that changes in stromatolite fortunes were a response to competition for substrates from other algae and metazoans and a result of increasing carbonate production rates in the Phanerozoic, through sediment generation by the breakdown of carbonate skeletons, leading to dilution of the contribution from stromatolites.

9.2.4 Mud mounds

A feature of mid-Palaeozoic outer-ramp sequences are mud mounds, consisting largely of micrite. Many have few skeletal organisms within them, whereas

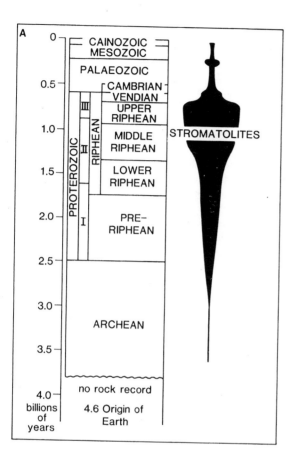

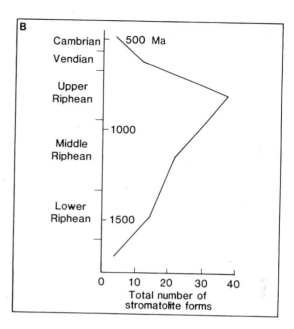

Fig. 9.7 *The geological record of stromatolites. (A) Abundance of stromatolites through time. After Awramik (1984). (B) Diversity of columnar stromatolites in the Proterozoic. After Awramik (1971).*

others have significant numbers of bryozoans, calcified algae or sponges. There is usually abundant evidence of synsedimentary cements, mostly radiaxial fibrous calcite, which fill cavities. These are commonly of the stromatactis type, and there has been much discussion of their origin (see Section 4.5.3a and Fig. 4.96; Bathurst, 1982; Pratt, 1982a; Wallace, 1987). The geological record of mud mounds does not appear to be continuous; they have been recorded from Early Cambrian through Upper Jurassic strata. The two major problems with mud mounds are the source of the micrite and the mechanism by which it is trapped (see Section 4.5.8e). A distinct vertical sequence of four facies types has been identified by Lees & Miller (1985) in Carboniferous Waulsortian-type mud mounds and this helps in our understanding of the variability between mud mounds of different areas and ages. Pratt (1982a) has proposed that the lime mud is trapped and bound by organic matter, largely of cyanobacteria, so that in effect the mud mounds are giant thrombolites (unlaminated stromatolites).

The reasons for the apparent absence of mud mounds since the Late Jurassic is not clear. Analogies have been made with mud banks occurring in Florida Bay and on the inner Florida Shelf (see Section 3.3.1g, and Bosence *et al.*, 1985), but these are shallow-water buildups, there is little syn-sedimentary cementation, and sea-grass, which evolved in the Cretaceous, plays a major role in trapping the mud.

9.2.5 Pelagic facies

One other carbonate facies which has a chequered history is pelagic limestone. At the present time, pelagic oozes are composed mostly of coccoliths and foraminifera, with pteropods less important, and these oozes cover vast areas of the ocean floors and deeper-water shelves (see Chapter 5). Coccoliths evolved in the Early Jurassic and the familiar Ammonitico Rosso of southern Europe is an extensively developed pelagic limestone largely composed of these algal platelets (see Section 5.7.1). The spread of relatively deep

(50–150 m) epeiric seas in the Cretaceous resulted in deposition of the coccolith chalks (see Section 5.7.3). Pelagic foraminifera evolved in the Cretaceous and contribute to these chalks. Thin-shelled bivalves are important in some Triassic and Jurassic pelagic facies. In the Palaeozoic, pelagic limestones are best known from the Devonian of western Europe and North Africa (see Section 5.7.2). They contain pelagic macro and microfossils, such as ammonoids, thin-shelled bivalves and conodonts, but the origin of much of the fine-grained carbonate in these limestones is still not known. Ordovician pelagic limestones, commonly with many orthocones, are well known from Scandinavia, but otherwise there are few other truely pelagic limestones in the Palaeozoic. The preservation potential of pelagic facies is not high of course, those deposited on the ocean floor are likely to be subducted or preserved in mountain belts along with ophiolites.

9.3 TEMPORAL VARIATIONS IN PHANEROZOIC MARINE PRECIPITATES

During the 1950s to mid-1970s when much petrographic work was undertaken on modern and ancient carbonates, it was tacitly assumed that most limestones began their life as a mixture of aragonite, high-Mg calcite and low-Mg calcite grains and were cemented in the marine realm by aragonite and high-Mg calcite, just like carbonate sediments today. There was generally little consideration of the possibility that the mineralogy of carbonate precipitates from seawater had changed through the Phanerozoic. The predominance of dolomites in the Precambrian was appreciated, however, and there were suggestions of primary dolomite precipitation then, from a seawater of slightly different composition. In 1975, Sandberg argued from the fabrics of ooids from the Great Salt Lake, Utah, the Pleistocene of Miami and a variety of older limestones, that some ancient ooids were originally composed of calcite, rather than aragonite, which is the mineralogy of the 'classic', normal marine ooids from the Bahamas and Arabian Gulf. One point that Sandberg made is that ooids behave in an identical way to bioclasts during diagenesis. If the ooids were originally composed of calcite (HMC or LMC), then like brachiopods and some bivalves such as oysters, during diagenesis the ooid fabric would generally be unaltered and the radial–fibrous structure, which is characteristic of calcite ooids, would be well preserved. Calcitic ooids may show poor preservation

through micritization of course (Section 1.2), and in rare cases they have suffered dissolution (Sandberg, 1983). Originally aragonitic ooids, on the other hand, may be (1) calcitized with neomorphic spar crystals replacing the aragonite but retaining relics of the original texture, or (2) preserved as oomoulds through complete aragonite dissolution, with or without a calcite spar cement filling, in just the same manner as aragonite bioclasts are preserved (see Section 1.2.2). Sandberg (1975) thus provided an explanation for the different fabrics of ancient ooids; they had different original mineralogies, and this did away with the need for special mechanisms of replacement, such as the template model of Shearman et al. (1970), to explain radial–fibrous calcite ooids as replacement of originally aragonite ooids. In fact, the contrast in fabric of modern and Jurassic ooids was known to Sorby, publishing in 1859, and he thought the calcite mineralogy was primary!

Sandberg (1983) surveyed the fabrics of Phanerozoic oolites and postulated that there had been a secular variation in their composition with aragonite (and high-Mg calcite) ooids occurring in the Late Precambrian–Cambrian, Mid-Carboniferous through Triassic, and Tertiary to Recent, and calcite ooids, with low Mg content, occurring in the Mid-Palaeozoic and Jurassic–Cretaceous (see Fig. 9.8). Wilkinson et al. (1985) documented a similar pattern. There are some exceptions to Fig. 9.8, but this is to be expected. For example, the Upper Jurassic Smackover Oolite of the Gulf Coast region, over which there has been much discussion and argument, has both formerly aragonitic ooids, in a nearshore belt, and calcitic ooids (Fig. 7.14), in a more offshore belt (Swirydczuk, 1988), and deposition was at a time when one would predict an absence of aragonite from the Sandberg curve. The Late Jurassic Purbeck strata of the Swiss Jura also contain ooids of both mineralogies (Strasser, 1986).

Sandberg (1983, 1985) has also considered the original mineralogy of marine carbonate cements through the Phanerozoic and come up with a similar pattern. One of the distinctive aragonite cement morphologies, botryoids (Section 7.4), do seem to be restricted to the Early Cambrian, Late Mississippian to Early Jurassic, and Mid- to Late Cainozoic.

Two particular types of calcite cement which have given rise to much discussion are radiaxial fibrous calcite (RFC) and syntaxial overgrowths on calcite. RFC, a common early cement in mid-Palaeozoic mud mounds, was widely believed to be a replacement in the 1960s and 1970s, with one of the arguments being

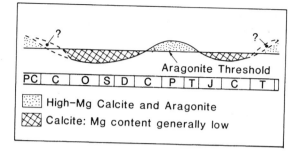

Aragonite Threshold

| PC | C | O | S | D | C | P | T | J | C | T |

▨ High-Mg Calcite and Aragonite

▧ Calcite: Mg content generally low

Fig. 9.8 *Mineralogy of marine carbonate precipitates through the Phanerozoic. Note the broad similarity of this curve to the global sea-level curve (Fig. 9.1B), suggesting calcite precipitation during high sea-level stands and calcite (high Mg) plus aragonite precipitation during low sea-level stands. After Sandberg (1983).*

that coarse calcite crystals of this type are not precipitated from modern marine waters. Re-examination of the fabrics of RFC (Kendall, 1985; Sandberg, 1985) has shown that they are primary (see Section 7.4.6) and near-Recent RFC has now been found at depth within Enewetak Atoll (Saller, 1986) and in Pleistocene limestones from Japan (Sandberg, 1985).

In many mid-Palaeozoic and Mesozoic grainstones, there are prominent syntaxial overgrowths of calcite on echinoderm debris, especially crinoids. In modern shallow-water carbonates, however, these cements are very rare indeed. In a similar fashion, equant calcite spar is the most common pore-filling cement in limestones of all ages and it has long been regarded as a typical meteoric or burial cement. However, in the Ordovician (Wilkinson *et al.*, 1982) and Jurassic (Wilkinson *et al.*, 1985) evidence has been presented from hardgrounds for seafloor precipitation of sparite (see Section 7.4.6). Another dogma hits the dust.

Fibrous calcites, some sparry calcites and echinoderm overgrowths, all of which probably had a range of Mg contents, can thus be regarded as another facet of the secular variation in marine carbonate precipitates. They were preferentially precipitated at times when calcite was the predominant marine precipitate.

The temporal variation in carbonate precipitates through the Phanerozoic (Fig. 9.8) is the consequence of subtle changes in the composition of ocean water. The two major controls on the mineralogy of carbonate precipitates are the P_{CO_2} and Mg/Ca ratio. With P_{CO_2}, low levels promote aragonite and high-Mg calcite precipitation, and with increasing P_{CO_2} calcite with a decreasing Mg content is precipitated (Mackenzie & Piggott, 1981). High P_{CO_2} could permit

dolomite precipitation, so they say. With Mg/Ca ratio, a molar value above 5 is thought to favour the precipitation of aragonite and high-Mg calcite, and a lower Mg/Ca ratio leads to low-Mg calcite precipitation. The precipitates out of modern low-latitude seawater, with a Mg/Ca molar ratio of 5.2, are thus aragonite and high-Mg calcite. There has been much discussion as to which factors determine whether aragonite or high-Mg calcite are formed (see Section 7.4.3). Carbonate precipitation systems are very complicated; apart from Mg/Ca ratio, carbonate ion concentration, SO_4^{2-}, orthophosphate, temperature and organic acids may all be important. It appears that the precipitation of aragonite from seawater is kinetically favoured over calcite at higher Mg^{2+} and SO_4^{2-} levels, lower PO_4^{3-} and increasing temperature (Mucci & Morse, 1983; Walter, 1986; Burton & Walter, 1987; Railsback & Anderson, 1987). Given & Wilkinson (1985) placed emphasis on the CO_3^{2-} supply rate, with high rates favouring aragonite and low rates high-Mg calcite, although this hypothesis has been questioned (Morse, 1985).

Sandberg (1975) suggested that an increase in the seawater Mg/Ca ratio through the preferential extraction of Ca^{2+} by coccoliths accounted for the change from calcite ooids in the Jurassic–Cretaceous to the aragonitic ooids in the Cainozoic. Sandberg (1983, 1985) later noted that such a mechanism would not explain the earlier occurrences of aragonite and he placed more emphasis on the P_{CO_2} of the atmosphere–hydrosphere system.

The pattern of marine precipitates (Fig. 9.8) correlates in a general way with the global sea-level curve. Wilkinson *et al.* (1985) found a better correlation of mean ooid cortical mineralogy to be with the per cent of continent flooded. Thus calcite seas correspond to periods of relative sea-level high and continents in submergent mode, and aragonite seas to periods of relative sea-level low and an emergent mode. Since the global position of sea-level is determined largely by plate tectonic processes, higher sea-level stands during periods of active plate movements and high rates of seafloor spreading, it would appear that there is a geotectonic control on seawater chemistry. Now both P_{CO_2} and Mg/Ca ratio of seawater are strongly affected by geotectonics. Oceanic Mg/Ca ratio is largely controlled by hydrothermal weathering of basalts at mid-ocean ridges; Mg is preferentially extracted from seawater as it is pumped through, to form clay minerals such as chlorite, and epidote. Thus lower Mg/Ca ratios can be expected during periods of high rates of seafloor spreading, as occur during periods of high sea-level stand. P_{CO_2} levels are affected by diagenetic and metamorphic reactions at subduction zones (Fig. 9.9; Mackenzie & Piggott, 1981; Wilkinson *et al.*, 1985), so that higher levels coincide with phases of active plate movements. It would thus appear that both factors, Mg/Ca ratio and P_{CO_2}, move in the appropriate direction to promote calcite precipitation during high sea-level stands when P_{CO_2} is likely to be higher and Mg/Ca ratio lower through increased rates of seafloor spreading and subduction, and aragonite and high-Mg calcite precipitation during low sea-level stands, when P_{CO_2} is lower and Mg/Ca ratio higher.

Another major factor in global geochemistry is the rate of continental weathering processes. This also controls atmospheric CO_2 levels, as well as the flux of Sr, Ca, C and other dissolved species to the oceans (Raymo *et al.*, 1988). Major phases of mountain building and uplift, which generally correlate with lower sea-level stands, have occurred over the past 20 million years with the Andes, Himalayas and Tibet, and these could have lowered atmospheric P_{CO_2} (leading to global cooling) and increased the input of dissolved salts to the oceans. Aragonite would be expected under these conditions and indeed that is what has been precipitated during the later Cainozoic.

A consensus is emerging that P_{CO_2} is more important than Mg/Ca ratio as the overriding control (Mackenzie & Piggott, 1981; Sandberg, 1983, 1985; Wilkinson *et al.*, 1985). It is generally easier to have local variations in the Mg/Ca ratio of seawater, through evaporative precipitation of sulphate minerals for example, or influxes of meteoric water. Thus the odd occurrences of aragonite which do not fit into the Sandberg curve, such as in the Late Jurassic, could reflect local conditions of higher Mg/Ca ratio.

9.4 SECULAR VARIATIONS IN BIOMINERALIZATION

Skeletal carbonates are composed of aragonite, high-Mg calcite and low-Mg calcite, or in some cases a mixture of two of these minerals (see Section 1.3 and 6.2). Important factors which appear to control the mineralogy of biogenic grains include the 'vital effect' (the organisms determining the mineralogy of their skeleton irrespective of environmental factors), seawater chemistry and water temperature. The last is particularly important with regard to the Mg content of calcitic skeletons; in cooler waters, the Mg content is lower (Section 6.3). Some organisms secrete aragonite preferentially over calcite in warmer water. A compilation of the mineralogy of important carbonate-forming organisms through the Phanerozoic (Fig. 9.2; Wilkinson, 1979) shows that during the Palaeozoic most benthic organisms, such as rugose corals, bryozoans, echinoderms and brachiopods produced calcite skeletons, with only molluscs secreting aragonite (although some of these had calcite shells). By contrast, in the Mesozoic and especially the Cainozoic, benthic aragonite-secreting molluscs, scleractinian corals and green algae were much more important. High-Mg calcite was precipitated by the red calcareous algae and echinoderms, and some molluscs, such as the oysters, had low-Mg calcite shells. Thus, broadly, in the Palaeozoic shallow-water bioclastic limestones were dominantly calcitic, in the Mesozoic a mixture of calcite and aragonite, and in

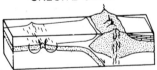

HIGH SEA-LEVEL STANDS
CALCITE SEAS

1. increased pCO₂ and lower oceanic CO_3^{2-} through production of CO_2 by diagenetic and metamorphic reactions at subduction zones.

2. lower Mg/Ca ratio of seawater through hydrothermal alteration of basalts at mid-ocean ridges (Mg extracted from seawater to form chlorite and epidote)

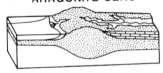

LOW SEA-LEVEL STANDS
ARAGONITE SEAS

1. decreased pCO₂ through increased photosynthesis.

2. increased Mg/Ca ratio through increased Mg²⁺ supply to oceans by rivers from subaerial weathering.

3. removal of Ca²⁺ by evaporite precipitation raising Mg/Ca ratio.

Fig. 9.9 *The processes affecting seawater chemistry and leading to the precipitation of calcite (low Mg) and aragonite plus calcite (high Mg) out of seawater during high and low sea-level stands respectively. After Wilkinson* et al. *(1985).*

the Cainozoic, dominantly aragonite. Contrasting with Mesozoic–Cainozoic shallow-water limestones, deeper-water limestones were (and are) largely calcitic (low Mg) composed of coccoliths and planktonic foraminifera, with aragonitic pteropods a minor component. It is difficult to decide whether this temporal change in skeletal carbonate is a function of fluctuations in seawater chemistry, with Pco_2 and Mg/Ca ratio the controls again (Fig. 9.9) or entirely a consequence of biological factors and the evolution of organism groups with a preference for a particular skeletal mineralogy. One of the most conspicuous and rather abrupt changes in the mineralogy of marine skeletons occurs across the Permian–Triassic boundary, when calcite-dominated Palaeozoic organisms gave way to Mesozoic groups mostly with aragonite skeletons (e.g. rugose corals replaced by scleractinian corals and calcitic brachiopods by aragonitic bivalves). There has been much discussion of the cause of the end-Permian extinction event, when some 90% of all marine species died out, but there is evidence for Triassic seawater having high SO_4^{2-}, elevated temperature and possibly high Mg/Ca ratio. These factors would have promoted the development of aragonitic skeletons when biological productivity returned to normal after the extinction event (Railsback & Anderson, 1987).

Variations in skeletal mineralogy through time (Fig. 9.2) have some significance for fine-grained limestones. There has always been much discussion over the original mineralgoy of micrite (Sections 1.4 and 7.6.4). In modern shallow-water carbonate environments, much lime mud is produced by the breakdown of carbonate skeletons, particularly those of the aragonitic green algae. A dominant calcite mineralogy of Palaeozoic bioclasts could indicate that Palaeozoic lime mudstones were also originally of calcite, if skeletal disintegration was as important then as now. In this context, from SEM studies, Lasemi & Sandberg (1984) have been able to recognize micrites which were originally aragonite dominated, although their distribution through the Phanerozoic does show departures from the trend of Fig. 9.8. This is attributed to temporal and/or local fluctuations in the contributions of material from inorganic and biogenic sources and/or in the mineralogy of skeletal debris available.

Cyanobacteria (blue–green algae) have a history which is relevant to this section. They have a long geological record, with many forms originating early in the Precambrian. However, calcified marine forms, such as *Renalcis* and *Girvanella*, appear rather suddenly close to the Precambrian–Cambrian boundary

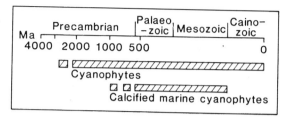

Fig. 9.10 *Geological record of calcified and non-calcified marine cyanobacteria ('blue–green algae'). After Riding (1982).*

and continue through until the Late Cretaceous–Early Tertiary (Fig. 9.10, Riding, 1982). Calcification of blue–green algae appears to be determined by environmental factors, rather than 'vital' effects, so that Riding postulated subtle changes in seawater chemistry, again in Mg/Ca ratio, to account for the temporal distribution of these calcified algae.

9.5 DIAGENESIS THROUGH TIME

Since there have been changes in the composition of carbonate sediments through time (Figs 9.2 and 9.8), then it follows that there should be variations in the diagenesis too, particularly in the early stages. This is because the carbonate minerals, aragonite, high-Mg calcite and low-Mg calcite, have different diagenetic potentials (Fig. 9.11). In particular, aragonite is metastable in meteoric environments and will dissolve to liberate $CaCO_3$ for cementation. In sediments largely composed of calcite, mid-Palaeozoic carbonate sands and muds for example, one would expect meteoric cementation to be less important than in sediments having a large component of aragonite. The latter would have extensive vadose cements and abundant evidence of grain dissolution/calcitization. Apart from the mineralogical control, climate is also important in determining the path of early diagenesis (Fig. 9.12). Basically, it is a humid versus arid influence: at times of more arid climate, marine cementation will be much more widespread through the development of hypersalinities over carbonate platforms and more rapid CO_2 degassing. Also evaporative and seepage–reflux dolomites are more likely to form during arid periods. With meteoric diagenesis, more cementation and dissolution can be predicted under humid climates, with differences between dominantly calcitic and dominantly aragonitic sediments (Fig. 9.12). Also during more humid times, one can expect mixing zones to be more active, so that the

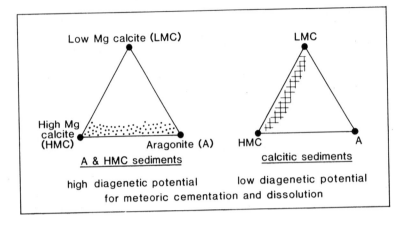

Fig. 9.11 *The diagenetic potential of marine carbonate sediments; those with high A plus HMC contents, such as modern tropical and Tertiary carbonates, have a much higher diagenetic potential for cementation and dissolution in meteoric environments compared to dominantly calcite sediments, such as many Lower-Mid Palaeozoic skeletal and oolitic grainstones.*

potential for dilute water dolomitization is increased. The effects of a climatic control on early diagenesis are well shown by the Lower Carboniferous oolites of south Wales (Fig. 9.13; Hird & Tucker, 1988). The Broviscin Oolite was deposited during a humid phase and shows vadose cement fabrics. The younger Gully Oolite, capped by an emergence horizon, has local marine cements (acicular calcite) in intraclasts, otherwise little early cement. The Gully suffered extensive grain–grain pressure dissolution during burial, as a result of the absence of near-surface cementation. Paleokarsts and paleosols on the Gully Oolite indicate

an arid/semi-arid climate. Although meteoric cements are present in the Broviscin Oolite, they are not pore filling, just grain coats and meniscus structures. This is attributed to the dominantly calcitic mineralogy of the original sediment (these ooids were calcitic), so that there were few aragonite grains to release $CaCO_3$ for cementation.

Dissolution and/or calcitization of aragonite grains and cements are widely assumed to be near-surface phenomena, as a result of meteoric diagenesis (Section 7.5). Much of this state of affairs is based on the common occurrence and extensive documentation

Fig. 9.12 *Diagram illustrating the effects of climate on diagenesis in the marine, meteoric and mixing-zone environments. After Hird & Tucker (1988).*

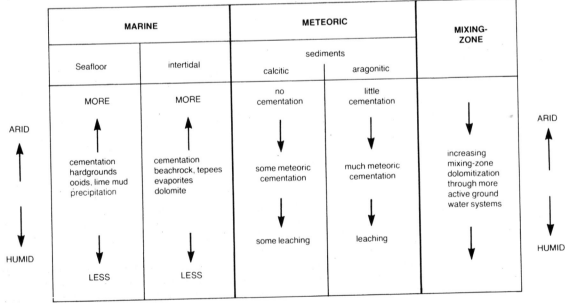

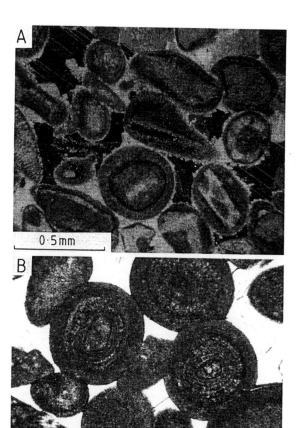

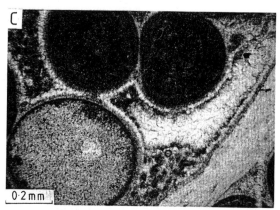

Fig. 9.13 *Contrasting early diagenesis of two Lower Carboniferous oolites from south Wales. (A) Brofiscin Oolite, showing early equant calcite spar patchily developed, but preferentially at grain contacts, indicating vadose meteoric diagenesis. A later poikilotopic calcite spar occludes the porosity. (B) Gully Oolite, showing no early cement, but grain–grain sutured and concavo-convex contacts through chemical compaction. Burial poikilotopic cement again occludes porosity. (C) Gully Oolite, intraclast with marine acicular calcite cement and some internal sediment. The early diagenesis of the Brofiscin Oolite took place under a more humid climate compared to a more arid/semi-arid climate for the Gully. The stratigraphic position of these oolites is shown in Fig. 2.20; also see Fig. 4.24 and Fig. 7.24. (A) crossed polars, (B) and (C) plane polarized light. For details see Hird & Tucker (1988).*

of near-surface meteoric aragonite dissolution in Pleistocene limestones, which is the result of rapid fluctuations in sea-level over the last 200 000 years (see Fig. 2.32). However, in the geological record, many thick platform limestone sequences were deposited at times of stable or rising sea-level and some of these are likely to have had marine porewaters throughout much of their diagenetic history. As discussed in Section 9.3 and shown in Fig. 9.8, there were times when calcite was the main precipitate from seawater, and during these periods, seawater may well have been undersaturated with respect to aragonite. *Sea-floor* dissolution in shallow depths should have occurred, especially in areas of slow sedimentation. This has now been documented from Jurassic and Ordovician hardgrounds (Palmer *et al.*, 1988). Where

carbonate sediments were buried with marine pore-waters supersaturated *re* aragonite, then the aragonite can be expected to have survived until substantial depths, into the burial, late diagenetic realm. Burial, as opposed to near-surface, calcitization of aragonite has been described by Kendall (1975), Sandberg & Hudson (1983) and Tucker (1985b). Meteoric waters can penetrate deep into the subsurface, notably where there are confined aquifers, and they may even emerge on to the seafloor as submarine springs. Enos (1988) has described such a situation, with attendant aragonite dissolution, from the Cretaceous of Mexico and Dorobek (1987) another from the Siluro-Devonian of the Appalachians. The topic of aragonite dissolution is another illustration of the danger in rigidly adhering to the principle of uniformitarianism, in this case assuming it is a product of near-surface meteoric diagenesis.

Another facet of limestone diagenesis is *dolomitization* and it has long been written that the abundance of dolomites, mostly on the basis of Mg/Ca ratios of carbonates, increases back into the geological record (Fig. 9.14) with dolomites dominating over limestones in the Early Palaeozoic and Precambrian (e.g. Chilingar, 1956; Vinogradov & Ronov, 1956; Ronov, 1964). This apparent trend has been interpreted as the result of either more extensive dolomitization environments in the Early Phanerozoic or the slow rate of dolomitization and older limestones having more chance and time to come into contact with dolomitizing fluids. The possibility of primary dolomite precipitation in the Precambrian has been raised (see Section 9.7). The tenet of increasing dolomite content with age has recently been questioned and shown to be erroneous by Given & Wilkinson (1987). They recalculated the published Ca/Mg ratios to dolomite percentages, used a finer time interval for plotting the data and included information from some periods unrepresented in the original compilations. The result is a graph which does not show an increase in dolomite abundance with increasing age, but one which shows two maxima of dolomite content that broadly correspond to the high sea-level stands of the Phanerozoic (Fig. 9.15). The paucity of dolomite in Cainozoic strata is apparent in both presentations (Figs 9.14 and 9.15). Given & Wilkinson (1987) see the correlation of dolomitization with high sea-level stands as an extension of the role of geotectonics in the hydrosphere–atmosphere system, discussed in Section 9.3: increased continental flooding, leading to higher atmospheric P_{CO_2}, lower oceanic carbonate concentration and/or lower calcite saturation resulting in increased rates of seawater dolomitization of carbonate sediments. The correlation could also reflect the increased flow of seawater through carbonate platforms by ocean current pumping, and Kohout convection, with the consequent dolomitization (Section 8.7.5); these processes should be more widespread at times of high sea-level stand.

9.6 SECULAR VARIATIONS IN CARBONATE GEOCHEMISTRY

9.6.1 Trace elements

If secular variations in the mineralogy of marine carbonate precipitates are accepted then it follows that there are likely to be secular variations in the

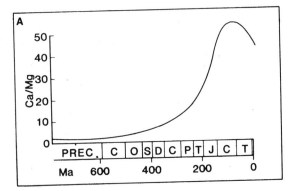

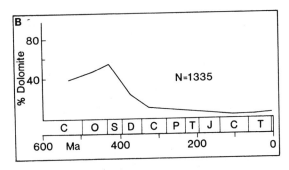

Fig. 9.14 *The geological record of dolomites through the Phanerozoic. (A) Ca/Mg ratio in carbonate rocks. (B) Percentage dolomite in carbonate sequences. After Chilingar (1956) and Given & Wilkinson (1987).*

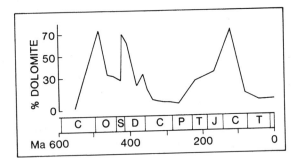

Fig. 9.15 *The geological record of dolomites through the Phanerozoic using data recalculated from Chilingar (1956) and new data for the Triassic–Jurassic. After Given & Wilkinson (1987).*

geochemistry of the limestones they form too. Marine aragonite, high-Mg and low-Mg calcite have quite different original trace element signatures (Sections 6.3 and 7.4.2) so that vestiges of these may be preserved after diagenetic mineral stabilization and cementation. Low-Mg calcite is the most stable, i.e. it has a low diagenetic potential, and generally retains its original geochemistry of low Mg, Na, Sr, Fe and Mn. High-Mg calcite is less stable; it generally loses some or even all of its Mg during diagenesis, and with this Na and Sr are likely to go too, but Fe and Mn may be picked up. Wholesale dissolution, however, is rare. Aragonite is generally replaced by calcite but this may involve wholesale dissolution and calcite filling of the voids, and thereby complete loss of original chemistry, or calcitization in which some of the original chemistry, notably high Sr, may be retained.

Thus, in plots of limestone trace element contents against time, one would predict that mid-Palaeozoic carbonates and Jurassic–Cretaceous oolites should show low Mg and Sr from a dominantly low-Mg calcite sediment, with the minor element contents being close to original values. Carbonates of the Late Palaeozoic–Early Mesozoic and Cainozoic, on the other hand, will show a range of Mg, Na and Sr contents, reflecting a higher proportion of aragonite and high-Mg calcite in the sediment and a variable diagenesis. Extensive late diagenetic calcite spar cement, with its depleted Mg, Sr and Na values, will lower bulk limestone concentrations of these elements. The point to make here is that in the interpretation of trace element contents of limestones, consideration must be given to the original mineralogy of the sediment, since this strongly controls the original trace element chemistry. However, the latter will be modified by diagenetic processes, especially where aragonite and high-Mg calcite were important constituents of the sediment.

9.6.2 **Stable isotopes**

Secular variations in the isotope geochemistry of carbonates are well documented (Veizer & Hoefs, 1976; Veizer et al., 1980; Mackenzie & Piggott, 1981). The *carbon isotope ratio*, $\delta^{13}C$, varies between −1 and +2‰ through the Phanerozoic, with lighter values characterizing the Early–Mid-Palaeozoic (400–500 Ma), more positive values occurring in the Late Precambrian–Cambrian and Permian, and intermediate values since the Triassic (Fig. 9.16). The carbon isotopic composition of limestone mostly reflects the $\delta^{13}C_{TDC}$ (Total Dissolved Carbonate) of the seawater from which it was precipitated (see Section 6.4). There are two main reservoirs of carbon, oxidized inorganic carbon, chiefly contained in carbonate sediments, and reduced organic carbon, contained in organic matter. Long-term changes in $\delta^{13}C_{TDC}$ are thought to be the result of variations in the ratio of

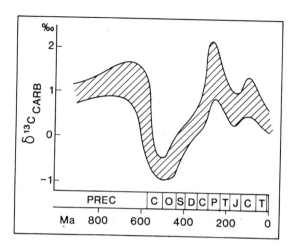

Fig. 9.16 *Generalized secular variations of* $\delta^{13}C$ *in marine carbonate rocks. After Veizer* et al. *(1980).*

organic carbon to carbonate carbon that is sedimented. An increase in the amount of organic C buried means that ^{12}C is preferentially taken out of seawater, so that the latter becomes heavier and carbonate precipitates are more positive. The secular trend (Fig. 9.16) is of a decrease of the organic C/carbonate C depositional ratio from the Late Precambrian through to the Ordovician, then an increase of this ratio through to the Permian, and a slow decrease again through to the Recent. Interestingly, the $\delta^{13}C$ curve correlates with the global sea-level curve: limestones with more negative $\delta^{13}C$ values correlate with high sea-level stands, and more positive $\delta^{13}C$ limestones correlate with low sea-level stands (Mackenzie & Piggott, 1981). The ratio of reduced C (in organic matter) to total C (in organic matter plus carbonate rocks) also correlates with the sea-level curve: a higher ratio at lower sea-level stands. What these trends in $\delta^{13}C$ and reduced C/total C show is that during progressive flooding of continental margins, as through the Lower Palaeozoic and Mesozoic, carbon was transferred from the reduced carbon reservoir (mostly organic C in mudrocks) to the oxidized C reservoir in carbonate sediments, and the reverse during the subsequent regressive phases, such as the Mid- to Late Palaeozoic. Extensive carbonate sedimentation during the high sea-level stands leads to a lower reduced C/total C ratio and to more negative $\delta^{13}C$ in limestones. Mackenzie & Piggott (1981) postulated that the forcing mechanism behind these secular patterns was variation in P_{CO_2} of the atmosphere–hydrosphere system brought about by tectonic processes, as explained in Section 9.3 and Fig. 9.9.

In addition to the long-term changes in $\delta^{13}C$ of total dissolved carbonate in seawater, short-term changes, referred to as isotope anomalies, excursions or perturbations, occur in carbonate rocks. They are well documented from Cretaceous and Jurassic limestones, especially those in pelagic facies, and most of these carbon isotope excursions are positive (e.g. Fig. 9.17, Scholle & Arthur, 1980; Jenkyns & Clayton, 1986). These $\delta^{13}C$ excursions occur on a global scale, at times of regional black 'shale' deposition and transgression, and during oceanic anoxic events (Jenkyns, 1980b). Much organic matter is deposited at these times of increased organic productivity, so that the ocean becomes depleted in ^{12}C and carbonate precipitates are thus more positive (^{13}C-rich).

A prominent negative $\delta^{13}C$ excursion occurs in pelagic chalks close to the Cretaceous–Tertiary boundary (Hsü et al., 1982), and this coincides with the now infamous iridium anomaly (Fig. 9.18). The latter is thought to be evidence of a meteorite impact (Alvarez L.W. et al., 1980), which killed off much of the biota, one way or another, and as a result of which the oceanic $\delta^{13}C$ became more ^{12}C-rich since there was little organic productivity for a time. Carbonates deposited during this interval thus show a negative $\delta^{13}C$ excursion. This scenario of a short-lived pause in organic productivity has been referred to as a 'Strangelove' effect. It should be stressed that although much evidence to support a catastrophe at the K/T

boundary has been gathered in recent years, there is a strong body of opinion arguing for a terrestrial causation of the end-Cretaceous mass extinction event (e.g. Hallam, 1987).

Carbon isotope excursions have been reported from several Precambrian–Cambrian boundary sections and in Morocco (Tucker, 1986a), Siberia (Magaritz et al., 1986) and the Himalayas (Aharon et al., 1987), they are positive ones, ascribed to increases in organic productivity and a higher rate of organic matter burial. Hsü et al. (1985) reported a negative $\delta^{13}C$ spike and an iridium anomaly just above the boundary in China and gave a predictable interpretation (although see Awramik et al., 1986). Carbon isotope excursions are being reported from other boundaries too, such as the Permian–Triassic and Ordovician–Silurian.

Thus in carbonate rocks, a long-term secular variation in $\delta^{13}C$ is well established and reflects subtle changes in the ratio of organic C to carbonate C sedimented. There are also short-term excursions of $\delta^{13}C$ in seawater, which mostly reflect geologically rapid changes in organic productivity and burial rates of organic carbon.

Secular variations also occur in the *oxygen isotope data* (Fig. 9.19). However, in this case there is much more scatter in the data since the $\delta^{18}O$ of a limestone (whole rock) is an average of sedimentary components and cements, and the latter are commonly very depleted relative to the former. Nevertheless, on a broad scale a trend is apparent, of

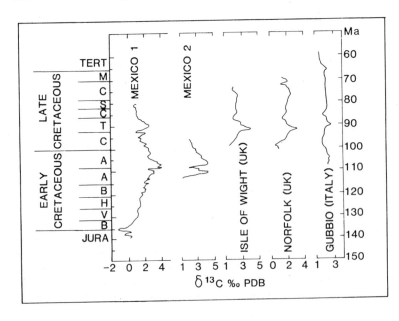

Fig. 9.17 *Carbon isotope anomalies in Cretaceous pelagic limestones from Mexico and western Europe. After Scholle & Arthur (1980).*

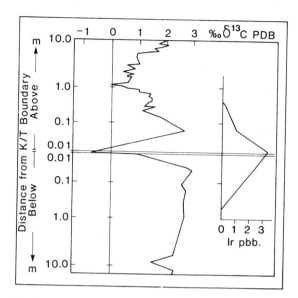

Fig. 9.18 *Carbon isotope stratigraphy and iridium anomaly in Cretaceous–Tertiary boundary strata. Note the negative $\delta^{13}C$ excursion at the K/T boundary as a result of the 'Strangelove' effect. After Hsü (1983).*

increasingly negative $\delta^{18}O$ limestone back through the Phanerozoic, but particularly from the Permian back to the Precambrian (Fig. 9.19A). There is less variation over the last 250 Ma; limestone values are generally +1 to −2‰, although data from molluscs show a negative to positive swing the Early Tertiary.

The temporal trend to more negatives values in older rocks can be explained in two ways. It could mean that seawater was more depleted in ^{18}O in the Palaeozoic, or that seawater had a similar isotopic composition to that now ($\delta^{18}O_{SMOW} \approx 0$) but its temperature was higher. The first possibility is more likely since there is no sedimentary evidence for 50°C seawater in the Cambrian (although chert phosphate $\delta^{18}O$ values suggest high temperatures too, Karhu & Epstein, 1986). The reason for lighter seawater is not clear, but it may be connected with the rate of seawater cycling through seafloor basalts at mid-ocean ridges.

There have been suggestions of very depleted or hot seawater in the Precambrian too (e.g. Perry & Ahmad, 1983; Karhu & Epstein, 1986, see Fig. 9.19B), and some support for this comes from very negative $\delta^{18}O$ values occurring in cherts. However, isotopic work on Precambrian carbonates has rarely taken into account the diagenesis and Tucker (1983b, 1986b) has shown that some Proterozoic limestones

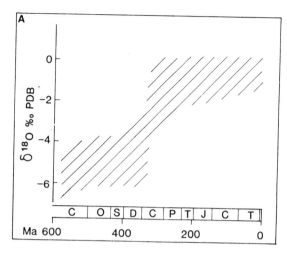

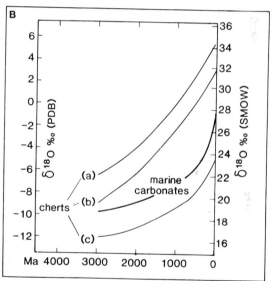

Fig. 9.19 *Secular variation of $\delta^{18}O$ in marine carbonate rocks. (A) Phanerozoic trend based mainly on brachiopod analyses. After Veizer et al. (1980). (B) Very generalized trend of $\delta^{18}O$ of carbonates through time compared with data from cherts. The upper two chert curves (a and b) are the best estimates for marine shallow-water cherts, whereas the lower chert line (c) is for cherts which have recrystallized in meteoric water. Note that both PDB and SMOW scales are given here. After Anderson & Arthur (1983), data from several sources.*

with very negative values (−10 to −17‰) were originally composed of aragonite. The latter was replaced at depth during burial diagenesis, so that the low $\delta^{18}O$ values can be interpreted as the product of high burial temperatures rather than a ^{18}O-depleted seawater.

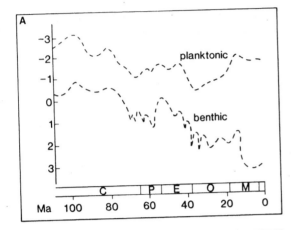

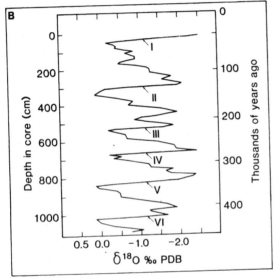

As with carbon isotopes, on a smaller time-scale there are commonly significant fluctuations in $\delta^{18}O$. These are best documented from the Cainozoic (Fig. 9.20B), the $\delta^{18}O$ values of deep sea calcareous oozes temperature and the isotopic composition of seawater, this time strongly controlled by the volume of continental ice sheets and the amount of meltwater entering the oceans, since meltwater is usually quite negative with respect to seawater. In the Quaternary (Fig. 9.20B), the $\delta^{18}O$ values of deep-sea calcareous oozes are strongly affected by the waxing and waning of the Pleistocene ice sheets, with the now well-documented astronomical control of orbital forcing and Milankovitch cycles (also see Section 2.10.3).

Secular variations in $\delta^{34}S$ of evaporites are well documented (Claypool *et al.*, 1980) and the trend correlates negatively with $\delta^{13}C$ of carbonate (Veizer *et al.*, 1980). In the similar way to $\delta^{13}C$, changes in $\delta^{34}S$ of seawater reflect variations in the masses of oxidized inorganic sulphate and reduced sulphide reservoirs, and a tectonic interpretation has been put forward by Mackenzie & Piggott (1981) to explain the carbon−sulphur coupling.

The strontium isotopic ratio $^{87}Sr/^{86}Sr$ of carbonates also varies through the Phanerozoic (Fig. 9.21; Veizer & Compston, 1974). In fact, the changes are sufficient in some instances that the $^{87}Sr/^{86}Sr$ value of a limestone can give an age estimate (e.g. Saller, 1984, used this ratio to date dolomitization in the Enewetak subsurface). The $^{87}Sr/^{86}Sr$ ratio of seawater appears to be strongly influenced by a range of global tectonic processes, including the variation in lithological composition of the Earth's crust exposed to weather-

Fig. 9.20 *Record of $\delta^{18}O$ from pelagic carbonates. (A) Cretaceous−Cainozoic isotope stratigraphies from tropical planktonic and benthic foraminifera. Fluctuations are the result of the preferential extraction of ^{16}O into ice to make $\delta^{18}O_{seawater}$ heavier and/or changes in tropical sea surface temperatures. (B) Oxygen isotope record from an 11 m long deep sea core from the Caribbean. The major fluctuations I to VI are on a 100 000 year scale, the Milankovitch eccentricity rhythm (see Section 2.10.3). Note that the cycles are asymmetric, with a rapid swing to more negative values followed by a more gradual trend to more positive values. This asymmetry is a reflection of rapid deglaciation (the influx of ^{16}O-depleted meltwater and temperature rise giving the more negative $\delta^{18}O$ signatures), as contrasted with a more gradual buildup of polar ice. Smaller-scale fluctuations in $\delta^{18}O$ within some of the 100 000 year cycles could reflect the Milankovitch precession/obliquity rhythms. From Hardie (1986b) after the work of W.S. Broecker and J. van Donk.*

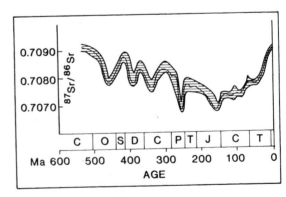

Fig. 9.21 *$^{87}Sr/^{86}Sr$ variation in seawater through the Phanerozoic. The line within the shaded band is the likely seawater trend; deviations are mostly the result of diagenesis. After Koepnick et al. (1985).*

ing, volcanic activity, rate of seafloor spreading, climate, global sea-level and palaeo-oceanography, so that the $^{87}Sr/^{86}Sr$ curve through time is more complicated than $\delta^{13}C$ or $\delta^{34}S$ (Koepnick *et al.*, 1985).

9.7 PRECAMBRIAN DOLOMITES: IS THERE A PROBLEM?

Available compilations of sedimentary rock composition through time (e.g. Fig. 9.22, Ronov, 1964) show that dolomites are much more common than limestones in the Precambrian compared with the Phanerozoic (although see Section 9.5 and Given & Wilkinson, 1987). To account for this, primary precipitation from seawater is often suggested for Precambrian dolomites *or* the existence of conditions (environmental/geochemical) more conducive to dolomitization. Age alone and a longer diagenetic history are not the answer.

As explained in Chapter 8, the majority of dolomites are of replacement origin, although dolomite cements are common in dolomitized limestones. However, with some Precambrian dolomites, it can be difficult from textural evidence alone to *prove* a replacement origin. This is because they commonly have very well-preserved original fabrics (e.g. Fig. 9.23), at least at the hand specimen and microscopic level, so that a primary dolomite mineralogy for the grains and cements appears a possibility. Such a hypothesis would require some modification to seawater chemistry to promote direct dolomite precipitation, since although seawater is supersaturated with respect to dolomite and the thermodynamic drive is towards dolomitization (Section 8.2), $CaCO_3$ minerals (A, HMC and LMC) are the principal marine precipitates because of the kinetic hindrances to dolomite

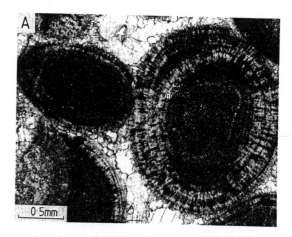

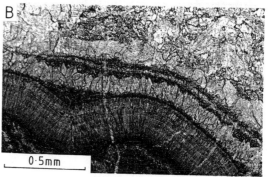

Fig. 9.23 *Good textural preservation in Precambrian dolomites. (A) Dolomitized ooid with radial–concentric structure and micritic peloidal nucleus, broken ooid with nucleus showing a replacement fabric and a superficial ooid, all in a dolospar cement with drusy fabric. Porsanger Dolomite, Late Proterozoic, Finnmark, Arctic Norway. (B) A range of cements (all dolomite) with internal sediment between: early isopachous acicular cement with fine growth banding, columnar cement with crystal terminations, and cavity-filling dolospar with coarse crystal silt. Beck Spring Dolomite, Late Proterozoic, California.*

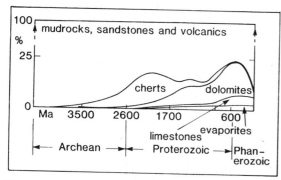

Fig. 9.22 *The relative abundance of limestones and dolomites through time. After Ronov (1964).*

precipitation. The factors likely to promote direct dolomite precipitation from seawater are a higher temperature, higher P_{CO_2}, higher Mg/Ca ratio, lower SO_4^{2-} and organic acid effects (Section 8.2). Indeed, there is evidence from a variety of sources, and with varying degrees of reliability, that in the Precambrian the Mg/Ca ratio was higher (data from mudrock geochemistry for example, Schwab, 1978), P_{CO_2} was higher (from current ideas of atmosphere evolution, e.g. Mizutani & Wada, 1982; Holland, 1984), temperatures were higher (isotope data, Fig. 9.19B, Perry & Ahmad, 1983) and that SO_4^{2-} was lower (Schopf,

1980). It is also plausible to suggest that the organic geochemistry of seawater was different in the Precambrian, when only alage, bacteria and fungi were available to produce organic matter, and metazoans were absent (pre-1400 Ma) or just evolving. Recently, Kazmierczak et al. (1985) have emphasized the important of the Ca^{2+} ion in biological processes and explained biocalcification as a form of calcium detoxification. They postulated a gradual increase of Ca^{2+} in seawater through time and the appearance of calcareous skeletons close to the Precambrian–Cambrian boundary as a function of the Ca^{2+} level passing a threshold. They further suggested that the sporadic development of heavily calcified organisms in the Phanerozoic, such as the rudistids in the Cretaceous, was a response to high and toxic Ca^{2+} concentrations. In the model of Kazmierczak et al. (Fig. 9.24) a 'soda ocean' with limiting Ca^{2+} content existed in the Precambrian, gradually giving way to a sodium chloride ocean into the Phanerozoic, with optimal to toxic levels of Ca^{2+}. They would explain Precambrian dolomites as a consequence of the Mg-rich, Ca-poor soda ocean.

With some Precambrian dolomites, the whole formation is dolomite, there are no lenses or beds of limestone in the field nor calcite relics in thin section. Some Phanerozoic carbonates are also 100% dolomite, but generally there is clear thin-section evidence for a replacement origin with dolomite crystals crosscutting the original grain fabrics. Nevertheless, as discussed in Section 8.4, good fabric preservation does occur in some dolomites and this can be attributed to early replacement of high-Mg calcite grains and/or the presence of numerous nucleation sites.

In spite of arguments to support a primary dolomite origin, the majority of Precambrian dolomites must be replacements just like Phanerozoic dolomites. Nevertheless, the temporal distribution of Fig. 9.22, showing more dolomites in the Precambrian, appears to be real and so must mean something. Accepting dolomitization, in most Precambrian cases it appears to be an early near-surface phenomenon, and this could mean that some chemical properties of surface and shallow groundwater were promoting dolomitization. The same factors mentioned above (P_{CO_2}, Mg/Ca ratio, SO_4^{2-}, temperature, organic effects) could have been instrumental here too. For example, $CaCO_3$ minerals could have been precipitated on the seafloor and then dolomitized there by seawater, if the latter had a higher dolomitizing potential in the Precambrian. Alternatively, dolomitization may have been the major near-surface meteoric and mixing-

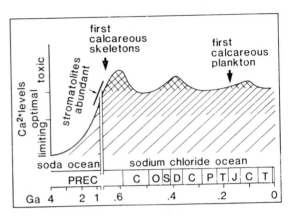

Fig. 9.24 *Possible secular variations in Ca^{2+} in seawater. It is suggested that toxic levels were reached in the Late Precambrian and that this led to the development of biocalcification. The Ca^{2+} levels during the Phanerozoic were controlled first by benthic calcareous skeletal organisms and then by calcareous plankton from the Jurassic. After Kazmierczak* et al. *(1985).*

zone diagenetic process, rather than calcite precipitation and calcitization, through some chemical differences in groundwaters. The lack of extensive soils in the Precambrian may have been a factor, although Beeunas & Knauth (1985) have presented isotopic evidence for some type of subaerial vegetative microbial cover from a 1200 dolomite. The good fabric preservation in many Precambrian dolomites could indicate a predominance of high-Mg calcite grains in the sediment (Tucker, 1982, 1983a), although there is evidence for Precambrian aragonite in several dolomite formations now (e.g. Grotzinger & Read, 1983; Ricketts, 1983; Zempolich et al., 1988). There are many Precambrian limestones, and these commonly contained aragonite and high-Mg calcite grains (e.g. Tucker, 1984) and show a pattern of diagenesis similar to Phanerozoic carbonates. The secular variations in marine precipitates suggested for the Phanerozoic are likely to extend back into the Proterozoic, although local variations are likely to have been common. It is possible that Precambrian dolomites are in some way a further reflection of these secular variations and the geotectonic control on carbonate sedimentation, mineralogy and diagenesis.

Back into the Archaean, carbonate rocks are really quite rare (Fig. 9.22, Ronov, 1964). In the greenstone belts of the Canadian Shield for example, they account for much less than 1% of the sedimentary rocks, which are dominantly siliciclastics derived from volcanic terrains. The paucity of carbonates is attributed to the thin curst of Archaean times, which did not allow the development of extensive shelves and platforms where much carbonate was deposited in later times (Cameron & Bauman, 1972). Some carbonates may well have been deposited in deep ocean basins, through microbial-induced precipitation, but subduction-type processes would have largely destroyed these sediments.

References

Abbott, D.P., Ogden, J.C. & Abbott, I.A. (1974) Studies on the activity pattern, behaviour and food of the echinoid *Echinometra lucunter* on beachrock and algal reefs and St Croix, US Virgin islands. *West Indies Lab. spec. Publ.* **4**, 1–111.

Abell, P.I., Awramik, S.M., Osborne, R.H. & Tomellini, S. (1982) Plio-Pleistocene lacustrine stromatolites from Lake Turkana, Kenya: morphology, stratigraphy and stable isotopes. *Sedim. Geol.* **32**, 1–26.

Acker, K.L. & Risk, M.J. (1985) Substrate destruction and sediment production by the boring sponge *Cliona caribbaea* on Grand Cayman Island. *J. sedim. Petrol.* **55**, 705–711.

Adams, A.E., Mackenzie, W.S. & Guilford, C. (1984) *Atlas of Sedimentary Rocks Under the Microscope*, pp. 104. Longman, Harlow.

Adams, J.E. & Rhodes, M.L. (1960) Dolomitization by seepage refluxion. *Bull. Am. Ass. petrol. Geol.* **44**, 1912–1920.

Adey, W.H. (1975) The algal ridges of St Croix: their structure and Holocene development. *Atoll Res. Bull.* **187**, 1–67.

Adey, W.H. (1978) Coral reef morphogenesis: a multidimensional model. *Science* **202**, 831–837.

Adey, W.H. & Burke, R.B. (1977) Holocene bioherms of the Lesser Antilles — geologic control of development. In: *Reefs and Related Carbonates — Ecology and Sedimentology* (Ed. by S.H. Frost, M.P. Weiss and J.B. Saunders) Am. Ass. petrol. Geol. Studies in Geol. 4, 67–81.

Adey, W.H., Macintyre, I.G., Stuckenrath, R. & Dill, R.F. (1977) Relict barrier-reef system off St Croix: its implications with respect to late Cenozoic coral-reef development in the western Atlantic. *Proc. 3rd Int. Coral Reef Symp. Miami* 2, 15–22.

Adey, W.H., Townsend, R.A. & Boykins, W.T. (1982) The crustose coralline algae of the Hawaiian Archipelago. *Smith. Contrib. mar. Sci.* **15**, 1–74.

Aharon, P., Schidlowski, M. & Singh, I.B. (1987) Chronostratigraphic carbon isotope record of the lesser Himalaya. *Nature* **327**, 699–702.

Ahr, W.M. (1973) The carbonate ramp: an alternative to the shelf model. *Trans. Gulf-Cst. Ass. geol. Socs* **23**, 221–225.

Ahr, W.M. & Standon, R.J. (1973) The sedimentologic and paleoecologic significance of *Lithotrya* a rock boring barnacle. *J. sedim. Petrol.* **43**, 20–23.

Aigner, T. (1982) Calcareous tempestites: storm-dominated stratification, Upper Muschelkalk limestones (Middle Trias, SW Germany). In: *Cyclic and Event Stratification* (Ed. by G. Einsele and A. Seilacher) pp. 180–198. Springer-Verlag, Berlin.

Aigner, T. (1984) Dynamic stratigraphy of epicontinental carbonates, Upper Muschelkalk (M. Triassic), South German Basin. *Neues Jb. Geol. Paläont. Abh.* **169**, 127–159.

Aigner, T. (1985) *Storm Depositional Systems. Lecture Notes in Earth Sciences*, Vol. 3, p. 174. Springer-Verlag, Berlin.

Aigner, T. & Reineck, H.E. (1982) Proximity trends in modern storm sands from the Heligoland Bight (North Sea) and their implication for basin analysis. *Senckenbergiana Marit.* **14**, 183–215.

Aissaoui, D.M. (1985) Botryoidal aragonite and its diagenesis. *Sedimentology* **32**, 345–361.

Aissaoui, D.M. (1988) Magnesian calcite cements and their diagenesis: dissolution and dolomitization, Mururoa Atoll. *Sedimentology* **35**, 821–841.

Aissaoui, D.M., Buigues, D. & Purser, B.H. (1986) Model of reef diagenesis: Mururoa Atoll, French Polynesia. In: *Reef Diagenesis* (Ed. by J.H. Schroeder and B.H. Purser), pp. 27–52. Springer-Verlag, Berlin.

Aissaoui, D.M. & Purser, B.H. (1983) Nature and origin of internal sediments in Jurassic limestones of Burgundy (France) and Fnoud (Algeria). *Sedimentology* **30**, 273–283.

Aissaoui, D.M. & Purser, B.H. (1985) Reef diagenesis: cementation at Mururoa Atoll, French Polynesia. *Proc. 5th Int. Coral Reef Congr. Tahiti* 3, 257–262.

Aitken, J.D. (1966) Middle Cambrian to Middle Ordovician cyclic sedimentation, southern Rocky Mountains of Alberta. *Bull. Can. petrol. Geol.* **14**, 405–441.

Aitken, J.D. (1978) Revised models for depositional Grand Cycles, Cambrian of the southern Rocky Mountains, Canada. *Bull. Can. petrol. Geol.* **26**, 515–542.

Alexandersson, T. (1974) Carbonate cementation in coralline algal nodules in the Skagerrak, North Sea: biochemical precipitation in under-saturated waters. *J. sedim. Petrol.* **44**, 7–26.

Alexandersson, T. (1978) Destructive diagenesis of carbonate sediments in the Eastern Skagerrak, North Sea. *Geology* **6**, 324–327.

Al-Hashimi, W. & Hemingway, J.E. (1974) Recent dolomitization and the origin of the rusty crusts of Northumberland: A Reply. *J. sedim. Petrol.* **44**, 271–274.

Ali, Y.A. & West, I.M. (1983) Relationships of modern gypsum nodules in sabkhas of loess to compositions of brines and sediments in northern Egypt. *J. sedim. Petrol.* **53**, 1151–1168.

Allan, J.R. & Matthews, R.K. (1977) Carbon and oxygen isotopes as diagenetic and stratigraphic tools: data from surface and subsurface of Barbados, West Indies. *Geology* **5**, 16–20.

Allan, J.R. & Matthews, R.K. (1982) Isotope signatures associated with early meteoric diagenesis. *Sedimentology* **29**, 797–817.

Allanson, B.R. (1973) The fine structure of the periphyton of *Chara* sp. and *Potamogeton natans* from Wytham Pond, Oxford, and its significance to the macrophyte–periphyton metabolic model of R.B. Wetzel and H.L. Allen. *Freshwater Biol.* **3**, 535–541.

Allen, J.R.L. (1982) Sedimentary Structures: their character and physical basis. *Developments in Sedimentology*, 30A & B. Elsevier, Amsterdam.

Allen, P.A. (1981) Wave generated structures in the Devonian lacustrine sediments of southeast Shetland and ancient wave conditions. *Sedimentology* **28**, 369–379.

Allen, P.A. (1985) Hummocky cross-stratification is not produced purely under progressive gravity waves. *Nature* **313**, 562–564.

Allen, P.A. & Collinson, J.D. (1986) Lakes. In: *Sedimentary Environments and Facies* (Ed. by H.G. Reading) pp. 63–94. Blackwells, Oxford.

Alvarez, L.W., Alvarez, W., Asaro, F. & Michel, H.V. (1980) Extraterrestrial cause of the Cretaceous–Tertiary extinction. *Science* **208**, 1095–1108.

Alvarez, W., Colacicchi, R. & Montanari, A. (1985) Syn-sedimentary slides and bedding formation in Apennine pelagic limestones. *J. sedim. Petrol.* **55**, 720–734.

Alvarez, W., Kent, D.V., Premoli-silva, I., Schweickert, R.A. & Larson, R.A. (1980) Franciscan complex limestone deposited at 17 degrees south paleolatitude. *Bull. geol. Soc. Am.* **91**, 476–484.

Anadón, P. & Zamarrēno, I. (1981) Paleogene nonmarine algal deposits of the Ebro Basin, northeastern Spain. In: *Phanerozoic Stromatolites* (Ed. by C.L.V. Monty) pp. 140–154. Springer-Verlag, Berlin.

Anderson, E.J., Goodwin, P. & Sobieski, T. (1984) Episodic accumulation and the origin of formation boundaries in the Helderberg Group of New York State. *Geology* **12**, 120–123.

Anderson, T.F. & Arthur, M.A. (1983) Stable isotopes of oxygen and carbon and their application to sedimentologic and paleoenvironmental problems. In: *Stable Isotopes in Sedimentary Geology* (Ed. by M.A. Arthur and T.F. Anderson) Soc. econ. Paleont. Miner. Short Course No. 10, 1.1–1.151.

Anderson, E.J. & Goodwin, P.W. (1978) Punctuated aggradational cycles (PACS) in Middle Ordovician and Lower Devonian sequences. *Proc. 50th Ann. Mtg. New York State Geol. Soc.* **50**, 204–224.

Andrews, J.E. (1986) Microfacies and geochemistry of Middle Jurassic algal limestones from Scotland. *Sedimentology* **33**, 499–520.

Arakel, A.V. (1982) Genesis of calcrete in Quaternary soil profiles. Hutt and Leeman Lagoons, Western Australia. *J. sedim. Petrol.* **52**, 109–125.

Arakel, A.V. & McConchie, D. (1982) Classification and genesis of calcrete and gypsite lithofacies in paleo-drainage systems of inland Australia and their relationship to carnotite mineralization. *J. sedim. Petrol.* **52**, 1149–1170.

Arthur, M.A., Anderson, T.F., Kaplan, I.R., Veizer, J. & Land, L.S. (1983) *Stable Isotopes in Sedimentary Geology*. Soc econ. Paleont. Miner. Short Course No. 10.

Ashton, M. (1977) *Stratigraphy and carbonate environments of the Lincolnshire Limestone Formation, eastern England*. Unpublished PhD Thesis, Hull University.

Asquith, G.B. (1979) *Subsurface Carbonate Depositional Models*, p. 121. Penwell Books, Tulsa.

Assereto, R.L. & Folk, R.L. (1980) Diagenetic fabrics of aragonite, calcite and dolomite in an ancient peritidal spelean environment: Triassic Calcare Rosso, Lombardia, Italy. *J. sedim. Petrol.* **50**, 371–394.

Assereto, R.L. & Kendall, C.G. St.C. (1977) Nature, origin and classification of peritidal tepee structures and related breccias. *Sedimentology* **24**, 153–210.

Astin, T.R. (1989) A revised stratigraphy of Devonian lacustrine sediments of Orkney. Scotland: implications for climatic cyclicity, basin structure and maturation history. *J. geol. Soc.* **146**, 550–565.

Austin, J.A., Schlager, W., Palmer, A.A. *et al.* (1986) Proc. Init. Repts (Pt.A), O D P, 101, p. 247.

Awramik, S.M. (1971) Precambrian columnar stromatolite diversity: reflection of metazoan appearance. *Science* **174**, 825–826.

Awramik, S.M. (1984) Ancient stromatolites and microbial mats. In: *Microbial Mats: Stromatolites* (Ed. by Y. Cohen, R.W. Castenholz and H.O. Halvorson) pp. 1–22. A.R. Liss Inc., New York.

Awramik, S.M., Morris, S.C., Bengston, S. & Hsü, K.J. (1986) The Precambrian–Cambrian boundary and geochemical perturbations. *Nature* **319**, 696–697.

Azeredo, A.C. & Galopim de Carvalho, A.M. (1986) Novos elementos sobre o 'Paleogénico' carbonatado dos Arredores de Lisboa. *Comun. Serv. Geol. Portugal* **72**, 111–118.

Babcock, J.A. (1977) Calcareous algae, organic boundstones and the genesis of the Upper Capitan Limestone (Permian, Guadalupian), Guadalupe Mts., West Texas & New Mexico. In: *Upper Guadalupian Facies, Permian Reef Complex, Guadalupe Mountains, New Mexico and West Texas* (Ed. by M.E. Hileman and S.J. Mazzullo) Permian Basin Sect., Soc. econ. Paleont. Miner. Publ. 77–16, 3–44.

Bacelle, L. & Bosellini, A. (1965) Diagrammi per la stima visiua della composizione percentuale nelle rocce sedimentaire. *Ann. Univ. Ferrara, N.S. Sez IX, Sci. Geol. Paleont.* **1**, 59–62.

Back, W., Hanshaw, B.B., Herman, J.S. & van Driel, J.N. (1986) Differential dissolution of a Pleistocene reef in the ground-water mixing zone of coastal Yucatan, Mexico. *Geology* **14**, 137–140.

Back, W., Hanshaw, B.B., Pyle, T.E., Plummer, L.N. & Weidie, A.E. (1979) Geochemical significance of ground-water discharge and carbonate solution to the formation of Caleta Xel Ha, Quintana Roo, Mexico. *Water Resources Res.* **15**, 1531–1535.

Back, W., Hanshaw, B.B. & van Driel, J.N. (1984) Role of groundwater in shaping the eastern coastline of the Yucatan Peninsula, Mexico. In: *Groundwater as a Geomorphic Agent* (Ed. by R.G. LaFleur) pp. 157–172. Allen & Unwin, Mass.

Badiozamani, K. (1973) The Dorag dolomitization model — application to the Middle Ordovician of Wisconsin. *J. sedim. Petrol.* **43**, 965–984.

Badiozamani, K., Mackenzie, F.T. & Thorstenson, D.C. (1977) Experimental carbonate cementation: salinity, temperature and vadose-phreatic effects. *J. sedim. Petrol.* **47**, 529–542.

Baker, P.A. & Burns, S.J. (1985) The occurrence and formation of dolomite in organic-rich continental margin sediments. *Bull. Am. Ass. petrol. Geol.* **69**, 1917–1930.

Baker, P.A., Gieskes, J.M. & Elderfield, H. (1982) Diagenesis of carbonates in deep sea sediments — evidence from Sr/Ca ratios and interstitial dissolved Sr^{++} data. *J. sedim. Petrol.* **52**, 71–82.

Baker, P.A. & Kastner, M. (1981) Constraints on the formation of sedimentary dolomites. *Science* **213**, 214–216.

Ball, M.M. (1967) Carbonate sand bodies of Florida and the Bahamas. *J. sedim. Petrol.* **37**, 556–591.

Ball, M.M., Shinn, E.A. & Stockman, K.W. (1967) The geologic effects of Hurricane Donna in South Florida. *J. Geol.* **75**, 583–597.

Bandel, (1974) Deepwater limestones from the Devonian–Carboniferous of the Carnic Alps, Austria. In: *Pelagic Sediments: on Land and under the Sea* (Ed. by K.J. Hsü and H.C. Jenkyns) Spec. Publ. int. Ass. Sediment. 1, 93–115.

Barbieri, M., Masi, U. & Tolomeo, L. (1979) Stable isotope evidence for a marine origin of ophicalcites from the north-central Apennines (Italy). *Mar. Geol.* **30**, 193–204.

Barker, C.E. & Halley, R.B. (1986) Fluid inclusion, stable isotope, and vitrinite reflectance evidence for the thermal history of the Bone Spring Limestone, Southern Guadalupe Mountains, Texas. In: *Roles of Organic Matter in Sediment Diagenesis* (Ed. by D.L. Gautier) Spec. Publ. Soc. econ. Paleont. Miner. 38, 189–203.

Barnes, D.J. (1973) Growth in colonial scleractinians. *Bull. Mar. Sci.* **23**, 280–298.

Barnes, R.S.K. & Hughes, R.N. (1982) *An Introduction to Marine Ecology*, p. 339. Blackwells, Oxford.

Barrett, T.J. (1982) Stratigraphy and sedimentology of Jurassic bedded chert overlying ophiolites in North Apennines, Italy. *Sedimentology* **29**, 353–373.

Barwis, J.H. (1978) Sedimentology of some South Carolina tidal-creek point bars, and a comparison with their fluvial counterparts. In: *Fluvial Sedimentology* (Ed. by A.D. Miall) Mem. Can. Soc. petrol. Geol. 5, 129–160.

Basan, P.B. (1973) Aspects of sedimentation and development of carbonate bank in the Barracuda Keys, South Florida. *J. sedim. Petrol.* **43**, 42–53.

Bassett, H.G. & Stout, J.G. (1967) Devonian of western Canada. *Proc. Int. Symp. Devonian System, Calgary*, **1**, 717–725.

Bathurst, R.G.C. (1966) Boring algae, micrite envelopes and lithification of molluscan biosparites. *Geol. J.* **5**, 15–32.

Bathurst, R.G.C. (1967) Oolitic films on low energy carbonate sand grains, Bimini Lagoon, Bahamas. *Mar. Geol.* **5**, 89–109.

Bathurst, R.G.C. (1970) Problems of lithification in carbonate muds. *Proc. geol. Ass.* **81**, 429–440.

Bathurst, R.G.C. (1975) *Carbonate sediments and their diagenesis*, pp. 658. Elsevier, Amsterdam.

Bathurst, R.G.C. (1980a) Deep crustal diagenesis in limestones. *Revista Inst. Investigaciones Geol. Univ. Barcelona*, **34**, 89–100.

Bathurst, R.G.C. (1980b) *Stromatactis* — origin related to submarine cemented crusts in Paleozoic mud mounds. *Geology* **8**, 131–134.

Bathurst, R.G.C. (1982) Genesis of stromatactis cavities between submarine crusts in Palaeozoic carbonate mud buildups. *J. Geol. Soc.* **129**, 165–181.

Bathurst, R.G.C. (1986) Carbonate diagenesis and reservoir development: conservation, destruction and creation of pores. *Quart. J. Colorado Sch. Mines* **81**, 1–25.

Bathurst, R.G.C. (1987) Diagenetically enhanced bedding in argillaceous platform limestones: stratified cementation and selective compaction. *Sedimentology* **34**, 749–778.

Bauld, J. (1981) Occurrence of benthic microbial mats in saline lakes. *Hydrobiologia* **81**, 87–111.

Bausch, W.M. (1968) Clay content and calcite crystal size of limestones. *Sedimentology* **10**, 71–75.

Bay, T.A. Jr. (1977) Lower Cretaceous stratigraphic models from Texas and Mexico. In: *Cretaceous Carbonates of Texas & Mexico: Applications to Subsurface Exploration* (Ed. by D.G. Bebout and R.G. Loucks) pp. 12–30. Bureau of Economic Geology, Rept. of Investigations, 89. Austin, Texas.

Bayly, I.A.E. & Williams, W.D. (1974) *Inland Waters and Their Ecology*. Longmans, London.

Beach, D.K. & Ginsburg, R.N. (1982) Facies succession of Pliocene–Pleistocene carbonates, northwestern Great Bahama Bank: reply *Bull. Am. Ass. petrol. Geol.* **66**, 106–108.

Beadle, L.C. (1974) *The Inland Waters of Tropical Africa*, pp. 365. Longmans, London.

Beales, F.W. (1958) Ancient sediments of Bahaman type. *Bull. Am. Ass. petrol. Geol.* **42**, 1845–1880.

Beales, F.W. & Hardy, J.W. (1980) Criteria for the recognition of diverse dolomite types with an emphasis on studies of host rocks for Mississippi Valley-type ore deposits. In: *Concepts and Models of Dolomitization* (Ed. by D.H. Zenger, J.R. Dunham and R.L. Ethington) Spec. Publ. Soc. econ. Paleont. Miner. 28, 197–214.

Becher, J.W. & Moore, C.H. (1976) The Walker Creek Field: a Smackover diagenetic trap. *Trans. Gulf-Cst Ass. Geol. Socs* **26**, 34–56.

Bechstädt, T., Hagemeister, A., Schweizer, T. & Zeeh, S. (1987) Asymmetric and symmetric cycles in Triassic carbonate and carbonate-clastic sequences of the Eastern Alps. *8th IAS Regional Meeting, Tunis (Abst).* 64–65.

Beeunas, M.A. & Knauth, L.P. (1985) Preserved stable isotopic signature of subaerial diagenesis in the 1.2-b.y. Mescal Limestone, central Arizona. Implications for the timing and development of a terrestrial plant cover. *Bull. geol. Soc. Am.* **96**, 737–745.

Begin, A.B., Ehrlich, A. & Nathan, Y. (1974) Lake Lisan, the Pleistocene precursor of the Dead Sea. *Bull. geol. Surv. Israel* **63**, 30.

Behrens, E.W. & Land, L.S. (1972) Subtidal Holocene dolomite, Baffin Bay, Texas. *J. sedim. Petrol.* **42**, 155–161.

Beier, J.A. (1985) Diagenesis of Quaternary Bahamian beachrock: petrographic and isotopic evidence. *J. sedim. Petrol.* **55**, 755–761.

Beier, J.A. (1987) Petrographic and geochemical analysis of caliche profiles in a Bahamian Pleistocene dune. *Sedimentology* **34**, 991–998.

Bellanca, A., Calderone, S. & Neri, R. (1986) Isotope geochemistry, petrology and depositional environments of the diatomite-dominated Tripoli Formation (Lower Messinian) Sicily. *Sedimentology* **33**, 729–744.

Belperio, A.P., Gostin, V.A., Cann, J.H. & Murray-Wallace, C.V. (1988) Sediment organism zonation and the evolution of Holocene tidal sequences in southern Australia. In: *Tide-Influenced Sedimentary Environments and Facies* (Ed. by P.L. de Boer *et al.*) pp. 475–497. D. Riedel, Dordrecht.

Belperio, A.P., Hails, J.R., Gostin, V.A. & Polach, H.A. (1984a) The stratigraphy of coastal carbonate banks and Holocene sea-levels of northern Spencer Gulf, South Australia. *Mar. Geol.* **61**, 297–313.

Belperio, A.P., Smith, B.W., Polach, H.A., Nitterhouse, C.A., DeMaster, D.J., Prescott, J.R., Hails, J.R. & Gostin, V.A. (1984b) Chronological studies of the Quaternary marine sediments of northern Spencer Gulf, South Australia. *Mar. Geol.* **61**, 265–296.

Benson, L.V. (1974) Transformation of a polyphase sedimentary assemblage into a single phase rock — a chemical approach. *J. sedim. Petrol.* **44**, 123–136.

Berger, W.H. (1970) Planktonic foraminifera: selective solution and the lysocline. *Mar. Geol.* **8**, 111–138.

Berger, W.H., Vincent, E. & Thierstein, H.R. (1981) The deep-sea record: major steps in Cenozoic ocean evolution. In: *The Deep Sea Drilling Project: a Decade of Progress* (Ed. by J.E. Warme, R.G. Douglas and E.L. Winterer) Spec. Publ. Soc. econ. Paleont. Miner. 32, 489–504.

Berger, W.H. & Winterer, E.L. (1974) Plate stratigraphy and the fluctuating carbonate line. In: *Pelagic Sediments: on Land and under the Sea* (Ed. by K.J. Hsü and H.C. Jenkyns) Spec. Publ. int. Ass. Sediment. 1, 11–48.

Berner, R.A. (1975) The role of magnesium in the crystal growth of calcite and aragonite from seawater. *Geochim. Cosmochim. Acta* **39**, 489–504.

Berner, R.A. (1980) *Early Diagenesis, a Theoretical Approach*, pp. 241. Princeton University Press, Princeton, NJ.

Berner, R.A., Westrich, J.T., Graber, J., Smith, J. & Martens, C.S. (1978) Inhibition of aragonite precipitation from supersaturated seawater: a laboratory and field study. *Am. J. Sci.* **278**, 816–837.

Bernoulli, D. & Jenkyns, H.C. (1974) Alpine, Mediterranean and Central Atlantic Mesozoic facies in relation to the early evolution of the Tethys. In: *Modern and Ancient Geosynclinal Sedimentation* (Ed. by R.H. Dott and R.H. Shaver) Spec. Publ. Soc. econ. Paleont. Miner. 19, 129–160.

Bertani, R.T. & Carozzi, A.V. (1985) Lagoa Feia Formation (Lower Cretaceous) Campos Basin, offshore Brazil: rift valley stage carbonate reservoirs, I & II. *J. petrol. Geol.* **8**, 37–58, 199–220.

Bertrand-Sarfati, J., Freytet, P. & Plaziat, J.C. (1966) Les calcaires concrétionnés de la limite Oligocène–Miocène des environs de St. Pourcain sur Sioule (Limagne d'Allier). Rôle des Algues dans leur édification, analogie avec les stromatolites et rapports avec la sédimentation. *Bull. Soc. géol. France* **8**, 652–662.

Bhattacharyya, A. & Friedman, G.M. (1984) Experimental compaction of ooids under deep-burial diagenetic temperatures and pressures. *J. sedim. Petrol.* **54**, 362–372.

Bice, D. (1988) Synthetic stratigraphy of carbonate platform and basin systems. *Geology,* **16**, 703–706.

Biddle, K.T. (1981) The basinal Cipit boulders: indicators of Middle to Upper Triassic buildup margins, Dolomite Alps, Italy. *Riv. Ital. Paleont.* **86**, 779–794.

Binkley, K.L., Wilkinson, B.H. & Owen, R.M. (1980) Vadose beachrock cementation along a southeastern Michigan marl lake. *J. sedim. Petrol.* **50**, 953–962.

Birke, L. (1974) Marine blue–green algal mats. In: *Coastal Ecological Systems of the United States,* 1 (Ed. by H.T. Odum, B.J. Copeland and E.A. McMahan) pp. 331–345. Conservation Foundation, Washington, DC.

Birnbaum, S.J. & Radlick, T.M. (1982) A textural analysis of trona and associated lithologies, Wilkins Peak Member, Eocene Green River Formation, southwestern Wyoming. In: *Depositional and Diagenetic Spectra of Evaporites — a Core workshop* (Ed. by C.R. Handford, R.G. Loucks and G.R. Davies) Soc. econ. Paleont. Miner. Core Workshop 3, 75–99.

Bischoff, W.D., Bishop, F.C. & Mackenzie, F.T. (1983)

Biogenically produced magnesian calcite: inhomogeneities in chemical and physical properties; comparison with synthetic phases. *Am. Mineral.* **68**, 1183–1188.

Bischoff, W.D., Sharma, S.K. & Mackenzie, F.T. (1985) Carbonate ion disorder in synthetic and biogenetic magnesian calcites: a Raman spectral study. *Am. Mineral.* **70**, 581–589.

Bissel, H.J. & Barker, H.K. (1977) Deep-water limestones of the Great Blue Formation (Mississippian) in the eastern part of the Cordilleran Miogeosyncline in Utah. In: *Deep-water Carbonate Environments* (Ed. by H.E. Cook and P. Enos) Spec. Publ. Soc. econ. Paleont. Miner. 25, 171–186.

Bissel, H.J. & Chilingar, G.V. (1967) Classification of sedimentary carbonate rocks. In: *Carbonate Rocks* (Ed. by G.V. Chilingar, H.J. Bissell and R.W. Fairbridge) pp. 87–168. Elsevier, Amsterdam.

Black, M. (1933) The algal sediments of Andros Island, Bahamas. *Phil. Trans. R. Soc. Lond. B* **222**, 165–192.

Blake, D.F. & Peacor, D.R. (1981) Biomineralization of crinoid skeletal elements using TEM and STEM microanalysis. *Scan. Elect. Microscopy* **III**, 321–328.

Blake, D.F., Peacor, D.R. & Wilkinson, B.H. (1982) The sequence and mechanism of low-temperature dolomite formation: calcian dolomites in a Pennsylvanian echinoderm. *J. sedim. Petrol.* **52**, 59–70.

Blatt, H. (1982) *Sedimentary Petrology*, pp. 564. W.H. Freeman & Co., San Francisco.

Blatt, H., Middleton, G.V. & Murray, R.C. (1980) *Origin of Sedimentary Rocks*, pp. 634. Prentice Hall, New Jersey.

Blendinger, W. (1986) Isolated stationary carbonate platforms: the Middle Triassic (Ladinian) of the Marmolada area, Dolomites, Italy. *Sedimentology* **33**, 159–184.

Bloom, A.L., Broecker, W.S., Chappell, J.M.A., Matthews, R.K. & Mesolella, K.J. (1974) Quaternary sea-level fluctuations on a tectonic coast. *Quat. Res.* **4**, 185–205.

Bloxsom, W.E. (1972) *A Lower Cretaceous (Comanchean) prograding shelf and associated environments of deposition, northern Coahuila, Mexico.* Unpubl. MSc thesis. Univ. Texas, Austin.

Bodine, M.W., Holland, H.D. & Borcsik, M. (1965) Coprecipitation of manganese and strontium with calcite. In: *Symposium on problems of postmagmatic ore deposition*, Prague II, 401–406.

Bogli, A. (1964) Mischungskorrosion — Ein Beitrag Zum Verkarstungsproblemm. *Erdkunde* **18**, 83–92.

Bogli, J. (1980) *Karst Hydrology and Physical Speleology*, pp. 285. Springer-Verlag, Berlin.

Boles, J.R. & Franks, S.G. (1979) Clay diagenesis in Wilcox Sandstones of southwest Texas: Implications of smectite diagenesis on sandstone cementation. *J. sedim. Petrol.* **49**, 55–70.

Bond, G.C. & Kominz, M.A. (1984) Construction of tectonic subsidence curves for the early Paleozoic miogeocline, southern Canadian Rocky Mountains: Implications for subsidence mechanisms, age of breakup and crustal thinning. *Bull. geol. Soc. Am.* **95**, 155–173.

Bosence, D.W.J. (1980) Sedimentary facies, models and production rates of coralline algal gravels, western Ireland. *Geol. J.* **15**, 91–111.

Bosence, D.W.J. (1983a) Description and classification of rhodoliths (rhodoids, rhodolites). In: *Coated Grains* (Ed. by T.M. Peryt) pp. 217–224. Springer-Verlag, Berlin.

Bosence, D.W.J. (1983c) The occurrence and ecology of recent rhodoliths — a review. In: *Coated Grains* (Ed. by T.M. Peryt), pp. 225–242. Springer-Verlag, Berlin.

Bosence, D.W.J. (1983c) Coralline algal reef frameworks. *J. geol. Soc.* **140**, 365–376.

Bosence, D.W.J. (1984) Construction and preservation of two modern coralline algal reefs, St. Croix, Caribbean. *Palaeontology* **27**, 549–574.

Bosence, D.W.J. (1985) Preservation of coralline algal frameworks. *Proc. 5th Int. Coral Reef Congress, Tahiti* **2**, 39–45.

Bosence, D.W.J. (1989a) Surface sublittoral sediments of Florida Bay. *Bull. Mar. Sci.* **49**.

Bosence, D.W. (1989b) Carbonate budgets for carbonate mounds, Florida USA. *Proc. 6th Int. Coral Reef Symp. Townsville 1988.*

Bosence, D.W.J., Rowlands, R.J. & Quine, M.L. (1985) Sedimentology and budget of a Recent carbonate mound, Florida Keys. *Sedimentology* **32**, 317–343.

Bosellini, A. (1984) Progradation geometries of carbonate platforms: examples from the Triassic of the Dolomites, northern Italy. *Sedimentology* **31**, 1–24.

Bosellini, A. (1989) *La Storia Geologica delle Dolomiti*, p. 149. Edizioni, Dolomiti.

Bosellini, A. & Ginsburg, R.N. (1970) Form and internal structure of recent algal nodules (rhodolites) from Bermuda. *J. Geol.* **79**, 669–682.

Bosellini, A. & Hardie, L.A. (1973) Depositional theme of a marginal marine evaporite. *Sedimentology* **20**, 5–27.

Bosellini, A., Masetti, D. & Sarti, M. (1981) A Jurassic 'Tongue of the Ocean' infilled with oolitic sands: the Belluno Trough, Venetian Alps, Italy. *Mar. Geol.* **44**, 59–95.

Bosellini, A. & Rossi, D. (1974) Triassic carbonate buildups of the Dolomites, Northern Italy. In: *Reefs in Time and Space* (Ed. by L.F. Laporte) Spec. Publ. Soc. econ. Paleont. Miner. 18, 209–233.

Bosellini, A. & Winterer, E.L. (1975) Pelagic limestone and radiolarite of the Tethyan Mesozoic: a genetic model. *Geology* **3**, 279–282.

Bottinga, Y. (1968) Calculation of fractionation factors for carbon and oxygen exchange in the system calcite–carbon dioxide–water. *J. Chem. Phys.* **72**, 800–808.

Bottinga, Y. (1969) Calculated fractionation factors for carbon and hydrogen isotope exchange in the system calcite–carbon dioxide–graphite–methane–hydrogen–water vapour. *Geochim. Cosmochim. Acta* **33**, 49–64.

Botz, R.W. & Von der Borch, C.C. (1984) Stable isotope study of carbonate sediments from the Coorong area, South Australia. *Sedimentology* **31**, 837–849.

Bouma, A.H., Normark, W.R. & Barnes, N.E. (Eds.)

(1986) *Submarine Fans and Related Turbidite Systems*, pp. 351. Springer-Verlag, New York.

Bourque, P.A. & Gignac, H. (1983) Sponge-constructed stromatactis mud mounds, Silurian of Gaspé, Québec. *J. sedim. Petrol.* **53**, 521–532; **56**, 459–463.

Bourrouilh-le Jan, F.G. (1980) Hydrologie des Nappes d'eau superficielles de L'Île Andros, Bahamas. Dolomitisation et diagenèse de plaine d' estran en climat tropical humide. *Bull. Cent. Rech. Explor-Prod. Elf-Aquitaine* **4**, 661–707.

Bourrouilh-le Jan, F.G. (1982) Geometrie et minéralogie des crops sédimentaires dans une mangrove carbonatée sous l'influence des ourangans Île Andros, Bahamas. *Mem. Soc. geol. Fr. N S* **144**, 77–92.

Boyer, B.W. (1982) Green River laminites: does the playa-lake model really invalidate the stratified-lake model? *Geology* **10**, 321–324.

Boyles, J.M. & Scott, A.J. (1982) A model for migrating shelf-bar sandstones in Upper Mancos Shale (Campanian), northwestern Colorado. *Bull. Am. Ass. petrol. Geol.* **66**, 491–508.

Braithwaite, C.J.R. (1973a) Reefs: just a problem of semantics. *Bull. Am. Ass. petrol. Geol.* **57**, 1100–1116.

Braithwaite, C.J.R. (1973b) Settling behaviour related to sieve analysis of skeletal sands. *Sedimentology* **20**, 251–262.

Brasier, M. & Donahue, J. (1985) Barbuda — an emerging reef and lagoon complex on the edge of the Lesser Antilles island arc. *J. geol. Soc.* **142**, 1101–1117.

Brennand, T.P. & van Veen, F.R. (1975) The Auk oilfield. In: *Petroleum and the Continental Shelf of NW Europe*, Vol. 1. Geology (Ed. by A.W. Woodland) pp. 275–281. Applied Science Publ., London.

Brice, S.E., Kelts, K.R. & Arthur, M.A. (1980) Lower Cretaceous lacustrine source beds from early rifting phases of South Atlantic. Abstract. *Bull. Am. Ass. petrol. Geol.* **64**, 680–681.

Bridges, P.H. (1982) The origin of cyclothems in the late Dinantian platform carbonates at Crich, Derbyshire. *Proc. Yorks. geol. Soc.* **44**, 159–180.

Bridges, P.H. & Chapman, A.J. (1988) The anatomy of a deep water mud-mound complex to the north of the Dinantian platform in Derbyshire, UK. *Sedimentology* **35**, 139–162.

Bromley, R.G. (1978) Bioerosion of Bermuda reefs. *Palaeogeog. Palaeoclimatol. Palaeoecol.* **23**, 169–197.

Bromley, R.G. (1981) Enhancement of visibility of structure in marly chalk: modification of the Bushinsky oil technique. *Bull. geol. Soc. Denmark* **29**, 111–118.

Bromley, R.G. & Ekdale, A.A. (1984) Trace fossil preservation in flint in the European chalk. *J. Paleont.* **58**, 298–311.

Bromley, R.G. & Ekdale, A.A. (1986) Composite ichnofabrics and tiering of burrows. *Geol. Mag.* **123**, 59–65.

Bromley, R.G. & Gale, A.S. (1982) The lithostratigraphy of the English Chalk Rock. *Cret. Res.* **3**, 273–306.

Brown, B.J. & Farrow, G.E. (1978) Recent dolomite concretions of crustacean burrows from Loch Sunart, west coast of Scotland. *J. sedim. Petrol.* **48**, 825–834.

Brown, R.E. & Wilkinson, B.H. (1981) The Draney Limestone: early Cretaceous lacustrine carbonate deposition in western Wyoming and southeastern Idaho. *Contrib. Geol., Univ. Wyoming* **20**, 23–31.

Brunskill, G.J. (1969) Fayetteville Green Lake, New York, II. Precipitation and sedimentation of calcite in a meromictic lake with laminated sediments. *Limnol. Oceanog.* **14**, 830–847.

Buchbinder, B. (1981) Morphology, microfabric and origin of stromatolites of the Pleistocene precursor of the Dead Sea, Israel. In: *Phanerozoic Stromatolites* (Ed. by C.L.V. Monty) pp. 181–196. Springer-Verlag, Berlin.

Buchbinder, L.G. & Friedman, G.M. (1980) Vadose, phreatic and marine diagenesis of Pleistocene–Holocene carbonates in a borehole: Mediterranean coast of Israel *J. sedim. Petrol.* **50**, 395–408.

Buchheim, H.P. & Surdam, R.C. (1977) Fossil catfish and the depositional environment of the Green River Formation, Wyoming. *Geology* **5**, 196–198.

Buchheim, H.P. & Surdam, R.C. (1981) Paleoenvironments and fossil fishes of the Laney Member, Green River Formation, Wyoming. In: *Communities of the Past* (Ed. by J. Gray, A.J. Boucot and W.B.N. Berry) pp. 415–452. Hutchinson Ross Publ. Co., Penn.

Budd, D.A. & Loucks, R.G. (1981) Smackover and lower Buckner Formation, South Texas; depositional systems on a Jurassic carbonate ramp. *Bur. econ. Geol. Rept.* **112**, 1–38.

Buddemeier, R.W. & Kinzie, R.A. (1976) Coral growth. *Oceanograph. Mar. Biol. Ann. Rev.* **14**, 183–225.

Buddemeier, R.W. & Oberdorfer, J.A. (1986) Internal hydrology and geochemistry of coral reefs and atoll islands: key to diagenetic variations. In: *Reef Diagenesis* (Ed. by J.H. Schroeder and B.H. Purser) pp. 91–111. Springer-Verlag, Berlin.

Bukry, D. (1981) Cenozoic coccoliths from the deep sea drilling project. In: *The Deep Sea Drilling Project: a Decade of Progress* (Ed. by J.E. Warme, R.G. Douglas and E.L. Winterer) Spec. Publ. Soc. Econ. Paleont. Miner. 32, 335–353.

Bullen, S.B. & Sibley, D.F. (1984) Dolomite selectivity and mimic replacement. *Geology*, **12**, 655–658.

Burchette, T.P. (1981) European Devonian reefs: a review of current concepts and models. In: *European Fossil Reef Models* (Ed. by D.F. Toomey) Spec. Publ. Soc. econ. Paleont. Miner. 30, 85–142.

Burchette, T.P. (1987) Carbonate-barrier shorelines during the basal Carboniferous transgression: the Lower Limestone Shale Group, South Wales and western England. In: *European Dinantian Environments* (Ed. by J. Miller, A.E. Adams and V.P. Wright) pp. 239–263. John Wiley, Chichester.

Burchette, T.P. (1988) Tectonic control on carbonate platform facies distribution and sequence development: Miocene, Gulf of Suez. *Sedim. Geol.* **59**, 179–204.

Burke, W.H., Denison, R.E., Hetherington, E.A., Koepnick, R.B., Nelson, H.F. & Otto, J.B. (1982) Variation of seawater $^{87}Sr/^{86}Sr$ throughout Phanerozoic time. *Geology* **10**, 516−519.

Burne, R.V. (1982) Relative fall of Holocene sea level and coastal progradation, north eastern Spencer Gulf, Australia. *BMR J. Aust. Geol. & Geophys.* **7**, 35−45.

Burne, R.V., Bauld, J. & deDecker, P. (1980) Saline lake charophytes and their geological significance. *J. sedim. Petrol.* **50**, 281−293.

Burne, R.V. & Colwell, J.B. (1982) Temperate carbonate sediments of Northern Spencer Gulf, South Australia: a high salinity 'foramol' province. *Sedimentology* **29**, 223−238.

Burne, R.V. & James, N.P. (1986) Subtidal origin of club-shaped stromatolites, Shark Bay (Abst.). Int. Ass. Sedim. Congress, Canberra, p. 49.

Burne, R.V. & Moore, L.S. (1987) Microbialites: organo-sedimentary deposits of benthic microbial communities. *Palaios* **2**, 241−254.

Burns, S.J. & Baker, P.A. (1987) A geochemical study of dolomite in the Monterey Formation, California. *J. sedim. Petrol.* **57**, 128−139.

Burruss, R.C., Cercone, K.R. & Harris, P.M. (1985) Timing of hydrocarbon migration: evidence from fluid inclusions in calcite cements, tectonics and burial history. In: *Carbonate Cements* (Ed. by N. Schneidermann and P.M. Harris) Spec. Publ. Soc. econ. Paleont. Miner. 36, 277−289.

Burton, E.A. & Walter, L.M. (1987) Relative precipitation rates of aragonite and Mg calcite from seawater: temperature or carbonate ion control? *Geology* **15**, 111−114.

Busenberg, E. & Plummer, L.N. (1985) Kinetic and thermodynamic factors controlling the distribution of SO_4^{2-} and Na^+ in calcites and selected aragonites. *Geochim. Cosmochim. Acta* **49**, 713−725.

Bush, P. (1973) Some aspects of the diagenetic history of the sabkha in Abu Dhabi, Persian Gulf. In: *The Persian Gulf* (Ed. by B.H. Purser) pp. 395−407. Springer-Verlag, Berlin.

Butler, G.P. (1969) Modern evaporite deposition and geochemistry of coexisting brines, the sabkha, Trucial Coast, Arabian Gulf. *J. sedim. Petrol.* **39**, 70−89.

Butler, G.P. (1970) Holocene gypsum and anhydrite of the Abu Dhabi sabkha, Trucial Coast: An alternative explanation of origin. *Third Symposium on Salt. N. Ohio geol. Soc.* **1**, 120−152.

Butler, G.P., Harris, P.M. & Kendall, C.G. St.C. (1982) Recent evaporites from the Abu Dhabi coastal flats. In: *Deposition and Diagenetic Spectra of Evaporites* (Ed. by C.R. Handford, R.G. Loucks and G.R. Davies) Soc. econ. Paleont. Miner. Core Workshop 3, 33−64.

Buxton, T.M. & Sibley, D.F. (1981) Pressure solution features in shallow buried limestone. *J. Sedim. Petrol.* **51**, 19−26.

Byers, C.W. & Stasko, L.E. (1978) Trace fossils and sedimentologic interpretation — McGregor Member of Platteville Formation (Ordovician) of Wisconsin. *J.*

sedim. Petrol. **48**, 1303−1309.

Cairns, S.D. & Stanley, G.C. (1981) Ahermatypic coral banks: living and fossil counterparts. *Proc. 4th Int. Coral Reef Symp., Manila* **1**, 611−618.

Calvet, F. (1979) Evolucio diagenetica en els sediment carbonates del Pleistoceno Mallorqui. Unpubl. PhD thesis, University of Barcelona, Spain.

Calvet, F. & Esteban, M. (1977) Evolucion diagenetica en los sedimentos carbonatados marinos del Pleistoceno de Mallorca. *Bol. Soc. Hist. Nat de Baleares* **22**, 96−118.

Calvet, F. & Julia, R. (1983) Pisoids in the caliche profiles of Tarragona (north east Spain). In: *Coated Grains* (Ed. by T.M. Peryt) pp. 456−473. Springer-Verlag, Berlin.

Calvet, F., Plana, F. & Traveira, A. (1980) La tendecia mineralogica de las eolianites de Pleistocene de Mallorca. Mediante la aplicacion del metedo de Chung. *Acta Geol. Hispanica* **15**, 39−44.

Calvet, F. & Tucker, M.E. (1988) Outer ramp carbonate cycles in the Upper Muschelkalk, Catalan Basin, NE Spain. *Sedim. Geol.* **57**, 185−198.

Cameron, E.M. & Bauman, A. (1972) Carbonate sedimentation during the Archean. *Chem. Geol.* **10**, 17−30.

Camoin, G., Bernet-Rollande, M.C. & Philip, J. (1988) Rudist−coral frameworks associated with submarine volcanism in the Maastrichtian of the Pachino area (Sicily). *Sedimentology* **35**, 123−138.

Cann, J.H. & Gostin, V.A. (1985) Coastal sedimentary facies and foraminiferal biofacies of the St Kila Fm. at Port Gawler, South Australia. *Trans. Roy. Soc. S. Austr.* **109**, 121−142.

Carballo, J.D., Land, L.S. & Miser, D.E. (1987) Holocene dolomitization of supratidal sediments by active tidal pumping, Sugarloaf Key, Florida. *J. sedim. Petrol.* **57**, 153−165.

Carlisle, D. (1983) Concentration of uranium and vanadium in calcretes and gypcretes. In: *Residual Deposits: Surface Related Weathering Processes and Materials* (Ed. by R.C.L. Wilson) Spec. Publ. geol. Soc. Lond. 11, 185−195.

Carlson, W.D. (1980) The calcite−aragonite equilibrium effects of Sr substitution and anion orientation disorder. *Am. Mineral.* **55**, 1252−1262.

Carpenter, A.B. (1976) Discussion. *J. sedim. Petrol.* **46**, 254−257.

Carpenter, A.B. (1980) The chemistry of dolomite formation I: the stability of dolomite. In: *Concepts and Models of Dolomitization* (Ed. by D.H. Zenger, J.B. Dunham and R.L. Ethington) Spec. Publ. Soc. econ. Paleont. Miner. 28, 111−121.

Carrasco, V.B. (1977) Albian sedimentation of submarine autochthonous and allochthonous carbonates, east edge of the Valles-San Luis Potosi Platform, Mexico. In: *Deep-water Carbonate Environments* (Ed. by H.E. Cook and P. Enos) Spec. Publ. Soc econ. Paleont. Miner. 25, 263−273.

Cerling, T.E. (1984) The stable isotopic composition of modern soil carbonate and its relationship to climate.

Earth Planet. Sci. Letts. **71**, 229–240.

Chafetz, H.S. (1986) Marine peloids: a product of bacterially induced precipitation of calcite. *J. sedim. Petrol.* **56**, 812–817.

Chafetz, H.S. & Butler, J.C. (1980) Petrology of recent caliche pisolites, spherulites and speleothem deposits from central Texas. *Sedimentology* **27**, 497–518.

Chafetz, H.S., McIntosh, A.G. & Rush, P.F. (1988) Freshwater phreatic diagenesis in the marine realm of Recent Arabian Gulf carbonates. *J. sedim. Petrol.* **58**, 433–440.

Chappell, J. (1980) Coral morphology, diversity and reef growth. *Nature* **286**, 249–252.

Chappell, J.M.A. & Shackleton, N.J. (1986) Oxygen isotopes and sea-level. *Nature* **324**, 137–140.

Chave, K.E. (1954) Aspects of the biochemistry of magnesium. 1. Calcareous and marine organisms. 2. Calcareous sediments and rocks. *J. Geol.* **62**, 266–283 and 587–599.

Chilingar, G.V. (1956) Relationship between Ca/Mg ratio and geological age. *Bull. Am. Ass. petrol. Geol.* **40**, 2256–2266.

Choi, D.R. & Ginsburg, R.N. (1982) Siliciclastic foundations of Quaternary reefs in the southernmost Belize lagoon, British Honduras. *Bull. geol. Soc. Am.* **93**, 116–126.

Choquette, P.W. & James, N.P. (1987) Diagenesis in limestones — 3, the deep burial environment. *Geoscience Canada* **14**, 3–35.

Choquette, P.W. & James, N.P. (1988) Introduction. In: *Paleokarst* (Ed. by N.P. James and P.W. Choquette) pp. 1–21. Springer-Verlag, New York.

Choquette, P.W. & Pray, L.C. (1970) Geologic nomenclature and classification of porosity in sedimentary carbonates. *Bull. Am. Ass. petrol. Geol.* **54**, 207–250.

Choquette, P.W. & Steinen, R.P. (1980) Mississippian nonsupratidal dolomite, Ste. Genevieve Limestone, Illinois Basin: evidence for mixed-water dolomitization. In: *Concepts and Models of Dolomitization* (Ed. by D.H. Zenger, J.B. Dunham and R.H. Ethington) Spec. Publ. Soc. econ. Paleont. Miner. **28**, 168–196.

Chow, N. & James, N.P. (1987a) Cambrian Grand Cycles: northern Appalachian perspective. *Bull. geol. Soc. Am.* **98**, 418–429.

Chow, N. & James, N.P. (1987b) Facies-specific calcitic and bimineralic ooids from the Middle and Upper Cambrian platform carbonates, western Newfoundland, Canada. *J. sedim. Petrol.* **57**, 907–921.

Chowdhury, A.N. & Moore, C.H. (1986) Diagenesis and dolomitization of Upper Smackover grainstones: East Texas Shelf (Abs.). Int. Ass. Sedim. Congress, Canberra, p. 59.

Cisne, J.L. (1986) Earthquakes recorded stratigraphically on carbonate platforms. *Nature* **323**, 320–322.

Clark, D.N. (1979) Patterns of porosity and cement in ooid reservoirs in Dogger (Middle Jurassic) of France: discussion. *Bull. Am. Ass. petrol. Geol.* **63**, 676–677.

Clark, D.N. (1980) The diagenesis of Zechstein carbonate sediments. *Contrib. Sediment.* **9**, 167–203.

Clausaude, M., Gavier, N., Picard, J., Pichon, M., Roman, M.L., Thomassin, B., Vasseur, P., Vivien, M. & Weydert, P. (1971) Morphologie des récifs coralliens de la région de Tuléar (Madagascar): Eléments de terminologie récifale. *Téthys* supplement 2, 1–74.

Claypool, G.E., Holser, W.T., Kaplan, I.R., Sakai, H. & Zak, I. (1980) The age curves of sulphur and oxygen isotopes in marine sulphate and their interpretation. *Chem. Geol.* **28**, 199–260.

Cloetingh, S., McQueen, H. & Lambeck, K. (1985) On a tectonic mechanism for regional sealevel variations. *Earth planet. Sci. Letts.* **75**, 157–166.

Cloud, P.E. (1962) Environment of calcium carbonate deposition west of Andros Island, Bahamas. *Prof. Pap. US geol. Surv.* **350**, 138.

Cloud, P. & Lajoie, K.R. (1980) Calcite impregnated defluidization structures in littoral sands of Mono Lake, California. *Science* **210**, 1009–1012.

Cluff, R.M. (1984) Carbonate sand shoals in the Middle Mississippian (Valmeyeran) Salem–St. Louis–Ste-Genevieve Limestones, Illinois Basin. In: *Carbonate Sands — a Core Workshop* (Ed. by P.M. Harris) Soc. econ. Paleont. Miner. Core Worshop 5, 94–135.

Cohen, A.D. & Spackman, W. (1980) Phytogenic organic sediments and sedimentary environments in the Everglades Mangrove Complex of Florida. *Palaeontogr. Abt. B* **172**, 125–149.

Cohen, A.S. & Nielsen, C. (1986) Ostracodes as indicators of paleohydrochemistry in lakes: a Late Quaternary example from Lake Elmenteita, Kenya. *Palaios* **1**, 601–609.

Cohen, A.S. & Thouin, C. (1987) Nearshore carbonate deposits in Lake Tanganyika. *Geology* **15**, 414–418.

Colacicchi, R., Passeri, L. & Pialli, G. (1975) Evidence of tidal environment deposition in the Calcare Massiccio Formation (Central Apennines, Lower Lias). In: *Tidal Deposits: a Casebook of Recent Examples and Fossil Counterparts* (Ed. by R.N. Ginsburg) pp. 345–353. Springer-Verlag, Berlin.

Collerson, K.D., Ullman, W.J. & Torgersen, T. (1988) Groundwaters with unradiogenic $^{87}Sr/^{86}Sr$ ratios in the Great Artesian Basin, Australia. *Geology* **16**, 59–63.

Collins, A.G. (1975) *Geochemistry of Oilfield Waters*, p. 496. Elsevier, New York.

Collins, S.E. (1980) Jurassic Cotton Valley and Smackover Reservoir Trends, East Texas, North Louisiana and South Arkansas. *Bull. Am. Ass. petrol. Geol.* **64**, 1004–1013.

Conaghan, P.J., Mountjoy, E.W., Edgecomb, D.R. *et al.* (1976) Nubrigyn algal reef (Devonian), eastern Australia: allochthonous blocks and megabreccias. *Bull. geol. Soc. Am.* **87**, 515–530.

Coniglio, M. & James, N.P. (1985) Calcified algae as sediment contributors to early Paleozoic limestones: evidence from deep-water sediments of the Cow Head Group, Western Newfoundland. *J. sedim. Petrol.* **55**, 746–754.

Coogan, A.H. (1969) Recent and ancient carbonate cyclic sequences. In: *Symposium on Cyclic Sedimentation in the*

Permian Basin (Ed. by I.G. Elan and S. Chubers) West Texas geol. Soc. pp. 5–16.

Cook, H.E., Hine, A.C. & Mullins, H.T. (Eds.) (1983) *Platform Margin and Deep Water Carbonates*. Soc. econ. Paleont. Miner. Short Course No. 12, p. 573.

Cook, H.E., McDaniel, P.N., Mountjoy, E.W. & Pray, L.C. (1972) Allochthonous carbonate debris flows at Devonian bank ('reef') margins, Alberta, Canada. *Bull. Can. petrol. Geol.* **20**, 439–497.

Cook, H.E. & Mullins, H.T. (1983) Basin margin environment. In: *Carbonate Depositional Environments* (Ed. by P.A. Scholle, D.G. Bebout and C.H. Moore) Mem. Am. Ass. petrol. Geol. **33**, 540–617.

Cook, H.E. & Taylor, M.E. (1977) Comparison of continental slope and shelf environments in the Upper Cambrian and lowermost Ordovician of Nevada. In: *Deep-water Carbonate Environments* (Ed. by H.E. Cook and P. Enos) Spec. Publ. Soc. econ. Paleont. Miner. 25, 51–81.

Craig, D.H. (1988) Caves and other features of Permian karst in San Andres Dolomite, Yates Field Reservoir, West Texas. In: *Paleokarst* (Ed. by N.P. James and P.W. Choquette) pp. 342–363. Springer-Verlag, New York.

Craig, H. (1961) Isotopic variations in meteoric waters. *Science* **133**, 1702–1703.

Craig, H. (1965) The measurements of oxygen isotope paleotemperatures. In: *Stable Isotopes in Oceanographic Studies and Paleotemperatures* (Ed. by E. Tongiorgi) pp. 1–24. Spoleto, July 26–27, Consiglio Nazionale delle Richerche, Laboratorio di Geologia Nucleare, Pisa.

Cram, J.M. (1979) The influence of continental shelf width on tidal range: paleo-oceanographic implications. *J. Geol.* **87**, 441–447.

Crame, J.A. (1980) Succession and diversity in the Pleistocene coral reefs of the Kenya coast. *Palaeontology* **23**, 1–37.

Crame, J.A. (1981) Ecological stratification in the Pleistocene coral reefs of the Kenya coast. *Palaeontology* **24**, 609–646.

Crevello, P.D. & Schlager, W. (1980) Carbonate debris sheets and turbidites, Exuma Sound, Bahamas. *J. sedim. Petrol.* **50**, 1121–1148.

Crevello, P.D., Wilson, J.L., Sarg, J.F & Read, J.F. (Eds) (1989) *Controls on Carbonate Platform and Basin Development*. Spec. Publ. Soc. econ. Paleont. Miner. 44, p. 405.

Cronan, D.S. (1980) *Underwater Minerals*, pp. 362. Academic Press, London.

Cros, P.G. (1979) Genesis of oolites and grapestones, the Bahama Platform (Joulters Cays and the Great Bank). *Bull. Cent. Rech, Explor-Prod. Elf Aquitaine* **3**, 63–139.

Cummings, E.R. (1932) Reefs or bioherms? *Bull. geol. Soc. Am.* **43**, 331–352.

Czerniakowski, L.A., Lohmann, K.C. & Wilson, J.L. (1984) Closed-system marine burial diagenesis: isotopic data from the Austin Chalk and its components. *Sedimentology* **31**, 863–877.

Dabrio, C.J., Esteban, M. & Martin, J.M. (1981) The coral reef of Nijar, Messinian (uppermost Miocene), Almeria Province, SE Spain. *J. sedim. Petrol.* **51**, 521–539.

Dahanayake, K., Gerdes, G. & Krumbein, W.E. (1985) Stromatolites, oncolites and oolites biogenically formed *in situ. Naturwissenschaften* **72**, 513–518.

Davies, G.R. (1970a) Carbonate bank sedimentation, eastern Shark Bay, Western Australia. *Mem. Am. Ass. petrol. Geol.* **13**, 85–168.

Davies, G.R. (1970b) Algal laminated sediments, Gladstone embayment, Shark Bay, Western Australia. *Mem. Am. Ass. petrol. Geol.* **13**, 169–205.

Davies, G.R. (1977) Former magnesian calcite and aragonite submarine cements in upper Paleozoic reefs of the Canadian Arctic: A summary. *Geology* **5**, 11–15.

Davies, P.J. (1983) Reef growth. In: *Perspectives on Coral Reefs* (Ed. by D.J. Barnes) pp. 69–106. Australian Institute of Marine Science, Townsville, Australia.

Davies, P.J., Bubela, B. & Ferguson, J. (1978) The formation of ooids. *Sedimentology* **25**, 703–730.

Davies, P.J. & Hopley, D. (1983) Growth fabrics and growth rates of Holocene reefs in the Great Barrier Reef. *B.M.R. J. of Austr. Geol. Geophys.* **8**, 237–251.

Davies, P.J. & Marshall, J.F. (1980) A model of epicontinental reef growth. *Nature* **287**, 37–38.

Davies, P.J. & Marshall, J.F. (1985) *Halimeda* bioherms — low energy reefs, northern Great Barrier Reef. *Proc. 5th Coral Reef Congress, Tahiti* **5**, 1–7.

Davies, P.J. & Martin, K. (1976) Radial aragonite ooids, Lizard Island, Great Barrier Reef. *Geology* **4**, 120–122.

Davies, T.A. & Worsley, T.R. (1981) Paleoenvironmental implications of oceanic carbonate sedimentation rates. In: *The Deep Sea Drilling Project: a Decade of Progress* (Ed. by J.E. Warme, R.G. Douglas and E.L. Winterer) Spec. Publ. Soc. econ. Miner. Paleont. 32, 169–179.

Davis, R.A. (Ed.) (1978) *Coastal Sedimentary Environments*, pp. 420. Springer-Verlag, New York.

Davis, R.L. & Wilkinson, B.H. (1983) Sedimentology and petrology of freshwater lacustrine carbonate: mid-Tertiary Camp Davis Formation, northwestern Wyoming. *Contrib. Geol., Univ. Wyoming* **22**, 45–55.

Davis, S.N. & DeWiest, R.J.M. (1966) *Hydrogeology*, p. 463. John Wiley, London.

Dean, W.E. (1981) Carbonate minerals and organic matter in sediments of modern north temperate hard-water lakes. In: *Recent and Ancient Non-marine Depositional Environments: Models for Exploration* (Ed. by F.G. Ethridge and R.M. Flores) Spec. Publ. Soc. econ. Paleont. Miner. 31, 213–231.

Dean, W.E. & Eggleston, J.R. (1975) Comparative anatomy of marine and freshwater algal reefs, Bermuda and central New York. *Bull. geol. Soc. Am.* **86**, 665–676.

Dean, W.E. & Fouch, T.D. (1983) Lacustrine environment. In: *Carbonate Depositional Environments* (Ed. by P.A. Scholle, D.G. Bebout and C.H. Moore) Mem. Am. Ass. petrol. Geol. **33**, 97–130.

De Boer, R.B. (1977) On the thermodynamics of pressure solution–interaction between chemical and mechanical forces. *Geochim. Cosmochim. Acta* **41**, 249–256.

Decandia, F.A. & Elter, P. (1972) La 'zona' ofiolitifera del Bracco nel settore compreso fra Levanto e la Val Graveglia (Appenino ligure). *Mem. Soc. geol. Ital.* **11**, 503–530.

Deckker, P. De & Last, W.M. (1988) Modern dolomite deposition in continental, saline lakes, western Victoria, Australia. *Geology* **16**, 29–32.

Deelman, J.C. (1978a) Experimental ooids and grapestones: carbonate aggregates and their origin. *J. sedim. Petrol.* **49**, 1269–1278.

Deelman, J.C. (1978b) Protodolomite redefined — Discussion. *J. sedim. Petrol.* **48**, 1004–1007.

Deffeyes, K.S., Lucia, F.J. & Weyl, P.K. (1965) Dolomitization of Recent and Plio-Pleistocene sediments by marine evaporite waters on Bonaire, Netherlands Antilles. In: *Dolomitization and Limestone Diagenesis — a Symposium* (Ed. by L.C. Pray and R.C. Murray) Spec. Publ. Soc. econ. Paleont. Miner. 13, 71–88.

Degens, E.T. (1976) Molecular mechanisms in carbonate, phosphate and silica deposition in the living cell. *Topics in Current Chemistry* **64**, 1–112.

Degens, E.T. & Epstein, S. (1964) Oxygen and carbon isotope ratios in coexisting calcites and dolomites from recent and ancient sediments. *Geochim. Cosmochim. Acta* **28**, 23–44.

Deines, P. (1980) The isotopic composition of reduced organic carbon. In: *Handbook of Environmental Isotope Geochemistry, 1. The Terrestrial Environment*, A, pp. 329–406. Elsevier, Amsterdam.

DeJong, E.W., Borman, A.H., Thierry, R., Westbroek, P., Gruter, M. & Kamerling, J.P. (1986) Calcification in the Coccolithophorides *Emiliania Huxleyi* and *Pleurochrysis carterae* II Biochemical aspects. In: *Biomineralization in Lower Plants and Animals* (Ed. by B.S.C. Leadbeater and R. Riding) The Systematics Association. Spec. vol. No. 30, pp. 205–217. Oxford University Press, Oxford.

Demaison, G.J. & Moore, G.E. (1980) Anoxic environments and oil source bed genesis. *Bull. Am. Ass. petrol. Geol.* **64**, 1179–1209.

Demicco, R.V. (1985) Platform and off-platform carbonates of the Upper Cambrian of western Maryland, USA. *Sedimentology* **32**, 1–22.

DePaolo, D.J. (1986) Detailed record of the Neogene Sr isotopic evolution of seawater from D.S.D.P. Site 590B. *Geology* **14**, 103–106.

Desborough, G.A. (1978) A biogenic–chemical stratified lake model for the origin of the oil shale of the Green River Formation: an alternative to the playa-lake model. *Bull. geol. Soc. Am.* **89**, 961–971.

Desrochers, A. & James, N.P. (1988) Early Palaeozoic surface and subsurface paleokarst: Middle Ordovician carbonates, Mingan Islands, Quebec. In: *Paleokarst* (Ed. by N.P. James and P.W. Choquette) pp. 183–210. Springer-Verlag, New York.

Devol, A.H., Anderson, J.J., Kuivila, K. & Murray, J.W. (1984) A model for coupled sulfate reduction and methane oxidation in the sediments of Saanich Inlet. *Geochim. Cosmochim. Acta* **48**, 993–1004.

Dickman, M. (1985) Seasonal succession and microlamina formation in a meromictic lake displaying varved sediments. *Sedimentology* **32**, 109–118.

Dickson, J.A.D. (1966) Carbonate identification and genesis as revealed by staining. *J. sedim. Petrol.* **36**, 491–505.

Dickson, J.A.D. (1983) Graphical modelling of crystal aggregates and its relevance to cement diagnosis. *Phil. Trans. R. Soc. Lond.* A **309**, 465–502.

Dickson, J.A.D. & Coleman, M.L. (1980) Changes in carbon and oxygen isotope composition during limestone diagenesis. *Sedimentology* **27**, 107–118.

Dill, R.F., Shinn, E.A., Jones, A.T., Kelly, K. & Steinen, R.P. (1986) Giant subtidal stromatolites forming in normal salinity waters. *Nature* **324**, 55–58.

Dix, G.R. & Mullins, H.T. (1988) Rapid burial diagenesis of deep-water carbonates: Exuma Sound, Bahamas. *Geology* **16**, 680–683.

Donahue, J. (1978) Pisolite. In: *The Encyclopaedia of Sedimentology* (Ed. by R.W. Fairbridge and J. Bourgeois), pp. 582–583. Stroudsburg, Pennsylvania.

Donaldson, J.A. & Ricketts, B.D. (1979) Beachrock in Proterozoic dolostone of the Belcher Islands, NWT, Canada. *J. sedim. Petrol.* **49**, 1287–1294.

Done, T.J. (1983) Coral zonation: its nature and significance. In: *Perspectives on Coral Reefs* (Ed. by D.J. Barnes), pp. 107–149. Australian Institute of Marine Science, Townsville, Australia.

Donovan, D.T. & Jones, E.J.W. (1979) Causes of worldwide changes in sea-level. *J. geol. Soc.* **136**, 187–192.

Donovan, R.N. (1975) Devonian lacustrine limestones at the margin of the Orcadian Basin, Scotland. *J. geol. Soc.* **131**, 489–510.

Donovan, R.N. & Collins, A. (1978) Mound structures from the Caithness Flags (Mid Devonian), northern Scotland. *J. sedim. Petrol.* **48**, 171–173.

Donovan, R.N. & Foster, R.J. (1972) Subaqueous shrinkage cracks from Caithness Flagstone Series (mid Devonian) of northeast Scotland. *J. sedim. Petrol.* **42**, 309–317.

Dorobek, S.L. (1987) Petrography, geochemistry and origin of burial diagenetic facies, Siluro-Devonian Helderberg Group (carbonate rocks), Central Appalachians. *Bull. Am. Ass. petrol. Geol.* **71**, 492–514.

Dott, R.H. Jr. & Bourgeois, J. (1982) Hummocky stratification: significance of it's variable bedding sequences. *Bull. geol. Soc. Am.* **93**, 663–680.

Dravis, J. (1979) Rapid and widespread generation of Recent oolitic hardgrounds on a high energy Bahamian Platform, Eleuthera Bank, Bahamas. *J. sedim. Petrol.* **49**, 195–208.

Dravis, J. & Yurewicz, D.A. (1985) Enhanced carbonate petrography using fluorescence microscopy. *J. sedim. Petrol.* **55**, 795–804.

Drever, J.L. (1982) *The Geochemistry of Natural Waters*, p. 388. Prentice Hall, New Jersey.

Droxler, A.W., Schlager, W. & Whallon, C.C. (1983)

Quaternary aragonite cycles and oxygen-isotope record in Bahamian carbonate ooze. *Geology* **11**, 235–239.

Druckman, Y. (1981) Sub-recent manganese-bearing stromatolites along shorelines of the Dead Sea. In: *Phanerozoic Stromatolites* (Ed. by C.L.V. Monty) pp. 197–208. Springer-Verlag, Berlin.

Druckman, Y. & Moore, C.H. (1985) Late subsurface secondary porosity in a Jurassic grainstone reservoir, Smackover Formation, Mt. Vernon field, southern Arkansas. In: *Carbonate Petroleum Reservoirs* (Ed. by P.O. Roehl and P.W. Choquette) pp. 369–384. Springer-Verlag, New York.

Duke, W.L. (1985) Hummocky cross-stratification, tropical hurricanes, and intense winter storms. *Sedimentology* **32**, 167–194.

Dunham, J.B. & Olson, E.R. (1978) Diagenetic dolomite formation related to Paleozoic paleogeography of the Cordilleran miogeocline in Nevada. *Geology* **6**, 556–559.

Dunham, J.B. & Olson, E.R. (1980) Shallow subsurface dolomitization of subtidally deposited carbonate sediments in the Hanson Creek Formation (Ordovician–Silurian) of central Nevada. In: *Concepts and Models of Dolomitization* (Ed. by D.H. Zenger, J.B. Dunham and R.L. Ethington) Spec. Publ. Soc. econ. Paleont. Miner. 28, 139–161.

Dunham, R.J. (1962) Classification of carbonate rocks according to depositional texture. In: *Classification of Carbonate Rocks* (Ed. by W.E. Ham) Mem. Am. Ass. petrol. Geol. 1, 108–121.

Dunham, R.J. (1969a) Early vadose silt in Townsend mound (reef), New Mexico. In: *Depositional Environments in Carbonate Rocks* (Ed. by G.M. Friedman) Spec. Publ. Soc. econ. Paleont. Miner. 14, 139–182.

Dunham, R.J. (1969b) Vadose pisolite in the Capitan reef (Permian), New Mexico and Texas. In: *Depositional Environments in Carbonate Rocks* (Ed. by G.M. Friedman) Spec. Publ. Soc. econ. Paleont. Miner. 14, 182–190.

Dunham, R.J. (1970a) Stratigraphic reefs versus ecologic reefs. *Bull. Am. Ass. petrol. Geol.* **54**, 1931–1932.

Dunham, R.J. (1970b) Keystone vugs in carbonate beach deposits (Abs.). *Bull. Am. Assoc. petrol. Geol.* **54**, 845.

Durand, J.P., Gaviglio, P., Gonzalez, J.F., & Montenau, R. (1984) Sediments fluvio-lacustres du Crétacé Supérieur du Paleocene et de l'Eocene dans le Synclinal de l'Arc (Region d'Aix en Provence). *Excurs. Guide, 5th Eur. Reg. Mtrg. Int. Ass. Sedimentol.*, pp. 44.

Eardley, A.J., Shuey, R.T., Gvosdetsky, V., Nash, W.P., Picard, D.M., Grey, D.C. & Kukla, G.J. (1973) Lake cycles in the Bonneville Basin, Utah. *Bull. geol. Soc. Am.* **84**, 211–216.

Easton, A.J. & Claugher, D. (1986) Variations in a growth form of synthetic vaterite. *Min. Mag.* **50**, 332–336.

Ebanks, W.J. (1975) Holocene carbonate sedimentation and diagenesis, Ambergris Cay, Belize. In: *Belize Shelf — Carbonate Sediments, Clastic Sediments and Ecology* (Ed. by K.F. Wantland and W.C. Pusey) Am. Ass. petrol.

Geol. Studies in Geology 2, 234–296.

Ebanks, W.J. & Bubb, J.N. (1975) Holocene carbonate sedimentation Matecumbe Keys Tidal Bank, South Florida. *J. sedim. Petrol.* **45**, 422–439.

Eberli, G.P. (1987) Carbonate turbidite sequences deposited in rift-basins of the Jurassic Tethys Ocean (eastern Alps, Switzerland). *Sedimentology* **34**, 363–388.

Eberli, G.P. & Ginsburg, R.N. (1987) Segmentation and coalescence of Cenozoic carbonate platforms, north-western Great Bahama Bank. *Geology* **15**, 75–79.

Eder, W. (1970) Genese Riff-naher Detritus-Kalk. *Verh. Geol. B.A.* **4**, 551–569.

Eder, W. (1982) Diagenetic redistribution of carbonate, a process in forming limestone–marl alternations (Devonian and Carboniferous, Rheinisches Schiefergebirge, W. Germany). In: *Cyclic and Event Stratification* (Ed. by G. Einsele and A. Seilacher) pp. 98–112. Springer-Verlag, Berlin.

Edmunds, W.M. & Walton, N.R.G. (1983) The Lincolnshire Limestone — hydrochemical evolution over a ten-year period. *J. Hydrol.* **61**, 201–211.

Edwards, D.C. & Riding, R. (1988) Permian reefs: aragonite cement or neomorphosed algal skeleton. *9th IAS Regional Meeting, Leuven, Abst.* 64–65. Abstr. Leuven, 1988, 64–65.

Eggleston, J.R. & Dean, W.E. (1976) Freshwater stromatolitic bioherms in Green Lake, New York. In: *Stromatolites* (Ed. by M.R. Walter) pp. 479–483. Elsevier, Amsterdam.

Ekdale, A.A. & Bromley, R.G. (1984) Comparative ichnology of shelf-sea and deep-sea chalk. *J. Paleont.* **58**, 322–332.

Ekdale, A.A., Ekdale, S.F. & Wilson, J.L. (1976) Numerical analysis of carbonate microfacies in the Cupido Limestone (Neocomian-Aptian), Coahuila, Mexico. *J. sedim. Petrol.* **46**, 362–368.

Elderfield, H. (1986) Strontium isotope stratigraphy. *Palaeogeog. Palaeoclimatol. Palaeoecol.* **57**, 71–90.

Elderfield, H., Gieskes, J.M., Baker, P.A., Oldfield, R.K., Hawkesworth, C.T. & Miller, R. (1982) $^{87}Sr/^{86}Sr$ and $^{18}O/^{16}O$ ratios, interstitial water chemistry and diagenesis in deep-sea carbonate sediments of the Ontong Java Plateau. *Geochim. Cosmochim. Acta* **46**, 2259–2268.

Eliuk, L.S. (1978) The Abenaki Formation, Nova Scotia Shelf, Canada — a depositional and diagenetic model for a Mesozoic carbonate platform. *Bull. Can. petrol. Geol.* **26**, 424–514.

Elliott, T. (1986) Siliciclastic shorelines. In: *Sedimentary Environments and Facies* (Ed. by H.G. Reading) pp. 155–188. Blackwells, Oxford.

Elliott, T.L. (1982) Carbonate facies, depositional cycles and the development of secondary porosity during burial diagenesis. In: *4th International Williston Basin Symposium* (Ed. by J.E. Christopher and J. Kaldi) Spec. Publ. Saskatchewan Geol. Soc. 6, 131–151.

Ellis, J.P. & Milliman, J.D. (1986) Calcium carbonate suspended in Arabian Gulf and Red Sea Waters: biogenic and detrital, not 'chemogenic'. *J. sedim. Petrol.* **55**, 805–808.

Ellis, P.M., Wilson, R.C.L. & Leinfelder, R.R. (1990) Controls on Upper Jurassic carbonate build up development in the Lusitanian Basin, Portugal. In: *Carbonate Platforms* (Ed. by M.E. Tucker *et al.*) Spec. Publ. int. Ass. Sedim. 9.

Embry, A.F. & Klovan, J.E. (1971) A late Devonian reef tract on northeastern Banks Island, Northwest Territories. *Bull. Can. petrol. Geol.* **19**, 730–781.

Emerich, G.H. & Wobber, F.H. (1963) A rapid visual method for estimating sedimentary parameters. *J. sedim. Petrol.* **33**, 831–843.

Enos, P. (1977a) Holocene sediment accumulations of the South Florida shelf margin. In: *Quaternary Sedimentation in South Florida* (Ed. by P. Enos and R.D. Perkins) Mem. geol. Soc. Am. 147, 1–130.

Enos, P. (1977b) Tamabra Limestone of the Poza Rica Trend, Cretaceous, Mexico. In: *Deep-water Carbonate Environments* (Ed. by H.E. Cook and P. Enos) Spec. Publ. Soc. econ. Paleont. Miner. **25**, 273–314.

Enos, P. (1983) Shelf. In: *Carbonate Depositional Environments* (Ed. by P.A. Scholle, D.G. Bebout and C.H. Moore) Mem. Am. Ass. petrol. Geol. 33, 267–296.

Enos, P. (1988) Evolution of pore space in the Poza Rica trend (Mid-Cretaceous), Mexico. *Sedimentology* **35**, 287–326.

Enos, P. & Moore, C.H. (1983) Fore-reef slope. In: *Carbonate Depositional Environments* (Ed. by P.A. Scholle, D.G. Bebout and C.H. Moore) Mem. Am. Ass. petrol. Geol. 33, 507–538.

Enos, P. & Perkins, R.D. (1979) Evolution of Florida Bay from island stratigraphy. *Bull. geol. Soc. Am.* **90**, 59–83.

Enos, P. & Sawatsky, L.H. (1981) Pore networks in Holocene carbonate sediments. *J. sedim. Petrol.* **51**, 961–985.

Epstein, S., Buchsbaum, R., Lowenstam, H.A. & Urey, H.C. (1953) Revised carbonate-water isotopic temperature scale. *Bull. geol. Soc. Am.* **64**, 1315–1326.

Esteban, M. (1979) Significance of Upper Miocene coral reefs of the Western Mediterranean. *Palaeogeog. Palaeoclimatol. Palaeoecol.* **29**, 169–188.

Esteban, M. & Klappa, C.F. (1983) Subaerial exposure environment. In: *Carbonate Depositional Environments* (Ed. by P.A. Scholle, D.G. Bebout and C.H. Moore) Mem. Am. Ass. petrol. Geol. 33, 1–54.

Esteban, M. & Pray, L.C. (1983) Pisoids and pisolite facies (Permian), Guadalupe Mountains, New Mexico and West Texas. In: *Coated Grains* (Ed. by T.M. Peryt) pp. 503–537. Springer-Verlag, Berlin.

Eugster, H.P. (1980) Lake Magadi, Kenya and its precursors. In: *Hypersaline Brines and Evaporites* (Ed. by A. Nissenbaum), pp. 195–232. Elsevier, Amsterdam.

Eugster, H.P. & Hardie, L.A. (1975) Sedimentation in an ancient playa-lake complex: the Wilkins Peak Member of the Green River Formation of Wyoming. *Bull. geol. Soc. Am.* **86**, 319–334.

Eugster, H.P. & Hardie, L.A. (1978) Saline lakes. In: *Lakes, Chemistry, Geology, Physics* (Ed. by A. Lerman) pp. 237–293. Springer-Verlag, Berlin.

Eugster, H.P. & Kelts, K. (1983) Lacustrine chemical sediments. In: *Chemical Sediments and Geomorphology – Precipitates and Residua in the Near-Surface Environment* (Ed. by A.S. Goudie and K. Pye) pp. 321–368. Academic Press, London.

Evamy, B.D. (1973) The precipitation of aragonite and its alteration to calcite on the Trucial Coast of the Persian Gulf. In: *The Persian Gulf* (Ed. by B.H. Purser) pp. 329–342. Springer-Verlag, Berlin.

Evamy, B.D. & Shearman, D.J. (1965) The development of overgrowths from echinoderm fragments. *Sedimentology* **5**, 211–233.

Evans, C.C. (1984) Development of ooid sand shoal complex: the importance of antecedent and syndepositional topography. In: *Carbonate Sands – a Core Workshop* (Ed. by P.M. Harris) Soc. econ. Paleont. Miner. Core Workshop, 5, 392–428.

Evans, G., Murray, J.W., Biggs, H.E.J., Bate, R. & Bush, P. (1973) The oceanography, ecology, sedimentology and geomorphology of parts of the Trucial Coast barrier island complex, Persian Gulf. In: *The Persian Gulf* (Ed. by B.H. Purser) pp. 233–277. Springer-Verlag, Berlin.

Evans, G., Schmidt, V., Bush, P. & Nelson, H. (1969). Stratigraphy and geologic history of the sabkha, Abu Dhabi, Persian Gulf. *Sedimentology* **12**, 145–159.

Fabricius, F.H. (1977) Origin of marine ooids and grapestones. *Contrib. Sediment.* **7**, 113.

Fagerstrom, J.A. (1987) *The Evolution of Reef Communities*, pp. 600. Wiley – Interscience, New York.

Fairchild, I.J. (1980a) Sedimentation and origin of a Late Precambrian 'dolomite' from Scotland. *J. sedim. Petrol.* **50**, 423–446.

Fairchild, I.J. (1980b) Stages in a Precambrian dolomitization, Scotland: cementing versus replacement textures. *Sedimentology* **27**, 631–650.

Fairchild, I.J. (1982) Discussion and reply. *Sedimentology* **29**, 299–302.

Fairchild, I.J. (1983) Chemical controls of cathodoluminescence of natural dolomites and calcites: new data and review. *Sedimentology* **30**, 579–583.

Fanning, K.A., Byrne, R.H., Breland, J.A., W.S. Elsinger, R.J. & Pyle, T.E. (1981) Geothermal springs of the west Florida continental shelf: evidence for dolomitization and radionuclide enrichment. *Earth planet. Sci. Letts.* **52**, 345–354.

Farrow, G.E., Allen, N.H. & Akpan, E.B. (1984) Bioclastic carbonate sedimentation on a high-latitude, tide-dominated shelf, Northeast Orkney Islands, Scotland. *J. sedim. Petrol.* **54**, 374–393.

Faucette, R.C. & Ahr, W.M. (1984) Depositional and diagenetic history of Upper Townsite Haynesville Formation, Teague Townsite Field, Freestone County, Texas. In: *Jurassic of the Gulf Rim* (Ed. by W.P.S. Ventress, D.G. Bebout, B.F. Perkins and C.H. Moore) Publ. Soc. econ.

Paleont. Miner. Gulf Coast Section, pp.103–129.

Faulkner, T.J. (1988) The Shipway Limestone of Gower: sedimentation on a storm-dominated early Carboniferous ramp. *Geol. J.* **23**, 85–100.

Faure, G. (1986) *Principles of Isotope Geology*, p. 589. John Wiley, New York.

Feazel, C.T., Keany, J. & Peterson, R.M. (1985) Cretaceous and Tertiary Chalk of the Ekofisk Field Area, Central North Sea. In: *Carbonate Petroleum Reservoirs* (Ed. by P.O. Roehl and P.W. Choquette) pp. 497–507. Springer-Verlag, Berlin.

Feist, M. & Grambast-Fessard, N. (1984) New Porocharaceae from the Bathonian of Europe: phylogeny and palaeoecology. *Palaeontology* **27**, 295–305.

Ferguson, J., Bubela, B. & Davies, P.J. (1978) Synthesis and possible mechanisms of formation of radial carbonate ooids. *Chem. Geol.* **22**, 285–308.

Ferguson, J., Burne, R.V. & Chambers, L.A. (1982) Lithification of peritidal carbonates by continental brines at Fisherman Bay, South Australia, to form a megapolygon/spelean limestone association. *J. sedim. Petrol.* **52**, 1127–1147.

Ferguson, J., Chambers, L.A., Donnelly, T.H. & Burne, R.V. (1988) Carbon and oxygen isotope composition of a recent megapolygon–spelean limestone, Fisherman Bay, South Australia. *Chem Geol.* **72**, 63–76.

Fischer, A.G. (1961) Stratigraphic record of transgressing seas in the light of sedimentation on Atlantic coast of New Jersey. *Bull. Am. Ass. petrol. Geol.* **45**, 1656–1666.

Fischer, A.G. (1964) The Lofer cyclothems of the Alpine Triassic. In: *Symposium on Cyclic Sedimentation* (Ed. by D.F. Merriam) Bull. geol. Surv. Kansas 169, 107–149.

Fischer, A.G. (1975) Tidal deposits, Dachstein Limestone of the North-Alpine Triassic. In: *Tidal Deposits: a Casebook of Recent Examples and Fossil Counterparts* (Ed. by R.N. Ginsburg) pp. 235–242. Springer-Verlag, Berlin.

Fischer, A.G. & Garrison, R.E. (1967) Carbonate lithification on the sea floor. *J. Geol.* **75**, 488–496.

Fisher, J.B. (1982) Effects of macrobenthos on the chemical diagenesis of freshwater sediments. In: *Animal–Sediment Relations – the Biogenic Alteration of Sediments. Topics in Geobiology*, Vol. 2 (Ed. by P.L. McCall and M.J.S. Tevesz) pp. 177–218. Plenum Press, New York.

Fisher, R. & Stueber, A.M. (1976) Strontium isotopes in selected streams within the Susquehanna River basin. *Water Resources Res.* **12**, 1061–1068.

Fisher, W.L. & Rodda, P.U. (1969) Edwards Formation (Lower Cretaceous), Texas: Dolomitization in a carbonate platform system. *Bull. Am. Ass. petrol. Geol.* **53**, 55–72.

Flügel, E. (1972) Microfazielle Untersuchungen in der Alpinen Trias. Methoden und Probleme. *Mitt. Ges. Geol. Bergbaustud.* **21**, 9–64.

Flügel, E. (1981) Paleoecology and facies of Upper Triassic reefs in Northern Calcareous Alps. In: *European Fossil Reef Models* (Ed. by D.F. Toomey) Spec. Publ. Soc.

econ. Paleont. Miner. 30, 260–291.

Flügel, E. (1982) *Microfacies Analysis of Limestones*, pp. 633. Springer-Verlag, Berlin.

Flügel, E., Franz, H.E. & Ott, W.F. (1968) Review on electron microscope studies of limestones. In: *Recent Developments in Carbonate Sedimentology in Central Europe* (Ed. by G. Müller and G.M. Friedman) pp. 85–97. Springer-Verlag, Berlin.

Folk, R.L. (1959) Practical petrographic classification of limestones. *Bull. Am. Ass. petrol. Geol.* **43**, 1–38.

Folk, R.L. (1962) Spectral subdivision of limestone types. In: *Classification of Carbonate Rocks* (Ed. by W.E. Ham) *Mem. Am. Ass. petrol. Geol.* **1**, 62–84.

Folk, R.L. (1965) Some aspects of recrystallization in ancient limestones. In: *Dolomitization and Limestone Diagenesis* (Ed. by L.C. Pray and R.C. Murray) Spec. Publ. Soc. econ. Paleont. Miner. 13, 14–48.

Folk, R.L. (1973) Carbonate petrography in the post-Sorbian age. In: *Evolving Concepts in Sedimentology* (Ed. by R.N. Ginsburg) pp. 118–158. Johns Hopkins University Press, Baltimore.

Folk, R.L. (1974) The natural history of crystalline calcium carbonate: effect of magnesium content and salinity. *J. sedim. Petrol.* **44**, 40–53.

Folk, R.L. & Assereto, R. (1976) Comparative fabrics of length-slow and length-fast calcite and calcitized aragonite in a Holocene speleothem, Carlsbad Caverns, New Mexico. *J. sedim. Petrol.* **46**, 486–496.

Folk, R.L. & Land, L.S. (1975) Mg/Ca ratio and salinity, two controls over crystallization of dolomite. *Bull. Am. Ass. petrol. Geol.* **59**, 60–68.

Folk, R.L. & Robles, R. (1964) Carbonate sands of Isla Perez, Alacran Reef complex, Yucatan. *J. Geol.* **72**, 255–292.

Folk, R.L & Siedlecka, A. (1974) The schizohaline environment: its sedimentary and diagenetic fabrics as exemplified by Late Paleozoic rocks of Bear Island, Svalbard. *Sedim. Geol.* **15**, 1–15.

Ford, D. (1988) Characteristics of dissolutional cave systems in carbonates. In: *Paleokarst* (Ed. by N.P. James and P.W. Choquette) pp. 25–57. Springer-Verlag, New York.

Forester, R.M. (1983) Relationship of two lacustrine ostracode species to solute composition and salinity: implications for paleohydrochemistry. *Geology* 11, 435–438.

Forester, R.M. (1986) Determinations of the dissolved anion composition of ancient lakes from fossil ostracodes. *Geology* 14, 796–798.

Francis, J.E. (1983) The dominant conifer of the Jurassic Purbeck Formation, England. *Palaeontology* 26, 277–294.

Frank, J.R., Carpenter, A.B. & Oglesby, T.W. (1982) Cathodoluminescence and composition of calcite cement in the Taum Sauk Limestone (Upper Cambrian), southeast Missouri. *J. sedim. Petrol.* **52**, 631–638.

Freeze, R.A. & Cherry, J.A. (1979) *Groundwater*, pp. 640.

Prentice-Hall, Englewood Cliffs, N.J.

Frey, R.W. & Bromley, R.G. (1985) Ichnology of American chalks: the Selma Group (Upper Cretaceous), western Alabama. *Can. J. earth Sci.* **22**, 801–828.

Freytet, P. (1984) Carbonate lacustrine sediments and their transformations by emersion and pedogenesis. Importance of identifying them for paleogeographical reconstructions. *Bull. Centres Rech. Explor.-Prod. Elf-Aquitaine* **8**, 223–247.

Freytet, P. & Plaziat, J.C. (1965) Importance des constructions algaires dues à des Cyanophycées dans les formations continentales, du Crétacé supérieur de l'Eocene du Languedoc. *Bull. Soc. géol. France* **7**, 679–694.

Freytet, P. & Plaziat, J.C. (1982) Continental carbonate sedimentation and pedogenesis — Late Cretaceous and Early Tertiary of southern France. *Contrib. Sediment.* **12**, 213.

Friedman, G.M. (1965) Terminology of crystallization textures and fabrics in sedimentary rocks. *J. Sedim. Petrol.* **35**, 643–655.

Friedman, G.M. (1975) The making and unmaking of limestones or the downs and ups of porosity. *J. Sedim. Petrol.* **45**, 379–398.

Friedman, G.M. (1985) The problems of submarine cement in classifying reefrock: an experience in frustration. In: *Carbonate Cements* (Ed. by N. Schneidermann and P.M. Harris) Spec. Publ. Soc. econ. Paleont. Miner. 36, 117–121.

Friedman, G.M., Amiel, A.J., Braun, M. & Miller, D.S. (1973) Generation of carbonate particles and laminites in algal mats — example from sea-marginal hypersaline pool, Gulf of Aqaba, Red Sea. *Bull. Am. Ass. petrol. Geol.* **57**, 541–557.

Friedman, I. & O'Neil, J.R. (1977) Compilation of stable isotope fractionation factors of geochemical interest. In: *Data of Geochemistry.* US Geol. Surv. Prof. Paper, 440-KK, 1–12.

Fritz, P. & Smith, D.G.W. (1970) The isotopic composition of secondary dolomites. *Geochim. Cosmochim. Acta* **34**, 1161–1173.

Frost, S.H. (1977) Cenozoic reef systems of the Caribbean — prospects for paleoecological synthesis. *Am. Ass. petrol. Geol. Studies in Geology* **4**, 93–110.

Frydl, P. & Stearn, C.W. (1978) Rates of bioerosion by parrotfish in Barbados reef environments. *J. sedim. Petrol* **48**, 1149–1158.

Frykman, P. (1986) Diagenesis of Silurian bioherms in the Klinteberg Formation, Gotland, Sweden. In: *Reef Diagenesis* (Ed. by J.H. Schroeder and B.H. Purser) pp. 399–423. Springer-Verlag, Berlin.

Füchtbauer, H. (1974) Sediments and sedimentary rocks — 1. In: *Sedimentary Petrology II* (Ed. by W. Engelhardt, H. Fuchtbauer and G. Müller) pp. 464. Schweizerbart, Stuttgart.

Füchtbauer, H. *et al.* (1977) Tertiary lake sediments of the Ries, research borehole Nördlingen 1973 — a summary. In: *Ergebnisse der Ries — Forschungsbohrung 1973: Struktur des Kraters und Entwicklung des Kratersees* (Ed. by H. Schmidt-Kaler) pp. 13–19. Geologica Bavarica, 75.

Fürsich, F.T. (1979) Genesis, environments and ecology of Jurassic hardgrounds. *Neues. Jb. Geol. Paläont. Abh.* **158**, 1–63.

Fütterer, D.K. (1974) Significance of the boring sponge *Cliona* for the origin of fine grained carbonate sediments. *J. sedim. Petrol.* **44**, 79–84.

Gaetani, M., Foise, E., Jadoul, F. & Nicora, A. (1981) Nature and evolution of Middle Triassic carbonate build-ups in the Dolomites (Italy). *Mar. Geol.* **44**, 25–57.

Gaffey, S.J. (1983) Formation and infilling of pits of marine ooid surfaces. *J. Sedim. Petrol.* **53**, 193–208.

Gaillard, C. & Jautée, E. (1987) The use of burrows to detect compaction and sliding in fine-grained sediments: an example from the Cretaceous of SE France. *Sedimentology* **34**, 585–593.

Gaines, A. (1977) Protodolomite redefined. *J. sedim. Petrol.* **47**, 543–546.

Gaines, A. (1978) Protodolomite redefined — Reply. *J. Sedim. Petrol.* **48**, 1009–1011.

Gaines, A. (1980) Dolomitization kinetics: recent experimental studies. In: *Concepts and Models of Dolomitization* (Ed. by D.H. Zenger, J.B. Dunham and R.L. Ethington) Spec. Publ. Soc. Econ. Paleont. Miner. 28, 139–161.

Gamboa, L.A., Truchan, M. & Stoffa, P.L. (1985) Middle and Upper Jurassic depositional environments at outer shelf and slope of Baltimore Canyon Trough. *Bull. Am. Ass. petrol. Geol.* **69**, 610–621.

Gardner, W.C. & Bray, E.E. (1984) Oils and source rocks of Niagaran Reefs (Silurian) in the Michigan Basin. In: *Petroleum Geochemistry and Source Rock Potential of Carbonate Rocks* (Ed. by J.G. Palacas) Am. Ass. petrol. Geol. Studies in Geology 18, 33–44.

Garrett, P. (1970) Phanerozoic stromatolites: non-competitive ecology restriction by grazing and burrowing animals. *Science* **169**, 171–173.

Garrett, P. & Hine, A.C. (1979) Probing Bermuda's lagoon and reefs (Abst.). *Bull. geol. Soc. Am.* **63**, 455.

Garrett, P., Smith, D.L., Wilson, A.O. & Patriquin, D. (1971) Physiography, ecology and sediments of two Bermuda patch reefs. *J. Geol.* **79**, 647–668.

Garrison, R.E. (1973) Space–time relations of pelagic limestones and volcanic rocks, Olympic Peninsula, Washington. *Bull. geol. Soc. Am.* **84**, 583–594.

Garrison, R.E. (1981) Diagenesis of oceanic carbonate sediments: a review of the DSDP perspective. In: *The Deep Sea Drilling Project: a Decade of Progress* (Ed. by J.E. Warme, R.G. Douglas and E.L. Winterer) Spec. Publ. Soc. econ. Paleont. Miner. 32, 181–207.

Garrison, R.E. & Fischer, A.G. (1969) Deep-water limestones and radiolarites of the Alpine Jurassic. In: *Depositional Environments in Carbonate Rocks* (Ed. by

G.M. Friedman) Spec. Publ. Soc. Econ. Paleont. Miner. 14, 20−56.

Garrison, R.E., Kastner, M. & Zenger, D.H. (Eds.) (1984) *Dolomites of the Monterey Formation and Other Organic-rich Units.* Publ. Soc. econ. Paleont. Miner. Pacific Section, 41, pp. 215.

Garrison, R.E. & Kennedy, W.J. (1977) Origin of solution seams and flaser structure in Upper Cretaceous chalks of southern England. *Sedim. Geol.* 19, 107−137.

Gartner, S. & Kearn, J. (1978) The terminal Cretaceous Event: a geologic problem with an oceanographic solution. *Geology* 6, 708−712.

Gasiewicz, A. (1984) Eccentric ooids. *Neues. Jb. Geol. Paläont. Mh.* 4, 204−211.

Gavish, E. (1974) Geochemistry and mineralogy of a recent sabkha along the coast of Sinai, Gulf of Suez. *Sedimentology* 21, 397−414.

Gavish, E. & Friedman, G.M. (1969) Progressive diagenesis in Quaternary to Late Tertiary carbonate sediments: sequence and time scale. *J. sedim. Petrol.* 39, 980−1006.

Gawthorpe, R.L. (1986) Sedimentation during carbonate ramp-to-slope evolution in a tectonically active area: Bowland Basin (Dinantian), N. England. *Sedimentology* 33, 185−206.

Gawthorpe, R.L. (1987) Burial dolomitization and porosity development in a mixed carbonate−clastic sequence: an example from the Bowland Basin, northern England. *Sedimentology* 34, 533−558.

Gebelein, C.D. (1969) Distribution, morphology and accretion rate of recent subtidal algal stromatolites, Bermuda. *J. sedim. Petrol.* 39, 49−69.

Gebelein, C.D. (1974a) *Guidebook for Modern Bahamian Platform Environments.* Geol. Soc. Am. Ann. Mtg. Fieldtrip Guide, pp. 93.

Gebelein, C.D. (1974b) Biologic control of stromatolite microstructure: implications for Precambrian time stratigraphy. *Am. J. Sci.* 274, 575−598.

Gebelein, C.D. (1975) Holocene sedimentation and stratigraphy southwest Andros Island, Bahamas. *9th Internat. Sediment. Congress, Nice.* Theme 5, T. I, 193−198.

Gebelein, C.D. (1976) Open marine subtidal and intertidal stromatolites (Florida, the Bahamas and Bermuda). In: *Stromatolites* (Ed. by M.R. Walter), p. 381−388. Elsevier, Amsterdam.

Gebelein, C.D. (1977a) *Dynamics of Recent Carbonate Sedimentation and Ecology: Cape Sable, Florida.* E.J. Brill Publ., Leiden.

Gebelein, C.D. (1977b) Mixing zone dolomitization of Holocene tidal flat sediments, south-west Andros Island, Bahamas (Abs.). *Bull. Am. Ass. petrol. Geol.* 61, 787−788.

Gebelein, C.D., Steinen, R.P., Garrett, P., Hoffman, E.J., Queen, J.M. & Plummer, L.N. (1980) Subsurface dolomitization beneath the tidal flats of Central West Andros Island, Bahamas. In: *Concepts and Models of Dolomitization* (Ed. by D.H. Zenger, J.B. Dunham and R.L. Ethington) Spec. Publ. Soc. econ. Paleont. Miner. 28, 31−49.

Geister, J. (1977) The influence of wave exposure on the ecological zonation of Caribbean coral reefs. *Proc. 3rd Int. Coral Reef Symp. Miami,* 1, 23−29.

Geno, K.R. & Chafetz, H.S. (1982) Petrology of Quaternary fluvial low-magnesian calcite coated grains from central Texas. *J. sedim. Petrol.* 52, 833−842.

Gidman, J. (1978) Protodolomite redefined − Discussion. *J. Sedim. Petrol.* 48, 1007−1008.

Gilbert, G.K. (1885) The topographic features of lakeshores. *Ann. Rept. US Geol. Surv.* 5, 75−123.

Gilbert, G.K. (1890) Lake Bonneville. *US Geol. Surv. Mongr.* 1, pp. 438.

Gill, D. (1977) Salina A-1 sabkha in the Late Silurian paleogeography of the Michigan Basin. *J. Sedim. Petrol.* 47, 979−1017.

Gill, W.D. & Ala, M.A. (1972) Sedimentology of Gachsaran Formation (Lower Fars Series), southwest Iran. *Bull. Am. Ass. petrol. Geol.* 56, 1965−1974.

Ginsburg, R.N. (1956) Environmental relationships of grain size and constituent particles in some south Florida carbonate sediments. *Bull. Am. Ass. petrol. Geol.* 40, 2384−2427.

Ginsburg, R.N. (1957) Early diagenesis and lithification of shallow water carbonate sediments in south Florida. In: *Regional Aspects of Carbonate Deposition* (Ed. by R.J. LeBlanc and J.C. Breeding) Spec. Publ. Soc. Econ. Paleont. Miner. 5, 80−99.

Ginsburg, R.N. (1964) *South Florida Carbonate Sediments.* Geol. Soc. Am. Ann. Mtg. Fieldtrip Guide, p. 72.

Ginsburg, R.N. (1975) *Tidal Deposits: A Casebook of Recent Examples and Fossil Counterparts,* p. 428. Springer-Verlag, New York.

Ginsburg, R.N. & Hardie, L.A. (1975) Tidal and storm deposits, northwestern Andros Island, Bahamas. In: *Tidal Deposits: A Casebook of Recent Examples and Fossil Counterparts* (Ed. by R.N. Ginsburg) pp. 201−208. Springer-Verlag, Berlin.

Ginsburg, R.N., Hardie, L.A., Bricker, O.P., Garrett, P. & Wanless, H.R. (1977) Exposure index: a quantitative approach to defining position within the tidal zone. In: *Sedimentation on the Modern Carbonate Tidal Flats of Northwest Andros Island, Bahamas* (Ed. by L.A. Hardie) pp. 7−11. Johns Hopkins University Press, Baltimore.

Ginsburg, R.N. & James, N.P. (1974) Holocene carbonate sediments of continental margins. In: *The Geology of Continental Margins* (Ed. by C.A. Burke and C.L. Drake) pp. 137−155. Springer-Verlag, New York.

Ginsburg, R.N. & James, N.P. (1976) Submarine botryoidal aragonite in Holocene reef limestones, Belize. *Geology* 4, 431−436.

Ginsburg, R.N. & Schroeder, J.H. (1973) Growth and submarine fossilization of algal cup reefs, Bermuda. *Sedimentology* 20, 575−614.

Ginsburg, R.N., Schroeder, J.H. & Shinn, E.A. (1971) Recent synsedimentary cementation in subtidal Bermuda reefs. In: *Carbonate Cements* (Ed. by O.P. Bricker) Johns Hopkins University Studies in Geology 19, 54−58.

Given, R.K. & Lohmann, K.C. (1985) Derivation of the original isotopic composition of Permian marine cements. *J. sedim. Petrol.* **55**, 430–439.

Given, R.K. & Lohmann, K.C. (1986) Isotopic evidence for the early meteoric diagenesis of the reef facies, Permian Reef Complex of West Texas and New Mexico. *J. sedim. Petrol.* **56**, 183–193.

Given, R.K. & Wilkinson, B.H. (1985) Kinetic control of morphology, composition and mineralogy of abiotic sedimentary carbonates. *J. sedim. Petrol.* **55**, 109–119.

Given, R.K. & Wilkinson, B.H. (1987) Dolomite abundance and stratigraphic age: constraints on rates and mechanisms of Phanerozoic dolostone formation. *J. sedim. Petrol.* **57**, 1068–1078.

Glass, S.W. & Wilkinson, B.H. (1980) The Peterson Limestone — early Cretaceous lacustrine carbonate deposition in western Wyoming and southeastern Idaho. *Sedim. Geol.* **27**, 143–160.

Gleason, P.J. (1972) *The origin, sedimentation and stratigraphy of a calcitic mud located in the southern freshwater Everglades.* Unpubl. PhD Thesis, Pennsylvania State University.

Gleason, P.J. & Spackman, W. (1974) Calcareous periphyton and water chemistry in the Everglades. In: *Environments of South Florida, Present and Past* (Ed. by P.J. Gleason) Mem. Miami Geol. Soc. 2, 146–181.

Goldhammer, R.K. (1988) Superimposed platform carbonate cycles: eustatic response of an aggradational carbonate buildup, Middle Triassic of the Dolomites (Abst.). *Bull. Am. Ass. petrol. Geol.* **73**, 190.

Goldhammer, R.K., Dunn, P.A. & Hardie, L.A. (1987) High frequency glacio-eustatic sea-level oscillations with Milankovitch characteristics recorded in Middle Triassic cyclic platform carbonates, northern Italy. *Am. J. Sci.* **287**, 853–892.

Goldsmith, I.R. & King, P. (1987) Hydrodynamic modelling of cementation patterns in modern reefs. In: *Diagenesis of Sedimentary Sequences* (Ed. by J.D. Marshall) Spec. Publ. geol. Soc. Lond. 36, 1–13.

Goldsmith, J.R. (1983) Phase relations of rhombohedral carbonates. In: *Carbonates: Mineralogy and Chemistry* (Ed. by R.J. Reeder) Reviews in Mineralogy, **11**, 49–96. Mineralogical Society of America, Washington.

Goldsmith, J.R. & Graf, D.L. (1958a) Relations between lattice constants and compositions of the Ca–Mg carbonates. *Am. Miner.* **43**, 84–101.

Goldsmith, J.R. & Graf, D.L. (1958b) Structural and compositional variations in some natural dolomites. *J. Geol.* **66**, 678–693.

Goldsmith, J.R. & Heard, H.C. (1961a) Subsolidus phase relations in the system $CaCO_3$–$MgCO_3$. *J. Geol.* **69**, 45–74.

Goldsmith, J.R. & Heard, H.C. (1961b) Lattice constants of the calcium–magnesium carbonates. *Am. Miner.* **46**, 453–457.

Golubic, S. & Fischer, A.G. (1975) Ecology of calcareous nodules forming in Little Connestaga creek near Lancaster, Pennsylvania. *Verh. Intern. Verein. Limnol.* **19**,
2315–2323.

Golubic, S., Perkins, R.D. & Lukas, K.J. (1975) Boring microorganisms and microborings in carbonate substrates. In: *Trace Fossils* (Ed. by R.W. Frey) pp. 229–259. Springer-Verlag, New York.

Gonfiantini, R. (1986) Environmental isotopes in lake studies. In: *Handbook of Environmental Isotope Geochemistry, 2. The Terrestrial Environment, B*, pp. 113–168. Elsevier, Amsterdam.

Gonzalez, L.A. & Lohmann, K.C. (1985) Carbon and oxygen isotopic composition in Holocene reefal carbonates. *Geology* **13**, 811–814.

Goodell, H.G. & Garman, R.K. (1969) Carbonate geochemistry of Superior deep testwell, Andros Island, Bahamas. *Bull. Am. Ass. petrol. Geol.* **53**, 513–536.

Goodwin, P.W. & Anderson, E.J. (1985) Punctuated aggradational cycles: a general hypothesis of episodic stratigraphic accumulation. *J. Geol.* **93**, 515–533.

Goreau, T.F. & Hartman, W.D. (1963) Boring sponges as controlling factors in the formation and maintenance of coral reefs. In: *Mechanisms of Hard Tissue Destruction* (Ed. by R.F. Sognnaes) Am. Ass. Adv. Sci. Publ. 75, 24–54.

Goreau, T.F. & Land, L.S. (1974) Fore-reef morphology and depositional processes, North Jamaica. In: *Reefs in Time and Space* (Ed. by L.F. Laporte) Spec. Publ. Soc. Econ. Paleont. Miner. 18, 77–89.

Gorin, G.E., Racz L.G. & Walter, M.R. (1982) Late Precambrian–Cambrian sediments of the Huqf Group, Sultanate of Oman. *Bull. Am. Ass. petrol. Geol.* **66**, 2609–2627.

Gostin, V.A., Belperio, A.P. & Cann, J.H. (1988) The Holocene non-tropical coastal and shelf carbonate province of southern Australia. Rept. 88/17 Dept of Mines & Energy, South Australia.

Gostin, V.A., Hails J.R. & Belperio, A.P. (1984) The sedimentary framework of northern Spencer Gulf, South Australia. *Mar Geol.* **61**, 111–138.

Goudie, A.S. (1983) Calcrete. In: *Chemical Sediments and Geomorphology: Precipitates and Residua in the Near Surface Environment* (Ed. by A.S. Goudie and K. Pye) pp. 93–131. Academic Press, London.

Gould, S.J. (1980) The promise of paleobiology as a nomothetic, evolutionary discipline. *Paleobiology* **6**, 96–118.

Gow, C.E. (1981) Unusual occurrence of biogenic (cyanobacterial) carbonate sediments from a sinkhole lake in the western Transvaal, South Africa. *S. Afr. J. Sci.* **77**, 564–565.

Graham, D.W., Bender, M.L., Williams, D.F. & Keighwin, L.D. (1981) Strontium to calcium ratio in Cenozoic planktonic foraminifera. *Geochim. Cosmochim. Acta* **46**, 1281–1292.

Graus, R.R., Chamberlain, J.A. & Boker, A.M. (1977) Structural modification of corals in relation to waves and currents. In: *Reefs and Related Carbonates — Ecology and Sedimentology* (Ed. by S.H. Frost, M.P. Weiss and J.B. Saunders) Am. Ass. petrol. Geol. Studies in Geol. 4, 135–153.

Graus, R.R. & Macintyre, I.G. (1982) Variation in growth forms of the reef coral *Montastrea annularis* (Ellis & Sollander): a quantitative evaluation of growth response to light distribution using computer simulation. In: *The Atlantic Barrier Reef Ecosystem at Carrie Bow Cay, Belize, I — Structure and Communities* (Ed. by K. Rützler and I.G. Macintyre) Smithsonian Contrib. to the Marine Sciences, 12, 441−464.

Graus, R.R., Macintyre, I.G. & Herghenroder, B.E. (1984) Computer simulation of reef zonation at Discovery Bay, Jamaica — Hurricane disruption and long-term physical oceanographic controls. *Coral Reefs*, 3, 59−68.

Gray, D.I. (1981) *Lower Carboniferous Shelf Carbonate Palaeoenvironments in North Wales.* Unpubl. PhD Thesis, University of Newcastle Upon Tyne.

Grayson, R.F. & Oldham, L. (1987) A new structural framework for the northern British Dinantian as a basis for oil, gas and mineral exploration. In: *European Dinantian Environments* (Ed. by J. Miller, A.E. Adams and V.P. Wright) pp. 33−60. John Wiley, Chichester.

Gregg, J.M. (1985) Regional epigenetic dolomitization in the Bonneterre Dolomite (Cambrian), southeastern Missouri. *Geology* 13, 503−506.

Gregg, J.M. & Sibley, D.F. (1984) Epigenetic dolomitization and the origin of xenotopic dolomite texture. *J. Sedim. Petrol.* 54, 908−931.

Griffith, L.S., Pitcher, M.G. & Rice, G.W. (1969) Quantitative environmental analysis of a Lower Cretaceous reef complex. In: *Depositional Environments in Carbonate Rocks* (Ed. by G.M. Friedman) Spec. Publ. Soc. Econ. Paleont. Miner. 14, 120−138.

Gross, M.G. (1964) Variations in the $^{18}O/^{16}O$ and $^{13}C/^{12}C$ ratios of diagenetically altered limestones in the Bermuda Islands. *J. Geol.* 72, 120−194.

Grossman, E.L. & Ku, Teh-Lung (1986) Oxygen and carbon isotope fractionation in biogenic aragonite: temperature effects. *Chem. Geol.* 59, 59−74.

Grotzinger, J.P. (1986a) Cyclicity and paleoenvironmental dynamics, Rocknest platform, northwest Canada. *Bull. Geol. Soc. Am.* 97, 1208−1231.

Grotzinger, J.P. (1986b) Upward shallowing platform cycles: a response to 2.2 billion years of low-amplitude, high frequency (Milankovitch band) sea-level oscillations. *Paleoceanography* 1, 403−416.

Grotzinger, J.P. & Read, J.F. (1983) Evidence for primary aragonite precipitation, Lower Proterozoic (1.9 Ga) Rocknest Dolomite, Northwest Canada. *Geology* 11, 710−713.

Grover, G. & Read, J.F. (1978) Fenestral and associated vadose diagenetic fabrics of tidal flat carbonates. Middle Ordovician, New Market Limestone, southwestern Virginia. *J. sedim. Petrol.* 48, 453−473.

Grover, G. & Read, J.F. (1983) Paleoaquifer and deep burial related cement defined by regional cathodoluminescent patterns, Middle Ordovician carbonates, Virginia. *Bull. Am. Ass. petrol. Geol.* 67, 1275−1303.

Gruszczynski, M. (1986) Hardgrounds and ecological succession in the light of early diagenesis (Jurassic, Holy Cross Mts, Poland). *Palaeontologica* 31, 163−212.

Guilcher, A. (1988) *Coral Reef Morphology*, pp. 248. John Wiley, Chichester.

Gunatilaka, A., Saleh, A., Al-Temeemi, A. & Nasser, N. (1984) Occurrence of subtidal dolomite in a hypersaline lagoon, Kuwait. *Nature* 311, 450−452.

Gutteridge, P. (1985) Grain size measurement from acetate peels. *J. sedim. Petrol.* 55, 595−596.

Guzzetta, G. (1984) Kinematics of stylolite formation and physics of the pressure-solution process. *Tectonophys.* 101, 383−394.

Gygi, R.A. (1975) *Sparisoma viride* (Bonnaterre), the stoplight parrotfish: a major sediment producer on coral reefs of Bermuda. *Eclogae Geol. Helv.* 68, 327−359.

Haas, J. (1982) Facies analysis of the cyclic Dachstein Limestone Formation (Upper Triassic) in the Bakony Mountains, Hungary. *Facies* 6, 75−84.

Haas, J. & Dobosi, K. (1982) Felsö — Triász ciklusos karbonatós közetek vizsgálata bakonyi alapszelvényenken. *M. Áll. Földt. Int. Evi. Jel. 1980 − Ról.*, 135−168.

Hagan, G.M. & Logan, B.W. (1974a) Development of carbonate banks and hypersaline basins, Shark Bay, Western Australia. *Mem. Am. Ass. petrol. Geol.* 22, 61−139.

Hagan, G.M. & Logan, B.W. (1974b) History of Hutchinson Embayment Tidal Flat, Shark Bay, Western Australia. *Mem. Am. Ass. Petrol. Geol.* 22, 283−355.

Håkanson, L. & Jansson, M. (1983) *Principles of Lake Sedimentology*, pp. 316. Springer-Verlag, Berlin.

Hallam, A. (1977) Secular changes in marine inundation of USSR and North America through the Phanerozoic. *Nature* 269, 762−772.

Hallam, A. (1986) Origin of minor limestone−shale cycles: climatically-induced or diagenetic? *Geology* 14, 609−612.

Hallam, A. (1987) End-Cretaceous mass extinction: argument for terrestrial causation. *Science* 238, 1237−1242.

Halley, R.B. (1976) Textural variation with Great Salt Lake algal mounds. In: *Stromatolites* (Ed. by M.R. Walter) pp. 435−445. Elsevier, Amsterdam.

Halley, R.B. (1977) Ooid fabric and fractures in the Great Salt Lake and the geologic record. *J. sedim. Petrol.* 47, 1099−1120.

Halley, R.B. (1978) Estimating pore and cement volumes in thin section. *J. sedim. Petrol.* 48, 642−650.

Halley, R.B. (1987) Burial diagenesis of carbonate rocks. *Quart. J. Colorado Sch. Mines 82*, 1−15.

Halley, R.B. & Beach, D.K. (1979) Porosity preservation and early freshwater diagenesis of marine sands (Abs.). *Bull. Am. Ass. petrol. Geol.* 63, 460.

Halley, R.B. & Evans, C.C. (1983) The Miami Limestone: a guide to selected outcrops and their interpretation. *Miami geol. Soc.* p. 67.

Halley, R.B. & Harris, P.M. (1979) Freshwater cementation of a 1000 year old oolite. *J. Sedim. Petrol.* 49, 969−988.

Halley, R.B., Harris, P.M. & Hine, A.C. (1983) Bank margin. In: *Carbonate Depositional Environments* (Ed.

by P.A. Scholle, D.G. Bebout and C.H. Moore) Mem. Am. Ass. Petrol. Geol. 33, 463–506.

Halley, R.B. & Rose, P.R. (1977) Significance of freshwater limestones in marine carbonate successions of Pleistocene and Cretaceous age. In: *Cretaceous Carbonates of Texas and Mexico: Applications to Subsurface Exploration* (Ed. by D.G. Bebout and R.G. Loucks) Reports of Investigations 89, 206–215. Bureau of Economic Geology, University of Austin, Texas.

Halley, R.B. & Schmoker, J.W. (1983) High-porosity Cenozoic carbonate rocks of South Florida: progressive loss of porosity with depth. *Bull. Am. Ass. Petrol. Geol.* 67, 191–200.

Halley, R.B., Shinn, E.A., Hudson, J.H. & Lidz, B. (1977) Recent and relict topography of Boo Bee Patch Reef, Belize. *Proc. 3rd Int. Coral Reef Symp. Miami*, 2, 29–35.

Hallock, P. (1981) Production of carbonate sediments by selected large benthic foraminifera on two Pacific coral reefs. *J. sedim. Petrol.* 51, 467–474.

Hallock, P. & Schlager, W. (1986) Nutrient excess and the demise of coral reefs and carbonate platforms. *Palaios* 1, 389–398.

Hancock, J.M. & Scholle, P.A. (1975) Chalk of the North Sea. In: *Petroleum and the Continental Shelf of Northwest Europe*, Vol. 1. Geology (Ed. by A.W. Woodland) pp. 413–427. Applied Science Publications, London.

Handford, C.R. (1982) Sedimentology and evaporite genesis in a Holocene continental sabkha playa basin — Bristol Dry Lake, California. *Sedimentology* 29, 239–253.

Handford, C.R. (1988) Review of carbonate sand-belt deposition of ooid grainstone and application to Mississippian Reservoir, Damme Field, Southwestern Kansas. *Bull. Am. Ass. Petrol. Geol.* 72, 1184–1199.

Handford, C.R., Kendall, A.C., Prezbindowski, D.R., Dunham, J.B. & Logan, B.W. (1984) Salina-margin tepees, pisoliths, and aragonite cements, Lake MacLeod, Western Australia: their significance in interpreting ancient analogs. *Geology* 12, 523–527.

Hanley, J.H. (1976) Paleosynecology of non-marine Mollusca from the Green River and Wasatch Formations (Eocene) southwestern Wyoming and northwestern Colorado. In: *Structure and Classification of Paleocommunities* (Ed. by R.W. Scott and R.R. West) pp. 235–261. Hutchinson & Ross, Stroudsbourg.

Hanor, J.S. (1983) Fifty years of development of thought on the origin and evolution of subsurface sedimentary brines. In: *Revolution in the Earth Sciences* (Ed. by S.J. Boardman) pp. 99–111. Kendall/Hunt, Dubuque.

Hanshaw, B.B. & Back, W. (1980) Chemical mass wasting of the northern Yucatan Peninsula by groundwater dissolution. *Geology* 8, 222–224.

Hanshaw, B.B., Back, W. & Dieke, R.G. (1971) A geochemical hypothesis for dolomitization by groundwater. *Econ. Geol.* 66, 710–724.

Haq, B.U., Hardenbol, J. & Vail, P.R. (1987) Chronology of fluctuating sea-levels since the Triassic. *Science* 235, 1156–1167.

Harbour, J.L. & Mathis, R.L. (1984) Sedimentation, diagenesis and porosity evolution of carbonate sands in the Black Lake field of Central Louisiana. In: *Carbonate Sands — a Core Workshop* (Ed. by P.M. Harris) Soc. Econ. Paleont. Miner. Core Workshop 5, 306–333.

Hardie, L.A. (1977) Introduction. In: *Sedimentation on the Modern Carbonate Tidal Flats of Northwest Andros Island, Bahamas* (Ed. by L.A. Hardie) pp. 1–3. Johns Hopkins Studies in Geology, 22. Johns Hopkins University Press, Baltimore.

Hardie, L.A. (1986a) Ancient carbonate tidal-flat deposits. *Quart. J. Colorado Sch. Mines* 81, 37–57.

Hardie, L.A. (1986b) Stratigraphic models for carbonate tidal flat deposition. *Quart. J. Colorado Sch. Mines* 81, 59–74.

Hardie, L.A. (1987) Dolomitization: a critical view of some current views. *J. Sedim. Petrol.* 57, 166–183.

Hardie, L.A., Bosellini, A. & Goldhammer, R.K. (1986) Repeated subaerial exposure of subtidal carbonate platforms, Triassic, Northern Italy: evidence for high frequency sea-level oscillations on a 10^4 year scale. *Paleoceanography* 1, 447–457.

Hardie, L.A. & Garrett, P. (1977) General environmental setting. In: *Sedimentation on the Modern Carbonate Tidal Flats of Northwest Andros Island, Bahamas* (Ed. by L.A. Hardie) pp. 12–49. Johns Hopkins Studies in Geology, 22. Johns Hopkins University Press, Baltimore.

Hardie, L.A. & Ginsburg, R.M. (1977) Layering: The origin and environmental significance of lamination and thin bedding. In: *Sedimentation on the Modern Carbonate Tidal Flats of Northwest Andros Island, Bahamas* (Ed. by L.A. Hardie) pp. 50–123. Johns Hopkins Studies in Geology, 22. Johns Hopkins University Press, Baltimore.

Hardie, L.A., Smoot, J.P. & Eugster, H.P. (1978) Saline lakes and their deposits: a sedimentological approach. In: *Modern and Ancient Lake Sediments* (Ed. by A. Matter and M.E. Tucker) Spec. Publ. int. Ass. Sediment. 2, 7–41.

Hardman, R.F.P. (1982) Chalk reservoirs of the North Sea. *Bull. Geol. Soc. Denmark* 30, 119–137.

Hardy, R. & Tucker, M.E. (1988) X-ray diffraction. In: *Techniques in Sedimentology* (Ed. by M.E. Tucker) pp. 191–228. Blackwells, Oxford.

Harland, T.L. (1981) Middle Ordovician reefs of Norway. *Lethaia* 14, 169–188.

Harrell, J. (1981) Measurement errors in the thin section analysis of grain packing. *J. sedim. Petrol.* 51, 674–676.

Harris, P.M. (1979) *Facies Anatomy and Diagenesis of a Bahamian Ooid Shoal*. Sedimenta 7, Comparative Sed. Lab. Miami, p. 163.

Harris, P.M. (Ed.) (1984) *Carbonate Sands — a Core Workshop*. Soc. Econ. Paleont. Miner. Core Workshop, 5, pp. 463.

Harrison, R.S. (1975) Porosity in Pleistocene grainstones from Barbados: some preliminary observations. *Bull. Can. Petrol. Geol.* 23, 383–392.

Hart, M.B. (1987) Orbitally-induced cycles in the Chalk facies of the United Kingdom. *Cret. Res.* **8**, 335–348.

Harwood, G.M. & Moore, C.H. (1984) Comparative sedimentology and diagenesis of Upper Jurassic ooid grainstone sequences, East Texas Basin. In: *Carbonate Sands – a Core Workshop* (Ed. by P.M. Harris) Soc. econ. Paleont. Miner. Core Workshop, 5, 176–232.

Harwood, G.M. & Towers, P.A. (1988) Seismic sedimentologic interpretation of a carbonate slope, north margin of Little Bahama Bank. Init. Rpts, ODP 101, 263–277.

Havard, C. & Oldershaw, A. (1976) Early diagenesis in back-reef sedimentary cycles, Snipe Lake reef complex, Alberta. *Bull. Can. Petrol. Geol.* **24**, 27–69.

Hays, J.D., Imbrie, J. & Shackleton, N.J. (1976) Variations in the Earth's orbit: pacemaker of the ice ages. *Science* **194**, 1121–1132.

Heath, K.C. & Mullins, H.T. (1984) Open-ocean, off-bank transport of fine-grained carbonate sediments in northern Bahamas. In: *Fine-grained Sediments: Deepwater Processes and Facies* (Ed. by D.A.V. Stow and D.J.W. Piper) Spec. Publ. Geol. Soc. Lond. 15, 199–208.

Heckel, P.H. (1972) Possible inorganic origin for stromatactis in calcilutite mounds in the Tully Limestone, Devonian of New York. *J. sedim. Petrol.* **42**, 7–18.

Heckel, P.H. (1974) Carbonate buildups in the geologic record: a review. In: *Reefs in Time and Space* (Ed. by L.F. Laporte) Spec. Publ. Soc. econ. Paleont. Miner. 18, 90–154.

Heckel, P.H. (1983) Diagenetic model for carbonate rocks in Mid-continent Pennsylvanian Eustatic Cyclothems. *J. sedim. Petrol.* **53**, 733–759.

Heijnen, W.M. (1986) Crystal growth and morphology of calcium oxalates and carbonates. *Geol. Ultraictina* **42**, 167–175.

Hein, F.J. & Risk, M.J. (1975) Bioerosion of coral heads: inner patch reefs, Florida Reef Tract. *Bull. Mar. Sci.* **25**, 133–138.

Heller, P.L., Komar, P.D. & Pevear, D.R. (1980) Transport processes in ooid genesis. *J. Sedim. Petrol.* **50**, 948–952.

Henning, N.G., Meyers, W.J. & Grains J.C. (1989) Cathodoluminescence in diagenetic calcites: the roles of Fe and Mn as deduced from electron probe and spectrophotometric measurements. *J. sedim. Petrol.* **59**, 404–411.

Hennebert, M. & Lees, A. (1985) Optimized similarity matrices applied to the study of carbonate rocks. *Geol. J.* **20**, 123–131.

Henrich, R. & Zankl, H. (1986) Diagenesis of Upper Triassic Wetterstein Reefs of the Bavarian Alps. In: *Reef Diagenesis* (Ed. by J.H. Schroeder and B.H. Purser) pp. 245–268. Springer-Verlag, Berlin.

Hernandez-Avila, M.L., Roberts, H.H. & Rouse, L.J. (1977) Hurricane generated waves and coastal boulder-rampart formation. *Proc. 3rd Int. Coral Reef Symp. Miami* 2, 71–78.

Hess, J., Bender, M.L. & Schilling, J.G. (1986) Evolution of the ratio of strontium-87 to strontium-87 in seawater from Cretaceous to Present. *Science* **231**, 979–984.

Hesselbo, S.P. & Trewin, N.H. (1984) Deposition, diagenesis and structures of the Cheese Bay Shrimp Bed, Lower Carboniferous, East Lothian. *Scott. J. Geol.* **20**, 281–296.

Hillaire-Marcel, C. & Casanova, J. (1987) Isotopic hydrology and paleohydrology of the Magadi (Kenya)–Natron (Tanzania) Basin during the Late Quaternary. *Palaeogeog. Palaeoclimatol. Palaeoecol.* **58**, 155–181.

Hine, A.C. (1977) Lily Bank, Bahamas: history of an active oolite sand shoal. *J. sedim. Petrol.* **47**, 1554–1581.

Hine, A.C. & Neumann, A.C. (1977) Shallow carbonate bank margin growth and structure, Little Bahama Bank, Bahamas. *Bull. Am. Ass. petrol. Geol.* **61**, 376–406.

Hine, A.C., Wilber, R.J. & Neumann, A.C. (1981a) Carbonate sand-bodies along contrasting shallow-bank margins facing open seaways, northern Bahamas. *Bull. Am. Ass. Petrol. Geol.* **65**, 261–290.

Hine, A.C., Wilber, R.J., Bane, J.M., Neumann, A.C. & Lorenson, K.R. (1981b) Offbank transport of carbonate sands along leeward bank margins, northern Bahamas. *Marine Geol.* **42**, 327–348.

Hird, K. (1986) *Petrography and Geochemistry of some Carboniferous and Precambrian Dolomites.* PhD Thesis, University of Durham.

Hird, K. & Tucker, M.E. (1988) Contrasting diagenesis of two Carboniferous oolites from South Wales: a tale of climatic influence. *Sedimentology* **35**, 587–602.

Hird, K., Tucker, M.E. & Waters, R.A. (1987) Petrography, geochemistry and origin of Dinantian dolomites from south-east Wales. In: *European Dinantian Environments* (Ed. by J. Miller, A.E. Adams and V.P. Wright) pp. 359–377. John Wiley, Chichester.

Hiscott, R.N. & James, N.P. (1985) Carbonate debris flows, Cow Head Group, Western Newfoundland. *J. sedim. Petrol.* **55**, 735–745.

Hobson, J.P. Jr., Fowler, M.L. & Beaumont, E.A. (1982) Depositional and statistical exploration models, Upper Cretaceous offshore sandstone complex, Sussex Member, House Creek Field, Wyoming. *Bull Am. Ass. petrol. Geol.* **66**, 689–707.

Hoefs, J. (1988) *Stable Isotope Geochemistry*, pp. 208. Springer-Verlag, Berlin.

Hoffman, A. & Narkiewicz, M. (1977) Developmental pattern of Lower and Middle Palaeozoic banks and reefs. *Neues Jb. Geol. Paläont. Mh.* **5**, 272–283.

Hoffman, P. (1976) Environmental diversity of Middle Precambrian stromatolites. In: *Stromatolites* (Ed. by M.R. Walter) pp. 599–611. Elsevier, Amsterdam.

Hoffmeister, J.E., Stockman, K.W. & Multer, H.G. (1967) Miami Limestone of Florida and its recent Bahamian counterpart. *Bull. geol. Soc. Am.* **78**, 175–190.

Holland, H.D. (1978) *The Chemistry of the Atmosphere and Oceans*, pp. 351. Wiley-Interscience, New York.

Holland, H.D. (1984) *The Chemical Evolution of the Atmosphere and Oceans*, pp. 582. Princeton University Press, Princeton, N.J.

Holland, H.D., Holland, H.J. & Munoz, J.L. (1984b) The coprecipitation of cations with CaCO₃ II. The coprecipitation of Sr^{2+} with calcite between 90° and 100°C. *Geochim. Cosmochim. Acta* **28**, 1287–1301.

Holland, H.D., Kirsipu, T.V., Huebner, J.S. & Oxburgh, U.M. (1964a) On some aspects of the chemical evolution of cave water. *J. Geol.* **72**, 36–67.

Hook, J.E., Golubic, S. & Milliman, J.D. (1984) Micritic cement in microborings is not necessarily a shallow-water indicator. *J. sedim. Petrol.* **54**, 425–431.

Hooke, R.L. & Schlager, W. (1980) Geomorphic evolution of the Tongue of the Ocean and the Providence Channels, Bahamas. *Mar. Geol.* **35**, 343–366.

Hopkins, J.C. (1977) Production of foreslope breccia by differential submarine cementation and downslope displacement of carbonate sands, Miette and Ancient Wall Buildups, Devonian, Canada. In: *Deep-water Carbonate Environments* (Ed. by H.E. Cook and P. Enos) Spec. Publ. Soc. econ. Paleont. Miner. 25, 155–170.

Hopley, D. (1982) *The Geomorphology of the Great Barrier Reef*, pp. 453. John Wiley, New York.

Horodyski, R.J. & Vonder Haar, S.P. (1975) Recent calcareous stromatolites from Laguna Mormona (Baja California) Mexico. *J. sedim. Petrol.* **45**, 894–906.

Horowitz, A.S. & Potter, P.E. (1971) *Introductory Petrography of Fossils*, pp. 302. Springer-Verlag, Berlin.

Hovland, M., Talbot, M.R., Quale, H., Olaussen, S. & Aasberg, L. (1987) Methane-related carbonate cements in pockmarks of the North Sea. *J. sedim. Petrol.* **57**, 881–892.

Howson, M.R., Pethybridge, A.D. & House, W.A. (1987) Synthesis and distribution coefficients of low-magnesium calcites. *Chem. Geol.* **64**, 79–87.

Hsü, K.J. (1967) Chemistry of dolomite formation. In: *Carbonate Rocks* (Ed. by G.V. Chilingar, H.J. Bissel and R.W. Fairbridge) pp. 169–191. Elsevier, Amsterdam.

Hsü, K.J. (1976) Paleo-oceanography of the Mesozoic Alpine Tethys. *Spec. Pap. geol. Soc. Am.* **170**, 44.

Hsü, K.J. (1983) Actualistic catastrophism. *Sedimentology* **30**, 3–9.

Hsü, K.J. & Kelts, K. (1978) Late Neogene chemical sedimentation in the Black Sea. In: *Modern and Ancient Lake Sediments* (Ed. by A. Matter and M.E. Tucker) Spec. Publ. int. Ass. Sediment. 2, 129–145.

Hsü, K.J., McKenzie, J.A. & He, Q.X. (1982) Terminal Cretaceous environmental and evolutionary changes. *Spec. Pap. geol. Soc. Am.* **190**, 317–328.

Hsü, K.J., Oberhansli, H., Gao, J.Y., Sun Shu, Chen, H. & Krahenbuhl, U. (1985) 'Strangelove ocean' before the Cambrian explosion. *Nature* **316**, 809–811.

Hsü, K.J. & Schneider, J. (1973) Progress report on dolomitization and hydrology of Abu Dhabi sabkhas, Arabian Gulf. In: *The Persian Gulf* (Ed. by B.H. Purser) pp. 409–422. Springer-Verlag, Berlin.

Hsü, K.J. & Siegenthaler, C. (1969) Preliminary experiments on hydrodynamic movement induced by evaporation and their bearing on the dolomite problem. *Sedimentology*

12, 11–25.

Hubbard, D.K., Burke, R.B. & Gill, I.P. (1986) Styles of reef accretion along a steep, shelf-edge reef, St Croix, US Virgin Islands. *J. sedim. Petrol.* **56**, 848–861.

Hubbard, J.A.E.B. & Swart, P.K. (1982) Sequence and style in scleractinian coral preservation in reefs and associated facies. *Palaeogeog. Palaeoclim. Palaeoecol.* **37**, 165–219.

Hubbard, R.J. (1988) Age and significance of sequence boundaries on Jurassic and early Cretaceous rifted continental margins. *Bull. Am. Ass. petrol. Geol.* **72**, 49–72.

Hubert, J.F. (1966) Sedimentary history of Upper Ordovician geosynclinal rocks, Girvan, Scotland. *J. sedim. Petrol.* **36**, 677–699.

Hubert, J.K., Suchecki, R.K. & Callahan, R.K.M. (1977) The Cowhead Breccia: Sedimentology of the Cambro-Ordovician continental margin, Newfoundland. In: *Deep-water Carbonate Environments* (Ed. by H.E. Cook and P. Enos) Spec. Publ. Soc. Econ. Paleont. Miner. 25, 125–154.

Hudson, J.D. (1962) Pseudo-pleochroic calcite in recrystallized shell limestones. *Geol. Mag.* **99**, 492–500.

Hudson, J.H. (1977) Long term bioerosion rates on a Florida reef: a new method. *Proc. 3rd Coral Reef Symp. Miami* 2, 491–497.

Humphrey, J.D. (1988) Late Pleistocene mixing-zone dolomitization, southeastern Barbados, West Indies. *Sedimentology* **35**, 327–348.

Humphrey, J.D., Ransom, K.L. & Matthews, R.K. (1985) Early meteoric diagenetic control of Upper Smackover production, Oaks Field, Louisiana. *Bull. Am. Ass. petrol. Geol.* **70**, 70–85.

Hutchings, P.A. (1986) Biological destruction of coral reefs: a review. *Coral Reefs* **5**, 239–252.

Hutchings, P.A. & Bamber, L. (1985) Variability of bioerosion rates at Lizard Island, Great Barrier Reef: preliminary attempts to explain these rates and their significance. *Proc. 5th Int. Coral Reef Congr. Tahiti* 2, 333–338.

Hutchinson, G.E. (1957) *A Treatise on Limnology, Vol 1: Geography, Physics and Chemistry*, pp. 1015. Wiley, New York.

Ichikuni, M. (1973) Partition of strontium between calcite and solution: effect of substitution by manganese. *Chem. Geol.* **11**, 315–319.

Ijirigho, B.T. & Schreiber, J.F. (1988) Composite classification of fractured and brecciated carbonate rocks — examples from the Ordovician Ellenburger Group, West Texas. *J. petrol. Geol.* **11**, 193–204.

Illing, L.V. (1954) Bahamian calcareous sands. *Bull. Am. Ass. petrol. Geol.* **38**, 1–95.

Illing, L.V., Wells, A.J. & Taylor, J.C.M. (1965) Penecontemporaneous dolomite in the Persian Gulf. In: *Dolomitization and Limestone Diagenesis* (Ed. by R.C. Murray and L.C. Pray) Spec. Publ. Soc. econ. Paleont. Miner. 13, 89–111.

Imbrie, J. & Imbrie, J.Z. (1980) Modelling the climatic

response to orbital variations. *Science* **207**, 943–953.

Inden, R.F. & Moore, C.H. (1983) Beach. In: *Carbonate Depositional Environments* (Ed. by P.A. Scholle, D.G. Bebout and C.H. Moore) Mem. Am. Ass. Petrol. Geol. 33, 211–265.

Irwin, H., Curtis, C.D. & Coleman, M. (1977) Isotopic evidence for source of diagenetic carbonates formed during burial of organic-rich sediments. *Nature* **269**, 209–213.

Irwin, M.L. (1965) General theory of epeiric clear water sedimentation. *Bull Am. Ass petrol. Geol.* **49**, 445–459.

Jackson, J.B.C. & Winston, J.E. (1982) Ecology of cryptic coral communities, 1. Distribution and abundance of major groups of encrusting organisms. *J. Expt. Mar. Biol. Ecol.* **57**, 135–147.

Jacobson, R.L. & Usdowski, H.E. (1976) Partitioning of strontium between calcite, dolomite and liquids: an experimental study under higher temperature diagenetic conditions and the model for the prediction of mineral pairs for geothermometry. *Contrib. Min. Pet.* **59**, 171–185.

James, N.P. (1983) Reefs. In: *Carbonate Depositional Environments* (Ed. by P.A. Scholle, D.G. Bebout and C.H. Moore) Mem. Am. Ass. petrol. Geol. 33, 345–462.

James, N.P. (1984a) Shallowing-upward sequences in carbonates. In: *Facies Models* (Ed. by R.G. Walker) pp. 213–228. Geoscience Canada.

James, N.P. (1984b) Reefs. In: *Facies Models* (Ed. by R.G. Walker) pp. 229–244. Geoscience Canada.

James, N.P. & Choquette, P.W. (1983) Diagenesis 6. Limestones — the sea floor diagenetic environment. *Geoscience Canada* **10**, 162–179.

James, N.P. & Choquette, P.W. (1984) Diagenesis 9. Limestones — the meteoric diagenetic environment. *Geoscience Canada* **11**, 161–194.

James, N.P. & Choquette, P.W. (1988) *Paleokarsts*, pp. 416. Springer-Verlag, New York.

James, N.P., Coniglio, M., Aissaoui, D.M. & Purser, B.H. (1988) Facies and geologic history of an exposed Miocene rift-margin carbonate platform: Gulf of Suez, Egypt. *Bull. Am. Ass. petrol. Geol.* **72**, 555–572.

James, N.P., Geldsetzer, H.H. & Tebbutt, G.E. (1985) Canadian reef inventory: case histories of Proterozoic and Phanerozoic buildups of Canada and adjacent areas: conference announcement. *Int. Assoc. Sedimentol. Newsletter* **82**, Dec. 1985, 6–8.

James, N.P. & Ginsburg, R.N. (1979) The seaward margin of Belize barrier and atoll reefs. *Spec. Publ. int. Assoc. Sedimentol.* **3**, 191.

James, N.P., Ginsburg, R.N., Marszalek, D.S. & Choquette, P.W. (1976) Facies and fabric specificity of early subsea cements in shallow Belize (British Honduras) reefs. *J. sedim. Petrol.* **46**, 523–544.

James, N.P. & Klappa, C.F. (1983) Petrogenesis of early Cambrian reef limestones, Labrador, Canada. *J. sedim. Petrol.* **53**, 1051–1096.

James, N.P. & Kobluk, D.R. (1978) Lower Cambrian patch reefs and associated sediments, southern Labrador. *Sedimentology* **25**, 1–35.

James, N.P. & Macintyre, I.G. (1985) Carbonate depositional environments, modern and ancient. Part 1 — Reefs, zonation, depositional facies and diagenesis. *Quart. J. Colorado Sch. Mines* **80**, 70.

James, N.P. & Mountjoy, E.W. (1983) Shelf-slope break in fossil carbonate platforms: an overview. In: *The Shelf-break: Critical Interface on Continental Margins* (Ed. by D.J. Stanley and G.T. Moore) Spec. Publ. Soc. Econ. Paleont. Miner. 33, 189–206.

James, N.P., Wray, J.L. & Ginsburg, R.N. (1988) Calcification of encrusting aragonitic algae (Peyssonneliaceae): implications for the origin of Late Paleozoic reefs and cements. *J. sedim Petrol.* **53**, 291–303.

Jansa, L.F. (1981) Carbonate platforms and banks of the eastern North American margin. *Mar. Geol.* **44**, 97–117.

Jansen, J.H.F., Woensdregt, C.F., Kooistra, M.J. & van der Gaast, S.J. (1987) Ikaite pseudomorphs in the Zaire deep-sea fan: an intermediate between calcite and porous calcite. *Geology* **15**, 245–248.

Jefferson, D.P. (1980) Cyclic sedimentation in the Holkerian (middle Visean) north of Settle, Yorkshire. *Proc. Yorks. Geol. Soc.* **42**, 483–503.

Jenkyns, H.C. (1970) Fossil manganese nodules from the west Sicilian Jurassic. *Eclog. Geol. Helv.* **63**, 741–774.

Jenkyns, H.C. (1971) The genesis of condensed sequences in the Tethyan Jurassic. *Lethaia* **4**, 327–352.

Jenkyns, H.C. (1972) Pelagic 'oolites' from the Tethyan Jurassic. *J. Geol.* **80**, 21–33.

Jenkyns, H.C. (1974) Origin of red nodular limestones (Ammonitico Rosso, Knollenkalke) in the Mediterranean Jurassic: a diagenetic model. In: *Pelagic Sediments: On Land and Under the Sea* (Ed. by K.J. Hsü and H.C. Jenkyns) Spec. Publ. Int. Ass. Sediment. 1, 249–271.

Jenkyns, H.C. (1977) Fossil nodules. In: *Marine Manganese Deposits* (Ed. by G.P. Glasby), pp. 85–108. Elsevier, Amsterdam.

Jenkyns, H.C. (1980a) Tethys: past and present. *Proc. geol. Ass.* **91**, 107–118.

Jenkyns, H.C. (1980b) Cretaceous anoxic events: from continents to oceans. *J. Geol. Soc.* **137**, 171–188.

Jenkyns, H.C. (1986) Pelagic environments. In: *Sedimentary Environments and Facies* (Ed. by H.G. Reading) pp. 343–397. Blackwells, Oxford.

Jenkyns, H.C. & Clayton, C.J. (1986) Black shales and carbon isotopes in pelagic sediments from the Tethyan Lower Jurassic. *Sedimentology* **33**, 87–105.

Jennings, J.N. (1968) Syngenetic karst in Australia. In: *Contributions to the Study of Karst* (Ed. by P.W. Williams and J.N. Jennings) Nat. Univ. Dept. Geogr. Publ. G/5, pp. 41–110. Canberra Univ. Press, Canberra.

Jindrich, V. (1969) Recent carbonate sedimentation by tidal channels in the Lower Florida Keys. *J. Sedim. Petrol.* **39**, 531–553.

Johnson, D.P., Cuff, C. & Rhodes, E. (1984) Holocene reef sequences and geochemistry, Britomart Reef, central

Great Barrier Reef, Australia. *Sedimentology* **31**, 515–529.

Johnson, D.P. & Searle, D.E. (1984) Post-glacial seismic stratigraphy, central Great Barrier Reef, Australia. *Sedimentology* **31**, 335–352.

Johnson, R.H. (1983) The saltwater–freshwater interface in the Tertiary limestone aquifer, southeastern Atlantic continental shelf of the USA. *J. Hydrol.* **61**, 239–251.

Jones, B. (1987) The alterations of sparry calcite crystals in a vadose setting, Grand Cayman Islands. *Can. J. earth Sci.* **24**, 2292–2304.

Jones, B. & Dixon, O.A. (1976) Storm deposits in the Read Bay Formation (Upper Silurian), Somerset Island, Arctic Canada (an application of Markov chain analysis). *J. sedim. Petrol.* **46**, 393–401.

Jones, B. & Goodbody, Q.H. (1985) Oncolites from a shallow lagoon, Grand Cayman Island. *Bull. Can. petrol. Geol.* **32**, 254–260.

Jones, B., Lockhart, E.B. & Squair, C. (1984) Phreatic and vadose cements in the Tertiary Bluff Formation of Grand Cayman Island, British West Indies. *Bull. Can. petrol. Geol.* **32**, 382–397.

Jones, B. & Pemberton, S.G. (1988) *Lithophaga* borings and their influence on the diagenesis of corals in the Pleistocene Ironshore Formation of Grand Cayman Island, British West Indies. *Palaios* **3**, 3–21.

Jones, B.F. & Bowser, C.J. (1978) The mineralogy and related chemistry of lake sediments. In: *Lakes: Chemistry, Geology and Physics* (Ed. by A. Lerman) pp. 179–235. Springer-Verlag, Berlin.

Jones, F.G. & Wilkinson, B.H. (1978) Structure and growth of lacustrine pisoliths from recent Michigan marl lakes. *J. sedim. Petrol.* **48**, 1103–1110.

Jorgensen, N.O. (1987) Oxygen and carbon isotope compositions of Upper Cretaceous chalk from the Danish sub-basin and the North Sea Central Graben. *Sedimentology* **34**, 559–570.

Kahle, C.F. (1974) Ooids from Great Salt Lake, Utah, as an analogue for the genesis and diagenesis of ooids in marine limestones. *J. sedim. Petrol.* **44**, 30–39.

Kahle, C.F. (1981) Origin of ooids in Pleistocene Miami Limestone, Florida Keys (abst.). *Bull. Am. Ass. petrol. Geol.* **65**, 943.

Kaldi, J. (1986) Sedimentology of sand waves in an oolite shoal complex in the Cadeby Magnesian Limestone Formation (Upper Permian) of eastern England. In: *The English Zechstein and Related Topics* (Ed. by G.M. Harwood and D.B. Smith) Spec. Publ. geol. Soc. Lond. **22**, 63–74.

Kaldi, J. & Gidman, J. (1982) Early diagenetic dolomite cements: examples from the Permian Lower Magnesian Limestone of England and the Pleistocene carbonates of the Bahamas. *J. sedim. Petrol.* **52**, 1073–1085.

Karhu, J. & Epstein, S. (1986) The implication of the oxygen isotope records in coexisting cherts and phosphates. *Geochim. Cosmochim. Acta* **50**, 1745–1756.

Karner, G.D. (1986) Effects of lithospheric in-plane stress on sedimentary basin stratigraphy. *Tectonics* **5**, 573–588.

Kastner, M. (1984) Control of dolomite formation. *Nature* **311**, 410–411.

Katz, A. (1973) The interaction of magnesium with calcite during crystal growth at 25–90°C and one atmosphere. *Geochim. Cosmochim. Acta* **37**, 1563–1586.

Katz, A. & Matthews, A. (1977) The dolomitization of $CaCO_3$: an experimental study at 252–295°C. *Geochim. Cosmochim. Acta* **41**, 297–308.

Katz, A., Sass, E., Starinsky, A. & Holland, H.D. (1972) Strontium behaviour in the aragonite–calcite transformation: an experimental study at 40–98°C. *Geochim. Cosmochim. Acta* **36**, 481–496.

Kaufman, J., Cander, H.S., Daniels, L.D. & Meyers, W.J. (1988) Calcite cement stratigraphy and cementation history of the Burlington–Keokuk Formation (Mississippian), Illinois and Missouri. *J. sedim. Petrol.* **58**, 312–326.

Kazmierczak, J., Ittekkot, V. & Degens, E.T. (1985) Biocalcification through time: environmental challenge and cellular response. *Paläont. Z.* **59**, 15–33.

Keith, B.D. & Friedman, G.M. (1977) A slope–fan–basin–plain model, Taconic sequence, New York and Vermont. *J. sedim. Petrol.* **47**, 1220–1241.

Kelts, K. (1982) Lacustrine environments of deposition for laminites with hydrocarbon source potential. *3rd IAS European Meeting, Copenhagan Abst.* 20–21.

Kelts, K. & Arthur, M.A. (1981) Turbidites after ten years of deep-sea drilling — wringing out the mop? In: *The Deep Sea Drilling Project: A Decade of Progress* (Ed. by J.E. Warme, R.G. Douglas and E.L. Winterer) Spec. Publ. Soc. econ. Paleont. Miner. **32**, 91–127.

Kelts, K. & Hsü, K.J. (1978) Freshwater carbonate sedimentation. In: *Lakes: Chemistry, Geology and Physics* (Ed. by A. Lerman) pp. 295–353. Springer-Verlag, Berlin.

Kelts, K. & McKenzie, J. (1982) Diagenetic dolomite formation in Quaternary anoxic diatomaceous muds of Deep Sea Drilling Project Leg 64, Gulf of California. *Init. Repts. DSDP* **64**, 553–570.

Kelts, K. & McKenzie, J. (1984) A comparison of anoxic dolomite from deep-sea sediments: Quaternary Gulf of California and Messinian Tripoli Formation of Sicily. In: *Dolomites of the Monterey Formation and Other Organic-rich Units* (Ed. by R.E. Garrison, M. Kastner and D.H. Zenger) Publ. Soc. econ. Paleont. Miner. Pacific Section **41**, 119–140.

Kendall, A.C. (1975) Post-compactional calcitization of molluscan aragonite in a Jurassic limestone from Saskatchewan Canada. *J. sedim. Petrol.* **45**, 399–404.

Kendall, A.C. (1979) Continental and supratidal (sabkha) evaporites. In: *Facies Models* (Ed. by R.G. Walker) pp. 145–157. Geoscience Canada.

Kendall, A.C. (1985) Radiaxial fibrous calcite: a reappraisal. In: *Carbonate Cements* (Ed. by N. Schneidermann and P.M. Harris) Spec. Publ. Soc. econ. Paleont. Miner. **36**, 59–77.

Kendall, A.C. & Tucker, M.E. (1973) Radiaxial fibrous calcite: a replacement after acicular carbonate. *Sedimentology* **20**, 365–389.

Kendall, C.G.St.C. & Schlager, W. (1981) Carbonates and relative changes in sea-level. *Mar. Geol.* **44**, 181–212.

Kendall, C.G.St.C. & Skipwith, P.A.d'E. (1968) Recent algal mats of a Persian Gulf lagoon. *J. sedim. Petrol.* **38**, 1040–1058.

Kendall, C.G.St.C. & Skipwith, P.A.d'E. (1969) Holocene shallow-water carbonate and evaporite sediments of Khor al Bazam, Abu Dhabi, southwest Persian Gulf. *Bull. Am. Ass. petrol. Geol.* **53**, 841–869.

Kendall, C.G.St.C. & Warren, J. (1987) A review of the origin and setting of tepees and their associated fabrics. *Sedimentology*, **34**, 1007–1027.

Kennedy, W.J. (1987) Late Cretaceous and early Palaeocene Chalk Group sedimentation in the Greater Ekofisk Area, North Sea Graben. *Bull. Centres Rech. Explor. Prod. Elf-Aquitaine* **11**, 91–126.

Kennedy, W.J. & Garrison, R.E. (1975) Morphology and genesis of nodular chalks and hardgrounds in the Upper Cretaceous of southern England. *Sedimentology* **22**, 311–386.

Kepper, J.C. (1981) Sedimentology of a Middle Cambrian outer shelf margin with evidence for syndepositional faulting, Eastern California and Western Nevada. *J. sedim. Petrol.* **51**, 807–822.

Kerans, C. (1985) Petrology of Devonian and Carboniferous carbonates of the Canning and Bonaparte Basins. West. Austr. Miner. Petrol. Res. Inst. Rept. 12.

Kerans, C. (1988) Karst-controlled reservoir heterogeneity in Ellenburger Group carbonates, West Texas. *Bull. Am. Ass. petrol. Geol.* **72**, 1160–1183.

Kerans, C., Hurley, N.F. & Playford, P.E. (1986) Marine diagenesis in Devonian reef complexes of the Canning Basin, Western Australia. In: *Reef Diagenesis* (Ed. by J.H. Schroeder and B.H. Purser) pp. 357–380. Springer-Verlag, Berlin.

Kharaka, Y.K., Carothers, W.W. & Rosenbauer, R.J. (1983) Thermal decarboxylation of acetic acid: implications for origin of natural gas. *Geochim. Cosmochim. Acta* **47**, 397–402.

Kier, J.S. & Pilkey, O.H. (1971) The influence of sea-level changes on sediment carbonate mineralogy, Tongue of the Ocean, Bahamas. *Mar. Geol.* **11**, 189–200.

Kinsman, D.J.J. (1964) The recent carbonate sediments near Halat al Bahrani, Trucial Coast, Persian Gulf. In: *Deltaic and Shallow Marine Deposits* (Ed. by L.M.J.U. van Straaten) Developments in Sedimentology, 1, 189–192.

Kinsman, D.J.J. & Holland, H.D. (1969) The co-precipitation of cations with $CaCO_3$ IV. The co-precipitation of Sr^{2+} with aragonite between 16° and 96°C. *Geochim. Cosmochim. Acta* **33**, 1–17.

Kinsman, D.J.J. & Park, R.K. (1976) Algal belt and coastal sabkha evolution, Trucial Coast, Persian Gulf. In: *Stromatolites* (Ed. by M.R. Walter) pp. 421–433.

Elsevier, Amsterdam.

Kitano, Y., Kanamori, N. & Oomori, T. (1971) Measurements of distribution coefficients of strontium and barium between carbonate precipitate and solution. Abnormally high values of distribution coefficients measured at early stages of carbonate formation. *Geochim. J.* **4**, 183–206.

Klappa, C.F. (1978) Biolithogenesis of Microcodium: elucidation. *Sedimentology* **25**, 489–522.

Klappa, C.F. & James, N.P. (1980) Small lithistid sponge bioherms, early middle Ordovician Table Head Group, western Newfoundland. *Bull. Can. petrol. Geol.* **28**, 425–451.

Klein, G.deV. & Ryer, T.A. (1978) Tidal circulation patterns in Precambrian, Paleozoic and Cretaceous epeiric and mioclinal shelf seas. *Bull. geol. Soc. Am.* **89**, 1050–1058.

Klein, G. de V. et al. (1987) Hummocky cross-stratification, tropical hurricanes, and intense winter storms (Discussion). *Sedimentology* **34**, 333–359.

Knox, G.J. (1977) Caliche profile formation, Saldanha Bay (South Africa). *Sedimentology* **24**, 657–674.

Kobluk, D.R. (1981a) The record of cavity dwelling (coelobiontic) organisms in the Palaeozoic. *Can. J. earth Sci.* **18**, 181–190.

Kobluk, D.R. (1981b) Lower Cambrian cavity-dwelling endolithic (boring) sponges. *Can. J. earth Sci.* **18**, 972–980.

Kobluk, D.R. (1984) Coastal paleokarst near the Ordovician–Silurian boundary, Mantioulin Island, Ontario. *Bull. Can. petrol. Geol.* **32**, 398–407.

Kobluk, D.R. & Kozelj, M. (1985) Recognition of a relationship between depth and macroboring distribution in growth framework reef cavities, Bonaire, Netherlands Antilles. *Bull. Can. petrol. Geol.* **33**, 462–470.

Kobluk, D.R. & Risk, M.J. (1977a) Micritization and carbonate grain binding by endolithic algae. *Bull. Am. Ass. petrol. Geol.* **61**, 1069–1082.

Kobluk, D.R. & Risk, M.J. (1977b) Calcification of exposed filaments of endolithic algae, micrite envelope formation and sediment production. *J. sedim. Petrol.* **47**, 517–528.

Kocurko, M.J. (1979) Dolomitization by spray-zone brine-seepage, San Andres, Columbia. *J. sedim. Petrol.* **49**, 209–214.

Koepnick, R.N., Burke, W.H., Denison, R.E., Hetherington, E.A., Nelson, H.F., Otto, J.B. & Waite, L.E. (1985) Construction of the seawater $^{87}Sr/^{86}Sr$ curve for the Cenozoic and Cretaceous: supporting data. *Chem. Geol.* **58**, 55–81.

Kohout, F.A. (1967) Groundwater flow and the geothermal regime of the Floridan plateau. *Trans. Gulf-Cst. Ass. Geol. Socs.* **17**, 339–354.

Komar, P.D. (1976) *Beach Processes and Sedimentation*, pp. 429. Prentice-Hall, New Jersey.

Krause, F.F. & Oldershaw, A.E. (1979) Submarine carbonate breccia beds — a depositional model for two-layer sediment gravity flows from the Sekwi Formation (Lower Cambrian), Mackenzie Mountains, Northwest Territories, Canada. *Can. J. earth Sci.* **16**, 189–199.

Krauskopf, K.B. (1979) *Introduction to Geochemistry*, pp. 617. McGraw-Hill, New York.

Krebs, W. (1969) Early void-filling cementation in Devonian fore-reef limestones (Germany). *Sedimentology* **12**, 279–299.

Kreisa, R.D. (1981) Storm-generated sedimentary structures in sub-tidal marine facies with examples from the Middle and Upper Ordovician of south-western Virginia. *J. sedim. Petrol.* **51**, 823–848.

Kreisa, R.D. & Bambach, R.K. (1982) The role of storm processes in generating shell beds on Paleozoic shelf environments. In: *Cyclic and Event Stratification* (Ed. by G. Einsele and A. Seilacher) pp. 200–207. Springer-Verlag, Berlin.

Kretz, R. (1982) A model for the distribution of trace elements between calcite and dolomite. *Geochim. Cosmochim. Acta* **46**, 1979–1981.

Krumbein, W.E. (1984) In: *Stromatolite Newsletter No. 11* (Ed. by C.L.V. Monty) pp. 47–48.

Kulm, L.D., Suess, E. & Thornburg, T.M. (1984) Dolomites in organic-rich muds of the Peru forearc basins: Analogue to the Monterey Formation. In: *Dolomites of the Monterey Formation and Other Organic-rich Units* (Ed. by R.E. Garrison, M. Kastner and D.H. Zenger) Publ. Soc. econ. Paleont. Miner. Pacific Section 41, 29–47.

Kumar, R. & Sanders, J.E. (1974) Inlet sequences: a vertical succession of sedimentary structures and textures created by the lateral migration of tidal inlets. *Sedimentology* **21**, 291–323.

Lahann, R.W. (1978) A chemical model for calcite growth and morphology control. *J. sedim. Petrol.* **48**, 337–344.

Lahann, R.W. & Siebert, R.M. (1982) A kinetic model for distribution coefficients and application to Mg-calcite. *Geochim. Cosmochim. Acta* **46**, 2229–2238.

Lancaster, N. (1979) The changes in lake level. In: *Lake Chilwa: Studies of Change in a Tropical Ecosystem* (Ed. by M. Kalk, A.J. McLachlan and C. Howard-Williams) pp. 41–58. W. Junk, Hague.

Land, L.S. (1967) Diagenesis of skeletal carbonates. *J. sedim. Petrol.* **37**, 914–930.

Land, L.S. (1973a) Contemporaneous dolomitization of middle Pleistocene reefs by meteoric water, north Jamaica. *Bull. Mar. Sci.* **23**, 64–92.

Land, L.S. (1973b) Holocene meteoric dolomitization of Pleistocene limestones, north Jamaica. *Sedimentology* **70**, 411–424.

Land, L.S. (1980) The isotopic and trace element geochemistry of dolomite: the state of the art. In: *Concepts and Models of Dolomitization* (Ed. by D.H. Zenger, J.B. Dunham and R.L. Ethington) Spec. Publ. Soc. econ. Paleont. Miner. 28, 87–110.

Land, L.S. (1983) The application of stable isotopes to studies of the origin of dolomite and to problems of diagenesis of clastic sediments. In: *Stable Isotopes in Sedimentary Geology* (Ed. by M.A. Arthur and T.F. Anderson) Soc. econ. Paleont. Miner. Short Course No. 10, 4.1–4.22.

Land, L.S. (1985) The origin of massive dolomite. *J. geol. Educ.* **33**, 112–125.

Land, L.S. (1987) The major ion chemistry of saline brines in sedimentary basins. In: *Physics and Chemistry of Porous Media* (Ed. by J.R. Banavar, J. Koplik and K.W. Winkler) Conf. Proc. Amer. Inst. Physc. 154, 160–179.

Land, L.S., Behrens, E.W. & Frishman, S.A. (1979) The ooids of Baffin Bay, Texas. *J. sedim. Petrol.* **49**, 1269–1278.

Land, L.S. & Epstein, S. (1970) Late Pleistocene diagenesis and dolomitization, north Jamaica. *Sedimentology* **14**, 187–200.

Land, L.S. & Goreau, T.F. (1970) Submarine lithification of Jamaican reefs *J. sedim. Petrol.* **40**, 457–462.

Land, L.S. & Hoops, G.K. (1973) Sodium in carbonate sediments and rocks: a possible index to the salinity of diagenetic solutions. *J. sedim. Petrol.* **43**, 614–616.

Land, L.S., MacKenzie, F.T. & Gould, S.J. (1967) Pleistocene history of Bermuda. *Bull. geol. Soc. Am.* **78**, 993–1006.

Land, L.S. & Moore, C.H. (1977) Deep forereef and upper island slope, north Jamaica. *Am. Ass. petrol. Geol. Studies in Geology* **4**, 53–65.

Land, L.S. & Moore, C.H. (1980) Lithification, micritization and syndepositional diagenesis of biolithites on the Jamaican Island slope. *J. sedim. Petrol.* **50**, 357–370.

Land, L.S., Salem, M.R.I. & Morrow, D.W. (1975) Paleohydrology of ancient dolomites: geochemical evidence. *Bull. Am. Ass. petrol. Geol.* **59**, 1602–1625.

Laporte, L.F. (1967) Carbonate deposition near mean sea-level and resultant facies mosaic: Manlius Formation (Lower Devonian) of New York State. *Bull. Am. Ass. petrol. Geol.* **51**, 73–101.

Laporte, L.F. (1969) Recognition of a transgressive carbonate sequence within an epeiric sea: Helderberg Group (Lower Devonian) of New York State. In: *Depositional Environments of Carbonate Rocks* (Ed. by G.M. Friedman) Spec. Publ. Soc. econ. Paleont. Miner. 14, 98–119.

Lasemi, Z. & Sandberg, P.A. (1984) Transformation of aragonite-dominated lime muds to microcrystalline limestones. *Geology* **12**, 420–423.

Last, W.M. (1982) Holocene carbonate sedimentation in Lake Manitoba, Canada. *Sedimentology* **29**, 691–704.

Leckie, D.A. & Walker, R.G. (1982) Storm and tide dominated shorelines in Cretaceous Moosebar-Lower Gates interval — outcrop equivalent of deep basin trap in western Canada. *Bull. geol. Soc. Am.* **66**, 138–157.

Leeder, M.R. (1975) Lower Border Group (Tournaisian) stromatolites from the Northumberland Basin. *Scott. J. Geol.* **11**, 207–226.

Leeder, M.R. (1988) Recent developments in Carboniferous geology: a critical review with implications for the British Isles and NW Europe. *Proc. geol. Ass.* **99**, 73–100.

Leeder, M.R. & Strudwick, A.E. (1987) Delta–marine interactions: a discussion of sedimentary models for Yoredale-type cyclicity in the Dinantian of Northern England. In: *European Dinantian Environments* (Ed. by

J. Miller, A.E. Adams and V.P. Wright) pp. 115–130. John Wiley, Chichester.

Leeder, M.R. & Zeidan, R. (1977) Giant late Jurassic sabkhas of Arabian Tethys. *Nature* **268**, 42–44.

Lees, A. (1975) Possible influences of salinity and temperature on modern shelf carbonate sedimentation. *Mar. Geol.* **19**, 159–198.

Lees, A. & Buller, A.T. (1972) Modern temperate water and warm water shelf carbonate sediments contrasted. *Mar. Geol.* **13**, 1767–1773.

Lees, A., Hallet, V. & Hibo, D. (1985) Facies variation in Waulsortian build-ups. Part 1. A model from Belgium. *Geol. J.* **20**, 133–158.

Lees, A. & Miller, J. (1985) Facies variation in Waulsortian buildups, Part 2; Mid-Dinantian buildups from Europe and North America. *Geol. J.* **20**, 159–180.

Leggett, J.K. (1985) Deep-sea pelagic sediments and palaeo-oceanography: a review of recent progress. In: *Sedimentology: Recent Developments and Applied Aspects* (Ed. by P.J. Brenchley and B.P.J. Williams) Spec. Publ. geol. Soc. Lond. 18, 95–121.

Leighton, M.W. & Pendexter, C. (1962) Carbonate rock types. In: *Classification of Carbonate Rocks* (Ed. by W.E. Ham) Mem. Am. Ass. petrol. Geol. 1, 33–61.

Lemoine, M. (1972) Eugeosynclinal domains of the Alps and the problem of past oceanic areas. *24th Int. geol. Cong. Montreal Sect.* **3**, 476–485.

Leonardi, P. (1967) *Le Dolomiti*, p. 1019. Cons. Naz. Ricerche, Rome.

Leutloff, A.H. & Meyers, W.J. (1984) Regional distribution of microdolomite inclusions in Mississippian echinoderms from southwest New Mexico. *J. sedim. Petrol.* **54**, 432–446.

Lighty, R.G. (1985) Preservation of internal reef porosity and diagenetic sealing of submerged early Holocene barrier reef, southeast Florida Shelf. In: *Carbonate Cements* (Ed. by N. Schneidermann and P.M. Harris) Spec. Publ. Soc. econ. Paleont. Miner. 36, 123–151.

Lighty, R.G., Macintyre, I.G. & Stukenrath, R. (1978) Submerged early Holocene barrier reef, southeast Florida Shelf. *Nature* **276**, 59–60.

Lindstrom, M. (1963) Sedimentary folds and the development of limestone in an Early Ordovician Sea. *Sedimentology* **2**, 243–292.

Lindstrom, M. (1974) Volcanic contribution to Ordovician pelagic sediments. *J. sedim. Petrol.* **44**, 287–291.

Link, M.H. & Osborne, R.H. (1978) Lacustrine facies in Pliocene Ridge Basin Group; Ridge Basin, California. In: *Modern and Ancient Lake Sediments* (Ed. by A. Matter and M.E. Tucker) Spec. Publ. int. Ass. Sediment. 2, 169–187.

Link, M.H., Osborne, R.H. & Awramik, S.M. (1978) Lacustrine stromatolites and associated sediments of the Pliocene Ridge Route Formation, Ridge Basin, California. *J. sedim. Petrol.* **48**, 143–158.

Lippmann, F. (1973) *Sedimentary Carbonate Minerals*, pp. 228. Springer-Verlag, Berlin.

Littler, M.M. & Doty, M. (1975) Ecological component structuring in the seaward edges of tropical Pacific reefs: the distribution, communities and productivity-ecology of *Porolithon*. *J. Ecol.* **63**, 117–129.

Logan, B.W. (1961) Cryptozoon and associated stromatolites from the Recent, Shark Bay, Western Australia. *J. Geol.* **69**, 517–617.

Logan, B.W. (1974) Inventory of diagenesis in Holocene–Recent carbonate sediments, Shark Bay, Western Australia. *Mem. Am. Ass. petrol. Geol.* **22**, 195–249.

Logan, B.W., Hoffman, P. & Gebelein, C.D. (1974) Algal mats, crypt-algal fabrics and structures, Hamelin Pool, Western Australia. *Mem. Am. Ass. petrol. Geol.* **22**, 140–193.

Logan, B.W., Rezaki, R. & Ginsburg, R.W. (1964) Classification and environmental significance of algal stromatolites. *J. Geol.* **72**, 68–83.

Logan, B.W. & Semeniuk, V. (1976) Dynamic metamorphism; processes and products in Devonian carbonate rocks, Canning Basin, Western Australia. *Spec. Publ. geol. Soc. Austr.* **16**, 1–138.

Logan, B.W. *et al.* (1969) Carbonate sediments and reefs, Yucatan shelf, Mexico. *Mem. Am. Ass. petrol. Geol.* **11**, 198.

Logan, B.W. *et al.* (1970) Carbonate sedimentation and environments, Shark Bay, Western Australia. *Mem. Am. Ass. petrol. Geol.* **13**, 223.

Logan, B.W. *et al.* (1974) Evolution and diagenesis of Quaternary Carbonate Sequences, Shark Bay, Western Australia. *Mem. Am. Ass. petrol. Geol.* **22**, 358.

Lohmann, K.C. (1976) Lower Dresbachian (Upper Cambrian) platform to deep shelf transition in eastern Nevada and western Utah: an evaluation through lithologic cycle correlation. *Brigham Young University, Studies in Geology* **23**, 111–122.

Lohmann, K.C. (1988) Geochemical patterns of meteoric diagenetic systems and their application to studies of paleokarst. In: *Paleokarst* (Ed. by N.P. James and P.W. Choquette) pp. 58–80. Springer-Verlag, New York.

Lohmann, K.C. & Meyers, W.J. (1977) Microdolomite inclusions in cloudy prismatic calcites — a proposed criterion for former high magnesium calcites. *J. sedim. Petrol.* **47**, 1078–1088.

Longman, M.W. (1977) Factors controlling the formation of microspar in the Bromide Formation. *J. sedim. Petrol.* **47**, 347–350.

Longman, M.W. (1980) Carbonate diagenetic textures from nearshore diagenetic environments. *Bull. Am. Ass. petrol. Geol.* **64**, 461–487.

Longman, M.W. (1981) A process approach to recognising facies of reef complexes. In: *European Fossil Reef Models* (Ed. by D.F. Toomey) Spec. Publ. Soc. econ. Paleont. Miner. 30, 9–40.

Longman, M.W. (1982) *Carbonate Diagenesis as a Control on Stratigraphic Traps*. Am. Ass. petrol. Geol. Educ. Course Note Series 21, pp. 159.

Longman, M.W. & Mench, P.A. (1978) Diagenesis of Cretaceous limestones in the Edwards aquifer system of south-central Texas: a scanning electron microscopy study. *Sedim. Geol.* **21**, 241–276.

Loreau, J.P. (1982) Sediments aragonitiques et leur genese. *Memoirs du Museum d'Histoire Naturelle, Serie C* **47**, 312.

Loreau, J.P. & Purser, B.H. (1973) Distribution and ultra-structure of Holocene ooids in the Persian Gulf. In: *The Persian Gulf* (Ed. by B.H. Purser) pp. 279–328. Springer-Verlag, Berlin.

Lorens, R.B. (1981) Sr, Cd, Mn and Co distribution coefficients in calcite as a function of calcite precipitation rate. *Geochim. Cosmochim. Acta* **45**, 553–561.

Loucks, R.G. & Anderson, J.H. (1980) Depositional facies and porosity development in Lower Ordovician Ellenburger Dolomite, Puckett Field, Pecos County, Texas. Soc. Econ. Paleontol. Miner. Core Workshop 1, 1–31.

Loucks, R.G. & Bebout, D.G. (1984) Shelf-interior carbonate grainstone shoals: Lower Cretaceous Pearsall Formation, South Texas. In: *Carbonate Sands: a Core Workshop* (Ed. by P.M. Harris) Soc. econ. Paleont. Miner. Core Workshop 5, 334–364.

Loucks, R.G. & Folk, R.L. (1976) Fan-like rays of former aragonite in Permian Capitan Reef pisolite. *J. sedim. Petrol.* **46**, 483–485.

Loucks, R.G. & Longman, M. (1982) Lower Cretaceous Ferry Lake Anhydrite, Fairway Field, East Texas: product of shallow-subtidal deposition. In: *Depositional and Diagenetic Spectra of Evaporites* (Ed. by C.R. Handford, R.G. Loucks and G.R. Davies) Soc. econ. Paleont. Miner. Core Workshop 3, 130–173.

Lowe, D.R. (1975) Water escape structures in coarse-grained sediments. *Sedimentolgy* **22**, 157–204.

Lowe, D.R. (1976) Grain flow and grain flow deposits. *J. sedim. Petrol.* **46**, 188–199.

Lowe, D.R. (1982) Sediment gravity flows: II. Depositional models with special reference to the deposits of high-density turbidity currents. *J. sedim. Petrol.* **52**, 279–297.

Lowenstam, H.A. (1981) Minerals formed by organisms. *Science* **211**, 1126–1131.

Lowenstam, H.A. & Weiner, S. (1983) Mineralization by organisms and the evolution of biomineralization. In: *Biomineralization and Biological Metal Accumulation* (Ed. by P. Westbroek and E.W. DeJong) pp. 191–203. Reidel Pub. Co., Dordrecht.

Lowenstein, T.K. & Hardie, L.A. (1985) Criteria for the recognition of salt pan evaporites. *Sedimentology* **32**, 627–644.

Lucia, F.J. (1968) Recent sediments and diagenesis of south Bonaire, Netherlands Antilles. *J. sedim. Petrol.* **38**, 845–858.

Ludlam, S.D. (1974) Fayetteville Green Lane, New York, VI. The role of turbidity currents in lake sedimentation. *Limnol. Oceanogr.* **19**, 656–664.

Ludlam, S.D. (1981) Sedimentation rates in Fayetteville Green Lake, New York, USA. *Sedimentology* **28**, 85–96.

Lumsden, D.N. (1979) Discrepancy between thin section and X-ray estimates of dolomite in limestone. *J. sedim. Petrol.* **49**, 429–436.

Lumsden, D.N. & Chimahusky, J.S. (1980) Relationship between dolomite nonstoichiometry and carbonate facies parameters. In: *Concepts and Models of Dolomitization* (Ed. by D.H. Zenger, J.B. Dunham and R.L. Ethington) Spec. Publ. Soc. econ. Paleont. Miner. 28, 123–137.

MacGeachy, J.K. & Stearn, C.W. (1976) Borings by macro-organisms in the coral *Montastrea annularis* on the Barbados reef. *Int. Rev. Ges. Hydrobiol.* **61**, 715–745.

Machel, H.G. (1985) Cathodoluminescence in calcite and dolomite and its chemical interpretation. *Geoscience Canada* **12**, 139–147.

Machel, H.G. & Mountjoy, E.W. (1986) Chemistry and environments of dolomitization — a reappraisal. *Earth Sci. Rev.* **23**, 175–222.

Machel, H.G. & Mountjoy, E.W. (1987) General constraints on extensive pervasive dolomitization and their application to the Devonian carbonates of western Canada. *Bull. Can. petrol. Geol.* **35**, 143–158.

Machette, M.N. (1985) Calcic soils of the southwestern United States. In: *Soils and Quaternary Geology of the Southwestern United States* (Ed. by D.L. Wiede) Spec. Pap. geol. Soc. Am. 203, 1–21.

Macintyre, I.G. (1984a) Growth, depositional facies and diagenesis of a modern bioherm, Galeta Point, Panama. In: *Carbonate Buildups: a Core Workshop* (Ed. by P.M. Harris) Soc. econ. Paleont. Miner. Core Workshop, 478–593.

Macintyre, I.G. (1984b) Preburial and shallow-subsurface alteration of modern scleractinian corals. *Paleontographica Americana* **54**, 229–244.

Macintyre, I.G. (1985) Submarine cements — the peloidal question. In: *Carbonate Cements* (Ed. by N. Schneidermann and P.M. Harris) Spec. Publ. Soc. econ. Paleont. Miner. 36, 109–116.

Macintyre, I.G., Burke, R.B. & Stuckenrath, R. (1977) Thickest recorded Holocene reef section, Isla Pérez core hole, Alacran Reef, Mexico. *Geology* **5**, 749–754.

Macintyre, I.G., Burke, R.B. & Stuckenrath, R. (1981) Core holes in the outer fore reef off Carie Bow Cay, Belize: a key to the Holocene history of the Belizean Barrier Reef Complex. *Proc. 4th Int. Coral Reef Symp. Manilla* 1, 567–574.

Macintyre, I.G. & Milliman, J.D. (1970) Physiographic features on the outer shelf and upper slope, Atlantic continental margin, southeastern US. *Bull. geol. Soc. Am.* **81**, 2577–2598.

Macintyre, I.G., Mountjoy, E.W. & D'Anglejan, B.F. (1968) An occurrence of submarine cementation of carbonate sediments off the west coast of Barbados, WI. *J. sedim. Petrol.* **38**, 660–664.

Mackenzie, F.T., Bischoff, W.D., Bishop, F.C., Loijens, M., Schoonmaker, J. & Wollast, R. (1983) Magnesian calcites: Low-temperature occurrence, solubility and solid-solution behaviour. In: *Carbonates: Mineralogy and Chemistry* (Ed. by R.J. Reeder) Reviews in Mineralogy 11, 94–144. Mineralogical Society of America, Washington.

Mackenzie, F.T. & Piggottt, J.D. (1981) Tectonic controls of Phanerozoic sedimentary rock cycling. *J. geol. Soc.* **138**, 183–196.

Magara, K. (1978) *Compaction and Fluid Migration – Practical Petroleum Geology*, p. 319. Elsevier, Amsterdam.

Magaritz, M., Holser, W.T. & Kirschvink, J.L. (1986) Carbon-isotope events across the Precambrian/Cambrian boundary on the Siberian Platform. *Nature* **320**, 258–259.

Majewske, O.P. (1969) *Recognition of Invertebrate Fossil Fragments in Rocks and Thin Sections*, p. 101. Brill, Leiden.

Major, R.P., Halley, R.B. & Lukas, K.J. (1988) Cathodoluminescent bimineralic ooids from the Pleistocene of the Florida continental shelf. *Sedimentology*, **35**, 843–855.

Majoube, M. (1971) Fractionnement en oxygene 18 et en deuterium entre l'eau et sa vapeur. *J. Chimie Phys. Phys-chimie Biol.* **68**, 1423–1436.

Manze, U. & Richter, D.K. (1979) Die Veränderung des $^{13}C/^{12}C$ – Verhaltnisser in Seeigelcoronen die der Umwandlung von Mg-Calcit in Calcit unter meteorischvadosen Bedingungen. *Neues Jb. Geol. Paläont. Abh.* **158**, 334–345.

Maragos, J.E., Baines, G.B.K. & Beveridge, P.J. (1973) Tropical cyclone creates a new land formation of Funafuti Atoll. *Science* **181**, 1161–1164.

Markello, J.R. & Read, J.F. (1981) Carbonate ramp to deeper shale shelf transitions of an upper Cambrian intrashelf basin, Nolichucky Formation, Southwest Virginia Appalachians. *Sedimentology* **28**, 573–597.

Marland, G. (1975) The stability of $CaCO_3.6H_2O$ (ikaite). *Geochim. Cosmochim. Acta* **39**, 83–91.

Marlowe, J.I. (1971) Dolomite, phosphorite and carbonate diagenesis on a Caribbean seamount. *J. sedim. Petrol.* **41**, 809–827.

Marsaglia, K.M. & Klein, G. de V. (1983) The paleogeography of Paleozoic and Mesozoic storm depositional systems. *J. Geol.* **91**, 117–142.

Marshall, J.D. & Ashton, M. (1980) Isotopic and trace element evidence for submarine lithification of hardgrounds in the Jurassic of England. *Sedimentology* **27**, 271–289.

Marshall, J.D., Paul, C.R.C. & Wright, V.P. (1988) Diagenesis in Tertiary palustrine carbonate paleosols. *Soc. econ. Miner. Paleont. Ann. Midyear Mtg. Abst.* **5**, 34.

Marshall, J.F. (1983a) Submarine cementation in a high energy platform reef: One Tree Reef, southern Great Barrier Reef. *J. sedim. Petrol.* **53**, 1133–1149.

Marshall, J.F. (1983b) Lithology and diagenesis of the carbonate foundations of modern reefs in the southern Great Barrier Reef. *B.M.R. J. Austr. Geol. Geophys.* **8**, 253–265.

Marshall, J.F. (1986) Regional distribution of submarine

cements within an epicontinental reef system: Central Great Barrier Reef, Australia. In: *Reef Diagenesis* (Ed. by J.H. Schroeder and B.H. Purser) pp. 8–26. Springer-Verlag, Berlin.

Marshall, J.F. & Davies, P.J. (1975) High magnesium calcite ooids from the Great Barrier Reef. *J. sedim. Petrol.* **45**, 285–291.

Marshall, J.F. & Davies, P.J. (1981) Submarine lithification on windward reef slopes: Capricorn-Bunker Group, southern Great Barrier Reef. *J. sedim. Petrol.* **51**, 953–960.

Marshall, J.F. & Davies, P.J. (1982) Internal structure and Holocene evolution of One Tree Reef, southern Great Barrier Reef. *Coral Reefs* **1**, 21–28.

Marshall, J.F. & Davies, P.J. (1984) Facies variation and Holocene reef growth in the southern Great Barrier Reef. In: *Coastal Geomorphology in Australia* (Ed. by B.G. Thom) pp. 123–134. Academic Press, Sydney.

Massari, F. (1981) Cryptalgal fabrics in the Rosso Ammonitico sequences of the Venetian Alps. In: *Proc. Rosso Ammonitico Symposium* (Ed. by A. Farinacci and S. Elmi) pp. 435–469. Edizioni Techoscienza, Rome.

Massari, F. (1983) Oncoids and stromatolites in the Rosso Ammonitico sequences (Middle and Upper Jurassic) of the Venetian Alps. In: *Coated Grains* (Ed. by T.M. Peryt) pp. 358–366. Springer-Verlag, Berlin.

Matter, A. (1974) Burial diagenesis of pelitic and carbonate deepsea sediments from the Arabian Sea. *Init. Repts. DSDP* **23**, 421–470.

Matter, A., Douglas, R.G. & Perch-Nielson, K. (1975) Fossil preservation, geochemistry and diagenesis of pelagic carbonates from Shatsky Rise, northwest Pacific. *Init. Repts. DSDP* **32**, 891–922.

Mattes, D.H. & Mountjoy, E.W. (1980) Burial dolomitization of the Upper Devonian Miette buildup, Jasper National Park, Alberta. In: *Concepts and Models of Dolomitization* (Ed. by D.H. Zenger, J.B. Dunham and R.L. Ethington) Spec. Publ. Soc. econ. Paleont. Miner. **28**, 259–297.

Matthews, A. & Katz, A. (1977) Oxygen isotope fractionation during dolomitization of calcium carbonate. *Geochim. Cosmochim. Acta* **41**, 1431–1438.

Matthews, R.K. (1966) Genesis of Recent lime mud in southern British Honduras. *J. sedim. Petrol.* **36**, 428–454.

Matthews, R.K. (1984) *Dynamic Stratigraphy*, pp. 489. Prentice-Hall, Englewood Cliffs, N.J.

Matthews, R.K. & Frohlich, C. (1987) Forward modelling of bank-margin carbonate diagenesis. *Geology* **15**, 673–676.

Maxwell, W.G.H. (1968) *Atlas of the Great Barrier Reef*, p. 258. Elsevier, Amsterdam.

Mazzullo, S.J. (1982) Stratigraphy and depositional mosaics of Lower Clear Fork and Wichita Groups (Permian), Northern Midland Basin, Texas. *Bull. Am. Ass. petrol. Geol.* **66**, 210–227.

Mazzullo, S.J. & Cys, J.M. (1977) Submarine cements in Permian boundstones and reef-associated rocks, Guadalupe Mountains, West Texas and south-eastern New Mexico. In: *Upper Guadalupian Facies, Permian Reef Complex, Guadalupe Mountains, New Mexico and West Texas* (Ed. by M.E. Hileman and S.J. Mazzullo). Publ. Soc. econ. Paleont. Miner. Permian Basin Section Publ. Soc. econ. Paleont. Miner. 77–116, 151–200.

Mazzullo, S.J. & Cys, J.M. (1979) Marine aragonite sea-floor growths and cements in Permian phylloid algal mounds, Sacramento Mountains, New Mexico. *J. sedim. petrol.* **49**, 917–936.

Mazzullo, S.J. & Friedman, G.M. (1975) Conceptual model of tidally influenced deposition on margins of epeiric seas; Lower Ordovician (Canadian) of eastern New York and south western Vermont. *Bull. Am. Ass. petrol. Geol.* **59**, 2123–2141.

Mazzullo, S.J. & Reid, A.M. (1988) Sedimentary textures of Recent Belizean peritidal dolomite. *J. sedim. Petrol.* **58**, 479–488.

Mazzullo, S.J., Reid, A.M. & Gregg, J.M. (1987) Dolomitization of Holocene Mg-calcite supratidal deposits, Ambergris Cay, Belize. *Bull. geol. Soc. Am.* **98**, 224–231.

McCall, P.L. & Tevesz, M.J.S. (1982) The effects of benthos on physical properties of freshwater sediments. In: *Animal–Sediment Relations – the Biogenic Alteration of Sediments*. Topics in Geobiology (Ed. by P.L. McCall and M.J.S. Tevesz) pp. 105–176. Plenum Press, New York.

McGillis, K.A. (1984) *Stratigraphy and Diagenesis of the Upper Jurassic, North East Texas*. Applied Carbonate Research Program, University of Louisiana, Baton Rouge, Technical Series Contrib. No. 18, p. 154.

McGraw, M.M. (1984) Carbonate facies of the Upper Smackover Formation (Jurassic), Paup Spur-Mandeville Fields, Muller County, Arkansas. In: *Jurassic of the Gulf Rim* (Ed. by W.P.S. Ventress, D.G. Bebout, B.F. Perkins and C.H. Moore) Publ. Soc. econ. Paleont. Miner. Gulf Coast Section, pp. 255–274.

McGregor, A.R. (1983) The Waitkere Limestone, a temperate algal carbonate in the lower Tertiary of New Zealand. *J. geol. Soc.* **140**, 387–400.

McHargue, T.R. & Price, R.C. (1982) Dolomite from clay in argillaceous or shale-associated marine carbonates. *J. sedim. Petrol.* **52**, 873–886.

McIlreath, I.A. (1977) Accumulation of a Middle Cambrian, deep-water limestone debris apron adjacent to a vertical submarine carbonate escarpment, southern Rocky Mountains, Canada. In: *Deep-water Carbonate Environments* (Ed. by H.E. Cook and P. Enos) Spec. Publ. Soc. econ. Paleont. Miner. 25, 113–124.

McIlreath, I.A. & James, N.P. (1984) Carbonate slopes. In: *Facies Models* (Ed. by R.G. Walker), pp. 245–257. Geoscience Canada.

McKee, E.D. & Gutschick, R.C. (1969) History of Redwall Limestone of northern Arizona. *Mem. geol. Soc. Am.* **114**, 726.

McKee, E.D. & Ward, W.C. (1983) Eolian environment.

In: *Carbonate Depositional Environments* (Ed. by P.A. Scholle, D.G. Bebout and C.H. Moore) Mem. Am. Ass. petrol. Geol. 33, 132–170.

McKenzie, J.A. (1981) Holocene dolomitization of calcium carbonate sediments from the coastal sabkhas of Abu Dhabi, UAE: a stable isotope study. *J. Geol.* **89**, 185–198.

McKenzie, J.A., Hsü, K.J. & Schneider, J.F. (1980) Movement of subsurface waters under the sabkha, Abu Dhabi, UAE, and it's relation to evaporative dolomite genesis. In: *Concepts and Models of Dolomitization* (Ed. by D.H. Zenger, J.B. Dunham and R.L. Ethington) Spec. Publ. Soc. econ. Paleont. Miner. 28, 11–30.

McLimans, R.K. & Videtich, P.E. (1986) Oil migration versus diagenesis in the Wealden Basin, a race against time (Abst.). *3rd Conference on Petroleum Geology of NW Europe. Barbican Centre, London*, p. T21.

Medwedeff, D.A. & Wilkinson, B.H. (1983) Cortical fabrics in calcite and aragonite ooids. In: *Coated Grains* (Ed. by T.M. Peryt) pp. 109–115. Springer-Verlag, Berlin.

Meischner, K.D. (1964) Allodapische Kalke, Turbidite in Riff-nahen sedimentations–Becken. In: *Turbidites* (Ed. by A.H. Bouma and A. Brouwer) pp. 156–191. Elsevier, Amsterdam.

Mengard, R.O. (1968) Planktonic photosynthesis and the environment of carbonate deposition in lakes. *Univ. Minnesota, Limnol. Res. Center Interim Rept.* **2**, 47.

Mesolella, K.J., Sealy, H.A. & Matthews, R.K. (1970) Facies geometries within Pleistocene reef Barbados, West Indies. *Bull. Am. Ass. petrol. Geol.* **54**, 1899–1917.

Meyers, J.H. (1987) Marine vadose beachrock cementation by cryptocrystalline magnesian calcite — Maui, Hawaii. *J. sedim. Petrol.* **55**, 558–570.

Meyers, W.J. (1974) Carbonate cement stratigraphy of the Lake Valley Formation (Mississippian), Sacramento Mountains, New Mexico. *J. sedim. Petrol.* **44**, 837–861.

Meyers, W.J. (1978) Carbonate cements: their regional distribution and interpretation in Mississippian limestones of southwestern New Mexico. *Sedimentology* **25**, 371–400.

Meyers, W.J. (1980) Compaction in Mississippian skeletal limestones, southwestern New Mexico. *J. sedim. Petrol.* **50**, 457–474.

Meyers, W.J. & Lohmann, K.C. (1978) Microdolomite-rich syntaxial cements: proposed meteoric–marine mixing zone phreatic cements from Mississippian limestones, New Mexico. *J. sedim. Petrol.* **48**, 475–488.

Meyers, W.J. & Lohmann, K.C. (1985) Isotope geochemistry of regionally extensive calcite cement zones and marine components in Mississippian Limestones, New Mexico. In: *Carbonate Cements* (Ed. by N. Schneidermann and P.M. Harris) Spec. Publ. Soc. econ. Paleont. Miner. 36, 223–239.

Michard, G. (1968) Coprecipitation de l'ion magnaeaux avec le carbonate de calcium. *Comptes Rendus Acad., Paris, ser. D* **267**, 1685–1688.

Middleton, G.V. & Hampton, M.A. (1976) Subaqueous

sediment transport and deposition by sediment gravity flows. In: *Marine Sediment Transport and Environmental Management* (Ed. by D.J. Stanley and D.J.P. Swift) pp. 197–218. John Wiley, New York.

Miller, J. (1986) Facies relationships and diagenesis in Waulsortian mud mounds from the Lower Carboniferous of Ireland and N. England. In: *Reef Diagenesis* (Ed. by J.H. Schroeder and B.H. Purser) pp. 311–335. Springer-Verlag, Berlin.

Miller, J.A. (1975) Facies characteristics of Laguna Madre wind-tidal flats. In: *Tidal Deposits: a Casebook of Recent Examples and Fossil Counterparts* (Ed. by R.N. Ginsburg) pp. 67–74. Springer-Verlag, Berlin.

Milliken, K.L. (1979) The silicified evaporite syndrome — two aspects of the silicification history of former evaporite nodules from southern Kentucky and northern Tennessee. *J. sedim. Petrol.* **49**, 245–256.

Milliken, K.L., Land, L.S. & Loucks, R.G. (1981) History of burial diagenesis determined from isotopic geochemistry, Frio Formation, Texas. *Bull. Am. Ass. petrol. Geol.* **65**, 1397–1413.

Milliman, J.D. (1969) Four southwestern Caribbean atolls: Courtown Cays, Albuquerque Cays, Roncador Bank and Serrana Bank. *Atoll Res. Bull.* **129**, 41.

Milliman, J.D. (1971) Examples of lithification in the deep sea. In: *Carbonate Cements* (Ed. by O.P. Bricker) pp. 95–102. Johns Hopkins University Press, Baltimore.

Milliman, J.D. (1974) *Marine Carbonates*, p. 375. Springer-Verlag, New York.

Milliman, J.D. & Barretto, H.T. (1975) Relic magnesian calcite oolite and subsidence of the Amazon shelf. *Sedimentology* **22**, 137–145.

Milliman, J.D., Hook, J.A. & Golubic, S. (1985) Meaning and usage of micrite cement and matrix — reply to discussion. *J. sedim. Petrol.* **55**, 777–784.

Milliman, J.D. & Müller, J. (1973) Precipitation and lithification of magnesian calcite in the deep-sea sediments of the eastern Mediterranean sea. *Sedimentology* **20**, 29–45.

Milliman, J.D., Ross, D.A. & Ku, T.U. (1969) Precipitation and lithification of deep-sea carbonates in the Red Sea. *J. sedim. Petrol.* **39**, 724–736.

Mitchell, J.T., Land, L.S. & Miser, D.E. (1987) Modern marine dolomite cement in a north Jamaican fringing reef. *Geology* **15**, 557–560.

Mitchell, R.W. (1985) Comparative sedimentology of shelf carbonates of the Middle Ordovician St Paul Group, central Appalachians. *Sedim. Geol.* **43**, 1–41.

Mitchum, R.M. Jr., Vail, P.R. & Thompson, III S. (1977) Seismic stratigraphy and global changes of sea-level Part 2: The depositional sequence as a basic unit for stratigraphic analysis. In: *Seismic Stratigraphy — Applications to Hydrocarbon Exploration* (Ed. by C.E. Payton) Mem. Am. Ass. petrol. Geol. 26, 53–62.

Mitterer, R.M. (1972) Biogeochemistry of aragonite mud and oolites. *Geochim. Cosmochim. Acta* **36**, 1407–1412.

Mizutani, H. & Wada, E. (1982) Effect of high atmospheric CO_2 concentration on $\delta^{13}C$ of algae. *Origin of Life* **12**, 377–390.

Moller, N.K. & Kvingan, K. (1988) The genesis of nodular limestones in the Ordovician and Silurian of the Oslo Region (Norway). *Sedimentology* **35**, 405–420.

Monty, C.L. (1967) Distribution and structure of recent stromatolitic algal mats, Eastern Andros Island, Bahamas. *Ann. Soc. geol. Belg.* **90**, 55–100.

Monty, C.L.V. (1971) An autoecological approach to intertidal and deep water stromatolites. *Ann. Soc. geol. Belg.* **94**, 265–276.

Monty, C.L. (1972) Recent algal stromatolitic deposits, Andros Island, Bahamas. *Geol. Rdsch.* **61**, 742–783.

Monty, C.L. (1973) Precambrian background and Phanerozoic history of stromatolitic communities, an overview. *Ann. Soc. geol. Belg.* **96**, 585–624.

Monty, C.L.V. (1976) The origin and development of cryptalgal fabrics. In: *Stromatolites* (Ed. by M.R. Walter) pp. 193–249. Elsevier, Amsterdam.

Monty, C.L.V. (1981) Spongiostromate vs. Porostromate stromatolites and oncolites. In: *Phanerozoic Stromatolites* (Ed. by C.L.V. Monty) pp. 1–4. Springer-Verlag, Berlin.

Monty, C.L.V. (1982) Cavity and fissure dwelling stromatolites (endostromatolites) from the Belgian Devonian mud mounds. *Ann. Soc. geol. Belg.* **105**, 343–344.

Monty, C.L.V., Bernet-Rollande, M.C. & Maurin, A.F. (1982) Re-interpretation of the Frasnian classical 'reefs' of the southern Ardennes, Belgium. *Ann. Soc. géol. Belg.* **105**, 339–341.

Monty, C.L.V. & Hardie, L.A. (1976) The geological significance of the freshwater blue–green algal calcareous marsh. In: *Stromatolites* (Ed. by M.R. Walter) pp. 447–477. Elsevier, Amsterdam.

Monty, C.L.V. & Mas, J.R. (1981) Lower Cretaceous (Wealden) blue–green algal deposits of the Province of Valencia, eastern Spain. In: *Phanerozoic Stromatolites* (Ed. by C.L.V. Monty) pp. 85–120. Springer-Verlag, Berlin.

Moore, C.H. (1979) Porosity in carbonate rock sequences, geology of carbonate porosity. *Am. Ass. petrol. Geol. Continuing Education Course Notes* 11, A1–A124.

Moore, C.H. (1980) Porosity in carbonate rock sequences. *Am. Ass. petrol. Geol., Fall Education Conference 1980*, pp. 67.

Moore, C.H. (1984) The Upper Smackover of the Gulf Rim: Depositional systems, diagenesis, porosity, evolution and hydrocarbon production. In: *Jurassic of the Gulf Rim* (Ed. by W.P.S. Ventress, D.G. Bebout, B.F. Perkins and C.H. Moore) Publ. Soc. econ. Paleont. Miner. Gulf Coast Section, pp. 283–308.

Moore, C.H. (1985) Upper Jurassic cements: a case history. In: *Carbonate Cements* (Ed. by N. Schneidermann and P.M. Harris) Spec. Publ. Soc. econ. Paleont. Miner. 36, 291–308.

Moore, C.H., Graham, E.A. & Land, L.S. (1976) Sediment transport and dispersal across the deep fore-reef and island slope (−55 to 305 m), Discovery Bay, Jamaica. *J.*

sedim. Petrol. **46**, 174–187.

Moore, C.H. & Shedd, W.W. (1977) Effective rates of sponge bioerosion as a function of carbonate production. *Proc. 3rd Int. Coral Reef Symp. Miami* **2**, 499–505.

Morner, N. (1976) Eustasy and geoid changes. *J. Geol.* **84**, 123–151.

Morner, N. (1983) Sea-levels. In: *Megageomorphology* (Ed. by R. Gardner and H. Scoging) pp. 74–91. Oxford University Press, Oxford.

Morrow, D.W. (1978) The influence of the Mg/Ca ratio and salinity on dolomitization in evaporite basins. *Bull. Can. petrol. Geol.* **26**, 389–392.

Morrow, D.W. (1982a) Diagenesis I. Dolomite — part I. The chemistry of dolomitization and dolomite precipitation. *Geoscience Canada* **9**, 5–13.

Morrow, D.W. (1982b) Diagenesis II. Dolomite — part II. Dolomitization models and ancient dolostones. *Geoscience Canada* **9**, 95–107.

Morrow, D.W. & Ricketts, B.D. (1986) Chemical controls on the precipitation of mineral analogues of dolomite: the sulphate enigma. *Geology* **14**, 408–410.

Morse, J.W. (1985) Kinetic control of morphology, composition and mineralogy of abiotic sedimentary carbonates. Discussion. *J. sedim. Petrol.* **55**, 919–921.

Morton, R.A. & Land, L.S. (1987) Regional variations in formation water chemistry, Frio Formation (Oligocene), Texas Gulf Coast. *Bull. Am. Ass. petrol. Geol.* **71**, 191–206.

Mount, J. (1985) Mixed siliciclastic and carbonate sediments: a proposed first-order textural and compositional classification. *Sedimentology* **32**, 435–442.

Mountjoy, E.W., Cook, H.E., Pray, L.C. & McDaniel, P.N. (1972) Allochthonous carbonate debris flows — worldwide indicators of reef complexes, banks or shelf margins. *24th Int. geol. Cong. Montreal Sect.* 6, 172–189.

M'Rabet, A. (1981) Differentiation of environments of dolomite formation, Lower Cretaceous of Central Tunisia. *Sedimentology* **28**, 331–352.

Mucci, A. (1987) Influence of temperature on the composition of magnesian calcite overgrowths precipitated from seawater. *Geochim. Cosmochim. Acta* **51**, 1977–1984.

Mucci, A. & Morse, J.W. (1983) The incorporation of Mg^{2+} and Sr^{2+} into calcite overgrowths: influences of growth rate and solution composition. *Geochim. Cosmochim. Acta* **47**, 217–233.

Mucci, A., Morse, J.W. & Kaminsky, M.S. (1985) Auger spectroscopy analysis of magnesian calcite overgrowths precipitated from seawater and solutions of similar composition. *Am. J. Sci.* **285**, 289–305.

Muir, M., Lock, D. & Von der Borch, C.L. (1980) The Coorong model for penecontemporaneous dolomite formation in the Middle Proterozoic McArthur Group, Northern Territory, Australia. In: *Concepts and Models of Dolomitization* (Ed. by D.H. Zenger, J.B. Dunham and R.L. Ethington) Spec. Publ. Soc. econ. Paleont. Miner. 28, 51–67.

Müller, G. (1971) Aragonite inorganic precipitation in a freshwater lake. *Nature* **229**, 18.

Müller, G., Irion, G. & Förstner, U. (1972) Formation and diagenesis of inorganic Ca-Mg carbonates in the lacustrine environment. *Naturwissenschaften* **59**, 158–164.

Müller, G. & Oti, M. (1981) The occurrence of calcified planktonic green algae in freshwater carbonates. *Sedimentology* **28**, 897–902.

Müller, G. & Tietz, G. (1971) Dolomite replacing 'cement A' in biocalcarenites from Fuerteventura. Canary Islands, Spain. In: *Carbonate Cements* (Ed. by O.P. Bricker) pp. 327–329. Johns Hopkins University Press, Baltimore.

Müller, G. & Wagner, F. (1978) Holocene carbonate evolution of Lake Balaton (Hungary): a response to climate and the impact of man. In: *Modern and Ancient Lake Sediments* (Ed. by A. Matter & M.E. Tucker). pp. 57–81. Spec. Publ. Int. Ass. Sedimentol. 2.

Muller-Jungblath, W.U. (1968) Sedimentary petrologic investigation of the Upper Triassic 'Hauptdolomit' of the Lechtaler Alps, Tyrol, Austria. In: *Recent Developments in Carbonate Sedimentology in Central Europe* (Ed. by G. Müller and G.M. Friedman) pp. 228–239. Springer-Verlag, Berlin.

Mullins, H.T. (1983) Modern carbonate slopes and basins of the Bahamas. In: *Platform Margin and Deepwater Carbonates* (Ed. by H.E. Cook, A.C. Hine and H.T. Mullins) Soc. econ. Paleont. Miner. Short Course No. 12, 4.1–4.138.

Mullins, H.T. & Cook, H.E. (1986) Carbonate apron models: alternatives to the submarine fan model for paleoenvironmental analysis and hydrocarbon exploration. *Sedim. Geol.* **48**, 37–79.

Mullins, H.T., Gardulski, A.F., Hinchey, E.J. & Hine, A.C. (1988) The modern carbonate ramp slope of central west Florida. *J. sedim. Petrol.* **58**, 273–290.

Mullins, H.T., Heath, K.C., Van Buren, H.M. & Newton, C.R. (1984) Anatomy of modern open-ocean carbonate slope: Northern Little Bahama Bank. *Sedimentology* **31**, 141–168.

Mullins, H.T., Hine, A.C. & Wilber, R.J. (1982) Facies succession of Pliocene–Pleistocene carbonates, north-western Great Bahama Bank: discussion. *Bull. Am. Ass. petrol. Geol.* **66**, 103–105.

Mullins, H.T. & Neumann, A.C. (1979) Deep carbonate bank margin structure and sedimentation in the northern Bahamas. In: *Geology of Continental Slopes* (Ed. by L.J. Doyle and O.H. Pilkey) Spec. Publ. Soc. econ. Paleont. Miner. 27, 165–192.

Mullins, H.T., Neumann, A.C., Wilber, R.J. & Boardman, M.R. (1980) Nodular carbonate sediment on Bahamian slopes: possible precursors to nodular limestones. *J. sedim. Petrol.* **50**, 117–131.

Mullins, H.T., Newton, C.R., Heath, K. & Van Buren, H.M. (1981) Modern deep-water coral mounds north of Little Bahama Bank: criteria for recognition of deep-water coral bioherms in the rock record. *J. sedim. Petrol.* **51**, 999–1013.

Mullins, H.T. & Van Buren, H.M. (1979) Modern modified

carbonate grain flow deposit. *J. sedim. Petrol.* **48**, 747–752.

Mullins, H.T., Wise, S.W., Gardulski, A.F., Hinchey, E.J., Masters, P.M. & Siegel, D.I. (1985) Shallow subsurface diagenesis of Pleistocene periplatform ooze: northern Bahamas. *Sedimentology* **32**, 473–494.

Multer, H.G. (1977) *Field Guide to some Carbonate Rock Environments, Florida Keys and Western Bahamas*, pp. 415. Kendall/Hunt, Dubuque.

Murat, A. & Got, H. (1987) Middle and Late Quaternary depositional sequences and cycles in the eastern Mediterranean. *Sedimentology* **34**, 885–900.

Murphy, D.H. & Wilkinson, B.H. (1980) Carbonate deposition and facies distribution in a central Michigan marl lake. *Sedimentology* **27**, 123–135.

Murray, R.C. (1964) Preservation of primary structures and fabrics in dolomite. In: *Approaches to Paleoecology* (Ed. by J. Imbrie and N. Newell) pp. 388–403. John Wiley, New York.

Murray, R.C. (1969) Hydrology of south Bonaire, Netherlands Antilles — a rock selective dolomitization model. *J. sedim. Petrol.* **15**, 987–1035.

Murray, R.C. & Lucia, F.J. (1967) Cause and control of dolomite distribution by rock selectivity. *Bull. geol. Soc. Am.* **78**, 21–35.

Murris, R.J. (1980) Middle East: stratigraphic evolution and oil habitat. *Bull. Am. Ass. petrol. Geol.* **64**, 597–618.

Mussman, W.J., Montanez, I.P. & Read, J.F. (1988) Ordovician Knox paleokarst unconformity, Appalachians. In: *Paleokarst* (Ed. by N.P. James and P.W. Choquette) pp. 211–228. Springer-Verlag, New York.

Mutti, E. & Ricci-Lucci, F. (1978) Turbidites of the northern Apennines; introduction to facies analysis. *Int. Geol. Rev.* **20**, 125–166.

Neev, D. & Emery, K.O. (1967) The Dead Sea — depositional processes and environments of evaporites. *Bull. Israel geol. Surv.* **41**, 147.

Nelson, C.S. (1978) Temperate shelf carbonate sediments in the Cenozoic of New Zealand. *Sedimentology* **25**, 737–771.

Nelson, C.S., Hancock, G.E. & Kamp, P.J.J. (1982) Shelf to basin, temperate skeletal carbonate sediments, Three kings plateau, New Zealand. *J. sedim. Petrol.* **52**, 717–732.

Nelson, C.S. & Lawrence, M.F. (1984) Methane-derived high Mg calcite submarine cement in Holocene nodules from the Fraser Delta, British Columbia, Canada. *Sedimentology* **31**, 645–654.

Nelson, H.F., Brown, C.W. & Brineman, J.H. (1962) Skeletal limestone classification. In: *Classification of Carbonate Rocks* (Ed. by W.E. Ham) Mem. Am. Ass. petrol. Geol. 1, 224–253.

Netterberg, F. (1980) Geology of southern African calcretes: 1. Terminology, description, macrofeatures, and classification. *Trans. geol. Soc. S. Afr.* **83**, 255–283.

Neugebauer, G. (1974) Some aspects of cementation in chalk. In: *Pelagic Sediments on Land and Under the Sea*

(Ed. by K.J. Hsü and H.C. Jenkyns) Spec. Publ. int. Ass. Sediment. 1, 149–176.

Neumann, A.C., Gebelein, C.D. & Scoffin, T.P. (1970) The composition, structure and erodability of subtidal mats, Abaco, Bahamas. *J. sedim. Petrol.* **40**, 274–297.

Neumann, A.C., Kofoed, J.W. & Keller, G.H. (1977) Lithoherms in the Straits of Florida. *Geology* **5**, 4–11.

Neumann, A.C. & Land, L.S. (1975) Lime mud deposition and calcareous algae in the Bight of Abaco, Bahamas: a budget. *J. sedim. Petrol.* **45**, 763–786.

Neumann, A.C. & Macintyre, I.G. (1985) Response to sea-level rise: keep-up, catch-up or give-up. *Proc. 5th Coral Reef Congr., Tahiti* **3**, 105–110.

Newell, N.D., Imbrie, J., Purdy, E.G. & Thurber, D.L. (1959) Organism communities and bottom facies, Great Bahama Bank. *Bull. Amer. Mus. Nat. Hist.* **117**, 177–228.

Newell, N.D. & Rigby, J.K. (1957) Geological studies on the Great Bahama Bank. In: *Regional Aspects of Carbonate Deposition: a Symposium with Discussion* (Ed. by R.J. Le Blanc and J.G. Breeding) Spec. Publ. Soc. econ. Paleont. Miner. 5, 15–72.

Newell, N.D., Rigby, J.K., Fischer, A.G., Whiteman, A.J., Hickox, J.E. & Bradley, J.S. (1953) *The Permian Reef Complex of the Guadalupe Mountains Region, Texas and New Mexico*, pp. 236. W.H. Freeman, San Francisco.

Newton, C.R., Mullins, H.T., Gardulski, A.F., Hine, A.C. & Dix, G.R. (1987) Coral mounds on the west Florida slope: unanswered questions regarding the development of deep-water banks. *Palaios* **2**, 359–367.

Nickel, E. (1983) Environmental significance of freshwater oncoids, Eocene Guarga Formation, southern Pyrenees, Spain. In: *Coated Grains* (Ed. by P.M. Peryt) pp. 308–329. Springer-Verlag, Berlin.

Niemann, J.C. & Read, J.F. (1988) Regional cementation from unconformity — recharged aquifer and burial fluids. Mississippian Newman Limestone, Kentucky. *J. sedim. Petrol.* **58**, 688–705.

Northrop, D.A. & Clayton, R.N. (1966) Oxygen isotope fractionation in systems containing dolomite. *J. Geol.* **74**, 174–196.

Nygaard, E., Lieberkind, K. & Frykman, P. (1983) Sedimentology and reservoir parameters of the Chalk Group in the Danish Central Graben. *Geol. Mijnb.* **62**, 177–190.

Oertel, G.E. & Curtis, C.D. (1972) Clay-ironstone concretions preserving fabrics due to progressive compaction. *Bull. geol. Soc. Am.* **83**, 2597–2606.

Ogden, J.C. (1977) Carbonate-sediment production by parrotfish and sea urchins on Caribbean reefs. In: *Reefs and Related Carbonates — Ecology and Sedimentology* (Ed. by S.H. Frost, M.P. Weiss and J.B. Saunders) Am. Ass. petrol. Geol. Studies in Geol. 4, 281–288.

Okumura, M. & Kitano, Y. (1986) Coprecipitation of alkali metal ions with calcium carbonate. *Geochim. Cosmochim. Acta* **50**, 49–58.

Olsen, P.E. (1986) A 40-million year lake record of early

Mesozoic orbital climate forcing. *Science* **234**, 842–848.

O'Neil, J.R., Clayton, R.N. & Meyeda, T.K. (1969) Oxygen isotope fractionation in divalent metal carbonate. *J. Chem. Phys.* **51**, 5547–5558.

O'Neil, J.R. & Epstein, S. (1966) Oxygen isotope fractionation in the system dolomite–calcite–carbon dioxide. *Science* **152**, 198–201.

Oomori, T., Kaneshima, H. & Maezato, Y. (1987) Distribution coefficient of Mg^{2+} ions between calcite and solution at 10–50°C. *Mar. Chem.* **20**, 237–336.

Orme, G.R. (1977) The coral sea plateau — a major reef province. In: *Biology and Geology of Coral Reefs* (Ed. by O.A. Jones and R. Endean) 4, 267–306. Academic Press, New York.

Orme, G.R. (1985) The sedimentological importance of *Halimeda* in the development of back-reef lithofacies, northern Great Barrier Reef (Australia). *Proc. 5th Coral Reef Congr. Tahiti* 5, 31–37.

Orme, G.R., Flood, P.G. & Sargent, G.E. (1978) Sedimentation trends in the lee of the outer (ribbon) reefs, northern region of the Great Barrier Reef Province. *Phil. Trans. Roy. Soc. Lond.* A 29, 85–99.

Osborne, R.H., Licari, G.R. & Link, M.H. (1982) Modern lake stromatolites, Walker Lake, Nevada. *Sedim. Geol.* **32**, 39–61.

Otsuki, A. & Wetzel, R.G. (1974) Calcium and total alkalinity budgets and calcium carbonate precipitation of a small hard-water lake. *Arch. Hydrobiol.* **73**, 14–30.

Oudin, E. & Constantinou, G. (1984) Black smoker chimney fragments in Cyprus sulphide deposits. *Nature* **308**, 349–353.

Palmer, T.J. (1979) The Hampen Marly and White Limestone formations: Florida-type carbonate lagoons in the Jurassic of central England. *Palaeontology* **22**, 189–228.

Palmer, T.J., Hudson, J.D. & Wilson, M.A. (1988) Palaeoecological evidence for early aragonite dissolution in ancient calcite seas. *Nature* **335**, 809–810.

Park, R.K. (1976) A note on the significance of lamination in stromatolites. *Sedimentology* **23**, 379–393.

Park, R.K. (1977) The preservation potential of some stromatolites. *Sedimentology* **24**, 485–506.

Patterson, R.J. & Kinsman, D.J.J. (1981) Hydrologic framework of a sabkha along the Persian Gulf. *Bull. Am. Ass. petrol. Geol.* **65**, 1457–1475.

Patterson, R.J. & Kinsman, D.J.J. (1982) Formation of diagenetic dolomite in coastal sabkhas along the Arabian (Persian) Gulf. *Bull. Am. Ass. petrol. Geol.* **66**, 28–43.

Pentecost, A. (1978) Blue–green algae and freshwater carbonate deposits. *Proc. Roy. Soc. Lond.* B **200**, 48–63.

Perkins, R.D. (1977) Depositional framework of Pleistocene rocks in South Florida. *Mem. geol. Soc. Am.* **147**, 131–198.

Perkins, R.D. & Enos, P. (1968) Hurricane Betsy in the Florida–Bahama area — geologic effects and comparison with hurricane Donna. *J. Geol.* **76**, 710–717.

Perry, E.C. & Ahmad, S.N. (1983) Oxygen isotope geochemistry of Proterozoic chemical sediments. *Mem. geol.*

Soc. Am. **161**, 253–263.

Peryt, T.M. (Ed.) (1983a) *Coated Grains*, p. 655. Springer-Verlag, Berlin.

Peryt, T.M. (1983b) Classification of coated grains. In: *Coated Grains* (Ed. by T.M. Peryt) pp. 3–6. Springer-Verlag, Berlin.

Pettijohn, F.J., Potter, P.N. & Siever, R. (1972) *Sand and Sandstone*, pp. 618. Springer-Verlag, Berlin.

Pfeil, R.W. & Read, J.F. (1980) Cambrian carbonate platform margin facies, Shady Dolomite, Southwestern Virginia, USA. *J. sedim. Petrol.* **50**, 91–115.

Phillips, S.E. & Self, P.G. (1987) Morphology, crystallography and origin of needle-fibre calcite in Quaternary pedogenic calcretes of South Australia. *Aust. J. Soil Res.* **25**, 249–264.

Phipps, C.V., Davies P.J. & Hopley, D. (1985) The morphology of *Halimeda* banks behind the Great Barrier Reef east of Cooktown, Queensland. *Proc. 5th Coral Reef Congr. Tahiti* 5, 27–30.

Picard, M.D. & High, L.R. (1974) Criteria for recognizing lacustrine rocks. In: *Recognition of Ancient Sedimentary Environments* (Ed. by J.K. Rigby and W.K. Hamblin) Spec. Publ. Soc. econ. Paleont. Miner. 16, 108–145.

Picard, M.D. & High, L.R. (1981) Physical stratigraphy of ancient lacustrine deposits. In: *Recent and Ancient Nonmarine Depositional Environments: Models for Exploration* (Ed. by F.G. Ethridge and R.M. Flores) Spec. Publ. Soc. econ. Paleont. Miner. 31, 233–259.

Picha, F. (1978) Depositional and diagenetic history of Pleistocene and Holocene oolitic sediments and sabkhas in Kuwait, Persian Gulf. *Sedimentology* **20**, 365–389.

Pichon, J.F. & Lys, M. (1976) Sur l'existence d'une serie du Jurassique superieur a Cretace inferieur, surmontant les ophiolites dans les collines de Krapa (massif du Vourinos, Grece). *C.R. hebd. seanc. Acad. Sci. Paris*, D **282**, 523–526.

Pickering, K.T., Stow, D.A.V., Watson, M.P. & Hiscott, R.N. (1986) Deep-water facies, processes and models; a review and classification scheme for modern and ancient sediments. *Earth Sci. Rev.* **23**, 75–174.

Pierre, C., Ortlieb, L. & Person, A. (1984) Supratidal evaporitic dolomite at Ojo de Liebre lagoon: mineralogical and isotopic arguments for primary crystallization. *J. sedim. Petrol.* **54**, 1049–1061.

Pierson, B.J. (1981) The control of cathodoluminescence in dolomite by iron and manganese. *Sedimentology* **28**, 601–610.

Pierson, B.J. & Shinn, E.A. (1985) Cement distribution and carbonate mineral stabilization in Pleistocene limestones of Hogsty Reef, Bahamas. In: *Carbonate Cements* (Ed. by N. Schneidermann and P.M. Harris) Spec. Publ. Soc. econ. Paleont. Miner. 36, 153–168.

Pilkey, O.H., Morton, R.W. & Lutenauer, J. (1967) The carbonate fraction of beach and dune sands. *Sedimentology* **8**, 311–327.

Piller, W. (1976) Fazies und Lithostratigraphic des gebankten Dachsteinkalkes (Obertrias) am Nordrand des Toten

Geberges (S. Grunau/Almtal, Oberosterreich). *Mitt. Gesell. Geol. Bergbaust. Ost.* **23**, 113−152.

Pingitore, N.E. (1976) Vadose and phreatic diagenesis: processes, products and their recognition in corals. *J. sedim. Petrol.* **46**, 985−1006.

Pingitore, N.E. Jr. & Eastman, M.P. (1986) The coprecipitation of Sr^{2+} with calcite at 25°C and 1 atm. *Geochim. Cosmochim. Acta* **50**, 2195−2203.

Platt, N.H. (1985) Freshwater limestones from the Cameros Basin, N. Spain. *6th IAS European Meeting, Leiden, Abst.* 366−369.

Playford, P.E. (1980) Devonian 'Great Barrier Reef' of Canning Basin, Western Australia. *Bull. Am. Ass. petrol. Geol.* **64**, 814−840.

Playford, P.E. (1984) Platform-margin and marginal-slope relationships in the Devonian reef complexes of the Canning Basin. In: *The Canning Basin, Western Australia* (Ed. by P.G. Purcell) pp. 189−214. Proc. Geol. Soc. Aust. & Petrol. Expl. Soc. Aust. Symp., Perth.

Playford, P.E. & Cockbain, A.E. (1976) Modern algal stromatolites at Hamelin Pool, a hyper-saline barred basin in Shark Bay, Western Australia. In: *Stromatolites* (Ed. by M.R. Walter) pp. 389−411. Elsevier, Amsterdam.

Playford, P.E. *et al.* (1976) Devonian stromatolites from the Canning Basin, Western Australia. In: *Stromatolites* (Ed. by M.R. Walter) pp. 543−563. Elsevier, Amsterdam.

Pleydell, S.M. & Jones, B. (1988) Boring of various faunal elements in the Oligocene−Miocene Bluff Formation of Grand Cayman, British West Indies. *J. Paleont.* **62**, 348−367.

Plummer, L.N. & Busenburg, E. (1987) Thermodynamics of aragonite−strontianite solid solutions: Results from stoichiometric solubility at 25°C and 76°C. *Geochim. Cosmochim. Acta* **51**, 1393−1411.

Plummer, L.N., Wigley, T.M.L. & Parkhurst, D.L. (1979) Critical review of the kinetics of calcite dissolution and precipitation. In: *Chemical Modelling of Aqueous Systems* (Ed. by R.F. Gould) Am. chem. Soc. Symp. Series 93, 537−577.

Plummer, P.S. & Gostin, V.A. (1981) Shrinkage cracks: desiccation or synaeresis. *J. sedim. Petrol.* **51**, 1147−1156.

Pomar, L., Fornos, J.J. & Rodriguez-Perea, A. (1985) Reef and shallow carbonate facies of the Upper Miocene of Mallorca. Excursion 11. In: *6th IAS European Regional Meeting. Lleida, Excursion Guidebook* (Ed. by M.D. Mila and J. Rosell) pp. 493−518. Institut d'Estudis Ilerdencs, Lleida.

Popp, B.N., Anderson, T.F. & Sandberg, P.A. (1986) Textural, elemental and isotopic variations among constituents in Middle Devonian limestones. *J. sedim. Petrol.* **56**, 715−727.

Popp, B.N. & Wilkinson, B.H. (1983) Holocene lacustrine ooids from Pyramid Lake, Nevada. In: *Coated Grains* (Ed. by T.M. Peryt) pp. 142−153. Springer-Verlag, Berlin.

Powers, D.W. & Easterling, R.G. (1982) Improved methodology for using embedded Markov chains to describe cyclical sediments. *J. sedim. Petrol.* **52**, 913−923.

Pratt, B.R. (1982a) Stromatolitic framework of carbonate mud mounds. *J. sedim. Petrol.* **52**, 1203−1227.

Pratt, B.R. (1982b) Stromatolite decline — a reconsideration. *Geology* **10**, 512−515.

Pratt, B.R. & James, N.P. (1986) The St George Group (Lower Ordovician) of western Newfoundland: tidal flat island model for carbonate sedimentation in shallow epeiric seas. *Sedimentology* **33**, 313−343.

Presley, M.W. & McGillis, K.A. (1982) Coastal evaporite and tidal flat sediments of the Upper Clear Fork and Glorieta Formations, Texas Panhandle. *Reports of Investigations*, **115**, pp. 50. Bureau of Economic Geology, University of Austin, Texas.

Prezbindowski, D.R. (1985) Burial cementation — is it important? A case study, Stuart City trend, south-central Texas. In: *Carbonate Cements* (Ed. by N. Schneidermann and P.M. Harris) Spec. Publ. Soc. econ. Paleont. Miner. 36, 241−264.

Purdy, E.G. (1963a) Recent calcium carbonate facies of the Great Bahama Bank. 1. Petrography and reaction groups. *J. Geol.* **71**, 334−355.

Purdy, E.G. (1963b) Recent carbonate facies of the Great Bahama Bank. II. Sedimentary facies. *J. Geol.* **71**, 472−497.

Purdy, E.G. (1974a) Reef configurations: cause and effect. In: *Reefs in Time and Space* (Ed. by L.F. Laporte) Spec. Publ. Soc. econ. Paleont. Miner. 18, 9−76.

Purdy, E.G. (1974b) Karst-determined facies patterns in British Honduras: Holocene carbonate sedimentation model. *Bull. Am. Ass. petrol. Geol.* **58**, 825−855.

Purser, B.H. (1969) Syn-sedimentary marine lithification of Middle Jurassic limestones in the Paris Basin. *Sedimentology* **12**, 205−230.

Purser, B.H. (Ed.) (1973a) *The Persian Gulf: Holocene Carbonate Sedimentation and Diagenesis in a Shallow Epicontinental Sea*, p. 471. Springer-Verlag, Berlin.

Purser, B.H. (1973b) Sedimentation around bathymetric highs in the southern Persian Gulf. In: *The Persian Gulf* (Ed. by B.H. Purser) pp. 157−177. Springer-Verlag, Berlin.

Purser, B.H. (1975) Tidal sediments and their evolution in the Bathonian carbonates of Burgundy, France. In: *Tidal Deposits: A Casebook of Recent Examples and Fossil Counterparts* (Ed. by R.N. Ginsburg) pp. 335−343. Springer-Verlag, Berlin.

Purser, B.H. (1978) Early diagenesis and the preservation of porosity in Jurassic limestones. *J. Petroleum Geol.* **1**, 83−94.

Purser, B.H. & Evans, G. (1973) Regional sedimentation along the Trucial Coast, SE Persian Gulf. In: *The Persian Gulf* (Ed. by B.H. Purser) pp. 211−232. Springer-Verlag, Berlin.

Purser, B.H. & Loreau, J.P. (1973) Aragonitic, supratidal encrustations on the Trucial Coast, Persian Gulf. In: *The Persian Gulf* (Ed. by B.H. Purser) pp. 343−376.

Springer-Verlag, Berlin.

Purser, B.H. & Schroeder, J.H. (1986) The diagenesis of reefs: a brief review of our present understanding. In: *Reef Diagenesis* (Ed. by J.H. Schroeder and B.H. Purser), pp. 424–446. Springer-Verlag, Berlin.

Purser, B.H., Soliman, M. & M'Rabet, A. (1987) Carbonate, evaporite, siliciclastic transitions in Quaternary rift sediments of the northwestern Red Sea. *Sedim. Geol.* 53, 247–268.

Pusey, W.C. (1975) Holocene carbonate sedimentation on northern Belize shelf. *Am. Ass. petrol. Geol. Studies in Geology* 2, 131–233.

Quine, M. (1988) *Sedimentology of the Upper Cretaceous Chalk of Normandy, France.* Unpubl. PhD Thesis, University of London (RHBNC).

Raaf, J.F.M. De, Boersma, J.R. & Gelder, A. Van (1977) Wave-generated structures and sequences from a shallow marine succession. Lower Carboniferous, County Cork, Ireland. *Sedimentology* 4, 1–52.

Racki, G. (1982) Ecology of primitive charophyte algae: a critical review. *Neues Jb. Geol. Paläont. Abh.* 162, 388–399.

Radke, B.H. & Mathis, R.L. (1980) On the formation and occurrence of saddle dolomite. *J. sedim. Petrol.* 50, 1149–1168.

Railsback, L.B. & Anderson, T.F. (1987) Control of Triassic seawater chemistry and temperature on the evolution of post-Paleozoic aragonite-secreting faunas. *Geology* 15, 1002–1005.

Raiswell, R. (1988) Chemical model for the origin of minor limestone–shale cycles by anaerobic methane oxidation. *Geology* 16, 641–644.

Raiswell, R. & Brimblecombe, P. (1977) The partition of manganese into aragonite between 30 and 60°C. *Chem. Geol.* 19, 145–151.

Ramsay, A.T.S. (1977) Sedimentological clues to palaeo-oceanography. In: *Oceanic Micropalaeontology* (Ed. by A.T.S. Ramsay), pp. 1371–1453. Academic Press, London.

Ramsay, A.T.S. (1987) Depositional environments in the Dinantian Limestones of Gower, South Wales. In: *European Dinantian Environments* (Ed. by J. Miller, A.E. Adams and V.P. Wright), pp. 265–308. John Wiley, Chichester.

Randazzo, A.F. & Cook, D.J. (1987) Characterization of dolomite rocks from the coastal mixing zone of the Floridan Aquifer, Florida, USA. *Sedim. Geol.* 54, 169–192.

Rao, C.O. & Naqvi, I.H. (1977) Petrography, geochemistry and factor analysis of a Lower Ordovician subsurface sequence, Tasmania, Australia. *J. sedim. Petrol.* 47, 1036–1055.

Rao, C.P. & Green, D.C. (1982) Oxygen and carbon isotopes of Early Permian cold-water carbonates, Tasmania, Australia. *J. sedim. Petrol.* 52, 1111–1125.

Rasmussen, K. & Brett, C.E. (1985) Taphonomy of Holocene cryptic biotas from St Croix, Virgin Islands: infor-mation loss and preservational bias. *Geology* 13, 551–553.

Raymo, M.E., Ruddiman, W.F. & Froelich, P.N. (1988) Influence of late Cenozoic mountain building on ocean geochemical cycles. *Geology* 16, 649–653.

Read, J.F. (1973) Carbonate cycles, Pillara Formation (Devonian), Canning Basin, Western Australia. *Bull. Can. petrol. Geol.* 21, 38–51.

Read, J.F. (1974a) Calcrete deposits and Quaternary sediments, Edel Province, Shark Bay, Western Australia. *Mem. Am. Ass. Petrol. Geol.* 12, 250–282.

Read, J.F. (1974b) Carbonate bank and wave-built platform sedimentation, Edel Province, Shark Bay, Western Australia. *Mem. Am. Ass. petrol. Geol.* 22, 1–60.

Read, J.F. (1975) Tidal flat facies in carbonate cycles, Pillara Formation (Devonian), Canning Basin, Western Australia. In: *Tidal Deposits: A Casebook of Recent Examples and Fossil Counterparts* (Ed. by R.N. Ginsburg) pp. 251–256. Springer-Verlag, Berlin.

Read, J.F. (1980) Carbonate ramp-to-basin transitions and foreland basin evolution. Middle Ordovician, Virginia Appalachians. *Bull. Am. Ass. petrol. Geol.* 64, 1575–1612.

Read, J.F. (1982) Carbonate platforms of passive (extensional) continental margins: types, characteristics and evolution. *Tectonophys.* 81, 195–212.

Read, J.F. (1985) Carbonate platform facies models. *Bull. Am. Ass. petrol. Geol.* 66, 860–878.

Read, J.F., Grotzinger, J.P., Bova, J.A. & Koerschner, W.F. (1986) Models for generation of carbonate cycles. *Geology* 14, 107–110.

Read, J.F. & Grover, G.A. (1977) Scalloped and planar erosion surfaces, middle Ordovician limestones, Virginia; analogues of Holocene exposed karst or tidal rock platforms. *J. sedim. Petrol.* 47, 956–972.

Read, J.F. & Pfeil, R.W. (1983) Fabrics of allochthonous reefal blocks, shady Dolomite (Lower to Middle Cambrian), Virginia Appalachians. *J. sedim. Petrol.* 53, 761–778.

Reading, H.G. (1986) Facies. In: *Sedimentary Environments and Facies* (Ed. by H.G. Reading) pp. 4–19. Blackwells, Oxford.

Reeckman, S.A. & Gill, E.D. (1981) Rates of vadose diagenesis in Quaternary dune and shallow marine calcarenites, Warrnambool, Victoria, Australia. *Sedim. Geol.* 30, 157–172.

Reed, J.K. (1980) Distribution and structure of deep water *Oculina varicosa* coral reefs off central eastern Florida. *Bull. Mar. Sci.* 30, 667–677.

Reeder, R.J. (1981) Electron optical investigation of sedimentary dolomites. *Contrib. Min. Pet.* 76, 148–157.

Reeder, R.J. (1983a) (Ed.) Carbonates, mineralogy and chemistry. *Min. Soc. Am. Rev. in Mineral.* 11, 450.

Reeder, R.J. (1983b) Crystal chemistry of the rhombohedral carbonates. *Min. Soc. Am. Rev. in Min.* 11, 1–48.

Reeder, R.J. & Prosky, J.L. (1986) Composition sector zoning in dolomite. *J. sedim. Petrol.* 56, 237–247.

Reeder, R.J. & Wenk, H.R. (1979) Microstructures in low temperature dolomites. *Geophys. Res. Letts.* **6**, 77—80.

Reid, R.P. (1987) Non-skeletal peloidal precipitates in Upper Triassic Reefs, Yukon Territory (Canada). *J. Sedim. Petrol.* **57**, 893—900.

Reid, R.P. & Ginsburg, R.N. (1986) The role of framework in Upper Triassic patch reefs in the Yukon (Canada). *Palaios* **1**, 590—600.

Reijers, T.J.A. & ten Have, A.H.M. (1983) Ooid zonation as indication for environmental conditions in a Givetian-Frasnian carbonate shelf-slope transition. In: *Coated Grains* (Ed. by T.M. Peryt) pp. 188—198. Springer-Verlag, Berlin.

Reinhardt, J. (1977) Cambrian off-shelf sedimentation, Central Appalachians. In: *Deepwater Carbonate Environments* (Ed. by H.E. Cook and P. Enos) Spec. Publ. Soc. Econ. Paleont. Miner. 25, 83—112.

Reinson, G.E. (1984) Barrier Island and associated strand-plain systems. In: *Facies Models* (Ed. by R.G. Walker) pp. 119—140. Geoscience Canada.

Rich, M. (1982) Ooid cortices composed of neomorphic pseudospar: possible evidence for ancient originally aragonitic ooids. *J. Sedim. Petrol.* **52**, 843—847.

Richter, D.K. (1979) Die Stufen der meteorisch-vadosen Umwandlung von Mg-Calcit in Calcit in rezenten die pliozanen Biogenen Griechen lands. *Neues Jb. Geol. Paläont. Abh.* **158**, 277—333.

Richter, D.K. (1983a) Classification of coated grains: discussion. In: *Coated Grains* (Ed. by T.M. Peryt) pp. 7—8. Springer-Verlag, Berlin.

Richter, D.K. (1983b) Calcareous ooids: a synopsis. In: *Coated Grains* (Ed. by T.M. Peryt) pp. 71—99. Springer-Verlag, Berlin.

Richter, D.K. & Füchtbauer, H. (1978) Ferroan calcite replacement indicates former magnesian calcite skeletons. *Sedimentology* **25**, 843—860.

Richter, D.K., Herforth, A. & Ott, E. (1979) Brackish blue—green algal bioherms composed of *Rivularia haematites* in Pleistocene deposits of the Perachora Peninsula, near Korinthos (Greece). *Neues Jb. Geol. Paläont. Abh.* **159**, 14—40.

Ricken, W. (1986) *Diagenetic Bedding. Lecture Notes in Earth Sciences, 6,* pp. 210. Springer-Verlag, Berlin.

Ricken, W. (1987) The carbonate compaction law: a new tool. *Sedimentology* **34**, 571—584.

Ricketts, B.D. (1983) The evolution of a Middle Precambrian Dolostone sequence — a spectrum of dolomitization regimes. *J. Sedim. Petrol.* **53**, 565—586.

Riding, R. (1977) Reef concepts. *Proc. 3rd Int. Coral Reef Symp. Miami* 2, 209—213.

Riding, R. (1979) Origin and diagenesis of lacustrine algal bioherms at the margin of the Ries Crater, Upper Miocene, southern Germany. *Sedimentology* **26**, 645—680.

Riding, R. (1981) Composition, structure and environmental setting of Silurian bioherms and biostromes in northern Europe. In: *European Reef Models* (Ed. by D.F. Toomey)
Spec. Publ. Soc. econ. Paleont. Miner. 30, 9—40.

Riding, R. (1982) Cyanophyte calcification and changes in ocean chemistry. *Nature* **299**, 814—815.

Riding, R. (1983) Cyanoliths (cyanoids): oncoids formed by calcified cyanophytes. In: *Coated Grains* (Ed. by T.M. Peryt) pp. 277—283. Springer-Verlag, Berlin.

Riding, R. & Wright, V.P. (1981) Paleosols and tidal-flat/lagoon sequences on a Carboniferous carbonate shelf: sedimentary associations of triple disconformities. *J. Sedim. Petrol.* **51**, 1323—1339.

Rigby, J.K. & Roberts, H.H. (1976) Grand Cayman Island: Geology, sediments and marine communities. *Spec. Publ. Brigham Young University, Studies in Geology* 4, 1—95.

Risacher, F. & Eugster, H.P. (1979) Holocene pisoliths and encrustations associated with spring-fed surface ponds, Pastos Grandes, Bolivia. *Sedimentology* **26**, 253—270.

Risk, M.J. & MacGeachy (1978) Aspects of bioerosion of modern Caribbean reefs. *Rev. Biol. Trop.* (Suppl.) **26**, 85—105.

Roberts, H.H., Phipps, C.V. & Effendi, L. (1987) *Halimeda* bioherms of the eastern Java Sea, Indonesia. *Geology* **15**, 371—374.

Robertson, A.H.F. (1976) Origin of ochres and umbers: evidence from Skouriotissa, Troodos Massif, Cyprus. *Trans. Inst. Min. Metall. B.* **85**, 245—251.

Robertson, A.H.F. (1977) The origin and diagenesis of cherts from Cyprus. *Sedimentology* **24**, 11—30.

Robertson, A.H.F. & Hudson, J.D. (1973) Cyprus umbers: chemical precipitates on a Tethyan ocean ridge. *Earth planet. Sci. Letts.* **18**, 93—101.

Robertson, A.H.F. & Hudson, J.D. (1974) Pelagic sediments in the Cretaceous and Tertiary history of the Troodos Massif, Cyprus. In: *Pelagic Sediments: on Land and Under the Sea* (Ed. by K.J. Hsü and H.C. Jenkyns) Spec. Publ. int. Ass. Sediment. 1, 403—436.

Robin, P. (1978) Pressure solution at grain-to-grain contacts. *Geochim. Cosmochim. Acta* **42**, 1383—1389.

Rodgers, K.A., Easton, A.J. & Downes, C.J. (1982) The chemistry of carbonate rocks of Niue Island, South Pacific. *J. Geol.* **90**, 645—662.

Roehl, P.O. & Choquette, P.W. (Eds) (1985) *Carbonate Petroleum Reservoirs*, p. 622. Springer-Verlag, New York.

Ronov, A.B. (1964) Common tendencies in the chemical evolution of the Earth's crust, ocean and atmosphere. *Geochem. Int.* **4**, 713—737.

Rosen, B.R. (1975) The distribution of reef corals. *Rep. Underwater Ass.* **1**, 1—16.

Rosenberg, P.E. & Foit, F.F. (1979) The stability of transition metal dolomites in carbonate systems, a discussion. *Geochim. Cosmochim. Acta* **43**, 951—955.

Ruiz-Ortiz, P.A. (1983) A carbonate submarine fan in a fault-controlled basin of the Upper Jurassic, Betic Cordillera, southern Spain. *Sedimentology* **30**, 33—48.

Runnels, D.D. (1969) Diagenesis, chemical sediments and the mixing of natural waters. *J. sedim. Petrol.* **39**, 1188—1201.

Runnels, D.D. (1970) Errors in X-ray analysis of carbonates due to solid solution variation in the composition of component minerals. *J. sedim. Petrol.* **40**, 1156–1166.

Ruppel, S.C. & Walker, K.R. (1984) Petrology and depositional history of a Middle Ordovician carbonate platform: Chickamanga Group, northeastern Tennessee. *Bull. Geol. Soc. Am.* **95**, 568–583.

Rutzler, K. (1975) The role of burrowing sponges in bioerosion. *Oceologia* **19**, 203–216.

Rutzler, K. & Macintyre, I.G. (1982) The habitat, distribution and community structure of the barrier reef complex at Carrie Bow Cay, Belize. In: *The Atlantic Barrier Reef Ecosystem at Carrie Bow Cay, Belize. 1 Structure and Communities* (Ed. by K. Rutzler and I.G. Macintyre) Smithsonian Contrib. Mar. Sci. 12, 9–45.

Ryder, R.T., Fouch, T.D. & Elson, J.H. (1976) Early Tertiary sedimentation in the western Uinta Basin, Utah. *Bull. Geol. Soc. Am.* **87**, 496–512.

Saller, A.H. (1984) Petrologic and geochemical constraints on the origin of subsurface dolomite, Enewetak Atoll: an example of dolomitization by normal seawater. *Geology* **12**, 217–220.

Saller, A.H. (1986) Radiaxial calcite in Lower Miocene strata, subsurface Enewetak Atoll. *J. Sedim. Petrol.* **56**, 743–762.

Salomons, W. & Mook, W.G. (1986) Isotope geochemistry of carbonates in the weathering zone. In: *Handbook of Environmental Isotope Geochemistry, 2. The Terrestrial Environment, B*, pp. 239–269. Elsevier, Amsterdam.

Sandberg, P.A. (1975) New interpretations of Great Salt Lake ooids and of ancient non-skeletal carbonate mineralogy. *Sedimentology* **22**, 497–538.

Sandberg, P.A. (1980) The Pliocene Glenns Ferry Oolite: Lake margin carbonate deposition in the southwestern Snake River Plain — discussion. *J. sedim. Petrol.* **50**, 997–998.

Sandberg, P.A. (1983) An oscillating trend in Phanerozoic non-skeletal carbonate mineralogy. *Nature* **305**, 19–22.

Sandberg, P.A. (1985) Aragonite cements and their occurrence in ancient limestones. In: *Carbonate Cements* (Ed. by N. Schneidermann and P.M. Harris) Spec. Publ. Soc. econ. Paleont. Miner. 36, 33–57.

Sandberg, P.A. & Hudson, J.D. (1983) Aragonite relic preservation in Jurassic calcite-replaced bivalves. *Sedimentology* **30**, 879–892.

Sander, B. (1951) *Einführung in die Gefügekunde als Geologischer Korpen 2, Teil. Die Korngefügemerkmale*, pp. 409. Springer-Verlag, Berlin.

Sanders, J.E. & Kumar, N. (1975) Evidence of shoreface retreat and in-place 'drowning' during Holocene submergence of barriers, shelf off Fire Island, New York. *Bull. geol. Soc. Am.* **86**, 65–76.

Santisteban, C. & Taberner, C. (1983) Shallow marine and continental conglomerates derived from coral reef complexes after desiccation of a deep marine basin: the Tortonian-Messinian deposits of the Fortuna Basin, south east Spain. *J. Geol. Soc.* **140**, 401–411.

Sarg, J.F. (1988) Carbonate sequence stratigraphy. In: *Sea-Level Changes: an Integral Approach* Spec. Publ. Soc. econ. Paleont. Miner. 42, 155–181.

Sarkar, S., Chanda, S.K. & Bhattacharyya, A. (1982) Soft sediment deformation fabric in the Precambrian Bhanda Oolite, Central India. *J. sedim. Petrol.* **52**, 95–107.

Sassen, R. (1988) Geochemical and carbon isotopic studies of crude oil destruction, bitumen precipitation and sulfate reduction in the deep Smackover Formation. *Org. Geochem.* **30**, 79–91.

Schäfer, A. & Stapf, K.R.G. (1978) Permian Saar-Nahe basin and Recent Lake Constance (Germany): two environments of lacustrine algal carbonates. In: *Modern and Ancient Lake Sediments* (Ed. by A. Matter and M.E. Tucker) Spec. Publ. int. Ass. Sediment. 2, 83–107.

Schäfer, K. (1969) Vergleichs-Schaubilder zur Bestimmung des Allochemgehalts bioklastischer Karbonatgesteine. *N. Jb. geol. Palaont. Mh.* **3**, 173–184.

Schäfer, P. & Senowbari-Daryan, B. (1981) Facies development and paleoecologic zonation of four Upper Triassic patch-reefs, Northern Calcareous Alps near Salzburg, Austria. In: *European Fossil Reef Models* (Ed. by D.F. Toomey) Spec. Publ. Soc. Econ. Paleont. Miner. 30, 241–259.

Schenk, P. (1969) Carbonate–sulfate redbed facies and cyclic sedimentation of the Windsorian Stage (M. Carboniferous) Maritime Provinces. *Can. J. earth Sci.* **6**, 1019–1066.

Scherer, M. (1977) Preservation, alteration and multiple cementation of aragonite skeletons from the Cassian Beds (U. Triassic, Southern Alps), petrographic and geochemical evidence. *Neues Jb. Geol. Paläont. Abh.* **154**, 213–262.

Schlager, W. (1974) Preservation of cephalopod skeletons and carbonate dissolution on ancient Tethyan sea floors. In: *Pelagic Sediments: on Land and Under the Sea* (Ed. by K.J. Hsü and H.C. Jenkyns) Spec. Publ. Int. Ass. Sediment. 1, 49–70.

Schlager, W. (1981) The paradox of drowned reefs and carbonate platforms. *Bull. geol. Soc. Am.* **92**, 197–211.

Schlager, W. & Chermak, A. (1979) Sediment facies of platform-basin transition, Tongue of the Ocean, Bahamas. In: *Geology of Continental Slopes* (Ed. by L.J. Doyle and O.H. Pilkey) Spec. Publ. Soc. econ. Paleont. Miner. 27, 193–207.

Schlager, W. & Ginsburg, R.N. (1981) Bahama carbonate platforms — the deep and the past. *Marine Geol.* **44**, 1–24.

Schlager, W. & James, N.P. (1978) Low-magnesian calcite limestones forming at the deep-sea floor, Tongue of the Ocean, Bahamas. *Sedimentology* **25**, 675–702.

Schlanger, S.O. (1964) Petrology of the limestones of Guam. *US Geol. Surv. Prof. Papers*, **403-D**, 1–52.

Schlanger, S.O., Arthur, M.A., Jenkyns, H.C. & Scholle, P.A. (1987) The Cenomanian–Turonian Oeanic Anoxic Event, 1. Stratigraphy and distribution of organic carbon-rich beds and the marine $\delta^{13}C$ excursion. In:

Marine Petroleum Source Rocks (Ed. by J. Brooks and A.J. Fleet) Spec. Publ. Geol. Soc. Lond. 26, 371–400.

Schlanger, S.O. & Douglas, R.G. (1974) The pelagic ooze–chalk–limestone transition and its implications for marine stratigraphy. In: *Pelagic Sediments: on Land and Under the Sea* (Ed. by K.J. Hsü and H.C. Jenkyns) Spec. Publ. Int. Ass. Sediment. 1, 117–148.

Schlanger, S.O. & Jenkyns, H.C. (1976) Cretaceous oceanic anoxic events: causes and consequences. *Geol. Mijnb.* 55, 179–184.

Schmoker, J.W. & Halley, R.B. (1982) Carbonate porosity versus depth — a predictable relation for south Florida. *Bull. Am. Ass. Petrol. Geol.* 66, 2561–2570.

Schmoker, J.W. & Hester, T.C. (1986) Porosity of the Miami Limestone (Late Pleistocene), lower Florida Keys. *J. Sedim. Petrol.* 56, 629–634.

Schneider, J. (1977) Carbonate construction and decomposition by epilithic and endolithic micro-organisms in salt and freshwater. In: *Fossil Algae* (Ed. by E. Flügel) pp. 248–260. Springer-Verlag, Berlin.

Schneider, J., Schröder, H.G. & Le Campion-Alsumard, Th. (1983) Algal micro-reefs — coated grains from freshwater environments. In: *Coated Grains* (Ed. by T. Peryt) pp. 284–298. Springer-Verlag, Berlin.

Schneider, J.F. (1975) Recent tidal deposits, Abu Dhabi, United Arab Emirates, Arabian Gulf. In: *Tidal Deposits: A Casebook of Recent Examples and Fossil Counterparts* (Fd. by R.N. Ginsburg) pp. 209–214. Springer-Verlag, Berlin.

Scnneidermann, N. & Harris, P.M. (Eds.) (1985) *Carbonate Cements.* Spec. Publ. Soc. Econ. Paleont. Miner. 36, pp. 379.

Schoell, M. (1983) Genetic characterization of natural gases. *Bull. Am. Ass. Petrol. Geol.* 67, 2225–2238.

Schoell, M. (1984) Recent advances in petroleum isotope geochemistry. *Org. Geochem.* 6, 645–663.

Scholle, P.A. (1975) Chalk diagenesis. In: *Petroleum and the Continental Shelf of North-West Europe* (Ed. by A.W. Woodland) pp. 420–425. Applied Science Publ., London.

Scholle, P.A. (1977) Chalk diagenesis and its relation to petroleum exploration: oil from chalks, a modern miracle? *Bull. Am. Ass. Petrol. Geol.* 61, 982–1009.

Scholle, P.A. (1978) A color illustrated guide to carbonate rock constituents, textures, cements and porosities. *Mem. Am. Ass. petrol. Geol.* 27, 241.

Scholle, P.A. & Arthur, M.A. (1980) Carbon-isotope fluctuations in Cretaceous pelagic limestones: potential stratigraphic and petroleum exploration tool. *Bull. Am. Ass. petrol. Geol.* 64, 67–87.

Scholle, P.A., Arthur, M.A. & Ekdale, A.A. (1983) Pelagic environments. In: *Carbonate Depositional Environments* (Ed. by P.A. Scholle, D.G. Bebout and C.H. Moore) Mem. Am. Ass. Petrol. Geol. 33, 620–691.

Scholle, P.A., Bebout, D.G. & Moore, C.H. (Eds) (1983) *Carbonate Depositional Environments.* Mem. Am. Ass. petrol. Geol. 33, pp. 708.

Scholle, P.A. & Halley, R.B. (1985) Burial diagenesis: out of sight, out of mind. In: *Carbonate Cements* (Ed. by N. Schneidermann and P.M. Harris) Spec. Publ. Soc. Econ. Paleont. Miner. 36, 309–334.

Scholle, P.A. & Kinsman, D.J.J. (1974) Aragonite and high magnesian calcite caliche from the Persian Gulf, a modern analogy for the Permian of Texas and New Mexico. *J. sedim. Petrol.* 44, 904–916.

Schopf, T.J.M. (1980) *Paleooceanography*, p. 341. Harvard University Press, Mass.

Schreiber, B.C., Tucker, M.E. & Till, R. (1986) Arid shorelines and evaporites. In: *Sedimentary Environments and Facies* (Ed. by H.G. Reading) pp. 189–228. Blackwells, Oxford.

Schröder, H.G., Windolph, H. & Schneider, J. (1983) Balance of the biogenic calcium carbonate production in an oligotrophic lake (Lake Attersee, Salzkammergut, Austria). *Arch. Hydrobiol.* 97, 356–372.

Schroeder, J.H. (1972a) Fabrics and sequences of submarine carbonate cements in Holocene Bermuda cup reefs. *Geol. Rundsch.* 61, 708–730.

Schroeder, J.H. (1972b) Calcified filaments of an endolithic alga in Recent Bermuda reefs. *Neues Jb. Geol. Paläont. Monatsh.* Vol. 16–33.

Schroeder, J.H. & Purser, B.H. (Eds.) (1986) *Reef Diagenesis*, pp. 455. Springer-Verlag, Berlin.

Schroeder, J.H. & Zankl, H. (1974) Dynamic reef formation: a sedimentological concept based on studies of recent Bermuda and Bahama reefs. *Proc. 2nd Int. Coral Reef Symp. Brisbane* 2, 413–428.

Schwab, F.L. (1978) Secular trends in the composition of sedimentary rock assemblages — Archean through Phanerozoic time. *Geology* 6, 532–536.

Schwarzacher, W. & Fischer, A.G. (1982) Limestone-shale bedding and perturbations of the Earth's orbit. In: *Cyclic and Event Stratification* (Ed. by G. Einsele and A. Seilacher) pp. 72–95. Springer-Verlag, Berlin.

Scoffin, T.P. (1970) The trapping and binding of subtidal carbonate sediments by marine vegetation in Bimini Lagoon, Bahamas. *J. sedim. Petrol.* 40, 249–273.

Scoffin, T.P. (1972) Cavities in the reefs of the Wenlock Limestone (Mid Silurian) of Shropshire, England. *Geol. Rundschau* 61, 565–578.

Scoffin, T.P. (1987) *An Introduction to Carbonate Sediments and Rocks*, pp. 274. Blackie, Glasgow.

Scoffin, T.P. & Garrett, P. (1974) Processes in the formation and preservation of internal structure in Bermuda patch reefs. *Proc. 2nd Int. Coral Reef Symp. Brisbane* 2, 429–448.

Scoffin, T.P. & Hendry, M.D. (1984) Shallow-water sclerosponges on Jamaican reefs and a criterion for recognition of hurricane deposits. *Nature* 307, 728–729.

Scoffin, T.P. & McLean, R.F. (1978) Exposed limestones of the northern province of the Great Barror Reef. *Phil. Trans. Roy. Soc. Lond.* A. 291, 119–138.

Scoffin, T.P., Stearn, C.W., Boucher, D., Frydl, P., Hawkins, C.M., Hunter, I.G. & MacGeachy, J.K. (1980)

Calcium carbonate budget of a fringing reef on the west coast of Barbados, 2 — erosion, sediments and internal structure. *Bull. Mar. Sci.* **30**, 475–508.

Scoffin, T.P. & Stoddart, D.R. (1983) Beachrock and intertidal cements. In: *Chemical Sediments and Geomorphology* (Ed. by A.S. Goudie and K. Pye) pp. 401–425. Academic Press, London.

Scoffin, T.P., Stoddart, D.R., McLean, R.F. & Flood, P.G. (1978) The recent development of the reefs in the Northern Province of the Great Barrier Reef. *Phil. Trans. R. Soc. Lond. B* **284**, 129–139.

Scoffin, T.P. & Tudhope, A.W. (1985) Sedimentary environments of the central region of the Great Barrier Reef of Australia. *Coral Reefs* **4**, 81–93.

Sears, S.O. & Lucia, F.J. (1980) Dolomitization of northern Michigan Niagara reefs by brine refluxion and freshwater/seawater mixing. In: *Concepts and Models of Dolomitization* (Ed. by D.H. Zenger, J.B. Dunham and R.L. Ethington) Spec. Publ. Soc. econ. Paleont. Miner. **28**, 215–235.

Sellwood, B.W. (1971) A *Thalassinoides* burrow containing the crustacean *Glyphaea udressieri* (Meyer) from the Bathonian of Oxfordshire. *Palaeontology* **14**, 589–591.

Sellwood, B.W. (1986) Shallow-marine carbonate environments. In: *Sedimentary Environments and Facies* (Ed. by H.G. Reading) pp. 283–342. Blackwells, Oxford.

Sellwood, B.W. & Netherwood, R.E. (1984) Facies evolution in the Gulf of Suez area: sedimentation history as an indicator of rift initiation and development. *Mod. Geol.* **9**, 43–69.

Sellwood, B.W., Scott, J., Mikkelsen, P. & Akroyd, P. (1985) Stratigraphy and sedimentology of the Great Oolite Group in the Humbly Grove oilfield, Hampshire, S. England. *Marine petrol. Geol.* **2**, 44–55.

Seminiuk, V. & Meagher, T.D. (1981) Calcrete in Quaternary coastal dunes in southwestern Australia: a capillary rise phenomenon associated with plants. *J. Sedim. Petrol.* **51**, 47–68.

Senowbari-Daryan, B. (1978) Anomuren-Koprolithen aus der Obertrias der Osterhorngruppe (Hintersee/Salzburg, Österreich). *Ann. naturhist. Mus. Wien* **82**, 99–107.

Shaver, R.H. (1977) Silurian reef geometry — new dimensions to explore. *J. sedim. Petrol.* **47**, 1409–1424.

Shawkat, M.G. & Tucker, M.E. (1978) Stromatolites and sabkha cycles from the Miocene of Iraq. *Geol. Rundsch.* **67**, 1–14.

Shearman, D.J. (1978) Evaporites of coastal sabkhas. In: *Marine Evaporites* (Ed. by W.E. Dean and B.C. Schreiber) Soc. econ. Paleont. Miner. Short Course 4, 6–42.

Shearman, D.J. & Smith, A.J. (1985) Ikaite, the parent mineral of jarrowite-type pseudomorphs. *Proc. geol. Ass.* **96**, 305–314.

Shearman, D.J., Twyman, J. & Karmi, M.Z. (1970) The diagenesis of oolites. *Proc. geol. Ass.* **81**, 561–575.

Sheppard, S.M.F. & Schwarcz, H.P. (1970) Fractionation of carbon and oxygen isotopes and magnesium between coexisting metamorphic calcite and dolomite. *Contrib.*

Miner. Pet. **26**, 161–198.

Shimmield, G.B. & Price, N.B. (1984) Recent dolomite formation in hemipelagic sediments off Baja, California, Mexico. In: *Dolomites of the Monterey Formation and Other Organic-rich Units* (Ed. by R.E. Garrison and D.H. Zenger) Publ. Soc. econ. Paleont. Miner. Pacific Section 41, 5–18.

Shinn, E.A. (1963) Spur and groove formation on the Florida Reef Tract. *J. Sedim. Petrol.* **33**, 291–303.

Shinn, E.A. (1968a) Practical significance of birds eye structures in carbonate rocks. *J. sedim. Petrol.* **38**, 215–223.

Shinn, E.A. (1968b) Selective dolomitization of recent sedimentary structures. *J. Sedim. Petrol.* **38**, 612–616.

Shinn, E.A. (1969) Submarine lithification of Holocene carbonate sediments in the Persian Gulf. *Sedimentology* **12**, 109–144.

Shinn, E.A. (1973) Carbonate coastal accretion in an area of longshore transport, northeast Qatar, Persian Gulf. In: *The Persian Gulf* (Ed. by B.H. Purser) pp. 179–191. Springer-Verlag, Berlin.

Shinn, E.A. (1980) Geologic history of Grecian rocks, Key Largo Coral Reef Marine Sanctuary. *Bull. Mar. Sci.* **30**, 646–656.

Shinn, E.A. (1983a) Tidal flat environment. In: *Carbonate Depositional Environments* (Ed. by P.A. Scholle, D.G. Bebout and C.H. Moore) Mem. Am. Ass. petrol. Geol. **33**, 173–210.

Shinn, E.A. (1983b) Birdseyes, fenestrae, shrinkage pores and loferites: a re-evaluation. *J. sedim. Petrol.* **53**, 619–629.

Shinn, E.A. (1986) Modern carbonate tidal flats: their diagnostic features. *Quart. J. Colorado Sch. Mines* **81**, 7–35.

Shinn, E.A., Ginsburg, R.N. & Lloyd, R.M. (1965) Recent supratidal dolomite from Andros Island, Bahamas. In: *Dolomitization and Limestone Diagenesis, a Symposium* (Ed. by L.C. Pray and R.D. Murray) Spec. Publ. Soc. econ. Paleont. Miner. **13**, 112–123.

Shinn, E.A., Halley, R.B., Hudson, J.H. & Lidz, B.H. (1977) Limestone compaction — an enigma. *Geology* **5**, 21–24.

Shinn, E.A., Hudson, J.H., Halley, R.B. & Lidz, B. (1977) Topographic control and accumulation rate of some Holocene coral reefs: south Florida and Dry Tortugas. *Proc. 3rd Int. Coral Reef Symp. Miami* **2**, 1–7.

Shinn, E.A., Hudson, J.H., Halley, R.B., Lidz, B., Robbin, D.M. & Macintyre, I.G. (1982) Geology and sediment accumulation rates at Barrie Bow Cay, Belize. In: *The Atlantic Barrier Reef Ecosystem at Barrie Bow Cay, Belize, 1 Structure and Communities* (Ed. by K. Rutzler and I.G. Macintyre) Smithsonian Contrib. Mar. Sci. 12, 63–75. Smithsonian Inst. Press, Washington.

Shinn, E.A., Hudson, J.H., Robbin, D.M. & Lidz, B. (1981) Spurs and grooves revisited: construction versus erosion, Looe Key Reef, Florida. *Proc. Fourth Internat. Coral Reef Symp. Manila* **1**, 475–483.

Shinn, E.A. & Lidz, B.H. (1988) Blackened limestone pebbles: fire at subaerial unconformities. In: *Paleokarst*

(Ed. by N.P. James and P.W. Choquette) pp. 117–131. Springer-Verlag, New York.

Shinn, E.A., Lloyd, R.M. & Ginsburg, R.N. (1969) Anatomy of a modern carbonate tidal flat, Andros Island, Bahamas. *J. Sedim. Petrol.* **39**, 1202–1228.

Shinn, E.A. & Robbin, D.M. (1983) Mechanical and chemical compaction in fine grained shallow-water limestones. *J. sedim. Petrol.* **53**, 595–618.

Shinn, E.A., Steinen, R.P., Lidz, B.H. & Swart, R.K. (1989) Whitings, a sedimentologic dilemma, *J. Sedim. Petrol.* **59**, 147–161.

Sibley, D.F. (1980) Climatic control of dolomitization, Seroe Domi Formation (Pliocene), Bonaire, N.A. In: *Concepts and Models of Dolomitization* (Ed. by D.H. Zenger, J.B. Dunham and R.L. Ethington) Spec. Publ. Soc. econ. Paleont. Miner. 28, 247–258.

Sibley, D.F. (1982) The origin of common dolomite fabrics. *J. sedim. Petrol.* **52**, 1987–1100.

Sibley, D.F. & Gregg, J.M. (1987) Classification of dolomite rock texture. *J. sedim. Petrol.* **57**, 967–975.

Simms, M. (1984) Dolomitization by groundwater-flow systems in carbonate platforms. *Trans. Gulf-Cst, Ass. geol. Socs* **34**, 411–420.

Simo, A. (1986) Carbonate platform depositional sequences, Upper Cretaceous, south-central Pyrenees (Spain). *Tectonophys.* **129**, 205–231.

Simone, L. (1974) Genesi e significato ambientale degli ooidi a struttura fibroso-raggiata di alcuni depositi mesozocici dell'area Appenino-dinarica e delle Bahamas merdionali. *Boll. Soc. Geol. Ital.* **93**, 513–545.

Simone, L. (1980) Ooids: a review. *Earth Sci. Rev.* **16**, 319–355.

Simpson, J. (1985) Stylolite-controlled layering in a homogeneous limestone: pseudo-bedding produced by burial diagenesis. *Sedimentology* **32**, 495–505.

Singh, U. (1987) Ooids and cements from the Late Precambrian of the Flinders Ranges, South Australia. *J. Sedim. Petrol.* **57**, 117–127.

Skelton, P. (1986) The tyranny of uniformitarianism: the present as the key to geofantasies. *Reef Encounter* **3**, 5–6.

Slingerland, R. (1986) Numerical computation of co-oscillating paleotides in the Catskill epeiric sea of eastern North America. *Sedimentology* **33**, 487–498.

Sly, P.G. (1978) Sedimentary processes in lakes. In: *Lakes: Physics, Chemistry and Geology* (Ed. by A. Lerman), pp. 166–200. Springer-Verlag, Berlin.

Smart, P.L., Palmer, R.J., Whitaker, F. & Wright, V.P. (1988a) Neptunean dykes and fissure fills; an overview and account of some modern examples. In: *Paleokarst* (Ed by N.P. James and P.W. Choquette) pp. 149–162. Springer-Verlag, New York.

Smart, P.L., Dawans, J.M. & Whitaker, F. (1988b) Carbonate dissolution in a modern mixing zone. *Nature* **335**, 811–813.

Smith, D.B. (1974) Origin of tepees in Upper Permian shelf carbonate rocks of Guadalupe Mountains, New Mexico.

Bull. Am. Ass. petrol. Geol. **58**, 63–70.

Smith, D.B. (1981) Bryozoan–algal patch reefs in the Upper Permian Magnesian limestones of Yorkshire, Northeast England. In: *European Fossil Reef Models* (Ed. by D.F. Toomey) Spec. Publ. Soc. econ. Paleont. Miner. 30, 187–202.

Smith, S.V. (1973) Carbon dioxide dynamics: a record of organic carbon production, respiration and calcification in the Enewetak windward reef flat community. *Limnol. Oceanogr.* **18**, 106–120.

Smoot, J.P. (1978) Origin of the carbonate sediments in the Wilkins Peak. Member of the lacustrine Green River Formation (Eocene), Wyoming, USA. In. Modern and Ancient Lake Sediments (Ed. by A. Matter & M.E. Tucker) pp. 109–127. Spec. Publ. int. Assoc. Sedim. 2.

Smoot, J.P. (1983) Depositional subenvironments in an arid closed basin: the Wilkins Peak Member of the Green River Formation (Eocene), Wyoming, USA. *Sedimentology* **30**, 801–827.

Smosna, R. (1987) Compositional maturity of limestones — a review. *Sedim. Geol.* **51**, 137–146.

Smosna, R. & Warshauer, S.M. (1979) A scheme for multi-variate analysis in carbonate petrology with an example from the Silurian Tonoloway Limestone. *J. Sedim. Petrol.* **49**, 257–272.

Smosna, R. & Warshauer, S.M. (1981) Rank exposure index on a Silurian carbonate tidal flat. *Sedimentology* **28**, 723–731.

Snedden, J.W. & Nummedal, D. (1990) Origin and geometry of storm-deposited sand beds in modern sediments of the Texas continental shelf. In: *Shelf Sand and Sandstone Bodies: Geometry, Facies and Distribution* (Ed. by D.J.P. Swift, G.F. Oertel, R.W. Tillman and M.E. Tucker) Spec. Publ. Int. Ass. Sediment. 11, In press.

Sneh, A. & Friedman, G.M. (1980) Spur and groove patterns on the reefs of the northern Gulf of the Red Sea. *J. sedim. Petrol.* **50**, 981–986.

Somerville, I.D. (1979) Minor sedimentary cyclicity in late Asbian limestones in the Llangollen district of North Wales. *Proc. Yorks. geol. Soc.* **42**, 317–341.

Sorby, H.C. (1879) The structure and origin of limestones. *Proc. geol. Soc. Lond.* **35**, 56–95.

Speer, J.A. (1983) Crystal chemistry and phase relations of orthorhombic carbonates. In: *Carbonates: Mineralogy and Chemistry* (Ed. by R.J. Reeder) Reviews in Mineralogy, 11, 145–190. Mineralogical Society of America, Washington.

Spencer-Davies, P., Stoddart, D.R. & Siege, D.C. (1971) Reef forms of Addu Atoll, Maldive Islands. In: *Regional Variations in Indian Ocean Coral Reefs* (Ed. by D.R. Stoddart and M. Yonge) pp. 217–259. Symp. Zool. Soc. Lond., 28.

Spencer, R.C., Eugster, H.R., Kelts, K., McKenzie, J., Jones, B.F., Baedecker, M.J., Rettig, S.L., Goldhaber, M.B. & Bowser, E.J. (1981) Late Pleistocene and Holocene sedimentary history of Great Salt Lake, Utah. Abstr. Am. Ass. petrol. Geol. Mtg. San Francisco.

Spencer, R.J., Baedecker, M.J., Eugster, H.P., Forester, R.M., Goldhaber, M.B., Jones, M.F., Kelts, K., McKenzie, J., Madsen, D.B., Rettig, S.L., Rubin, M. & Bowser, C.J. (1984) Great Salt Lake, and precursors, Utah: the last 30 000 years. *Contrib. Miner. Petrol.* **86**, 321–334.

Sperber, C.M., Wilkinson, B.H. & Peacor, D.R. (1984) Rock composition, dolomite stoichiometry and rock/water reactions in dolomitic carbonate rocks. *J. Geol.* **92**, 609–622.

Squires, D.F. (1964) Fossil coral thickets in Warrarapa, New Zealand. *J. Paleont.* **38**, 904–915.

Stahl, W.J. (1979) Carbon isotopes in petroleum geochemistry. In: *Lectures in Isotope Geology* (Ed. by E. Jager and J.C. Hunziker) pp. 274–282. Springer-Verlag, New York.

Stearn, C.W. (1982) The shapes of Paleozoic and modern reef builders: a critical review. *Paleobiology* **8**, 228–241.

Stearn, C.W. & Scoffin, T.P. (1977) The carbonate budget of a fringing reef, Barbados. *Proc. 3rd Int. Coral Reef Symp. Miami* **2**, 471–476.

Stearn, C.W., Scoffin, T.P. & Martindale, W. (1977) Calcium carbonate budget of a fringing reef on the west coast of Barbados. Part 1. Zonation and productivity. Bull. Mar. Sci. **27**, 479–510.

Steinen, R.P. (1974) Phreatic and vadose diagenetic modification of Pleistocene limestone: petrographic observations from subsurface of Barbados, West Indies. *Bull. Am. Ass. Petrol. Geol.* **58**, 1008–1024.

Steinen, R.P. (1978) On the diagenesis of lime mud: scanning electron microscopic observations of subsurface material from Barbados. WI. *J. sedim. Petrol.* **48**, 1139–1147.

Steinen, R.P. (1982) SEM observations on the replacement of Bahaman aragonitic mud by calcite. *Geology* **10**, 471–475.

Stockman, K.W., Ginsburg, R.N. & Shinn, E.A. (1967) The production of lime mud by algae in south Florida. *J. sedim. Petrol.* **37**, 633–648.

Stoddart, D.R. (1969) Post-hurricane changes in the British Honduras reefs and cays. *Atoll Res. Bull.* **131**, 1–25.

Stoddart, D.R., McLean, R.F., Scoffin, T.P., Thom, B.G. & Hopley, D. (1978) Evolution of reefs and islands, northern Great Barrier Reef: synthesis and interpretation. *Phil. Trans. R. Soc. Lond.* B **284**, 149–159.

Stoffers, P. & Hecky, R.E. (1978) Late Pleistocene–Holocene evolution of Kiva–Tanganyika Basin. In: *Modern and Ancient Lake Sediments* (Ed. by A. Matter and M.E. Tucker) Spec. Publ. int. Assoc. Sedim. **2**, 43–55.

Stow, D.A.V. (1985) Deep-sea clastics: where are we and where are we going? In: *Sedimentology: Recent Developments and Applied Aspects* (Ed. by P.J. Brenchley and B.J.P. Williams) Spec. Publ. geol. Soc. Lond. **18**, 67–93.

Stow, D.A.V. (1986) Deep clastic seas. In: *Sedimentary Environments and Facies* (Ed. by H.G. Reading) pp. 399–444. Blackwells, Oxford.

Stow, D.A.V., Howell, D.G. & Nelson, C.H. (1984) Sedimentary, tectonic and sea-level controls on submarine fans and slope-apron turbidite systems. *Geo. Mar. Letts.* **3**, 57–64.

Strasser, A. (1984) Black pebble occurrence and genesis in Holocene carbonate sediments (Florida Keys, Bahamas, and Tunisia). *J. Sedim. Petrol.* **54**, 1097–1109.

Strasser, A. (1986) Ooids in Purbeck limestones (lowermost Cretaceous) of the Swiss and French Jura. *Sedimentology* **33**, 711–728.

Strasser, A. (1988) Shallowing-upward sequences in the Purbeckian peritidal carbonates (lowermost Cretaceous, Swiss and French Jura Mountains). *Sedimentology* **35**, 369–383.

Strasser, A. & Davaud, E. (1983) Black pebbles of the purbeckian (Swiss and French Jura): lithology, geochemistry and origin. *Eclogae geol. Helv.* **76**, 551–580.

Strasser, A. & Davaud, E. (1986) Formation of Holocene limestone sequences by progradation, cementation and erosion: Two examples from the Bahamas. *J. sedim. Petrol.* **56**, 422–428.

Stricklin, F.L. & Smith, C.I. (1973) Environmental reconstruction of a carbonate beach complex, Cow Creek (Lower Cretaceous) Formation of central Texas. *Bull. Geol. Soc. Am.* **84**, 1349–1368.

Strong, A. & Eadie, B.P. (1978) Satellite observations of calcium carbonate precipitation in the Great Lakes. *Limnol. Oceanogr.* **23**, 115–130.

Stross, R.G. (1979) Density and boundary regulations of the *Nitella* meadow in Lake George, New York. *Aquatic Botany* **6**, 285–300.

Stueber, A.M., Pushkar, P. & Hetherington, E.A. (1984) A strontium isotope study of Smackover brines and associated solids, southern Arkansas. *Geochim. Cosmochim. Acta* **48**, 1637–1649.

Sturm, M. & Matter, A. (1978) Turbidites and varves in Lake Brienz (Switzerland): deposition of clastic detritus by density currents. In: *Modern and Ancient Lake Sediments* (Ed. by A. Matter & M.E. Tucker) pp. 145–166. Spec. Publ. int. Ass. Sedim. **2**.

Suess, E. & Fütterer, D. (1972) Aragonitic ooids: experimental precipitation from seawater in the presence of humic acid. *Sedimentology* **19**, 129–139.

Supko, P.R. (1977) Subsurface dolomites, San Salvador, Bahamas. *J. sedim. Petrol.* **47**, 1063–1077.

Surdam, R.C. & Stanley, K.O. (1979) Lacustrine sedimentation during the culminating phase of Eocene Lake Gosiute, Wyoming (Green River Formation). *Bull. geol. Soc. Am.* **90**, 93–110.

Surdam, R.C. & Wolfbauer, C.A. (1975) Green River Formation, Wyoming: a playa lake complex. *Bull. Geol. Soc. Am.* **86**, 335–345.

Surdam, R.C. & Wray, J.L. (1976) Lacustrine stromatolites, Eocene Green River Formation, Wyoming. In: *Stromatolites* (Ed. by M.R. Walter) pp. 535–541. Elsevier, Amsterdam.

Swift, D.J.P. (1968) Coastal erosion and transgressive stratigraphy. *J. Geol.* **76**, 444–456.

Swift, D.J.P. (1975) Barrier island genesis: evidence from the Middle Atlantic Shelf of North America. *Sedim. Geol.* **14**, 1–43.

Swirydczuk, K. (1988) The original mineralogy of ooids in the Smackover formation, Texas. *J. Sedim. Petrol.* **58**, 339–347.

Swirydczuk, K., Wilkinson, B.H. & Smith, G.R. (1979) The Pliocene Glenns Ferry Oolite: Lake margin carbonate deposition in the southwestern Snake River Plain. *J. Sedim. Petrol.* **49**, 995–1004.

Swirydczuk, K., Wilkinson, B.H. & Smith, G.R. (1980) The Pliocene Glenns Ferry Oolite II: sedimentology of oolitic lacustrine terrace deposits. *J. Sedim. Petrol.* **50**, 1237–1247.

Swirydczuk, K., Wilkinson, B.H. & Smith, G.R. (1981) Synsedimentary lacustrine phosphorites from the Pliocene Glenns Ferry Formation of southwest Idaho. *J. sedim. Petrol.* **51**, 1205–1214.

Symonds, P.A., Davies, P.J. & Parisi, A. (1983) Structure and stratigraphy of the central Great Barrier Reef. *B.M.R. J. Austr. Geol. Geophys.* **8**, 277–291.

Szulczewski, M. (1968) Slump structures and turbidites in Upper Devonian limestones of the Holy Cross Mts. *Acta Geol. Pol.* **18**, 303–324.

Tada, R., Maliva, R. & Siever, R. (1987) A new mechanism for pressure solution in porous quartzose sandstone. *Geochim. Cosmochim. Acta* **51**, 2295–2301.

Tan, F.C. & Hudson, J.D. (1971) Carbon and oxygen isotopic relationships of dolomites and co-existing calcites, Great Estuarine Series (Jurassic), Scotland. *Geochim. Cosmochim. Acta* **35**, 755–767.

Tarutani, T., Clayton, R.N. & Mageda, T.K. (1969) The effects of polymorphism and magnesium substitution on oxygen isotope fractionation between calcium carbonate and water. *Geochim. Cosmochim. Acta* **33**, 987–996.

Taylor, J.C.M. & Illing L.V. (1969) Holocene intertidal calcium carbonate cementation, Qatar, Persian Gulf. *Sedimentology* **12**, 69–107.

Taylor, T.R. & Sibley, D.F. (1986) Petrographic and geochemical characteristics of dolomite types and the origin of ferroan dolomite in the Trenton Formation, Ordovician, Michigan Basin, USA. *Sedimentology* **33**, 61–86.

Tebbut, G.E., Coneley, C.D. & Boyd, D.W. (1965) Lithogenesis of a distinctive carbonate rock fabric. *Contrib. Wyoming geol. Surv.* **4**, 1–13.

Ten Have, T. & Heijnen, W. (1985) Cathodoluminescence activation and zonation in carbonate rocks: an experimental approach. *Geol. Mijn.* **64**, 297–310.

Terlecky, M.P. (1974) The origin of a Late Pleistocene and Holocene marl deposit. *J. Sedim. Petrol.* **44**, 456–465.

Thorne, J.A. & Watts, A.B. (1989) Quantitative analysis of North Sea subsidence. *Bull. Am. Ass. Petrol. Geol.* **73**, 86–116.

Thrailkill, J. (1968) Chemical and hydrologic factors in the excavation of limestone caves. *Bull. Geol. Soc. Am.* **87**, 19–46.

Todd, D.K. (1980) *Groundwater Hydrology*, pp. 535. John Wiley & Sons, London.

Toomey, D.F. & Babcock, J.A. (1983) *Precambrian and Paleozoic Algal Carbonates, West Texas–southern New Mexico*. Prof. Contrib. Col. Sch. Mines, 11, pp. 343. Colorado School Mines, Golden, Colorado.

Towe, K.M. & Hembleben, K.M. (1976) Diagenesis of magnesian calcite: evidence from miliolacean foraminifera. *Geology* **4**, 337–339.

Treese, K.L. & Wilkinson, B.H. (1982) Peat-marl deposition in a Holocene paludal-lacustrine basin — Sucker Lake, Michigan. *Sedimentology* **29**, 375–390.

Trewin, N.H. (1986) Palaeoecology and sedimentology of the Achanarras Fish bed of the Middle Old Red Sandstone, Scotland. *Trans. Roy. Soc. Edin. Earth Sci.* **77**, 21–46.

Trichet, J. (1968) Étude de la composition de la fraction organique des oolites. Comparison avec celle des membranes des bactéries et des cyanophycées. *C.R. Acad. Sci. Paris.* **267**, 1392–1494.

Trudgill, S. (1985) *Limestone Geomorphology*, pp. 196. Longmans, London.

Tsien, H.H. (1981) Ancient reefs and reef carbonates. *Proc. 4th Int. Coral Reef Symp. Manilla* **1**, 601–609.

Tsien, H.H. (1985a) Origin of stromatactis — a replacement of colonial microbial associations. In: *Paleoalgology: Contemporary Research and Applications* (Ed. by D.F. Toomey and M.H. Nitecki) pp. 274–289. Springer-Verlag, Berlin.

Tsien, H.H. (1985b) Algal–bacterial origin of micrites in mud mounds. In: *Paleoalgology: Contemporary Research and Applications* (Ed. by D.F. Toomey and M.H. Nitecki) pp. 290–296. Springer-Verlag, Berlin.

Tucker, M.E. (1969) Crinoidal turbidites from the Devonian of Cornwall and their palaeogeographic significance. *Sedimentology* **13**, 281–290.

Tucker, M.E. (1973) Sedimentology and diagenesis of Devonian pelagic limestones (Cephalopodenkalk) and associated sediments of the Rhenohercynian geosyncline, West Germany. *Neues Jb. Geol. Paläont. Abh.* **142**, 320–350.

Tucker, M.E. (1974) Sedimentology of Palaeozoic pelagic limestones: the Devonian griotte (Southern France) and Cephalopodenkalk (Germany). In: *Pelagic Sediments: on Land and Under the Sea* (Ed. by K.J. Hsü and H.C. Jenkyns) Spec. Publ. int. Ass. Sediment. 1, 71–92.

Tucker, M.E. (1977) Stromatolite biostromes and associated facies in the Late Precambrian Porsanger Dolomite Formation of Finnmark, Arctic Norway. *Palaeogeog. Palaeoclimatol. Palaeoecol.* **21**, 55–83.

Tucker, M.E. (1981) *Sedimentary Petrology: an Introduction*, pp. 252. Blackwells, Oxford.

Tucker, M.E. (1982) Precambrian dolomites: petrographic and isotopic evidence that they differ from Phanerozoic dolomites. *Geology* **10**, 7–12.

Tucker, M.E. (1983a) Diagenesis, geochemistry and origin of a Precambrian dolomite: the Beck Spring Dolomite of eastern California. *J. Sedim. Petrol.* **53**, 1097–1119.

Tucker, M.E. (1983b) Sedimentation of organic-rich limestones from the Late Precambrian of Southern Norway. *Precamb. Res.* **22**, 295–315.

Tucker, M.E. (1984) Calcite, aragonite and mixed calcitic–aragonitic ooids from the mid-Proterozoic Belt supergroup, Montana. *Sedimentology* **31**, 627–644.

Tucker, M.E. (1985a) Shallow-marine carbonate facies and facies models. In: *Sedimentology: Recent Developments and Applied Aspects* (Ed. by P.J. Brenchley and B.P.J. Williams) Spec. Publ. Geol. Soc. Lond. **18**, 139–161.

Tucker, M.E. (1985b) Calcitized aragonitic ooids and cements from the Late Precambrian of southern Norway. *Sedim. Geol.* **43**, 67–84.

Tucker, M.E. (1986a) Carbon isotope excursions in the Late Precambrian–Cambrian boundary beds, Anti-Atlas, Morocco. *Nature* **319**, 48–50.

Tucker, M.E. (1986b) Formerly aragonitic limestones associated with tillites in the Late Proterozoic Kingston Peak Formation of Death Valley, California. *J. Sedim. Petrol.* **56**, 818–830.

Tucker, M.E. (Ed.) (1988) *Techniques in Sedimentology*, pp. 394. Blackwells, Oxford.

Tucker, M.E. & Bathurst, R.G.C. (Eds) (1990) *Carbonate Diagenesis*. Reprint Series, Int. Ass. Sedim. 1.

Tucker, M.E. & Hollingworth, N.T.J. (1986) The Upper Permian Reef Complex (EZ) of North East England: Diagenesis in a marine to evaporitic setting. In: *Reef Diagenesis* (Ed. by J.H. Schroeder and B.H. Purser), pp. 270–290. Springer-Verlag, Berlin.

Tucker *et al.* (Eds) (1990) *Carbonate Platforms, facies, sequences and Evolution.* Spec. Publ. int. Ass. Sedim. 9. Blackwells, Oxford.

Tudhope, A.W. & Risk, M.J. (1985) Rate of dissolution of carbonate sediments by microboring organisms, Davies Reef, Australia. *J. sedim. Petrol.* **55**, 440–447.

Tudhope, A.W. & Scoffin, T.P. (1984) The effects of *Callianasa* bioturbation on the preservation of carbonate grains in Davies Reef lagoon, Great Barrier Reef, Australia. *J. sedim. Petrol.* **54**, 1091–1096.

Tudhope, A.W., Scoffin, T.P., Stoddart, D.R. & Woodroffe, C.D. (1985) Sediments of Suwarrow atoll. *Proc. 5th Int. Coral Reef Congr. Tahiti*, 6, 611–616.

Tunnicliffe, V.J. (1982) The role of boring sponges in coral fracture. *Colloque internationaux du C.N.R.S.* 291, Biologie des spongiaires, 309–315.

Turi, B. (1986) Stable isotope geochemistry of travertines. In: *Handbook of Environmental Isotope Geochemistry, 2. The Terrestrial Environment, B*, pp. 207–238. Elsevier, Amsterdam.

Turmel, R.J. & Swanson, R.G. (1976) The development of Rodriguez Bank, a Holocene mudbank in the Florida reef tract. *J. sedim. Petrol.* **46**, 497–518.

Turner, J.V. (1982) Kinetic fractionation of carbon-13 during calcium carbonate precipitation. *Geochim. Cosmochim. Acta* **46**, 1183–1191.

Urey, H.C. (1947) The thermodynamic properties of isotopic substances. *J. Chem. Soc. (London)* 562–581.

Vacher, H.L. (1978) Hydrology of Bermuda — significance of an across-the-island variation in permeability. *J. Hydrol.* **39**, 207–226.

Vail, P.R., Mitchum, R.M. & Thompson, S. (1977) Seismic stratigraphy and global changes of sea-level, part 4: global cycles of relative changes of sea-level. *Mem. Am. Ass. Petrol. Geol.* **26**, 83–97.

Van Andel, T.H., Thiede, J., Sclater, J.G. & Hay, W.W. (1977) Depositional history of the South Atlantic ocean during the last 125 million years. *J. Geol.* **85**, 651–698.

Van den Bark, E. & Thomas, O.D. (1981) Ekofisk: first of the giant oil fields in western Europe. In: *Giant Oil and Gas Fields of the Decade 1968–1978* (Ed. by M.T. Halbouty) Mem. Am. Ass. Petrol. Geol. 30, 195–224.

Van Laer, P. & Monty, C.L.V. (1984) The cementation of mud mound cavities by microbial spars. *5th Europ. IAS Meeting, Marseilles*, Abst. 30–31.

Veizer, J. (1983a) Chemical diagenesis of carbonates: theory and application of trace element technique. In: *Stable Isotopes in Sedimentary Geology* (Ed. by M.A. Arthur and T.F. Anderson) Soc. econ. Paleont. Miner. Short Course No. 10, 3.1–3.100.

Veizer, J. (1983b) Trace elements and isotopes in carbonate minerals. *Mineral. Soc. Am. Rev. in Mineral.* **11**, 265–299.

Veizer, J. & Compston, W. (1974) $^{87}Sr/^{86}Sr$ composition of seawater during the Phanerozoic. *Geochim. Cosmochim. Acta* **38**, 1461–1484.

Veizer, J. & Hoefs, J. (1976) The nature of $^{18}O/^{16}O$ and $^{13}C/^{12}C$ secular trends in sedimentary carbonate rocks. *Geochim. Cosmochim. Acta* **40**, 1387–1395.

Veizer, J., Holser, W.T. & Wilgus, C.K. (1980) Correlation of $^{13}C/^{12}C$ and $^{34}S/^{32}S$ secular variations. *Geochim. Cosmochim. Acta* **44**, 579–587.

Veizer, J., Lemieux, J., Jones, B., Gibling, M.R. & Savelle, J. (1978) Paleosalinity and dolomitization of a Lower Paleozoic carbonate sequence, Somerset and Prince of Wales Islands, Arctic Canada. *Can. J. earth Sci.* **15**, 1448–1461.

Ventress, W.P.S., Bebout, D.G., Perkins, B.F. & Moore, C.H. (Eds.) (1984) *The Jurassic of the Gulf Rim.* Publ. Soc. econ. Paleont. Miner. Gulf Coast Section, pp. 408.

Vera, J-A., Ruiz-Ortiz, P.A., Garcia-Hernandez, M. & Molina, J.M. (1988) Paleokarst and related pelagic sediments in the Jurassic of the Subbetic Zone, southern Spain. In: *Paleokarst* (Ed. by N.P. James and P.W. Choquette) pp. 364–384. Springer-Verlag, New York.

Vernon, J.E.N. & Hudson, R.C.L. (1978) Ribbon reefs of the Northern Region. *Phil. Trans. R. Soc. Lond. B* **284**, 3–21.

Vernon, P.D. (1969) The geology and hydrology associated with a zone of high permeability ('Boulder Zone') in Florida. *Soc. of Mining Engineers, Am. Inst. Min. Eng.* 69–AG–12.

Viau, C. (1983) Depositional sequences, facies and evolution of the upper Devonian Swan Hills Reef buildup, Central Alberta, Canada. In: *Carbonate Buildups: a Core Work-*

shop (Ed. by P.M. Harris) Soc. Econ. Paleont. Miner. Core Workshop, 112–143.

Videtich, P.E. & Matthews, R.K. (1980) Origin of discontinuity surfaces in limestones: isotopic and petrographic data, Pleistocene of Barbados, West Indies. *J. sedim. Petrol.* **50**, 971–980.

Vinogradov, A.P. & Ronov, A.B. (1956) Compositions of the sedimentary rocks of the Russian platform in relation to the history of its tectonic movements. *Geokhimiya* **6**, 3–24.

Vogel, J.C., Grootes, P.M. & Mook, W.G. (1970) Isotope fractionation between gaseous and dissolved carbon dioxide. *Z. Phys.* **230**, 255–258.

Von der Borch, C.C. (1976) Stratigraphy and formation of Holocene dolomitic carbonate deposits of the Coorong area, South Australia. *J. sedim. Petrol.* **46**, 952–966.

Von der Borch, C.C. & Jones, J.B. (1976) Spherular modern dolomite from the Coorong area, South Australia. *Sedimentology* **23**, 587–591.

Von der Borch, C.C. & Lock, D. (1979) Geological significance of Coorong dolomites. *Sedimentology* **26**, 813–824.

Wagner, C.W. & Van der Togt, C. (1973) Holocene sediment types and their distribution in the Southern Persian Gulf. In: *The Persian Gulf* (Ed. by B.H. Purser) pp. 123–156. Springer-Verlag, Berlin.

Wagner, P.D. & Matthews, R.K. (1982) Porosity preservation in the Upper Smackover (Jurassic) carbonate grainstone, Walker Creek Field, Arkansas: response of paleophreatic lenses to burial processes. *J. sedim. Petrol.* **52**, 3–18.

Walkden, G.M. (1974) Paleokarstic surfaces in the upper Visean (Carboniferous) limestones of the Derbyshire Block, England. *J. sedim. Petrol.* **44**, 1232–1247.

Walkden, G.M. (1987) Sedimentary and diagenetic styles in Late Dinantian carbonates of Britain. In: *European Dinantian Environments* (Ed. by J. Miller, A.E. Adams and V.P. Wright) pp. 131–155. John Wiley, Chichester.

Walkden, G.M. & Walkden, G.D. (1990) Cyclic sedimentation in carbonate and mixed carbonate/clastic environments: four simulation programs for a desktop computer. In: *Carbonate Platforms*. Spec. Publ. int. Ass. Sedim. 9 (in press).

Walker, K.R. & Alberstadt, L.P. (1975) Ecological succession as an aspect of structure in fossil communities. *Paleobiology* **3**, 238–257.

Walker, R. & Moss, B. (1984) Mode of attachment of six epilithic crustose Corallinaceae (Rhodophyta). *Phycologia* **23**, 321–329.

Walker, R.G. (1984) General introduction: facies, facies sequences and facies models. In: *Facies Models* (Ed. by R.G. Walker) pp. 1–9. Geoscience Canada.

Walker, R.G., Duke, W.L. & Leckie, D.A. (1983) Hummocky stratification: significance of its variable bedding sequences: discussion and reply. *Bull. geol. Soc. Am.* **94**, 1245–1251.

Wallace, M.W. (1987) The role of internal erosion and sedimentation in the formation of stromatactis mudstones

and associated lithologies. *J. sedim. Petrol.* **57**, 695–700.

Wallace, R.J. & Schafersman, S.D. (1977) Patch reef ecology and sedimentology of Glovers Reef Atoll, Belize. In: *Reefs and Related Carbonates — Ecology and Sedimentology* (Ed. by S.H. Frost, M.P. Weiss and J.B. Saunders) Am. Ass. petrol. Geol. Studies in Geol. 4, 37–52.

Walls, R.A. (1983) Golden Spike Reef Complex, Alberta. In: *Carbonate Depositional Environments* (Ed. by P.A. Scholle, D.G. Bebout and C.H. Moore) Mem. Am. Ass. petrol. Geol. 33, 445–453.

Walls, R.A. & Burrowes, G. (1985) The role of cementation in the diagenetic history of Devonian reefs, Western Canada. In: *Carbonate Cements* (Ed. by N. Schneidermann and P.M. Harris) Spec. Publ. Soc. Econ. Paleont. Miner. 36, 185–220.

Walls, R.A., Mountjoy, E.W. & Fritz, P. (1979) Isotopic composition and diagenetic history of carbonate cements in Devonian Golden Spike reef, Alberta, Canada. *Bull. geol. Soc. Am.* **90**, 963–982.

Walter, L.M. (1985) Relative reactivity of skeletal carbonates during dissolution: implications for diagenesis. In: *Carbonate Cements* (Ed. by N. Schneidermann and P.M. Harris) Spec. Publ. Soc. econ. Paleont. Miner. 36, 3–16.

Walter, L.M. (1986) Relative efficiency of carbonate dissolution and precipitation during diagenesis: a progress report on the role of solution chemistry. In: *Roles of Organic Matter in Sediment Diagenesis* (Ed. by D.L. Gautier) Spec. Publ. Soc. econ. Paleont. Miner. 38, 1–11.

Walter, L.M. & Morse, J.W. (1984) A re-evaluation of magnesian calcite stabilities. *Geochim. Cosmochim. Acta* **48**, 1059–1069.

Walter, M.R. (Ed.) (1976) *Stromatolites*, pp. 790. Elsevier, Amsterdam.

Wanless, H.R. (1969) Sediments of Biscayne Bay distribution and depositional history. *Tech. Rept. 62–2. Inst. Marine & Atmospheric Sciences, Univ. of Miami*, pp. 260.

Wanless, H.R. (1979) Limestone response to stress: pressure solution and dolomitization. *J. sedim. Petrol.* **49**, 437–462.

Wanless, H.R. (1981) Fining upwards sedimentary sequences generated in sea grass beds. *J. sedim. Petrol.* **51**, 445–454.

Wanless, H.R. (1982) Sea-level is rising — so what? *J. sedim. Petrol.* **52**, 1051–1054.

Wanless, H.R. (1983) Burial diagenesis in limestones. In: *Sediment Diagenesis* (Ed. by A. Parker and B.W. Sellwood) pp. 379–417. Reidel Pub. Co., Dordrecht.

Wanless, H.R., Burton, E.A. & Dravis, J. (1981) Hydrodynamics of carbonate fecal pellets. *J. sedim. Petrol.* **51**, 27–36.

Wanless, H.R., Tyrell, K.M., Tedesco, L.P. & Dravis, J.J. (1988) Tidal-flat sedimentation from Hurricane Kate, Caicos Platform, British West Indies. *J. sedim. Petrol.* **58**, 724–738.

Ward, W.C. (1975) Petrology and diagenesis of carbonate

eolianites of Northeastern Yucatan Peninsula, Mexico. In: *Belize Shelf-carbonate Sediments, Clastic Sediments and Ecology* (Ed. by K.F. Wantland and W.C. Pusey), Am. Ass. petrol. Geol. Studies in Geology, 2, 500−571.

Ward, W.C. (1978a). Collectors guide to carbonate cement types, northeastern Yucatan Peninsula. In: *Geology and Hydrogeology of Northeastern Yucatan* (Ed. by W.C. Ward and A.E. Weidie), pp. 147−171. New Orleans Geol. Soc., New Orleans.

Ward, W.C. (1978b) Indicators of climate in carbonate dune rocks. In: *Geology and Hydrogeology of Northeastern Yucatan* (Ed. by W.C. Ward and A.E. Weidie), pp. 191−208. New Orleans Geol. Soc., New Orleans.

Ward, W.C. & Brady, M.J. (1979) Strandline sedimentation of carbonate grainstones, Upper Pleistocene, Yucatan Peninsula, Mexico. *Bull. Am. Ass. petrol. Geol.* **63**, 362−369.

Ward, W.C. & Halley, R.B. (1985) Dolomitization in a mixing zone of near-seawater composition, late Pleistocene, northeastern Yucatan Peninsula. *J. sedim. Petrol.* **55**, 407−420.

Ward, W.C., Weidie, A.E. & Back, W. (1985) *Geology and Hydrology of the Yucatan*, pp. 160. New Orleans Geol. Soc.

Wardlaw, N.C. (1979) Pore systems in carbonate rocks and their influence on hydrocarbon recovery efficiency. *Am. Ass. Petrol. Geol. Continuing Education Course Notes* **11**, E1−E24.

Wardlaw N.C. (1980) The effects of pore structure as displacement efficiency in reservoir rocks and in glass micromodels. *Proc. First Joint SPE/DOE Symp. on Enhanced Oil Recovery*, pp. 345−352.

Wardlaw, N.C. & Cassan, J.P. (1978) Estimation of recovery efficiency by visual observation of pore systems in reservoir rocks. *Bull. Can. Petrol. Geol.* **26**, 572−585.

Warme, J.E. (1977) Carbonate borers − their role in reef ecology and preservation. In: *Reefs and Related Carbonates − Ecology and Sedimentology* (Ed. by S.H. Frost, M.P. Weiss and J.B. Saunders) Am. Ass. Petrol. Geol. Studies in Geol. 4, 261−279.

Warren, J.K. (1982a) The hydrological significance of Holocene tepees, stromatolites, and boxwork limestones in coastal salines in South Australia. *J. sedim. Petrol.* **52**, 1171−1201.

Warren, J.K. (1982b) The hydrological setting, occurrence and significance of gypsum in late Quaternary salt lakes in South Australia. *Sedimentology* **29**, 609−637.

Warren, J.K. (1983) Tepees, modern (South Australia) and ancient Permian (Texas and New Mexico) − a comparison. *Sedim. Geol.* **34**, 1−19.

Warren, J.K. & Kendall, C.G. St. C. (1985) Comparison of marine (subaerial) and salina (subaqueous) evaporites: ancient and modern. *Bull. Am. Ass. petrol. Geol.* **69**, 1013−1023.

Watts, N.L. (1983) Microfractures in chalks of Albuskjell Field, Norwegian Sector, North Sea: possible origin and distribution. *Bull. Am. Ass. Petrol. Geol.* **67**, 201−234.

Watts, N.L., Lapré, J.F., Schijndel-Goester, F.S. Van & Ford, A. (1980) Upper Cretaceous and Lower Tertiary chalks of the Albuskjell area, North Sea: deposition in a slope and base-of-slope environment. *Geology* **8**, 217−221.

Watts, N.R. (1981) *Sedimentology and diagenesis of the Hogklint reefs and their associated sediments, Lower Silurian, Gotland, Sweden*. Unpubl. PhD Thesis, University of Wales, Cardiff.

Watts, N.R. (1988) Carbonate particulate sedimentation and facies within the Lower Silurian Hogklint patch reefs of Gotland, Sweden. *Sedim. Geol.* **59**, 93−113.

Weber, J.N. (1968) Fractionation of the stable isotopes of carbon and oxygen in calcareous marine invertebrates − the Asteroidea, Ophiuroidea and Crinoidea. *Geochim. Cosmochim. Acta* **32**, 33−70.

Weber, J.N. & Woodhead, P.M.J. (1970) Carbon and oxygen isotopic fractionation in the skeletal carbonate of reef-building corals. *Chem. Geol.* **6**, 93−117.

Weedon, G.P. (1986) Hemipelagic shelf sedimentation and climatic cycles: the basal Jurassic (Blue Lias) of south Britain. *Earth planet. Sci. Letts.* **76**, 321−335.

Weiner, S. & Traub, W. (1984) Macromolecules in mollusc shells and their functions in biomineralization. *Phil. Trans. R. Soc. Lond. B*, **304**, 425−434.

Wells, N.A. (1983) Carbonate deposition, physical limnology and environmentally controlled chert formation in Paleocene−Eocene Lake Flagstaff, central Utah. *Sedim. Geol.* **35**, 263−296.

Wendt, J. (1969) Foraminiferen 'Riffe' im Karnischen Hallstatter Kalk des Feuerkogels (Steiermark, Osterreich). *Paläont. 2.* **43**, 177−193.

Wendt, J. (1971) Genese und fauna submariner sedimentarer spaltenfullengen im mediterranean Jura. *Palaeontographica, A*, **136**, 122−192.

Wendt, J. & Aigner, T. (1985) Facies patterns and depositional environments of Palaeozoic cephalopod limestones. *Sedim. Geol.* **44**, 263−300.

Wendt, J., Aigner, T. & Neugebauer, J. (1984) Cephalopod limestone deposition on a shallow pelagic ridge: the Tafilalt platform (Upper Devonian, eastern Anti-Atlas, Morocco). *Sedimentology* **31**, 601−625.

Wenk, H.R., Barber, D.J. & Reeder, R.J. (1983) Microstructures in carbonates. *Mineral. Soc. Am. Rev. in Mineral.* **11**, 301−368.

West, I.M. (1975) Evaporites and associated sediments of the basal Purbeck Formation (Upper Jurassic) of Dorset. *Proc. Geol. Ass.* **85**, 205−225.

West, I.M. (1979) Review of evaporite diagenesis in the Purbeck Formation of southern England. In: *Symposium de Sédimentation jurassique Européan*. Spec. Publ. Ass. Sedim. Francais 1, 407−416.

West, I.M., Ali, Y.A. & Hilmy, M.E. (1979) Primary gypsum nodules in a modern sabkha on the Mediterranean coast of Egypt. *Geology* **7**, 354−358.

West, I.M., Brandon, A. & Smith, M. (1968) A tidal flat evaporitic facies in the Visean of Ireland. *J. Sedim.*

Petrol. **38**, 1079–1093.

Westbroek, P., Van der Wal, P., Van Emburg, P.R., DeJong, E.W. & De Bruijn, W.C. (1986) Calcification in the Coccolithophorids *Emiliania huxlei* and *Pleurochrysis carterae* 1. Ultrastructural aspects. In: *Biomineralization in Lower Plants and Animals* (Ed. by B.S.C. Leadbeater and R. Riding) The Systematics Association Special Volume No. 30, 189–203. Oxford University Press Oxford.

Wetzel, R.G. (1975) *Limnology*, pp. 743. W.B. Saunders Co., Philadelphia.

Weyl, P.K. (1959) Pressure solution and the force of crystallization – a phenomenological theory. *J. Geophys. Res.* **64**, 2001–2025.

White, A.H. & Youngs, B.C. (1980) Cambrian alkali playa-lacustrine sequence in the northeastern Officer Basin, South Australia. *J. Sedim. Petrol.* **50**, 1279–1286.

Whiticar, M.J. & Faber, E. (1986) Methane oxidation in seawater and water column environments – Isotopic evidence. *Org. Geochem.* **10**, 759–768.

Whiticar, M.J., Faber, E. & Schoell, M. (1986) Biogenic methane formation in marine and freshwater environments: CO_2 reduction vs. acetate fermentation – Isotopic evidence. *Geochim. Cosmochim. Acta* **50**, 693–709.

Wiggins, W.D. (1986) Geochemical signatures in carbonate matrix and their relation to deposition and diagenesis, Pennsylvanian Marble Falls Limestone, Central Texas. *J. sedim. Petrol.* **56**, 771–783.

Wiggins, W.D. & Harris, P.M. (1984) Cementation and porosimetry of shoaling sequences in the subsurface Pettit limestone, Cretaceous of east Texas. In: *Carbonate Sands: a Core Workshop* (Ed. by P.M. Harris), Soc. Econ. Paleont. Miner. Core Workshop, 5, 263–305.

Wigley, P.L., Bouvier, J.D. & Dawans, J.M. (1988) Karst and mixing-zone porosity in the Amposta-Marino Field, off-shore Spain (Abst.). *Bull. Am. Ass. Petrol. Geol.* **72**, 1031.

Wigley, T.M.L. & Plummer, L.N. (1976) Mixing of carbonate waters. *Geochim. Cosmochim. Acta* **40**, 989–995.

Wilgus, C.K. *et al.* (Eds) (1989) *Sea-level Changes – an Integrated Approach*. Spec. Publ. Soc. Econ. Paleont. Miner, **42**, pp. 407.

Wilkinson, B.H. (1979) Biomineralization, paleoceanography and the evolution of calcareous marine organisms. *Geology* **7**, 524–527.

Wilkinson, B.H. (1982) Cyclic cratonic carbonates and Phanerozoic calcite seas. *J. geol. Educ.* **30**, 189–203.

Wilkinson, B.H., Buczynski, C. & Owen, R.M. (1984) Chemical control of carbonate phases: implications from Upper Pennsylvanian calcite–aragonite ooids in southeastern Kansas. *J. sedim. Petrol.* **54**, 932–947.

Wilkinson, B.H., Janecke, S.U. & Brett, C.E. (1982) Low-magnesian calcite marine cement in Middle Ordovician hardgrounds from Kirkfield, Ontario. *J. sedim. Petrol.* **52**, 47–57.

Wilkinson, B H. & Landing, E. (1978) 'Eggshell-diagenesis' and primary radial fabric in calcite ooids. *J. sedim. Petrol.* **48**, 1129–1138.

Wilkinson, B.H., Owen, R.M. & Carroll, A.R. (1985) Submarine hydrothermal weathering, global eustacy, and carbonate polymorphism in Phanerozoic marine oolites. *J. sedim. Petrol.* **55**, 171–183.

Wilkinson, B.H., Popp, B.N. & Owen, R.M. (1980) Near-shore oolite formation in a modern temperate-region marl lake. *J. Geol.* **88**, 697–704.

Wilkinson, B.H., Smith, A.L. & Lohmann, K.C. (1985) Sparry calcite marine cement in Upper Jurassic limestones of southeastern Wyoming. In: *Carbonate Cements* (Ed. by N. Schneidermann and P.M. Harris) Spec. Publ. Soc. econ. Paleont. Miner. 36, 169–184.

Williams, L.A. (1980) Community succession in a Devonian patch reef, Conondaga Formation, New York – physical and biotic controls. *J. sedim. Petrol.* **50**, 1169–1186.

Williamson, C.R. & Picard, M.D. (1974) Petrology of carbonate rocks of the Green River Formation (Eocene). *J. sedim. Petrol.* **44**, 738–759.

Wilson, J.L. (1974) Characteristics of carbonate platform margins. *Bull. Am. Ass. petrol. Geol.* **58**, 810–824.

Wilson, J.L. (1975) *Carbonate Facies in Geologic History*, pp. 471. Springer-Verlag, Berlin.

Winland, H.D. & Matthews, R.K. (1974) Origin and significance of grapestone, Bahama Islands. *J. sedim. Petrol.* **44**, 921–927.

Wise, S.W. (1977) Chalk formation: early diagenesis. In: *The Fate of Fossil CO_2 in the Oceans* (Ed. by N.R. Anderson and A. Malahoff) pp. 717–739. Plenum Press, New York.

Wolf, K.H. (1965) Gradational sedimentary products of calcareous algae. *Sedimentology* **5**, 1–37.

Wong, P.K. (1979) Sequential cementation in the Upper Devonian Kaybob Reef, Alberta. *Can. Soc. petrol. Geol. Symp in Honour of A.D. Baillie*, pp. 24–30.

Wong, P.K. & Oldershaw, A.E. (1980) Causes of cyclicity in reef interior sediments, Kaybob reef, Alberta. *Bull. Can. petrol. Geol.* **28**, 411–424.

Wong, P.K. & Oldershaw, A.E. (1981) Burial cementation in the Devonian Kaybob reef complex, Alberta, Canada. *J. sedim. Petrol.* **51**, 507–520.

Wood, G.V. & Wolfe, M.J. (1969) Sabkha cycles in the Arab/Darb formation off the Trucial Coast of Arabia. *Sedimentology* **12**, 165–191.

Woronick, R.E. & Land, L.S. (1985) Late burial diagenesis, Lower Cretaceous Pearsall and Lower Glen Rose formations, South Texas. In: *Carbonate Cements* (Ed. by N. Schneidermann and P.M. Harris) Spec. Publ. Soc. econ. Paleont. Miner. 36, 265–275.

Wray, J.L. (1977) *Calcareous Algae*, p. 185. Elsevier, Amsterdam.

Wright, R.F., Matter, A., Schweingruber, M. & Siegenthaler, U. (1980) Sedimentation in Lake Biel, an entrophic, hard water lake in northwestern Switzerland. *Schweiz. Z. Hydrol.* **42**, 101–126.

Wright, V.P. (1981) Algal aragonite encrusted pisoids from a Lower Carboniferous schizohaline lagoon. *J. sedim. Petrol.* **51**, 479–489.

Wright, V.P. (1982) The recognition and interpretation of

paleokarsts: two examples from the Lower Carboniferous of South Wales. *J. Sedim. Petrol.* 52, 83–94.

Wright, V.P. (1983) Morphogenesis of oncoids in the Lower Carboniferous Llanelly Formation of South Wales. In: *Coated Grains* (Ed. by T.M. Peryt) pp. 424–434. Springer-Verlag, Berlin.

Wright, V.P. (1984) Peritidal carbonate facies models: a review. *Geol. J.* 19, 309–325.

Wright V.P. (1985) Algal marsh deposits from the Upper Jurassic of Portugal. In: *Paleoalgology: Contemporaneous Research and Applications* (Ed. by D.F. Toomey and M.H. Nitecki) pp. 330–341. Springer-Verlag, New York.

Wright, V.P., (1986a) Facies sequences on a carbonate ramp: the Carboniferous Limestone of South Wales. *Sedimentology* 33, 221–241.

Wright, V.P. (1986b) The polyphase karstification of the Carboniferous Limestone in South Wales. In: *New Directions in Karst. Proceedings of the Anglo-French Karst Symposium* (Ed. by K. Paterson and M.M. Sweeting) pp. 569–580 Geoabstracts, Norwich.

Wright, V.P. (1986c) The role of fungal biomineralization in the formation of early Carboniferous soil fabrics. *Sedimentology* 33, 831–838.

Wright, V.P. (1987) The evolution of the early Carboniferous limestone province in southwest Britain. *Geol. Mag.* 124, 477–480.

Wright, V.P. (1988) Paleokarsts and paleosols as indicators of paleoclimate and porosity evolution: a case study from the Carboniferous of South Wales. In: *Paleokarst* (Ed. by N.P. James and P.W. Choquette) pp. 329–341. Springer-Verlag, New York.

Wright, V.P. (1989) A micromorphological classification of fossil and Recent calcic and petrocalcic microstructures. *Proc. Int. Wkg. Mtg. Sol Micromorphol.*, San Antonio, 1988. Elsevier, Amsterdam.

Wright, V.P. & Mayall, M. (1981) Organism–sediment interactions in stromatolites: an example from the Upper Triassic of southwest Britain. In: *Phanerozoic Stromatolites: Case Studies* (Ed. by C.L.V. Monty) pp. 74–84. Springer-Verlag, Berlin.

Wright, V.P., Platt, N.H. & Wimbledon, W.A. (1988) Biogenic laminar calcretes: evidence of root-mat horizons in paleosols. *Sedimentology* 35, 603–620.

Wright, V.P., Ries, A.C. & Munn, S.G. (1989) Intraplatformal basin fill sequences from the Infracambrian of central Oman. In: *The Geology and Tectonics of the Oman Region* (Ed. by A.H.F. Robertson and A.C. Ries) Spec. Publ. geol. Soc. Lond. 48, 603–618.

Wright, V.P. & Wilson, R.C.L. (1984) A carbonate submarine-fan sequence from the Jurassic of Portugal. *J.*

sedim. Petrol. 54, 394–412.

Wright, V.P. & Wilson, R.C.L. (1985) Lacustrine carbonates and source rocks from the Upper Jurassic of Portugal. *6th IAS Regional Meeting, Leida, Abst.* 487–490.

Wright, V.P. & Wilson, R.C.L. (1987) A terra rossa-like paleosol complex from the Upper Jurassic of Portugal. *Sedimentology* 34, 259–273.

Wright, V.P. & Wilson, R.C.L. (1987) Paralic carbonate facies and source rocks, Upper Jurassic, Portugal. *Soc. econ. Paleont. Miner. Abs.*, 4, 93.

Wu, Xian-Tao (1982) Storm-generated depositional types and associated trace fossils in Lower Carboniferous shallow marine carbonates of Three Cliffs Bay and Ogmore-by-sea, South Wales. *Palaeogeogr. Palaeoclimatol. Palaeoecol.* 39, 187–202.

Yurewicz, D.A. (1977a) The origin of massive facies of the Lower and Middle Capitan Limestone (Permian) Guadalupe Mountains, New Mexico and West Texas. In: *Upper Guadalupian Facies, Permian Reef Complex, Guadalupe Mountains, New Mexico and West Texas* (Ed. by M.E. Hileman and S.J. Mazzullo) Publ. Soc. econ. Palcont. Miner. Permian Basin Section 77–16, 45–92.

Yurewicz, D.A. (1977b) Sedimentology of Mississippian basin carbonates, New Mexico and West Texas — the Rancheria Formation. In: *Deep-water Carbonate Environments* (Ed. by H.E. Cook and P. Enos) Spec. Publ. Soc. econ. Paleont. Miner. 25, 203–219.

Yurtsever, Y. (1976) Worldwide survey of stable isotopes in precipitation. Rep. Sect. Isotope Hydrol., IAEA, Nov. p. 40.

Zankl, H. (1969) Structural and textural evidence of early lithification in fine-grained carbonate rocks. *Sedimentology* 12, 241–256.

Zeff, M.L. & Perkins, R.D. (1979) Microbial alteration of Bahamian deep-sea carbonates. *Sedimentology* 26, 175–201.

Zempolich, W.G., Wilkinson, B.H. & Lohmann, K.C. (1988) Diagenesis of Proterozoic carbonates: the Beck Spring Dolomite of eastern California. *J. sedim. Petrol.* 58, 656–672.

Zenger, D.H. (1972) Dolomitization and uniformitarianism. *J. geol. Educ.* 20, 107–124.

Zenger, D.H. (1983) Burial dolomitization in the Lost Burro Formation (Devonian), east-central California, and the significance of late diagenetic dolomitization. *Geology* 11, 519–522.

Zhao, X. & Fairchild, I.J. (1987) Mixing-zone dolomitization of Devonian carbonates, Guangxi, South China. In: *Diagenesis of Sedimentary Sequences* (Ed. by J.D. Marshall) Spec. Publ. Geol. Soc. Lond. 36, 157–170.

Index

Page numbers in *italics* refer to figures.

Printed and bound by CPI Group (UK) Ltd, Croydon, CR0 4YY